Institution of Civil Engineers

International conference on

Ground Anchorages and Anchored Structures in Service

Proceedings of the two day international conference organised by the Institution of Civil Engineers and held in London on 26th and 27th November 2007

Edited by Stuart Littlejohn

Sponsored by:

DSI GMGE ritchies KELLER

BACHY SOLETANCHE JAPAN ANCHOR ASSOCIATION SAMWOO-AnchorTec

Co-sponsored by:

BTS ADSC The International Association of Foundation Drilling The British Dam Society

Engineering Group of the Geological Society of London

GEO INSTITUTE of ASCE

Organising committee:

Stuart Littlejohn, University of Bradford (*Chairman*)
Raymond Coe, Black and Veatch Ltd
Mike Crilly, British Geotechnical Association
John Graham, Keller Ground Engineering Ltd
Roger Margerison, AMEC Group Ltd
Caesar Merrifield, Coffey Geotechnics
Devon Mothersille, Geoserve Global Ltd

Technical advisory group:

Stuart Littlejohn, University of Bradford (*Chairman*)
Tony Barley, SBMA Ltd
Donald Bruce, Geosystems L.P.
John Finnis, Dywidag Systems International
Gareth John, CAPCIS Ltd
Hiroaki Kubo, Japan Anchor Association
Ursula Lawrence, Engineering Group of the Geological Society of London
Yves Legendre, Bachy Soletanche
Paul McBarron, Austress Menard

Published for the Organising Committee by Thomas Telford Publishing, Thomas Telford Ltd, 1 Heron Quay, London E14 4JD. www.thomastelford.com

Distributors for Thomas Telford books are
USA: ASCE Press, 1801 Alexander Bell Drive, Reston, VA 20191-4400, USA
Japan: Maruzen Co. Ltd, Book Department, 3–10 Nihonbashi 2-chome, Chuo-ku, Tokyo 103
Australia: DA Books and Journals, 648 Whitehorse Road, Mitcham 3132, Victoria

First Published 2008

A catalogue record for this book is available from the British Library
ISBN 978-07277-3561-4

Flash drive manufactured by Golding Products Ltd, Telford, Shropshire
Printed and bound in Great Britain by MPG Books, Bodmin, Cornwall

Preface

Millions of ground anchorages have been installed over the past 80 years with relatively few recorded failures. However, as permanent anchorages in service become older, the subject is of growing importance, particularly for anchorages installed over 30 years ago, many of which have been designed with corrosion protection considered inadequate by today's standard.

Water levels reached 150-year highs in 2002 with associated flooding across Central and Eastern Europe, so the importance of anchored river and seawalls cannot be overemphasised. Winter storms are now judged by the insurance industry to represent the largest potential event loss in Europe. Elsewhere in the world, the highest potential losses are associated with earthquakes and typhoons, where anchored slopes are vital in resisting landslides and rock avalanches.

Routine programmes of inspection and monitoring can extend the service life of anchored structures, such as post-tensioned dams, dry docks, bridges, sports stadiums and multi-storey buildings, stabilised slopes, earth-retaining walls, river and seawall coastal defences, reinforced tunnels and underground caverns that represent key elements of a country's infrastructure.

Alternatively, where inspection highlights unacceptable tendon corrosion or over-stressing, the results provide early warning of the need for precautionary or remedial measures, in order to safeguard the integrity of the anchored structure.

In spite of these benefits, insufficient attention is currently paid to routine maintenance inspection and service behaviour monitoring. As a consequence, the papers selected for this conference are designed to illustrate the condition and service performance of anchorages in practice, provide guidance on appropriate monitoring and maintenance procedures, introduce recent developments in non-destructive testing and describe the latest corrosion monitoring and corrosion protection systems.

It is hoped that readers will enjoy these proceedings and benefit from the knowledge and experience of the authors. It is also important to review the associated discussions for each session, since they highlight some controversial issues, as well as future challenges relating to the management and monitoring of the condition and behaviour of ground anchorages and anchored structures in service.

I wish to thank personally members of the organising committee and the technical advisory group who acted as referees, authors, session chairs who also acted as rapporteurs, members of the ICE Conference office and the delegates, all of whom made an important contribution. As for the last ICE conference on ground anchorages in 1997, I much appreciate the willingness of authors and delegates to share hard won experience and knowledge, and their natural enthusiasm for an exciting subject.

Collaboration among old friends and new colleagues with a common interest has made the conference a great success and helped to create this landmark publication for the benefit of all who use ground anchorages.

Stuart Littlejohn
(*Editor and Chairman of the Organising Committee*)

Contents

Keynote presentation

Maintenance testing and service behaviour monitoring of permanent ground anchorages
Stuart Littlejohn and Devon Mothersille — i

Session one: Inspection procedures and physical conditions recorded in service

A case study of inspection, integrity investigation and repair of anchors used to stabilise slopes around dam lake
Minoru Okuno and Takahiro Takemata — 1

Inspection of anchorages supporting waterside structures
Devon Mothersille and Tony Barley — 12

Inspection, integrity investigation and repair of 30-year-old anchors
Yukio Fuseya and Hiroyuki Takeda — 23

Performance and condition assessment of 30-year-old anchorages, River Clyde, Glasgow
Devon Mothersille, Stuart Jackman and Jim Ferrier — 33

Some examples of the poor installation of electrically isolated prestressed ground anchors in Slovenia
Iztok Klemenc and Viljem Kuhar — 43

Maintenance control of ground anchors constructed in Japanese Expressways
Masahiro Yoshimura, Masaru Takemoto and Kiyoyuki Amano — 53

Session two: Service behaviour monitoring procedures and performance in service

Long term monitoring and routine maintenance of ground anchorages at Devonport Royal Dockyard, Plymouth UK
Michael Whitworth and Steve Parrish — 63

The long-term performance of permanent ground anchors forming part of the Thames Barrier project
Mike Turner and David Richards — 72

25 years of monitoring of rock anchors installed for landslide stabilisation at Nantgarw, South Wales
Hugo Wood and Peter Martin 82

Anchored slope monitoring and test for periodic inspection in Japanese Expressways
Katsumi Okubo, Masaru Takemoto and Hiroshi Yamada 92

Monitoring of multi-strand ground anchors at the City of Manchester stadium
Lee Jordan 92

Repeated lift-off testing of single bore multiple anchors for dam retaining wall over a 5-year period
Mary Ellen Bruce, Jesús Gomez and Robert Traylor 111

Session three: North American Dam rehabilitation

Rock anchors for North American dams: the national research program bibliography and database
Donald Bruce and John Wolfhope 121

Post-tensioned rehabilitation applied to hydropower dams: continuously improving on an anchor program
John Wolfhope, Donald Bruce, Gregor Forbes and Leslie Boyd 132

Case histories of high capacity anchors for two dams in Canada involving long term monitoring
Horst Aschenbroich 144

The stabilization of Gilboa Dam, New York using high capacity rock anchors: addressing service performance issues
Kessi Zicko, Donald Bruce and Robert Kline 155

The performance of Pacoima Dam in two major earthquakes, with and without post-tensioned anchors
Mark Schultz, Shawn Jones and Jeff Howard 166

Session four: Rock bolts in tunnels and mines

A review of glass reinforced polymer (GFRP) tendon for rock bolting in tunnels
Alison Littlejohn 177

Design and construction of large rock caverns supported by ground anchorages
Kenji Aoki 188

Monitoring and testing of mine anchor systems: cases studies and applications to the civil industry
Peter Altounyan, David Bigby and Lorraine Kent 199

Session five: Non-destructive integrity testing

Common anchorage issues addressed by numerical modelling
Ana Ivanović, Richard Neilson, Andrew Starkey and Albert Rodger 209

Non-destructive testing of deep mine anchorages
Andrew Starkey, Richard Neilson, Ana Ivanović and Albert Rodger 220

Ultrasonic technique for rock bolt and rock mass bond quality control
Takayuki Mori, Makoto Nakajima and Tohru Tsutsui 231

The development of the method of detecting the anchor tensile force to impact elastic wave
Kuniyasu Tanaka, Hidenobu Okada and Masayasu Izakura 238

Development of loading apparatus and test procedure of rapid load test for ground anchors
Tatsuro Sueyoshi and Kazuo Tani 250

Session six: Corrosion monitoring and corrosion protection systems

Durability of buried and encased ground anchorages and structural components; requirements for corrosivity assessment and protection to ensure service life
Gareth John and Stuart Littlejohn 263

Application of linear polarisation resistance measurements and system modelling for corrosion assessment of high strength rock anchorages
Gareth John and Petra Ernst 273

Corrosion protection options for permanent ground anchorages
Chris Irvin and Devon Mothersille 283

Replenishment of underhead protection (below bearing plate)
Kouichi Suga, Akira Yonemura and Yoshio Iwaida 293

Durability of ground anchors and other geotechnical stabilization applications in concrete reinforcement steel quality subject to new European regulations
Dieter Jungwirth and Susanne Adler 302

A case study on corrosion protection effect of using field-mixed resin as tie-back anchors' grout material for retaining wall and its inspection method
Chihping Kuo and Hung-Jiun Liao 312

Advancements in ground anchors: carbon fibre reinforced polymer (CFRP) strands
Matthew Sentry, Abdelmalek Bouazza, Riadh Al-Mahaidi, Darren Loidl, Chris Bluff and Len Carrigan 321

Ground anchor tendons in carbon fibre-reinforced polymers
Tony Barley and John Hartley 331

Ground anchor made of new materials applicable for highly corrosive environments
Tsuyoshi Enomoto, Toshio Sugisaki, Kouichi Suga and Tatsuro Sueyoshi 341

New development of low relaxation epoxy coated strand
Tiger Kido, Takayuki Yamagiwa, Toshihiko Niki and Yoshihiko Touda 351

Session seven: Case histories of satisfactory performance, shortcomings and failures in service

Assessment of cement grout cover as sole protection of permanent tensioned steel tendons
Stuart Littlejohn 357

The failure of anchorage B160 at Devonport Royal Dockyard, Plymouth; reasons for failure and lessons learnt
Michael Whitworth and Steve Parrish 368

Service performance of multiple anchors in difficult ground conditions for both permanent anchors and temporary removable anchor usage
Tony Barley and Mark Edwards 380

Investigation of the failure of permanent anchorages adjacent to a tidal surge barrier
Tony Barley, Devon Mothersille and Peter Chamley 387

Reliability-based failure probability and resistance factor for flush-drilled anchors in Taipei Basin
Hung-Jiun Liao, Jianye Ching and Chia-Wei Sue 397

Field surveys of anchored slopes after 2004 Niigataken Chuetsu earthquake
Jiro Takemura, Hiroaki Kubo and Junichi Yamazaki 407

Rehabilitation of a failing anchored retaining wall
Thomas Richards and Daniel Thome 417

Predicted and observed performance of an anchored retaining wall in granite
Abid Adekunte, Douglas Madsen, Neil Margetson and Liam Quinn 427

The design, installation & monitoring of high capacity antiflotation bar anchors to restrain deep basements in Dublin
Jim Martin, Philip Daynes, Cathal McDonnell and Martin Pedley 438

Session eight: Recommendations and standards for inspection, monitoring and repair

A new manual for ground anchor inspection, integrity investigation and remedial measures
Hiroaki Miyatake, Takeshi Oshita, Hiroaki Kubo and Mitsunari Takeyama 451

Structural integrity and durability of ground anchors – general programme of research in Japan
Junichi Yamazaki, Tatsuro Sueyoshi and Nobuyuki Urakawa 461

Engineering assessment of the grease-filled annulus of sheathed prestressing strand for use in permanent ground anchorages
Tony Barley and Devon Mothersille 471

Rock anchors for North America dams: the development of the national recommendations (1974–2004)
Donald Bruce and John Wolfhope 481

Author index 493

Discussions

Session one: Inspection procedures and physical conditions recorded in service 495

Session two: Service behaviour and monitoring procedure and performance in service 511

Session three: North American Dam rehabilitation 518

Session four: Rock bolts in tunnels and mines 525

Session five: Non-destructive integrity testing 530

Session six: Corrosion monitoring and corrosion protection systems 544

Session seven: Case histories of satisfactory performance, shortcomings and failures in service 571

Session eight: Recommendations and standards for inspection, monitoring and repair 587

KEYNOTE PRESENTATION

MAINTENANCE TESTING AND SERVICE BEHAVIOUR MONITORING OF PERMANENT GROUND ANCHORAGES

Stuart Littlejohn, University of Bradford, UK
Devon Mothersille, Geoserve Global Ltd, UK

Abstract
Maintenance and monitoring of permanent anchorages are important, particularly for those installed over 30 years ago that may have been designed with inferior or inadequate corrosion protection. Following a review of tendon corrosion failures and observed anchor head conditions after 28 to 33 years of service in aggressive environments, guidelines on good practice for inspection, testing, maintenance and service behaviour monitoring are presented. Topics include designer responsibility, anchor head inspection, component and material testing, load monitoring, extent and frequency of inspections, and associated records. Current national, European and international recommendations are summarised. Case histories of permanent anchorage corrosion are reported, where the conditions of the ground or groundwater and anchorage components are quantified. These cases illustrate the value of detailed metallurgical examination and testing to investigate the nature, intensity and extent of the corrosion, in order to determine the risk of failure and the most appropriate remedial measures.

Introduction
The purpose of this paper is to provide guidelines on good practice in maintenance testing and service behaviour monitoring of individual ground anchorages.

As permanent anchorages in service become older, the subject is of growing importance, particularly for anchorages installed over 30 years ago that may have been designed with corrosion protection considered inferior or inadequate by today's standards, e.g. reliance on cement grout cover alone, where the grout has been injected remotely under pressure or by gravity displacement to encase the tendon in situ. Even more recently installed anchorages are not immune due to the employment of inappropriate protective materials and occasional shortcomings in the quality of workmanship, combined with an absence of rigorous QA/QC procedures in the field.

Routine programmes of inspection and monitoring, where satisfactory condition and service performance are confirmed, can extend the service life of anchored structures, such as post-tensioned dams, dry docks, bridges, sports stadiums and multi-storey buildings, stabilised slopes, earth-retaining walls, river and seawall coastal defences, reinforced tunnels and underground caverns, that represent key elements of a country's infrastructure.

Alternatively, in cases where an inspection highlights unacceptable tendon exposure to corrosion, or service behaviour monitoring confirms tendon over-stressing, the results can provide an early warning of the need to carry out precautionary or remedial measures, in order to safeguard the integrity and performance of the anchored structure.

In spite of these benefits, insufficient attention is paid to routine maintenance inspections and service behaviour monitoring in current practice, and the potential consequences should not be ignored.

Water levels reached 150-year highs during August 2002 with associated flooding across central and eastern Europe, so the importance of anchored river and seawalls cannot be overemphasised. Winter storms are now judged by the insurance industry to represent the largest potential event loss in Europe (Swiss Re, 2003). The Association of British Insurers has identified that in London alone, 500,000 houses, 200 schools, 16 hospitals and eight power stations are located in floodplains (ABI, 2006).

Elsewhere in the UK there are numerous examples of waterside infrastructure incorporating anchored quay and wall structures. These include ports and harbours along the east coast of Scotland and England, the Thames Barrier, structures along the Thames Estuary, ports and harbours along the south coast of England, the Bristol Channel, the Welsh coast and west Scotland.

Elsewhere in the world, the highest potential losses are associated with earthquakes and typhoons. Given torrential rainfall, anchored slopes are vital in resisting landslides and rock avalanches. In Japan, for example, where the ground anchorage market is $650 million per annum, 64% of an average of 2277km of anchorages installed each year is devoted to landslide prevention and slope stabilisation (Kubo, 2006).

Failures of anchorages in service

In the 1980s, the Fédération Internationale de la Précontrainte (FIP) carried out a world-wide search of anchorage failures and published a study of 35 case histories of failure, primarily by tendon corrosion, relating to ground anchorages installed over the period 1934 to 1980. The detailed findings and records are published in FIP (1986) and Littlejohn (1987).

The findings may be summarised as follows.

- Corrosion is localised and appears irrespective of type of tendon in that 9 incidents involved bar, 19 involved wire and 8 involved strand.
- The period of service before failure ranged from a few weeks to many years for each tendon type; short term failures after a few weeks of service were due to stress corrosion cracking or hydrogen embrittlement.
- In terms of failure location, 19 incidents occurred at, or within 1m of the anchor head, 21 incidents in the free length, and 2 incidents in the fixed length.

Both fixed anchor problems were caused by inadequate cement grouting of the multi-strand tendon bond length exposing the tendon to corrosion. In Switzerland, where the groundwater contained chlorides and sulphides, tendon failures led to collapse in 1981 of a 60m span pipeline bridge over the River Thur at Dietfurt after 5 years (Figure 1). In South Africa, unacceptable pitting up to 1mm in depth was discovered after 2 years on a poorly grouted tendon (cement grout cover range = 0-6mm) founded in fill and led to rejection of all anchorages restraining a cantilevered grandstand in 1974.

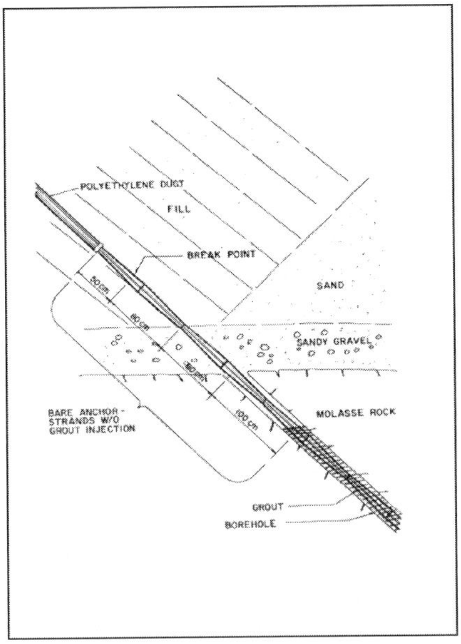

Figure 2. Absence of protective sheathing immediately below stressing head

Figure 1. Section of fixed anchor zone showing part exposure of tendon

Failures in the free length occurred due to factors such as tendon over-stressing, inadequate or no cement grout cover in an aggressive environment, and inappropriate choice of protective material, e.g. chemical grout containing nitrate ions, and hygroscopic mastic.

Failures at or immediately under the anchor head were due to various causes ranging from absence of protection to inadequate cover due to incomplete filling initially or slumping of the protective filler during service (Figure 2).

In one example in Hong Kong, delays on site between completion of stressing and encasement of the anchor head in concrete exposed the tendon to a humid and slightly saline atmosphere and caused up to 2.7% and 12% loss of wire diameter after delays of up to 8 and 36 months, respectively. Brian-Boyes and Howells (1984) have also recorded problems with anchorages specific to Hong Kong and installed in the 1970s when site supervision and quality controls on site were generally poor.

Parry-Davies and Knottenbelt (1997) studied anchorages installed in the 1970s and 1980s in South Africa when different techniques were employed reflecting the variable knowledge of contractors. Based on observed field performance, aspects of ground anchorage corrosion protection and installation requiring particular attention to avoid long-term service behaviour problems are highlighted, and include (i) assessment of environmental conditions during service, (ii) the need for meticulous care in applying protection to both the inner and outer anchor head, and (iii) the use of skilled operatives to carry out the works coupled with close supervision.

Recent examples of anchorage problems in service leading to progressive failure of sea and river walls, include the Langkawi seawall in Thailand, the Kiunga loading wharf in Papua New Guinea, the 21-year old sheet pile quay wall on the River Thames in London and the 19-year old Chef de Baie Quay in La Rochelle harbour, France.

On the River Thames, progressive failure of the quay occurred in 1990 over a 142m length (Figure 3) due to over-stressing and corrosion of the multi-strand tendon at the anchor head caused by inadequate under head grouting and failure of the epoxy resin coating on the outer head (Barley, 1997). At La Rochelle harbour, tie bar over-stressing during backfilling behind the front wall led to horizontal displacements up to 400mm over a 100m length of wall in 2001 (Plaxis, 2005). Regular visual inspections and service load monitoring would have highlighted corrosion and over-stressing at an early stage and permitted timely remedial measures to avoid progressive failure.

Figure 3. Anchored quay wall failure on River Thames (after Barley, 1997)

Most recently, failed structural steel tie bars up to 75mm in diameter have been found in 2005 at the 490m long East Arm Port Stage 1 Darwin Wharf in Australia, where the wharf has been in service since 1999. In essence, a brittle heat affected zone from welding combined with severe bending of the tie bar created a crack and led to a brittle "single event" fracture of the bar.

Subsequent to discovery of three bar failures, non-destructive ultrasonic testing located a further two failures. On-site acceptance testing of the tie bars (in accordance with the recommendations of BS8002 and BS8081) after backfilling behind the seawall would have permitted early recognition of the nature and potential extent of the problem.

General considerations for maintenance testing and service behaviour monitoring

Maintenance testing involves inspection of the condition of materials and components of the anchorage, and where appropriate, testing to determine the nature and severity of the condition, e.g. metallurgical and tension tests to assess the type and significance of a corroded tendon unit. The prime purpose of maintenance testing is to establish if the anchorage has suffered corrosion or mechanical damage, and whether the conditions recorded are within acceptable limits.

Service behaviour monitoring focuses on the performance of

(i) the anchored structure with respect to overall movement and local deformation, and

(ii) individual anchorages with particular reference to residual load and anchor head displacement.

The prime purpose of service behaviour monitoring is to establish if the anchored structure has recorded any trend in terms of overall movements and local deformations, and whether or not individual anchorages have maintained their design load in compliance with acceptance criteria.

The essential elements relating to routine maintenance testing and service behaviour monitoring include:

- Visual survey of physical condition of anchored structure
- Measurement of overall movement and local deformation of anchored structure
- Visual survey of physical condition of all protected individual anchorages
- Inspection of the anchor head above and below bearing plate of selected anchorages
- Measurement of residual load (and preferably load-displacement behaviour) of selected anchorages

The extent and frequency of maintenance testing and service behaviour monitoring depends on the type and scale of the anchored structure under consideration. Critical structures where excessive deformations or failure might lead to loss of life and/or wide-scale damage to property such as adjacent buildings and service networks will demand the greatest attention during service. In addition, the effect of 1, 2 and 3 adjacent anchorages failing should be quantified with particular reference to the risk and consequences for service behaviour of the anchored structure and its overall stability.

All operations involving maintenance testing and service behaviour monitoring should be in accordance with a specification created for the project; the works should be carried out by a competent specialist contractor and be supervised by an engineer experienced in this type of investigation.

Responsibility of the designer

For any permanent anchored structure where the consequence of anchorage failures is serious, the designer should indicate clearly at the outset the value and necessity of maintenance testing and service behaviour monitoring.

The designer should provide a method statement or performance specification for such works. In the specification, the designer should prescribe the inspection/monitoring programme in terms of anchorage number, location, frequency, acceptance criteria specific to both the condition and service behaviour of the anchored structure and individual anchorages, and reporting procedures for both maintenance testing and service behaviour monitoring.

For residual load lift-off checks, the designer should ensure an adequate protruding length of tendon at time of installation to permit subsequent stressing. Alternatively, restressable or detensionable heads can be specified, augmented where appropriate by load cells. Where the importance of the structure warrants it, consideration may be given to the provision of extra anchor head locations at the design stage, e.g. base slab of a dry dock.

In maintenance testing, the designer should determine the type and level of damage and corrosion that can be tolerated and what is unacceptable leading to de-rating of the anchorage, e.g. tendon unit kinking (Figure 4), or anchorage rejection e.g. environmentally assisted stress corrosion cracking of the tendon unit(s). In service behaviour monitoring, the designer should determine the maximum loss or gain of load that can be tolerated in any anchorage during its service life, taking into account the design of the individual anchorages and anchored structure.

Figure 4. Protruding strands above anchor head with one kinked strand that may be judged redundant

For all programmes related to maintenance testing and service behaviour monitoring, it is highly desirable to obtain detailed records covering the original anchorage design, construction and testing, in order to appreciate, for example, the corrosion protection provided and the original design assumptions. Preferably, records should include:

- site investigation data
- records of any pre-construction anchorage tests on site
- anchorage number and location
- anchorage design and geometry, including design free length
- design working load and load safety factor(s)
- tendon type, elastic modulus and characteristic tensile strength
- tendon corrosion protection
- grout design, including material constituents
- anchorage construction details, e.g. drilling, water testing, tendon installation, grout injection
- specifications for all anchorage materials and components
- initial mode of stressing and procedure
- initial on-site suitability and/or on-site acceptance test results
- lock-off load, i.e. initial residual load in service
- anchor head type and its corrosion protection after stressing
- all relevant QA/QC records.

Subsequent results of on-site inspections, surveys and individual anchorage loads, together with the type, accuracy and calibration data of monitoring equipment should be recorded using standard approved forms.

Maintenance testing and service behaviour monitoring records, in accordance with the designer's specification, should be copied in a timely manner to persons with the knowledge and authority to instruct nil, precautionary, supplementary or remedial actions, as appropriate.

Inspection and maintenance testing of individual anchorages

Where individual anchor heads are inspected, the condition of all components and materials should be recorded. If feasible, a sample number of anchor heads should be exposed to permit examination of the tendon, particularly in the region of the inner head immediately below the bearing plate, e.g. via fibrescope camera. This is particularly important where anchorage load monitoring is not possible.

All tendon units within the tensioned free length exhibiting localised corrosion, e.g. stress corrosion cracking/hydrogen embrittlement and significant pitting, should be rejected. General corrosion should only be accepted where measurements of tendon unit cross sectional dimensions confirm losses no greater than the allowable manufacturing defects of the tendon unit, e.g. longitudinal cracks or depressions <4% of the nominal diameter of the tendon component or 1mm maximum depth of depression, whichever is the lesser (BS DD82:1982, FIP:1986, BS8081:1989).

Where doubt exists, selected anchorages can be partially exhumed to provide access to the upper free length and the area immediately beneath the anchor head. In such circumstances, liaison with the designer is recommended before deciding on the appropriate anchorages for intrusive testing. In this regard, it is noteworthy that on important projects a few redundant anchorages may be considered at the design stage to facilitate this type of extensive test work during service.

Performance tests on the gripping efficiency of barrel and wedge components for multi-unit tendons can be carried out, together with metallurgical, tensile and bend tests, on representative tendon unit samples to determine if there is any loss of strength. The results can determine whether the tendon unit is accepted, de-rated or rejected.

Where wedges gripping strands or wires within the stressing head have experienced significant corrosion, particularly on the serrated faces (Figure 5), individual wedge units should be replaced (assuming available access) by jacking over a chair to increase the strand load and thereby release the existing wedges. In this regard, new wedge/tapered hole interfaces should be oiled or greased, to lubricate the gripping mechanism and provide a seal to prevent corrosion and ingress of deleterious materials such as cementitious products that can reduce grip efficiency.

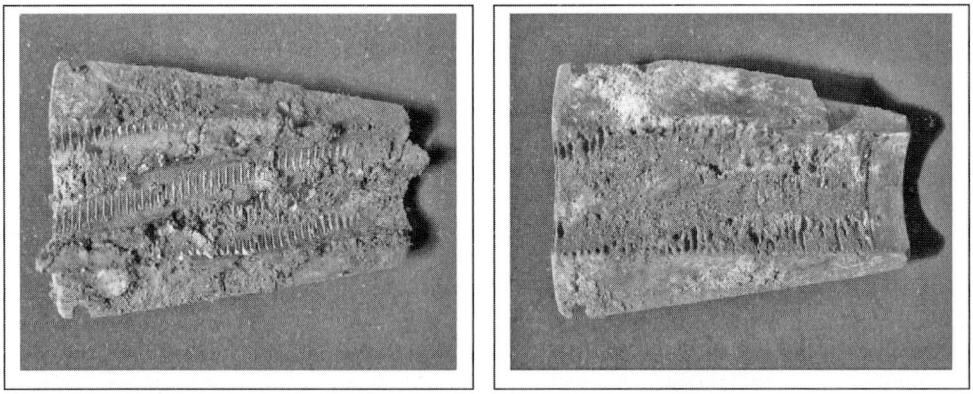

Figure 5. Serrations of wedges infilled with corrosion products that inhibit gripping efficiency after 33 years of service in a marine environment in the UK

Severe corrosion of the serrated face of wedges can lead to strand slippage or loss in service (Figure 6). Where the stressing head is encased in cement grout, strand slippage has been recorded within 12 months of anchorage installation (Figure 7). Severed tendons at the anchor head can often be reinstated using replacement tendon unit lengths, barrels, wedges, couplers and corrosion protection.

A visual survey of protected individual anchor heads should include assessment of corrosion, distortion or damage. Where protective caps have been damaged mechanically or by corrosion (Figure 8), they should be removed to permit inspection of the outer head and protruding tendon, in order to establish if the anchor head has been affected.

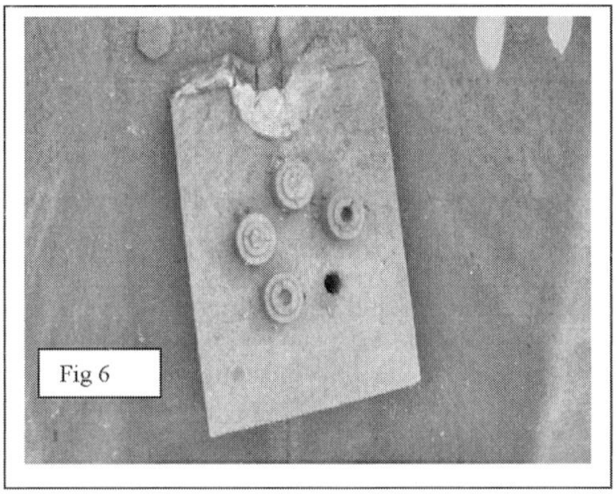

Figure 6. Strand loss and slippage at exposed anchor head after service of 33 years in a marine environment in the UK

Figure 7. Removal of cement grout cover above the stressing head to permit measurement of protruding strand lengths for slippage and to facilitate restressing, if appropriate

Figure 8. Severe corrosion up to 10mm deep recorded on protective steel cap after 28 years in a marine environment in the UK

For steel outer caps, DIN50 929 Part 3 (1985) provides guideline values for estimating mean corrosion rates depending on the corrosivity of the environment. For example, unalloyed and low-alloy steel caps subjected to marine conditions and particularly tidal effects, can suffer removal rates as high as 0.1mm per annum and a maximum penetration of 0.5mm per annum.

Given this evidence, consideration should be given to the use of fibre glass protective caps fabricated to recognised standards, e.g. ISO 12215-1(2000). Such anchor head caps were introduced to the UK anchorage industry in the early 1990s to improve protection against corrosion.

Grease checks should be carried out every 5 years on a selection of anchorages. Where corrosion inhibiting grease is emulsified by the presence of water or exhibits signs of ageing due to oxidation, the grease should be replaced with fresh material of the same specification. Alternatively, grease sampling and testing of properties should be carried out to check for compliance with the specified requirements, e.g. grease properties in British Standard BS8081 (1989) or European Standard EN1537 (2000). This approach is recommended when grease is located in zones that cannot easily be replaced, without the aid of steam jetting.

Where water ingress is encountered, sampling and testing for corrosivity are required, e.g. pH, chloride, sulphide and sulphate content, and the source/entry point and rate of ingress should be recorded. If the entry point for ingress is above ground level and outwith the anchor head, the point of ingress should be sealed or isolated immediately. The condition of the tendon and anchor head will determine if the anchorage should be accepted, de-rated or rejected. Where the ingress comes from within the anchorage, e.g. sheathing/grease or grease/tendon interface, the anchorage should be rejected unless the rate of ingress and corrosivity of the water are judged by corrosion specialists to be passivating, or sufficiently benign to accommodate the design life of the structure.

When the inspection of the outer head is complete, a new base seal should be fitted before replacing the outer protective cap, and any contaminated, weathered or lost grease should be replaced with fresh material.

Inspection records for each identified anchorage should include the following.

- Observations before removal of protective cap with particular regard to condition of adjacent ground/structure, anchor pad, protective cap, base seal and bearing plate.

- Observations after removal of protective cap with respect to condition of inside of cap, base seal, grease, anchor head, locking-off components, e.g. wedges, bearing plate and protruding tendon.

- Close-up photographs of anchor head assembly before and after removal of protective cap and grease.

- The results of any measurements and tests.

Monitoring service behaviour of anchored structure

The overall performance of the anchored structure should be monitored as a whole involving regular surveys of selected points on the anchored structure to check for general movement and local deformations. The information should be plotted and analysed on each occasion such that cumulative plots with time are produced and the results compared, in order to highlight trends and patterns.

It is important to ascertain the movement of the anchored structure in order to help determine the cause of any change in the anchorage load. Permissible limits for total movement and local deformation should be established that may be related to the original design assumptions. Where ground water levels are a critical factor in design, levels should be monitored by means of piezometers. Data loggers can be used at reasonable cost to monitor continuously or at a suitable frequency. Regular sampling for chemical analysis may also be prudent, particularly on sites where the ground water may be aggressive.

General considerations for monitoring service behaviour of individual anchorages

Service load monitoring should ideally be carried out via readings from load cells, installed at the time of construction, or lift-off measurements using a multi-unit jack (Figure 9) or mono-unit jack (Figure 10). The use of load cells is preferred since there is no disturbance to the anchorage and its corrosion protection system, particularly at the anchor head (Figure11). The use of load cells should also be considered where difficult access would present problems in attaching stressing equipment for lift-off tests. Where specified, load cells should be robust hydraulic electronic pressure transducer types, with capacity greater than the nominal ultimate capacity of the tendon and calibrated to cope with changes in environmental conditions such as large temperature variations.

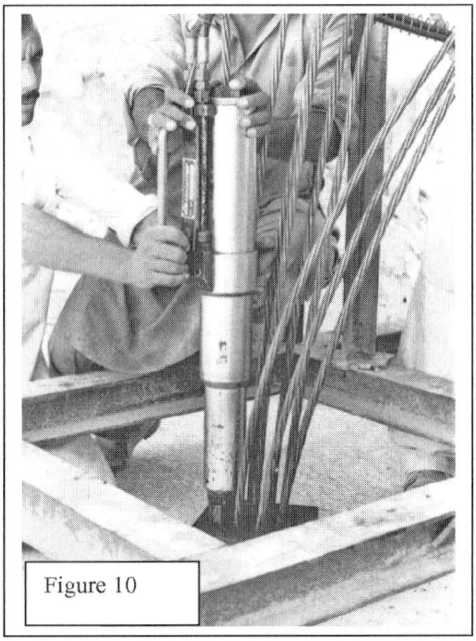

Figure 9 Multi-strand stressing with measurement of displacement of tendon and settlement of bearing pad

Figure 10. Mono-strand stressing

Figure 11. Load monitoring using annular load cell at anchor head inside protective cap

Service behaviour monitoring using check-lifting

Where monitoring takes the form of measuring the residual lift-off load on a tendon by applying a stressing jack to the anchor head, care must be taken to ensure minimal disturbance to the anchor tendon and that the act of monitoring does not jeopardise the safety of the anchorage, either by forming weak links in the corrosion protection, or by overloading any part of the system. In this regard, the lift-off load is the minimum load monitored during a re-stressing operation that permits a locking nut to turn on a bar tendon or provides a nominal clearance or lift in the case of a multi-wire or multi-strand tendon.

The distance the anchor head is raised is normally in the range 0.5mm to 1mm for minimal disturbance, and should be fixed using a feeler or dial gauge to provide a consistent procedure. The method of measurement should ensure that all sides of the stressing head are clear of the load distribution bearing plate. For load lift-off checks without a load cell, an absolute accuracy less than 5% is unlikely unless at least three lift-off observations confirm a consistent load, in which case the accuracy of the measurement is related to the accuracy of the calibrated jack and associated pressure gauge of 1-2%. It is recommended that gauges incorporate a large diameter face of at least 150mm and where possible some form of dual electronic readout to facilitate accurate measurement.

Initial lock-off loads for ground anchorages range typically from 1.0 to 1.15 x design working load (Tw), although there is a case for locking off at higher loads, e.g. 1.2Tw, where this can be justified by the measurement of a high cohesive/frictional resistance within long free tendon lengths during the initial cyclic loading phase of an on-site suitability or on-site acceptance test. This case can be applicable to high post-tensioned dams.

In certain circumstances, ground anchorages are locked off at only a nominal load, e.g. 10%Tw, where the permanent load in service is only developed as backfilling is completed behind the anchored retaining wall. It should be noted that it may be necessary to wait a specific time period to permit the full development of passive resistance behind the restraining element such as a sheet pile or concrete "deadman". On completion of such anchored structures, it is essential to carry out at least load lift-off checks, and preferably on-site acceptance tests, incorporating proof load and load-displacement monitoring, in order to ensure that the anchorages comply with design assumptions.

Subsequently, for anchorages in service, no attempt is made to alter the loads being monitored, unless significant load changes are recorded. The maximum load loss or gain that can be tolerated in any anchorage during service should be indicated, taking into account the design of the works.

Load changes (e.g. 3-5%Tw) within a short time frame, e.g. between monitoring cycles of 6-12 months, should be double-checked for accuracy and if the change confirms a trend, its probable cause should be determined. Long-term variations up to ±10%Tw do not cause concern in routine practice. For variations greater than 10%Tw, unless the reasons are known, e.g. structural movement, the causes and consequences should be analysed.

Supplementary lift-off checks, coupled with inspection of anchor heads, should be carried out when significant changes in load behaviour are recorded via load cells. In the event of a significant impact on the anchored structure or a seismic event, all anchorages fitted with load cells should be monitored daily for at least 10 days to confirm no significant change in load, or until the loads have sensibly stabilised.

Replacement criteria during service, i.e. permissible load loss or gain, or limiting range of movements, should be indicated at the design stage. By way of example, remedial action, which may involve partial de-stressing or additional anchorages, is recommended where load gains approach 40%Tw for anchorages, where the original proof load factor and load safety factor are 1.5Tw and 2.0Tw for the tendon, respectively. If the original proof load factor and load safety factor are 1.25Tw and 1.6Tw, respectively, remedial action, which may involve partial de-stressing or additional anchorages, is recommended where load gains approach 20%Tw. If de-stressing is anticipated then shims should be placed under the stressing head at the time of anchorage installation.

Lift-off checks should be carried out in accordance with published recommendations. In this regard, parameters such as tendon relaxation, consolidation of the ground between structure and fixed anchor zone, movement of the anchored structure, and seasonal temperature changes should be taken into account. In addition, due allowance should be made for external conditions, where appropriate, such as tidal levels, surcharges and local excavations that might affect stress conditions in the ground.

Service behaviour monitoring using on-site acceptance testing

Where high load losses are judged to be due to consolidation of the ground, then a remedial stressing programme is considered advisable, whereby the initial residual load, e.g. 1.1Tw is regained, and on-site acceptance test service load monitoring repeated (i.e. residual load v time or displacement v time at residual load, in accordance with recommended national procedures) to determine if the original design and on-site acceptance criteria can be restored.

Where the requirement during service monitoring is to determine or reconfirm a measured margin of safety (via a proof load) and the efficiency of load transfer (via apparent free length check), cyclic loading (residual load → proof load → residual load) with tendon displacement measurements should be carried out. Proof loading above 1.25Tw is not often recommended in Europe, bearing in mind the minimum proof loading requirements of 1.25Tw in EN 1537 (2000).

Cyclic loading over the range residual load → proof load → residual load is preferred to de-tensioning (where this is feasible) in the case of multi-strand or multi-wire tendons, since de-tensioning permits the indented tendon units (indentations caused initially by serrations of the gripping wedges located within the permanent stressing head) to enter the free tendon length that is subsequently tensioned.

At least two complete load-unload cycles should be carried out, in order to confirm consistent reproducible behaviour that can be relied upon. The number of cycles may be increased to three where the observed behaviour during the first cycle is not reproduced, e.g. where the first application of additional load creates some fixed anchor displacement or has to overcome friction within the stressing head or free tendon length.

Residual load checks will not detect the onset of corrosion in tendons. However, they will reveal failure of the tendons, where this is not otherwise obvious, and they may also indicate failures of individual strands or wires. Where bond failure is suspected in multi-unit tendons, both multi-unit jack and mono-unit jack stressing operations (Figures 9 and 10) should be carried out to permit assessment of the location of the bond failure, i.e. ground/grout or grout/tendon interface.

Where a defective or failed anchorage is encountered, consideration may be given to the possibility of gleaning safely additional information, such as tendon displacement over a nominal load increment and decrement cycle, in order to check the elastic and permanent displacement components of the anchorage. In this way it may be possible to justify a reduced safe load in service. Where an anchorage fails at a fixed anchor interface, a first estimate of the new de-rated working load may generally be taken as the maximum load at failure divided by a specified load safety factor, e.g. 2.0 for permanent anchorages.

Thereafter, consideration should be given (wherever feasible) to a standard on-site acceptance test using an appropriate proof load factor, e.g. 1.25 x de-rated working load, in order to determine whether to accept or reject the anchorage at its de-rated value. When a de-rated anchorage is accepted, the residual load should be checked thereafter at each routine monitoring period for the project.

For closely-spaced anchorages, e.g. 1 to 3m centres, simultaneous stressing operations on individual anchorages should only be carried out, where the anchorages are remote from each other and preferably restrain a separate element of the anchored structure, otherwise there is the potential for one stressing operation to influence the results of another during the monitoring programme.

Anchor head design to facilitate service behaviour monitoring

Where specified, individual anchor heads should be designed to facilitate accurate service behaviour monitoring. Appropriately designed protective caps should be provided on all standard, restressable and detensionable anchor heads to facilitate service load monitoring and increase or, when required, decrease the tendon load at any time after installation.

Standard stressing heads that may require controlled detensioning need protruding strand lengths equivalent to the recorded strand extension plus approximately 300mm. Extended protective caps are required in these circumstances.

On occasions where inadequate protruding strands lengths are encountered beyond the stressing head, (Figure 12) consideration can be given to the use of a wedge-secured circular externally threaded stressing head block (Figure 13). This allows the fitting of stressing equipment comprising a hub puller, stressing jack and stressing stool (Figure14). Load checks can subsequently be carried out by jacking the whole tendon. If the requirement is to recover losses in load, then the stressing head can be lifted and purpose-made shims fitted to maintain the increased anchorage load.

If at least 80mm of strand has been left beyond the stressing head, beneath the protective cap, then load checks may be carried out on individual strands using a coupler, stressing stool and calibrated monojack.

Figure 12. Standard stressing head with short protruding strands

Figure13. Externally threaded stressing head permitting the attachment of threaded hub puller

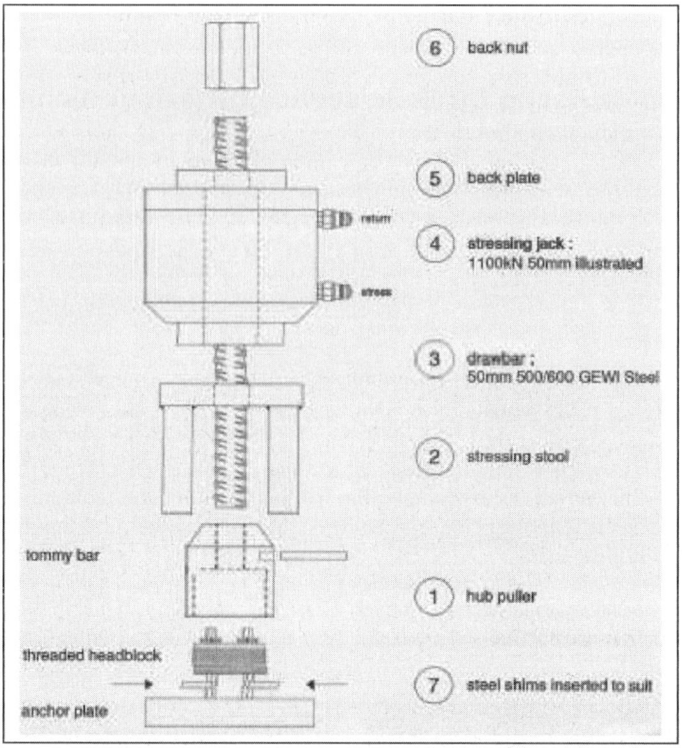

Figure 14. Assembly sequence for stressing up to seven strand tendons where the protruding strand length is not less than 55mm (courtesy of Dywidag-Systems International)

Records of anchorage service behaviour

The maintenance of accurate service behaviour monitoring records represents good practice and should include the following.

- Anchorage number, type, characteristic tensile strength of tendon, design free tendon length, elastic modulus of tendon, maximum proof load, design working load and initial residual load immediately after lock-off.
- Identification numbers and up-to-date calibration tables/graphs of jacks, gauges, pumps and load cells.
- Residual load in service (and lift-off movement, e.g. 1mm, if stressing employed).
- Total change in load as a percentage increase or decrease of initial residual load.

Detailed records using standard approved forms are necessary when assessing the condition and service performance of individual anchorages in a consistent manner.

Extent and frequency of maintenance inspections

For important anchored structures, where the environment is aggressive or failure of the anchored structure would risk life and cause significant damage or disruption, visual inspections of the anchored structure and the exterior of the individual anchor head caps (wherever practicable) should be carried out every year. Where cap damage is observed or the cap is lost (Figure 15), the anchor head and protruding tendon units should be exposed, examined and condition recorded. In such circumstances, all exposed anchor head units should be cleaned, the bearing plate provided with a protective coating, e.g. bitumen, then a grease filled repaired or new cap should be fitted with a new base seal and the likely cause of the cap damage or loss reported, in order to avoid a recurrence.

Figure 15. Sequence of remedial measures for corroded anchor head after loss of protective cover in service

Extent and frequency of monitoring anchorage service behaviour

The recommended proportion of permanent anchorages to be monitored for service behaviour ranges in practice from 5 to 15%, the vast majority lying within the range 5-10% for projects involving more than 100 anchorages.

The frequency of service behaviour monitoring over the long term ranges generally from 1 to 5 years. In some cases, monitoring may be required to ascertain ground or structural movements in the early life of the anchored structure and then to detect failures (potentially due to corrosion) during its later life. In these circumstances, tests should commence at short intervals of 3 to 6 months, with later tests at longer intervals depending on results, becoming gradually less frequent until they are carried out at not less than 5-year intervals for the life of the anchored structure.

For important anchored structures, where failure would risk life and cause significant damage or disruption, current good practice suggests that 10% of the anchorages are tested each year. Where anchor heads are detensionable or restressable, there is a growing trend to recommend monitoring of anchorages in important structures, e.g. nuclear-licensed naval dockyards, within a set period of time (e.g. 10 years) or at least within the design life of the anchored structure.

Where a significant number of representative anchorages (e.g. 5-15%) are fitted with load cells, more regular monitoring (e.g. every six months) is recommended and even shorter intervals of one month, if remote data logging is installed. The more frequent monitoring permits assessment of the effect of seasonal variations such as temperature, particularly over the first 2 to 3 years of service.

National recommendations

For readers interested in long-term monitoring and observed service behaviour of rock anchorages installed in the 1960s and 1970s, reference can be made to Littlejohn and Bruce (1976). More recent individual case histories of ground anchorage behaviour in service can be found in the proceedings of the international conference on ground anchorages and anchored structures in London (ICE, 1997).

National recommendations that are in use today are summarised below for a selection of countries where topics such as maintenance inspections and service behaviour monitoring are addressed. Whilst the recommended proportion of anchorages and the frequency of inspection and monitoring vary from country to country, the basic objective on any anchorage project is to investigate a representative number of each type of anchorage employed, such that it is possible to assess the probable condition and behaviour of all anchorages at the site.

Australia

In New South Wales, all anchorages must have provision to enable the anchorage load to be monitored, detensioned and restressed at any time. In addition, production anchorages (at least 1 in 50) must be fitted with a load cell, plus any critical anchorages detailed in the design, or as agreed with the client's representative (RTA, 1997).

Up to the end of the maintenance period (not specified in the RTA standard), all anchorages subjected to proving tests and on-site suitability tests, those fitted with a load cell, and a further 10% of the remaining anchorages must be monitored by the Contractor. Where concerns exists about the reliability of load cell readings, greater reliance has been placed on the use of confirmatory lift-off testing. RTA, 2007 (Draft in progress) states '*Measure by Lift-off Test the residual load where readings from a load cell are not reliable.*'

The monitoring frequency is 7 days, 14 days, 1 month, 3 months, 6 months and at 6-monthly intervals thereafter.

France

Since the issue of French Recommendations T.A.72 (Bureau Securitas, 1972), the extent of monitoring is typically 10% of anchorages installed (1-50 anchorages), 7% (51-500 anchorages) and 5% (> 500 anchorages).

The service behaviour of permanent anchorages is monitored for at least 10 years, where generally, anchorages are inspected every 3 months in year 1, every six months in year 2, and annually thereafter.

Hong Kong

Based on GEOSPEC1 (1989), the minimum number of permanent anchorages and frequency of monitoring by visual survey, anchor head inspections and residual load measurements during service are illustrated in Table 1. A visual survey includes inspection of all visible components of individual anchorages, the anchored structure and any associated drainage system, in order to identify any areas of actual or potential distress, e.g. cracks in bearing pads or rust stains from the anchor head.

Inspections of selected anchorages comprise a survey of anchor pad, adjacent anchored structure, protective cap and the visible components of the outer anchor head after removal of the grease. Where grease is removed, it should be sampled and its properties tested for compliance with the requirements of GEOSPEC 1. Grease properties are similar to those required in BS8081:1989.

When the protective cap is removed, the residual load should also be measured to verify that it remains acceptable.

Frequency of visual survey	**Monitoring of individual anchorages**		
	No. of anchorages[1]	Frequency of anchorage inspection and residual load measurement	Extent and frequency of special grease checks[2]
High risk category meaning risk to life where failures affect occupied buildings and economic risk relating to failures affecting urban trunk roads, essential services or excessive structural damage to buildings			
Weekly (up to end of maintenance period) and every 6 months thereafter	15% of first 50 anchorages 12% of second 50 anchorages 10% of additional anchorages	2 weeks, 1 month, 3 months, 6 months, 9 months, 1 year, 18 months, 2 years, and every year thereafter	Three anchorages 2 years, 5 years and every 5 years thereafter
Low risk category meaning risk to life where failures affect densely used open spaces and recreational facilities, roads with high traffic density and public waiting areas, and economic risk relating to failures affecting rural or primary distributor roads that are not sole accesses and temporary loss of essential services			
Fortnightly (up to end of maintenance period) and annually thereafter	10% of first 50 anchorages 7% of second 50 anchorages 5% of additional anchorages	2 weeks, 1 month, 3 months, 6 months, 9 months, 1 year, 18 months, 2 years, and every 2 years thereafter	Two anchorages 2 years, 5 years and every 5 years thereafter
Negligible risk category meaning risk to life where failures affect country parks, lightly used open recreational areas, roads with low traffic density and storage compounds for non-dangerous goods, and economic risk relating to failures affecting country parks, open air car parks, rural feeder and local distributor roads that are not sole accesses			
Monthly (up to end of maintenance period) and annually thereafter	7% of first 50 anchorages 3% of additional anchorages	2 weeks, 1 month, 3 months, 6 months, 9 months, 1 year, 18 months, 2 years, 5 years and every 5 years thereafter	One anchorage 2 years, 5 years and every 5 years thereafter
1. The same anchorages are to be monitored each time. The number may be increased if necessary to provide a representative sample. 2. Different anchorages are to be selected each time for special grease checks so that the original undisturbed grease can be sampled.			

Table 1. Guide to extent and frequency of monitoring of anchorages (after GEOSPEC 1, 1989)

Japan

Based on draft guidelines of the Japan Anchor Association (JAA, 2006), regular visual inspections should be carried out on 10% of anchorages (or more than 3, whichever is the greater number) within the first three years of completion. The recommended frequency of visual inspections is annual during the first three years and thereafter every 3-5 years. If the anchorages are important or old (e.g. not corrosion protected in accordance with the 1988 Standard that is reflected in the Guidebook for Design and Application of Ground Anchorages published in 1990 by the Japanese Geotechnical Society), annual visual inspections are maintained.

Given the results of such visual inspections, integrity investigations are carried out. The protective cap is removed to permit a closer examination of the stressing head on 20% of anchorages (or more than 5 anchorages, whichever is the greater number). The examination comprises visual checking of grease, protruding tendon, locking unit(s) and bearing plate and is carried out once within 5 years from completion and once every 2-3 years for extremely important ground anchorages.

For service behaviour monitoring, load lift-off tests are carried out on 10% of anchorages (or more than 3, whichever is the greater number). The frequency of lift-off tests is once within 5 years of completion, and for extremely important anchorages, every 2-3 years.

Detailed examination of the anchor head plinth, bearing plate, stressing head, wedges and tendon units for corrosion and damage, is carried out on 5% of anchorages (or more than 3, whichever is the greater number). The examination also checks the volume of grease and the sealing condition to prevent water penetration. In addition, a series of tests are carried out to confirm that the tensile strength of the tendon and the resistance to withdrawal of the fixed anchor exceed the design specification. The frequency of these examinations and tests is once within 5 years of completion, and for very important anchorages, every 2-3 years.

South Africa

In the case of permanent anchorages that are not fully grouted over their free length, the South African code SAICE (1989) recommends as a prudent measure load monitoring of 10% of anchorages over the first year of their effective life, followed by a further 10% at the end of another year, after which the degree of monitoring should be determined by an assessment of the risk from failure of any of the anchorages.

Should anchorage load gains be recorded at any stage of lift-off testing, monitoring should continue to ensure stabilisation of anchorage service load within a load increment of 10%Tw. In the event of a larger increase, the cause should be investigated and the overall anchored structure monitored.

UK

Where the purpose of monitoring is the detection of corrosion, at least 10% of anchorages should be monitored on projects with less than 100 anchorages. On larger projects at least a further 5% of the excess over 100 should be monitored (BS8081,1989).

Testing should be carried out at not greater than 6-month intervals for a period of 3 years and thereafter at longer intervals of not greater than 5 years throughout the service life of the anchored structure.

USA

Depending on the number of anchorages and the importance of the measurements, typically 3% to 10% of the anchorages, or more if desired, are monitored for service behaviour on any given project (PTI, 2004).

In general, monitoring commences at short intervals of 1-3 months and later at intervals not greater the 2 years, depending on the results. When an anchorage load gain is measured, monitoring should continue until the load stabilises. If the load in the anchorage approaches the original proof load, the anchorage is destressed to the design working load (Tw), additional anchorages are installed and the overall anchored structure is monitored until the overall system stabilises.

European Union (25 countries)

EN1537 (2000) contains no specific advice on maintenance testing or service behaviour monitoring other than to confirm that ground anchorages should be installed with a monitoring facility. Where a structure is sensitive to changes in load or ground movement, the number of anchorages to be monitored and the intervals between measurements must be specified. In addition, the corrosion protection of all accessible parts of the anchor heads must be inspected periodically and renewed, if necessary.

Fédération Internationale de la Précontrainte

International recommendations for type, duration, frequency and extent of monitoring the service behaviour of anchorages (FIP, 1996) are identical to those of the UK, i.e. BS8081 (1989).

General remarks

In spite of the published national, European and international recommendations that provide excellent guidance, there is little evidence that these recommendations are put into practice, to the obvious detriment of anchored structures in service.

Case histories of maintenance testing and service behaviour monitoring of anchorages

Detailed case histories, where anchorage failure or significant tendon corrosion has been observed, are relatively rare. The following projects are described to illustrate the type of observations made, the nature of testing and monitoring carried out, the lessons learned, and the precautionary or remedial measures implemented.

Grundfors Hydropower Station, Sweden

118 prestressed rock anchorages comprising 12 to 17m long, 26mm diameter prestressing bars were installed at 1m centres and post-tensioned to 300kN in 1955 to support overhead travelling crane beams and reinforce the abutments of the arch roof for a 15m wide and 60m long machine hall (Figure 16).

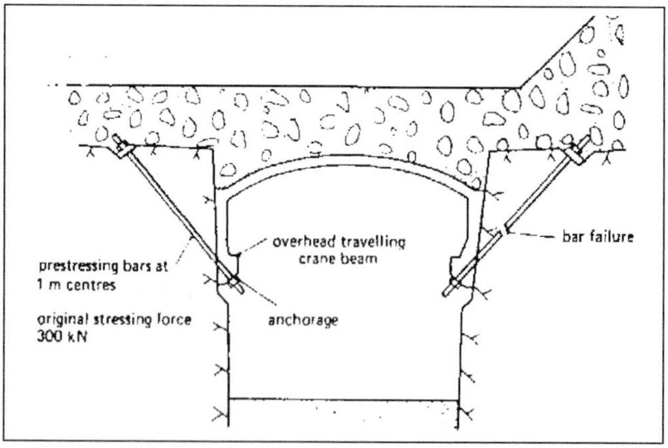

Figure 16. Section through machine hall at Grundfors hydropower station (after Jurell, 1985)

In 1981, one of the 26-year old bars failed unexpectedly when a 2.5m length of bar flew with great force through a light fitting and landed some 8m away on the floor. On inspection of this length, cement grout encasement was generally deficient and the entire length was superficially corroded and bent 24mm out of true (Jurell, 1985).

Rust-coloured water was also observed leaking from at least ten other anchorages in the machine hall, a fact that had been reported in the late 1950s, although no action appears to have been taken as a result. Examination of the construction records established that the bars were bent when delivered to site and subsequently straightened during post-tensioning. This attempt at straightening does not reflect prudent practice, albeit that the bar would exhibit stable behaviour where the steel had suffered strain hardening and would not yield. In such circumstances there would still be high local stresses at the locations of the bends where the bars would be most vulnerable in a corrosive environment.

Non-destructive boltmeter testing (where elastic compression and flexural waves are transmitted along the bar and the reflections measured) was carried out on the lower head of 96 bars. 10% of the tests gave very large echoes and 49% one or more minor echoes. Given this information, it was concluded that more than half the bars were unsatisfactorily grouted or damaged in some respect. The upper heads of 56 bars were then exposed and although coated originally with bitumen, there was hardly a trace left. Boltmeter readings were again taken and wherever possible, probe wires were inserted through slits in the anchor nuts. These tests confirmed the earlier picture of virtual non-existence of grout and often the presence of rusty water, although this water when tested for chloride and sulphur contents was not found to be specially aggressive (pH = 7-8).

One bar 17m long was completely removed and aside from the virtual absence of any trace of grout, the bar was covered by a moist superficial rust that had attacked the bar uniformly, leaving a close pattern of pitting. Overall, the reduction in cross-sectional area was found to be 6.8%. When this bar was tensile tested to failure, superficial corrosion cracks up to 1.3mm deep were observed.

Based on close examination of the fracture face and adjacent area of the broken bar, it was concluded that failure was triggered by a 5mm deep crack originating from the bottom of a corrosion pit 0.8mm deep (Figure 17). The primary crack showed clear signs of corrosion attack. Several minor superficial cracks of an intergranular nature were found parallel to the main crack implying that the primary crack occurred as a result of stress corrosion. Metallurgical investigations of the bar showed that its ductility at failure was low (possibly due to strain hardening) and that cracks deeper than a few millimetres created a serious risk of brittle failure.

The overall conclusions were that

- the anti-corrosion protection of the bars was deficient and unsatisfactory
- the prestressing bar material was sensitive to cracking
- the primary cause of failure was probably intergranular stress corrosion
- the factor of safety against repeated brittle failures was judged to be low

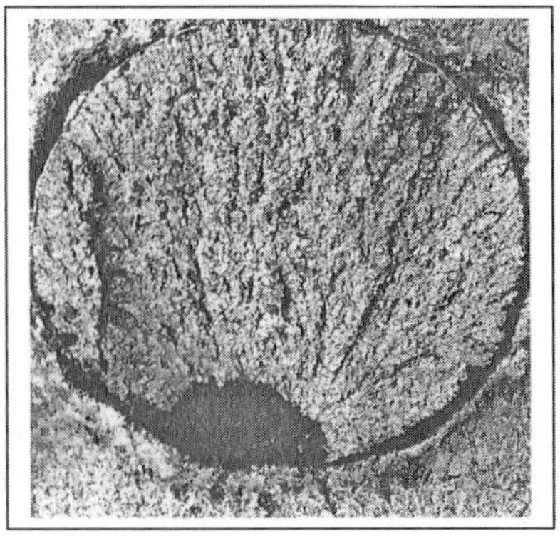

Figure 17. Magnified view of fractured face illustrating small depth of initial crack leading to brittle failure of bar (after Jurell, 1985)

On the basis of these results and the potential serious consequences of a further bar failure, it was decided to carry out extensive strengthening measures, comprising the installation of 232 deformed reinforcing steel bars through the crane beam and at least 4m into solid rock (Figure 18). Each bolt was grouted carefully with cement mortar. To prevent ejection of the existing prestressing bars in the event of any new failures, the original bars were de-tensioned and allowed to remain in position, in order to act as auxiliary reinforcing bars.

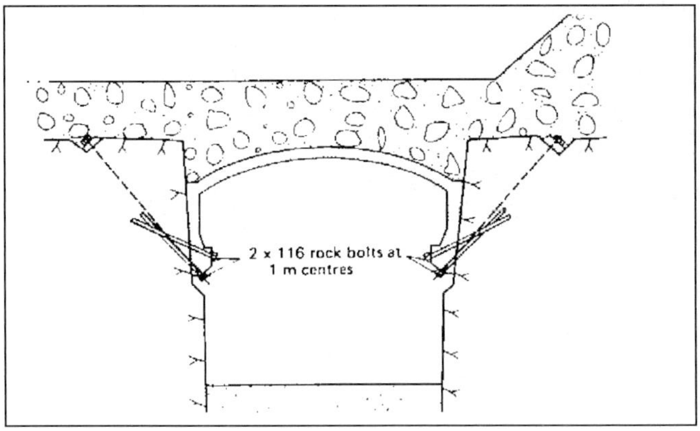

Figure 18. Section through machine hall showing locations of new rock bolts (after Jurell, 1985)

It is fortunate that dynamic failure of a single bar caused no injury and only limited damage. Routine annual maintenance inspections, including reports recording the engineering investigation and assessment of the rust-coloured water seeping out from the anchor heads, could have highlighted the initial corrosion and absence of cement grout encapsulation of the bars, thereby permitting more economical remedial measures within a few years of completion.

First Ring Road, Kuwait City

On this project, anchorages were installed in dense partly cemented sands to resist hydrostatic uplift of a major roadway constructed below groundwater level. However, construction was interrupted by the invasion of Kuwait in 1990 resulting in work being halted for a period of around five years. During this time, partially installed multi-strand anchorages (i.e. unprotected tendons immediately beneath unprotected anchor heads) were subjected over a period of 18 months to several metres of groundwater inundation, when the dewatering system was out of action.

The groundwater contained high levels of chlorides (up to 10,500mg/litre), nitrates and sulphates. Jones (1997) describes the metallurgical measurements and tests carried out on anchorages considered to be in a high risk category due to corrosion, i.e. (i) installed but not proof tested, (ii) installed, proof tested but not locked off, and (iii) locked off but protection of anchor head not complete.

The purpose of the tests was to study the degree and significance of tendon corrosion in the field as visual appearance alone suggested that the level of corrosion was unacceptable (Figure 19). Several options were considered, e.g. remove and replace, cut and couple, clean and retain, and accept with no remedial works.

Figure 19. Exposed protruding tendon after exposure to groundwater inundation for 18 months (after Jones, 1997)

Following careful exhumation of six anchorages by overcoring with fresh water flushing, corrosion was found on the exposed sheathed and greased strands but confined to a depth of 1.5m measured from the top of the polyethylene sheath surrounding each standard 15.2mm diameter strand. Although each strand was greased and sheathed, the exposed strand ends permitted water to travel down through the interstices around the king wire. Microscopic examination found pits at the contact points between the wires or with the sheathing. The maximum pit depth found was 0.25mm with pit depths of 0.1mm to 0.15mm appearing regularly at intervals not exceeding 150mm along the strand (Figure 20).

Figure 20. Magnified view of corroded strand showing pit (after Jones, 1997)

It was postulated that the corrosion was initiated by a crevice mechanism in the minute gap between the sheathing and the strand with moisture entering the crevice by capillary action.

> Note for information. Sheathing used up to the 1980s often incorporated a smaller grease filled annulus surrounding the strands resulting in a tighter fit than is supplied in current systems that provide more grease cover and reduce cohesive frictional resistance between the strand and sheath.

Given these observations and bearing in mind that EN1537 (2000) stipulates complete annulus filling, it is recommended that efficiency of grease infilling should be assessed by testing routinely sample lengths of greased and sheathed strand, as supplied by the manufacturer.

Essentially, the testing comprises full immersion of the sample in a cylinder (preferably graduated), in order to determine: (i) volume of greased strand withdrawn from the sheath Vgs; (ii) volume of strand after degreasing with solvent Vs; (iii) internal volume of plastic sheath, complete with any retained grease by sealing the basal end and measuring volume of water to fill void Vpsg; and (iv) volume of water to fill sheath, having cleaned sheath with solvent Vps.

Given Vps – Vs (volume of annulus Va), Vgs – Vs (grease volume around strand Gs), Vps - Vpsg (grease volume remaining within the plastic sheath Gps), then (Gs + Gps)100/Va = percentage grease infilling of the annulus %Gia.

Where specialised equipment and factory controlled conditions are employed, the efficiency of grease infilling ranges from 76% to 100%, but in general practice field measurements indicate a range of 0% to 100% where 1m sample lengths were tested in each case. A practical acceptance criterion ≥ 80% is recommended and all sheathed strand suppliers should provide factual data confirming compliance with this criterion that can also be checked routinely in the field. Sample lengths can vary from 0.3m to 1m, and it is important to establish the accuracy of testing, depending on method used (see also Barley and Mothersille, 2007).

It is also recommended that protruding greased and sheathed strands prior to stressing should be sealed at the exposed ends by greased tape or caps, in order to prevent moisture ingress, and after stressing the protruding strands should be immediately sealed by encasement in corrosion inhibiting grease. Figures 21-24 illustrate an appropriate sequence of applying corrosion protection to an anchor head with protruding strands.

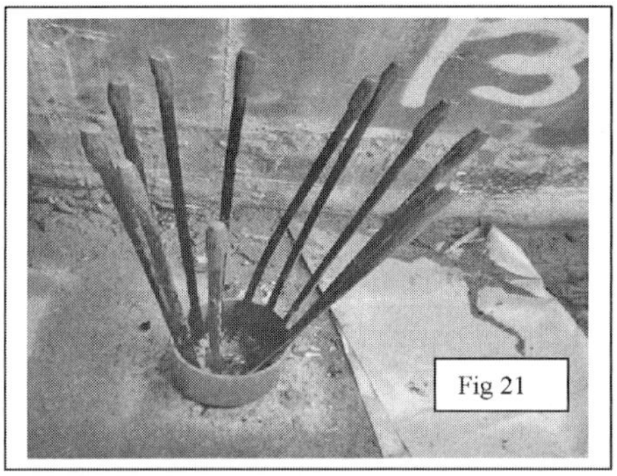

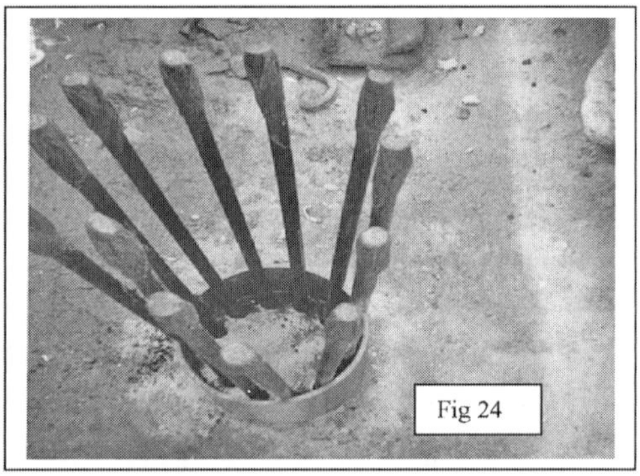

Figure 21. Greased and sheathed protruding strands with upper end seals to prevent ingress of moisture, plus plastic base tube to contain grease protection of stressing head

Figure 22. Heating of corrosion inhibiting grease to give pumpable self-levelling consistency

Figure 23. Placement of self-levelling grease to fully encapsulate stressing head and base of greased and sheathed protruding strands

Figure 24. Finished corrosion protection of stressing head and protruding strands

In spite of the high chloride content of the groundwater, chloride content measured along corroded strand was only 0.7mg/m and regarded as insignificant in terms of corrosion potential. The effect of chloride would have been worthy of study had the content exceeded 4mg/m. However, during the investigations, a faint smell of hydrogen sulphide was also detected when peeling back the sheathing which indicated the possibility that the corrosion could have been microbiological influenced corrosion (MIC).

An exposed sample of corroded strand was tested with hydrochloric acid which released hydrogen sulphide and confirmed the presence of iron sulphide corrosion deposit. Bacterial enumeration was also carried out on swabs from the strand which confirmed the sulphate reducing bacteria (SRB). The combination of iron sulphide corrosion product and presence of sulphate reducing bacteria confirmed that the corrosion was due to MIC. Where MIC occurs the corrosion process generates hydrogen which, due to the iron sulphide, preferentially enters the steel. Bacterial attack is often characterised by the formation of shallow pits (generally less than 0.1mm) that were a feature of the pitting found on the strands.

In order to determine if the anchorages had lost strength, proof testing of eight selected anchorages to 1.5Tw (Tw = 625kN) was undertaken. The proof load of 938kN was held for 3 days without failure, but in order to detension the anchorages, the load was increased further to release the wedges and failure occurred on two anchorages at loads of 970kN up to 990kN (1.58Tw).

On close examination, the strands were fairly heavily corroded and failures had initiated close to the root of the lowest serration of the gripping wedge where the indentation (notch) depth in the outer wires was 0.3mm (Figure 25). The crack was opened up and scanning electron microscopy revealed evidence of a brittle fracture over the full depth of a 40 micron crack that was present prior to failure and caused by hydrogen. Failure was judged to be due to the notch acting as a point of stress concentration in an area of pre-existing hydrogen damage. Where a proof load of 1.5Tw was applied at Kuwait, the typical indentation (notch) depth formed by the wedges was around 0.3mm.

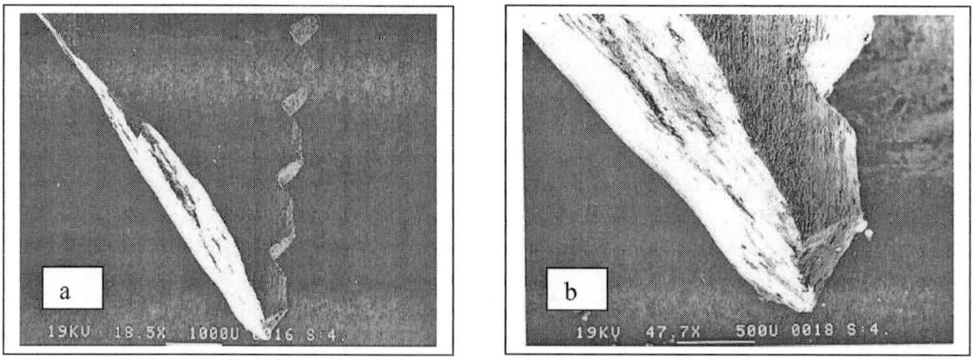

Figure 25. Magnified views of wire fracture located at lower serration of the gripping wedge (after Jones, 1997)

Standard laboratory tensile and bending tests were carried out on samples of strand and wire and indicated no loss of strength. On review, it was judged that the fast rate of strain during the tests would not necessarily indicate evidence of stress corrosion cracking (SCC) as both the rate of generation and diffusion of hydrogen to the point of high stress can be slow compared with the times for standard testing.

As a consequence, slow strain rate testing (10^{-6}/s) was carried out on individual 5mm diameter wires with machined notches of different depths, namely 0.3mm, 0.35mm, 0.5mm and 1.0mm in air and also in a 3% sodium chloride solution, in order to establish if there was a critical depth of pit by comparing failure loads and notch depth. Thereafter, constant load testing at 60% and 70%UTS (ultimate tensile strength of wire) was carried out for approximately 6 days in a 3% sodium chloride solution.

Overall, the results indicated that for strain controlled tests, fast fracture is not likely to occur from notches or pits $\leq$ 0.3mm in depth, and that notches of 0.5mm are only likely to cause fast fracture of wire at loads in excess of 70%UTS (> 30kN).

Where SCC associated with hydrogen embrittlement is the concern, key considerations are what depth of pitting related to applied tensile stress and the environment promotes this mechanism. In fracture mechanics, a critical value known as the threshold stress intensity factor K_{ISCC} is established, above which stress corrosion cracking will occur.

$K = \sigma\sqrt{(\pi a)}$, where σ = applied tensile stress (MN/m^2) and a = crack depth (m)

For a 0.5mm deep notch in 3% sodium chloride solution, a constant load of 30kN (σ = 1500MN/m^2) created failure by SCC after 110 hours (K = 60$MN/m^{1.5}$). For the same test with a 0.5mm notch and an applied load of 24kN (σ = 1200MN/m^2), no failure occurred during the 144-hour test (K = 48$MN/m^{1.5}$), although an 11mm long crack at the foot of the notch was recorded. For a 0.35mm notch and an applied load of 20kN (σ = 1000MN/m^2), no failure occurred during the 144-hour test (K = 33$MN/m^{1.5}$) with an absence of cracking. Overall, these results suggest a K_{ISCC} value between 33 and 48$MN/m^{1.5}$.

Given these test results, it was judged appropriate to accept pits in the outer wires up to 0.3mm in depth without any change in the applied tensile load, once the strands had been thoroughly cleaned and a dry corrosion free environment re-established. Nevertheless, it is important to record that depths of 0.46mm for the manufactured wedge serrations led to indented notches of 0.3mm on the outer wires of the strand. Whilst acceptable, this value is relatively close to the critical defect size and emphasises the importance of corrosion protection within and immediately below the stressing head. The wedge/tapered hole interface should always be oiled so that the wedge can move freely within the tapered hole and limit the ingress of contaminants, otherwise the grip efficiency is impaired.

The information obtained on this project illustrates the value of a detailed investigation involving tensile and metallurgical testing that permitted acceptance of in situ anchorages, once a corrosion-free environment had been re-established, as opposed to a simple visual assessment of the level of corrosion.

Devonport Royal Dockyard, England

333 multi-strand anchorages with free lengths up to 35m were installed through mass concrete retaining walls into Upper Devonian slate for two naval dry docks. The tendon bond length was encased in cement grout whilst each free length of strand was greased and sheathed in polyethylene. The anchorages were designed with a working load of 2000kN and lock-off loads ranged from 2200-2250kN.

Following stressing, the anchor heads were either coated in bitumen or cement grouted. The protruding 15.2mm diameter compact strands above the head were either greased and enclosed in polyethylene sheath or wrapped in grease impregnated tape. The anchor heads and their protruding strands were located within a services trench that was subsequently backfilled with sand by the main works contractor. The design, installation and testing of these rock anchorages are described by Littlejohn and Truman-Davies (1975).

The early load behaviour of ten anchorages was monitored using load cells at hourly intervals during the first 24 hours, weekly for the following six months and thereafter at monthly intervals up to a period of 46 months. The objective of the service behaviour monitoring was to measure load loss with time, load changes during the initial basin de-watering phase, and eventually confirm an overall stabilising trend. The maximum recorded load loss was 7% before stabilisation was confirmed, after which regular monitoring was discontinued on site (Littlejohn and Bruce, 1979).

Subsequently, Weerasinghe and Anson (1997) described an investigation into the condition and performance of 222 anchorages existing within two walls of Docks 14 and 15 after a period of 22 years in the marine environment.

The testing programme involved the following.

- Corrosion inspection above and below the anchor heads.
- Mono-strand jack lift-off testing to determine the residual loads in the strands and assess grout/strand bond.
- Cyclic load-extension tests of individual strands to determine apparent free length.
- Metallurgical and environmental investigations above and below the anchor heads to determine nature and severity of corrosion of steel components and corrosivity of the environment.

74 anchorages (33.3%) were inspected for corrosion above the head and 48 (21.6%) were inspected for corrosion below the head. 69 anchorages (31.0%) were subjected to load lift-off tests and 8 (3.6%) were subjected to cyclic load-displacement tests.

During testing, 4 strands (0.5% of strands tested) failed above the head during lift-off tests. A total of 3 strands (0.4% of strands tested) failed above the head during cyclic load testing. Only 2 strands (0.3% of strands tested) were found to have failed during service due to severe corrosion immediately below the anchor head. These strands were extracted over their full length and cut into 2m lengths and samples from below the anchor head, free length and bond length were then subjected to tensile and metallurgical testing. In addition, samples of the groundwater in boreholes, water in the services trench and sandy fill around the anchor head, were taken for chemical analysis.

Within the tendon bond length, only general corrosion was recorded on the strands but with negligible loss of strand diameter. Although the protection was limited to cement grout cover, the rock mass had been pre-grouted during anchorage installation, where appropriate, to eliminate grout loss during injection and ensure a low rock mass permeability. Within the free length, corrosion was absent where grease and polyethylene sheathing were in position to provide protection.

Although not quantified in terms of corrosion depth, where corrosion of the anchor head or strands was observed, corrosion was graded by Weerasinghe and Anson to ensure a consistent approach to the field inspection (Table 2). The term "flaking" describes loss in layers due to general corrosion.

Corrosion Grade	Description
1	Lightly corroded, light brown in colour, no flaking corrosion
2	Moderately corroded, deeper brown in colour, no/light flaking corrosion
3	Heavily corroded, variation of colours (brown/orange), flaking corrosion that can be removed or brushed off

Table 2. Corrosion grades used for corrosion inspection

A wire brush was used to remove any remnants of flaking corrosion from the portion of each strand immediately above the wedges. Vernier calipers were used to measure the diameter of this section of strand (Figure 26). It is noteworthy that digital readout vernier callipers used commonly today.

Figure 26. Measurement of strand diameter by vernier calipers

In addition, the lengths of protruding strands were measured with notes such as "kinked at wedge" or "strand bent". Such observations are vital in the UK because mechanically damaged strands (depending on severity of damage and location with respect to the wedges), are often considered redundant, in accordance with BS8081 (1989), when assessing a safe anchorage capacity

The nature of corrosion is also vital to record since stress corrosion cracking and hydrogen embrittlement are not normally tolerated because there is currently no predictive capacity for estimating the remaining life of the stressed tendon. For these localised corrosion mechanisms, very small losses of metal can lead to failure. Tendon units (i.e. bar, wire or strand) exhibiting these mechanisms of corrosion are usually rejected

Where grade 2 or 3 general corrosion was recorded, the loss of section of the tendon unit was measured and compared with acceptable manufacturing defects, in order to judge whether or not the loss of section could be accepted. The type and depth of manufacturing defects can be provided by the tendon supplier.

Above the stressing head the strand diameter ranged from 12.90mm to 16.77mm (expansion due to slight unwinding and corrosion of individual wires) for a manufactured diameter range of 15.0 to 15.6mm, thereby indicating a loss of diameter of 2.1mm for the minimum manufactured diameter due to flaking of the wires caused by general corrosion.

Whilst the stressing heads were mostly lightly to moderately corroded (grades 1 and 2), the internal surface of the serrated faces of wedges (removed during cyclic loading) ranged from serrations partly filled with corrosion products to serrations appearing smooth due to the complete filling of the notches with flaked surface rust. This clearly reduces the ability of the wedges to grip the strand efficiently.

Below the anchor head, the condition of the strands was investigated by fibrescope attached to a video camera. Weerasinghe and Anson graded the condition of the strands according to Table 3 and Figure 27.

Grade	Description
A	Strand sheathed and sheathing not damaged
B	Strand unsheathed, no corrosion obvious
B/C	Strand unsheathed, presence of corrosion uncertain
C	Strand unsheathed, corrosion observed

Table 3. Grading system used for fibrescope investigation

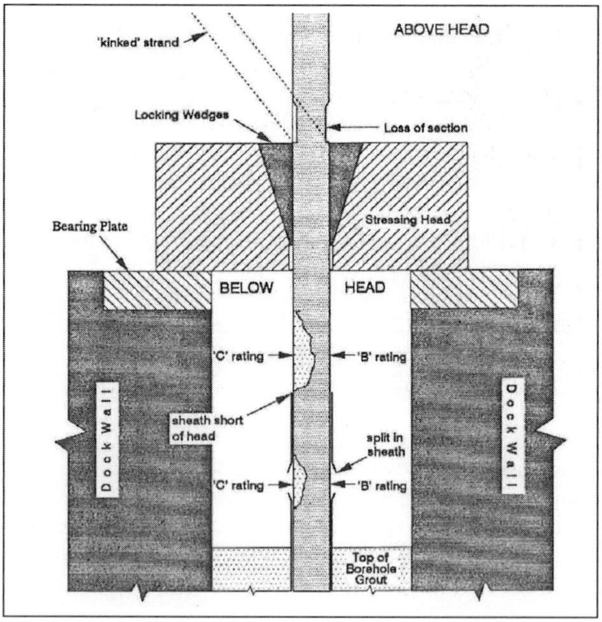

Figure 27. Typical section of an anchor head, strand and corrosion ratings (after Weerasinghe and Anson, 1997)

The strand condition varied from completely sheathed, i.e. abutting the underside of the stressing head, to exposed over a distance of 5 to 210mm below the head. Where the sheathing was absent, two types of corrosion were encountered, namely

(i) general corrosion over the local area exposed, and

(ii) environmentally assisted stress corrosion cracking (SCC).

It was concluded that the probable cause of the SCC was fresh water sulphate reducing bacteria that were found in the sandy fill surrounding the anchor heads.

The most severe general corrosion below the anchor head resulted in a loss of strand diameter of 1-2mm, but there was no evidence of pitting and no failures were encountered due to this level of corrosion after a period of 22 years. Had pitting been observed, it was judged from general experience that a 0.4mm pit would be acceptable, implying a minimum acceptable strand diameter of 14.6mm.

The cold drawn strands were not considered to be highly susceptible to SCC, but given the exposed sections of some strands below the stressing head, it was decided that strands with applied tensile stresses above 60%fpu (fpu = characteristic tensile strength) might be at risk and such strands were not relied upon when assessing the residual loads for each anchorage.

Since all the strands in an anchor head could not be tested due to lack of access and strand damage above the stressing head, a conservative approach was taken in estimating the reliable residual load in service for each anchorage, namely, the sum of the residual loads tested on a tendon minus the load in those strands found to have

- an average diameter of less than 14.6mm immediately above the wedges
- grade C or B/C below the stressing head
- grade B and have a residual load greater than 60%fpu

Given use of 15.2mm diameter compact strands with a characteristic tensile strength of 300kN and original design working load of 55%fpu, the residual loads on individual strands ranged from

- 14-68%fpu (41-204kN) with an average of 52%fpu (156kN) at 14 Dock, and
- 4-66%fpu (12-200kN) with an average of 47%fpu (141kN) at 15 Dock.

Following this maintenance inspection and programme of service behaviour monitoring, and bearing in mind the exposure of the stressed strands to a marine environment, particularly in the location of the anchor head, it is encouraging that these 1970's anchorages continued to hold a significant, albeit a de-rated residual load.

Final remarks

Anchored structures represent key elements of a country's built environment and more attention should be paid to programmes of maintenance inspection and service behaviour monitoring in current practice, particularly where failure would risk life and cause significant damage or disruption. Such programmes can provide confidence in relation to ongoing behaviour and potentially extend the working life of the anchored structure. Alternatively, the results can highlight the need for precautionary or remedial measures in order to ensure the ongoing integrity and performance of the individual anchorages and overall structure.

For new anchored structures, the designer should indicate clearly at the outset the value and necessity of regular inspections and testing, Annual visual inspections of anchor head caps are recommended and where significant mechanical damage or corrosion is observed, the protective cap should be removed to permit assessment of the condition of the anchor head, and the need for further investigations, as appropriate.

It should also be borne in mind that many anchored structures were designed and installed over thirty years ago when the corrosion protection provided might be considered inferior or inadequate by today's standards. For these older structures, reliance should not be placed on purely visual inspections of the outer protective caps. The condition of the stressing head should be assessed and load lift-off tests carried out on a representative sample of anchorages to check that the residual loads in service are in accordance with the original design criteria.

Out of the millions of ground anchorages that have been installed around the world over the past 80 years, it is reassuring that there have been relatively few failures recorded to date. With the passage of time, lessons have been learned and standards improved which augers well for the future. However, there is no room for complacency and engineers must continue to apply high standards in design and construction. In the future, rigorous maintenance inspections and service behaviour monitoring should be included to complete sensibly the engineering requirements for important permanent ground anchorages.

Acknowledgements

The authors wish to thank Mr Tony Barley (Director, SBMA Ltd.) for providing photographs of anchor heads exposed during 2005 in the UK, where the anchorages were installed in the 1970s prior to the establishment of the draft British Standard DD81 in 1982. Special thanks are also due to Dr Gareth John (Lead Consultant, CAPCIS Ltd.) for assessment of the corrosion mechanisms.

References

ASSOCIATION OF BRITISH INSURERS (2006). News release, Tuesday 10th October 2006 ref.83/06

BARLEY, A.D. (1997) "The failure of a 20 year old anchored sheet pile quay wall on the Thames". Ground Engineering, Vol.30 (2), 42-45.

BARLEY, A.D. & MOTHERSILLE, D.K.V. (2007) "Engineering assessment of the grease-filled annulus of sheathed pre-stressing strand for use in permanent ground anchorages" Proc. Int. Conf. 'Ground anchorages and anchored structures in service', Institution of Civil Engineers, London.

BRIAN-BOYES, K.C & HOWELLS, D.J. (1984) "Prestressed ground anchor practice in Hong Kong". Proc. Int. Conf. on "In Situ Soil and Rock Reinforcement", Vol. 1, 231-236. Paris.

BS8002 (1994) "Code of practice for earth-retaining structures". British Standards Institution, London.

BS DD81 (1982) "Recommendations for ground anchorages". British Standard Draft for Development British Standards Institution, London.

BS8081 (1989) "Code of practice for ground anchorages". British Standards Institution, London.

BS ISO 12215-1 (2000) "Small craft. Hull construction and scantlings. Materials. Thermosetting resins, glass-fibre reinforcement, reference laminate" British Standards Institution, London.

DIN50 929 (1985) "Probability of corrosion of metallic materials when subject to corrosion from the outside". Part 3. Deutsche Norm. Deutsches Institut fur Normung e.v. Berlin.

EN1537 (2000) "Execution of special geotechnical works – Ground anchors". European Committee for Standardisation, Central Secretariat, rue de Stassart, 36, B-1050 Brussels, Belgium.

FIP (1986) "Corrosion and corrosion protection of prestressed ground anchorages". State of the art report, Fédération Internationale de la Précontrainte, Institution of Structural Engineers, 11 Upper Belgrave Street, London SW1X 8BH.

FIP (1996) "Design and construction of prestressed ground anchorages". Fédération Internationale de la Précontrainte, Institution of Structural Engineers, 11 Upper Belgrave Street, London SW1X 8BH.

GEOSPEC (1989) "Model specification for prestressed ground anchors". GEOSPEC 1, Geotechnical Control Office, Civil Engineering Services Department, Hong Kong.

ICE (1997) "Ground anchorages and anchored structures". Proc. Int. Conf. organised by the Institution of Civil Engineers (editor G S Littlejohn), Thomas Telford, London (644p)

JAA (2006) "Manual of maintenance testing and monitoring" Draft guidelines. Japan Anchor Association, Yasaka Building 9-12, 2-chrome, Misakicho, Chiyoda-ku, Tokyo, 101-0061, Japan

JAPANESE GEOTECHNICAL SOCIETY (1990) "Guidelines for the Design and Application of Ground Anchors, including Interpretation". Japanese Geotechnical Society, 4-38-2, Sengoku, Bunkyo-ku, Tokyo, 112-0011, Japan

JONES, D.L. (1997) "The performance of prestressed ground anchorages in sands". PhD Thesis, Department of Civil Engineering, University of Bradford, England.

JONES, D.L. (1997) "Corrosion – establishing the limits of acceptability". Proc. Int. Conf. "Ground anchorages and anchored structures", Institution of Civil Engineers, London, 363-370.

KUBO, H. (2006) "Ground anchors in Japan". International Forum on Ground Anchors, Japan Anchor Association, Tokyo (18.10.06).

LITTLEJOHN, G.S. and TRUMAN-DAVIES, C. (1975) "Ground Anchors at Devonport Nuclear Complex". Ground Engineering, 7 (6), 19-24.

LITTLEJOHN, G.S. and BRUCE, D.A. (1976) "Rock Anchors - State-of-the-art Part 3: Stressing and Testing". Ground Engineering, 9 (2), 20-29, 9 (3), 55-60, 9 (4), 33-44 (see also

LITTLEJOHN, G.S. and BRUCE, D.A. (1977) "Rock Anchors: State-of-the-art". Foundation Publications Ltd., 7, Ongar Road, Brentwood, Essex CM15 9AU, England.

LITTLEJOHN, G.S. (1987) "Ground anchorages : corrosion performance". Proc. Instn. Civ. Engrs, Part 1, 82, 645-662.

PARRY-DAVIES, R. and KNOTTENBELT, E.C (1997) "Investigations into long-term performance of anchors in South Africa with emphasis on aspects requiring care". Proc. Int. Conf. "Ground anchorages and anchored structures", Institution of Civil Engineers, London, 384-392.

PLAXIS (2005) "How a distressed quay wall could be moved back in place....using Plaxis". Plaxis Bulletin, Issue 18, 15-18.

RTA (1997) "Permanent Rock Anchors". Roads and Traffic Authority, New South Wales, Australia. QA Specification DCM B114. Bridge Engineering Section, Roads & Traffic Authority, PO Box 3035, Parramatta, NSW 2124, Australia (see also draft in progress 2007).

SAICE (1989) "Code of practice on lateral support in surface excavations". Geotechnical Division, South African Institution of Civil Engineers, PO Box 94345, Yeoville 2143, South Africa.

SWISS RE (2003) "Natural catastrophes and reinsurance". Sigma No. 3. Swiss Reinsurance Company, Mythenquai 50/60, PO Box CH-8022, Zurich, Switzerland.

WEERASINGHE, R.B. and ANSON, R.W.W. (1997) "Investigation of the long term performance and future behaviour of existing ground anchorages". Proc. Int. Conf. "Ground anchorages and anchored structures", Institution of Civil Engineers, London, 353-362.

Session one:

Inspection procedures and physical conditions recorded in service

A CASE STUDY OF INSPECTION, INTEGRITY INVESTIGATION AND REPAIR OF ANCHORS USED TO STABILISE SLOPES AROUND DAM LAKE

Minoru Okuno and Takahiro Takemata, Japan Anchor Association, Tokyo, Japan

Abstract

This paper considers anchors constructed over twenty years ago for the stabilization of the slopes around Dam Lake in Tohoku Japan. The slopes were designed with 358 anchors and construction took place before the establishment of an Anticorrosion Standard in 1988 (in Japan). Visual inspections raised concerns and confirmed variation in the quality of some anchors. In view of this, further inspection, integrity investigation and repairs have been conducted over the past three years.

The Public Works Research Institute that is compiling a "maintenance manual for anchors" with input from the Japan Anchor Association monitored the implementation of this work.

Inspection, integrity investigation and result

Specification of anchors and preliminary survey

Figure 1 shows the location of the inspection and integrity investigation and the concrete slope beam and anchor head. In this preliminary survey, after reviewing visual data of the ground and road in the area, site investigations such as inspecting slopes of ground and road pavement were conducted. As for the anchors, visual inspections of all anchors, inspection of soundness by shock and integrity investigation by lift-off test were conducted.

Figure 1. Panoramic view of site and proximal anchors

The area for integrity investigation was selected from the standpoint of recognizing behavior of the whole slope.

Table 1. Specification of anchors

Item	Specification・Dimension	Shape of anchor head
Tendon	φ12.7mm×9	 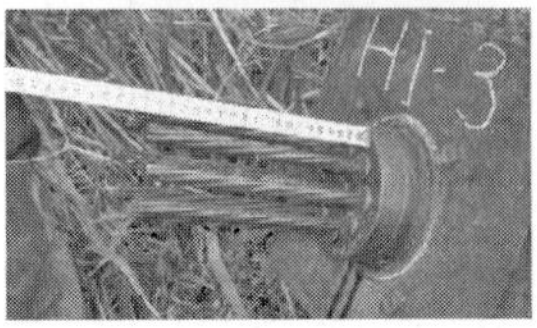
Designed anchor strength	892kN (91 ton)	
Fixed tension load	294kN (30 ton)	
Anchor length	Anchor fixed length: 8.0m Anchor free length: 14.5m to 44.0m	

The specification of anchors and the protruding length of steel wire measured after removal of the head cap are shown in Table1.

Visual inspection and survey result of anchor head

Ground was assumed to have moved in the direction of the arrow shown in Figure 2 from the deformation of ground, road and slope. In the visual inspection of anchor covers, corrosion protection compound had deteriorated or varied and was effected by water coming from below the bearing plates, rain water and also by outside temperature up to 40^0C in the summer.

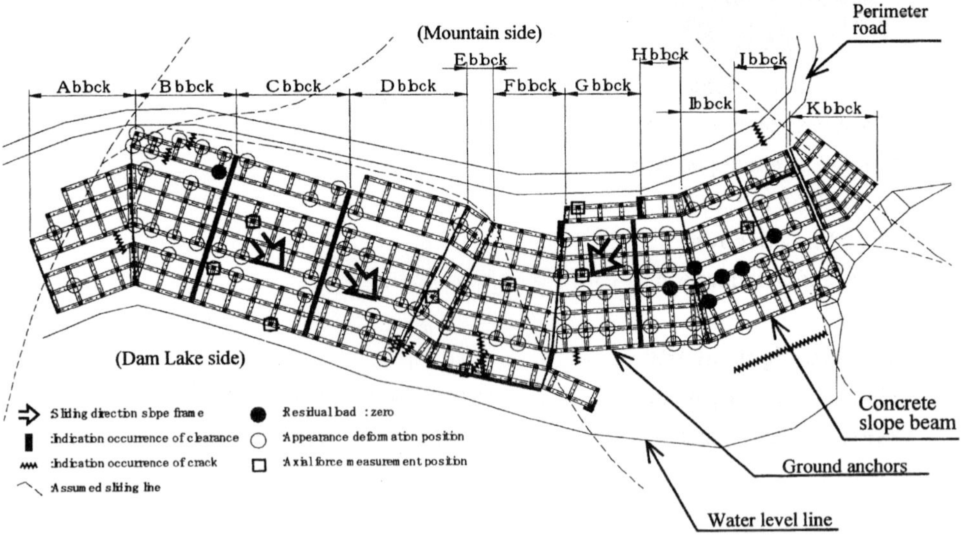

Figure 2. Result of visual inspection

Deformation of the anchor head and the concrete slope beam were found from visual inspection. Furthermore, the corrosion protection compound had been carried away drastically from the anchor cover as a result of deterioration of the anchor cover seal and from the gap of the wedges. Corrosion occurred on the surface of the prestressed stranded wire in the anchor head (Figure3).

In view of these results, enhancement of durability was judged to be problematic. Therefore, repair was conducted in order to improve the anticorrosion function of head cover and below anchor head.

Figure 3. Situation of visual inspection

① Displacement of concrete slope beam joint

② Settlement of upper part of slope of road

③ Corroded load cell

④ Damaged anchor cover

⑤ Leakage and deterioration of corrosion inhibiting compound

Integrity investigation and result

For the integrity investigation, a lift-off test, visual measurement of below the bearing plate after releasing tension, and a 1-cycle capacity verification test (up to design anchor load) were conducted.

The lift-off test was conducted as illustrated in Figures 4 and 5.

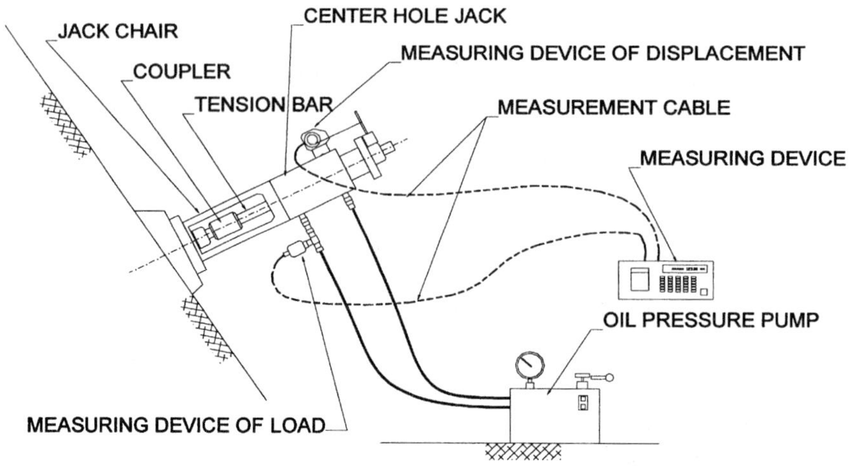

Figure 4. Outline of measurement

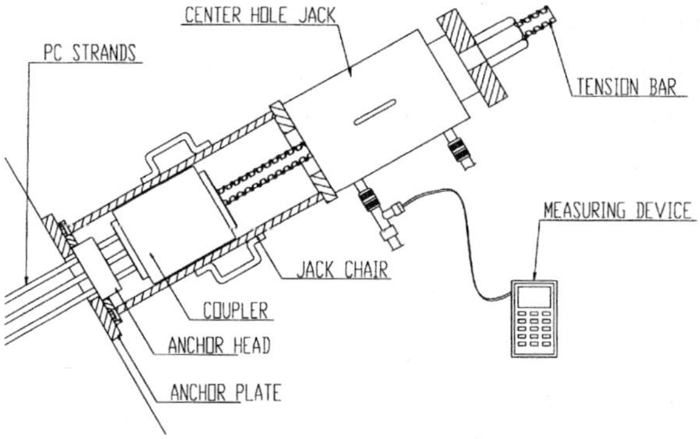

Figure 5. Drawing of test equipment for 500kN

Fluctuation of the residual load along with deformation of whole ground (clearance of frame joint in each block and settlement of road surface) was shown in the lift-off test (Figure 7).

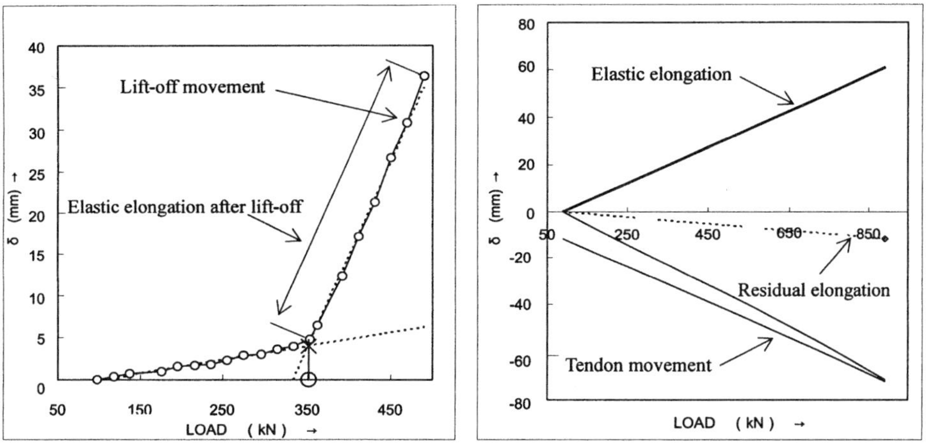

Figure 6. Lift-off test and result of 1 cycle recognition test

The apparent tendon free length was calculated from elastic elongation after lift-off load of the lift-off test graph and elastic elongation after the 1-cycle recognition test (Figure 6).

There were various observational results gained such as blocks where anchors were effective, blocks with stable sliding displacement while effectiveness of the anchor was reduced, blocks with a uniform load both in the upper and lower layer, and blocks with decreased load in the upper layer and increased load in the lower. Table 2 shows the summary of results for investigation items.

Table 2. Integrity investigation results

Test item	Result
Investigation in anchor cover	Deterioration, reduction of corrosion protection compound and corrosion of steel wire were confirmed.
Lift-off test	Anchor load of upper slope decreased and lower slope increased, while load varies due to ground movement
Investigation below the bearing plate (Figure 8)	Corrosion protection compound was deteriorated and decreased drastically. Water and soil from below the bearing plates and boreholes were accumulated. Corrosion of steel wire was observed.
1 cycle recognition test	Residual load was zero, however capacity of anchor to the planned maximum test load could be confirmed. Failure of steel was confirmed in one place. Some anchors were confirmed with rupture of steel wire near the boundary of free length and fixed length.

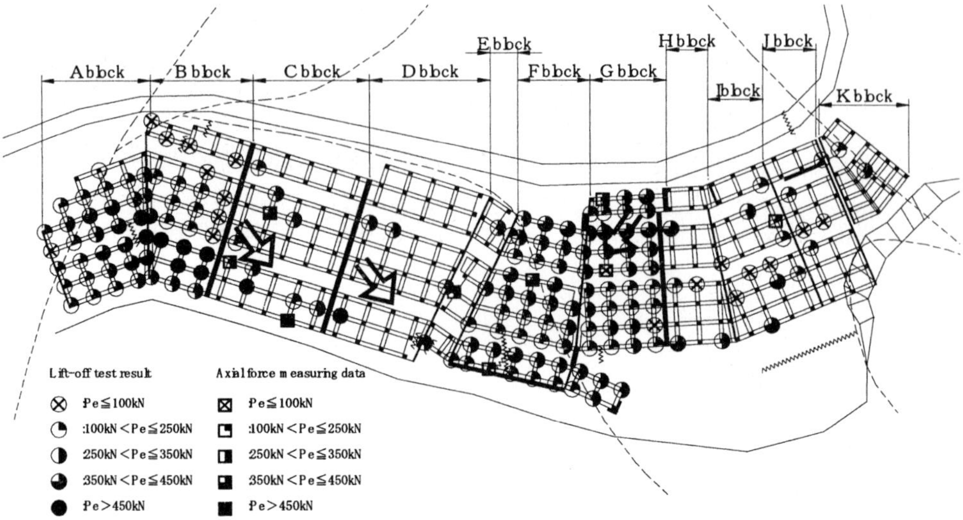

Figure 7. Lift-off test result

In order to investigate the function and durability of the deteriorated corrosion protection compound, the test stipulated (Table 3) was conducted on the extracted sample of the deteriorated corrosion protection compound.

The laboratory test results showed that performance of the corrosion protection compound varied drastically from the specified value at the time of manufacture although the effectiveness of corrosion protection remained.

Table 3. Analysis of corrosion protection compound

Test item	Unit	Standard value	New oil	No.1	No.2	No.3	No.4
Humidity test		A class	A class	A class	A class	A class	A class
Oxidation stability	kPa	Below 69.0	9.0	90.0	---	105.0	250.0
Oil separation	Mass %	Below 5.0	2.0	---	96.0	---	95.2
Evaporation	Mass %	Below 1.0	0.3	---	---	---	1.6

※ "---" could not measure because of lack of specimen.

Figure 8. Below the bearing plate

Drastic deterioration of corrosion protection compound by water and heat (construction site was located on a south facing slope and exposed to temperature differences of 50^{0}C to 60^{0}C) were confirmed before repair.

According to the 1 cycle recognition test result, anchors with zero load influenced by variation of ground had no problem for capacity within test load and anchors considered to not be working from their appearance could be recovered by re-tensioning.

On the other hand, during the 1 cycle recognition test, one anchor with one broken and protruded strand wire was found. The location of wire rupture was assumed to be near the boundary of tendon free length and tendon bond length from the apparent tendon free length calculated.

The rupture is thought to have been caused by a stress corrosion crack and corrosion has gradually progressed for a long time Pitting corrosion of large and small size was confirmed approximately 2m from the rupture point.

An investigation was conducted on the ruptured steel wire (Figure 9). Abnormality of metallic structure was not found near the rupture but severe pitting corrosion was confirmed. Hardness, measured by Micro Vickers hardness test, showed no abnormality. A strength test was conducted for each wire (Table 4).

Table 4. Strength test of ruptured steel wire

Wire	Main corrosion area		Lesser corrosion area	
	Rupture load	Elongation	Rupture load	Elongation
Circumference-1	19.5kN	0.2%	28.2kN	6.3%
Circumference-2	11.7kN	1.5%	28.5kN	5.8%
Circumference-3	12.8kN	1.0%	28.6kN	5.8%
Circumference-4	19.4kN	---	28.4kN	5.8%
Circumference-5	20.2kN	1.7%	28.6kN	5.8%
Circumference-6	22.1kN	1.1%	28.5kN	6.8%
Average	17.6kN	0.9%	28.5kN	6.1%
Core wire	29.0kN	4.5%	29.3kN	6.0%

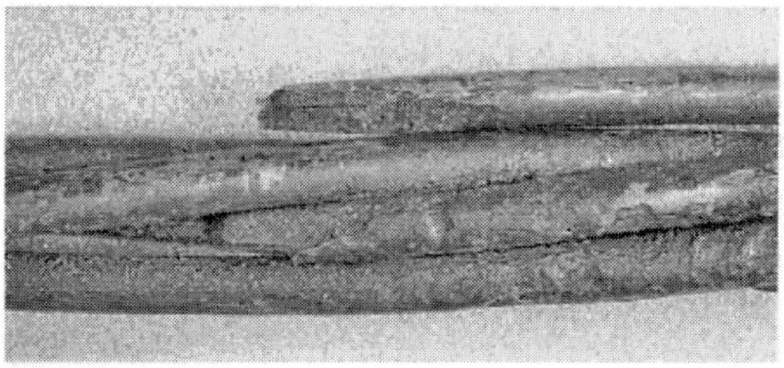

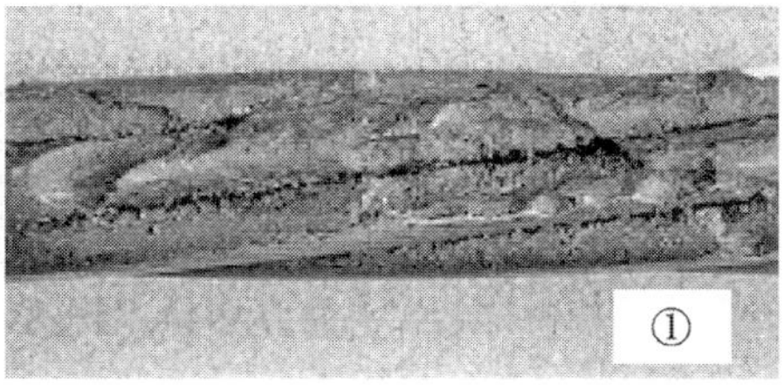

① Corroded area
② Ruptured area

Figure 9. Ruptured steel wires

Repair below the bearing plate

It is difficult to provide anticorrosion treatment for the whole length of existing anchor tendons. Therefore, the following method to enhance the anticorrosion function of the anchor head and below the bearing plate was adopted. In order to enhance the integrity of anchors, the locking unit and anchor cover were renewed and the corrosion protection compound in the anchor head and below the bearing plate was refilled with new compound (Figures 11 and 12). Also, the load cell was exchanged.

For treatment below the bearing plate, a special anchor plate with slide pipe was connected (Figure 10). The anticorrosion function was enhanced by preventing water ingress by connecting the anchor plate and slide pipe to the sheath around the tendon free length. The anchors cover and inside of the slide pipe below the anchor head was then filled with corrosion protection compound.

The structure below the bearing the plate was enhanced for corrosion protection effectiveness (Figure 10). For the ageing investigation of the corrosion protection compound, color change due to emulsification was confirmed after one year. However, changes such as loss in weight, leakage of corrosion protection compound and water and soil were not found.

This repair method was confirmed to enhance the effectiveness of the corrosion protection compound. However, deterioration and loss in weight of corrosion inhibiting compound will naturally progress and, therefore, regular investigation and maintenance should be implemented for the life of the anchorages and safety of the anchored structures. An ageing investigation of corrosion protection compound in the anchor cover showed color changing by emulsification 1 year after exchanging (Figure 13).

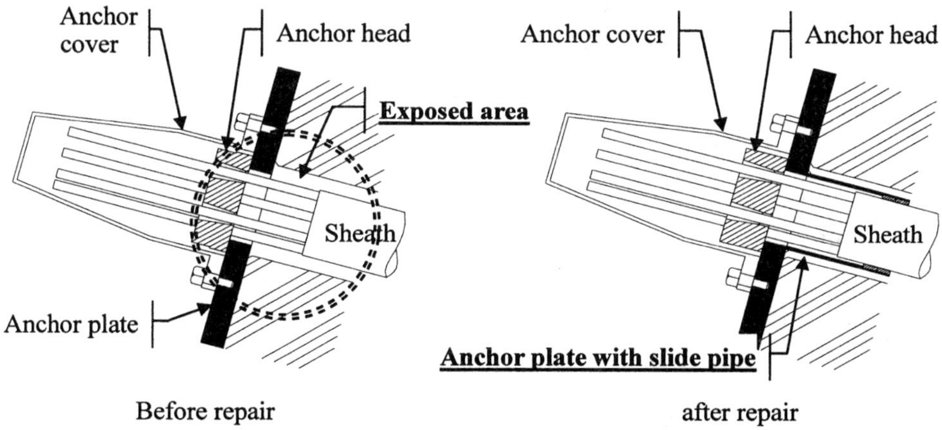

Figure 10. Corrosion protection below the bearing plate

Figure11. Corrosion protection below the bearing plate

before repair after repair

Figure12. Anticorrosion treatment of load cell

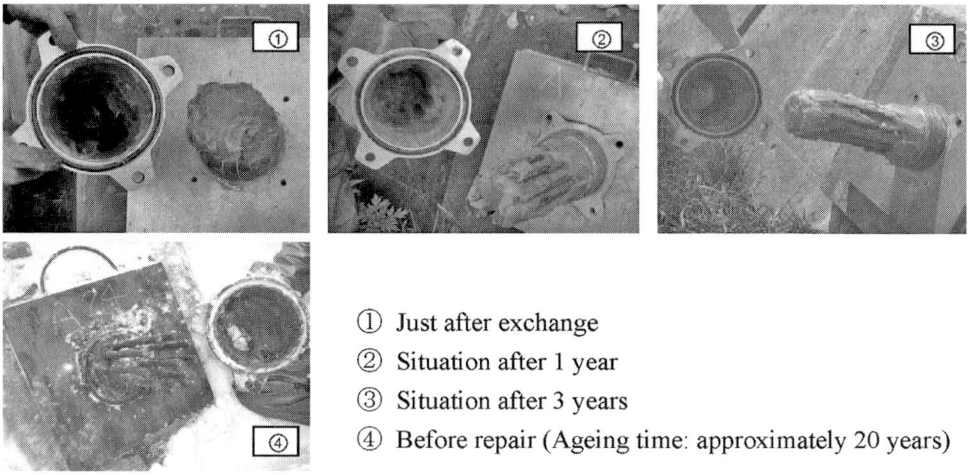

① Just after exchange

② Situation after 1 year

③ Situation after 3 years

④ Before repair (Ageing time: approximately 20 years)

Figure13. Ageing investigation of corrosion inhibiting compound in the anchor head cover

Future agenda

In case that there is any problem about the integrity of anchors, the anchor head and below the bearing plate can be repaired directly. However, it will be tough construction work due to operational conditions and environmental perspective and the maintenance of each anchor should be conducted carefully. It is important to remember not to overlook the warning signs indicating anchor deterioration and thereby lose a great opportunity for enhancing longevity of anchors.

Finally, a reliable corrosion protection method for existing anchors, development of new anchors that are easy to maintain, development of a reliable anticorrosion system and materials and investigation and monitoring technology for anchored structures should be expected.

Inspection of Anchorages Supporting Waterside Structures

Devon Mothersille, Geoserve Global Ltd, England
Tony Barley, Single Bore Multiple Anchor Ltd, England

Abstract

From 1969 ground anchorages were installed at Aberdeen Harbour to accommodate increases in quay wall depth brought on by accommodating a fully tidal harbour. For the majority of these anchorages, the corrosion protection measures employed would be judged inadequate by today's standards. The paper assesses the results of visual inspections and metallographic examinations, undertaken in 1991 and aimed at identifying the nature of corrosion on samples of strand that failed during testing.

Observations are compared with a similar programme of inspections, carried out on 35 year old anchorages on the River Thames in London, and undertaken in 2005. Observations confirmed that, the anchor head protection on both contracts was not fit for extended use and were all in various states of corrosive degradation. The paper also describes the rating system, developed to create objective consistency in the inspection process, and the procedure used to extend the service life of the existing anchorages.

Introduction

The effective maintenance of waterside infrastructure is a matter of increasing concern particularly when considered in the context of climatic change and associated flood risks. Over the last 40 years UK engineers have routinely used ground anchorage technology to provide structural support to various types of waterside structure such as quay walls (Weerasinghe and Anson, 1997, Barley, 1992 and 1997, Littlejohn and Truman-Davies, 1974,). Despite the success of these applications in providing a cost effective solution, questions have been raised about the sustainability of the early systems. This is particularly relevant when new developments are planned that rely on the integrity of the existing infrastructure or when changes in environmental conditions increase the demand on the existing systems. For these reasons inspections and testing programmes are crucial activities in assessing the condition of existing anchorages and, where appropriate, to establish suitable remedial measures. This paper describes experiences with anchorage inspections at two UK sites; Aberdeen Harbour, Scotland and Greenhithe, London.

Aberdeen Harbour, Scotland

Between 1969 and 1982 over 360No ground anchorages were installed at Aberdeen Harbour to accommodate changes in quay wall loading brought on by changes in tidal conditions. These anchorages supported nine quay walls within the harbour complex and were routinely exposed to conditions that could create extensive corrosion of steel components in a highly aggressive marine environment. The anchorages were installed in two phases of development of anchorage corrosion requirements; between 1969-1976 and 1976-1982. During both phases of construction, contract specifications called for corrosion protection measures that would be judged inadequate or *borderline* by today's standards. The following sections focus on some key aspects of the inspection and testing works associated with this study.

Anchorage construction

Fixed anchor length protection

The fixed anchor construction for the Aberdeen anchorages were of two types. Generally, the anchorages constructed during the first phase comprised fixed anchor protection that relied on the cement grout only and the cement's alkalinity in providing a corrosion inhibiting environment.

The fixed anchor lengths for the anchorages in the second phase of construction were encapsulated. Although these encapsulations were in early stages of development, the strands were protected by polyester resin within a galvanized cabco steel duct, and later by cement grout within a single corrugated plastic duct.

Free anchor length protection

Although it was not common practice prior to 1970, the free anchor lengths were protected, from that date, generally by greased sheathing. This was confirmed by inspection below the bearing plates in all the inspected quay walls.

Underhead and Overhead protection

In the underhead area protection was initially provided by cement grout and some partially effective ducting or sheathing. Between 1976-1982 a combination of cement grout and injected resin in the anchorages in combination with quality sheathing was developed. In the overhead areas, the anchors installed between 1969-1976 were either left unprotected or protected by means of bituminous paint applied to head plates, barrel and wedges and protruding strands or in some cases wrapped in grease impregnated tape. Anchors installed between1976-1982 were protected using screw on grease filled steel caps.

Metallurgical examination of strand samples

To improve understanding of the extent of corrosive damage, sections of free length strand with barrel and wedge attached were subjected to detailed examination. The samples were extracted from an anchor that had failed during routine proof load testing (Figure 1). Proof testing was undertaken to investigate debonding in the fixed anchor and to check that the fixed anchor was performing as designed. The sample was removed from a section of sheathed strand and visual inspection had confirmed that the sheathing had sustained damage exposing the steel strand beneath. The location of the failure was just below the anchor head which was known to be a vulnerable location for corrosion related failures.

Visual examination showed the failed strand to be extensively corroded with areas covered in shades of red and brown corrosion deposits, indicating localised contact with water, probably as a result of voids in the cement grout or poor quality construction operations. After removal of the corrosion product using inhibited acid, it was found that four of the seven wires showed evidence of pre existing cracks on the fractures faces (Figure 2a and 2b). The points of origin of the various cracks were found to be in the strand interstices, not in the outside facing regions. Observed distortion of the king wire suggests that the strand may have been loaded for some time with at least one of the peripheral wires broken. A longitudinal metallographic section, taken through one of the wire fractures, revealed the presence of several partially propagated cracks close to the fracture. The investigative report suggested that failure was due to a combination of corrosive attack in an area where corrosion protection had been damaged, exposing the strand, and fatigue mechanisms.

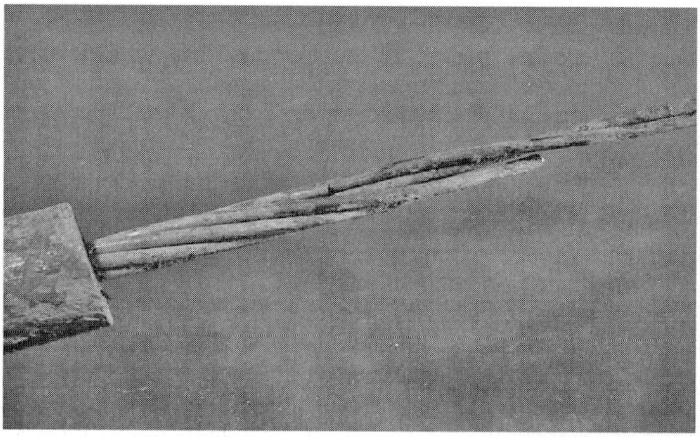

Figure 1 Failed 15mm diameter strand sample submitted for examination

During routine proof load testing, to loads between working load and 1.33 times working load on another quay wall, failures occurred in 2 of the 15 strands tested. One failure occurred in the upper free length and the other above the barrel. The failed samples were visually examined with the aid of a low power binocular microscope. They were then cleaned in solvent and re-examined. Metallographic sections were taken from two of the wires and, after preparation, were examined using conventional optical microscope.

Figure 2a Fractured wire showing the presence of several cracks on the wire surface

Figure 2b As Figure 2a but cracks appear at a point some distance from the fracture face

Some slight corrosion pitting was observed on the Dyform face in close proximity to the corroded areas. After cleaning it was found that even in these areas only the outward facing surfaces had corroded and that the inside of the strand was free from corrosion. It was also noted that the grease was partially emulsified with water. All six wire fractures were found to be ductile tensile overload breaks (Figure3).

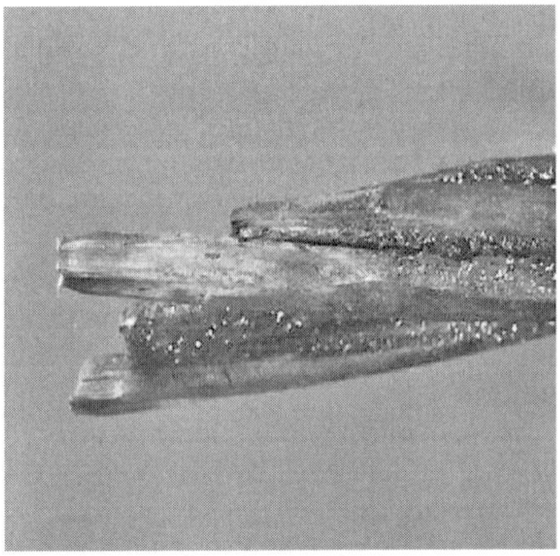

Figure 3 Sample of failed strand showing failure face

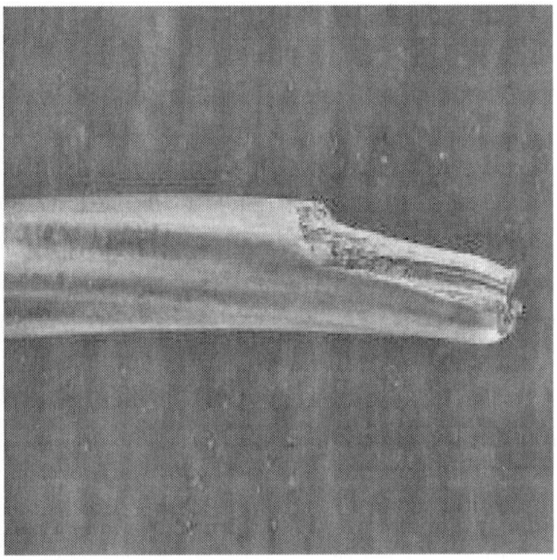

Figure 4 Localised corrosion in failed wire

Metallographic examination confirmed the non-existence of pre-existing surface defects and the corrosion appeared to have taken place by general dissolution of the surface rather than preferential crack formation and propagation. Conclusions from the specialist test report were that the tensile failures resulted from loss of section due to corrosion (Figure 4) which itself was caused by the absence of protective grease from that section of strand.

Figure 5 Seaweed covered caps confirm highly aggressive marine environment for corrosion of steel in the lower tidal range

Figure 6 Correctly protected anchor head barrel/wedges exposed after 16 years in service are corrosion free

However, the effective use of a protective cap filled with corrosion protection compound was clearly demonstrated at another location on this site. Figure 5 shows the conditions under which the anchors were subjected and Figure 6 shows the condition of the protruding strands and barrel/wedges beneath the protective cap after some 16 years in service. Clearly, the strands and barrel and wedges are in bright steel condition having suffered no corrosive degradation at all.

Greenhithe, London, UK

The anchors heads for anchorages supporting retaining walls on the River Thames, at Greenhithe, were inspected and photographed as a preliminary phase to allow grading of the anchor head corrosion. In view of the fact that there is no method established in the anchorage industry for predicting the future performance of corroded anchor components reliance was initially placed on a qualitative assessment of the anchor heads based on the visual inspection. In a subsequent phase of the works, selected anchorages will be subjected to lift-off tests to ascertain residual load levels.

General Observations

Circa 1976, the Greater London Council issued the first Ground Anchorage Specification in the UK demanding a precise corrosion protection system around the full length of the anchor tendon. It is noteworthy that no British Standard or related document existed until 1981 (DD81 – Draft for Development). The first British Standard for Ground Anchorages (BS8081:1989) was not published until 1989. With this in mind the quality of post 1976 anchorages on the Thames should be superior to those pre 1976 in Aberdeen.

Overhead and Underhead Condition

From visual inspection of the anchorages on the site there was every indication that the anchor head corrosion protection was consistent with that recommended in BS8081:1989. The protective caps, installed circa 1978 were bitumen coated steel caps threaded onto short male thread length of a steel tube itself welded to the anchor head plate. These anchor head caps and threaded bases have generally served their purpose for a thirty year period but in the majority of cases require replacement since it is essential that the anchor head components maintain this protection for a further lengthy period of possibly 50 years.

In all cases the anchor head components including protective caps, bearing plate, gusset plates and welds all show visible signs of surface corrosive damage with bitumen coating either removed or flaking off (Figure 7). This included section loss, laminated expansion of the corroded anchor cap steel and crevices and pitting up to 10mm on the 40mm thick bearing plate. In marine environments the effective life of the protective cap is judged to be limited in but the over design of bearing plate may tolerate some pitting.

Figure 7 Severe deterioration of bituminous paint on corroded protective cap

Figure 8 Exposed strands and barrel and wedges containing remnant grease after loss of protective cap

On site every anchor head and cap was visually inspected and its state, with regard to current and future performance *graded.* In 10% of the anchor heads inspected protective caps had corroded so severely that they had fallen off leaving the strands, barrel and wedges and remnant grease exposed (graded 'D' see Table 1) as shown in Figure 8. The raised barrel shown in this figure was due to damage sustained in the free length by subsequent, non-anchor related, building operations.

Table1 Grading system adopted for inspections

Grade category	Description
A	Good condition with superficial surface rust
B	Satisfactory condition with more extensive corrosion evident. Anticipated future lifespan of steel component in aggressive marine environment could be limited.
C	Poor condition with evidence of deep corroded surfaces. Some section loss and small crevices also visible. Anticipated lifespan of the steel component is short.
D	Very poor condition. Severe and significant section loss evident with deep pitting and crevices. Immediate replacement would need to be considered to prevent corrosive deterioration of steel component.

On the basis that corroded anchor head protection is required to be appropriately uprated to provide an extension to service life, the following procedure is recommended (Figure 9):

1. Uncover all anchor head recesses within the sheet pile wall
2. Clear and dry all recesses and anchor head components
3. Investigate presence of voids below anchor head plates by probing through grout holes. Advance rigid plastic tube, probably aided by air flush to maximum depth possible. Grout up to free length void with net cement grout, changing to 1:1 sand:cement grout should take become excessive. Withdraw nominal volume of grout nominal volume of grout from immediately below head plate to allow later injection of grease. *Note that this procedure was adopted specifically for the anchorages at Aberdeen harbour where underhead tendon failure had occurred.*
4. Blast clean all exposed steel surfaces (bearing plates, barrel, wedges and strand) to remove all existing coating and other contaminants
5. If required drill and tap holes for new protective cover bolts
6. Apply two coats of appropriate epoxy resin coal tar or approved bituminous paint to bearing plate
7. With suitable gasket seal fit glass reinforced plastic protective cap filled with corrosion inhibiting grease and secure with galvanised bolts.

The use of fibre glass protective caps fabricated to recognised standards, e.g. ISO 12215-1(2000) was pioneered in the UK in the 1980s.

Figure 9 Sequence of remedial measures for corroded anchor head after loss of protective cover in service as constructed by Keller Ground Engineeering (After Littlejohn and Mothersille, 2007)

Summary and Closing Remarks

A comprehensive visual inspection has shown that whilst the majority of anchor head caps were in place, at Greenhithe and on one wall in Aberdeen, and were still making some contribution as a retention unit for the corrosion inhibiting compound. However, they were all, without exception, in a state of corrosive degradation.

A site trial on cap replacement has proved successful and will provide a basis for the consideration of a more extensive program of cap replacement, particularly on the Thames Bank Raising Anchorages and other flood defenses.

The bearing plates and gusset plates were also included in the anchor head inspections. Generally, the condition of these components is judged to be medium to poor. The marine environment has accentuated the degree of corrosive degradation to an extent that section loss has occurred on many components. Although it would appear that the gusset plates and bearing plates are still functioning as load bearing elements, the longevity of the anchor head systems, particular that of the inaccessible underheads, could be questioned. In order to confirm the extent of corrosive degradation detailed metallurgical tests are recommended.

In the case of ground anchorages installed with limited or no corrosion protection, it is not possible to inspect or assess the extent of corrosion over the full tendon. Thus it is not possible to guarantee the ultimate capacity of the steel components in the current or future state, even though in the recommended environment further corrosion may be retarded. In Aberdeen Harbour, anchorages fall into this category having served their useful purpose for over thirty years. It is judged that the remedial measures employed at both sites will serve to prolong the effective service life of the anchorages.

Recently on the River Clyde an investigation was carried out involving inspections and the testing to failure of removed sections of strand, barrel and wedges (Mothersille *et al*, 2007). The extent of corrosion on the River Clyde anchorages was of a similar order to that observed at Greenhithe and the unprotected anchorages at Aberdeen Harbour and provides an indication as to what can be expected on sites which depend on anchorages installed around the same time and in a similar environment.

In essence this paper provides in insight into important aspects of ground anchorage maintenance and describes the procedures and approaches to testing, inspection, assessment and provides guidance to those embarking upon similar engineering assessments in the future.

Acknowledgements

The authors wish to thank Keller-Ground Engineering for their assistance with the site based operations at the Greenhithe site.

References

BARLEY, A.D. (1992) "The use of Ground Anchorages in Waterfront Structures" Presented at the International Shipping and Marine Technology Market with Congress, Hamburg, September – October.

BARLEY, A.D. (1997) "The failure of a 20 year old anchored sheet pile quay wall on the Thames". Ground Engineering, March issue, 42-45.

BS DD81 (1982) "Recommendations for ground anchorages". British Standard Draft for Development British Standards Institution, London.

BS ISO 12215-1 (2000) "Small craft. Hull construction and scantlings. Materials. Thermosetting resins, glass-fibre reinforcement, reference laminate" British Standards Institution, London.

BS8081 (1989) "Code of practice for ground anchorages". British Standards Institution, London.

ICE (1997) "Ground anchorages and anchored structures". Proc. Int. Conf. organised by the Institution of Civil Engineers (editor G S Littlejohn), Thomas Telford, London (644p)

LITTLEJOHN, G.S. and MOTHERSILLE, D.K.V. (2008) "Maintenance Testing and Service Behaviour Monitoring of Permanent Ground Anchorages". Geotechnical Engineering, (to be published).

LITTLEJOHN, G.S. AND TRUMAN-DAVIS, C (1974), "Ground Anchors at Devonport Nuclear Complex", Ground Engineering, 7 (6) pp 19-24

MOTHERSILLE, D.K.V., JACKMAN, S and FERRIER, J (2007) "Performance and condition assessment of 30 year old anchorages, River Clyde, Glasgow" Proc. Int. Conf. "Ground anchorages and anchored structures in service", Institution of Civil Engineers, London.

WEERASINGHE, R.B and ANSON, R.W.W (1997) "Investigation of the long term performance and future behaviour of existing ground anchorages" Proc. Conf. on Ground Anchorages and Anchored Structures. ICE. Thomas Telford, London.

Inspection, Integrity Investigation And Repair Of 30-Year Old Anchors

Yukio Fuseya and Hiroyuki Takeda
Shikoku Branch, Nittoc Co., Ltd. Takamatsu city, Japan

Abstract

In Japan, before the establishment of a corrosion prevention standard (established in 1988), many permanent anchors were constructed without adequate consideration for corrosion protection of pre-stressing steel. As a result, corrosion occurred in some anchors and many accidents resulting from rupture of anchors have been reported.

This paper reports on the inspection, integrity investigation and repair of 30-year-old anchors and the temporary repair that was carried out referring to the investigation and evaluation of integrity for each anchor.

Though the project site was not large, this investigation can be a future reference for similar cases.

Outline of construction

In this project, anchors were utilized for protection against landslide at a dam pond located in the country city. A landslide of 140m width x 185m length and soil (mass) volume of 64,750m^3 occurred after being affected by rising groundwater after torrential rain and increasing dam water level. 73 anchors, named “SEEE” and “pre-stressing steel bar” anchors, were installed from 1975 to 1976.

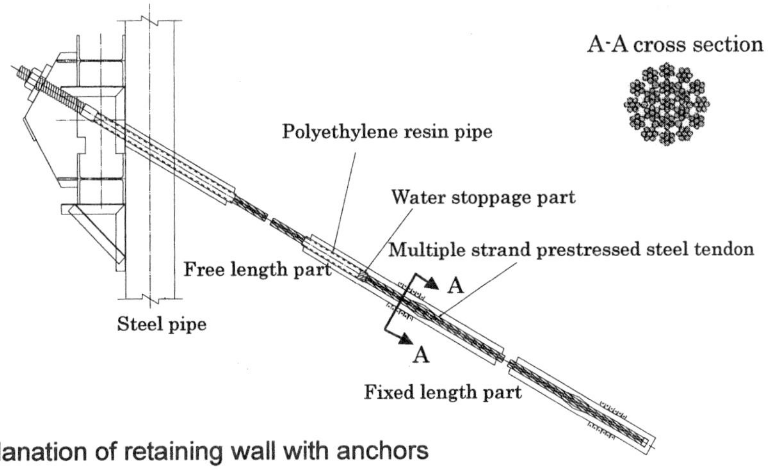

Figure 1 Explanation of retaining wall with anchors

Ground anchorages and anchored structures in service, Thomas Telford, London

SEEE anchors and a retaining wall are shown in Figure 1, anchors before repair are shown in Figure 2, and the steel wire after removal of the damaged corrosion inhibiting pipe is shown in Figure 3. These anchors are installed at the head of steel pipe piles via H-type steel beams. Therefore, 1.0m of exposed anchor free length exists. The anti-corrosion measures for this exposed free anchor length are insufficient and partial damage to some of the corrosion protection materials used can be observed. Corrosion can be found in the exposed steel wire where the protection is damaged and integrity of anchors is considered to be low. Cracks have developed in the retaining wall at the backside of the anchors and the surface of the road affected by movement of soil mass. A landslide is currently observed. With future soil movement there is a strong possibility of disaster by rupture of anchors with low integrity. Therefore, integrity investigation and repair are urgently required.

Figure 2 Retaining wall and anchors

Figure 3 Steel wire after removing damaged corrosion inhibiting pipe

Flow of integrity investigation

The integrity investigation comprised a preliminary and a main investigation the content of which is shown in Table 1. A flow chart of the integrity investigation is shown in Figure 4. The results of the integrity investigation are recorded for each anchor for reference for repair work in the future.

Table 1 Integrity investigation

	Subject	Procedure and item	Purpose of investigation
Preliminary investigation	Document research	· Research of basic document at the time of anchor installation · Repair record	· Confirming details of anchors · Confirming details of landslide · Damaged anchors
	Field work	· Geological research in the vicinity	· Confirming of current situation of landslide
Main investigation	Structure survey in the vicinity	· Variation of structure (visual inspection) · Repair record of structure	· Confirming of current situation of landslide
	Anchor head survey	· Detailed investigation by visual inspection	· Propagation of corrosion · Anchors with accident
	Physical investigation	· Investigation of hitting sound · Lift-off test	· Dashing out · Confirming residual anchor strength · Anchors with accident

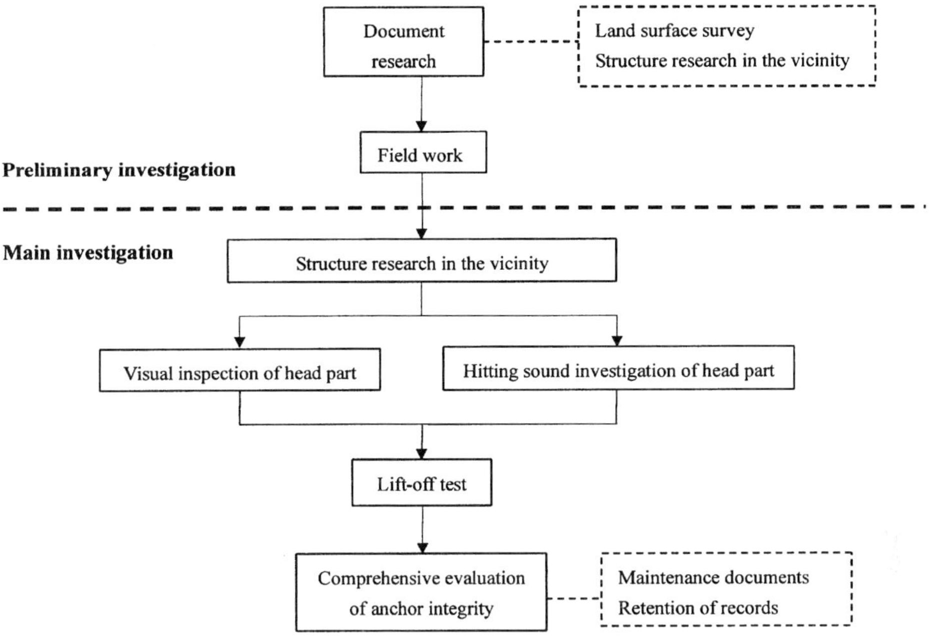

Figure 4 Flow chart of integrity investigation

Preliminary investigation

Document research

The results of the investigation into published documents are summarized as below: -

Details of the landslide are shown in Table 2.

Table 2 Outline of landslide

Geological situation	Weathered crystalline schist (mainly consisting of pelitic schist)
Strike, Dip	N82-73°E, 14-19°N
Depth of basement complex	2.00m - 4.50m
Type of landslide	Sliding of weathering rock layer
Angle of sliding surface	29°
Maximum width (W), Maximum length (L)	(W) 140m, (L) 185m
Moved soil mass volume	64,750m^3
Assumed depth of sliding surface	5.00m

Details of the installed anchors are shown in Table 3.

Table 3 Details of installed anchors

Method	Type	Fixed length part (m)	Free length part (m)	Design load (kN)	Initial load (kN)	No.	Installed location
SEEE	F170	8.00	9.00	Unknown	400	52	In the landslide block
PC steel bar	φ32 mm	7.00	8.00	Unknown	Unknown	20	Border of landslide block

Record of variation of anchors
No rupture was found in SEEE anchors. However, in the PC steel bar anchors three ruptures in 1996, one rupture in 1999 and one rupture in 2004 were discovered. As for structures in the vicinity, variation occurred in the area of the rock filled baskets and cracking occurred on the retaining wall at the back side of the pile.

Anchor repair records
The three anchors that ruptured in 1996 were renewed immediately. The new type of SEEE anchor with dual corrosion inhibiting measures was utilized, incorporating an anchor head cover to become a complete corrosion inhibiting structure. The anchor that ruptured in 1999 was renewed during the same year with an SEEE anchor as used for the previous failures. Corrosion inhibiting compound was provided to all 73 pits of anchor head. One anchor ruptured in 2004 and was renewed as above. Tendons were the same as used in 1996.

Repair record of structures in the vicinity
Repair works were conducted for some sites where a large-scale movement occurred in 2000 in the area of the rock filled baskets. In 2001, repair work was conducted for the remaining sites with movement in the area of the rock filled baskets.

Fieldwork

Fieldwork results for one of the largest blocks are shown in Figure 5. From the result of fieldwork, multiple new landslide blocks and new surface layer collapse blocks were found to be formed in the former landslide block. A cliff with gap difference of 1.5 to 4.0 m was formed in these new blocks and a new progressive crack had developed intermittently. In the central area of the new block, tree roots had deformed and a creep slope had formed. It was therefore concluded that landslide activity is continuing and a new landslide inside or a surface layer collapse of unstable soil mass occurred locally with a high possibility of inducing a large-scale landslide.

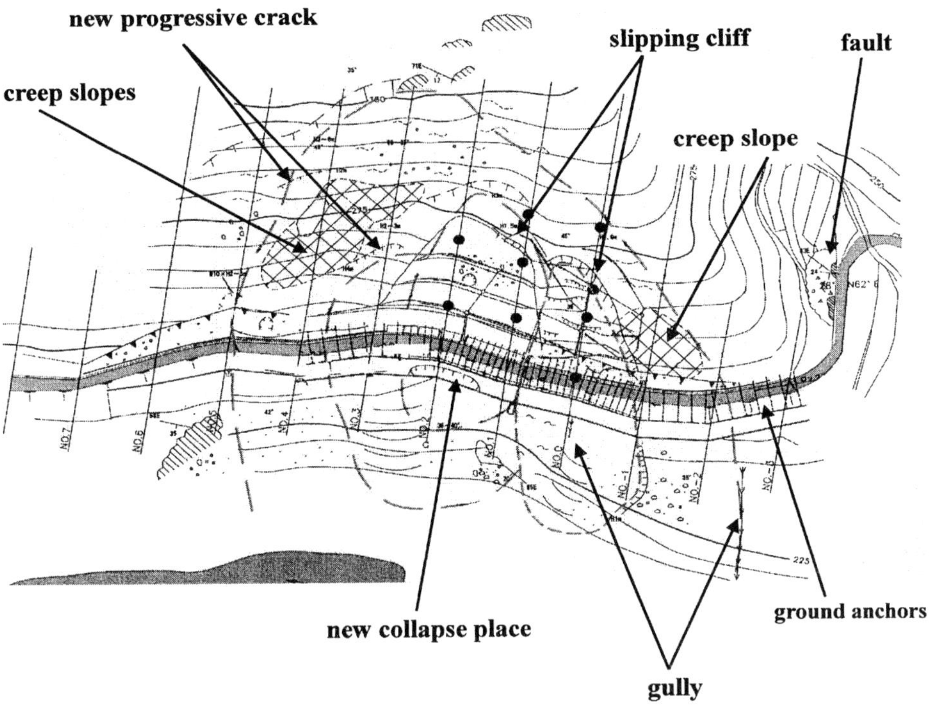

Figure 5 Fieldwork result

Main survey

Structure survey in the vicinity

Apart from the crack in the retaining wall found in the preliminary survey, the crack in the road at the back side of the pile was progressing and surface variations were judged to be increasing.

Visual inspection of anchor head

The noticeable points of the visual inspection are shown in Table 4. Visual inspection was implemented after exposed part of land surface installed anchors were divided into i. locking unit (fixed nut, anchor plate and bearing plate) and protruding tendon ii. fixed base part (pedestal metal, supporting unit) and iii. exposed free anchor length part (anti-corrosion cover, steel wire). The judgment standard for visual inspections of the locking unit and fixed base part are shown in Table 5 and judgment standard for visual inspections of the exposed free anchor length are shown in Table 6. Inspection results were divided into ranks as in Tables 5 & 6 and anchor integrity was evaluated. As shown in Figure 6, more than 60% of the fixed base parts and locking units were evaluated as rank I and few anchors were judged to require repair work. In the fixed base parts, 2 anchors were deformed severely in supporting unit (H type steel) and these were judged to be rank III requiring immediate repair work. In the exposed free anchor length, damage of corrosion inhibiting material had progressed in all pits and leakage of corrosion inhibiting grease was found. Deterioration of remaining corrosion inhibiting grease had progressed and solidified into ash white. In all the anchors investigated, corrosion of the steel was found to have progressed as a result of deterioration of the corrosion protection. In this investigation 34 anchors (65%) were judged to be rank IIIa that could be maintained by immediate repairing work and 18 anchors (35%) were judged to be rank IIIb that could not be recovered due to cross-section damage. As a result, current status of anchor function was maintained by immediate repair work for old SEEE anchors and it was decided to take drastic countermeasures for the 18 anchors judged as III b.

Table 4 Noticeable points of visual inspection

Part	Observation item
i. Fixed base part	Deformation. Corrosion status
ii. Locking unit	Harmful scratch. Deformation. Floating of nut, bearing plate and anchor plate. Flat sound of hitting by hammer. Corrosion of material
iii. Free anchor length exposed part	Damage of corrosion inhibiting material. Filling amount of corrosion inhibiting grease. Deterioration of corrosion inhibiting grease. Corrosion of steel wire.

Table 5 Judgment standard of visual inspection result (Locking unit, fixed base part)

Rank	Judgment rank
I	It is OK as it is.
II	Repair work is required.
III	Replacement is required becausc of severe deterioration.

Table 6 Judgment standard of visual inspection result (Free length exposed part)

Rank	Judgment rank
I	There is no problem because it is fresh.
II	Corrosion is progressed a little
III a	Corrosion is progressed but current status can be maintained by repair
III b	Corrosion is severely progressed, rupture of wire and damage of cross-section are progressed

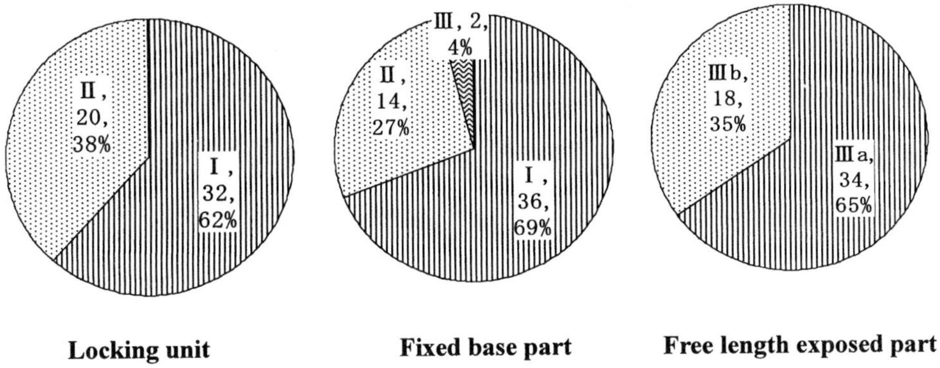

Figure 6. Rank of integrity by visual inspection

Lift-off test

In cases where integrity was judged to be low by visual inspection, a lift-off test was conducted to confirmed integrity by application of 400kN initial fixed load. Details of lift-off test are shown in Table-7. During the tests, steel wires ruptured in 2 anchors during tensioning, 5 anchors were drawn out and the initial fixed load was confirmed in 34 anchors. The 2 anchors that ruptured were evaluated as low integrity IIIb by visual inspection of the exposed free length. The rupture load was between 420kN and 470kN and rupture occurred immediately after surpassing the initial tension. The position of the rupture was not clarified but assumed to be in the vicinity of the boundary of the fixed length part and free length part. Stranded wire at the free length part was loosened in both anchors and anchor function completely lost. The ruptured steel wire is shown in Figure 7. Loosening of multiple stranded wire was observed. Drawn out anchors were confirmed from variation of head part after lift-off test. Drawn out load were gathered around 400 to 450kN as well as ruptured anchors. From the lift-off tests, 22 of the 34 anchors evaluated by visual inspection were confirmed as IIIa and 12 of the 18 anchors evaluated by visual investigation were confirmed as IIIb.

Table 7 Details of lift-off test

	Condition	Reference
Initial load	100kN	
Maximum load	600kN	Initial fixed load +200kN
Pressure rising level	20 to 30kN	
Load keeping time	3 minutes	
Variation measuring time	0, 1, 3 minutes	

47 anchors (90%) were confirmed as having an anchor force above the initial fixed load and 16 anchors (31%) were confirmed as having an anchor force more than 20% above the initial fixed load. Lift-off test results are shown in Table-8. The increase in the anchor forces was assumed to be caused by movement of landslide block. Drawing out load was 400 to 500 kN. If the increase in anchor forces continues in the future, rupture of steel wire in severely deteriorated anchors will increase, anchor function will be lost and instability of slopes will occur.

Figure-7 Ruptured steel wire (Loosening in multiple stranded wire is observed.)

Table-8 Lift-off test result

	Remained anchor load (kN) /Initial introduced load (kN)					
	>120%	119 to 110%	109 to 100%	99 to 90%	89 to 80%	<79%
Degree	16	24	7	4	1	0

Effect of repair

After the integrity investigation, anchor repair work was conducted. Repair of all anchors is impossible and, therefore, maintenance of anchor function at current status through application of corrosion inhibiting measures around anchor head was implemented to extend anchor life.

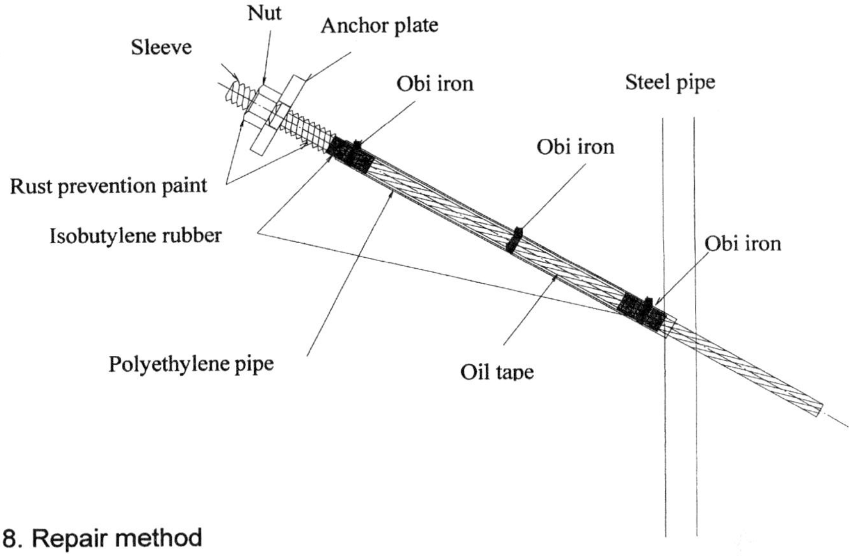

Figure 8. Repair method

For repair of the exposed free length part, protection against further corrosion propagation was provided by preventing contact with air by (1) Removal of corrosion inhibiting material in the damaged exposed free anchor length part, (2) Removal of rust, (3) Re-application of corrosion inhibiting grease and tape, (4) Covering of the repaired part by polyethylene pipe and butyl tape. This repair method to the exposed free anchor length part is shown in Figure 8. For repair of the fixed base part and locking unit, removal of rust and re-application of corrosion inhibiting treatment were conducted considering economy. In order to achieve effectiveness of the corrosion inhibition treatment, triple coatings of epoxy resin corrosion inhibiting compound and chlorinated rubber corrosion inhibiting compound were provided.

Repairs were conducted on 45 SEEE anchors. One year after completion of the repairs there has been no further deterioration of the anchors. No further variations in the structure at the back of the steel piles were observed and cracking of the road surface had not progressed. Therefore, it was considered that anchor force had not changed. The effect of repairing in this time was perfectly proceeded and each anchor maintains function at the time of repair.

Summary

Notable points learned from the inspection, integrity investigation and repair work of the aging anchors are as follows: -

1. During the integrity investigation of the aging anchors, documents and records were both are scattered and lost. Therefore, it is important to search as many documents as possible.
2. A corrosion inhibiting standard for tendons had not been established. Therefore,

corrosion had progressed around and below the anchor head and ruptures had occurred.

3. During implementation of the integrity investigation and repair of anchors aimed at preventing landslides, a survey of all the installed anchor positions and the land surface in the vicinity should be conducted carefully. From these surveys, it is important to identify abnormalities of topography and structures in the vicinity accurately and to prospect remained anchor force.
4. Lift-off tests should be conducted at a load level based on visual inspections results in order to avoid ruptures in tendon.
5. It is important to sum up integrity investigation result for each anchor and to utilize this information for repair, maintenance and inspection in the future.
6. It is impossible to repair whole anchors by current technology. Therefore repair work is conducted temporarily and is aimed at extending the life of anchors by maintaining current function. For anchors with severely low integrity, drastic measures such as renewal should be taken.
7. After completing repair work, regular inspections should be taken by the person in charge of technology. In case that repair work should be done again, immediate countermeasures should be conducted.

Integrity investigation and repair work is necessary for many anchors in the future. This investigation of integrity and repair work of 30 years old anchors will be a good reference for similar cases in the future, especially small projects.

References

1) The Japanese Geotechnical Society: "Ground anchors design and construction standard with explanation" pp.145-198, 2000

2) The Japanese Geotechnical Society: "Earth anchors construction method" pp.23-30, pp33, 1976

3) Toshiyuki Nagao: "Maintenance, management and repairing of ground anchors" ("Civil engineering technology" pp.25-31 Vol. 60, No.8)

4) Seiichi Onodera et al: "Integrity investigation of ground anchors" ("Civil engineering technology" pp.32-37 Vol.60, No.8)

Performance and Condition Assessment of 30 Year Old Anchorages, River Clyde, Glasgow

Devon Mothersille, Geoserve Global Ltd, London, UK
Stuart Jackman, Ritchies, Glasgow, UK
Jim Ferrier, Glasgow City Council, Glasgow, UK

Abstract

To assess the condition and service performance of 30 year old anchorages located within a tidal zone, a programme of inspections and testing was commissioned. In the first phase anchorages were subjected to detailed visual inspection and classification according to the extent of corrosion. In the second phase selected anchorages were subjected to lift-off tests and some strands were tested to failure. To further investigate the severity of corrosion, exhumations were carried out behind the anchor head and within the free anchor length where samples were extracted and subjected to metallurgical examination. In addition, samples of strand, with the original barrel and wedges still attached, were tested to establish strand capacity and the gripping efficiency of the existing barrel and wedge units.

The information derived from this inspection and testing programme provided justification for the down-rating of a proportion of the working anchorages and the recommendation that, in some cases, the anchorages were no longer fit for their intended purpose.

Introduction

A walk along the banks of the Clyde, through the city centre of Glasgow, reveals the many forms that the river containment has taken; gravity masonry walls, rockfill revetments, steel sheet pile walls and reinforced concrete structures. Many of the more recent examples have employed ground anchorages to resist lateral earth pressures imposed by the retained soil which was typically sands and gravels.

In 2005, Glasgow City Council awarded contracts to ascertain the condition and service performance of the anchorages, involving a programme of inspections and testing carried out in two phases. In the first phase anchorages were subjected to detailed visual inspection where each unit was photographed and rated to classify the extent of corrosive degradation. In the second phase selected anchorages were subjected to lift-off tests to confirm residual load capacity and ultimate load tests where strands were loaded to failure. To further investigate the severity of corrosion, exhumations were carried out behind the anchor head and within the free anchor length where samples were extracted and subjected to metallurgical examination.

Phase I – Visual inspection, categorisation and test classification

A typical general arrangement of the quay walls showing the location of walls 16, 18 and 20 is shown in Figure 1. The anchorages in quay wall 16 were constructed in 1972 and those in quay walls 18, 20 and 23 in 1974. Installation techniques and anchor tendon materials used were consistent with the limited technology available at that time and it should be noted that no British Standard Code of Practice for Ground Anchorages was published until 1989 (BS 8081:1989). The majority of the anchor heads were located within the tidal zone and installed with corrosion protection considered inadequate by current standards.

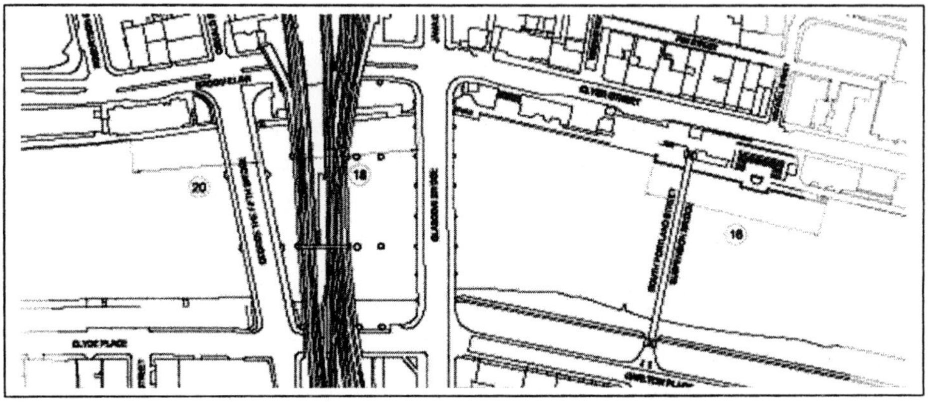

Figure 1 Location plan showing position of River Clyde quay walls 20,18 and 16

Visual inspections

During Phase I anchors heads were inspected and photographed as a preliminary phase to allow categorisation of the anchorages (Tables 1 and 2), establish length of protruding strands, general accessibility and identification of the anchorages that were to be subjected to more onerous load testing and metallurgical examination.

Grade category	Category definition
A	Good condition with superficial surface rust
B	Satisfactory condition with some corrosion evident. However, future lifespan of unprotected anchor head in aggressive marine environment limited.
C	Poor condition with evidence of deep corroded surfaces. Some section loss and small crevices also visible. Anticipated lifespan of corrosion protection to the anchor steel extremely short. Remedial action recommended.
D	Very poor condition. Severe and significant section loss evident. Immediate remedial action/replacement necessary to prevent further corrosive deterioration of anchor steel.

Table 1 Visual observation grading with definitions

Anchor No.	No. of strands	Strand length	Strand	Barrel/ wedges	Bearing plate	Gusset plate	Test type (see Table 3)
16/1	4	500	D	B	B	B	1 & 2
16/5	4	430	D	C/B	C/B	C/B	1
16/7	4	450	D	C/D	C	C	1
16/8	4	430	D	C	C	C	1
16/27	4	600	D	B	B	B	1
16/34	4	600	D	B	B	B	1 & 2
18/5	4	0	n/a	B	B	B	3 & 4
18/40	5	0	n/a	C/B	C/B	C/B	3 & 4
20/1	5	0	D	C/B	C/B	C/B	3 & 4
20/18	5	50	D	B	B	B	3 &4
23/1	5	0	D	B	B	B	3 & 4
23/22	5	0	D	B	B	B	3 & 4
23/39	5	0	D	B	B	B	3 &4

Table 2 Extract from field observations on anchorages selected for testing condition grading for anchor head components

Test classification

Generally, anchor head inspections, are limited to observing the integrity of the anchor head corrosion protection. Such inspections do not necessarily address the condition of the anchor components lying beneath the anchor head (i.e. the tendon and the internal corrosion protection system). It is noteworthy that there is currently no method established in the ground anchorage industry for predicting the future performance of corroded anchor components, consequently destructive tests on some in situ anchor components were undertaken and the results used to qualitatively assess the performance of the anchored walls (Table 3).

Test type	Test Description
1	Lift-off tests
2	Strand tested to failure through the anchor head
3	Exhumation inspection and removal of strand (minimum 1m length) for remote testing of strand to failure
4	Exhumation inspection and removal of strand (minimum 1m length) through the anchor head with barrel and wedge attached for remote testing of strand head to failure

Table 3 Summary of test categories

The assessment of the load performance of the 30 year old anchorages could only be carried out where adequate length of strands protruded from the anchor head. *Lift-off tests* were carried out only on quay wall 16 anchorages. Owing to the lack of protruding tendons no load checks were possible on anchorages in quay walls 18, 20 and 23. On these walls inspection of corrosion of the anchor tendon, within the wall, using exhumations, where access was available. Extracted samples of strand were subjected to metallurgical examination which included; image analysis of cross section macrographs to estimate remnant cross sectional areas of recovered strand samples and comparative metallurgical examination of recovered strand samples, using nital etching to determine strand microstructure and extent of corrosion.

The nature and configuration of the existing anchorages

Before discussing the results derived from the test programme it is important to appreciate the physical construction of the existing anchorages.

Tendon bond length within the fixed anchor length – all walls

Contrary to current UK practice the anchor tendons fabricated and installed in the quay walls were not encapsulated in the fixed anchor length but, as was the trend at the time, the bare steel strands were surrounded by grout within the borehole. At the depth of the fixed anchors, the grouting of the bare strands within the cementitious grout would undoubtedly provide some degree of protection and isolation of the strands, albeit some cracking of the cement grout column during anchor stressing is considered likely which could provide exposure of the steel to the environment. Furthermore in the absence of suitable centralisers, some of the steel strands may lie on the bottom of the borehole and be in direct contact with the soil.

Tendon debonded length within free anchor lengths – Walls 18, 20 and 23

Exhumations behind the retained face confirmed the existence of grease filled sheathing around each of the strands but contrary to the requirements of BS8081:1989, no group sheath was present. On each occasion where the grease filled polypropylene sheathing was exposed, a visual inspection was carried out. Generally, the sheathing was found to be intact but at three locations (out of 23 strands inspected) damage due to scuffing, abrading or tearing was apparent (Figure 2). It is also significant that under this sheathing damage, generally the strands appeared to have maintained their grease coating. There was, however, a risk element in the loss of water-tightness of the polypropylene sheath and for this reason, where damaged sheathing was encountered local repairs were carried out.

Underhead protection of the tendons – Walls 18, 20 and 23

The underhead is known to be a particularly vulnerable area owing to transition in the corrosion protection systems surrounding the strands, and the possible tendon contact or deviation through adjacent structures, and due to the frequent existence of variable environmental conditions i.e. ground water flow, tidal variations, transient contact with aggressive materials, combination of oxygen and chlorides and anaerobic conditions in the presence of sulphates. Trial pit inspection of the strands within each anchor tendon indicated that the grease-filled sheathing had been extended through the steel pile (through a short plastic duct) and actually passed through the hole in the head plate in an effort to extend the corrosion protection to the underside of the prestressing barrel.

Figure 2 Damaged polypropylene sheathing in free anchor length

Figure 3 Exposed anchor head on Wall 16 (Graded C/D)

Overhead protection of the strands /barrel wedges - Walls 16, 18, 20 and 23

The condition of the external faces of the barrels varied between moderately corroded Class B to extensively corroded Class D – (Table 1). For those barrels identified as Class B there was no evidence of deep pitting or the inability of the component to perform in its existing condition. An example of the more extensively corroded units was seen in the barrels on eight anchorages at the up stream end of wall 16. These were identified as class C/D with visible signs of deep corrosion and loss of section loss (Figure 3).

Evidence of strand slippage was also observed. In some cases the strand completely slipped through the anchor head leaving the barrel and wedges held in place on the bearing plate by corrosive deposits (Figure 4). In other cases some strands were still visible in the wedges but had slipped down to positions almost flush with the top surface of the wedges (Figure 4). Contributory factors to the slippage of strands is the accumulation of corrosive deposits within the wedges and loss grease or oil lubrication without which the performance of the wedge grip system is grossly reduced (Figure 5).

Testing of strand capacity on Walls 18, 20 and 23

Eight strand lengths of approximately 1.5m were exhumed and removed from within the grease filled sheath located in the upper tendon unbonded length. Six tensile tests produced failure capacities that averaged 6% greater than the 227 kN characteristic strength for the 15.2 diameter prestressing strand used. Two failure capacities were 2% and 5% less than the characteristic strength. Investigation works carried out on characteristic strength of prestressing strand produced and tested in 1973 (Barley, 1992), provides supportive evidence that the unexpectedly high capacities were achieved by the use of exceptionally high quality strand supplied by manufacturers.

Figure 4 Strand slippage on Wall 18

Figure 5 Post-failure sample of strand and wedges from Wall 20

Testing of strand capacity integral with the barrel and wedge system -Walls 18, 20 and 23

Following exhumation, strand lengths incorporating the locked-on barrel and wedges were withdrawn from nine anchor heads on walls 18, 20 and 23 and subjected to tensile and gripping efficiency tests. Five of the nine tests exhibited strand failure capacities within the wedges at average values 2% greater anticipated characteristic strength of 15.2mm diameter prestressing strand. The reasons for this unexpected occurrence are understood and are explained above.

One strand failure occurred within the wedges at a value of 220 kN equating to 97% of the strand characteristic strength. Two strand failures occurred within the under head below the wedge location, at 220 and 237 kN. Failure of strand below the wedges is an indication of loss of section due to strand wire corrosion either within the head plate or within the underhead zone. One of the nine tested strands failed by strand pull out through the wedges at a load of 182 kN. Failure of strand by slippage through the wedges can also result from loss of section of the strand steel due to corrosion. A similar mechanisms of failure was reported by Barley (1997) during the investigation of an anchored quay wall failure at Greenwich, UK. It was estimated that corrosive degradation accounts for a section loss in the peripheral wires of between 2% and 16% and a barely detectable loss of area of the king wire (see Figure 6). Further test results show losses of 11.25% and 16.1% of peripheral wire area in wall 16 anchorages.

On completion of the extractions new lengths of strand were coupled onto the existing strand, protected and stressed with replacement barrel and wedges.

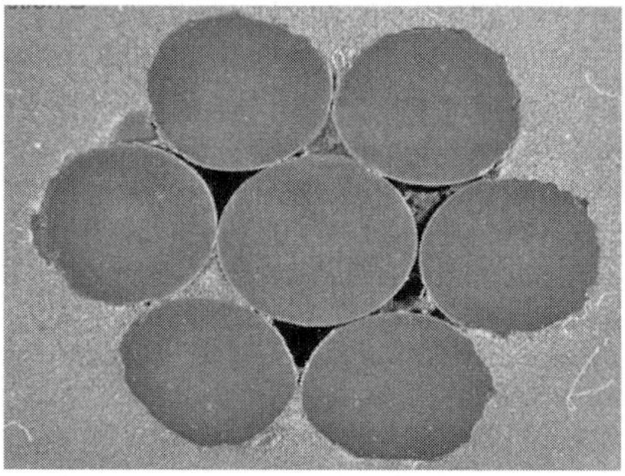

Figure 6 Macrograph showing section loss in peripheral wires (extracted from Wall 23)

Figure 7 Post-failure sample of strand and barrel extracted from Wall 16

Tendon dedonded length within the free anchor length – Wall 16

In 1972 the fabrication of greased-filled sheaths around prestressing strand was in its infancy and not generally available in the anchorage market. Inspections confirmed that no greased sheathing was used to protect the strand during the construction of wall 16 anchorages. Protection took the form of wrapping of each strand with Densotape, a system frequently used in protecting tie-rod works. However, the Densotape wrapped around the strand is susceptible to damage during anchor installation and to drying out over longer periods. The degree of protection of the steel strands against corrosion is compatible with that utilised at Greenwich where collapse of the anchored wall took place (Barley, 1997).

Underhead protection of tendon strands

The underhead protection applied to the anchorages in wall 16 would, at best, be similar to that applied to the tendon debonded length. However it is probably that considerable lengths of strand without Densotape wrapping and without underhead grout would be present at the point of strand passage through the sheet piles. Owing to the lack of provision of any grout containment, no protection of any sort would exist around the strand below the head plates. The high grade steel strands were completely exposed to corrosive attack in the tidal environment.

Testing of strand capacity - Wall 16

No access was available behind wall 16 therefore exhumation and sampling was not possible to assess strand capacity. However, during a previous investigation, carried out in 1996, a protruding length of bent strand, was cut off above the barrel and despatched to a laboratory for analysis, the resulting report (Bridon Report, 1996) stated that *the reduction in the sectional area of the king wire and outer wires was 28% and 9% respectively*. Although this investigation was carried out on a length of strand cut from above the anchor head, the exposure to the environment above and immediately below the head plate may be considered comparable, and strand corrosion of a similar order, in view of the absence of a protective cap.

During the execution of Lift-offs (Test type 1) loads in individual strands ranged from 35 to 85 kN with no strand failures occurring during the procedure. However, the test on two strands (No's 16/1 and 16/34) were specifically extended to establish the failure capacity of this strand (Test type 2). These strands failed at applied loads of 172 and 114 kN respectively. Failure occurred at 50 and 150mm, respectively, below the base of the barrels (i.e. in the underhead zone). These destructive tests were performed on anchorages at each end of the quay wall.

Investigation of loads in existing anchorages

Anchorages to Walls 18, 20 and 23

Based on four or five strands making up an anchor tendon, the probable tendon capacity can be evaluated. Considering a percentage of strand capacity utilised, it is possible to estimate the original anchor working loads.

Characteristic strand capacity = 227 kN

Percentage of capacity used = 50% x 227 = 113.5kN

Maximum working load of 4 or 5 strand anchorages = 454, 567 kN (say 400kN, 500kN)

It was common practice in 1974, as it frequently is now, to load the anchorages to a "lock-off" load at 110% working load. This overload is provided to accommodate a nominal 10% load loss due to anchor creep and or strand relaxation i.e. Lock-off loads were probably in the order of 440kN and 550 kN.

Where adequate length of strand is left protruding through the barrels and wedges (circa 200mm) then it is possible to investigate the actual existing load locked-in to each strand. This was possible only on anchorages to wall 16. The preliminary observations confirmed that the barrels and wedges generally remain in contact with the anchor head plate. This is an indication that some tensile force remains in each anchor strand.

Observation in the five trial pits confirmed that all but one of all the strands inspected were in tension. The remaining but major supportive factor in demonstrating that the in-situ anchorages were currently able to sustain the required load is the apparent stability of the existing walls. There was no evidence of the presence of tensile cracks in the paving behind the wall. In the event of anchorage failure this is frequently an early indication of distress. Furthermore there have been visual inspections of the anchored walls, which showed no sign that any local loaded areas of sheet piling have moved towards the river. This would probably be observed in the event of individual or progressive failure of anchorages along the quay (as recorded at Greenwich prior to wall failure – Barley, 1997). However, based on these observations it was not possible to assess what, if any, factor of safety is provided by the in-situ anchorages to resist wall failure.

Anchorages to Wall 16

Six anchorages present in wall 16 exhibited adequate lengths of individual protruding strand to allow lift-off of checks (Type 1, Table 3). The average recorded loads in strand tendons based on 3 lift-offs are as follows:

74, 80, 75, 71, 70, 37, 67, 66 and 67 kN.

The four 15.2mm diameter prestressing strands would probably have had a lock-off load of 440 kN. This would suggest that the 110 kN strand lock off loads have reduced by approximately 40% to an average lift off test value of 67 kN.

Should these losses be entirely due to the inward movement of the sheet piles the (assuming a 20m free length) pile movement of approximately 48mm would have been required. Alternatively load loss in the fixed anchor due to anchor creep or load loss due to increase in debonded length (progressive debonding) or direct relaxation of the prestressing strand could have contributed. These load loss factors could have been supplemented by the presence of reduced section of the prestressing strand, particularly in the underhead zone. Some strand slippage within the wedges could have taken place but owing to the variation in lengths of protruding strands it is not possible to establish precisely where slippage had occurred.

In addition to the six lift off tests, two extended tests, which involve loading anchor strand to failure, were carried out. Explosive failure loads at approximately 50mm and 150mm below the barrel base occurred at loads of 172 and 114 kN respectively (see Wall 16 – post failure samples, Figure 7).

Summary and closing remarks

Investigation has been carried out to determine the probable weakest component of the anchorages. There is every indication that the capacity of the prestressing strand within the barrel and wedge arrangement or immediately below the anchor head plate, or the barrel and wedge itself contribute a weakness in the existing system. Such potential weakness results from the omission of protection against steel corrosion to any or all of these components.

However, tests on 8 out of 458 working strands demonstrated that despite the exposure of the parts to corrosion for a 32 year period, the tensile capacity of these removed components equated closely to the characteristic strength of the strand.

BS 8081 recommends the application of a factor of safety of 2.0 to the guaranteed breaking capacity of the strand or barrel/wedge assembly or a factor of safety greater than 3.0 where the design utilises some component of capacity which is not proven, or is known to exhibit variation in performance. Despite 8 satisfactory strand results, the occurrence of slippage at loads varying from 183kN to approximately 100 kN on 4 strands prevents an accurate assessment of the failure capacity of the 32 year old anchor head components. Thus it would be inappropriate to utilise a factor of safety of 2 in a design check.

The application of the principles applied by BS 8081 demand that a factor of safety of at least 3 be applied to the original breaking capacity of the strand or to other load transfer mechanisms. This approach requires the working capacity of the 480 strands not to exceed 75 kN. Even with this interpretation it must be appreciated that excessive loss of section to a number of strands in a single anchor or in proximal anchorages still provides a risk element of local wall failure.

This general consideration recommends that working loads in 4 or 5 strand anchorages does not exceed 300 or 375 kN. It is necessary for the Engineer responsible for the review of the stability of the anchored wall to ensure that adequate factors of safety against wall failure are provided with the application of these reduced anchor capacities. It should be borne in mind that accommodation of these proposals only allows assessment of the current stability in the existing state of component exposure to corrosion.

Mothersille and Barley (2007) report on remedial measures applied to the anchor head of anchorages installed in 1970s. Such an approach should extend the useful life of the anchorages installed in walls 18, 20 and 23. Consideration should also be given to the long term monitoring of these 3 walls by observation and measurement with particular attention placed on Wall 18 in the areas where strand slippage occurred (e.g. anchorages 18/7, 18/9 and 18/29).

The 154 anchor strands (37 anchorages) installed in 1972 to wall 16 are provided with inferior corrosion protection to those in other walls. Lift-off tests on 9 strands indicated existing working loads of 66 to 80 kN, albeit one strand was known to be *slack*. Owing to the variation in length of strand protruding above the anchor head it is not possible to detect the number of strands that may already have slipped through the wedges, in the mode observed in wall 18.

The estimated loss of section on strands and barrels and wedges is severe (as much as 28% area loss on one king wire). Furthermore on 2 strands, tests were extended and failure loads occurred at 75% and 50% original strand breaking capacity. The open exposure of the steel in the strand and barrel and wedge components in wall 16 is identical to that encountered by the anchorages at Greenwich prior to failure. However, it is appreciated that the composition of the corrosive environment surrounding the anchor heads on the two sites has not been compared. The Greenwich anchorages, installed in 1969, failed in 1990 and total wall collapse took place. In view of this, it was recommended that immediate consideration be given to the observation and monitoring of Wall 16 followed by replacement of the anchorages.

Acknowledgements

The authors wish to thank Mr A.D. Barley for his assistance with the technical interpretation of the data derived from this project. The engineering management and project logistics provided by Mr Alistair Walker is also acknowledged with thanks.

References

BARLEY, A.D. (1997) "The failure of a 20 year old anchored sheet pile quay wall on the Thames". Ground Engineering, March issue, 42-45.

BARLEY, A.D. (1992) "The use of Ground Anchorages in Waterfront Structures" Presented at the International Shipping and Marine Technology Market with Congress, Hamburg, September – October.

BRIDON REPORT (1996) Confidential report produced by Bridon Wire Technical Centre, Doncaster, UK

BS8081 (1989) "Code of practice for ground anchorages". British Standards Institution, London.

ICE (1997) "Ground anchorages and anchored structures". Proc. Int. Conf. organised by the Institution of Civil Engineers (editor G S Littlejohn), Thomas Telford, London .

MOTHERSILLE, D.K.V and BARLEY, A.D (2007) "Inspection anchorages supporting waterside structures" Proc. Int. Conf. "Ground anchorages and anchored structures in service", Institution of Civil Engineers, London.

Some examples of the poor installation of electrically isolated prestressed ground anchors in Slovenia

Iztok Klemenc and Viljem Kuhar, Slovenian National Building and Civil Engineering Institute, Ljubljana, Slovenia

Abstract

Since 1994, when Slovenia's National Motorway Construction Programme began, a total of 85 retaining structures with permanent prestressed ground anchors have been built. Even though the design of all the used anchoring systems was thoroughly reviewed, and the performance of typical installed anchors was verified by means of preliminary field and laboratory tests, some cases of poor installation practice, indicating the possibility of significant corrosion damage to the elements of anchor heads, were discovered. The paper provides a description of the findings of inspections of the most critical anchors of several anchored structures. The poor condition of the investigated anchors can be ascribed mainly to the presence of water beneath the anchor caps and covers as a result of non-watertight fixing. The findings of laboratory tests of samples of the corrosion protection compounds used, i.e. grease and vaseline, are also described.

Introduction

Since the start of the implementation of Slovenia's National Motorway Construction Programme, in 1994, a total of nearly 370 km of motorways and dual carriageways have been built. Their routes frequently run across hilly and geologically difficult ground, so that so far a total of 85 anchored retaining structures have had to be constructed, with a total of more than 9000 permanent prestressed ground anchors (PPGAs), which were mostly standard anchors, in addition to which there were some instrumented anchors, equipped with an electronic load cell, as well as some control anchors for later verification of load-bearing capacity. As is well-known, the most sensitive part of any PPGA is the area around the anchor head [1],[2], with the transition from the factory-manufactured encapsulated anchor tendon to the anchor head elements. In the case of a non-watertight transition between the free length and the anchor head, the high tensile stresses in the tendons could result in stress corrosion cracking (SCC) and hydrogen embrittlement.

In the early 1990s there were no adequate regulations about the design and installation of PPGAs, so in 1996 the Ministry of Transport of RS decided to adopt the Swiss recommendation SIA V 191 [3], which states that only electrically isolated anchors with comprehensive protection against corrosion can be counted as permanent prestressed ground anchors. Soon after this it was agreed that complete tests of each type of permanent anchor, manufactured by an individual producer, should be carried out, in accordance with the requirements of the client: DARS, the "Motorway Company in the Republic of Slovenia". On the basis of a corresponding certification scheme, where the requirements of the SIA V 191 recommendation and other technical regulations [1],[4] as well as our own experience were taken into account, the key components of the different types of PPGAs were tested, and surveillance of their production was implemented. The suitability of different PPGA anchoring systems was also verified on a test field. The procedure was completed, for the first type of PPGA, in January 1998.

In connection with the installation of PPGAs, a surveillance system which would suit the particular characteristics of such anchors was developed, which included:

- continuous preliminary laboratory testing of the components used to manufacture PPGAs (i.e. the steel strands, bearing plates, wedge plates, wedges, smooth and corrugated polyethylene (PE) ducts, and the corrosion protection compounds – vaseline and grease),
- periodic inspections of the production of PPGAs,
- testing of the load-carrying capacity of installed PPGAs, including, in the case of the majority of structures, an investigation test to failure of a typical PPGA in order to determine its ultimate load-carrying capacity, on the basis of which criteria were defined for the performance of comprehensive acceptance tests on about 10 % of the anchors, and simple acceptance tests on the remaining anchors of each retaining structure,
- for each installed PPGA, testing of the integrity of the encapsulation of the anchor, i.e. the electrical resistance between the anchor and the surrounding ground.

In the years which followed the construction of anchored structures regular routine inspections and maintenance of the anchor heads were performed only exceptionally, in cases where particularly poor installation practice had been observed. Each such inspection consisted of:

- a general visual inspection of the installed anchors,
- visual assessment of the corrosion protection condition of the anchor's steel parts,
- electrical resistance measurements (ERM I) according to SIA V 191,
- in some cases, additional low-voltage resistance and capacitance measurements, and electrochemical impedance spectroscopy (EIS),
- laboratory tests of samples of vaseline and grease, taken from the area of the anchor heads.

The text which follows contains a description of the results of such inspections, carried out within the framework of a research project [5], at two different locations: on two anchored piled retaining walls and on a vertical anchored beam next to the portals of a tunnel, and on an anchored retaining wall next to an exit road from the motorway.

Damage to anchors supporting two anchored piled retaining walls, located next to the portals of a tunnel

In the year 2000, in the area in front of the portals of two tunnel tubes (see Figure 1 L), two anchored piled retaining walls were constructed, with lengths of 75 and 66 m respectively, into which a total of 178 PPGAs were installed, on six levels. In the same area a number of drainage boreholes had been constructed, which were designed in such a way that the water drained directly over the reinforced concrete horizontal beams with the embedded anchor heads of PPGAs, so that the concrete surfaces were subject to a more or less constant flow of water (see Figure 1 R). Taking into account the externally visible state of the anchored structure (the degree of wetting of individual parts of the piled walls, and the state of the concrete cover of the anchor heads), several typical anchors were selected for inspection, where the most severe corrosion damage was anticipated.

In the case of the piled wall next to the left tunnel tube, the concrete cover and metal anchor caps of three standard anchors were removed, whereas in the case of the piled wall next to the right tunnel tube the anchor cap of an instrumented anchor was removed. All four inspected anchors were of the 5-strand type, each having a total length of 24 m, with a 10 m long bond length. These anchors had been locked-off at a load of 600 kN (i.e. at a load equal to 48 % of P_{tk} - the characteristic load capacity of the tendon). In the case of the three standard anchors, it was found that (1) water had penetrated through the porous concrete which was used to embed the anchor cover, and (2) the embedding concrete itself had not bonded well with the RC beam. As a result, since there was no adequate gasket, water was able to penetrate beneath the caps of all the inspected anchors. The presence of water under the anchor caps caused wetting and consequently degradation of the grease. There was also quite a lot of water "trapped" in the grease. It had partly degraded into fatty acids, and had become potentially corrosive and hygroscopic. Taking into account the observed poor adhesion of the grease onto the steel parts of the inspected anchor heads, it is possible that they were greased in unsuitable (i.e. damp) weather conditions.

In the case of all four of the inspected anchors, the measured values of ERM I did not satisfy the electrical resistance criterion according to SIA V 191, i.e. $R_I \geq 0.1$ MΩ, which means that they did not satisfy the requirements for "comprehensive corrosion protection".

Figure 1. L: View of the two anchored retaining walls next to the portals of the two, not yet constructed tunnel tubes
R: Traces of water action on an embedded anchor cap

Inspection of embedded standard anchors

In the case of all three inspected standard anchors, water and the dark products of degraded grease could be seen to have leaked out through the lowest joint between the concrete cover of the anchor heads and the RC beams (see Figure 2 L). It was observed that no gasket had been fitted at the joint between the metal cap and the concrete (see Figure 2 M), and the grease, which had filled up approximately one-third of the cap, was superficially wetted, so that its original dark-green colour had changed, on its surface and to a depth of up to several millimetres, to a colour which was lighter than the original. The grease had also partially slipped towards the bottom of the cap. In all three cases corrosion damage to the bearing plate was visible (see Figure 2 R). Crevice corrosion might have begun at the joint between the bearing plate and the underside of the wedge plate, and water might have penetrated by capillary rise up inside the anchor head, to the wedges and strands. On the other hand, it appeared that the grease had been able to provide sufficient corrosion protection to the wedge plate and strand ends of all three inspected anchor heads, since there was no observable corrosion damage to these elements. It was also observed that the strands had too short cut-off lengths of less than 1 cm (see Figure 2 R), which were probably not due to strand slippage. In May 2004, in the case of the inspected anchors the measured electrical resistance R_I amounted to between 0.008 and 0.013 MΩ.

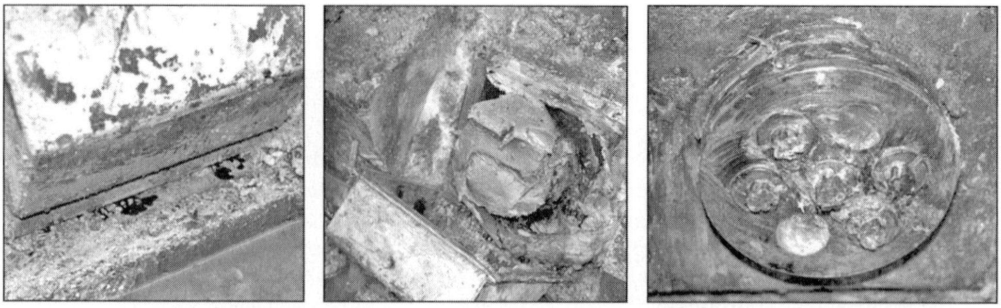

Figure 2. L: Discharge of the degraded products of grease from underneath the metal cap of one of the inspected embedded standard anchors
M: The heavily wetted grease beneath the cap of the same standard anchor
R: Corrosion damage to the most badly affected standard anchor

Inspection of an instrumented anchor

The corrosion protection of the anchor cap of the instrumented anchor was insufficient: (1) the cap was only protected by an anti-corrosion primer, and (2) no gasket had been fitted so that water had entered the cap from above, and then flowed out through the joint at the bottom of the cap (see Figure 3 L). The layer of grease covering the wedges and the strand-ends was approximately 1.0 cm thick, and the anchor head and load cell were wet. The colour of the surface layer of the wetted grease had changed to a lighter colour, and some grease had slipped towards the bottom of the cap (Figure 3 M). Two samples of grease were taken for laboratory tests, one from the anchor head the other from the cap. At the joint between the bearing plate and the load cell more severe corrosion damage was observed, whereas initial corrosion could be seen on the wedge plate (see Figure 3 R). The cut-off length of the strands was less than 1 cm. The measured electrical resistance R_I of the anchor was 0.004 MΩ.

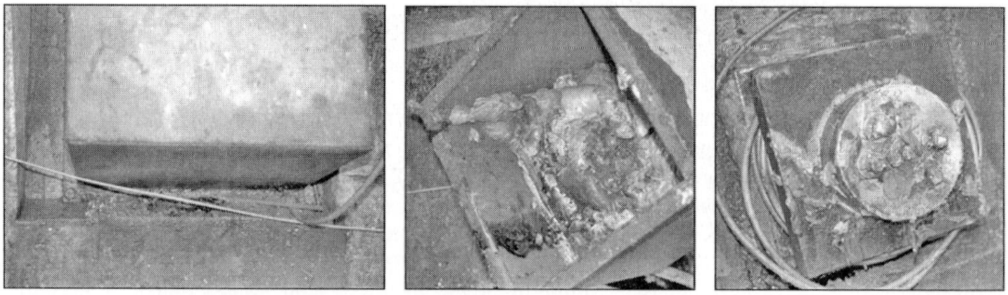

Figure 3. L: Discharge of water and the degraded products of grease through the joint between the instrumented anchor cap and the recess in the RC beam
M: The wet grease had degraded and slipped downwards
R: Corrosion damage had occurred to the anchor's bearing plate

Inspection of a standard anchor of a vertical anchored beam

In January 2000 eight 5-stranded standard PPGAs, each with a total length of 24 m and a 10 m long bond length, which had been locked-off at a load of 600 kN (i.e. at a load equal to 48 % of P_{tk}), were installed in a vertical RC beam next to the portal of a tunnel tube. This beam, when it was inspected in May 2004, was in a very poor state (see Figure 4 L): its bottom part was cracked, and in this area the sprayed concrete had fallen off, taking with it, in all probability, the cap of one of the anchors (see Figure 4 M). This PPGA, which was the second to bottom one, was selected for detailed inspection. The anchor head was found to be severely damaged by corrosion, so that only two of the initially five strands could be seen in the anchor head (see Figure 4 R). Water was observed to drip out from one of the holes in the wedge plate.

Figure 4. L: View of the anchors in the bottom part of the vertical RC beam, showing the most severely damaged anchor (in the middle)
M: The anchor cap of the most severely damaged anchor had fallen off due to too weak fixing of the cap
R: The anchor head of the most severely corroded anchor

Damage to anchors of an anchored retaining wall, located next to an exit road from the motorway

In autumn 2000 a total of 48 PPGAs were installed in 4 rows in an anchored retaining wall located next to the exit road from a motorway (see Figure 5). These anchors were of the 3-strand type, with total lengths of 15 and 19 m, and bond lengths of 7 m. The anchors were locked-off at loads of 360 and 380 kN respectively (i.e. at 46 % and 49 % of P_{tk}).

Figure 5. View of the anchored retaining wall next to the exit road from a motorway

Inspection of a control anchor

In November 2003 a visual inspection was made of the anchorage of the control anchor (in the second row of anchors). It was found that polyurethane (PU) foam had been used to fill the empty space between the anchor head, which had been covered in vaseline, and the anchor cover, and that there was no gasket, so that this joint was not watertight. The inner surface of the anchor cover was corroded (see Figure 6 L), as was the anti-corrosion coating on the outside of the cover. The surface of the PU foam inside the recess, which filled up about three-quarters of the void under the cover (see Figure 6 M) was dry to the touch but inside (particularly in the lower part of the foam) it was very damp. During the inspection a small piece of the PU foam was removed so that it was possible to access the layer of protective vaseline. This vaseline was compact and visually in its original state, and it was bonded strongly to all the metal parts of the anchor head. A sample of this vaseline was removed for laboratory tests. Corrosion products were not, however, visible on the steel parts of the anchor head (see Figure 6 R). The measured values of electrical resistance R_I of all three strands of the anchor amounted, in November 2003, to between 0.98 and 1.01 MΩ.

Figure 6. L: Corrosion damage to the inside of the cover of the control anchor
M: Hardened polyurethane foam on the anchor head of the control anchor
R: No corrosion damage was observed on the bearing plate and bush of one strand of the control anchor when the layer of vaseline was removed

Inspection of an instrumented anchor and some embedded standard anchors

A further inspection of the anchored retaining wall was performed in May 2004, including a repeated inspection of the control anchor and an inspection of an instrumented anchor in the lowest row. The cover of the instrumented anchor (with no gasket) was similar to that of the control anchor (Figure 7 L). It could be observed that, during expansion, the PU foam had pushed away part of the protective layer of vaseline. Some of the anchor head parts were damp, and initial corrosion damage was visible on the bearing plate and one of the bushes (see Figure 7 M). At the time of this inspection the electrical resistance R_I of the instrumented anchor amounted to 0.002 MΩ. During the visual inspection of the embedded heads of some of thc standard anchors of the retaining wall, it was observed that cracks had occurred in the protective layer of concrete on top of the anchor caps, particularly in the case of the embedded anchors in the bottom row, which were closest to the exit road from the motorway. Even in the dry summer period, the concrete next to some of these cracks was damp (see Figure 7 R).

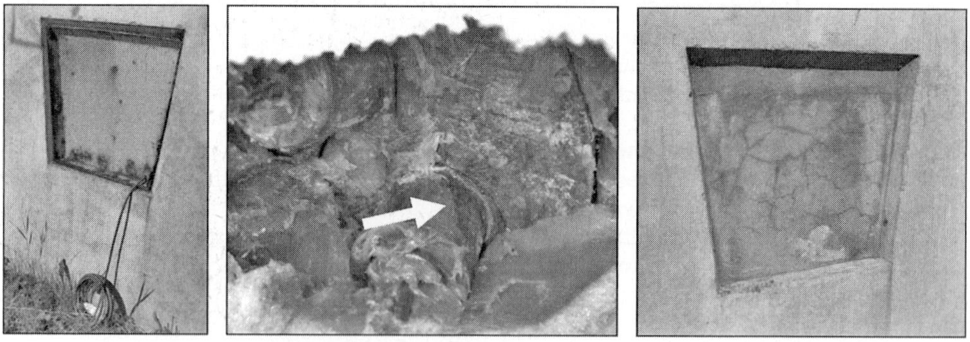

Figure 7. L: Corrosion damage to the cover of the inspected instrumented anchor
M: Initial corrosion damage to the instrumented anchor's head parts
R: Cracked concrete on top of the cap of an embedded standard anchor

Additional measurements on the control and instrumented anchors

The integrity of the encapsulation of various types of anchors is usually verified by measuring their electrical resistance R_I. However, additional information about the condition of the anchor's corrosion protection can be obtained by measuring the anchor's impedance (Z), its polarisation resistance (R_p) and its capacitance (C_f) [6],[7]. These values are obtained by performing measurements on the installed anchor using low voltages. In the case of very good electrical isolation the capacitance of the anchor will be high, whereas in the case of damage to the protective PE duct (cracks or perforations) the capacitance will be markedly less. In Table 1 the results of measurements of electrical resistance R_I have been compared with the measured values of resistance obtained by low-voltage measurements, and with the results of electrochemical impedance spectroscopy (EIS).

Table 1. Comparison of the results of measurements performed on the inspected control and instrumented anchors of the anchored retaining wall

	Resistance according to SIA V 191	Resistance*	Electrochemical impedance spectroscopy (EIS)**		
	R_I [MΩ]	R [MΩ]	R_p [MΩ]	R_u [Ω]	C_f [μF]
Control anchor	0.712	2.95	0.957	0.0004	0.032
Instrumented anchor	0.002	1.74	0.002	0.0026	0.048

* Obtained by using a Model 189 Fluke Multimeter, with alternating voltage < 5 V at 1 kHz.

** Obtained by using a Gamry PC3/750 Potentiostat, with a frequency spectrum ranging from 0.1 to 5000 Hz, and an amplitude 50 mV.

The relatively low values of R_I were probably due to electrical contact between the anchor head and the concrete surface of the retaining wall, via the damp PU foam, and not to damage to the protective PE ducts for the anchor tendons. It should, however, be noted that the most objective assessment of the integrity of the encapsulation of an anchor can be obtained by using a combination of several measuring methods, as described above.

Results of laboratory tests of the samples of grease and vaseline

At the same time that the anchors were inspected, three samples of the corrosion protection compounds were taken:

- In the case of the anchored piled retaining walls next to the tunnel portals, grease had been used to protect the anchor heads. As noted above, two samples were taken: one from the anchor cap of the instrumented anchor, which consisted partly of a water-saturated layer of grease with a light green colour and partly of grease which, visually, had not been affected by water, and one from the anchor head (this layer of grease was thin, and was water-saturated throughout its thickness, up to the steel parts).
- In the case of the anchored retaining wall which had been constructed next to the exit road, vaseline had been used to protect the anchor heads, whereas the empty space between the vaseline and the anchor covers was filled with PU foam. Due to the low temperature of the environment at the time of the inspection, in November 2003, the sample of vaseline was hard, and firmly attached to the steel parts of the control anchor and the PU foam, so it was not possible to obtain a sample of vaseline that did not contain particles of PU foam. The larger particles of PU foam were removed to the greatest possible extent in the laboratory at room temperature.

The IR spectrum of all three investigated samples of corrosion protection compounds indicated the presence of water. There was relatively little water in the sample of vaseline but a considerable amount in both of the samples of grease. A standard corrosion test was performed on all three samples: some of each of the samples of grease and vaseline was applied, to form a layer of thickness 0.5 mm, to the surface of plates made of low-carbon steel of quality S235JR. These plates were then exposed for 14 days in a salt spray chamber, according to the standard SIST ISO 9227:1999.

In the case of the two samples of grease, the corrosion process started at those locations where there was a higher water content, and where a change in the colour of the grease could be observed, which proves that water has an effect on the latter's corrosion protection properties. Individual larger damaged areas, with sizes up to 1.5 cm^2, were observed on the steel plates, and estimates according to the ASTM D610-95 corrosion scale varied from 5 to 9 (the sample of grease from underneath the anchor cap is shown in Figure 8 L), and from 6 to 10 (the sample of grease from the anchor head).

The results of the tests of the sample of vaseline showed that the small water-saturated particles of PU foam in the vaseline had, acting as initiators of the corrosion process, caused a deterioration in the latter's corrosion inhibiting properties. Some corrosion products of steel and some pits were observed on the surface of the steel plates, resulting in an assessment of between 7 and 9 on the ASTM D610-95 corrosion scale (see Figure 8 R).

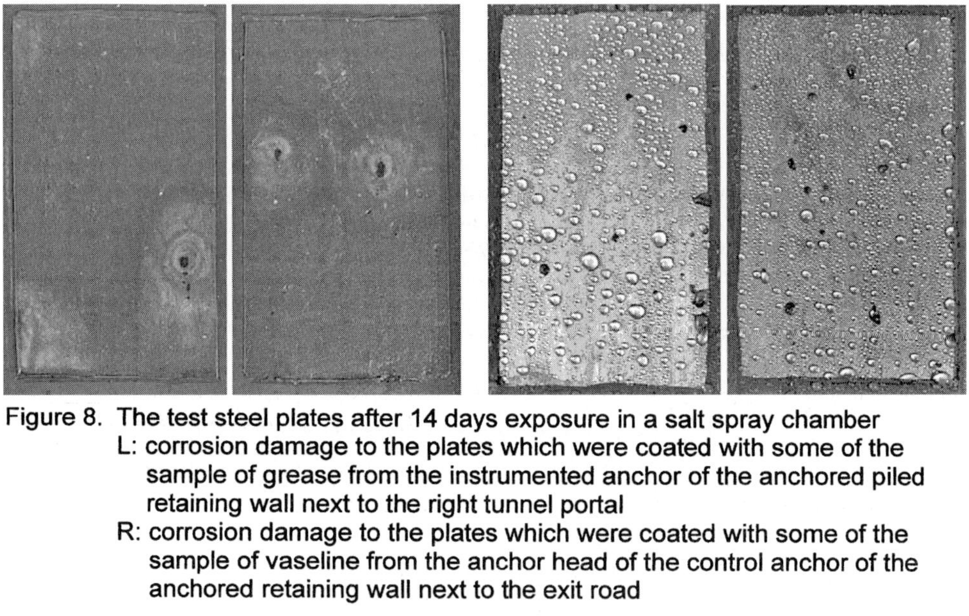

Figure 8. The test steel plates after 14 days exposure in a salt spray chamber
L: corrosion damage to the plates which were coated with some of the sample of grease from the instrumented anchor of the anchored piled retaining wall next to the right tunnel portal
R: corrosion damage to the plates which were coated with some of the sample of vaseline from the anchor head of the control anchor of the anchored retaining wall next to the exit road

Using ion-chromatography, the quantity of water soluble chlorides that could be extracted from the sample of hardened PU foam (which did not contain any flame-retardants) was determined by means of tests according to EN ISO 10304:1998. This quantity was found to amount to between 50 and 116 $\mu g\ Cl^-/g$ of foam, which is above the permissible limit for chloride content in a corrosion protection compound (the limit is set at less than 50 $\mu g\ Cl^-/g$). The results of the tests which were performed in order to determine the absorption of water in thermally aged and non-aged samples of the hardened PU foam showed that the PU foam was porous, and contained between 4 and 7 % of open pores, which, in the case of dampening, act as water reservoirs. This means that PU foam is not suitable for the filling of the hollow space underneath anchor caps or covers, since it can come into direct contact with the metal parts of the anchor heads, and the water which is present can extract chlorides from the PU foam. With the evaporation of this water the local concentration of the chlorides could, in the case of higher temperatures, be higher, and thus initiate the corrosion process.

Conclusions and recommendations

Based on the results of the measured ERM I values after lock-off, which were obtained on the installed anchors of the anchored structures which have been constructed since 1994, when Slovenia's National Motorway Construction Programme began, it can be concluded that, in the case of at least 90 % of these anchors, the integrity of their encapsulation was, at the time of installation, satisfactory ($R_I \geq 0.1\ M\Omega$). However, up until the present time periodic systematic inspection of these installed anchors has not been performed, except in the case when visible signs of the action of water have been observed, potentially causing degradation of the corrosion protection of the anchor heads, or in cases when it has been known in advance that the anchors were not been installed in accordance with the prescribed design and installation conditions.

In the case of the two inspected anchored piled retaining walls, whose anchor heads were protected by grease, the following potentially hazardous deviations from the prescribed design and installation conditions were observed:

- All the inspected anchor caps and covers were installed without gaskets, so that water could get underneath them. Apart from this, the metal caps of the two inspected instrumented anchors were only protected by a coating of anti-corrosion primer.
- In some cases there was not enough bond between the concrete of the RC beams and the embedding layer of concrete on top of the anchor caps. Particularly next to the edges of the cap this layer of concrete was too thin, and in some cases cracks had occurred. Locally, small pieces of the concrete had flaked off.
- As well as this, poorly arranged drainage systems behind the anchored retaining walls had led to constant water flow over all the inspected types of anchors and their elements. Due to all of the above-described effects, it is possible that crevice corrosion had occurred between the bearing plates and the underside of the wedge plates, so that water could, by means of capillary rise, have got inside the anchor heads. In combination with the high anchor loads (at the time of inspection amounting to approximately 70% of P_{tk}, which was probably due to the fact that the actual geological conditions were worse than the design expectations) conditions for the occurrence of SCC and hydrogen embrittlement may have occurred (the strands were made of steel grade Y1770S7 according to prEN10138-3:2005). Since the measured values of ERM I were, in the case of the inspected anchors, lower than the valid criterion (i.e. $R_I \geq 0.1\ M\Omega$) according to SIA V 191, the decision was made to extend both tunnel tubes almost as far as the end of the retaining walls, and to cover them with the fill which had been originally excavated before the retaining walls could be constructed.

An additional kind of deviation from the prescribed design and installation conditions was observed in the case of the anchored retaining wall next to the exit road, whose anchor heads had been protected by vaseline: unsuitable PU foam had been used to fill most of the empty space underneath the anchor covers.

As a result of the above-described deviations, the corrosion protection compounds which had been used to protect the anchor heads of the described retaining structures were degraded, to different degrees (more in the case of grease, and less in the case of vaseline), by the ingress of water and its constant presence beneath the anchor cap and covers.

To avoid the above-described damage phenomena, the following recommendations are made:

- When anchors are being installed, there should be strict and continuous supervision in order to verify whether the corrosion protection procedures are being performed in accordance with the prescribed design and installation conditions. Effective sealing of anchor caps and covers by means of corresponding gaskets is of key importance.
- Special attention should be paid to the execution of the corrosion protection of anchor heads located in the vicinity of carriageways which, in the winter period, are exposed to saline mist, or covered by snow that has been contaminated with chlorides.
- Regular visual inspections of completed anchored structures should be performed sufficiently frequently, e.g. once every two or three years, and, in the case that degradation of individual anchors is observed, more detailed inspections of these anchors should be immediately performed. Such regular inspections could then be used as a basis for the planning and implementation of maintenance and repair works.
- Within the framework of the maintenance plan, regular renewal should be performed of the corrosion protection of anchor heads, i.e. of any deteriorated gaskets beneath anchor caps, as well as of the protective coatings on anchor caps. Monitoring of the fixing of anchor caps and of the state of corrosion protection compounds beneath these caps is also essential. At the same time cleaning and maintenance of drainage systems in the vicinity of anchored structures needs to be ensured.

The question also arises as to whether the anchor caps of standard anchors should be embedded in concrete, in which case the areas around the anchor heads are better protected against negative environmental effects and possible vandalism, the conditions beneath the caps are more stable, and the corrosion protection compounds degrade at a slower rate.

Acknowledgements

The authors are grateful to DARS, which provided funds for the research project. The careful translation of this paper into English by Peter Sheppard is much appreciated.

References

[1] Corrosion and corrosion protection of prestressed ground anchorages (1986). FIP State of the art report, Thomas Telford Ltd, London.

[2] Hunkeler F. et al. (2005). Prestressing Tendons, Stay Cables and Ground Anchors – Description of the Systems and Lessons Learnt from Corrosion Damages. Swiss Federal Roads Authority, Department of the Environment, Transport, Energy and Communications, Switzerland (in German).

[3] Empfehlung SIA V 191 (1995). Vorgespannte Boden- und Felsanker. Schweizerischer Ingenieur- und Architekten-Verein, Zürich (in German).

[4] SIST EN 1537 (2002). Execution of special geotechnical works: Ground anchors. Comité Européen de Normalisation.

[5] Legat A., Leban M., Klemenc I., Trajkovski S., Bajt Ž., Čavlek K., Kuhar V. (2004). Final Report on the Research and Development Project: "Completion of Criteria for the Installation of Permanent Prestressed Ground Anchors" for the client: DARS – Motorway Company in the Republic of Slovenia (in Slovenian).

[6] Elsener B. (2004). Monitoring of electrically isolated post-tensioning tendons, IBWK, Swiss Federal Institute of Technology, Faculty of Engineering, University of Cagliari.

[7] Elsener B. (2004). Electrical Isolation as Enhanced Protection for Post-tensioning Tendons in Concrete Structures, Faculty of Engineering, University of Cagliari, Institute for Building Materials, ETH Hönggerberg, Zurich.

MAINTENANCE CONTROL OF GROUND ANCHORS CONSTRUCTED IN JAPANESE EXPRESSWAYS

Masahiro YOSHIMURA, Masaru TAKEMOTO and Kiyoyuki AMANO
NEXCO Expressway Research Institute, Tokyo, JAPAN

Abstract

Cut slopes on expressways are under extraordinarily severe conditions in Japan because approximately 70% of land is composed of unstable soils and thus frequently influenced by typhoons and earthquakes. Ground anchors have recently been adopted on numerous occasions for slope stabilization. Ground anchors have been in service for slope stabilization on expressways in Japan for approximately 40 years. Cases of fracture of ground anchors with inadequate corrosion protection (old-type ground anchors) on expressways in 1987 revealed the deformation of ground anchors due to corrosion. At present, ground anchors are applied with better corrosion protection. In the future, however, appropriate maintenance of ground anchors including those with inadequate corrosion protection, and the identification of life-cycle cost will be required.

This paper describes the functional deterioration of old-type ground anchors due to corrosion that was revealed as a result of investigations of anchors collected at six locations to detect the functional deterioration of old-type anchors. The paper also presents ground anchor maintenance standards on expressways. For investigations of the integrity of ground anchors, refer to another paper.

Investigation outline

Investigation sites were selected in places where tendons could be pulled out for major types of tendons adopted in old-type anchors, i.e. prestressing strand, composite prestressing tendon and prestressing bar (Table 1). In order to investigate the conditions for soil corrosion, a borehole was drilled diagonally adjacent to the point of tendon pull-out. Samples were taken at three levels, one, five and ten meters below the ground surface, to check hydrogen ion, sulfide ion, chloride ion, sulfate ion, specific resistivity value, sulfate-reducing bacteria and oxidation reduction potential. To test tendons, tension tests were conducted for the corroded section of the tendon, and the loss of cross-sectional area due to corrosion was measured.

Table 1. List of investigation points

District	Age (years)	Tendon specification	Number of tendons pulled out	Anchor length (m)	Free length (m)	Fixed length (m)	Length of tendon collected
A	17	PC bar/ 26 mm diameter	3	9.00	3.20	5.50	Three 9m
B	29	PC multi-strand/ 19-multi-9.5mm diameter strands with filler wires/45.7mm diameter	3	10.00	4.33	5.50	Two 10m and one 5m
C	25	PC bar/ 32 mm diameter	3	11.00	7.00	4.00	Two 6m and one 11m
D	18	PC 7-wire strand/ 5-12.7mm diameter strands	3	40.88	26.68	12.53	One 0.1m, one 29.7m and one 22.8m
E	17	PC bar/ 23 mm diameter	3	13.00	9.00	4.00	One 6m, one 4.9m and one 6.8m
F	32	PC 7-wire strand/ 7-12.7mm diameter strands	3	10.00	6.50	3.50	Three 0.2m

Investigation results

Soil corrosiveness at investigation sites

Identifying soil corrosiveness requires comprehensive evaluation of numerous environmental factors that contribute to corrosion in combination. Soil corrosiveness was evaluated at investigation sites using four methods that had already been proposed (Table 2). Figure 1 shows the ratings at respective points of investigation. Corrosiveness was rated higher in all of the districts by Baeckmann's method. In evaluation by Baeckmann's method, corrosiveness at the anchor head was at an intermediate level in districts A and B. Soil corrosiveness was low in other districts. Evaluation by the other three methods found low corrosiveness in all of the districts.

Table 2. Rating of soil corrosiveness

Item \ Rating index	The German Technical and Scientific association for gas and water (DVGW GW9-1971)[1] Classifications	Rating	Baeckmann (1971)[2] Classifications	Rating	American National Standards Institute (ANSI A21.5-1972)[3] Classifications	Rating	German Industrial Standards (DIN-50 929-1985)[4] Classifications	Rating
Soil	Calcareous soil, sand Loam Clay, corrosive soil Peat	2 0 -2 -4	Limestone, sand Loam, cohesive soil Sand Clay, corrosive soil	2 0 -2 -4				
Soil resistivity (Ωcm) (measured using a soil box)	>10,000 5,000~10,000 2,300~5,000 1,000~2,300 <1,000	0 -1 -2 -3 -4	>10,000 5,000~10,000 2,300~5,000 1,000~2,300 <1,000	0 -1 -2 -3 -4	>2,000 1,500~2,000 1,200~1,500 1,000~1,200 700~1,000 <700	0 1 2 5 8 10	>50,000 20,000~50,000 5,000~20,000 2,000~5,000 1,000~2,000 <1,000	4 2 0 -2 -4 -6
Water content (%)	<20 >20	0 -1	<20 >20	0 -1	Dry Wet Moist	0 1 2	≦25 >25	0 -1
pH	>6 <6	0 -1	>6 <6	0 -1	>8.5 7.5~8.5 6.5~7.5 4~6.5 2~4 0~2	3 0 0 0 3 5	>9 5.5~9 4~5.5 <4	2 0 -1 -3
Oxidation reduction potential	>400 200~400 0~200 <0	2 0 -2 -4	>400 200~400 0~200 <0	2 0 -2 -4	>100 50~100 0~50 <0	0 3.5 4 5		
Sulfide	None Stain Exists	0 -2 -4	None <0.5 >0.5	0 -2 -4	None Stain Detected	0 2 3.5	<5 5~10 >10	0 -3 -6
Chloride ion	<100 >100	0 -1	<100 >100	0 -1				
Sulfate ion	<200 200~500 500~1000 >1000	0 -1 -2 -3	<200 200~500 500~1000 >1000	0 -1 -2 -3				
Groundwater	None Exists Groundwater table varies	0 -1 -2	None Exists, groundwater table varies	0 -1			None Exists Groundwater table varies	0 -1 -2
Evaluation standards	Non-corrosive Slightly corrosive Corrosive Highly corrosive	>0 0~-4 -5~-10 <-10	Nearly non-corrosive Slightly corrosive Intermediately corrosive Highly corrosive	>0 0~-4 -5~-10 <-10	More than 10 indicates corrosiveness		Non-corrosive Slightly corrosive Corrosive Highly corrosive	≧0 -1~-4 -5~-10 <-10

Functional deterioration of tendon

Tendons were collected at investigation sites and corroded cross sectional areas were measured behind the head, in the free length section and in the fixed anchor section. Then, the rate of corrosion was obtained based on the years that had passed since installation and on the corroded cross sectional area (Figure 2). The rate ranged from 0.001 to 0.22 mm/year except for the fixed anchor section in district A. Figure 3 shows the rate of corrosion in the soil. The rate of corrosion was near the lower end of the range of uniform corrosion rate except in the fixed anchor section, and soil exhibited low corrosiveness at all the investigation sites (Figure

1). Judging from the rate of corrosion, therefore, the tendons collected during the investigation suffered from uniform corrosion.

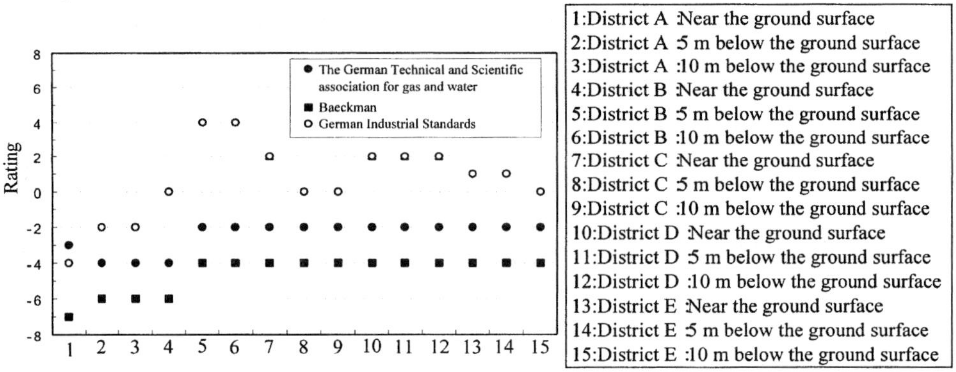

Figure 1. Evaluation of soil corrosiveness

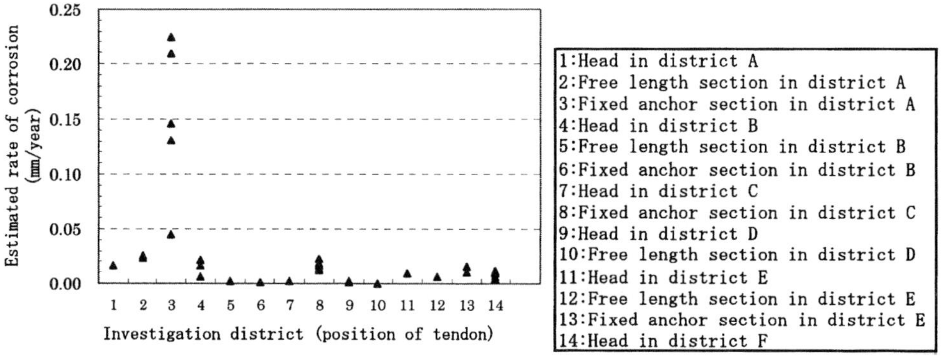

Figure 2. Rate of corrosion at different positions of tendon

The rate of corrosion in the fixed anchor section in district A was in the range of the maximum rate of corrosion and at the lower end of macrocell corrosion in Figure 3. In the fixed anchor section of an old-type anchor, the tendon is embedded in the grout in which the anchor is fixed. The tendon is, therefore, highly likely to be in contact with the ground because of poor grouting or unless the tendon is equipped with a centralizer. There is a high probability of potential difference between the tendon in the grout and that in the soil. In view of the soil corrosiveness at the location of installation, the corrosion in the fixed anchor section in district A may be highly likely to be macrocell corrosion.

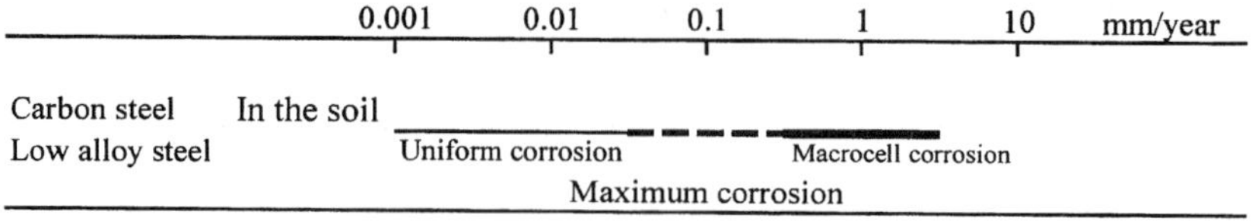

Figure 3. Rate of corrosion in the soil

Figure 4 shows a relationship between the loss of cross section of the tendon collected in the field and the ratio of actual tensile load to JIS tensile load. If the lower limit of functional deterioration of an old-type anchor is set to be 60% of standard tensile load (design anchorage force), the allowable loss of tendon cross section will be 22.5%. The standard tensile load of tendon generally reaches 60% at a loss of cross section of 40%. The standard tensile load reaches 60% at a lower loss of cross section probably because corrosion causes fracture with little plastic deformation. Figure 5 shows a relationship between the years passed since construction and the loss of cross section of tendon behind the head of anchors with nearly the same corrosion protection. It is assumed that tensile load is reduced below the design anchorage force behind the head of an old-type anchor about 38 years after construction.

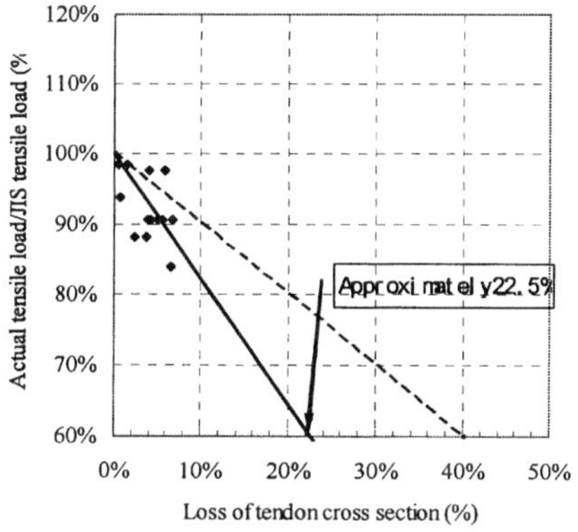

Figure 4. Loss of tendon cross section and tensile load ratio

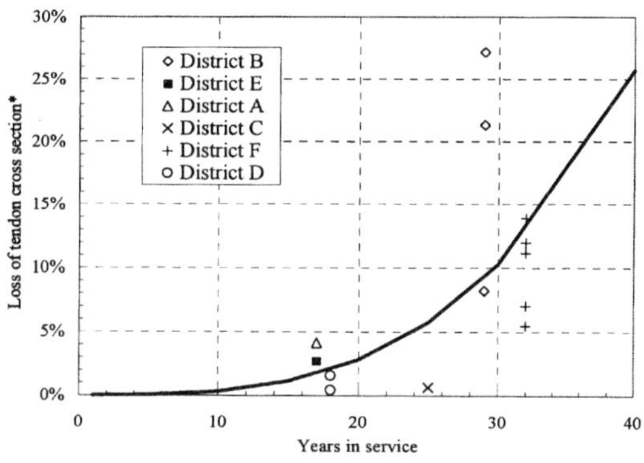

Figure 5. Years passed after construction and loss of tendon cross section

*: Loss of tendon cross section is based on the square of the loss

Maintenance

It was assumed that old-type anchors were likely to be no longer functional after approximately 40 years of service. Then, appropriate maintenance of ground anchors was considered necessary. The anchors with enhanced corrosion protection have been in service nearly 20 years and also require adequate maintenance.

Flow of anchor maintenance steps

Cuts, fills and bridges on expressways are checked by daily (inspect three days per week), annual and detailed inspections in Japan. Figure 6 shows a flow of anchor maintenance steps. An example of anchor reinforcement by the procedure shown in Figure 6 is given in Figure 7. Irregular conditions were found in the anchor and investigations were made to check the integrity of the anchor. The existing anchor was found to be overtensioned. Then, slope stability was examined, and a new anchor was additionally applied for reinforcement.

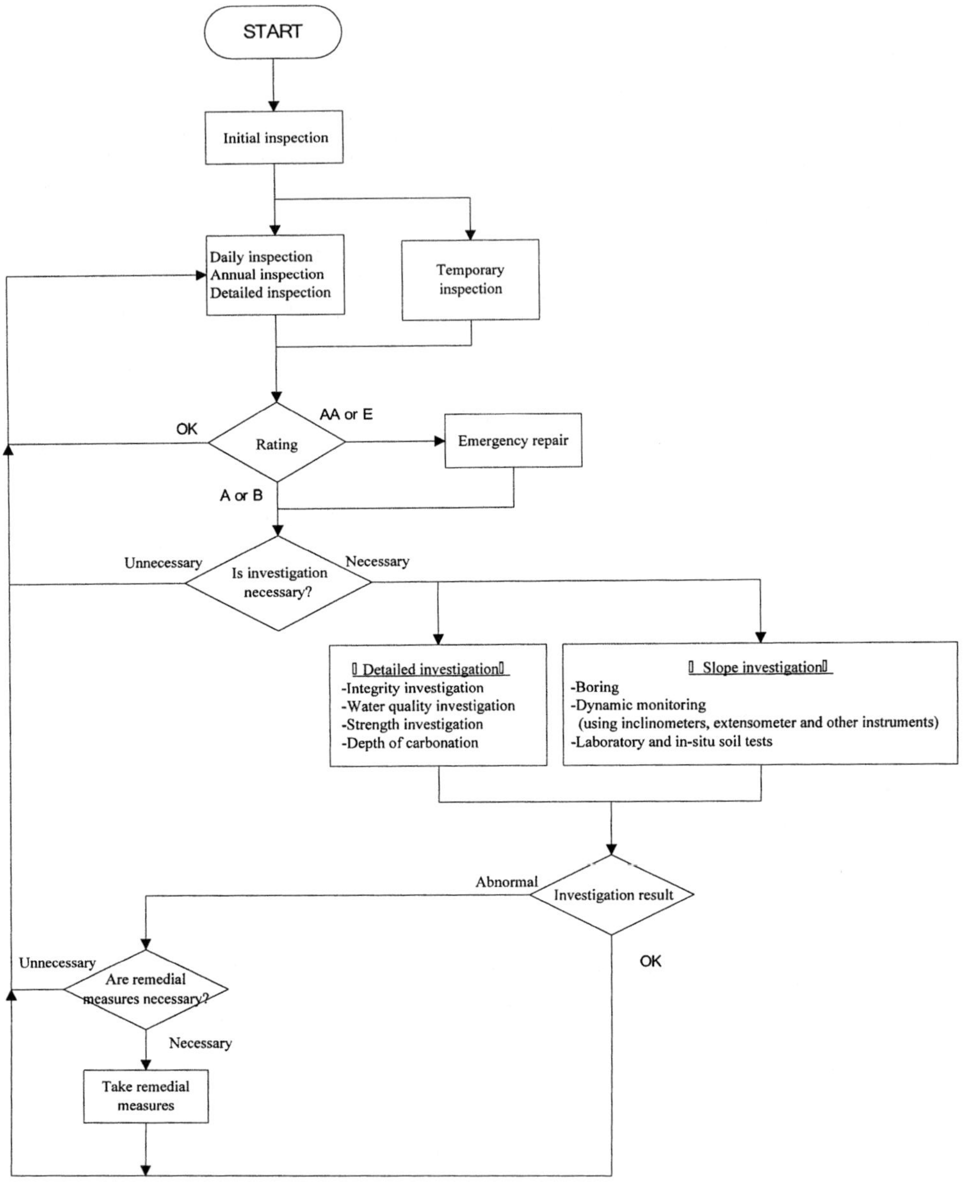

Figure 6. Flow of maintenance steps

Inspection

Inspections are classified into initial, daily, annual or detailed inspection as shown in the flow of maintenance steps in Figure 6. Initial inspection is conducted after the structure is completed and before it is placed into service to identify the initial conditions. "Daily inspection" is actually conducted approximately three days per week. Annual and detailed inspections are carried out once a year and once in five years, respectively.

Figure 7. An example of anchor reinforcement by the procedure

Anchor inspection involves visual check of the slope and the verification of tensile load of the anchor to determine whether the anchor is working satisfactorily or not. In visual inspection, the engineer sees and touches the structure. Most parts of an anchor are embedded in the earth and only the anchor head can be inspected visually. Visual inspection can therefore be conducted relatively easily. Table 3 lists ratings obtained in anchor inspection. Rating AA represents a case where there is serious damage or deformation and urgent functional repair is required. In the case with rating A, damage or deformation has caused functional deterioration and repair is required although not urgently. In rating B, no functional deterioration is found despite damage or deformation and continuous monitoring is required. Table 4 gives an example of inspection. Protection concrete was inspected for uplift or delamination. A photograph of the site shows uplift resulting in rating A.

Table 3. Sample rating of the result of inspection

Inspection element	Inspection item	Rating		
		AA	A	B
Protection concrete	Uplift and delamination	–	Uplift or dull sound in hammering test at two or more locations or in more than 20% of concrete inspected	Uplift or dull sound in hammering test
	Breakdown and falling	Concrete breaks up into fragments and falls at several positions.	–	–
	Deterioration	–	Isolation of lime in more than 50% of concrete inspected	Isolation of lime.

	Water leakage	–	–	Water leakage at anchor head
	Material deterioration	–	Deterioration found in more than 50% of concrete inspected	Deterioration
Protection cap	Deterioration of seal	–	Deterioration found in more than 50% of concrete inspected	Water leakage at anchor head
	Breakdown, deformation and falling	Concrete breaks up into fragments and falls at several positions.	Partial breakdown or deformation is found at least one location.	–
	Water leakage	–	–	Water leakage at anchor head

Conclusions

Old-type ground anchors mainly suffer from uniform corrosion in view of the rate of corrosion. Corrosion progresses at a rate of 0.001 to 0.022 mm per year even in low corrosive soils.

Macrocell corrosion is likely to occur in the fixed anchor section of an old-type anchor.

The tendon behind the head of an old-type anchor is likely to have functionally deteriorated below the design anchorage force after approximately 38 years of service.

The progress of functional deterioration has been identified and described. Ground anchor maintenance standards on expressways have been developed.

Table 4. Sample rating

Type of deformation	
Element	Inspection item
Protection concrete	Uplift

Photograph of the site	
Rating of inspection result	
Cause	Rating
Fracture of tendon due to corrosion	A

Closing remark

Investigations will be required for the corrosion of ground anchors applied with better corrosion protection. Investigations of corrosion will enable the introduction of the concept of life-cycle cost. Life-cycle cost is a combined total of the initial, maintenance and upgrade costs, and is determined by the service life of the structure. The total cost over a sufficiently long period of time can therefore be evaluated. Applying the concept of life-cycle cost in economical assessment is expected to encourage the adoption of structures that provide for greater durability and easier maintenance. Economical assessment that has been made based on the initial and maintenance costs will be made throughout the life cycle of the structure. Development of more effective new design, construction and maintenance technologies may be considered. In the future, performance enhancing technologies will be developed including anchor maintenance technologies.

References

1) DVGW, GW9 (1971)

2) W.Baeckmann, W.Schwenk: "Handbuch des Kathodisher Korrosion Schutzes", 55, Velag Chemie GmgH (1971)

3) ANSI A21.5 (1972)

4) DIN 50929 (1985-9)

Session two:

Service behaviour monitoring procedures and performance in service

Long Term Monitoring and Routine Maintenance of Ground Anchorages at Devonport Royal Dockyard, Plymouth U.K.

Michael Whitworth Devonport Royal Dockyard Ltd, Plymouth, United Kingdom
Steve Parrish Jacobs, Exeter, United Kingdom

Abstract

Devonport Royal Dockyard, located in Plymouth, SW England, currently carries out the maintenance of the UK's fleet of Nuclear Submarines. To allow this to take place, between 1995 and 2002 the existing docks and associated structures were upgraded to modern nuclear standards of seismic qualification. Part of this upgrade required the installation of 247 multistrand anchorages with a working load of 3150kN, and 531 Bar anchorages with a working load of 2040kN.

This paper presents the findings from the current maintenance regime, discusses how such a regime can provide confidence in the anchorage performance and thus the structures fitness for purpose.

The continued monitoring and maintenance at Devonport Dockyard has provided a high level of confidence in the anchorages and thus the structures that rely on them. Lift off checks have confirmed the general findings of load cell trends although some anomalies mean that periodic lift off checks are still required. Corrosion monitoring via visual inspections has proved successful for identifying corrosion and carrying out remedial action. It is noted that potential corrosion below the anchor head is possible, however research into non-destructive testing and trials of some methods have been fruitless. Confidence in the double corrosion protection and endoscope surveys where possible should provide enough confidence in the corrosion protection system.

Background

Devonport Royal Dockyard, located in Plymouth, SW England, currently carries out the maintenance of the UK's fleet of Nuclear Submarines. To allow this to take place, between 1995 and 2002 the existing docks and associated structures were upgraded to modern nuclear standards of seismic qualification. Part of this upgrade required the installation of 247 multistrand anchorages with a working load of 3150kN, and 531 bar anchorages with a working load of 2040kN within 5 different areas within the dockyard..

The bar anchorages range between 20-30m in length anchoring the dock floor to the slate bedrock. The bar anchorages are located within individual grease filled pits and periodically exposed to 15m hydrostatic load when the dock is flooded. The strand anchorages some in excess of 60m in length are used to anchor the dock walls into the slate bedrock. The strand anchorages are located within trenches along the dock walls.

Figure 1 shows a schematic through a dock, with an outline of the geological strata, with the location of strand and bar anchorages.

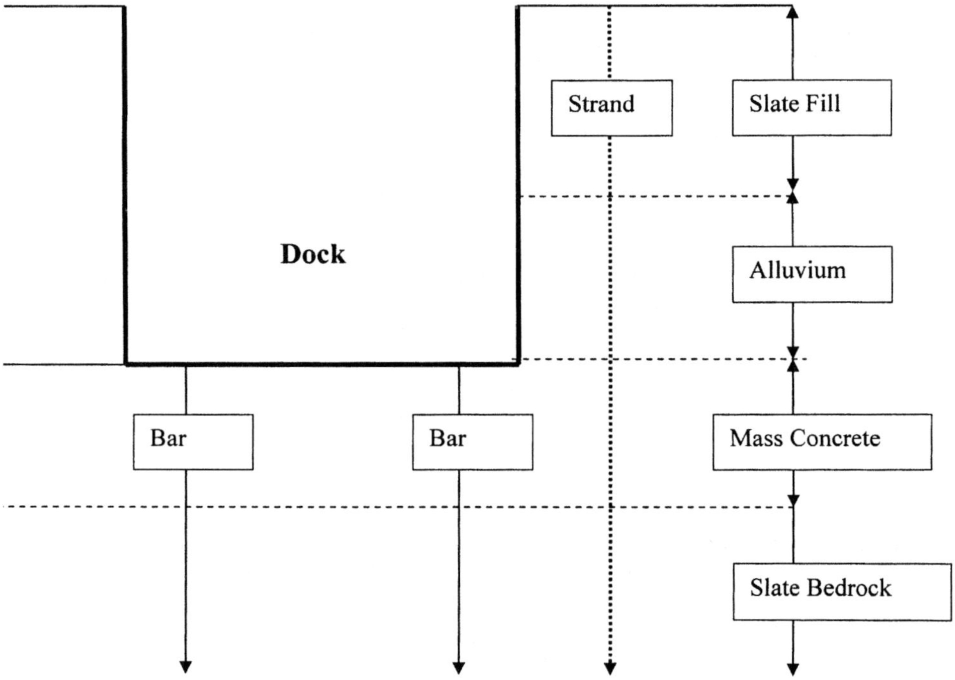

Figure 1 Schematic showing the position of bar anchorages through the dock floor and strand anchorages through the dock walls. The geology is shown, however depths are not given as these vary from facility to facility.

Ensuring fitness for purpose of the facilities requires a rigorous maintenance regime for the anchorages of visual inspections, check lifts, deformation surveys and continual load cell monitoring.

Purpose of Routine Maintenance

The aims of rock anchorage maintenance at Devonport Dockyard is twofold:

1. To demonstrate that the rock anchorages continue to apply the loads to the structure required to maintain structural performance within the design basis.
2. To ensure that the physical condition of the rock anchorages remain fit for purpose

Current practice at Devonport is to use a combination of permanently installed load cells and periodic jacking (lift-off) checks to achieve the first of these aims whilst the second is addressed by visual inspections of the anchor head

Review of Standards

Load Cells

BS8081 is considered to provide the most onerous requirement for the number of anchorages to be monitored at 10% for the first hundred anchorages and 5% for anchorages above this number. Other standards recommend monitoring in the range of 2 – 10%. The only exception to this is Hong Kong, which requires on safety critical structures, 15% of the first 50 anchorages, 12% of the next 50 and 10% of the remainder.

The Hong Kong standard, as it addresses safety critical structures, provides perhaps the best benchmark for Devonport facilities. The number of load cells provided at Devonport is in the range of 14 – 24% and is thus considered conservative in comparison.

All standards recommend increasing periods between reading of load cells as information is built up showing satisfactory performance. Typically long term monitoring is being carried out every 2 to 5 years. The Hong Kong standard specifies annual long term monitoring. Devonport has no relaxation of maintenance periodicity with age.
The Australian standard indicates that after initial monitoring, ground anchorages are periodically monitored every 6 months. However the case study from Burrinjuck Dam, Australia stated that anchorages were initially tested at 3 months, then 12 months and thereafter every 5 Years.

Lift Off Checks

All standards permit lift off checks as an alternative to load cell monitoring although the The industry's procedures for lift off checks means that they have the potential to compromise the integrity of the corrosion protection system. For strands there is a significant potential for damage due to operator error. For bar anchorages where lift off checks only extend the anchorage by a millimetre this is less likely, consensus is that load cells should be used in preference to lift off checks. The implication of this is that re-stressing above the lift off check level should be limited to the investigation of problems or the reinstatement of design margin when load losses fall outside the acceptance criteria.

Corrosion Monitoring

Corrosion monitoring methods are generally limited to the visual inspection of anchor heads. BSEN1537 does include an electrical resistance test to check for breaches of the corrosion protection system. The electrical test is considered unreliable and is not recommended by Littlejohn, 2005. An alternative electrical potential measurement test is available and is being investigated further, this method was previously considered and rejected by DML, 2002

Inspection periodicity varies from 6 monthly, in the case of Hong Kong standards to several years with most standards leaving the determination of the extent and frequency of the inspection to the discretion of the designer.

Table 1. Summary of anchorage type, number of load cells and periodicity of readings and analysis

Facility		**No. Strand Anchorages**	**No of Load Cells**	**No of |Bar anchorages**	**No of Load Cells**	**Periodicity of load Cell Readings**	**Load Cell analysis**
10 Dock		147	20	0	0	Yearly	Yearly
15 Dock	Floor	0	0	206	43	6 Months	Yearly
	Wall	6	6	0	0	6 Months	Yearly
14 Dock	Floor	0	0	50	12	Weekly	Yearly
	Wall	94	20	0	0	Weekly	Yearly
LLRF	Floor	0	0	112	12	Weekly	Yearly
	Wall	0	0	34	8	6 Months	Yearly
PRT Berth	Wall	0	0	129	22	Weekly	Yearly
Total		247	46	531	97		

Table 2. Corrosion monitoring acceptance criteria.

Corrosion Grade	**Definition**	**Accept/Reject**
1	No signs of corrosion or damage to paint Work	Accept
2	Lightly corroded, light brown in colour, no flaking of corrosion	Accept
3	Moderately corroded, deeper brown in colour, light flaking of corrosion	Accept but take action to remove and protect
4	Heavily corroded, variation in colour (brown/orange), flaking of corrosion which can be removed with brush	Reject

Maintenance Techniques Employed

Load Cell Analysis

Approximately 20% of anchorages at Devonport Dockyard are monitored using load cells. The most rigorous monitoring being automated weekly readings taken via a data acquisition system and analysed every 6 months. The type of anchorage, number of load cells and data collection frequency are given in Table 1. Due to the difference in readings of the actual installation lock off load read via a jack versus the readings observed via load cells at lock off up to 10% difference, load cells are used to indicate load loss trends over time on the anchorage systems rather than absolute load. These trends are then used to provide information on the load on the remaining anchorages confirming whether the system is performing within the design intent. Anchorages are deemed to be within acceptable limits if two criteria are satisfied; firstly that load loss is below 2% over a year; secondly that load loss is less than 8% since installation. Since anchorages are locked off at 110% working load this provides an early warning system, allowing for enhanced intrusive maintenance i.e. check lifts and visual inspections.

Check Lifts
All bar anchorages are check lifted once over a 10 year rolling programme, throughout the design life to confirm trends of load cells. Currently this has been interpreted as 10% of anchorages per year. Strand anchorages are not subjected to check lifts, unless load cells indicate load loss outside acceptable limits. This is due to concern of reseating the wedges during repeated check lifts, potentially leading to failure. This problem arose during early installation of strand anchorages. Three sacrificial strand anchorages within 10 dock are used to monitor load loss via periodic check lifts.

Corrosion Monitoring
All anchorages are visually inspected over a 10 year period for bar anchorages the same anchorages subjected to check lift are inspected. Inspection comprises the removal of top hats, the removal, visual inspection of the base plates, bar/strands and nut/wedges for signs of corrosion. Concrete bases and pit/trench walls are inspected for signs of deformation and cracking. Photographic records are taken at set stages and of any defects to allow monitoring over time. Endoscope surveys were access is available are employed to inspect below the head detail. The anchorages are then reinstated, with new gaskets and grease after all surfaces have been cleaned and protective painting. Acceptance criteria for corrosion are given in Table 2.

DRDL independently researched methods, the most comprehensive research to date was carried out by the American Transportation Research Board (NCHRP Report 477 Recommended Practice for Evaluation of Metal Tensioned Systems in Geotechnical Applications).

Using this study as a guide DRDL ground anchorages would be given a low vulnerability and low-medium corrosion hazard, due to the age of the anchorages, double corrosion protection and the minimal corrosive environment. This gives a maximum rank of 2 out of 6. meaning DRDL would be a low priority for assessing corrosion using NDT methods.

Results of Maintenance and Monitoring
Continual load cell monitoring for each facility with anchorages has provided an early indication prior to invasive maintenance of any load loss outside established acceptance criteria. The acceptance criteria are twofold. Anchorages are deemed acceptable if there has been less than 2% load loss per year and if the cumulative load loss is less than 8%. Outside of these criteria requires further investigation. Figures 2 shows load loss for one facility since installation, showing a rapid loss, followed by a steadying off over time. Figure 3 shows more recent results showing a relatively stable trend. Slight fluctuations have been found to be linked to tidal levels (tidal range of 5.5m), temperature differences (+/- 5°c) and the flooding of the dock structures up to 15m. On average load cells over a 12 month period, relax by 0.5%. Load cells have been found to be outside the criteria on two occasions prompting remedial action and investigation. Load cells analysis for some facilities has been compromised by the failure of load cells. The investigation led to the conclusion that the large hydrostatic head and the lack of water proofing of the load cells was the root cause of failure. Replacement was carried out, with improved water proofing.
Annual lift off checks of a sample of anchorages has found that generally anchorages have remained above the working load, with overall load losses from installation being comparative, within 2% of load cells. This provides a good level of confidence in what the

load cells are observing. On the rare occasion an anchorage has been found to be below the working load they are re-stressed.

Throughout the Dockyard where facilities have anchorages, significant effort has been made to establish the level of redundancy within the anchorage system, meaning that some load loss over anchorages is acceptable without the need for remedial action.

Visual inspections to monitor corrosion, have found varying levels of corrosion, although none gave concern for the immediate life of the anchorage, the corrosion un-checked would potentially have had a detrimental effect. Therefore remedial action is taken to clean up the anchorage and re-paint. Depending on the nature of the corrosion dictates whether more anchorages are inspected. In The majority of cases corrosion can be pinned down to poor quality control and inspections on installation. Several non-destructive corrosion monitoring methods have been considered within Devonport with some methods trialled with limited success. The most effective method remains visual inspections, with where possible endoscope surveys to visually inspect below the base plate.

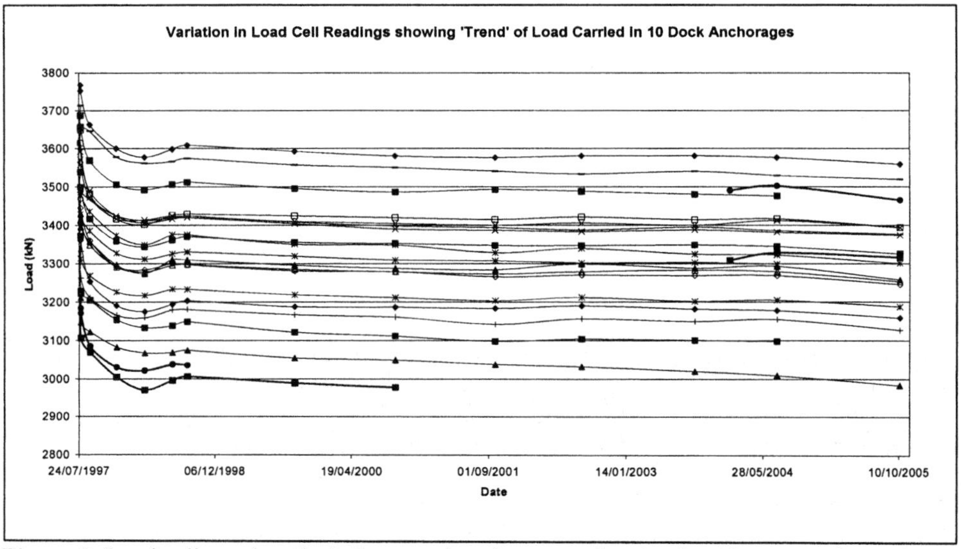

Figure 2. Load cell trend analysis for strand anchorages, showing load loss over time. The rapid loss after installation can clearly be seen with a steadying over time.

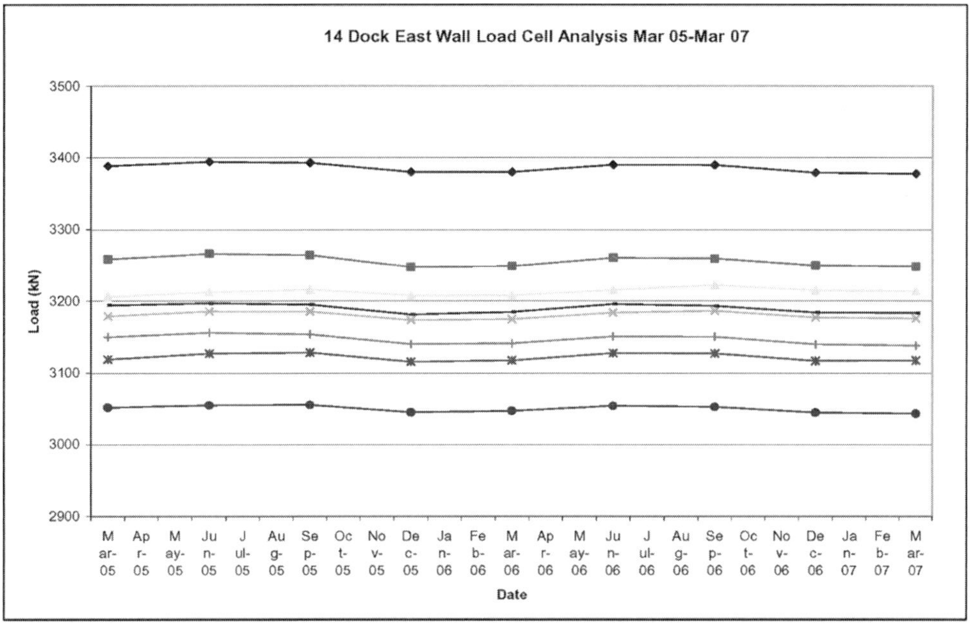

Figure 2. Load cell results for a two year period, between March 05 and March 07 showing a relatively stable trend. Fluctuations can be attributed to tidal and temperature variations.

Figure 3 Visual Inspection of a Multi-strand (left) and Bar (right) Anchorage after removal of top hat and void filler. The condition of the strands and wedges can be seen to show no signs of corrosion. The bar anchorage shows some levels of deterioration.

Figure 4. Lift off check on a sacrificial strand anchorage, a mono-strand jack is required to be used due to the limited length of strand available

Conclusions

The continued monitoring and maintenance at Devonport Dockyard has provided a high level of confidence in the anchorages and thus the structures that rely on them. Lift off checks have confirmed the general findings of load cell trends although some anomalies mean that periodic lift off checks are still required. Corrosion monitoring via visual inspections has proved successful for identifying corrosion and carrying out remedial action. It is noted that potential corrosion below the anchor head is possible, however research in to non-destructive testing and trails of some methods have been fruitless. Confidence in the double corrosion protection and endoscope surveys where possible should provide enough confidence in corrosion protection system.

Failure to have carried out these works would have meant that potentially the anchorages would have been compromised, leading to a reduced design life. The results from the maintenance have provided sufficient confidence in the anchorage system to allow 10% annual inspections and check lifts to be changed to 30% every 3 years.

Recommendations

To allow continued justification for ongoing use of nuclear related structures, requires a high level of confidence in the anchorages installed. Load cell trend analysis is not sufficient with periodic lift off checks required to confirm the findings and a rigorous regime of visual inspection to monitor corrosion. The level of confidence required is such that the regime at Devonport is extremely rigorous and employs several methods. Before dictating the level of maintenance, the consequence associated with failure of the structure has to be established with a level of maintenance and monitoring commensurate with the consequence is put in place.

References

BS 8081:1989 Code of practice for ground anchorages

BS EN 1537:2000 Execution of special geotechnical work. Ground anchors

Devonport Management Ltd. 2002. Ground Anchorage Overview Report.

GEOSPEC 1. 1997. Model Specification for Prestressed Ground Anchors. Geotechnical Engineering Office, Hong Kong

Littlejohn, G.S. 2005. (Correspondence). A Technical Review of Non-Destructive Techniques for Monitoring Post-Tensioned Steel Tendons.

NCHRP Report 477 Recommended Practice for Evaluation of Metal Tensioned Systems in Geotechnical Applications American Transportation Research Board

Roads and Traffic Authority, 1997. Permanent Rock Anchors. New South Wales, Australia

The long-term performance of permanent ground anchors forming part of the Thames Barrier Project

Mike J Turner, Applied Geotechnical Engineering Ltd, Buckingham, UK
David Richards, Ken Rush Associates, Braintree, UK

Abstract

This paper describes monitoring work undertaken on a site where permanent ground anchors had been installed in 1979 to support a new sheet piled river wall, on the south bank of the Thames.

In 1999/2000 the site was redeveloped for residential use. The building loads were supported on piled foundations close to, and in places threaded through, the permanent anchors. The condition and residual loads in the anchors were checked before, during and after the foundation construction works. At the same time, the response of the anchored river wall was checked by optical survey techniques to confirm that the new works were not adversely affecting the wall: particularly by any disturbance to the integrity or performance of the ground anchors.

The monitoring demonstrated that the anchored wall behaved in a resilient, robust and controllable manner, even when isolated anchors were destroyed by the piling works.

Matters of concern beyond the particular project are also highlighted. These include a general lack of maintenance and inspection of such anchored structures, and the disturbing fact that as-built and construction records, even of such relatively young and vital structures, appear to have rapidly been lost.

Introduction

From the mid-1970s to the early 1980s, major flood-defence works were constructed along the banks of the River Thames. These defence works stretched from the mouth of the Thames estuary, westwards to their culmination at the Thames Barrier, at Silvertown, just west of Woolwich. These works included raising the height of the flood defences flanking the river, utilising a variety of techniques depending upon the land available and the characteristics of the properties on the banks. Where existing river walls and working wharves immediately abutted the river, permanent ground anchors were often used to provide the additional restraint necessary to support the heightened walls.

The high technical specifications required of the ground anchors associated with the bank-raising project led directly to the development of sophisticated corrosion protection systems: mainly pioneered by a variety of innovative specialist geotechnical companies. The experiences gained in developing and utilising these systems contributed greatly to the development of the UK ground anchorage Codes, DD81(1982) and BS8081(1989).

A generation on, a decrease in industrial activity along the Thames, and an increasing realisation of the amenity value of such river-side sites has rendered an increasing number of these properties attractive for redevelopment. Where such a site incorporates permanent ground anchors to the river walls, it is normal practice that an appraisal of the condition of these anchors is undertaken.

This current paper outlines the inspection and evaluation procedures adopted for the ground anchors at such a site, at Erith on the southern bank of the river. The site measures some 250m long and 50m wide, and is illustrated on Figure 1. It is bounded on three sides by anchored sheet piled walls. Load checks and load monitoring of the ground anchors were undertaken before, during and after the construction of the new development. Measurements of river-wall movement were also taken.

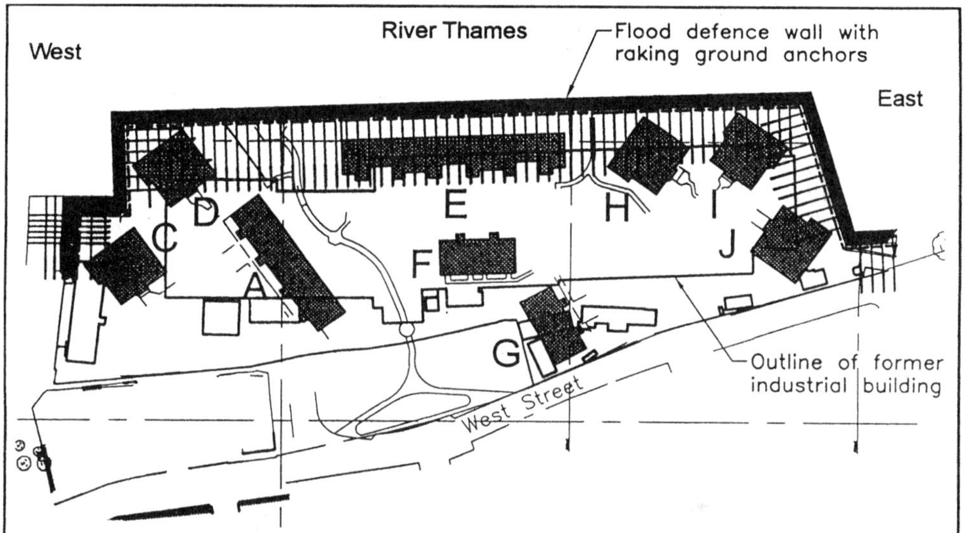

Figure 1: Site plan

Background.

The original bank-raising and anchoring works at Erith were undertaken in 1979. The river wall was formed from steel sheet piles, supported by ninety-five permanent strand anchors. The anchors were installed into chalk beneath a cover of fill, river alluvium and flood plain gravels. Overall anchor lengths varied between 28-30m at the eastern end of the site, and 36-40m at the western end.

The ground anchors were designed and installed by Messrs. Ground Anchors Ltd (a specialist contractor owned at that time 50/50 by PSC Freyssinet and Colcrete Ltd). Each anchor tendon was formed from 'Dyform' low-relaxation pre-stressing strand, with its lower end encased in a 70mm diameter OD high strength polyester resin encapsulation, which transferred the anchor loads into the founding strata along the grouted fixed anchor length. Above the encapsulated section, each strand was individually covered by a grease filled high-density polyethylene sheath. Anchor working loads of up to 600kN were required, and up to five 15.2mm diameter Dyform strands were used. At the anchor head, grease filled spigot tubes, screwed into the back of the anchor block, were slid over the strand sheaths to provide continuity of corrosion protection. The interior of the load bearing angle bracket assembly was back-grouted with a bentonite/cement mix to provide protection at the back of the block. Aspects of the ground anchor system were described in Turner (1980).

In 1999/2000 the site was redeveloped with nine residential apartment blocks. The piled foundations for some of these blocks had to be installed around and between the existing ground anchors. Part of the enabling works for the development involved checks upon the condition and behaviour of the ground anchors, and these are described in this current paper. Further details of the research and remediation works undertaken to facilitate the installation of the new foundations at the site are described by Richards and Turner (2005).

Preliminary load/condition checks

A detail of the anchor head assembly as used at Erith is illustrated in Figure 2. The ground anchor head system utilised a rectangular steel anchor head block, which was drilled with taper holes to accept standard 3-part strand wedge grips. This was located on a load distribution/bearing plate and gusset assembly, which transferred the anchor load onto the sheet pile wall. To provide a re-stressable facility, the rectangular anchor head block was drilled and tapped at each corner to accept four high tensile bolts. A prefabricated lifting bar was bolted onto the anchor head block to allow check-lifts to be undertaken and the anchor head to be shimmed if load adjustments had to be made. The anchor head was encased within a rectangular grease-filled cast-steel anchor head cap and gasket, which was bolted onto the load distribution/bearing plate. All exposed steel parts of the anchor head assembly were painted with a proprietary epoxy paint system.

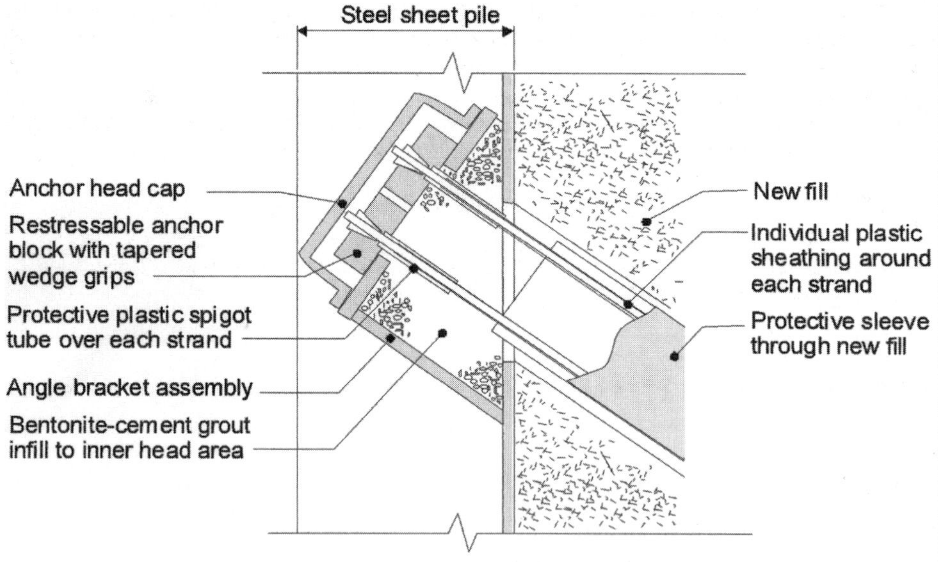

Figure 2: Details of anchor head assembly used at Erith

As part of the initial evaluation work, condition surveys and load checks were undertaken by Messrs Keller Ground Engineering on twenty-seven anchors along the wall fronting the first phase of development. Of these, one had a damaged cap, with one bolt hole broken off and the anchor head block was "badly" corroded; one cap had only two of its four bolts fitted. A third anchor head recorded no major deficiencies, but no grease within the cap, resulting in corrosion of the anchor head plate (and four bolts were corroded, hence replaced). In eleven further cases the anchor head cap bolts were serviceable, but were replaced due to corrosion.

However, it should be appreciated that the ground anchor heads had been subjected to tidal immersion twice a day for over twenty years, without, so far as was known, being inspected or tested. As a result, the head assemblies could be expected to be weed-covered and suffering at least to some extent from surface corrosion and degradation.

The condition of the wedge-grips and the strands within the anchor head plates was noted as being generally good, with little corrosion being observed. In only three cases were wedges noted as corroded. Two of the cases were associated with damaged caps or missing bolts to caps. The third was recorded on a fully grease-filled cap, and there was no obvious reason for such rusting. It was possibly associated with rusting before final capping of the anchor at the initial construction stage. However, in no case was the reported corrosion of sufficient import to preclude check-lifting of the anchor.

Twenty five of the twenty-seven accessible anchors were subsequently successfully check-lifted. The two unsuccessful anchors could not be checked because the stressing stool/jacking combination could not be fitted under the reinforced capping beam above the anchors.

The selected ground anchors had specified working loads of 550kN and were constructed using five 15.2mm 7-wire drawn strands with a specified characteristic strength of 300kN per strand. The lift-off checks of the 25 anchors recorded residual loads of between 170 and 317kN, with an average of around 280kN, as illustrated on Figure 3. Subsequent research of contemporaneous records indicated that, as a principal, the ground anchors at the site had been locked-off at 80% of their design working loads, namely 440kN. In some cases, however, because of movements of the newly constructed sheet pile wall (particularly over the eastern part of the site: near the present-day Blocks H and I), anchors were locked off at even lower loads, after proof-testing. It is not unusual, also, that such anchors continue to relax and lose load over the wall-construction period, as the land-ward wall movements continue: until the system either reaches equilibrium, or is stabilised by backfilling behind the new wall and the completion of capping works. Unfortunately records of the final as-built anchor loads had been lost. Hence, it was not possible to make an assessment of anchor load losses since construction.

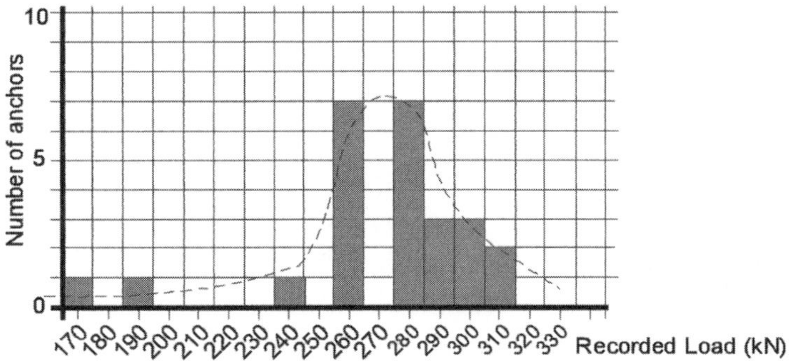

Figure 3: Measured residual loads recorded for the initial 25 anchors

Construction load checks

A source of discussion at the preliminary stages of the work revolved around the liability or otherwise of the developer for the ground anchors. If, as in the current case, the developer finds the residual anchor loads to be substantially less than their design loads: what is to be done about this? A decision was taken on the Erith site that, since there was documentary evidence that the anchors had been originally set at lower loads, then the loads in the anchors at the time of the first check represented an acceptable as-is condition of the system.

Subsequent to the initial check-loading and condition reporting on the first twenty-seven anchors, all accessible ground anchors along the complete river frontage of the site were examined, their condition recorded, and their residual loads checked, before piling work commenced in their vicinity. After piling of a section was completed, the anchor loads were again checked, to demonstrate that the anchor system had not been adversely affected by the works.

Load fluctuations recorded by the monitoring are summarised below:

Block D: Initial anchor loads in front of Block D varied between 380 and 550kN, with an average of 425kN for 17 anchors. It is not known if the loads higher than the nominal 440kN were due to wall movement or due to a higher lock-off load. Two months later, post-piling checks showed anchor loads varying between 402 and 548kN, with an average of 462kN, for 13 anchors (several anchors had been replaced by tie-bars and one had been destroyed by piling operations). Six months after initial checks, recorded anchor loads were found to be varying between 375 and 525kN, with an average of 462kN, in 13 anchors. Eight months after the initial checks, average loads in 4 anchors were 461kN.

Block E: Initial anchor loads in front of Block E varied between 350kN and 500kN, with an average anchor load of 387kN for the 21 anchors that could be checked. Ten months later, after Block E had been piled, recorded anchor loads varied between 265 and 504kN, with an average of 391kN for 19 anchors. As part of the piling works, four anchors had been replaced by deadman anchors. Seventeen months after the initial checks, rechecks on nine anchors recorded loads of between 320 and 549kN, with an average of 422kN.

Blocks H and I: Initial anchor loads in front of Blocks H and I varied between 262 and 300kN, with an average of 257kN for 20 anchors. Seventeen months later, after piling works had been completed the recorded loads varied between 247kN and 394kN, with an average of 317kN. To facilitate the piling works to these blocks, five anchors had been replaced by deadman anchors. In addition two anchors had been destroyed by the piling works and replaced by tie bars.

Block J: Initial anchor loads varied between 170 and 281 kN for 5 anchors, averaging 237kN per anchor. Fourteen months later after the completion of piling works, recorded loads varied between 170 and 280kN, with an average of 232kN. Seventeen months after the initial checks, loads varied between 247 and 302kN in four anchors, with an average of 275kN per anchor.

Wall movements

As indicated in Figure 1, the Erith site was roughly rectangular in shape, with its 250m long side parallel to the River Thames. This is termed the 'long' wall in the subsequent discussions. At each end of the site, the river walls turned through a right angle to return to the river bank. In the following discussion, these are termed the 'return' walls.

Wall movements were monitored by targets placed along the reinforced concrete river wall topping the sheet pile wall, as illustrated on Figure 4. The targets were monitored by precision survey techniques to record movements at right angles to and parallel to the line of the wall (termed 'lateral' and 'longitudinal' movements). These are also designated as the 'X-X' and 'Y-Y' directions, respectively. In addition, vertical movements ('Z-Z') were also monitored. A total of 58 survey targets were installed along the river wall.

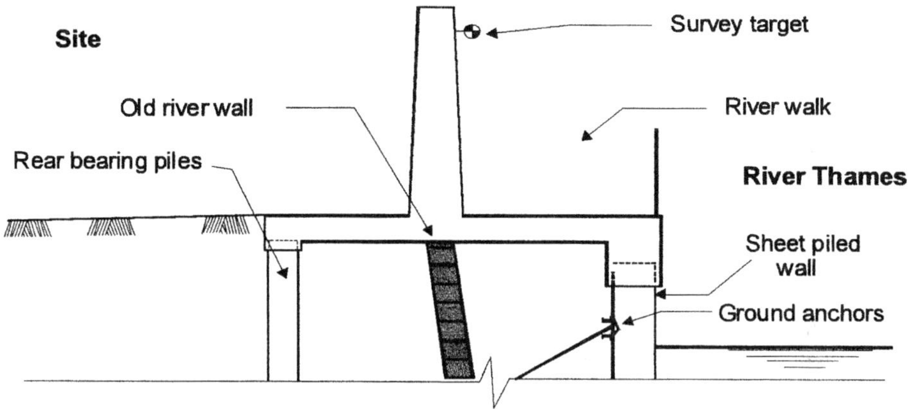

Figure 4: Location of survey targets on the river wall

Movement monitoring commenced up to six months before the piling works themselves commenced. Over the intervening period, enabling works, such as the detensioning and replacement of selected ground anchors by deadman ties, and the construction of piling platforms were undertaken.

Lateral (X-X) movements

Figure 5(a) illustrates the typical lateral wall movements measured by a monitoring point near the centre of the long wall, in front of Block E. Monitoring in this location extended over almost a full year, although the particular construction cycle of concern (anchor replacement and piling) lasted only two months.

The typical overall range of lateral movement measured over the whole centre section varied between 25 and 30mm over the year. The movements appeared to be cyclic, and there appeared to be a general trend of landward movement over the winter months and riverward movement in the summer: although it should be borne in mind that site construction activities were ongoing over all this time period.

Figure 5(b) shows a plot of a typical monitoring point close to the corners of the long wall. The lateral X-X movements are less than 50% of those shown in Figure 5(a). This reduced movement was noted to be about 20m from the corner points. X-X movements on the return walls were similarly reduced to 10-15mm over the year.

Piling-related movements: Figures 5(a) and (b) both show an increase in the 'peakiness' or amplitude of X-X wall movements during the anchor replacement and piling works. These were more noticeable on the Figure 5(b) plot, where the range of lateral movements increased to about 30mm: similar to those on the long wall. At least part of these observed increases were considered to be attributable to the increased frequency of readings during critical construction activities, where, at times, readings were taken twice a day. Other factors that could affect the short-term readings were perceived to be the diurnal effects of daily temperature and tidal ranges. A further effect of temperature was that difficulties were also noted in reading targets over long distances during hot weather: so that, again, some of the movements recorded may have been reading-related.

No correction or allowance was made for these effects: the main concern of the monitoring was to identify any major deviation from the observed range of wall movements, which might be indicative of a potential failure.

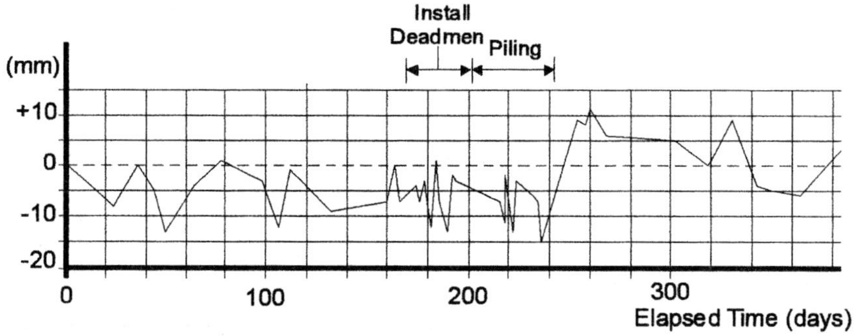

Figure 5(a) Typical lateral (X-X) movement of sheet piled wall near centre of 'long wall'

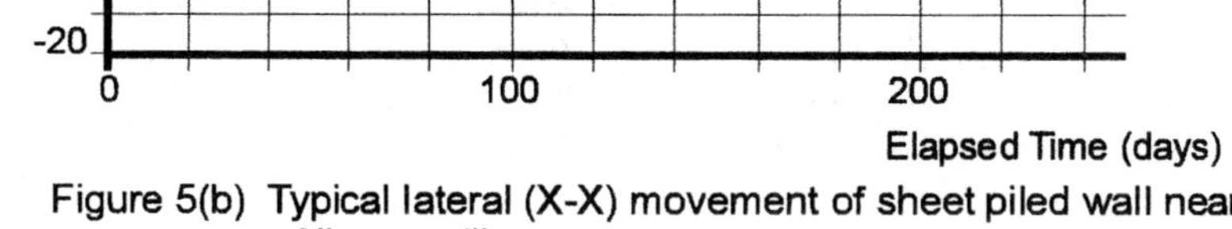

Figure 5(b) Typical lateral (X-X) movement of sheet piled wall near end of 'long wall'

Longitudinal Y-Y movements

In contrast to the lateral movements, the recorded longitudinal, Y-Y, movements of the long wall were greater towards the ends of the wall. The range of movement in the centre of the sheet piled wall typically varied between 2 and 6mm, with an average of 3-4mm. Towards the northern and southern corners, the range of recorded Y-Y movements increased to around 6-9mm over a year.

On the return walls, longitudinal movements typically varied between a range of 2mm and 6-9mm: with the higher values at the river end and the lower values at the landward end of the walls.

Again all movements appeared to be cyclic, and representative of the normal everyday motion of the anchored river wall. There was no obvious permanent effect from construction activities.

Vertical Z-Z movements

In a similar manner to the X-X movements, the amplitude of the vertical movements in the centre of the long wall, in front of Block E, was larger than the ends, with a range typically around 12-14mm. Towards the northern and southern corners, movements were typically reduced to 4-10mm.

Overview of the survey data

As a general overview, the survey system was simple and robust and had a good degree of redundancy within it. If some points were obscured by construction works or damaged, others nearby were serviceable. A baseline for the survey was set up well outside the confines of the site. It was clearly to be expected that there would be errors of repeatability in the optical surveying, but these appeared to be well within acceptable bounds. No attempt was made to review or calibrate the magnitude of such errors, other than the finding that adjacent targets gave consistent results, capable of allowing a coherent and understandable view of the performance of the site to be formulated.

Load changes compared with wall movements

Based upon the available information, the estimated free tendon lengths of the anchors would be expected to vary between 25 and 27m near Blocks H to J, increasing to 33 to 37m near Block D.

The elastic extension of a five strand anchor can be calculated to be around 1.5mm per 10kN of load change for the former case, and 2.0mm per 10kN of load for the latter. Hence, the estimated lateral pile movements that would produce the recorded average load changes in the Erith ground anchors can be calculated as between 6 and 9mm.

Such displacements are well within the range of lateral movements recorded for the river wall over the duration of the monitoring: both before, during and after the tie-bar replacement and piling works. They therefore seem to be consistent in the way that a flexible wall such as a sheet piled wall might be expected to respond to diurnal, monthly and seasonal changes in its environment.

Within this comfortable "envelope" of response, the loads recorded for individual anchors appeared at times to fluctuate quite widely from the average. Exceptionally, maximum load increases or decreases of 140 to 160kN were recorded on individual anchors: which would equate to wall displacements of around 30mm. On balance these seemed to be load measurement errors, since no corresponding changes in wall displacement or the loads of other nearby anchors were noted. With this caveat, all the load measurement readings were taken into account in the calculation of average loads for the particular sections of the wall, rather than excluding particular suspect results.

Anchor failures

As described in Richards and Turner (2005), three of the ground anchors were destroyed by the pile installation works. The survey data indicated no adverse wall movements due to these incidents, and the anchor loads in the vicinity remained unaffected. The monitoring gave confidence that remedial works could be designed and scheduled within a reasonable time window, rather than having to be viewed as an emergency situation.

Summary

A knowledge of the condition of the permanent ground anchors at the site was essential to the practicable and economic development of the site. Equally, it was essential to be confident that the anchors could continue to fulfil their design function of supporting the river wall after the new development had been completed. In addition to these basic questions, it was important to know whether the foundation piling works were having a detrimental affect upon the ground anchors and the anchored river wall. A corollary to this was that it was clearly most important that any unavoidable damage to the anchored system during construction was rapidly identified and appropriate measures were taken.

These requirements were achieved by instituting a comprehensive condition monitoring programme of every accessible anchor. This included load checks before, during and after the construction works, and, in addition, providing an extensive site survey system to monitor any movement of the anchored wall.

The condition monitoring of the anchors demonstrated that the anchor head corrosion protection system had shown an encouragingly high level of performance over its then twenty-two year life (but bearing in mind it has a required sixty-year design life). A weak point was mechanical damage to, or poor fitting of, the anchor head caps. Some of the latter problems probably dated back to the site operations themselves.

A depressing feature of the project, and this is not an uncommon feature of such works, was the clear lack of any maintenance and refurbishment programme up to the time of the redevelopment works. The companion paper by Richards and Turner (2005) on the construction aspects of the development of the site, also highlighted how little of the construction records of the bank-raising project had been preserved over a relatively short time-span. It was only the efforts of a single individual who had ensured that even these had survived in a traceable form.

The wall movement and anchor check-loading operations gave a clear insight into the flexibility and resilience of the anchored wall system, and its ability to absorb and resist the effects of the manifold foundation construction activities being undertaken around it. The measurements also gave an insight into the time-based cyclical load changes that can be experienced by such anchors.

Conclusions

It is considered that the wall survey monitoring fulfilled its essential purpose as a warning device, serving to indicate potential damage to the anchored wall caused by the works. This was important due to the significance of the flood defence wall.

The condition monitoring and check-loading of the ground anchors gave assurance of their capability to continue to provide support to the wall.

Appreciable cyclical, and recoverable, movements of the wall were observed, together with fluctuations in the anchor loads, which appear to be due to tidal, seasonal, temperature and other effects.

There did seem to be some increase in the magnitude of the movements during major construction disturbance (such as the de-stressing and replacement of anchors, or the piling works themselves), but these rapidly returned to the background values.

The extensive background research and observations associated with the ground anchor aspects of this project highlighted several areas which are considered should be of general concern to the engineering community at large. These may be summarised as follows: -

- The Erith work raises concerns for sites, along the Thames and elsewhere, which are not being developed, and where anchors could be deteriorating because of a lack of ongoing maintenance and inspection. The authors have been involved in several projects or potential projects where riverside sites featuring river walls supported by post-tensioned ground anchors have been considered for redevelopment. The authors have yet to meet a site where routine inspection has been undertaken, despite almost 30 years having passed since installation.
- Almost invariably, records are not readily available to indicate the types of anchor used, their positions, original prestress loads or procedures for maintenance, checks and inspections. This means that there are no benchmarks against which the condition and ongoing performance of the anchors can be checked. Sometimes records can be traced, but this is invariably very time-consuming, involves numerous enquiries and often relies on good luck.
- The need for records is also increasingly important now that complex forms of procurement are used. As responsibilities are handed over, or private companies amalgamate or split up, records may not always be passed on to those who need them.

Such problems do not only concern the bank-raising project. Undoubtedly, similar concerns could be raised from major infrastructure projects such as highways, harbour works, railways, and other major structures.

References

BS8081: 1989 British Standard Code of Practice for Ground Anchorages.

DD81: 1982 British Standard Draft for Development: Recommendations for Ground Anchorages.

Richards, D and Turner, M. J (2005) "Redevelopment of a riverside site in the proximity of existing ground anchors" *Ground Engineering*. Vol 38, No. 8. August, 2005.

Turner, M J (1980) "Rock anchors: An outline of some current design, construction and testing practices in the United Kingdom". International Conference on Structural Foundations in Rock. Sydney, 1980, pp87-103 (Pub. Balkema)

25 years of monitoring of rock anchors installed for landslide stabilisation at Nantgarw, South Wales

Hugo Wood and Peter Martin, High-Point Rendel, London, UK

Abstract

Investigation and monitoring of cracks in the A470 trunk road at Nantgarw identified landslide movement of the hillside below the road as the primary cause. A scheme comprising a row of anchored pads was designed and constructed in 1980 to stabilise the landslide. Particular consideration was given in the design to ensure the durability of the anchorage system whilst allowing monitoring and inspection of the anchor heads.

The loads in the anchors have been regularly checked over the period 1981 to 2006 since installation and where necessary the anchors have been re-stressed. On each occasion, the total lift-off load of all the anchors has been found to be below the total notional design load with individual anchors having generally lost rather than gained load with time. It would appear therefore that the landslide has demanded less restoring force from the anchors than the design landslide stabilisation load. This confirms site observations that the landslide has been adequately stabilised by the anchor scheme.

This paper describes the installation and monitoring of the anchor system, discusses the results of the load measurement and re-stressing and comments on the effectiveness of the stabilisation scheme.

Introduction

The A470 trunk road at Nantgarw is on steep ground sloping down to the River Taff. The ground conditions comprise glacial deposits overlying Coal Measures bedrock. Cracking was observed in the then recently constructed trunk road improvement and investigation and monitoring of cracks was commenced in 1972. These investigations identified the cause of cracking to be gradual landslide movement of the hillside downslope of the road (Kelly and Martin, 1985). Figure 1 shows a typical section of the landslide.

Detailed review and interpretation of the initial monitoring results was carried out to identify the best stabilisation option (Martin and Kelly, 2000). The form of the landslide suggested that the optimum remedial measures would be anchored bearing pads. These would provide a retaining force to the upper section of the landslide directly affecting the A470. The anchors were designed to be relatively flexible and the anchor pads were sized so that additional anchors could be installed if the force applied had been insufficient.

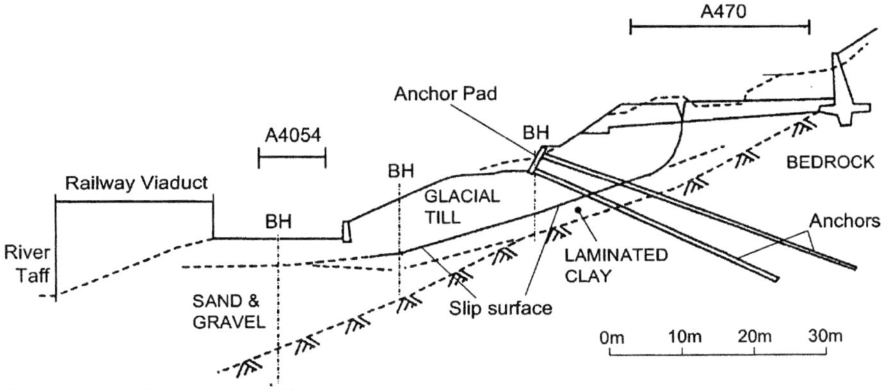

Figure 1 Section of Landslide

The stabilisation scheme was constructed in 1980 under a contract let by Mid Glamorgan County Council. South Wales Trunk Road Agency (SWTRA) is the authority currently responsible for the maintenance and monitoring of the stabilisation scheme and associated instrumentation on behalf of the Welsh Assembly Government as the trunk road Highway Authority. They have commissioned and overseen the most recent round of monitoring of the anchor loads and carried out the monitoring of the instrumentation installed at the site.

Design of Anchor System

The scheme consists of a total of 32 reinforced concrete anchor pads each 4.8m wide and 3m high with four rock anchors. The anchors were designed for a maximum test load of 900kN and a nominal working load of 600kN. The actual design landslide stabilisation load assumed and applied to each working anchor during commissioning was 500kN. This is equivalent to a stabilising force of approximately 400kN per metre width of the landslide across 160m width of landslide resulting in a total stabilising force of 64000kN.

Figure 2 shows a plan showing the location of the anchor pads relative to the carriageway. Pads 1 to 26 were installed on the hillside below the road. A gap along the line of the pads was required between pads 12 and 13 due to the presence of a 14" gas main and a 12" drainage pipe. Pads 27 to 32 were constructed against the rockfill embankment supporting the carriageway and were installed at a slightly lower elevation and steeper inclination compared with the other pads. In this paper, pads 1 to 12 are referred to as the southern pads, 13 to 26 as the central pads and 27 to 32 as the northern pads.

Details of Anchors

Figure 3 shows a general arrangement of the anchor. The anchors were based on a proprietary system with four 15.2mm diameter Dyform strand tendons and a 3.3m long reinforced resin capsule fixed bond length in bedrock. A series of five trial anchors were installed with eight strand tendons with the 3.3m fixed bond length tested to 1800kN prior to installation of the working anchors. The bond length is formed from a steel wire helix fitted over 3m lengths of unsheathed strand bound using purpose made spacers. This assembly was cast into an 88mm diameter corrugated plastic capsule using a high strength polyester resin.

Each tendon strand was individually greased with a bituminous lime based grease designed for long term corrosion protection and sheathed in 1.5mm thick high density polythene. A 3mm diameter flexible wire strand was attached to the tendon for the purpose of monitoring movement of the bond length relative to the anchor pad. This would be monitored by measuring the length of the strand relative to a fixed datum on the anchor head. The anchors were installed by homing into grout filled holes. The grout comprising neat SRPC cement with a water cement ratio of 0.4 was placed by tremie to the full depth of the hole. A 125mm diameter UPVC duct was then placed over the tendons from rockhead level to the anchor pad.

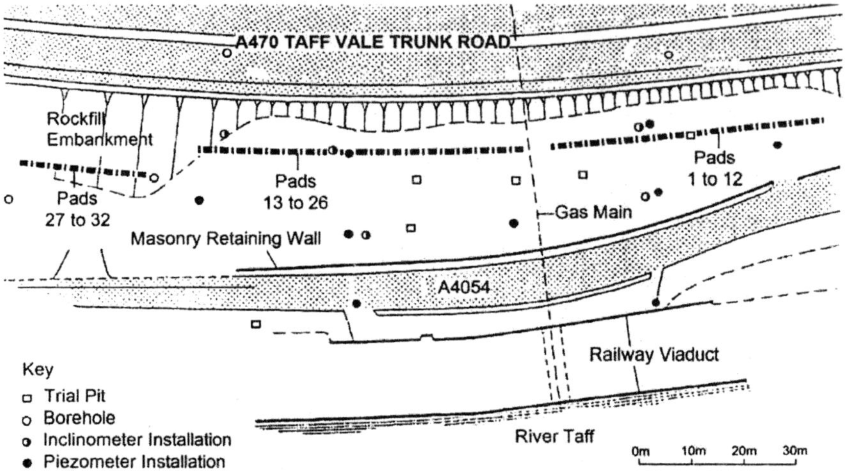

Figure 2 Plan of site

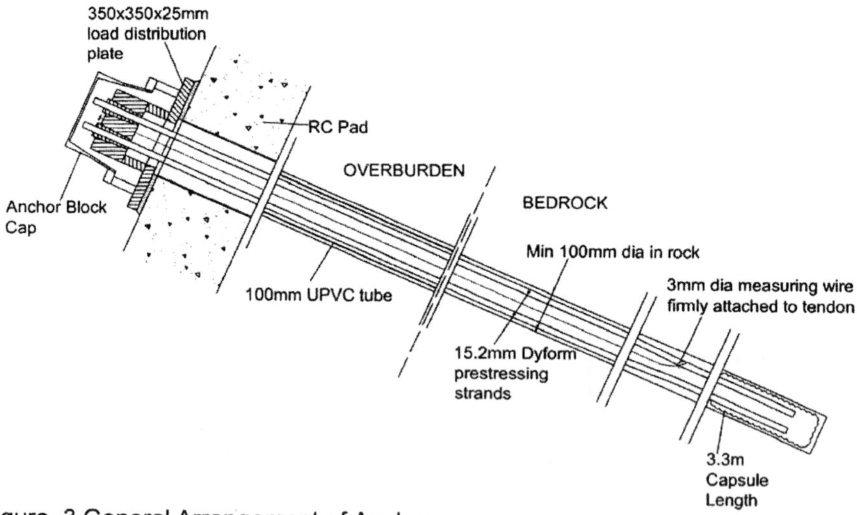

Figure 3 General Arrangement of Anchor

Anchor Head

The anchor head is assembled on a square galvanised steel load distribution plate. The tendons pass through a 150mm diameter duct in the anchor pad and are retained by wedges in the steel anchor block as shown in Figure 3. The anchor block is separated from the distribution plate by steel spacers and any additional shims added during re-stressing. The entire assembly is sealed within a grease-filled galvanised cast iron cover bolted to the load distribution plate.

Monitoring Regime

Construction of the works was completed in 1980 and following the maintenance period recommendations for monitoring and maintenance were made. The designers recommended regular monitoring of the anchor loads to confirm the performance and behaviour of the anchors and that where necessary, re-stressing of the anchors should be carried out. In conjunction with the stress checking, it was recommended that the pads should be inspected and examination of the anchor head assembly should be carried out to check for corrosion.

The loads in the anchors were checked on two occasions during the maintenance period (1980-81) and have been checked on six subsequent occasions in the period 1981-2006. After checking the load in each anchor, re-stressing was carried out where necessary to restore the design landslide stabilisation load of 500kN. On each occasion the monitoring has been carried out on behalf of SWTRA by the same specialist contractor thus providing continuity of experience in the monitoring procedure.

Monitoring and Re-stressing Procedure

To enable anchor loads to be measured following installation of the anchors a purpose-made puller bar was designed. This is bolted to the anchor head with a saddle block supporting the feet of the stressing jack against the load distribution plate and preventing the shims from dropping out on lift off the anchor head. Figure 4 shows a schematic of the stressing equipment.

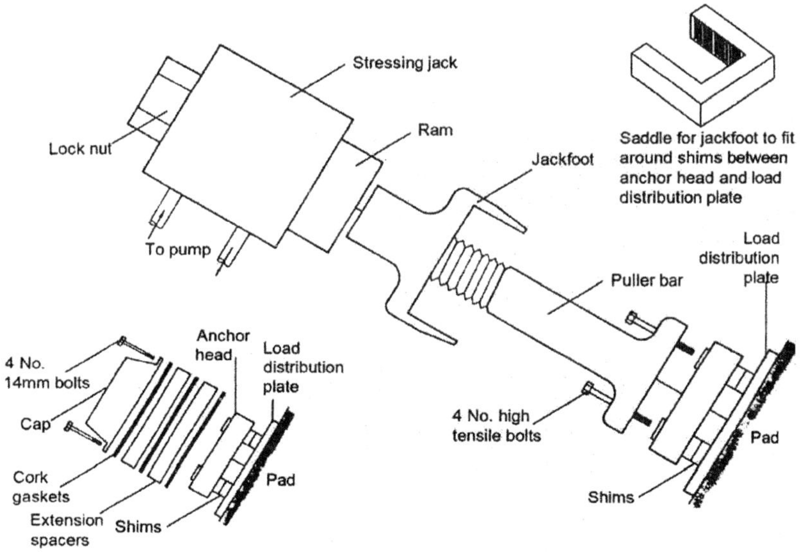

Figure 4 Schematic View of Load Monitoring Equipment

The load at which the anchor head is observed to lift off the load distribution plate is taken as the check lift load in the tendons. The length of the wire line attached to the fixed bond length of the anchor is measured to monitor movement of the fixed anchor length relative to the anchor head. Following the measurement of the lift-off load re-stressing is carried out when the load exceeds 510kN or is less than 495kN by the addition or removal of steel shims between the anchor head and load distribution plate followed by a further check lift measurement of the lock off load.

Results of Measurement and Re-stressing

On each occasion when the loads in the anchors have been monitored it has been found that most anchors have shown a loss of load with time while no anchor has shown an overall increase in load with time since installation. However, occasionally individual anchors have been found to be carrying a load slightly higher than the design load. Table 1 provides a comparison of the number of anchors found to have increased or significantly decreased in load between rounds of monitoring.

Table 1 Variation in loss of load between routine re-stressing operations.

Year	Anchors Gaining Load (>510kN lift off load)	Anchors Losing Load (<450kN lift off load)
2006	2	40
2002	0	54
1998	17	42
1991	0	98
1982	0	93
1981	3	12

An increase in anchor load is likely to reflect a locally increased demand to ensure stability of the landslide (for example reduced stability as a result of an increased imposed load or higher groundwater level). However, during any one re-stressing operation the lift-off load of all the anchors has been below the notional design load, as shown in Figure 5. The continuing loss of anchor load with time was anticipated and is likely to be a function of a number of factors. It would appear that the landslide has demanded slightly less restoring force from the anchors than the design load. This suggests that the landslide has been adequately stabilised by the anchor scheme.

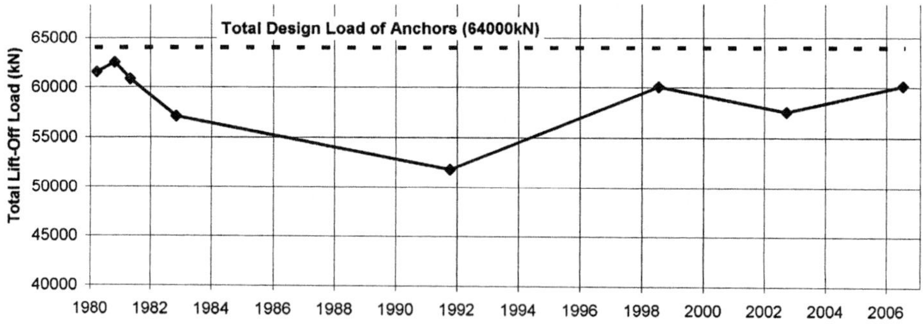

Figure 5 Comparison of total anchor lift off loads

Instrumentation Monitoring

Instrumentation monitoring is carried out on a 6-monthly programme by SWTRA and the results are reviewed for evidence of adverse readings. The monitoring of inclinometers and piezometers and site observations confirm that the stabilisation measures have adequately stabilised the landslide.

Prior to installation of the anchors, rates of movement of up to 15mm/yr were being recorded on shear surfaces within the landslide from inclinometer measurements. Immediately following the installation of the stabilisation measures, recorded movement reduced to less than 2mm/yr with most inclinometers indicating negligible movement. The most recent data plots of deflection for the inclinometers indicate that there is no significant movement on the shear surface or elsewhere within the landslide mass. One inclinometer has shown very limited shear movement (up to 1.5mm of movement) since 2002 at approximately 6m depth on the shear surface between the glacial till and soft silty clay. However, this rate of shear movement is significantly less than that recorded in this instrument prior to or immediately following the installation of the anchors.

All recent piezometer readings are within the previously recorded range and show groundwater levels lower than the highest previously recorded levels. The piezometers have recently been installed with buckets to provide a more accurate measure of variations in groundwater levels. It is not considered that the piezometer results show any cause for concern.

Prior to the installation of the stabilisation measures there was significant cracking and settlement in the A470 carriageway. The rate of movement across the cracks reached up to 10mm/month during the initial stages of the installation of the anchor pads although this reduced significantly following completion of the stabilisation scheme. There is no current evidence of significant cracking or settlement within the carriageway and verges of the A470 or of the anchor pads.

Loss of Anchor Load

It was anticipated that there would be loss of anchor load with time as a result of settlement of the pads, creep within the fixed bond length, relaxation and creep of the strand anchors. Any reduction in the load 'demand' from the landslide (i.e. a reflection of an apparent increase in stability of the landslide with time) could also result in loss of load.

Distribution of Anchor Load Loss

A comparison of average cumulative loss of load per anchor with time in each of the three main areas of anchors is shown as Figure 6. This shows that after an initial relatively high rate of loss of load due to pad settlement etc. there is now a relatively uniform loss of load with time although there are variations in the loss of load in different areas. It is noticeable that there is a lower loss of load in the northern pads. This may be attributable to the installation of the pads onto the rockfill embankment which is likely to have a stiffer response in respect of pad settlement. Alternatively it may be that the northern part of the landslide has a marginally higher load demand than the southern part.

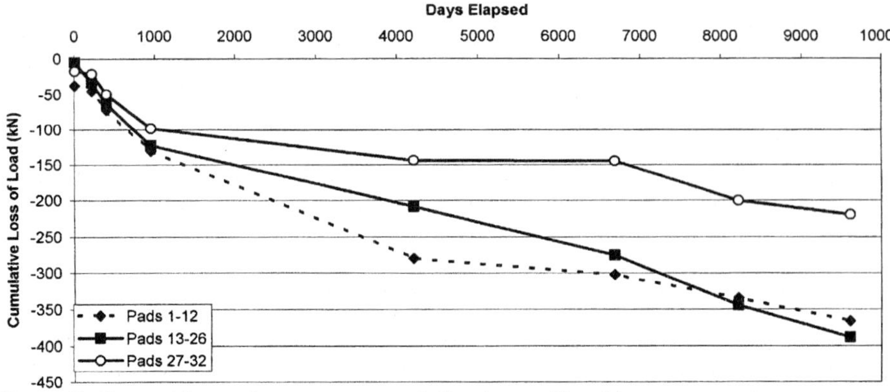

Figure 6 Comparison of average cumulative loss of load by area

Considering the observed behaviour since 1991 it is evident that the northern and southern pads have shown a lower loss of load (average loss per anchor 76kN and 87kN respectively) compared to the central pads which have lost an average of 180kN per anchor. This is indicated in Figure 7 which shows the average cumulative loss of load since 1991. This variation in loss of load since 1991 could be envisaged to be a function of arching of the body of the landslide onto the outermost anchor pads relieving the load on the central pads to a greater extent. However, this is speculative and it is not possible to confirm from the currently available instrumentation results.

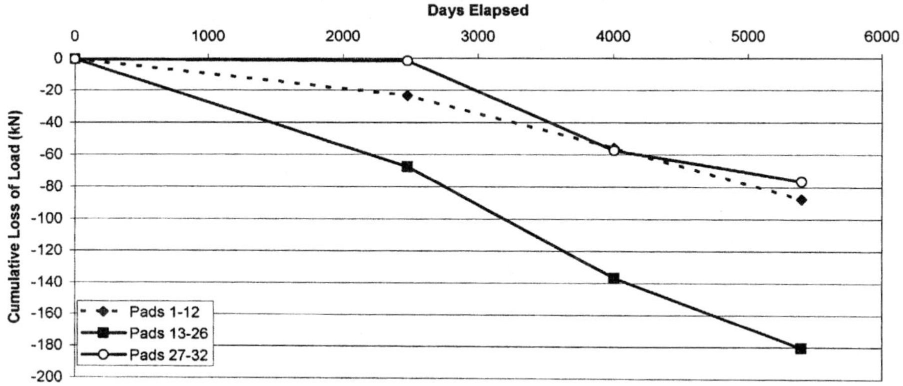

Figure 7 Comparison of average cumulative loss of load by area since 1991

Local Variation of Anchor Loads

Further to the comparison indicated above between the total loss of load in the northern, central and southern pads, a diagram of the anchor pads has been produced showing relatively high or low average loss of anchor load for the four anchors in each pad. This attempts to identify any patterns in the distribution of variation of anchor loads across the landslide. Figure 8 shows the findings for each of the central pads.

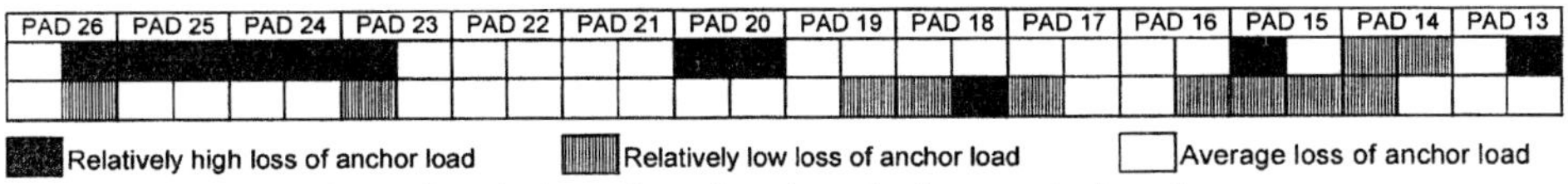

Figure 8 Distribution of variation of anchor loss in the central pads

These results are presented for all the routine monitoring operations since 1981 and highlights individual anchors where the average lift off-load from the monitoring operations is more than ±3% from the mean anchor load.

The variation in loss of load across the anchor pads may be due to a number of causes, including variations in the length of the anchors, the characteristics of the founding strata for the anchor or founding conditions for the pads. In practice the variation is likely to be due to a combination of all of these factors. However, in all cases the monitoring has shown the variation in load to be small (less than ±10% of anchor load) and is not considered to be a cause for concern.

Anchor Re-stressing

A review of the consistency of anchor behaviour has been made by looking at the relationship between the thickness of shims added and the increases in anchor load required during re-stressing. This is presented in Figure 9 for re-stressing operations from 1991. This figure shows cumulative thickness of shims added to restore the design load plotted against cumulative loss of anchor load measured as difference between successive lift-off and check lift loads. This has been plotted with the thickness of shims normalised by a factor L_1/L_2 where L_1 is the theoretical effective length of the anchor (from the drilling records, length of borehole less half the bond length) and L_2 is the average effective length of all anchors (21.3m). This ensures that the small differences in the length of the anchors are taken into account in the plot.

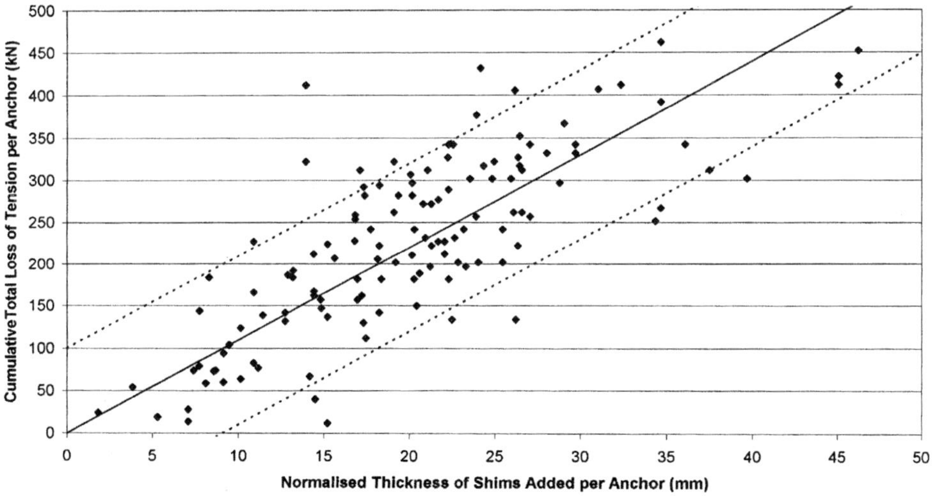

Figure 9 Total loss of load against net thickness of shims added per anchor

Figure 9 confirms, as would be anticipated, a linear relationship between the total increase in load required to restore the design load and the total additional strain in the anchor as determined by the thickness of shims added. However, it is apparent that there is considerable scatter about the mean although it is believed that this may be anticipated from the relatively crude determination of load in the anchors and the strain. Nevertheless, the form of the relationship between the two measurements does appear to indicate that the anchor system is working in a consistent and satisfactory manner.

No measurements are available for the strain recorded during initial stressing and as such it is difficult to extrapolate the anchor stiffness directly from this derived plot. However, the average anchor stiffness indicated by this plot is approximately 11MN/m. This is somewhat stiffer than would be suggested from the physical anchor properties.

This stiffer response indicated by analysis of the re-stressing data may possibly be due to a transfer of load over the free section of the anchor below rockhead. However, the polythene sheaths surrounding the tendons were greased where the anchor passes through the borehole and as such it is considered likely that transfer of load will be small. Alternatively, the higher apparent stiffness may be due to the assessed loss of load (i.e. the difference between the lift-off load and check load) having been over-recorded during the re stressing operation measurements. Additional investigation would be needed to identify the actual reasons for the deduced higher stiffness but this is not considered necessary for confirmation of the behaviour of the anchors.

Figure 10 View of Southern Pads View of Typical Anchor Head

Anchor Condition

Figure 10 shows a view of the southern pads following the re-stressing operations in 2006. The anchor heads all remain in good condition after 25 years of service with no signs of significant corrosion or pitting of the anchor assembly. Specially fabricated new rubber gaskets have been used to replace any original cork gaskets between the anchor head plate and the head cover which have deteriorated. New 'spacer' blocks have also been fabricated which would be used to lift the head covers off the plate if the repositioned head assembly (with shims in place) would prevent the head cover being tied down flush to the plate. These spacer blocks would be installed with a new gasket above and below to ensure watertightness. These have not been required to date, however, with the continuing loss of load and need for re-stressing it is likely that these will be required in future.

There is some localised spalling/loss of the mortar placed around the anchor bearing head plate (i.e. filling the edges of the sunken recess in the pad) although this does not appear deleterious to the function of the anchors. The anchor pads and bearing plinths are all in good condition with no sign of cracking or relative movement of the pads.

Conclusions

The operation of the rock anchors has been monitored over a period of 25 years and shows successful behaviour of the stabilisation scheme. Instrumentation demonstrates that landslide movement is continuing to be arrested by the anchor scheme. The benefits of regular monitoring and maintenance have been recognised by SWTRA to provide a safeguard against potential future disruption to the infrastructure.

No significant increase in load in the anchors has been recorded during the monitoring of the anchors and the anchors remain in good condition indicating the effectiveness of the design of the anchor head protection.

For the future management of the anchorage system, it is only considered necessary to continue the routine monitoring of the ground anchors and instrumentation to ensure the safe and efficient working of the stabilisation scheme.

Acknowledgements

The authors wish to thank the South Wales Trunk Road Agency and Transport Wales for permission to publish this paper and for their assistance and provision of the monitoring data. The authors would also like to thank Ritchies for their review of the re-stressing data.

References

Kelly, J.M.H and Martin, P.L, *Construction Works on or near Landslides*, Proc. Symp. Landslides in the South Wales coalfields, Polytechnic of Wales, 1985.

Martin, P.L and Kelly, J.M.H, *Design and Performance of Rock Anchored Bearing Pads Installed in 1980 for the Stabilisation of an Unstable Soil Slope at Nantgarw, South Wales, UK*, Proc. Conf. Landslides in research, theory and practice, Thomas Telford, London, 2000.

ANCHORED SLOPE MONITORING AND TEST FOR PERIODIC INSPECTION IN JAPANESE EXPRESSWAYS

Katsumi Okubo, Masaru TAKEMOTO, NEXCO Expressway Research institute and **Hiroshi YAMADA**, Nittoc Construction Co., Ltd, Sapporo, JAPAN

Abstract

Ground anchors and retraining piles have been installed on cut slopes on expressways in Japan to ensure slope stability. Ground anchors have been adopted on numerous slopes because of the ease of construction and short construction period required. Some of the ground anchors were constructed some time ago and have been in service for approximately 40 years. Some have fractured and are no longer functional as originally intended due to deterioration or ground deformation.

Objective

Ground anchors are available in two types in Japan. The old-type anchors constructed in or before 1988 have not been fully protected from corrosion at the head. The new-type anchors installed later have been applied with enhanced anticorrosive protection. Damage by corrosion is of less concern for new-type anchors. The new types still require appropriate maintenance because more than 20 years have passed since their installation. Maintenance is discussed in a separate paper. This paper focuses on the control of tensile load of ground anchors. For load measurement, load cell installation was made mandatory in 1992 to provide for load management. The life of a load cell is shorter than the service life of permanent anchors, so problems remain with load management. The paper describes load measurement evaluation methods and load measurement equipment for evaluating ground anchor integrity.

Inspection method

Cuts, fills and bridges on expressways are checked by regular, annual and detailed inspections in Japan. Ground anchor loads are measured in detailed inspections that are conducted once in five years. It has been made mandatory to measure loads for more than five percent of the anchors installed on a slope and more than three anchors. Multiple ground anchors are generally installed to stabilize a slope. Insufficient residual load of a ground anchor therefore does not necessarily mean instability of the entire slope. In cases where no urgent repair is required despite damage or deformation, continuous monitoring or investigating the progress of damage and deformation and examining the stability of the entire slope are necessary.

Anchor integrity evaluation

The integrity of a ground anchor should be evaluated through comprehensive consideration of the rust or damage of the locking unit and tension member. This paper describes technical standards for load measurement that provides for quantitative evaluation. Anchor loads may be measured either by using an existing load cell or by lift-off check. The residual load measured should be evaluated considering the load-displacement characteristics as well. Approximate guidelines are shown in Table 1. For symbols I, II and OK, refer to Table 2. Type A is a sound anchor. Type-B anchor has pulled out. In Type-C anchor, the wedge and tendon have slipped after lift-off. Type D is overloaded.

In the case where load-displacement characteristics are rated as I or II, a review meeting of engineers should be held immediately to discuss the cause and corrective measures.

Whether the scar at the bite of a wedge that is likely to lead to fracture or breakdown is on the free or extra length of the anchor after re-tensioning or member replacement results in large difference in the evaluation result. Great care should therefore be exercised at the time of re-tensioning or member replacement.

Table 1. Rating of load-displacement characteristics

Type	Symbol	Classification of characteristics	
A		$\theta_1 = \theta_2 = \theta$	
	OK	$\dfrac{E \cdot A}{(\ell f + 0.5 \cdot \ell a)} \leq \tan\theta \leq \dfrac{E \cdot A}{0.9 \cdot \ell f}$: Normal	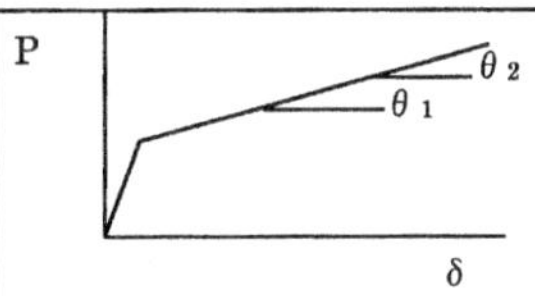
	II	$\dfrac{E \cdot A}{0.9 \cdot \ell f} < \tan\theta$: *Abnormal*	
		$\dfrac{E \cdot A}{(\ell f + 0.5 \cdot \ell a)} < \tan\theta$: *Abnormal*	
B	I	Load never increases after lift-off. Load decreases as displacement increases.	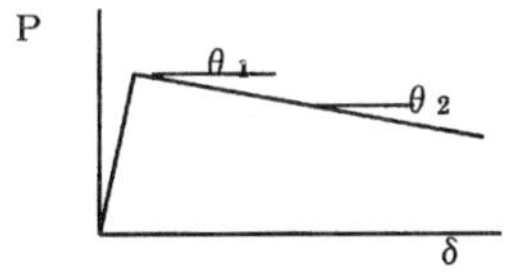
C	I or II	Load suddenly decreases and immediately starts increasing. -The wedge and tendon slip after lift-off.	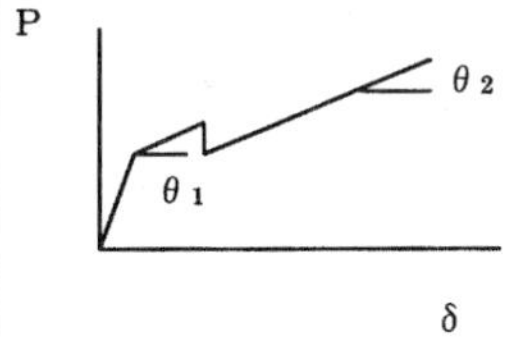

D	I or II	No lift-off can be explicitly confirmed. -The free length section may have been strained because grout entered the sheath in the free length or for other reasons. -The anchor is overloaded.	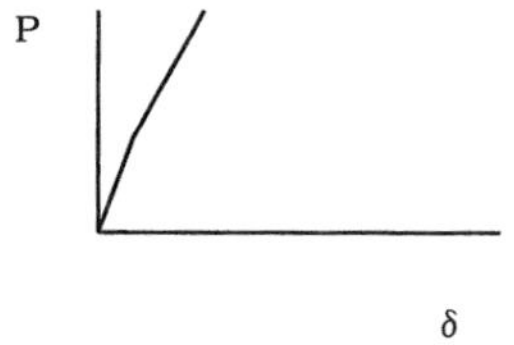

P: Load, δ: Displacement, lf: Free length of anchor, la: Locking length of anchor

E: Modulus of elasticity of anchor tendon, A: Effective cross section of anchor tendon

Table 2. Rating of anchor integrity

Rating	Description
I	The anchor does not work at all or has its capacity has deteriorated considerably. Neither corrective nor repair measures are expected to restore the capacity.
II	Capacity has deteriorated. Corrective or repair measures can restore and maintain the capacity.
OK	The anchor is working properly. Excellent conditions may continue without any corrective measures.

A new load measurement method

Outline

At present, anchor loads are generally measured either (i) by using an existing load cell or (ii) by lift-off check. For method (i), there is a concern about the durability of the load cell. Method (ii) requires much time and cost. A method is therefore required for measuring the load over a long period of time in a simpler manner. A non-destructive monitoring device using magnetostriction was developed jointly by Nippon Expressway Company Limited (NEXCO) and Mitsubishi Heavy Industries, Ltd. Figure 1 provides an outline of the device.

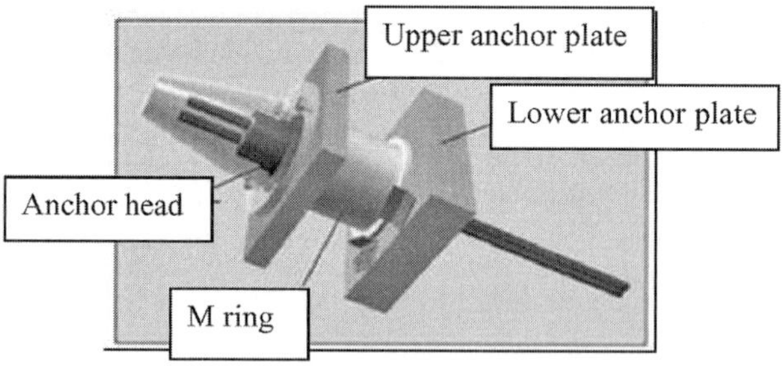

Figure 1. Equipment outline

Principle of measurement

The measurement equipment is designed to measure the stress in the M ring produced by residual load. A probe is used for measurement.

When steel undergoes a strain, the magnetic permeability changes according to the direction and extent of the strain (Villar effect). A probe with a combination of magnetized and detecting coils can detect the strain of magnetic field due to the anisotropy of permeability as voltage. Analyzing voltage leads to identification of main stress and its direction (Figure 2).

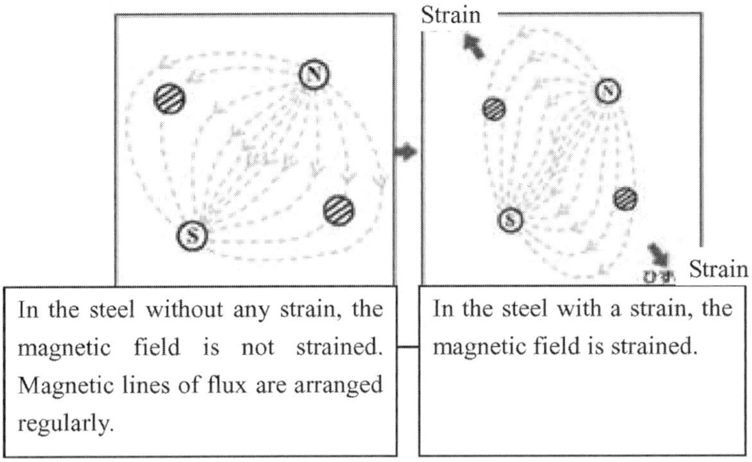

Figure 2. Principle of measurement

Measurement system

The measurement system of the measurement equipment is composed of a ring-shaped member that is inserted into the locking unit and a dedicated measurement device (Figure 3). In the field, only the ring-shaped member is installed. There is no electric wiring including electric signal cables or power source equipment. The measurement system is durable for decades like the anchor head. The measurement system is installed between the anchor head and bearing plate, so it is applicable to any type of anchor. Tensile load is measured by having a dedicated probe touch eight points on the outer surface of the ring-shaped member. Figure 4 shows how measurement is made. Measurement throughout the outer surface enables the identification of eccentric loading, and the monitoring of changes in tensile load balance due to partial anomaly of the strand as well as tensile load variation (Figure 5). The measurement system uses no electric wiring, so it is capable of long-term stable measurement. Without wiring, the system can be installed easily. Data collection never becomes impossible due to disconnection. Measurement in about five minutes is another advantage. Long-term durability and the short measurement time required sometimes make the system advantageous in terms of life-cycle cost. The system also provides for the monitoring of eccentric loading that cannot be handled by other equipment.

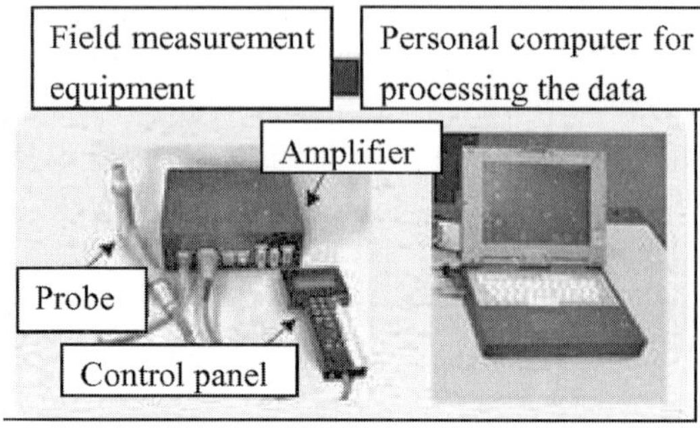

Figure 3. Measurement system

Figure 4. Measurement in the field

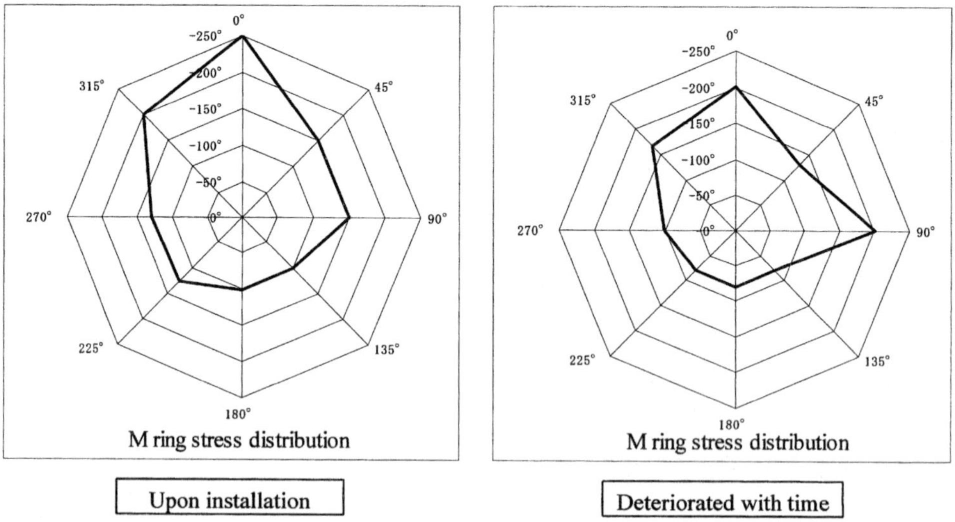

Figure 5. Monitoring of eccentric loading

Method of dynamic slope monitoring

Outline

As described earlier, insufficient residual load of a ground anchor found in load measurement does not necessarily mean instability of the slope. In cases where no urgent repair is required despite damage or deformation, continuously monitoring or investigating the progress of damage and deformation and examining the stability of the entire slope are necessary. Measurement at a few points is insufficient. Measurement should be made at numerous points to identify displacement in a plane. Using measurement equipment for measurement at numerous points requires much time and cost. The authors then took notice of digital cameras and personal computers that have recently been available at lower cost and have been undergoing rapid technological enhancement. A system has been developed that can easily measure slope displacement based on the image data obtained by a digital camera using a theory for calculating three-dimensional coordinate values of an actual object from two-dimensional coordinate values.

Measurement and analysis

A target is installed on the slope (Figure 6) for photographing from dozens of locations. The images obtained are analyzed using a personal computer to obtain the three-dimensional co-ordinate values of the target. At present, measurement is possible with millimeter precision at a photographing distance of less than 100 m.

Figure 6. Target

Advantages

The system requires some effort for installation of the target. Once the target has been installed, however, a digital camera and a personal computer suffice to take measurements. Thus, the measurement system is highly convenient. The key advantages of the system are listed below.

-Measurement can be made throughout a slope. Deformation in parts of a slope can be detected while no conventional technologies have provided such a capacity.

-A digital camera and a personal computer suffice to make measurement.

-Measurements can be taken easily without any expert knowledge.

-Analysis results can be obtained in a short time.

-Data can be stored in a system, so the measurement system is advantageous when identifying the slope deformation with time or managing measurement data.

Summary

(i) Technical standards have been developed for evaluating the integrity of ground anchors.

(ii) New load measurement equipment has been developed.

(iii) A technology has been developed for efficiently measuring the displacement throughout a slope.

Closing remark

High-quality equipment has recently been developed using dynamic monitoring technology including the technology for measuring the displacement of an entire slope. In the future, research will be required on the control standard values for determining whether the slope is likely to fail or remain safe based on the displacement values obtained. It will be necessary to evaluate the integrity not only of an anchor on the slope reinforced with anchors but also of the entire slope.

Monitoring of Multi-Strand Ground Anchors at the City of Manchester Stadium

Lee Jordan, Arup Manchester, UK

Abstract

The City of Manchester Stadium is a world class venue which formed the centre piece for the 2002 Commonwealth Games. Arup were employed by Client Manchester City Council for the Multi-disciplinary architectural and engineering design of the project. Following the Games it underwent a transformation in to its current function as home ground to Manchester City Football Club. This paper describes the use of permanent multi-strand anchors to support permanent loads from the stadium roof. The anchors were installed in 2001 and given their importance, a monitoring strategy was developed and has been implemented between 2001 and 2007. The methodology and findings of this monitoring are discussed as well as some key lessons learned.

Project Background

The City of Manchester Stadium project turned a derelict and contaminated former industrial site in East Manchester into a 48,000 capacity world class multifunctional football stadium, with an interim single usage as a 38,000 capacity athletics stadium, as centrepiece to the 2002 Commonwealth Games. Immediately after the Games, the athletics track and field were removed and the stadium underwent further modification which raised the seating capacity to 48,000 (Figure 1a). This new stadium became the home for Manchester City Football Club (MCFC) at the start of the 2003/04 season and has also been used for rugby events and music concerts.

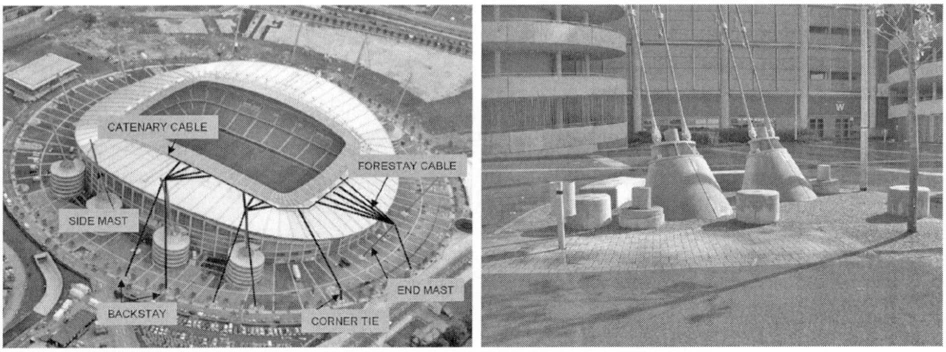

(a) Stadium and roof components **(b) Roof Backstays and anchor plinths**

Figure 1. Completed City of Manchester Stadium.

Requirement for and Function of the Ground Anchors

The Stadium structure comprises four principal components, namely the pre-stressed cable net roof, eight spiral ramp access towers, the seating bowl and the foundations, which included 1150 conventionally bored piles (Ref 1). There is a two level basement below the west stand that is used for football training, catering and office space.

The roof is a lightweight cable stayed structure suspended from twelve masts connecting to the 28 specialist backstay foundation elements, located about 35m outside the building footprint of the bowl (See Fig 1 a and b). The roof rafters are supported at one end of the bowl structure and near the tip by a series of forestays. Loads (inclined in the direction of forestay) are transferred from the forestays to the top of these masts. They are reacted by compression at the base of the mast and a tensile force which is transferred via galvanized steel rods to a set of 24 backstay foundations, external to the main structure, each containing two ground anchors (see Fig 2).

In order to avoid uplift of the roof, a grounded catenary tension ring is employed to preload the forestay tension cables and backstay rods such that they can never go slack. The catenary lies in the plane of the forestays and runs between the forestays near the leading edge of the roof. The catenary is then grounded by four corner tie down cables, one at each corner between the end and side stands. The pre-stress forces of the catenary ring are transferred to four corner ties (making up the 28 specialist backstay foundation elements), which are connected to a total of eight ground anchors.

Each foundation unit contains 2 permanent ground anchors, giving a total of 56 anchors. The anchor tensile working loads (Tw) are between 1000 and 1950 kN per anchor. The design life of the backstay elements, including anchors, is 120 years.

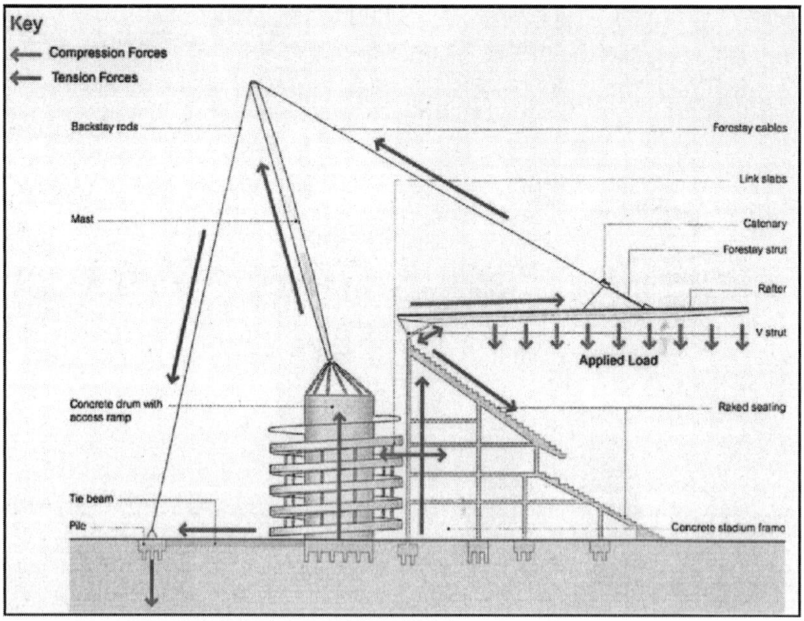

Figure 2. Structural Configuration and Force paths.

Arup designed the backstay structural elements and produced a performance specification for the anchors, initially based on BS 8081 (Ref 2). However the anchors were designed and were installed by a specialist contractor, Keller Ground Engineering, acting under a subcontract to Stent Foundations, who carried out the piling.

Ground Conditions

The site had an industrial history spanning over two hundred years, leaving a legacy of buried foundations and chemically contaminated soils. The Made Ground deposits, 2-3m thick, are underlain by Glacial Till (primarily clay), 7 – 15m thick. The top of bedrock varies around the Stadium footprint, at depths of between 12 and 18m. It comprises Carboniferous Coal Measures, predominantly mudstone but with thinner siltstones and sandstones and containing six shallow coal seams, the thickest being 1.2m. The Coal Measures were generally weak to moderately weak (Unconfined Compressive Strength (UCS) 1.25 to 12.5 MN/m^2) and typically highly weathered and fractured, particularly close to rockhead. The sandstone bands, occupying about 10% of the site footprint, were typically at the upper end of this strength range.

The coal seams had been worked both by partial extraction techniques and longwall techniques. To reduce the risks to the project, the shallow seams were in-filled as part of a site enabling works, using consolidation grouting.

Anchor Installations

Each of the 56 post tensioned, multi-strand ground anchors contains between 7 and 13 No. 15.2mm greased and sheathed dyform 7 wire strands, installed within a common 165mm corrugated duct. Each strand has a Guaranteed Ultimate Tensile Strength (GUTS) of 300kN. The anchors were double corrosion protected and had a factory grouted encapsulation. A further internal 110mm corrugated duct contained the opened and noded unsheathed strands in the encapsulation, and formed the second layer of protection. Within the free length, the protection was formed by the polypropylene strand sheath (17mm i.d./19mm o.d) and the 165mm corrugated polypropylene duct, which had a minimum wall thickness of 1mm.

The target lock off load was 110 to 115% of the working load (Tw), to allow for a degree of future load loss due to ground creep or strand relaxation whilst still remaining serviceable. The design was based on a working ground to grout bond stress of 260kN/m^2 (with a Factor of Safety of 3 on Ultimate Bond Stress) acting on a 240mm diameter drill hole, resulting in fixed anchor lengths of 6 to 10m. The anchors were installed within the mudstone and sandstone strata, with the distal end of the anchor between 25 and 30m below ground level.

Earlier during the works it had become apparent to Arup that strand anchors manufactured in a certain way have an inherent susceptibility to water ingress and premature corrosion. In particular, the anchor head region is often exposed to the elements during both temporary and permanent conditions and therefore if inadequately protected there is a potential for the development of a corrosive water and air environment. The anchor head detail plays a very important role in protecting the anchor. A considerable effort was spent on developing a head which would both prevent a moist gaseous corrosive environment from developing whilst allowing routine head lift off testing to be undertaken. The adopted head detail is illustrated as Figure 3.

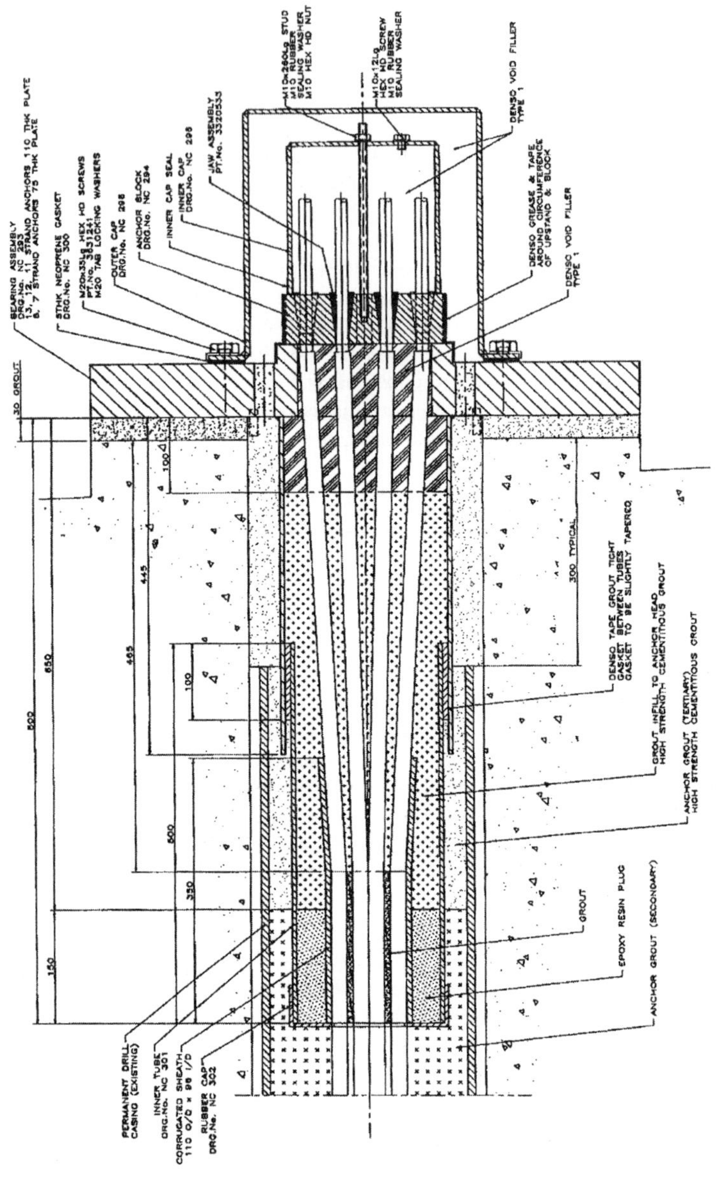

Figure 3. Anchor Head Detail – (After Freyssinet Ltd)

The head includes an outer and inner cap sized such that for routine lift off testing, only the outer cap needs to be removed, whilst the inner cap, which is completely grease filled, remains in place protecting the strands and wedges from rain ingress. A machined lip on the galvanised head plate provides protection from any rain water entering the outer cap, in addition to layers of Denso tape wrapped on the outer interfaces of the inner cap. Raised reference points were built in to the outside of the head plate to allow future level monitoring.

Anchor Monitoring Requirements

As the anchors are elements which are critical to the support of the roof structure, a monitoring /maintenance strategy was developed in order to identify any loss in pre-stress due to ground relaxation, strand corrosion, creep/relaxation of the steel strand or structural problems with the support. The objective of this is to enable early action to be taken in the event of a gradual loss of functionality of the anchors. The monitoring forms part of the Stadium public licensing, which falls under the remit of Manchester City Council (MCC), and is therefore mandatory.

The BS 8081 appeared to be limited in terms of definitive guidance on anchor maintenance. The BS EN 1537 (Ref 3) appeared even more limited, therefore a detailed strategy was developed and written in to an Operating and Maintenance Manual for the backstays. A copy of this is held by the tenant MCFC and a copy by MCC Building Control. This document provides a briefing on the history of the backstays and describes their structural function in a manner that will allow future personnel to quickly familiarise themselves with the force paths, testing procedures and the trigger and action levels associated with the monitoring.

Given potential problems with continuity of staff and therefore knowledge over time, the author's view is that this is a key document and means of passing safety and procedural knowledge on to future parties involved, including Stadium Operator, Consultant, testing contractor and Local Authority Building Control. Due to the high levels of pre-stress energy in the anchors, there are inherent risks and any future maintenance parties need to be fully aware of how the backstays operate before they can safely monitor them.

There are three components to the monitoring regime:

Anchor Lift off Testing

This monitoring allows the anchor residual load to be determined and hence any load loss to be detected. The load losses could be indicative of the following:

- Corrosion of the strands resulting in failure of single or multiple strands
- Slippage of the fixed length i.e. movement on the anchor/ground interface
- Structural creep movements

All anchors were intended to be locked off at about 110% of design load Tw such that over the design life, some load loss of up to 10% could be accommodated with the anchor remaining fully operational. In reality, most of the anchors were locked off at between 110 and 125% of working load due to practicalities involved in the stressing and lock-off procedures.

The anchor head is lifted using a specially adapted threaded puller coupled to a low friction hydraulic jack (typical setup is illustrated in Figure 4). By conducting this testing at periodic intervals, any gradual losses in load can be detected and remedial actions implemented.

Head lift off testing is carried out on all 56 of the anchors. Thus far there have been 9 rounds of monitoring, initially on 6 monthly intervals then annually. This monitoring was written in to the performance specification for the anchor installations:

- The baseline readings were carried out after installation in December 2001 (Visit 1).
- Monitoring rounds at 6 monthly intervals in April and December of 2002 (Visits 2 and 3), in June/July and December of 2003 (Visits 4 and 5), in June/July of 2004 (Visit 6).

- Monitoring at 12 monthly intervals in September 2005 (Visit 7), in July 2006 (Visit 8) and in July 2007 (Visit 9).

Figure 4. Anchor Lift off Testing

From July 2007, the maintenance schedule currently proposes that lift off testing be conducted every 5 years, although this is currently under review. Following July 2007, components of the apparatus are provided to MCFC and lift off testing continued until the end of the design life.

It was expected that a greater frequency of monitoring would be required in the early life of the anchors as this would detect any bedding in losses due to fixed length creep. By conducting a winter and summer monitoring it was envisaged that this could also build up knowledge of any seasonal fluctuations that might occur.

As part of this work, the head levels are recorded using a precise level referenced back to benchmarks which have been established on fixed points on the main stadium structure. The temperature is also recorded to ascertain any effects.

It is suggested that these works are supervised by an independent person in addition to the contractor's own staff.

Anchor Head Inspections

This involves the removal of both the outer and inner grease caps and inspection of the key visible components which transfer the tensile force from the structure in to the strands including the strand overlength, the portion of the wedges which are above the head, and the head block itself for visible evidence of metal corrosion or grease degradation (i.e. discoloration, emulsification, dessication and oxidation).

It is considered essential that the inspections are conducted in dry conditions. In the event of a risk of rainfall during the inspection period, shelter was provided to prevent water falling on to the exposed tendons. A typical head condition is indicated in Figure 5b.

Records were to include observations and photographs of the following on each strand/wedge:

- Presence of moisture droplets of any form, from a fine film to free water.
- Presence of rust staining on the strands/wedges/head.
- Presence of pitting corrosion on the strands/wedges/head.
- Check on the tightness of the wedges by hand.
- Presence of any strands where the top of strand is significantly lower than the others.

(a) Anchor Head Location **(b) Typical Head Condition**

Figure 5. Anchor Head Inspections

Once inspected, the grease is replaced and the caps and grease protection reinstalled.

Anchor head inspections were conducted by Arup/Keller in June 2003 and July 2007. The maintenance requirements have now become the responsibility of MCFC and it is proposed to inspect the heads at a rate of 25% every 5 years, to coincide with the 'lift off' visit (eg inspect Stadium North Stand anchors on visit 1, then East Stand on visit 2 etc). This allows any problems with regard to the overhead corrosion protection to be identified at an early stage from this sample rather than waiting for the full 20 years and inspecting all anchors at once.

Inspection of Backstay Structure

This is an annual visual structural inspection conducted by Arup, of the exposed metallic and concrete structures forming the backstay elements. It includes an inspection of the steel baseplates, tie down bolts, locknuts, flexible seals and the concrete plinths for signs of significant corrosion, bending, looseness and cracking which could indicate distress in the support system.

Practical Issues and Operational Errors in Lift off Testing

One of the earlier monitoring rounds resulted in sporadic high load results which the monitoring contractor could not explain. It later became apparent that the alignment of the stressing jack at setup is critical as any extension under slightly misaligned conditions can result in twisting of the jack cylinder and greatly increased internal jack friction giving artificially high loads. In order to minimise this on future visits, a centralising plate was fabricated to locate the top of the jack cylinder and keep the line of pull normal to the head.

It has been established that the heads do not lift evenly and parallel to the head plates. The head plate can be off by 1mm on one side whilst still being in contact on the other. This is due in some cases to uneven strand distributions in the head plate and in others due to variations in the individual strand loads due to differing wedge draw in at the time of lockoff. This is the reason for checking the full perimeter of the plate for lift off, to ensure all the plate is off by 1mm at the lowest point.

It should be noted that jacks often have different calibrations at different ram extensions. It is important that the ram extension used in the field at lift off matches as closely as possible that which was used in the calibration test bed.

The sensitivity and calibration of the jack and pressure gauges are critical in terms of obtaining accurate results. Class 1 pressure gauges were used which have subdivisions of 5 Bar pressure, allowing an estimate of pressure to the nearest 2 Bars to be made by the recorder. The jack has to be sensitive enough in terms of its capacity, in order to produce accurate results. The K350 jacks used on this project had a calibration of approximately 5kN per Bar.

It is very important that the lift off testing is done in as repeatable manner as possible in terms of the setup and ensuring that readings are always taken with the jack in an expanding mode, ensuring that the pressure is sufficiently reduced to allow seal reversal before commencing each subsequent test. The jack should also be expanded in a gradual and steady manner using small increases in pump pressure.

The lift off test can be subject to human error and errors in perception of movement and this can be minimised by repeating the test. In addition, the head block can on occasion become slightly bonded to the head plate and the first lift off test can be artificially higher until this bond is broken. For this project, a minimum of three lift off tests are performed providing these are reasonably consistent and averaged to form the full lift pressure. Further testing is conducted where the results are inconsistent.

With the form of lift off testing used, it is considered that the accuracy of the test is in the order of +/- 3 to 4% and possibly up to 6% for the lowest anchor loads. This is a combination of the ability to read the pressure gauges accurately, the quoted accuracy of the pressure gauge itself and selection of the correct calibration constant. The recorded pressures were generally very repeatable, providing further assurance on the reliability of the test method.

Should the anchor loads indicate significant losses compared to the previous visit, there is provision in the specification to conduct mono-strand lift off testing if necessary. However, there are further risks with adopting this approach such as risk of wedge slippage and strand weakening due to re-gripping which would need careful consideration. This has not been required during the monitoring to date.

Assessment of the Results

Following each monitoring visit, there is some Contractor processing of the data and a factual report is produced. Arup correlate this data with independent site records then carry out further assessments including the comparison with the working load (Tw), the original lock off loads, and the previous recorded loads. Due to symmetry in the structure, there are 7 types of backstays which contain anchors with the same specified working loads. These results are collated so a comparison can be made. There is also an assessment of the trend in overall system load, this being the sum of the anchor set and assessment of any spatial patterns in load changes or magnitudes which could be related to ground conditions. As assessment is also made of the backstay structural inspections, to identify features such as hairline cracking or corrosion that are degrading when compared to earlier visits. Arup then produce an interpretive report highlighting any issues where closer scrutiny, further testing or inspection is required and present this to MCFC and Building Control.

Findings to Date

Lift off Testing

There have been 9 rounds of monitoring so far. The anchors are generally performing at loads within the permitted range. A typical plot of load versus time, complete with trigger and action levels of 110% Tw and 100% Tw is presented as Figure 6.

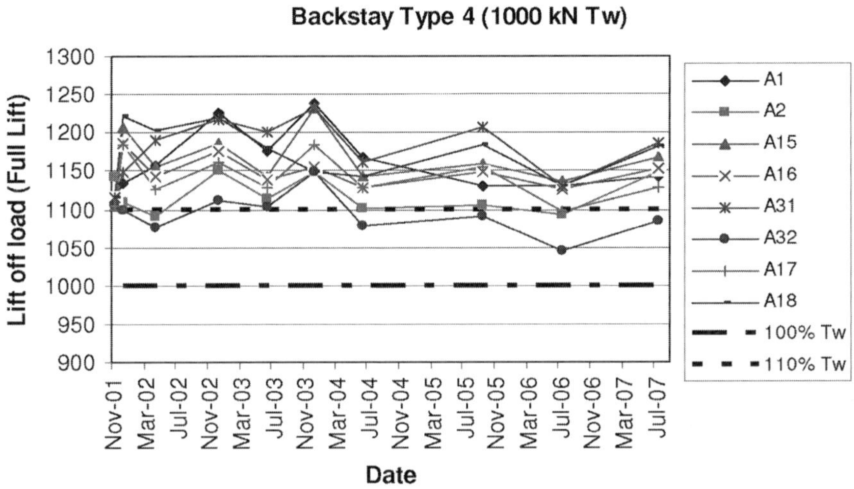

Figure 6. Typical Lift off Test Load versus Time

The results currently indicate that all the anchors are operating within a range of 104% to 119% Tw, with 48 of the anchors above 110% Tw and 8 anchors below. Closer attention is paid to any anchors below 110% Tw on each visit. In terms of the strand loads, they are currently operating at between 51 and 56% of the GUTS.

By plotting the collated results, it is possible to identify anchors which are outside the general trends in load. Figure 7 for example indicates 2 anchors which have been below 105% Tw and although there is no concern at present, they will require close scrutiny on subsequent visits.

An increase in air temperature is likely to cause a lengthening of the steel anchor tendon relative to its fixed anchorage and hence a slight drop in pre-stress. The average air temperature during each visit is recorded and for the monitoring to date has varied between 6 °C and 28 °C. Depending on the depth of temperature penetration assumed, this could account for between 0.5 and 2% of the change in working load recorded. If it is assumed that the full air temperature gradient affects the upper 5m of the anchor this gives a theoretical extension of about 1mm which would equate to a load decrease of the order of 1% Tw.

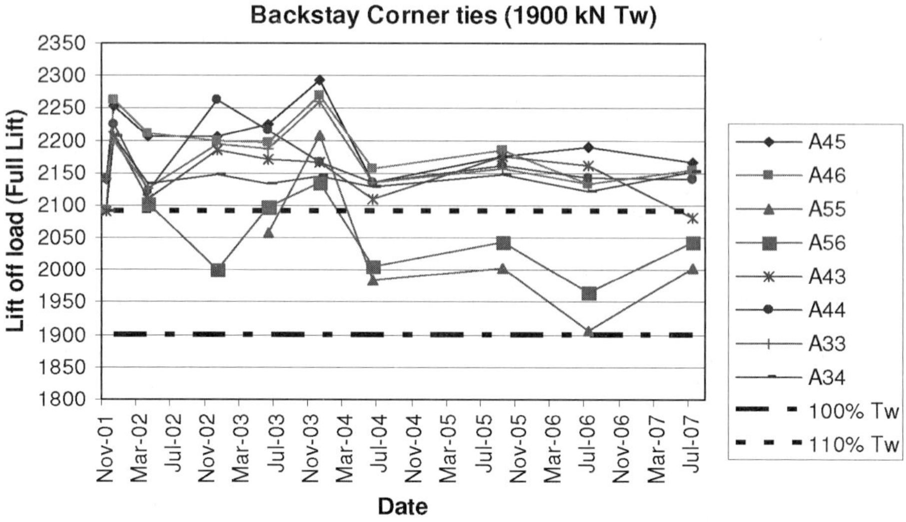

Figure 7. Corner Tie Loads versus Time

Head levels are generally recorded to be within +/- 2mm of the previous visit and it is considered that this is mainly due to the operational accuracy of the surveying and benchmarks used. There is no visible sign of vehicular impact or other head disturbance at the backstays.

In terms of the overall sum of system loads, the specified minimum lock off loads, representing 110% TW, was 96,800kN. The actual lock off loads achieved was 101,700kN, representing almost 116% Tw. Since lock off, the recorded system load has varied between 97,500kN and 103,000kN, a range of 111 to 117% Tw. The current system load is about 113% Tw. There is no evidence to suggest any significant problems with the anchor system as a whole.

Anchor Head Inspections

There have been 2 rounds of anchor head inspections. An example of the most recent inspection in July 2007 is shown on Figure 5b. There is no evidence of significant corrosion on the head block, strands and wedges and all components appear physically undamaged. There is evidence of some light surface staining on the head blocks and some of the strand overlength but this is not considered to be significant. The inspections are providing assurance that the protection detail is providing very good protection to the head.

Backstay Inspections

The backstay concrete and steel structures are generally performing well. There are minor issues relating to damage to the seals between the head plates and the concrete structure, scuffs and small patches of flaking galvanising on some of the metallic components. There have been some very minor concrete cracks to the backstay plinths which have mostly self healed, and a few minor scrapes which need localised repair. There is no evidence of unacceptable corrosion or weld damage to the metallic backstay components.

Conclusions

There are a number of key conclusions which have arisen out of this work:

- Monitoring of ground anchors in service is often overlooked when designing the anchorage system. The BS 8081 is not explicit in the requirements in terms of types of monitoring and methodology required and the Eurocode 1537 is extremely limited in its advice on the subject. Therefore the Consultant or specifier has to develop a monitoring strategy suitable for each particular project.

- The stadium permanent anchor system is only entering the 7th year of its 120 year design life but the regular monitoring has provided assurances that it is generally in a highly serviceable condition at present. There are several anchors where some load loss has occurred but loads remain above the required levels, and the monitoring has highlighted these for future close observation. The head detail has ensured that the degree of corrosion of the head components is very low. The backstay structures are also performing well.

- The testing has indicated a slight seasonal variation in loads, with those recorded in Summer typically lower than the Winter tests, probably due to slight expansion effects. Therefore if annual visits are specified, the Summer visit will usually provide a more conservative indication of the lock in loads. The typical variation between Summer and Winter testing represents about 2 % of working load.

- Given the length of time the anchors are in service and the spacing of the visits, there can be problems with testing procurement. Continuity of knowledge throughout the design life of the anchors is a major issue which can be addressed in a number of ways, in particular by the development of a concise and well written Operating and Maintenance Manual. This can be used both to inform future maintenance staff of the structural requirements and monitoring procedures and can be used as a basis for monitoring procurement. Also by ensuring that all parties, including client and local authority Building Control witness at least part of the testing each visit, so becoming familiar with the practical application, and reducing the knowledge lost if personnel leave companies between visits. It is critical that results are interpreted and appended to the O&M information to enable any long term trends in load to be detected.

- By integrating the initial monitoring requirements into the contractor's specification it is possible to get the benefits of using a consultant/contractor team that are familiar with the anchor system and performance requirements conducting the initial rounds of monitoring. These are critical in identifying any creep losses as the anchors 'bed in', or early problems with respect to corrosion.

- The monitoring procedures need to strike a balance between practicality and accuracy. In order to produce consistent and comparable test results, and minimise the unavoidable operational errors, the method of lift off testing must remain consistent on each visit. Small variations in method can greatly affect the accuracy of the results. As the testing arguably needs to be done to a higher accuracy than the original installation stressing, it is recommended that the equipment should be calibrated immediately before each visit.

- None of the monitoring must expose the anchors to either the risk of water ingress or the risk of overstressing. This can be overcome by the development of the head detail, the use of temporary shelters and the setting of safe limits on stress levels during the testing.

- The accuracy of the test method must be examined before attempting to draw conclusions on the long term behaviour. Movement of the anchor supports should also be checked on each visit in order to validate any apparent changes in load recorded.

Acknowledgements

The Author would like to thank other colleagues at Arup who were involved in the ground anchor installation and monitoring works at the Stadium. The works were particularly demanding at times and thanks in particular go to Colin Curtis, Mike King, Mark Boyle (ex. Arup), Martin Simpson and David Hughes for their support.

Thanks also to the anchor contractors Keller Ground Engineering and Freyssinet Ltd. who manufactured and installed the anchors and conducted the lift off testing.

In particular, the author would also like to thank Mr Asit Sarkar and Mr Rodney Burrows of Manchester City Council and Mr Paul Kenyon of Manchester City Football Club for their permission to publish this paper.

References

(1) CURTIS, LAW and JORDAN, 2002. Piling in Coal Measures at the City of Manchester Stadium for the Commonwealth Games 2002, 9th Int. DFI Conf. on Piling and Deep Foundations p387 to 394.
(2) BS 8081 (1989), British Standard Code of Practice for Ground Anchorages.
(3) BS EN 1537 (2000), Execution of Special Geotechnical Work – Ground Anchors.

Repeated Lift-off Testing of Single Bore Multiple Anchors for Dam Retaining Wall over a 5-Year Period

Mary Ellen C. Bruce, geotechnica, s.a., Inc., Venetia, PA, USA
Jesús Gómez, Schnabel Engineering, West Chester, PA, USA
Robert P. Traylor, Traylor LLC, Ellicott City, MD, USA

Abstract
The right downstream retaining wall of Hodenpyl Dam in Michigan underwent significant lateral displacement due to global instability of the in situ soils. After a failed remediation attempt with traditional tiebacks in 1996, a remediation effort was implemented in 2002 to limit the rate of movement of the wall while additional studies were conducted. The remediation design called for installation of additional anchors bonded within a high plasticity, stiff clay layer. The length of the bond zone of the anchors was limited by the presence of an artesian aquifer underlying the clay layer; therefore, the design unit bond stress was relatively large. To reduce the potential for creep and increase the unit bond strength, the contractor selected the Single Bore Multiple Anchor (SBMA) system with comprehensive, selective post grouting.

The anchors were effective in reducing significantly the rate of movement of the wall. Presently, the wall is being monitored for displacement and periodic lift off tests of the 2002 SBMA tiebacks are being performed while possible additional remediation efforts are devised for future implementation as needed. The monitoring and lift off test data provided a unique opportunity to correlate the response of the anchors with the wall movement. This paper describes the design and construction of the SBMA anchors, testing of a sacrificial anchor, lift-off testing procedures and results, and conclusions.

Background
Hodenpyl Dam, owned and operated by Consumers Electric Company of Cadillac, MI, is located in Wexford and Manistee Counties, Michigan, USA, and is one of a series of dams along the Manistee River. The facility was constructed between 1923 and 1925 and consists of an earthen embankment running roughly in the north-south direction. The powerhouse, located on the southern end of the embankment, discharges water to a tailrace channel excavated in natural soils. Cantilever retaining walls of varying height were originally constructed on the east and west sides of the tailrace channel. The right (east) downstream wall is approximately 250 feet long and 30 to 50 feet high (Figure 1).

After significant displacements of the wall were first noticed, a repair attempt was carried out in 1996. The repair consisted of conventional strand anchors bonded into the high plasticity clays existing behind the wall backfill. In addition, steel sheeting was installed bearing on the wall footing and attached to the wall face. The sheeting was intended to provide structural reinforcement and was not driven into the ground. Subsequent monitoring of the wall revealed that lateral displacement of the wall continued to develop, and that additional stabilization measures were required. The monitoring data collected show that the wall is rotating around a vertical axis located in upstream end of the shorter height section seen in Figure 1.

The Engineer specified a total of 13 restressable, regroutable anchors to improve the wall stability. The remediation was designed to provide immediate improvement of the wall stability while controlling construction costs. The Engineers also specified a detailed monitoring/action plan, the results of which were to be used to determine whether subsequent treatments were necessary. Full details of the investigation and analysis are discussed by Gómez et al. (2004).

Figure 1. View of the tailrace channel and right downstream retaining wall. Note the sheeting and tiebacks installed in 1996.

Bond Zone Materials

The soils behind the retaining wall comprise a layer of poorly graded sand, underlain by an interval of overconsolidated clay (bond zone materials). The clay typically contains a 10- to 30-percent fraction of fine to coarse sand (passing #4 sieve and retained in #200 sieve as per US practice) and traces of gravel-size particles (retained in #4 sieve). Sporadic lenses and/or seams of sand were also detected at some locations within the clay mass. The plasticity of the clay materials increases with depth. The 2002 tiebacks were bonded to high plasticity clay with a plasticity index ranging between 32 and 50 percent and a liquid limit of 50 to 72 percent. The clay stratum is underlain by a confined aquifer consisting of fine to coarse sands. The presence of the aquifer induces upward flow within the stiff clay and above-hydrostatic pore pressures around the tiebacks.

During design of the 2002 remediation, the Engineer realized that the available space for bonding the tiebacks was limited. The existing tiebacks were located near the top of the wall and inclined at angles steeper than 25 degrees with horizontal (see Figure 2). The Engineer also recognized that the new tiebacks should not penetrate the underlying aquifer. The available space for the new tiebacks did not allow for bond zone lengths longer than 40 ft. Finally, the number of tiebacks was also limited due to the configuration of the reinforcing sheeting installed in 1996.

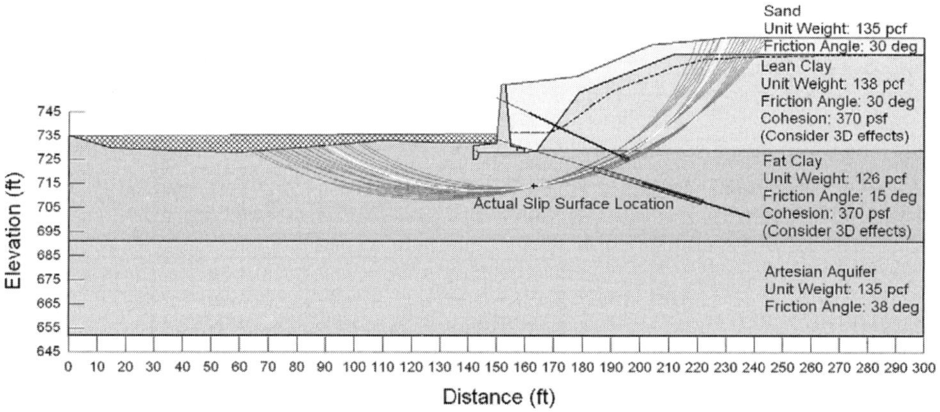

Figure 2. Typical section through the wall showing the 1996 (upper) tiebacks, the 2002 (lower) SBMAs, and the geotechnical conditions.
1 pcf = 0.157 kN/m^3; 1 psf = 0.048 kN/m^2

Because of the limitation on the number and maximum bond length of the tiebacks, there was a concern over the bond strength capacity of the tiebacks. Also, because the tiebacks were to be bonded to stiff, high plasticity clay, the potential for creep of the tiebacks was considered during design. The available proof testing data of the 1996 tiebacks showed that creep was observed during at least one of the tests.

Single Bore Multiple Anchor System

It is fully acknowledged by researchers who have investigated grout/ground load transfer that the distribution of stress along the fixed anchor is non-uniform due to general incompatibility between elastic moduli of the anchor tendon, anchor grout, and the ground (Littlejohn and Bruce, 1977; Barley, 2000). The top portion of Figure 3 depicts the "progressive debonding" that occurs due to this incompatibility in a traditional tieback. The area under the bond stress distribution line is representative of the ultimate load in the anchor. It can be seen that the load does not increase uniformly with increasing length. In an SBMA, load transfer from the strand to the grout takes place over a series of short lengths at staggered intervals along the borehole, thereby reducing or eliminating the occurrence of progressive debonding and substantially increasing the efficiency of the overall anchor. Full details on the SBMA system are discussed by Barley and Windsor (2000).

A typical SBMA (Figure 4) consists of several unit anchors (single or double strand) with varying lengths installed in a borehole (4- to 8-inch diameter (nominally 100 to 200 mm) such that their respective load transfer lengths are located at predetermined positions within the total fixed length.

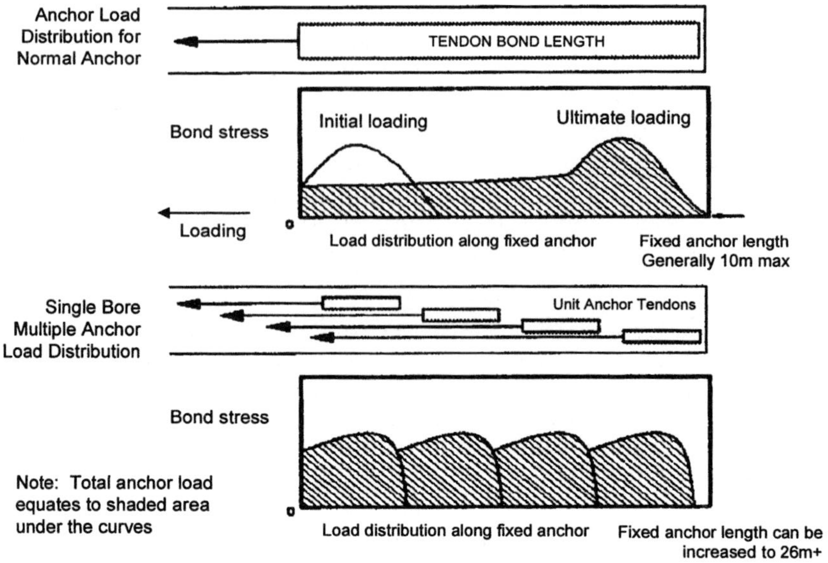

Figure 3. Comparison of load distribution along a conventional anchor and an SBMA (Barley, 2000).

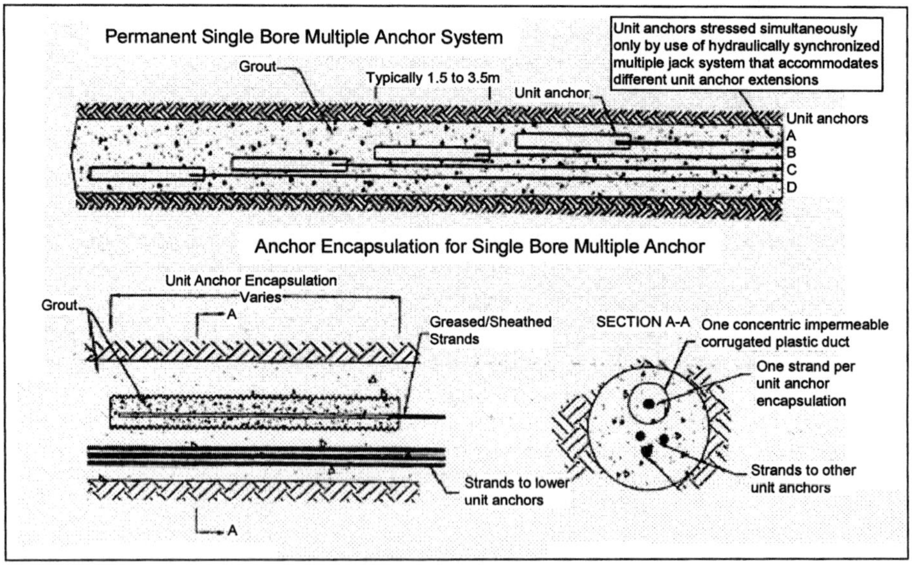

Figure 4. Elevations and cross section of a typical 4-unit SBMA.

Sacrificial Anchor Installation and Testing

The Engineer required the installation and testing of a sacrificial SBMA anchor prior to construction of production anchors. This anchor was installed using identical construction methods and materials (see next section for construction details) and bonded in the same soils as those proposed during production with the following exceptions a) the sacrificial anchor was installed vertically, as opposed to being inclined at an angle of 20° below horizontal; b) three 10-foot-long unit encapsulations were installed as opposed to four to avoid penetration into the sand aquifer; and c) the upper two unit anchors contained two strands in each encapsulation to allow load testing to very high grout/ground bond stress.

The jack arrangement for a three-unit SBMA includes three hydraulic rams that are synchronized by coupling to the same hydraulic powerpack, so that the same load is applied simultaneously to each unit anchor. The jacking arrangement is shown in Figure 5. A primary gauge and a reference gauge were calibrated with one of the jacks. The ram extensions were recorded using a stiff steel rule, and during creep testing by using a vernier caliper. Measurements were corrected for reaction pad movement measured by dial gauges mounted on an independent reference beam.

Two tests were conducted on the sacrificial anchor: 1) Performance/Extended Creep Tests (PTI, 1996), and 2) a high bond capacity test. Since creep was a concern on this project, the load-hold period was extended overnight (total load-hold time 810 minutes) at maximum test load. The linearity of the load/movement plots indicated that virtually no progressive debonding occurred into the unit fixed anchor. The maximum test load for the high bond capacity test was limited by the steel capacity, and so an ultimate bond stress was not measured. The average mobilized bond stress was 5.0 kip/ft^2 (239 kPa) at the maximum test load with no creep or failure.

Figure 5. Sacrificial test anchor stressing jack.

Details of Anchor Construction

The production SBMAs were installed between July 29 and August 23, 2002. Testing and lock off were conducted August 29 and 30, 2002. Anchor holes were drilled using rotary duplex with water flush and 7-inch (nominally 175 mm) casing with an approximate 1-inch (25 mm) overcut from the casing shoe teeth, resulting in an 8- to 8½-inch (nominally 200 to 216 mm) hole diameter. Casing was advanced to the target tip elevation for each anchor.

The specified working load for each anchor was 135 kips (600 kN). Anchor geometry and design requirements dictated a minimum free length of 50 feet (15 m); a minimum bond length of 40 feet (12 m); a minimum drill hole diameter of 5 inches (127 mm); and an anchor inclination of 20° (or less). Class I corrosion protection (PTI, 2004) was required for the anchors. A summary of "unit" anchor lengths and SBMA geometry is provided in Table 1.

Table 1. Summary of unit anchor lengths and SBMA geometry for each anchor.

UNIT ANCHOR	FREE LENGTH	FIXED LENGTH	TOTAL TENDON LENGTH*
A	55 ft (16.7 m)	11.5 (3.5 m)	71.5 (21.8 m)
B	66.5 (20.2 m)	10 (3.1 m)	81.5 (24.8 m)
C	76.5 (23.3 m)	10 (3.1 m)	91.5 (27.9 m)
D	86.5 (26.3 m)	10 (3.1 m)	101.5 (31.0 m)

*Total tendon length = 5-foot tail + free length + fixed length

All anchors were subjected to two phases of post grouting. Two performance and extended creep tests were conducted on production anchors. All other production anchors were proof-tested. The maximum test load was 180 kip (805 kN), or 44.9 kips (201 kN) per strand (see Bruce et al. 2004).

All unit anchors were locked off after testing at 34 kips (152 kN) using a mono-strand jack. The strands were then trimmed to approximately 8 inches (203 mm) above the gripping wedges. The trumpets were grouted, and the steel caps were filled with grease and installed as designed.

Lift-Off Testing Procedures

Lift-off tests were conducted in April 2004, April 2005, and May 2007. The steel anchor head caps were removed and the grease stripped from the strands. A mono-strand jack was used to lift-off each of the 4-strand SBMA individually. Lift-off was defined as the load at which the gauge pressure increase hesitated during loading and at which a 0.020 inch-thick (0.5 mm) feeler gauge could first be fitted below the barrel (Figure 6).

During the test it was observed that the strands and wedges did not show evidence of corrosion as a consequence of removing the greased-filled cap in each of the lift-off events. Indentations of the stressing wedges on the strands occurred during the 2004 and 2005 lift-off events were visible. However, these did not pose significant concerns for the last testing event. Should future monitoring of the tieback load be necessary, it is recommended that load cells be used which can be read remotely to avoid potential risks associated with loss of capacity of the strand or loss of grip due to increasing size and frequency of the indentations.

Figure 6. Measuring lift-off using feeler gauge.

Lift-Off Testing Results

Lift-off loads ranged from 29.55 to 38.27 kips (131 to 170 kN) per strand. The highest load measured was 113% of the generalized lock off load, which is equivalent to 65.6% GUTS. The lift-off loads measured are shown graphically in Figure 7. It is interesting to note that the lift off load increases in the downstream direction along the wall. This correlates well with the pattern of wall displacement measured, which suggests that the wall is rotating around a vertical axis located near the tieback located further upstream.

Figure 8 shows a comparison between the increase in tieback load measured during each lift-off event, and the theoretical increase in tieback load calculated from the measured wall displacements at each tieback location. The 45-degree line shown corresponds to full agreement between the theoretical and measured load increases. The data in the figure show that there is good agreement between the data and the 45-degree line, thus suggesting that the tiebacks have not undergone yielding or creep since installation.

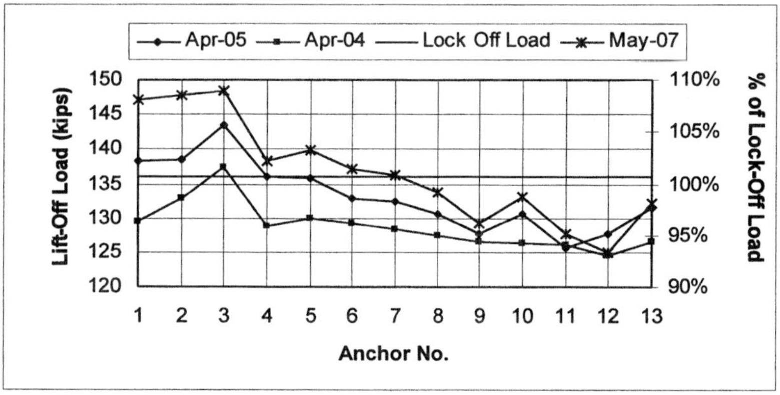

Figure 7. Lift-off loads (total of all strands). 1 kip = 4.448 kN

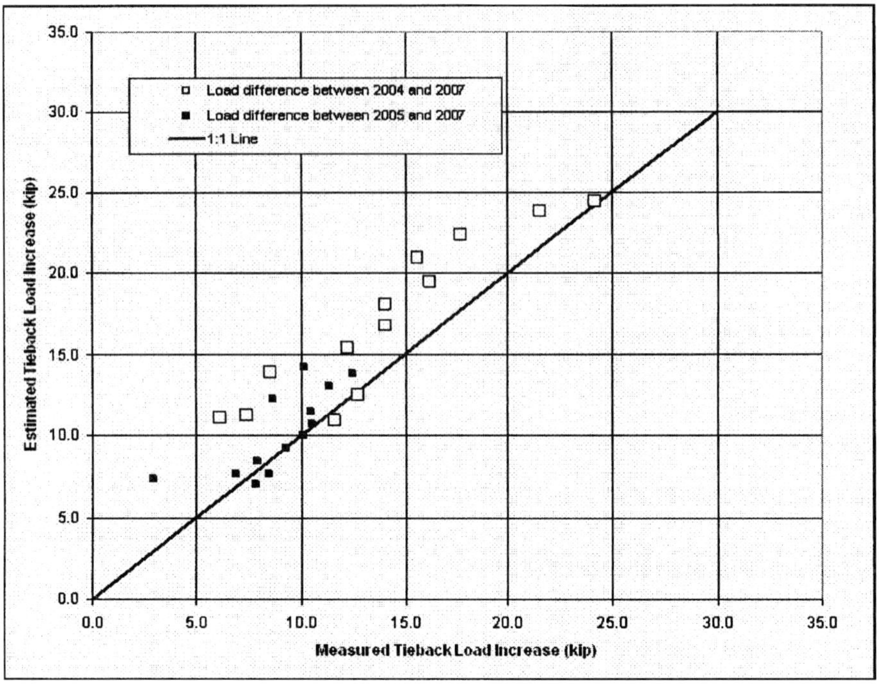

Figure 8. Comparison of estimated anchor load (based on wall movement) and anchor load measured during lift-off testing (2007). 1 kip = 4.448 kN

Conclusions

Based on the results obtained during load testing and lift-off testing of the SBMA elements, we may conclude the following:

- The sacrificial test showed that the SBMA installed in high plasticity clay may develop significant capacity.
- The sacrificial load and production tests showed that creep in the high plasticity clay was insignificant, even though the design bond stresses were relatively high.
- Comparison between the lift-off test data and the wall monitoring data shows that the production tiebacks have not undergone significant creep during the five years elapsed since their installation in 2002
- Repeated lift-off testing of the tiebacks did not appear to have a detrimental effect on the tiebacks.
- The SBMA technology is a valuable tool in projects on which it is necessary to install large-capacity tiebacks in high-plasticity clay.

Acknowledgements

We thank Consumers Energy Company and especially Mr. Stuart Johnson for kindly allowing the publication of the information contained in this paper. Additional thanks must also go to Gerace Construction Company for the cooperation during the construction and lift-off testing phases. Tony Barley (SBMA, LLC) provided important insight during design. We also thank Schnabel Engineering, Applied Geosciences and Engineering, and geotechnica, s.a., Inc. for their effort and time during preparation of this paper.

References

Barley, A.D. (2000). "Trial soil nails for tunnel face support in London Clay and the detected influence of tendon stiffness and bond length on load transfer." Proceedings of the Third International Geosystems, London, UK, June.

Barley, A.D., and Windsor, C.R. (2000). "*Recent advances in ground anchor and ground reinforcement technology with reference to the development of the art*," GeoEng 2000, International Conference on Geotechnical and Geological Engineering, Melbourne, November 12-19, pp, 1084-1094.

Bruce, M.E., R.P. Traylor, A.D. Barley, D.A. Bruce and J. Gómez. (2004). "Post Grouted Single Bore Multiple Anchors at Hodenpyl Dam, Michigan," *GeoSupport 2004 Drilled Shafts, Micropiling, Deep Mixing, Remedial Methods, and Specialty Foundation Systems*, ADSC: International Association of Foundation Drilling and American Society of Civil Engineers, Geotechnical Special Publication No. 124, Orlando, FL, January 29-31, pp. 361-373.

Gómez, J., M.E. Bruce, D.A. Bruce, D.L. Basinger and A.W. Cadden. (2004). "Stabilization of Retaining Wall along Tailrace Channel of Hodenpyl Dam," *GeoSupport 2004 Drilled Shafts, Micropiling, Deep Mixing, Remedial Methods, and Specialty Foundation Systems*, , ADSC: International Association of Foundation Drilling and American Society of Civil Engineers, Geotechnical Special Publication No. 124, Orlando, FL, January 29-31, pp. 346-360.

Littlejohn, G.S. and Bruce, D.A. (1977). "Rock Anchors - State of the Art." Foundation Publications, Essex, England, 50 pp. (Previously published in Ground Engineering in 5 parts, 1975-1976.)

Post Tensioning Institute (PTI) (1996). "Recommendations for prestressed rock and soil anchors." Post Tensioning Manual. Fourth Edition. Phoenix, Arizona. 41 p.

Session three:

North American Dam rehabilitation

Rock Anchors for North American Dams: The National Research Program Bibliography and Database

Donald Bruce, Geosystems, L.P., Venetia, PA, U.S.A.
John Wolfhope, Freese and Nichols, Inc., Austin, TX, U.S.A.

Abstract
This paper describes two of the three major goals of Phase 1 of a National Research Program dealing with rock anchors for North American dams. The study commenced in 2005 and has, to date, found approximately 400 case histories spanning four decades, and 230 technical papers. A review of the bibliography is provided with special emphasis on the evolution of the philosophies and details of corrosion protection. This focus is maintained when analyzing the information in the database, while data are also provided on costs and prices.

Introduction
During the period 2005-2006, Phase 1 of the National Research Program into the use of rock anchors for North American dams was undertaken. This work had three goals:
(i) complete a bibliography of all technical papers published on the subject of dam anchoring in North America;
(ii) create a database containing as much information as possible on each dam anchored in North America; and
(iii) conduct a comparative review of each of the five successive versions of the national "recommendations" documents which have been published in the U.S. since 1974;

This paper describes the results of the first two research goals with a particular focus on the evolution of corrosion protection concepts and details. The findings of the third research goal are presented in the companion paper, "Rock Anchors for North American Dams: The Development of the National Recommendations (1974-2004)," also published in these Conference Proceedings. The current authors were the Co-Principal Investigators for this project, funded by a consortium of American and Japanese interests. They relied heavily on the cooperation of specialty contractors and specialist post-tensioning suppliers who provided access to historical records.

The Bibliography
General
A comprehensive literature survey was completed to identify published dam anchoring case studies and various publications documenting the evolution of North American dam anchoring practices and construction methods. A total of 230 technical papers have been compiled relating to North American post-tensioned rock anchoring projects. Hard copy and electronic versions of each have been collected. Figure 1 shows the number of publications by year indicating that over the first five years of the twenty-first century, industry has been publishing at a rate of about 13 papers per year. These papers relate to over 200 dams.

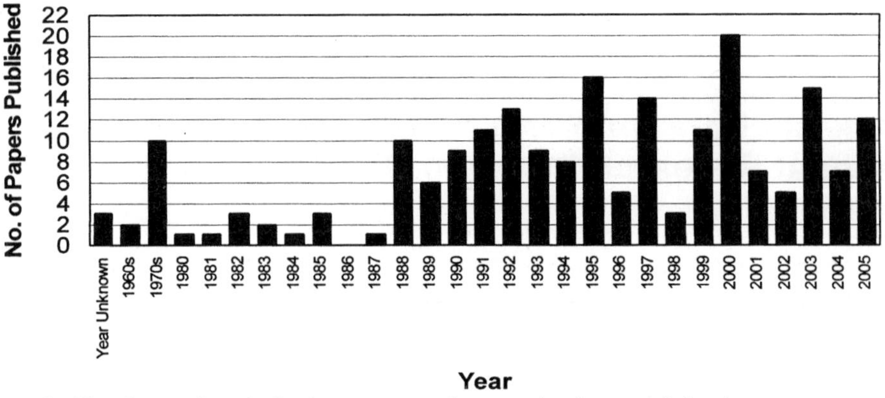

Figure 1. Numbers of technical papers on dam anchoring published per year

Observations on Corrosion Protection

When analyzing these technical papers, one occasionally wins fascinating insights into contemporary thinking regarding corrosion protection issues. The following is provided by way of illustration: no commentary is necessary and no claim is made that this is anything other than a sampling. The papers are presented in date order of project completion.

- Thompson (1969) (John Hollis Bankhead Dam, AL, 1968). Based on chemical analyses of the groundwater "...the laboratory concluded that the risk of corrosion would probably be no more than that of a coal free environment." "During the 12 months when the anchors were loaded [i.e., primary grouted and stressed only] their wires and the grout pipe developed corroded areas just below the anchor head. The wires subjected to flowing water had lost considerably more weight per unit of length than the wires in the still water." "...it was concluded that a good cover of cement grout would adequately protect the anchor wires."
- ENR (1971) (Ryan Dam, MT, 1970). "...the stressing length of the anchor was filled with grout to protect against corrosion."
- Buro (1972) (Libby Dam, MT, 1971). "After all the stressing procedures were completed, the anchor was secondary grouted — that is, the space between the top of the bond length and the stressing anchorage was filled with grout to provide the steel with positive and complete protection against corrosion." The concept of providing a greased and sheathed free length for "measuring anchors" is introduced.
- Bruce (2003) (Rock Island Dam WA, 1975). This dam had been anchored with fully bonded multiwire tendons in 1975. With considerable foresight on the part of the designers, four anchors were equipped with greased and sheathed "minitendons," each comprising 4 wires, which could be lifted-off in the future.
- Sharma and Sasaki (1985) (Pacoima Dam, CA, 1976). Eight of 35 anchors "were equipped with hydraulic lift-off load cells to monitor the working force in the stressed tendons. The monitored anchors were greased and sheathed with plastic tubing in the stressed secondary zone to insure free movement. All anchors were grouted in the secondary zone to provide corrosion protection."
- ENR (1977) Conowingo Dam, VA (1977-1978). "To finish the job, each hole receives a lean grout mixture to prevent corrosion of the tendon..."
- Troka and Lane (1988) (Bagnell Dam, MO, 1982). "Also, since second stage grouting would arrest any further corrosion..."

- Standig (1984) (Delta Dam, NY, 1983). "A double corrosion system includes a continuous plastic duct enclosing the strands, in addition to grout. Single stage grouting was specified...around the strands and between the duct and the drill hole simultaneously. In the free length of the anchor, the smooth duct and a grease and plastic coating on the strands prevents their being bonded to the grout. This permits tensioning later." The "VSL Double Corrosion Protected Rock Anchor," which was specified, was in fact identical in all respects to the anchors installed from 1978-1980 in Tarbela Dam, Pakistan, where Mr. Standig had worked before engineering the Delta Dam remediation.
- Corns and Jansen (1988) (Elkhart Dam, IN, 1986). The strands were individually greased and sheathed in the free length, and a full length corrugated tube was provided. However, "Openings at the bottom of the corrugated pipe were provided to allow the grout to flow around the pipe." Secondary grouting comprised "a weak grout (cement and bentonite) to provide corrosion protection."
- McWhorter and Meeks (1990) (Nacoochee Dam, GA, 1987). The purpose of the second stage grouting in the free length was described thus: "The grout not only functions as a load-carrying structural component but it also serves as the corrosion protection for the bare strand."
- Bruce (1988) General Commentary on Anchors for Dams. Notes Graber's (1981) report detailing cracks (radial and longitudinal) of up to 2 mm aperture at 100 mm centers in the grout of a Tarbela Dam test anchor installed in 1980. Plastic sheathing was found to be effective in arresting crack propagation to the steel. "Anchorages for dams can nearly always be regarded as permanent. By all international standards, such anchors must be properly protected against corrosion." "Corrosion protection of the fixed anchor length by applying an outer corrugated plastic sheath is becoming increasingly more common."
- Mishalanie (1990) (Newton Falls Upper Dam, NY, 1989). "Due to the corrosive environment of the reservoir water, all anchors were fully encapsulated with corrugated polyethylene tubing."
- Paolini and Petrovsky (1991) (Lighthouse Hill Dam, NY, 1992). "A chemical analysis of the reservoir water showed that a water aggressivity to steel was low (pH = 7.4, resistivity = 2,600 ohm-cm, chlorides = 2.4 mg/l, and nitrates = 0.87 mg/l). As a result, a single corrosion protection system was chosen for the tendon anchorage." (This is assumed to mean that no protective sheathing was specified around the bond zone.)
- Bruce and Nicholson (1994) General Review. "Virtually every rock anchorage installed in a dam is regarded as permanent. Corrosion protection is therefore a vital and integral part of anchorage design and construction. On the global stage, it is perhaps only in this aspect that U.S. practice is perceived as being deficient, even though considerable advances have been made in the last few years following the works of FIP (1986) and Littlejohn (1990) in particular. The major point of difference between U.S. and foreign practice is in the concept of double corrosion protection. Foreign engineers, following their national codes, do not regard cement grout as an acceptable barrier to corrosion, in that it carries the potential for microfissuring under load: the protective alkaline environment can then be depassivated quickly in the presence of aggressive anions, notably chloride. An acceptable barrier is one which can be inspected prior to installation. Therefore, a tendon incorporating a plastic sheath, and grouting in place with a normal cement grout is regarded as a singly protected tendon overseas, but a doubly protected tendon in the U.S. The least protected part of the tendon defines the class of protection, and joints or transitions are particularly vulnerable.

  American engineers may argue, with a certain justification, that most dams are founded on 'good', impermeable rock which is then further grouted, if necessary, prior to anchorage installation. In short, the real danger of water percolating through possible microfissures in both rock and grout – and then a flow through the plastic protection is generally regarded as a tolerable risk.

Within the last few years, attitudes toward long multistrand tendon protection have undergone the following chronological progression:

a) bare strand in bond zone, individual sheaths on the free length steel;
b) as a) except for a full length, outer 'group' sheath of corrugated plastic (polypropylene or polyethylene);
c) epoxy coated strand (and two phase grouting);
d) epoxy coated strand, with individual sheaths in the free length, permitting one phase of grouting.

In the current absence of a national policy towards corrosion protection, individual owners are responsible for specifying the degree of hole corrosion protection they want to pay for."

- Bogdan (2001) Review of Epoxy Protection. This is an excellent perspective of epoxy protected tendons, clearly listing the advantages and disadvantages of the technology.
- Bruce (2002) Review of Protection Evolution. "Although ground anchor practice in the United States has enjoyed a long, successful and internationally acclaimed reputation (Bruce, 1997) one area in which it differed from European concepts was in its somewhat more relaxed approach to corrosion protection. For example, what British practice (BS8081, 1989) regarded as single corrosion protection (i.e., the use of a protective corrugated sheath, grouted in situ) U.S. specialists typically referred to as double corrosion protection. The difference lay in the interpretation of the reliability of the grout in the bond zone as an acceptable layer of corrosion protection. Thus while the British tended not to count the grout as a reliable and permissible layer of corrosion protection since it could crack during stressing due to its strain differential with the far more elastic steel it encased, others disagreed. It was argued that any stress fractures would be of very small aperture, and that the highly alkaline environment of the grout would prevent corrosion of the steel – should it actually be exposed to direct contact with continually aggressive groundwater in any case. No case has been reported, nevertheless, of failure resulting from bond length corrosion in a properly grouted anchor.

  Around the same period in the late 1980s, U.S. contractors installing permanent ground anchors began to realize that the use of a corrugated plastic duct as corrosion protection over the bond length required special attention to construction detail during the grouting operation (e.g., tremie tubes inside and outside the sheath, grouted in careful sequence to avoid structural distress to the sheath due to differential fluid grout pressures); as well as demanding larger diameter drill holes to accommodate the tendon, the corrugated sheath, and the multiple tremie tubes.

  It was logical, therefore, that epoxy protected strand should become considered for strand tendons: it removed the necessity for a separate tendon protective encapsulating sheath, allowed hole diameters to be minimized, and simplified the grouting operation. Such construction efficiencies would have the potential to offset the far higher material costs of such strand."

The Database

General

Given that anchoring is conducted in dams other than earth embankments, Figure 2 presents a histogram of U.S. dam construction involving concrete and masonry structures.

Phase 1 of the database preparation has revealed 239 projects whose anchor details are essentially "complete," a further 50 projects classified as "nearly complete," and a further

104 case histories regarded as "incomplete." (This will clearly drive the nature of the Phase 2 studies.) Nevertheless, a total of 318 anchor projects do have sufficient data to allow year of commencement to be plotted (Figure 3). Over the 40-year period, well in excess of 20,000 tendons were installed, with the peak years, driven by Federal Regulatory demands, being in the period 1988-2002. Further details of individual projects are provided in Bruce and Wolfhope (2006).

Corrosion Protection

An analysis of the database provides the information upon which Figure 4 is based. The following comments pertain to the different types of tendons and protections used over the years.

- Wire tendons, comprising eventually up to 212 each 6 mm diameter wires, giving a Working Load of around 6,800 kN at 60% GUTS, were the original choice for dam remediation, but were used only very infrequently in subsequent years. Apparently, the last example was in 1982 at Bagnell Dam, MO. Such tendons had to be grouted in two stages, had no corrosion protection on the steel (other than cement grout) and were challenging and tedious to assemble, handle and stress. However, they were promoted to dam engineers by certain post-tensioning suppliers as being the state of practice in the wider world of post-tensioning and therefore would be appropriately suited to dam work also. According to Bogdan (2007), there were numerous examples of "improper" corrosion protection using wire tendons and he cites the case of Railroad Canyon Dam, California. Here, lift-off testing confirmed that the existing wire tendon anchors had to be replaced due to corrosion. This was accomplished in 1991 by using restressable anchors with epoxy coated strand.
- From the early days, post-tensioning suppliers also offered a variety of tendon types based on their standard 12.7 mm strand system which had been used for up to 55 strand tendons in nuclear facilities. A few strands in each tendon were omitted in favor of grout tubes and vents for rock anchor applications (Crigler, 2007). Early applications involved bare steel and two stage grouting, but by 1976 such tendons were typically used only where a "fully bonded" design was in fact to be implemented. By then, a plastic sheathing was used on individual strands, primarily as a bond breaker to thereby permit single stage grouting. This, itself, began to be superseded by the "Double Corrosion Protected Anchor" by 1978 where designers judged the conditions to be particularly aggressive, and an exterior "group sheath" (corrugated in the bond length and often smooth in the free length) was added. This was a Canadian-German concept which was first widely used for the Waldeck Cavern anchors in Germany in 1969, following work in Calgary, Alberta, in 1968.
- By the mid-1980s, 12.7 mm strand (190 kN GUTS) had been superseded by 15.2 mm strand (266 kN GUTS) which was by then more cost effective and readily available. Also, by this time, geotechnical contractors had become the prime movers in the dam remediation market and were pushing post-tensioning companies to provide dam specific solutions for tendon configurations. Tendons made from 15.2 mm strand were almost always greased and sheathed in the free length (Lang's extruded coating "polystrand" system, had been patented in 1972 but not widely used until 1982, except where full bonding was a design requirement). By 1986, full length secondary protection also became equally popular, partly as a result of marketing to DOTs, leading to the elimination of bare strand in the bond zone by 1998. One may note that in the early days of outer corrugated sheathing, only the bond zone was so covered. However, contractors found this an awkward detail to handle and to grout effectively in the field, and soon began to install the corrugated sheathing full length to facilitate constructability.

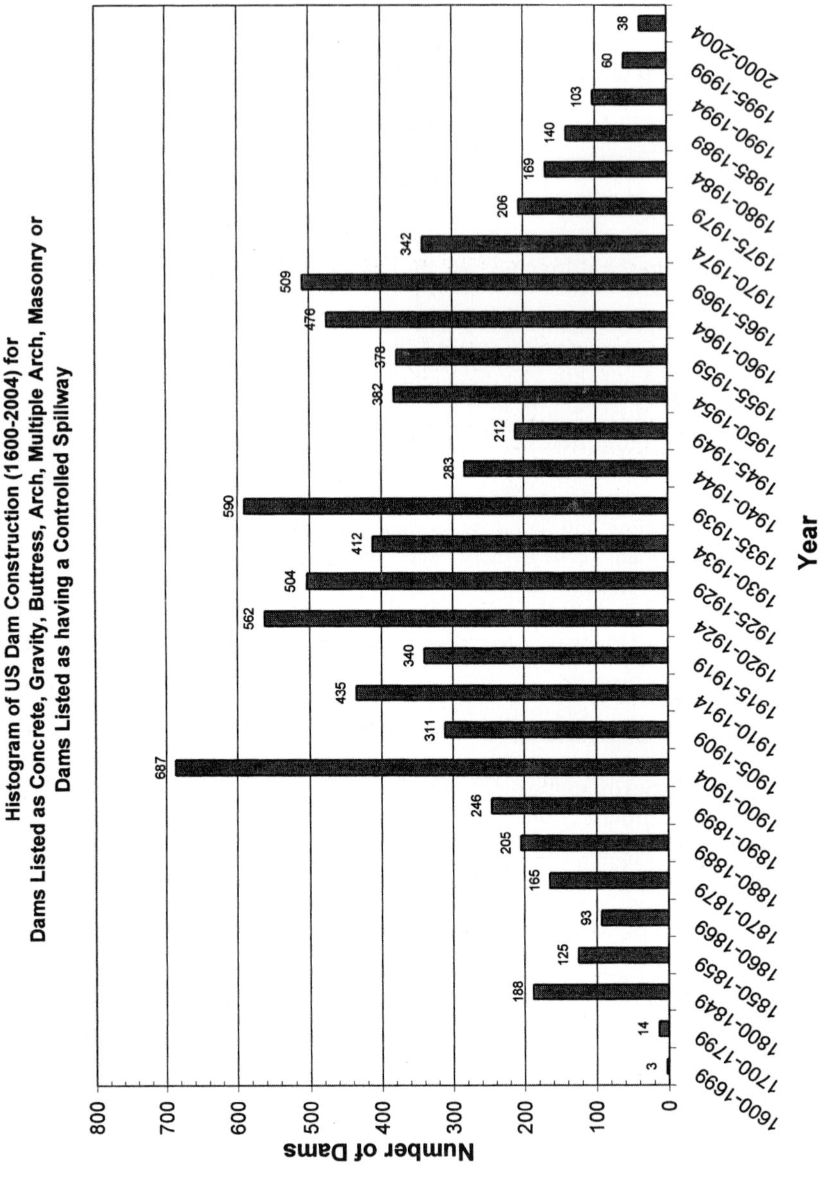

Notes: 1) Source of Data - National Inventory of Dams, USACE, 79777 Dams Total
2) Does not include 9500 dams where the year construction completed is not reported or invalid
3) Total Number of Dams (not including 9500 with unreported/invalid data) = 8178

Figure 2. Histogram of U.S. Dam Construction by Type of Dam

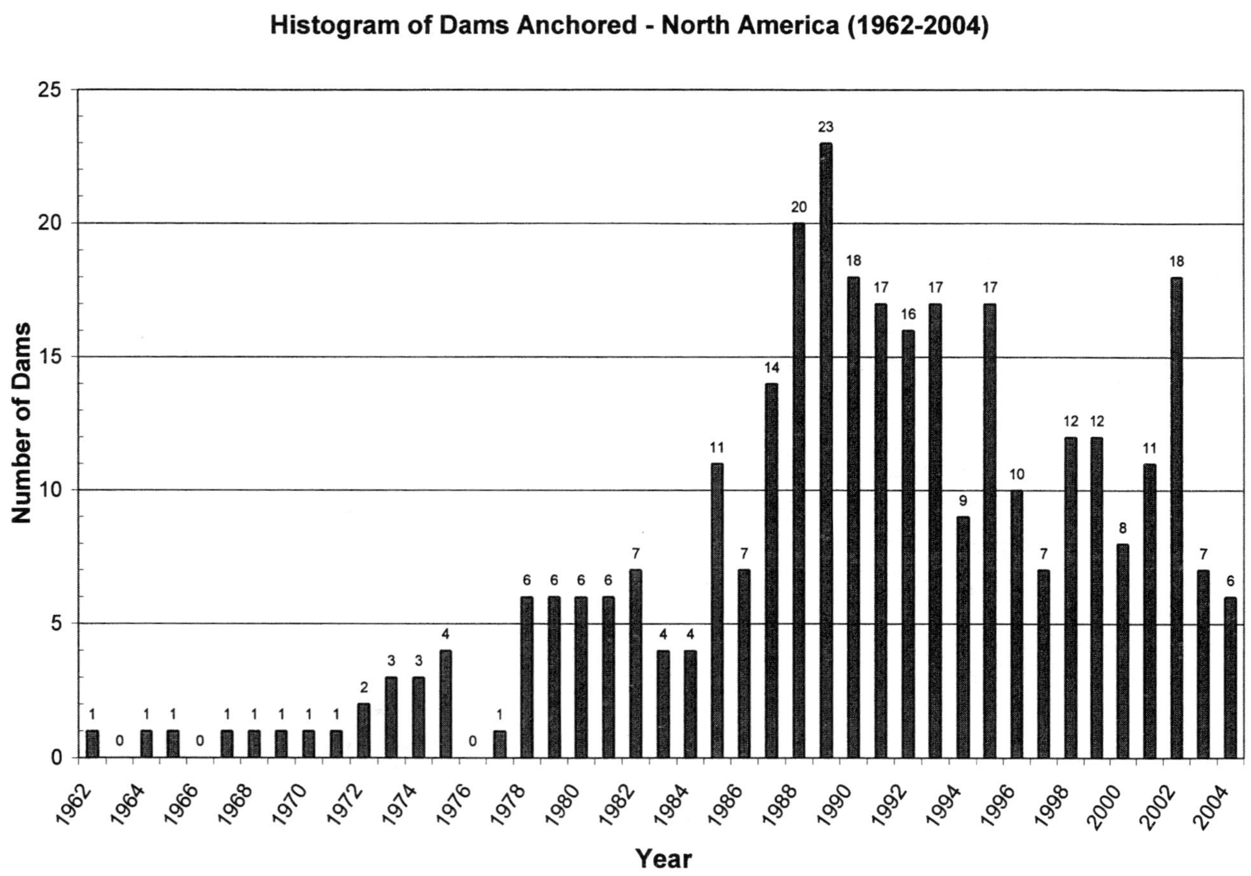

Notes: 1) Total Number of Dams Shown = 323
2) Does not include 70 anchor case studies where year anchored not reported or as yet ascertained.

Figure 3. Histogram of Dams Anchored by Year (1962-2004)

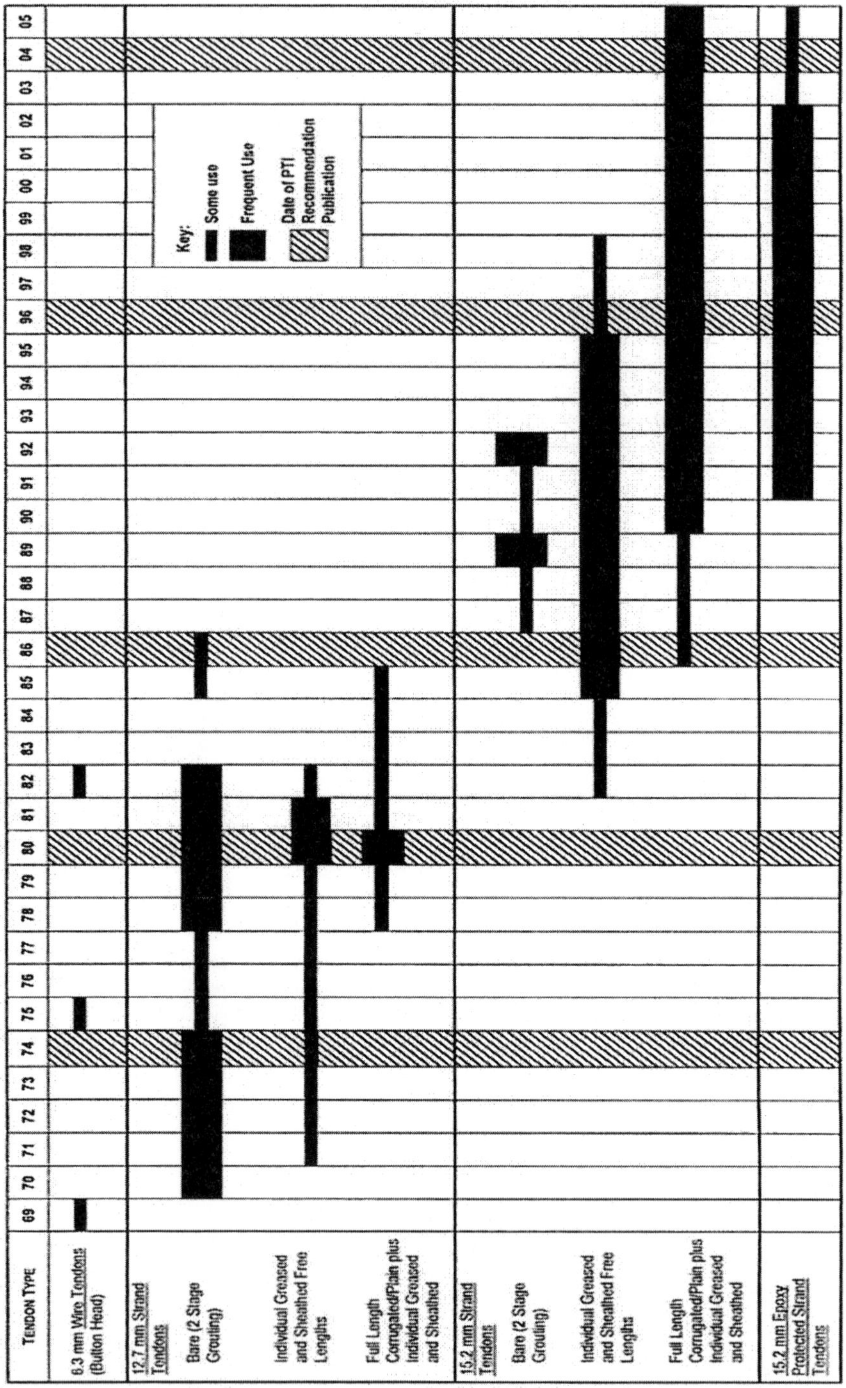

Figure 4. Illustrating how the Types of Corrosion Protection have Evolved (Data Drawn from the Authors' Research Database)

- On certain projects, the free length outer sheath is now plain and is joined very carefully to the bond zone corrugated sheath with heat-shrink bands and other defenses. In recent years, the outer protection has become so large (over 300 mm o.d.) that it has become common practice to grout the duct in place in the waterproofed drill hole before placing, grouting and stressing the tendon. This is a very intricate operation involving many water testing and grouting steps. It was first performed at Cross River Dam, NY in 1997. The method allows the acceptability of the corrosion protection to be verified at various stages prior to anchor completion (e.g., before placing, after placing, after exterior grouting and after tendon placement).
- According to Bogdan (2007) the current state of practice to provide individual strand sheaths is to have the plastic sheathing hot-melted and extruded in a controlled thickness over the greased strand. This method, imported from unbonded monostrand practice, assures that no air will be entrapped between sheathing and grease and that water will not penetrate inside. The traditional "stuffing" method, wherein a plastic tube is forced over a greased strand, is still acceptable for epoxy protected strand.
- A most interesting recent case history is the 2004 anchoring of Seven Mile Dam, a BC Hydro structure. Aschenbroich (2007) recounts that the owner researched corrosion protection systems in considerable detail since longevity for these tendons — at 92 strands the largest in the world with design working loads up to 14,000 kN – was critical. The decision was made to use petrolatum wax in lieu of the strand post-tensioning grease on the steel, inside the individual strand sheathing on the free length. This has now become standard practice in many quarters. Incidentally, these 130-m-long anchors had to be assembled on site and required a 20,000 kN jack with 1 m stroke since "stage stressing" was disallowed by Specification. The Seven Mile Dam achievement was indeed remarkable, but the massive size of the tendon verged on the impractical. It would seem that, in general, present practice is to limit individual tendon capacity to 61 strands (either bare in a pregrouted corrugated duct, or epoxy protected: each in a "waterproofed" hole).
- Epoxy protected strand made its dam debut in 1991 and, following an early surge of national popularity, has since accounted for less than 20% of dam projects and typically for one particular client. Of particular note is the "triple corrosion protection" selected by the designers for Pardee Dam, CA (Freitas et al., 1997) in 1995 when epoxy protected strand was encapsulated inside a full length corrugated pipe, while the free length strands also had individual greased and sheathed protection: "The California DSOD had concerns regarding long term corrosion resistance and performance of the anchors." Obviously!
- Bars have been used as relatively low capacity tendons (up to 800 kN) since the mid-1970s especially on smaller Canadian dams, and invariably were installed in a protective corrugated sheath, which gave rise to the term "Double Corrosion Protection," i.e., sheath plus grout, as noted above.
- As a final point of detail, it is widely believed that part of the tendon assembly at and just below the anchor head is most susceptible to corrosion. In accordance with PTI (2004), it is now common practice to provide a steel pipe trumpet that prevents water from penetrating behind the plate. The space inside the trumpet is usually filled with cementitious grout ("topping up"). Anchorages are placed inside blockouts and embedded in concrete. There are special cases when the owner requests some or all of the tendons to be retressable. In such cases, the trumpet will be filled with post-tensioning grease, and a restressable wedge plate and permanent load cell is added to the anchorage. This entire assembly is covered with a removable cover cap, also filled with grease. Such restressable systems are not recommended for anchors which may have to act in a

submerged condition, e.g., in an active spillway or plunge pool: the corrosion risk is simply too high.

Prices and Costs

In compiling the anchor case study database of over 400 dam anchoring projects, meaningful construction cost information has to date been obtained for over 100 projects. The database captures the overall project construction cost and the more specific value of the anchoring aspects of the project. Graphs have been developed to identify a general relationship between the cost of the anchoring construction versus the overall length of drilled hole, and versus the overall length of pre-stressing steel installed in the project. All cost data were adjusted to a common baseline of 2007 construction costs using cost indices published by the U.S. Bureau of Reclamation (USBR) for concrete dam construction. Figure 5 provides a comparison of overall project drilling length to the adjusted cost of the anchoring construction. Figure 6 provides a similar graph based on the length of the pre-stressing steel: it only includes data for multi-strand tendon anchors, excluding wire and bar anchors, to reduce the variation in the data. Since the available historical information on projects ranges from a simple reference to project costs in a published journal to detailed bid sheets and final payment estimates identifying installed quantities, there is a large degree of variation in the data as to what is included (or not included) in the costs.

Final Observations

The Phase 1 research has allowed a very detailed picture to emerge of the evolution of dam anchoring technology in North America over a 40-year period. Especially interesting points relate to developments in corrosion protection, costs, and to the overall size of the market. The authors hope that their colleagues in other countries which have also enjoyed a long history of dam anchoring will emulate the North American study.

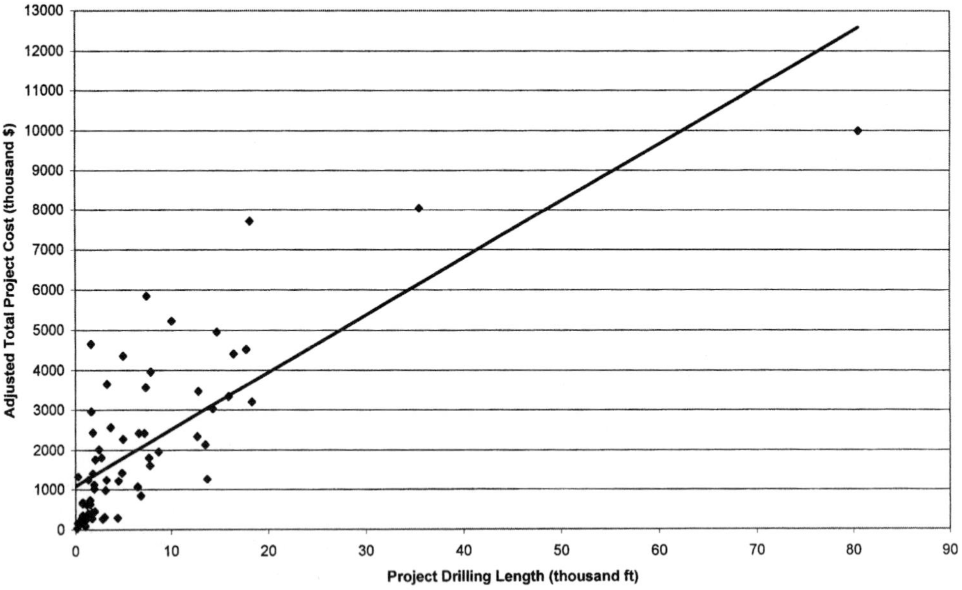

Figure 5. Drilling Length vs. Adjusted Cost

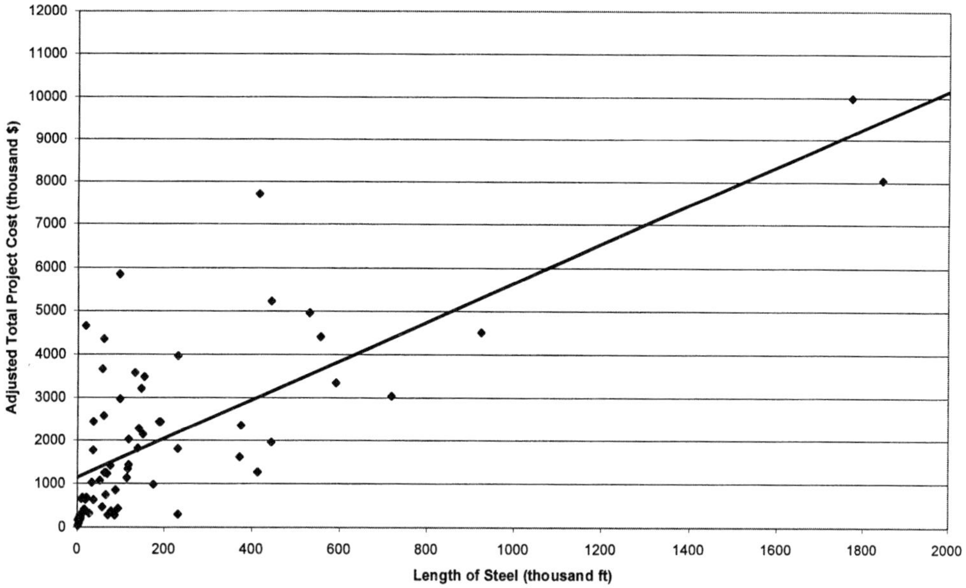

Figure 6. Length of Steel vs. Adjusted Cost

Acknowledgements

The "Rock Anchors for Dams National Research Program" was sponsored by The Association of Drilled Shaft Contractors (ADSC), Boart Longyear, Sumitomo (SEI) Steel Wire Corporation, Sumiden Wire Products Corporation, Freese and Nichols, Inc. and Geosystems, L.P.

References

A complete list of references for this paper is provided in the companion paper "Rock Anchors for North American Dams: The Development of the National Recommendations (1974-2004)," also published in these Conference Proceedings.

POST-TENSIONED REHABILITATION APPLIED TO HYDROPOWER DAMS: CONTINUOUSLY IMPROVING ON AN ANCHOR PROGRAM

John Wolfhope, Freese and Nichols, Inc., Austin, TX, U.S.A.
Donald Bruce, Geosystems, L.P., Venetia, PA, U.S.A.
Gregor Forbes, Lower Colorado River Authority, Austin, TX, U.S.A.
Leslie Boyd, Freese and Nichols, Inc., Austin, TX, U.S.A.

Abstract
As part of a $50 Million modernization program, post-tensioned anchors were used to stabilize four large high-hazard hydropower dams in central Texas, USA. The use of anchors for the rehabilitation of each structure evolved by incorporating lessons learned on the previous dams. The anchoring program was tailored for the unique challenges of each facility. The analysis of the difficult conditions and material quality issues encountered during rehabilitation of the first dam provided an opportunity to improve the stabilization design of the remaining dams. Through the use of extensive field investigations, test anchor programs, and peer review, the post-tensioned anchor details and specifications were progressively improved for each rehabilitation project. This paper discusses the lessons learned installing over 300 anchors on four high-hazard dams and developments made during the anchoring program. Although each application is unique and requires investigation and testing to establish the details and specifications, the model followed in this paper can be applied in response to unique hydropower configurations and site specific challenges to build a successful post-tensioned anchor rehabilitation program.

Background
In the early 1990s the Lower Colorado River Authority (LCRA) and Freese and Nichols, Inc. completed preliminary studies that determined four of the six hydropower dams on the lower Colorado River of Texas needed improvements to meet modern design standards. As a result of these studies, the LCRA began a $50 million, 10-year program to perform detailed investigations, design modernization improvements, and complete construction of improvements for four of the dams. Under the State of Texas dam safety program, all six of the Highland Lakes Dams are classified as large, high hazard dams. State regulations require all high hazard dams to safely pass the probable maximum flood (PMF). Wirtz Dam, Buchanan Dam, Inks Dam, and Tom Miller Dam required stabilization to withstand flood events as low as the 500 year event for certain failure modes. All four of the reservoirs are multipurpose reservoirs providing hydroelectric power to cities located in Central Texas, municipal water supply, and recreational uses for the area. Risk and failure mode analyses were used to prioritize the dams for modernization, with the modernization of Alvin Wirtz Dam being completed in 1998, Buchanan Dam in 2000, Inks Dam in 2001, and Tom Miller Dam in 2004. The use of custom tailored post-tensioning programs at each of the four dams allowed LCRA to complete the modernization of the Highland Lakes Dams within budget and ahead of

schedule. The post-tensioning projects cost a fraction of the other available stabilization methods while preserving the historic appearance of the structures.

Description of Projects Sites

Originally constructed from 1949 to 1951, **Wirtz Dam** is a concrete and earthen gravity dam with two gated spillways impounding Lake LBJ. The dam is approximately 5,580 feet long and includes a concrete spillway controlled by a single radial gate, a second concrete spillway controlled by nine radial gates, a powerhouse and intake section with two hydroelectric generation units, a roller compacted concrete plated embankment section, and interconnecting concrete bulkhead sections. The entire dam is founded on a massive granite rock formation. Prior to the recent modernization, the dam was last upgraded in 1970 to add a single-gated spillway and repair scour damage adjacent to the powerhouse. Figure 1 provides an aerial photograph of Wirtz Dam.

Figure 1 – Aerial view of Wirtz Dam

Buchanan Dam, is an impressive structure approximately two miles long combining elements of concrete and earthen gravity dams and arch dams. The dam impounds Lake Buchanan, the first lake in the chain of Highland Lakes. Buchanan Dam was constructed from 1931 to 1936 and consists of two multiple buttressed arch sections, three separate concrete gravity spillways controlled by 7, 14, and 16 radial gates respectively, a uncontrolled ogee-shaped concrete gravity spillway, a powerhouse with four hydroelectric generation units, interconnecting non-overflow concrete gravity bulkhead sections, and an earthen embankment section with roller-compacted concrete overtopping protection. The dam is also founded on a massive granite rock formation. Figure 2 provides an aerial photograph of Buchanan Dam.

Figure 2 – Aerial view of Buchanan Dam

Located just downstream of Buchanan Dam, **Inks Dam** is approximately 1550 feet long and impounds Inks Lake. The dam consists of an uncontrolled ogee shaped concrete gravity spillway, a powerhouse with a single hydroelectric generation unit, and a non-overflow concrete gravity bulkhead section. Inks Dam was constructed from 1936 to 1938, immediately following the completion of Buchanan Dam. The dam is founded on granitic-gneiss and a large rock outcrop forms the left bank of the dam. As a result of the rock outcrop, the foundation of the spillway rises nearly 75 feet in elevation from the center of the river channel to the left abutment. Figure 3 provides an aerial photograph of Inks Dam.

Figure 3 – Aerial view of Inks Dam

Tom Miller Dam is a unique structure with an interesting history. The last in the series of six dams, the dam was originally constructed in 1893 by the City of Austin, failed catastrophically in 1900, was re-built and failed again between 1912 and 1915, and was re-constructed in 1938 to 1941 by the LCRA. The dam is approximately 1440 feet long and combines elements of all three construction efforts. The major features of the dam include intake structure (built on top of portions of the original masonry dam and 1912 flat slab-and-buttress dam) with two hydroelectric generation units, a concrete flat slab-and-buttress spillway controlled by nine radial gates, and an uncontrolled ogee-shaped concrete gravity spillway comprised of the original cyclopean limestone masonry dam overlaid by a reinforced concrete cap. Tom Miller Dam is located within a very complex geologic setting known as the Balcones Fault Zone. The dam is founded on weathered limestone with severely fractured and highly permeable zones.

Tom Miller Dam serves the primary purpose of passing water released from the large upstream hydro-generation facilities to maintain a water supply impoundment. Figure 4 provides an aerial photograph of Tom Miller Dam.

Figure 4 – Aerial view of Tom Miller Dam

Table 1 summarizes the key features of the four Highland Lakes dams stabilized using post-tension anchors.

Table 1: Features of Stabilized Dams

Name of Dam	Post-Tensioned Anchor Stabilized Sections	Type of Structure (Stabilized Sections)	Combined Length of Stabilized Sections	Primary Foundation Type
Wirtz Dam	Gated Spillways and Non-Overflow Bulkhead Sections	Concrete Gravity	1200 Feet	Granite
Buchanan Dam	Non-Overflow Bulkhead Sections	Concrete Gravity	3700 Feet	Granite
Inks Dam	Uncontrolled Overflow Spillway	Concrete Gravity	880 Feet	Granitic-Gneiss
Tom Miller Dam	Uncontrolled Overflow Spillway and Powerhouse Intake Section	Cyclopean Masonry Overlain with Concrete	560 Feet	Weathered Limestone

Anchor Stabilization Program

Post-tensioning programs were customized for each of the four dams to be stabilized. Multiple-strand tendon anchors were installed at all four dams. All four projects used 0.6" (15.2mm) diameter 7-wire low relaxation strand. The first project used epoxy coated, filled, grit impregnated strand conforming to ASTM A 882; the remaining three projects used bare steel strand in accordance with ASTM A 416. The number of strands per anchor and anchor spacing were selected based on the required stabilization forces, geometric limitations, structural configuration, and foundation characteristics of each dam. Each of the anchor programs was specified according to the Post-Tensioning Institute's (PTI) published *Recommendations for Prestressed Rock and Soil Anchors (1996)*. The stabilization designs were based on a design service load of 60% GUTS (guaranteed ultimate tensile strength) of the tendons, with a lock-off load of 70% GUTS (1.16 x Design Load) and a test load of 80% GUTS (1.33 x Design Load). Site specific requirements were developed for each dam including corrosion protection details, treatment of the foundation, protection of the structural elements, restrictions on drilling methods, and various other design details. For each dam, lessons learned from the previous dams were incorporated into the subsequent designs. Strict qualification requirements were used in bidding each project to select specialized anchor contractors experienced in similar post-tensioned anchoring projects. To supplement the quality control programs implemented by the contractors, each individual anchor test was witnessed by the Owner/Engineer representatives.

Wirtz Dam Stabilization

While Wirtz Dam was the last of the modernized dams to be constructed, risk assessment prioritized Wirtz Dam as the first dam to be modernized. Wirtz Dam was also the dam owner's first endeavor into post-tensioned anchoring. At Wirtz Dam, 78 multiple-strand tendon anchors were used to stabilize the gated spillways and interconnecting gravity non-overflow sections, including:

- 72 twenty-seven strand anchors with a lock-off load of 1110 kips (70% GUTS)
- 6 eighteen strand anchors with a lock-off load of 740 kips (70% GUTS)

The post-tensioned anchors consisted of a bonded length in the foundation bedrock and an unbonded or free length extending from the top of the bond zone to the anchorage at the top of the dam. The anchor tendons ranged from 56 to 148 feet in total length with 30 feet bond zones. The multi-strand tendon anchors were installed in a drilled hole approximately 9-inches in diameter extending up to 50 feet into the bedrock. The drilled hole inclination ranged from 69 to 90 degrees from horizontal. Drilling restrictions and tolerances were specified to reduce the potential for damaging the existing foundation drain system during drilling and grouting operations.

Corrosion protection was provided through the use of epoxy coated strands. In addition to the epoxy coating, corrosion protection was provided by a minimum 1" grout cover between the strand and the borehole wall using a neat cement grout mixture with a water/cement ratio of 0.45 by weight. This provided what is known in the U.S. anchoring industry as Class I (encapsulated) corrosion protection, often referred to in North America as double corrosion protection. The unbonded length of the anchor was encapsulated in a grease-filled polypropylene sheath and grouted along with the bond zone in a single stage. The anchors placed in the gated spillways were recessed in pockets to avoid interfering with the discharge flow. The design provided for the anchor head and wedges to be protected by a grease-filled metal vault. Concrete was then placed flush with the spillway surface following stressing of the tendons.

The anchor system was designed using a finite element model of the structure and foundation rock to confirm that loads imparted on the dam from the post-tensioning forces would not cause binding of the radial gates preventing proper operation during flood releases. The structural deformation analyses considered a fault passing diagonally to the axis of the dam with variations in foundation properties on each side of the fault. The anchors were spaced from a minimum spacing of 8 feet on center to a maximum of 21 feet on center to avoid differential settlement of the structure occurring during stressing. Extensometers and precision surveys were used to verify that excessive deformations did not cause adverse affects on the structure and gates. Fifteen permanent load cells were installed on representative anchors in each section of the dam to provide post-construction performance monitoring during the 1-year warranty correction period. Each load cell was comprised of six individual vibrating wire elements and are read using a portable electronic readout device.

During construction, issues related to manufacturing quality control of the anchor strand surfaced. In response, careful re-engineering of the contractor's quality control program reduced the occurrence of wedge slippage after testing and lock-off of the post-tensioned anchors. Also, in response to wedge slippage, the anchorage was modified to add a secondary gripping mechanism achieved by fully grouting the strand tails in high strength epoxy grout. Despite difficulties with the anchor strand, the project was completed on-time and within budget. Following construction, the load cells were monitored for the one year warranty period on a quarterly basis. All load cells indicated load loss to be within the ASTM limits for seven-wire strand relaxation based on readings at 72-hours and 1000-hours following load transfer. Analysis of the load cell data using logarithmic regression based on instrumentation readings obtained during the first 10,000 hours of service indicates that all of the anchors are performing in accordance with the design requirements and will continue to provide the minimum required stabilization force based on 60% GUTS for the assumed 50-year service life of the post-tensioned anchors.

The primary lessoned learn from the experiences at Wirtz Dam is the importance of clearly specifying and implementing an effective quality control and quality assurance program for each aspect of the post-tensioning program including tendon fabrication, drilling, insertion, grouting, stressing, and acceptance testing. Strict qualification requirements should be provided in all bid documents to provide for the selection of qualified post-tensioning specialty contractors employing staff with demonstrated experience in similar project conditions. The active involvement of a qualified full-time on-site quality control engineer is paramount to a successful project. A properly engineered quality control program can allow the project team to respond to adverse material and installation conditions and still provide a successful project meeting the design requirements.

Buchanan Dam Stabilization

The lessons learned at Wirtz Dam were incorporated into the anchor program for the next dam, Buchanan Dam. A test anchor program was developed and conducted during the design phase. This program established the rock strength parameters used in sizing and configuring the anchors. Four test anchors were installed and performance tested in accordance with PTI criteria to evaluate the constructability of the proposed anchor system. Extended creep tests were performed on each of the test anchors and anchor loads were monitored for 1000 hours to determine that the relaxation of the tendons was within the 2.5% load loss limit prescribed by PTI when the tendons are initially loaded to 70% GUTS. Data from the performance tests were used to determine that the load carrying capacity of the test anchors was sufficient for the design loads, the observed extension of the tendon was within limits prescribed by PTI, and the rate of tendon creep would stabilize within the time limits prescribed by PTI. The performance tests demonstrated average rock/grout bond strengths in the granite foundation up to 370 psi without any signs of failure. Due to the competent nature of the foundation at Buchanan Dam, none of the tendons could be tested to failure within the 80%GUTS limit of the steel.

At Buchanan Dam, 138 multiple-strand tendon anchors were used to stabilize the non-overflow concrete gravity bulkhead sections, including:

- 42 twenty-one strand anchors with a lock-off load of 861 kips (70% GUTS)
- 96 five-strand anchors with a lock-off load of 205 kips (70% GUTS)

Seven-wire bare strand was selected for use in the Buchanan anchors. Post-tensioned bar anchors were also bid as an alternate to the smaller five strand anchors. Due to the length of the anchors, all bidders provided lower pricing for the multiple strand anchors than bar anchors. The anchor tendons ranged from 70 to 130 feet in total length and were installed with a vertical orientation. The multi-strand tendon anchors were installed in a drilled hole approximately 7-inches and 11-inches in diameter for the five-strand and twenty-one strand anchors respectively.

Corrosion protection at Buchanan Dam was developed to provide protection of the bare strand. Heavy-wall corrugated HDPE pipe provided an encapsulation around the tendon for Class I (encapsulated) corrosion protection. The HPDE pipe was 8" diameter for the 21-strand anchors and 4" diameter for the 5-stand anchors, with a minimum wall thickness of 60 mils. The annulus of the corrugated pipe was pre-grouted into the anchor hole prior to installing the anchor with a neat cement grout mixture with a water/cement ratio of 0.45 by weight. Greased sheaths and grease filled anchor head vaults were not used at Buchanan Dam. Instead, the tendons were installed and only the bond zone grouted prior to stressing. The specifications required the contractor to use power seating to ensure uniform seating of the strand wedges. Following testing and acceptance, the entire tendon was fully grouted with a neat cement grout mixture with a water/cement ratio of 0.45 by weight. Quality control tests confirmed that grout strengths were consistently greater than 5000 psi at 28 days.

The contractor had a severely restricted work area on the top of the dam requiring the use of drilling equipment smaller than four feet in width. Project specifications required that the contractor demonstrate the adequacy of his drilling, installation, and quality control procedures on one 5-strand and one 21-strand test anchor prior to proceeding with any other anchoring work. The contractor demonstrated that the selected equipment could provide efficient drilling rates and maintain the one degrees specified drilling tolerance. The stressing and testing of the pre-production test anchors identified improvements to the load-extension testing apparatus to accommodate the tight workspace constraints and enhance the efficiency and reliability of the post-tensioning process. As a result of the demonstration, the contractor refined his procedures and successfully installed 137 anchors without a single failure.

Some of the lessons learned at Buchanan Dam include the following:

- Design-phase anchor test programs provide important data in custom engineering the post-tensioned anchor system to the site specific conditions and demonstrating constructability of the proposed anchor system.
- Pre-production test anchor programs allow the contractor to refine fabrication, drilling, grouting, installation, stressing, and testing procedures prior to proceeding

with anchor production. The successfully installation and testing of 138 anchors at Buchanan Dam reinforced that spending extra attention in the early stages of the project to establish and refine quality control practices provides for significant improvements in production efficiency and overall project performance.

- Simplifying the anchor system design as much as possible and avoiding numerous variations in anchor configuration on a single project allows the contractor to concentrate on effective quality control practices and efficient production.
- The pre-grouting of the HDPE corrugated pipe into the borehole prior to tendon insertion provided for an effective corrosion barrier system that is straightforward to install and test. Efficient production rates can be achieved despite the specification of discrete installation and testing steps including installation and pre-grouting of corrugated pipe and the requirement for two stages for primary grout.

Inks Dam Stabilization

Having similar foundation conditions to Buchanan Dam, the same corrosion protection system, anchor configuration, and foundation strength parameters from the Buchanan test anchor program were used at Inks Dam. The primary challenge at Inks Dam was that the concrete crest of the uncontrolled overflow spillway was in poor shape, highly fractured in irregular patterns. The only significant difference in the Inks Dam and Buchanan Dam programs was developing a method for transferring the stabilization forces into the Inks structure in the form of a highly reinforced stressing beam.

Approximately 2000 cubic yards of concrete were excavated from the dam's crest to facilitate construction of the stressing beam and replacement of the spillway crest. Steel sleeves were placed in the stressing beam to facilitate drilling the 11-inch diameter anchor holes and to eliminate interference with the beam's reinforcing steel. At Inks Dam, 46 twenty-four strand tendon anchors with a lock-off load of 985 kips (70% GUTS) were used to stabilize the uncontrolled ogre-shaped concrete spillway. The anchor tendons ranged from 60 to 130 feet in total length and were installed with drilled hole inclination of three degrees from vertical to avoid interfering with the existing foundation underdrain system. The number of strands was increased from the 21 used at Buchanan to 24 strands based on the required stabilization loads. Figure 5 shows the anchoring approach including the stressing beam, anchor configuration, and cap replacement.

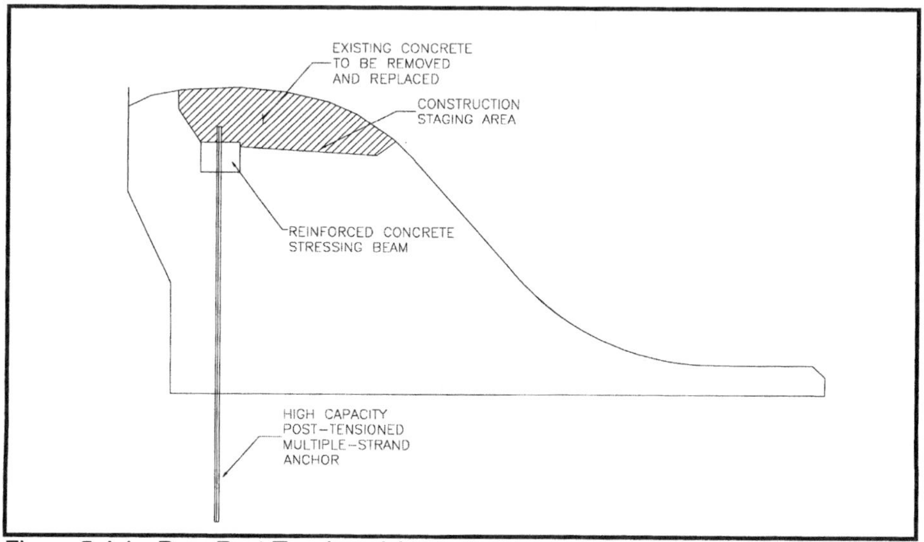

Figure 5: Inks Dam Post-Tensioned Anchoring Approach – Typical Cross Section

The successful program at Inks Dam demonstrated that a standard post-tensioned anchor system can be readily adapted for efficient installation and testing despite adverse site conditions such as highly fractured and deteriorated concrete structures.

Tom Miller Dam Stabilization

The last to be stabilized, Tom Miller Dam certainly presented the most challenging site conditions. Special design studies were required to overcome the significant challenges posed by the complicated geology and composite construction of the dam. The moderately weathered limestone foundation has several highly weathered zones in horizontal planes below the structure. Large voids and highly permeable zones were noted in the drill logs throughout the geotechnical investigation program. The stabilized portions of the dam consisted of mortared irregular limestone blocks capped with concrete. The limestone was quarried from a site adjacent to the right abutment of the dam, thus the limestone is of a similar weathered quality as the foundation rock beneath the dam. The powerhouse intake section is further complicated by portions of the flat slab-and-buttress dam that extend upstream of the section.

Fifty-five 18-strand tendon anchors were installed at Tom Miller Dam to stabilize the uncontrolled ogee-shaped spillway and the powerhouse intake structure. Each anchor had a lock-off load of 740 kips (70% GUTS). The anchor tendons ranged from 130 to 150 feet in total length and were installed in vertical, eight-inch diameter drilled holes. The anchor design was carefully developed to handle the difficult geologic and structural conditions that would often be considered detrimental to construction of post-tensioned anchor systems, including zones of poor quality, severely weathered, fractured, decomposed, cavitated, or highly permeable rock showing evidence of significant water movement through the core of the dam and foundation rock formation.

As with Buchanan and Inks Dams, a test anchor program was used during design to provide data for sizing and configuration of the production anchors. Two anchors were tested to failure and two were performance tested. The performance tests demonstrated average rock/grout bond strengths in the limestone foundation exceeding 200 psi without any signs of failure. The anchors tested to failure showed ultimate rock/grout bond stresses up to 430 psi. The bond length for the production anchors was set at 30 feet based on a 2.0 safety factor and a 1.5 adjustment factor for expected random zones of highly weathered rock as observed in the geotechnical sampling across the dam.

Based on the observations made during the test anchor program, a program was developed for the production anchors to pre-treat each anchor hole in the dam and bedrock by gravity grouting using a sand cement mixture comprised of one part sand to one part cement by weight and a water/cement ratio of 0.72 by weight. The holes were then re-drilled, watertightness tested, and neat cement grouted (where necessary) to enable the anchors to be installed in accordance with the PTI watertightness criterion for the bond zone (2 ½ gallons loss in 10 minutes at 5 psi excess head). To provide Class I corrosion protection, a corrugated heavy wall HDPE pipe with a minimum wall thickness of 60 mils was then pregrouted into the hole prior to tendon insertion. The grouting of the annulus between the corrugated pipe and borehole wall was conducted in 30-feet maximum stages to prevent collapse of the pipe. A polypropylene grease-filled sheath was used on each of the anchor strands to provide an unbonded length of the anchor to make certain that the post-tensioning forces were transferred into the foundation in a zone of relatively competent limestone and away from the masonry core of the dam. A length of bare strand was provided at the top of the anchor beneath the anchor head to provide fully grouted tendons bonded in the concrete cap.

A three-dimensional finite element model was used to verify that the post-tensioning forces would not have an adverse affect on the masonry core of the dam considering the irregular character of the masonry core and to determine the range of anticipated structural deformations. Restrictions were also specified for the contractor's drilling methods to make sure the drilling did not damage the masonry core. Vertical extensometers were installed adjacent to the test anchors to verify that the application of the post-tensioning forces did not adversely affect the structure.

By following a disciplined quality control program, the contractor successfully demonstrated that the holes could be drilled within the specified alignment tolerances, the foundation rock treated to meet watertightness criteria, the tendons effectively installed and grouted, and the anchors tested and locked-off in accordance with the acceptance criteria. Based on observations of the site specific field conditions, several improvements were made to the fabrication and installation program to enhance the construction of the remaining 53 anchors. The two test anchors were accepted as production anchors and the remaining anchors were released for hole drilling and tendon fabrication. To gain a familiarity with the groutability of the masonry core and foundation, grouting procedures were refined based on a series of trial mix batches to enhance the effectiveness of the remaining grouting operations. Sand-cement grout takes ranged up to 369% of the theoretical hole volume. In general, the volume of takes decreased as the project proceeded and the dam and its foundation progressively tightened. Subsequent neat cement grouting was only required in 17 of the anchor holes, highlighting the effectiveness and benefits of the routine sand-cement grouting operation. Takes were generally moderate, with only four holes requiring more than one neat cement grout treatment. Figure 6 presents a cross section of the anchor configuration at Tom Miller Dam.

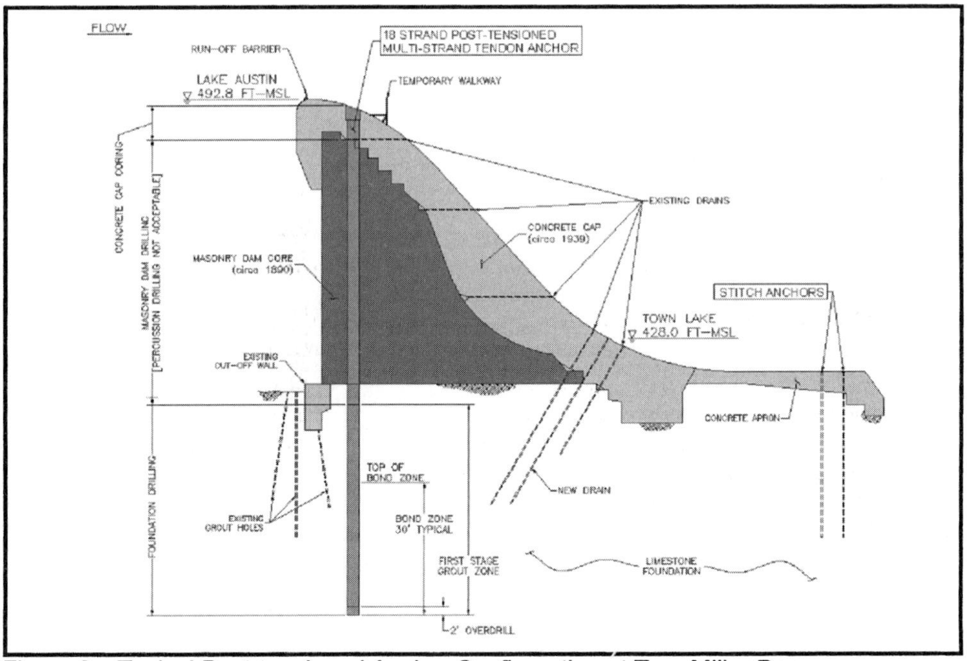

Figure 6 – Typical Post-tensioned Anchor Configuration at Tom Miller Dam

All other aspects of anchor construction (such as hole verticality, sheath testing and grouting, and tendon insertion and grouting) provided consistent, compliant, and acceptable results. Despite the challenging site conditions, the anchors were drilled, grouted, tested, and stressed efficiently without any anchor failures, allowing the project to be completed on schedule and within budget. The primary lesson learned at Tom Miller Dam is that rock foundations prone to significant water movement such as weathered Karstic limestone can be effectively pre-treated using grout mixtures to reduce permeability and improve grouting conditions.

Comparison of Anchor Programs

The anchor program for each dam was refined by incorporating the lessons learned on the previous projects and site specific requirements of each dam. Table 2 summarizes the various anchor configurations used at the four anchored dams. Table 3 provides a comparison of the anchor program features.

Table 2: Summary of Anchor Configurations

Dam Name	Number of Anchors	Number of Strands per Anchor	Average Tendon Length (feet)	Lock-Off Load at 70% GUTS (kips)	Hole Diameter (inches)	Inclination (degrees from vertical)
Wirtz Dam	72	27	120	1110	9	0 to 21
Wirtz Dam	6	18	120	740	9	0 to 21
Buchanan Dam	42	21	110	861	11	Vertical
Buchanan Dam	96	5	110	205	7	Vertical
Inks Dam	46	24	90	985	11	3
Tom Miller Dam	55	18	133	740	8	Vertical

Note: All tendons used 0.6" diameter (15.2mm) 7-wire low relaxation steel strand

Table 3: Features of Anchor Programs

Dam Name	Anchor Test Programs	Corrosion Protection	Free Zone Encapsulation	Foundation Treatment	Primary Grouting Stages
Wirtz Dam	None	Grout Encapsulated Epoxy-Coated And Filled Strand	Grease-Filled Polyethylene Sheath	None	One
Buchanan Dam	Design Phase and Pre-Production Demonstration	Bare Strand in Pre-Grouted Heavy-Wall Corr. HDPE Pipe	None	None	Two
Inks Dam	Design Phase and Pre-Production Demonstration	Bare Strand in Pre-Grouted Heavy-Wall Corr. HDPE Pipe	None	None	Two
Tom Miller Dam	Design Phase and Pre-Production Demonstration	Bare Strand in Pre-Grouted Heavy-Wall Corr. HDPE Pipe	Grease-Filled Polyethylene Sheath	Sand-Cement Gravity Pre-Grout	Two

Conclusions

Post-tensioned anchors provide a cost-effective and adaptable method for stabilizing structures to safely withstand increased loadings. When post-tensioned anchors are selected as the stabilization method, it is imperative that the unique characteristics and configuration of each structure control the design of the anchor system. With appropriate design, qualified contractors, and adequate quality control, post-tensioned anchor systems can be efficiently installed and tested with minimal chance of failure. Despite the challenging geologic, structural configuration, and site access conditions associated with four unique dam sites, post-tensioned anchors were successfully installed within budget and on schedule. Throughout the installation of over 300 anchors, the anchors have consistently demonstrated that they can carry the design load and provide necessary stabilization forces even after all creep or relaxation losses are considered. As a result of these successful custom-tailored anchor programs, the service life of four large high-hazard hydropower dams was extended by more than 50 years.

References

1. Post Tensioning Institute (PTI). (1996). "Recommendations for prestressed rock and soil anchors." Post Tensioning Manual. Fourth Edition. Phoenix, Arizona. 41 p.
2. Post Tensioning Institute (PTI). (2004). "Recommendations for prestressed rock and soil anchors." Post Tensioning Manual. Fifth Edition.
3. Bruce, D.A. (2002, 2003). "Epoxy Protected Strand: A Historical Review of its Use for Prestressed Rock Anchors, Part I and Part 2." Foundation Drilling. ADSC: The International Association of Foundation Drilling, Vol. XXII, No. 8, November, pp. 14-18 and No. 1, December/January, pp. 22-31.
4. Wolfhope, J.S., Scanlon, J.W., Forbes, G., Baldwin, R.A. (2001). "Post-Tensioned Stabilization of Roy Inks Dam." Association of State Dam Safety Officials (ASDSO), 2001 Annual Meeting Proceedings.
5. Boyd, M.L., Wolfhope, J.S., Scanlon, J.W. (1999). "Post-Tensioned Anchoring of Alvin Wirtz Dam." American Society of Civil Engineers (ASCE), 1999 Annual Conference Proceedings.
6. Bruce, D.A., Wolfhope, J.S., Jensen, J.J. (2005). "High Capacity Rock Anchors for Dams: Some Fundamental Observations on the Analysis of Stressing Data." United States Society of Dams (USSD), 2005 Annual Conference Proceedings.
7. Littlejohn, G.S. and D.A. Bruce. (1977). "Rock Anchors - State of the Art." Foundation Publications, Essex, England, 50 pp. (Previously published in Ground Engineering in 5 parts, 1975-1976.)

Case Histories of High Capacity Anchors for two Dams in Canada involving Long Term Monitoring

Horst Aschenbroich, Con-Tech Systems Ltd.
Delta BC, Canada

Abstract
Since the 1960's, in Canada and the USA, existing dams, powerhouses and other structures for hydropower, water supply, irrigation and flood water control have been repaired, retrofitted or upgraded[2]. For most structures, because of very high loads and limited accessibility, flexible and high capacity post-tensioned strand anchors with multiple corrosion protection (Class I) are now specified. This paper describes two recent case histories of significance.

Introduction
Testing and permanent monitoring is becoming a major requirement for the approval and Dam Safety Certification of such anchored structures in service. The two dams, (Figures 1, 2, 3, 4 and 17), described in this paper have been upgraded and retrofitted using high capacity, Class I corrosion protected, strand anchors. They are equipped and instrumented for future testing, long-term monitoring and possible restressing.

Pointe du Bois Spillways, Dam and Powerhouse (Manitoba Hydro)

Figure 1. The two spillways

Figure 2. The adjacent old Power house

The Spillways (Figure 1) were upgraded with Class I corrosion protected bar anchors without instrumentation but were designed to be re-stressable for possible future testing.

The Dam Anchors

Figure 3. The dam with the joint power house

Figure 4. Location of dam crest (Multi-Strand anchors)

The dam was built in 1909 for the powerhouse owned at that time by the city of Winnipeg, in the province of Manitoba, Canada (Figure 3). As per the latest Canadian dam safety requirements (minimum safety factor 2.0), the measured safety factor was under 2.0. Permanent Class I strand, (5-15mm strands), had to be installed to retrofit and upgrade the dam (Figure 4). They are permanently monitored with peak load memory strain gauges, (Figures 5 and 6), and load cells (Figure 7). The strain gauges do not require a data logger. After a seismic or extreme flood-overload event, the peak load which happened during an event as well as the real time load can be read with a standard read-out unit. The gauge is placed immediately under the anchor head or wedge plate on one of the strands, Figure 6. The gauges have been checked without any significant load changes recorded, (from 913 KN to 905 KN). The anchors did not have to be re-stressed. By reading the loads on the 6 vibrating wire strain gauges inside the load cell, differences were noticed due to the eccentric loading of the asymmetric location of the strands in the anchor head, see Figure 7. The average load however was similar to the anchors monitored with the gauges and were accepted as well

Figure 5. Gauge

Figure 6. Head assembly

Figure 7. Load cell

The gauge is connected to one strand inside a heavy wall open pipe chair. The chair is located between anchor-wedge plate and bearing plate, as shown in Figures 5 and 6.

The Powerhouse anchors.

The powerhouse was originally constructed in 1911 with 18 3.5 MW Turbine Generators. One was replaced with a new 8.4 MW Straflo Turbine Generator. It was found that an AAR (Alkali-Aggregate Reaction) in the old concrete caused differential movements between turbines (wet part) and generators (dry part), resulting in overheating and burnout of the generator-turbine shaft bearings. Also, if the apron gallery flooded, the increased uplift pressure beneath the powerhouse would reduce the stability of the powerhouse and increase the risk of failure.

A cost comparison indicated that placing new apron infill concrete and installing vertical post-tensioned anchors, was more economical than reconstruction (Figure 8). The introduced post-tensioning forces were designed to counteract the differential forces.

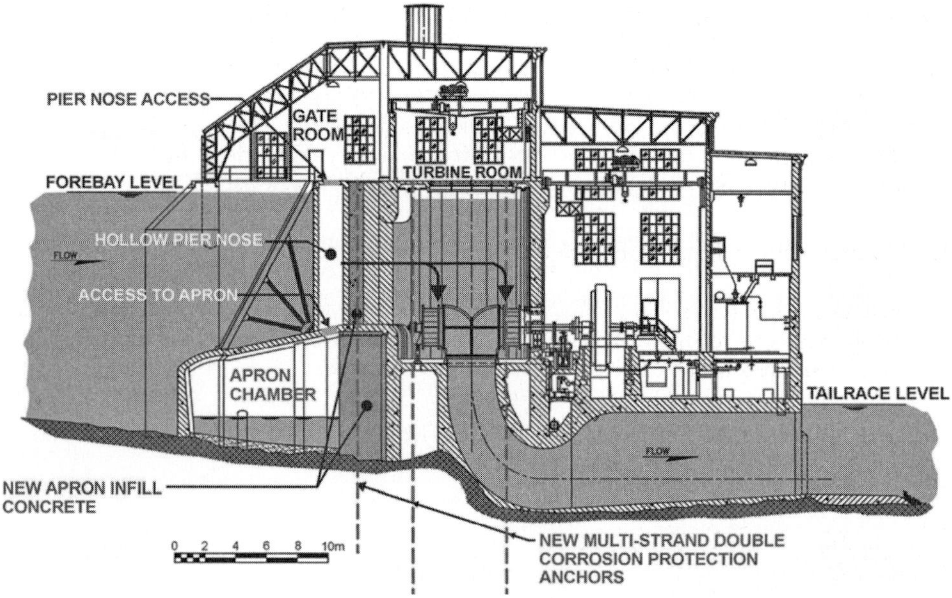

Figure 8. Cross-section of the powerhouse showing the location of the Class I, anchors.

48 Class I corrosion protected anchors, (18-15mm diameter strands with 4,694.0 KN ultimate load capacity each locked off at 3,150.0 KN), were installed with a special low headroom off-coiler (Figure 9) between the operating turbines and stressed simultaneously in groups of 3 each, using 3 gauges on a manifold at the pump, see (Figure 10). The elongations were measured with a dial gauge on each jack, see (Figure 11). The holes were drilled 180 mm in diameter and over 33 meters in length through the concrete intake piers of the powerhouse and into the granite bedrock below.

Figure 9. Using a special low headroom off-coiler to install the vertical anchors

Figure 10.
Three pressure gauges for the three jacks

Figure 11.
Dial gauge to measure elongation

For future testing and restressing, a short, 250 mm long strand tail was left exposed to receive an externally threaded re-stressing wedge plate. In case of a load increase, shims were placed under the anchor head which can be removed to lower the load.

A grease filled cover cap was placed over the assembly (Figure 12).

Figure 12 . Grease cap with connection bolt

After 5 years in service, the anchors were tested and checked for a possible load increase which could result in further movements. The cover cap was removed and the exposed wedge plate and strand tail cleaned to receive an externally threaded wedge plate (Figure 13). A sleeve with 3 extension rods was threaded onto the wedge plate (Figures 14 and 15), to receive the stressing jack (Figure 16).

Figure 13. Threaded wedge plate

Figure 14. Sleeve threaded on to wedge plate

Figure 15. Extension rods with chair

Figure 16. Stressing jack

During the test program, the anchors were stressed individually up to the lift-off force of the anchor head from the bearing plate. The typical anchor initial lock-off load was 3150.0 KN = 67.1% of the ultimate load, and the tested lift-off load was 3021.6 KN which represented a 2.7% loss in pre-stressed load. A dial gauge, which is normally used to measure tendon elongation during stressing, was used. Lift-off was considered at the moment the dial gauge needle started to move. The anchors have been successfully checked. The resulting anchor forces were actually less than predicted, (predicted was a small load increase) and no further movements took place. The integrity of the corrosion protection was also inspected. The Engineers report stated: “All of the anchor heads were well preserved with no indication of any rust or corrosion. All of the grippers were well seated and had no signs of any corrosion. The 300mm strand tails were in good condition with no signs of rust. The corrosion protection at the exposed heads was performing well, even inside pockets filled with water”.

Seven Mile Dam (BC Hydro) British Columbia, Canada

Figure 17. Dam, Spillway and Power house

The Seven Mile Dam (Figure 17) is up to now the most interesting dam retrofit and upgrade project in Canada. The dam is located on the Pend D'Oreille River near the city of Trail in the south east corner of British Columbia, Canada, 7 miles upstream from the Columbia River. The dam is 427 m long and 80 m high. The powerhouse with 4 generators delivers over 800 MW. It was originally constructed in 1980. The dam was raised in 1988 by 4.0 m. This extension and the now higher seismic loads required upgrading with post-tensioned anchors[3]. 57 vertical restressable anchors, 92 strands each, 23,990.8 KN ultimate load capacity, Class I corrosion protected, up to 130 m long with 100 m free length, were installed in the upstream side of the dam (Figure 18) to achieve the largest possible overturning moment resistance.

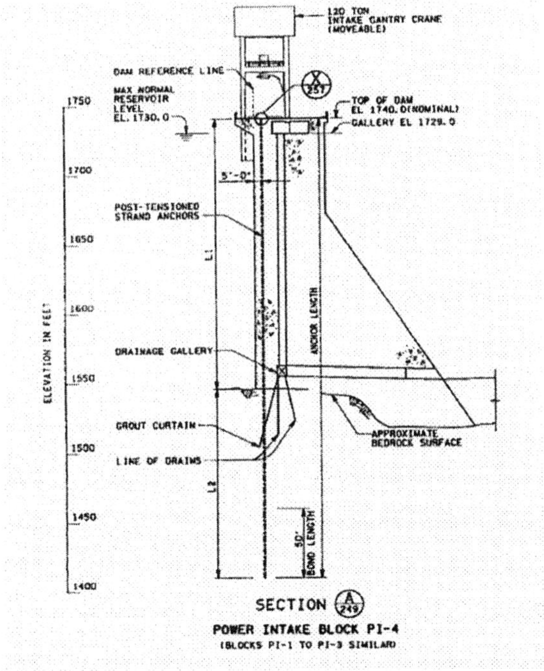

Figure 18. Location of anchors near the upstream face of the dam

These are the highest capacity and longest anchors in the world.
Because of the large size, the anchors had to be assembled at the jobsite (Figures 19) but still transported 1.6 Km to the dam (Figure 20).

Figure 19. On site fabrication plant

Figure 20. Transportation wagon

Over 457,000 m of 15mm Ø strand was required to fabricate all 57 anchors with 92 strands each.

Several special and first time fabrication features had to be addressed:

1. The use of the specified Petrolatum wax instead of PT grease required a special extrusion system to be developed (Figure 21).

Figure 21. Extrusion head for the white HDPE encapsulation over waxed strand

2. The large diameter anchor heads with external threads had to be pre-tested to assure future easy threading of the restressing system (Figure 22).

Figure 22. Checking threading-ability of anchor heads after stressing with a gauge nut

3. A 2000 Metric Ton Hydraulic Stressing Jack, with 90 cm stroke, had to be fabricated, tested and calibrated (Figure 23).

Figure 23. Fabrication of 2000 Metric Ton hollow ram jack with long stroke

4. The re-stressing system had to be designed and fabricated (Figure 24).

Figure 24. Re-stressing assembly, 4-500 Metric Ton solid ram jacks, pulling sleeve and hydraulic pump

The General Contractor used a rotary drill rig (Figure 25) with a 400mm down-hole hammer and a 6.0 m stabilizing rod to achieve the required straightness tolerance of the drill hole.

Figure 25. Hydraulic rotary drill rig with down hole hammer on top of dam

The specified linear tolerance, 1 in 300, was measured every 6.0 m with a digital magnetic inclinometer. An average tolerance of 1 in 800 was achieved.

Immediately after drilling of each hole, a water-pressure test was conducted and the hole was inspected for cracks with an in-hole video camera. Based on the results of the two tests, the amount of consolidation grouting with cement grout was determined. The holes were then re-drilled and re-tested. 8534m of hole had to be re-drilled. The corrugated HDPE sheath, (295 mm Ø, 4 mm wall) over the bond length, was pre-installed and welded to the smooth sheath (325mm Ø, 10 mm wall) over the free length. The long tendons were guided and lowered into the HDPE sheath over a 15 m diameter special roller-guided structure (Figure 26).

Figure 26. Guide structure

For the anchor grout, a neat cement grout (W/C 0.45 by weight) was pumped simultaneously inside the bore hole and inside the HDPE sheath using three staggered grout hoses. The grout plant consisted of a high speed, high shear colloidal mixer, agitating holding tank and double plunger pump developing up to 100 bar pressure.

Tensioning and Testing

The anchors were stressed as per PTI (Post-Tensioning Institute) Recommendations for Prestressed Rock and Soil Anchors Fourth Edition 2004[1], up to 80% of the Guaranteed Ultimate Tensile Strength (GUTS) = 19196.7 KN and locked of at 72% GUTS = 17,273.4 KN for anticipated 2% creep and relaxation losses. Because of the long free-stressing length and consequently long elongation, over 75cm, the 2000 Metric Ton Jack was fabricated with a 90cm stroke (Figure 28). Intermediate lock-off was not permitted.

Figure 28. Stressing jack, notice long elongation

All anchors were installed, tensioned, locked off and tested without any problem.
Since the completion in 2003 and starting 1 year after completion, the anchors have been re-tested several times using the re-stressing system. The four 500 Ton solid ram jacks were pressurized simultaneously against the stressing plate which was connected to the pulling sleeve threaded onto the anchor head. A dial gauge was placed between the anchor bearing plate and the pulling sleeve (Figure 29). The lift-off force was recognized on the pressure gauge at the time the dial gauge needle started to move.

Figure 29. The re-stressing system setup

The lift-off force on all tested anchors was very similar ≈17,000 KN. The tension losses were lower than the anticipated 2%. Therefore no formal data was recorded. The dam is now in conformance with all Dam Safety requirements. BC Hydro purchased the equipment for further re-stressing and also testing other large capacity anchors in future projects.

Conclusions and Recommendations

Post-tensioned high capacity multiple strand anchors with Class I corrosion protection have become a preferable solution to upgrade or retrofit hydro structures. As demonstrated in this paper the advantages are clearly visible:

1. Small required equipment to work in tight areas.
2. Flexibility of large anchors, which can be installed without any mechanical splices.
3. Very good corrosion protection of the individual strands inside a common sheath (Class I protection) gives them a larger redundancy and increases the safety of the structure.
4. Being an active anchor, monitoring and re-tensioning can easily be achieved to prove the lifetime performance of the anchors.
5. Anchors are more economical than other methods of re-constructing.

Recent experiences in connection with new concrete dam construction have shown that anchoring rather than increasing weight with mass concrete is more economical and also a speedier way. It is therefore a good recommendation that sensors or gauges are installed inside the anchors for permanent monitoring and inspection possibilities to reassure the owners and Engineers in their confidence of their design and confidence in their anchored structures in service at later dates. The additional cost of monitoring is relatively very minimal. The anchors shall also be equipped to allow for testing and re-tensioning if required.

References

[1] Post Tensioning Institute (PTI) "Recommendations for Prestressed Rock and Soil Anchors." Forth Edition, 2004

[2] Aschenbroich H., Dipl. Ing. "Ertuechtigungen von Staumauern in Kanada und USA" 13tes Deutsches Talsperrensymposium, Weimar, Oktober 13-15, 2004, (Upgrading of Dams in Canada and USA, 13th German Dam Symposium)

[3] Sutton, P.Eng., Fenwick, P.Eng., Stephenson, P.Eng. "Seven Mile Dam: Dam Stability Improvements, Canadian Dam Association Bulletin, Spring 2004"

The Stabilization of Gilboa Dam, New York, Using High Capacity Rock Anchors: Addressing Service Performance Issues

Kessi E. Zicko, Gannett Fleming, Harrisburg, PA, U.S.A.
Donald A. Bruce, Geosystems, L.P., Venetia, PA, U.S.A.
Robert A. Kline, Jr., Gannett Fleming, Harrisburg, PA, U.S.A.

Abstract
Prior to the proposed reconstruction of Gilboa Dam in upstate New York, 79 very high capacity rock anchors were installed from the dam crest and downstream face to improve the interim stability of the dam. Given concerns over potentially compromising corrosion protection at the head, and other logistical reasons, it was elected not to install load cells in the permanent anchors. In order to satisfy potential concerns regarding the long-term performance of the anchors, many of which were installed in argillaceous rocks of variable properties, several "defenses" were put in place. These included a conservative design process; preproduction pull-out tests; the concept of off-site "sentinel" anchors (with load cells); stringent installation and testing procedures; and Performance Testing (i.e., progressive cyclic) on every anchor (not just on a limited number).

The paper describes each step in the assessment, design, construction and testing/ evaluation process and, thereby, provides a comprehensive case history of a contemporary large dam stabilization using high capacity rock anchors. Units are provided in the actual, Imperial style used in the project. A conversion table is provided at the end of the paper.

Project Overview

Gilboa Dam is a major component of the New York City water supply system and is located in the Catskill Mountains approximately 120 miles north of New York City. Completed in 1927, the 180-foot-high dam consists of a 700-foot-long earth embankment and a 1,324-foot-long cyclopean concrete spillway. The Schoharie Reservoir, which is impounded by Gilboa Dam, can store up to 17.6 billion gallons and provides the City with a large percentage of its drinking water. The spillway was built on nearly horizontally interbedded layers of sandstone, mudstone, siltstone, and shale, with shale being the pre-dominantly weak rock unit. The spillway is composed of cyclopean concrete and has a stepped downstream face (Figure 1). A cutoff wall of varying depth was constructed near the upstream face.

Figure 1. Gilboa Dam in summer 2003.

In early fall 2005, during the preliminary design phase for dam reconstruction, preliminary analyses showed that sliding stability of the spillway structure did not meet current New York State dam safety criteria and was marginal for the 1996 record flood (JV 19Dec 2005). Given the critical nature of the reservoir both in terms of public safety to over 8,000 residents living downstream and dependability for New York City's water supply, an interim stability improvement project was implemented for completion before the end of 2006, years prior to the major reconstruction. The design and construction phases of this interim project were completed in an unprecedented time frame of twelve months.

To help ensure successful completion of the job given the tight time frame, bid packages for rock anchor installation were distributed to three specialty contractors judged to have the appropriate experience in high capacity rock anchors for dams. In addition to conservative design assumptions, several construction measures were implemented to ensure the long-term performance of the anchors including pre-production anchors, "sentinel" anchors, corrosion protection of tendons, water-tightness testing of the anchor hole and sheathing, performance testing of all production anchors, extended lift-off tests of select anchors, and anchor head encapsulation.

Design Approach

The rock anchor system was designed in general accordance with criteria and guidance provided in the Post-Tensioning Institute Recommendations for Prestressed Rock and Soil Anchors (PTI 2004) for Class I encapsulation.

Geotechnical Investigation

The schedule for interim improvements did not permit the completion of a full-scale geotechnical field and laboratory investigation prior to selection of the foundation shear strength parameters for use in the rock anchor design. Investigations by others (Chas T. Main 1977 and GZA 2003) showed that the site is underlain by relatively flat lying sedimentary rocks primarily consisting of sandstone, siltstone, shale, and mudstone. Only four unconfined compressive strength tests had been conducted on rock specimens; two on sandstone samples, one on a shale sample, and one on an interbedded shale and sandstone sample (GZA 2003). Results indicated that the shale and sandstone have unconfined compressive strengths (UCS) ranging from approximately 15,000 to 25,000 psi.

To supplement and confirm the existing data, a subsurface and laboratory investigation program was performed concurrently with anchor design (JV 16May 2006). This investigation consisted of five borings located immediately downstream of the spillway structure at locations that were accessible without regard to, or hindrance of, active flow within the spillway. Figure 2 shows the location of one of these new downstream borings relative to the spillway location at one cross-section. Because the site stratigraphy is relatively flat, samples from these borings were considered to be indicative of the foundation rock conditions beneath the adjacent spillway structure.

UCS testing was performed on rock specimens obtained from the five borings. Representative specimens were selected from the sandstone, siltstone and shale rock units. To supplement the 4 UCS tests by GZA and to verify the ultimate strength used in the bond zone design, 15 UCS tests were conducted on the rock samples. The results of these UCS tests indicated compressive strengths ranging from approximately 6,000 to 13,000 psi for shale/siltstone samples and from 11,000 to 20,000 psi for sandstone samples.

Construction Considerations

Since anchor installation was conducted along the crest and downstream face of the spillway, measures were required to minimize the potential for lost construction time due

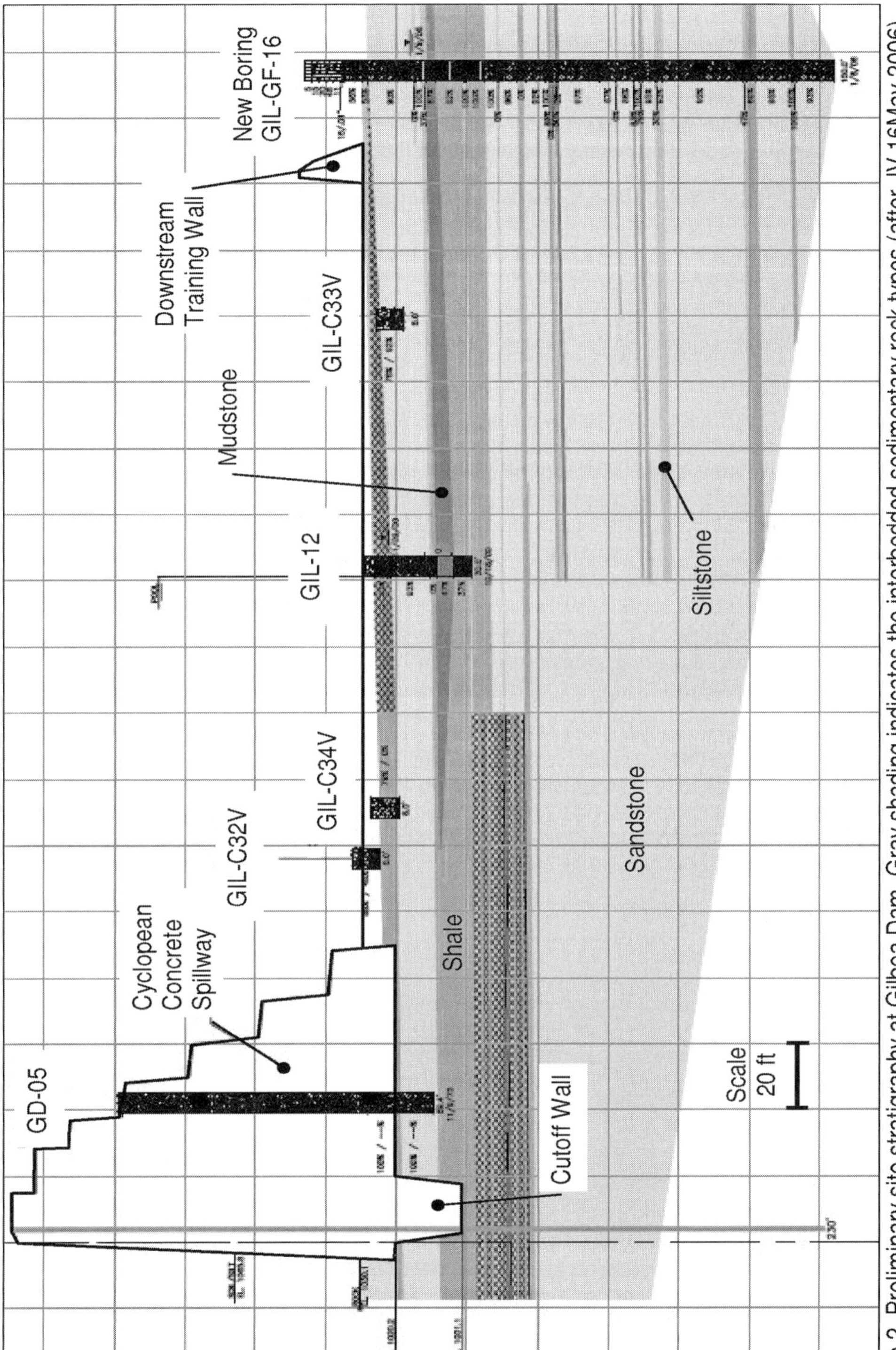

Figure 2. Preliminary site stratigraphy at Gilboa Dam. Gray shading indicates the interbedded sedimentary rock types (after JV 16May 2006).

to active flows from the Schoharie Reservoir. As provided in Kline and Cordell (2007), the following three measures directly impacted the anchor work:

1) Installation of a log boom across the reservoir to prevent debris from encroaching on the spillway crest.
2) Construction of a 5.5 feet deep by 220 feet long notch in Monoliths M15, M16, and M17 in order to isolate discharges, to the degree practical, from the majority of the spillway and side channel.
3) Installation of four siphons over the spillway structure to regulate the reservoir level during construction of the crest notch, and to slow rising reservoir levels prior to flood events that resulted in overtopping of the entire spillway and ultimately emergency demobilization.

The temporary notch and siphon system were considered in the anchor layout and design and also impacted construction efforts (see Figure 3).

Anchor Layout

Spillway stability analyses were conducted at seven cross-sections across the site to identify the required post-tensioned anchor loads (JV 19Dec 2005). Based on these results, vertical anchors were required along the entire length of the spillway crest, from Monolith M1 to Monolith M17. Furthermore, inclined anchors were needed along the downstream face in the central portion of the spillway. These central monoliths are taller than the eastern monoliths, but they have shallower cutoff walls resulting in shorter failure surfaces compared to the higher western monoliths with deeper cutoff walls. For Monoliths M6 through M11, inclined anchors were angled 48° from horizontal and were located at the corner of step numbers 3-4; for Monoliths M12 through M14, the anchors were inclined 45° from horizontal and were located at the corner of step numbers 4-5. A plan view of the anchor layout is shown in Figure 3, and a schematic of the general anchor configuration is depicted in Figure 4. A total of 79 anchors were installed including 47 vertical and 32 inclined anchors.

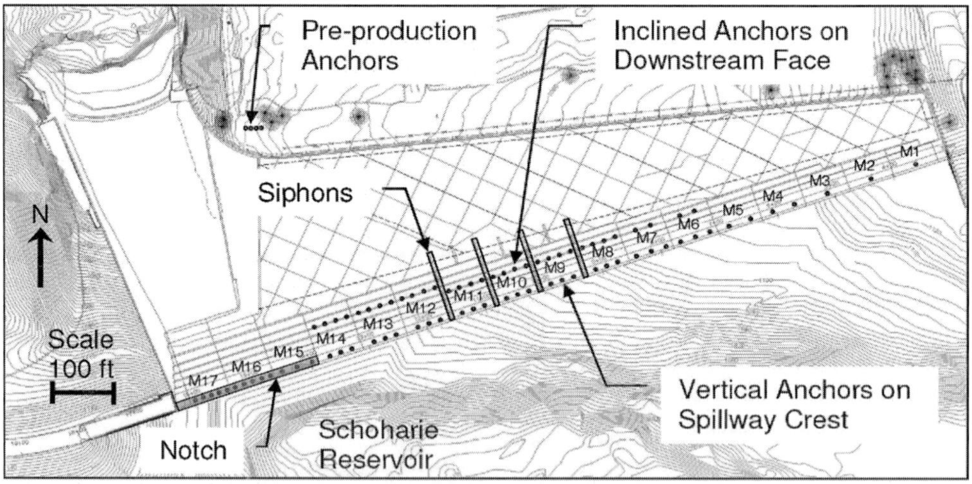

Figure 3. Plan view of Gilboa Dam showing the anchor layout.

The anchor demands were grouped into four ranges of anchor size, Groups A through D, based on their capacities and engineering judgment (Table 1). By grouping the anchors based on maximum required capacities and locking-off all the anchors to the same Design Load (DL) equal to 60% Guaranteed Ultimate Tensile Strength of the steel tendons (GUTS), some additional load was provided greater than that required to meet the minimum stability requirements.

Anchor Design

Computations were performed to determine the minimum required corrugated sheathing diameter, hole diameter, and bond length (JV 30May 2006). Based on the computed minimum diameters, commercially available products were selected to utilize readily available products minimizing lead times for material deliveries. Table 2 provides the design parameters as presented in the contract documents and also shows changes (in bold) implemented by the contractor. These changes were reportedly made to minimize tool sizes and to ensure satisfactory performance during stressing.

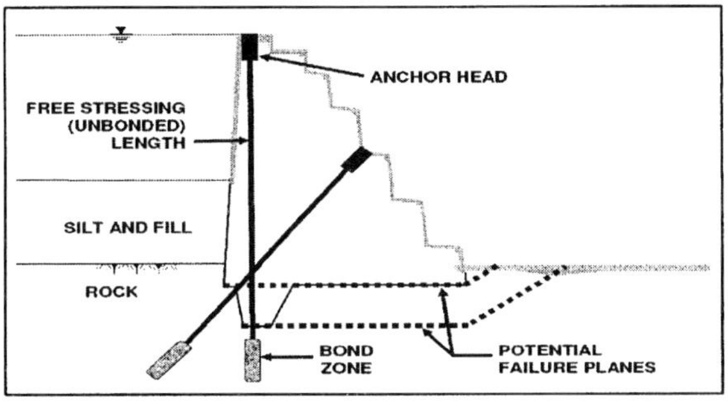

Figure 4. Typical section of the spillway with rock anchors.

Table 1. Anchor Groups

Group ID	Range of Number of Strands	Design Capacity Range (kips)	Selected Number of Strands	Individual Anchor Design Capacity (kips)
A	33-39	1160-1371	39	1371
B	40-45	1406-1582	45	1582
C	46-52	1617-1828	52	1828
D	53-58	1863-2039	58	2039

Table 2. Summary of Minimum Design Parameters

Number of Strands	Design Load (kips)	Drill Hole Diameter (inch)		Sheathing Diameter (inch)		Bond Length (ft)	
39	1371	12	**15**	8	**10**	31	**41**
45	1582	12	**15**	8	**10**	35	**45**
52	1828	14	**15**	8	**10**	35	**45**
58	2039	14	**15**	10	**10**	39	**49**

The bond length was calculated based on the selected drill hole diameter and the ultimate bond zone strength. To establish the ultimate bond stress, the interbedded site strata were presumed to be governed by sandstone and shale, which were the predominant strata in the bond zone. Typical ultimate bond strength values for preliminary design published in PTI (Table 6.1, 2004) for shale range from 30 to 200 psi and for sandstone range from 100 to 250 psi. These values were significantly less than the ultimate bond stresses of 1,500 to 2,500 psi based on UCS testing by GZA, which exceed the typical allowable bond stress in grout of 600 psi. Therefore, an ultimate bond stress of 200 psi was selected for design of the rock anchors, which is the upper bound for shale given in PTI. A working bond stress of 100 psi was selected providing a factor of safety of 2.0. These bond stresses were confirmed by the previously mentioned UCS testing on rock samples downstream of the spillway, by conducting a site specific pre-production test program early in construction, and by testing each installed anchor to a 33% overload to verify its load carrying capacity.

Free-Stressing Length

The free stressing length was selected to locate the top of the bond zone at a depth at least 10 feet below the base of the existing cutoff wall. The distance of 10 feet was intended to account for uncertainty associated with the location of the actual concrete/rock interface. For design purposes, the location of this interface was based upon the original construction as-built (record) drawings (NYCDEP 1927). The concrete/rock interface was encountered within 5 feet of its predicted location in preliminary borings drilled through the crest, thus documenting with reasonable accuracy the as-built drawings. PTI (2004) recommends that the free stressing length of anchors should extend a minimum distance of 5 feet beyond potential failure planes (see Figure 4). Therefore, 10 feet was selected to provide additional assurance that the bond zone was at least 5 feet beyond the bottom of the cutoff wall. This additional anchor footage also allowed the contractor to order anchor materials prior to confirming the concrete/rock interface location during drilling.

Anchor Group Effects

Anchor group effects were evaluated to ensure that the interaction between anchors would not decrease the overall capacity of the anchored system. The pull-out resistance of an anchor was equated to the weight of an inverted cone of rock as presented in Littlejohn & Bruce (1977). This method assumes that a vertical plane develops where adjacent cones overlap and decreases the cone volume accordingly, and it also ignores the rock shear strength along the edges of the cone. Although vertical and nearly vertical fractures were encountered in the borings drilled concurrently with the anchor design, these fractures were typically discontinuous and had surface roughness, or undulations, such that shear failure of the rock would be required to form a continuous vertical failure surface. Furthermore, grouting was completed in the dam foundation during the original construction (NYCDEP 1927), which would have further "locked" the rock together. Therefore, despite the presence of vertical fractures, the inverted cone method was considered appropriate since the rock shear strength was ignored.

Corrosion Protection of Tendons

Corrosion protection of the anchor tendons was addressed by specifying permanent PTI Class I encapsulation. This protection included a grout filled sheathing extending the full length of the strand, the trumpet welded to the bearing plate, and an overlap of the trumpet by the sheathing. The anchor holes and sheathing were subjected to extensive testing to ensure water tightness during the successive phases of anchor construction.

After drilling, each anchor hole was required to pass a water test that limited water loss to 5.5 gallons per 10 minutes under a pressure head of 10 psi. If required due to failure of the water test, pregrouting and redrilling were performed until the anchor hole was sufficiently watertight. Furthermore, each anchor hole was videotaped to "see" the sidewalls of the anchor hole. Consequently, the infiltration of water through rock joints into the anchor hole could be observed, whether as a trickle or a jet. Jets of water were more concerning since they would more likely wash anchor grout from the hole reducing the degree of corrosion protection and the area of grout to rock contact within the bond zone.

Water testing was required for all sheathing, which consisted of full length corrugated sheathing for vertical anchors and of smooth sheathing in the unbonded zone and corrugated sheathing in the bond zone for inclined anchors. Regardless of the sheathing configuration, the criterion for water testing was water loss less than 2.75 gallons per 10 minutes under a pressure head of 5 psi. For the vertical anchors, the corrugated sheathing was testing both prior to and after installation into the anchor hole. For the inclined anchors, the smooth and corrugated sheathing were heat welded and then water tested prior to installation into the anchor hole. The sheathing was further tested after its placement into the hole both prior to and after tendon installation. This third additional water test was required because of the increased potential for damaging the sheathing during tendon installation on an angle since the annulus between the sheathing and wall of the anchor hole had not been grouted.

Anchor Heads

Future dam reconstruction at Gilboa Dam will include removal of the deteriorated concrete and overly stone masonry façade of the spillway up to a depth of 6.5 feet. To minimize anchor constructability problems caused by setting the top of the encased anchor at this depth, a steel reinforced concrete column was constructed prior to installation of the anchors. The column was built by removing the existing concrete to a depth of 10 feet and then refilling with steel reinforced, high strength concrete to the proposed bearing plate elevation approximately 2.5 feet below the existing face. This column is intended to protect the anchor during future construction and will transmit the anchor load to the structure below the limit of scaling. To facilitate removal of the deteriorated materials around the columns during reconstruction, a bond breaker was applied between existing and new concretes.

To provide a high level of corrosion protection of the anchor heads, the bearing plate, wedge plate, and tendon tails were coated with bitumastic material and then directly encased in concrete at all locations, except in the notch. Restressable anchors were considered in the design of the anchorage but were not utilized due to the increased potential for corrosion of restressable anchor heads as compared to traditional anchor heads. This is an especially important consideration at Gilboa Dam where the anchor heads are within the active spillway structure.

Although originally designed to be directly encased in concrete, questions regarding the final dam reconstruction instigated revision of the anchor heads in the notch during construction. The possibility of decommissioning anchors was realized, so access to the wedge plate was required in the 13 vertical anchors in the notch. The anchor heads in the 5.5 feet deep notch were, therefore, modified to include a steel cap in-filled with corrosion inhibiting grease immediately around the wedge plate. This steel cap was encased in a larger steel cap bolted to the bearing plate that was filled with expandable closed-cell foam. Finally, the entire anchor head was encased in concrete. This configuration was used for the anchor heads to provide access, if necessary, without damaging the anchors during reconstruction. Lastly, due to construction restraints and to ensure adequate concrete cover, the anchor tails were cut short thereby not allowing for future restressing.

Long-Term Anchor Performance

Although load cells are frequently incorporated into anchor heads to provide long-term monitoring of anchor loads, this was not feasible at Gilboa Dam due to the location of the anchors (i.e. in an active spillway) and future reconstruction activities (i.e. scaling of deteriorated concrete). In addition to the previously discussed design conservatism and anchor installation testing, extensive test methods were put in place to ensure satisfactory anchor performance including pre-production anchor tests, installation of "sentinel" anchors, Performance Testing all production anchors, and extended lift-off testing.

Pre-Production Anchors ("Sentinel" Anchors)

At the beginning of the anchor contract, an extensive pre-production anchor test program was performed to verify the bond stress and factor of safety used in design or establish the actual bond stress of the site strata, evaluate creep susceptibility of the site strata, and provide instrumented "sentinel" anchors at the site. Four pre-productions anchors were installed downstream of the spillway (see Figure 3) utilizing the same construction techniques and materials used in the production anchors but were smaller in size. Table 3 summarizes the test conditions, including anchor sizes and rock types, and the test results. The pre-production anchors were incrementally loaded and unloaded until an ultimate bond stress of 200 psi was exceeded. As shown, the required 200 psi bond stress was achieved in each anchor, so the factors of safety were greater than the value of 2.0 used in design. It is important to note that no anchor was taken to failure.

Table 3. Pre-Production Anchor Test Conditions and Results

Test Anchor Number	Number of Strands	Test Load (kips)	Bond Length (ft)	Bond Zone Lithology	Tested Average Bond Stress (psi)	Tested Factor of Safety
PP1	10	468	4.9	Shale	282	2.8
PP2	15	701	9.9	Shale	209	2.0
PP3	10	468	5.0	Sandstone	276	2.7
PP4	15	701	9.8	Sandstone	211	2.1

To evaluate creep susceptibility of the site strata, a constant load equivalent to 80% GUTS was applied to each pre-production anchor. The anchors were monitored for up to 75 hours (4500 minutes), which is significantly greater than typical creep tests. Elongation of the tendons indicated creep less than 0.04 inches per log cycle of time, which is acceptable.

Two pre-production anchors were equipped with permanent load cells after completion of the performance and creep testing and were locked-off at 70% GUTS. Load cell readings for these "sentinel" anchors were recorded intermittently for one year, or approximately 6 months beyond completion of the anchor contract. The results are shown in Figure 5, a semi-log plot of the anchor loads over time. Assuming a 100-year service life for the dam, it is apparent that 91% to 92% of the anchor lock-off load will be available, which corresponds closely to the design loads of the anchors within the spillway.

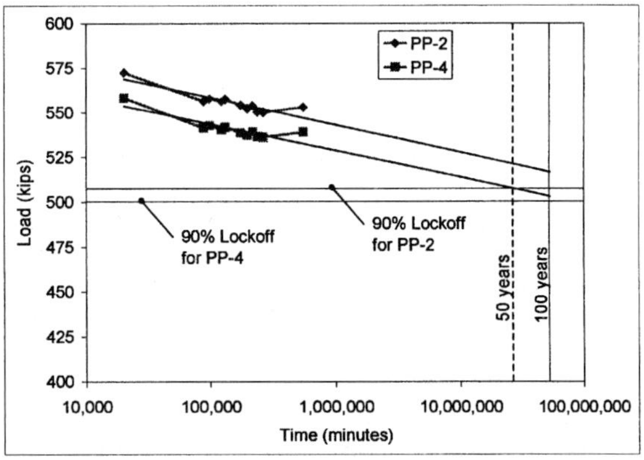

Figure 5. Results of load-cell monitoring of "sentinel" anchors.

Performance Testing

All 79 production anchors, vertical and inclined, installed in the spillway were subjected to Performance Testing. This exceeds standard practice of Proof Testing the majority of the anchors and Performance Testing only the first 2 or 3 anchors and 2% thereafter. Due to the limited amount of subsurface information available during design, it was deemed prudent to apply this additional level of testing to ensure satisfactory anchor performance. The Performance Testing was in accordance with PTI (2004), which included application of an Alignment Load to each strand (10% Design Load per strand), cyclic loading and unloading to a maximum load of 133% Design Load, and then creep testing at the maximum load. During stressing, both dial gauges and load cells were utilized to measure the applied load and tendon elongation.

With the exception of one anchor, which is discussed below, all of the rock anchors were successfully stressed per the contract documents. Of these 78 anchors, 73 anchors performed adequately during Performance Testing and passed the 10 minute creep test, which required the strand elongation not to exceed 0.04 inches. Due to excessive elongation during the 10 minute creep test, the 5 remaining anchors were subjected to the 60 minute creep test, which they subsequently passed with strand elongation not exceeding 0.08 inches. Typical Performance Test results are shown in Figure 6 in terms of total, elastic and residual movement.

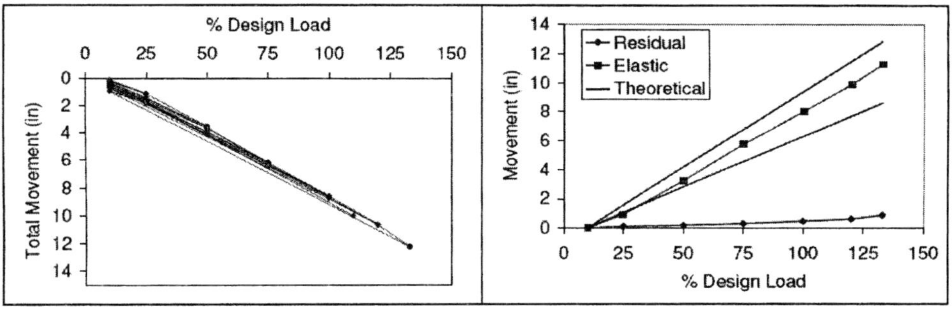

Figure 6. Movements recorded during Performance Testing of Anchor A40, a vertical 58-strand anchor.

One 58-strand inclined anchor had wires on 8 strands break during Performance Testing. The anchor had been successfully loaded to 120% Design Load and experienced strands breaking as loading was cycled to 133% Design Load, at which time stressing was ceased. The cause of failure was most likely either an uneven Alignment Load on individual strands or misalignment of the strands and jack. The remaining undamaged 50 strands were subsequently (and successfully) restressed using a Design Load of 50% GUTS rather than the contract specified 60% GUTS. This lesser Design Load was selected because the total load required for structural stability of the monolith was still satisfied while the maximum load (133% the new Design Load) during restressing was less than the previously achieved load. Prior to restressing, the load cell used during Performance Testing was removed so the strands could be regripped closer to the lock-off plate, thereby eliminating the potential of regripping already pinched strands.

Lift-off Testing

Following locked-off of each anchor, an initial lift-off test was conducted to verify that load was successfully transferred to the anchor bond zone. Each of the anchors was within the contract limits of 5% of the specified lock-off load, which was 110% Design Load. Approximately 30 days after the initial lift-off test, 6 vertical and 4 inclined anchors were subjected to an extended lift-off test. The term "extended" simply refers to the time between the initial and second lift-off tests and does not imply that load was applied and subsequently held for an extended period of time. The lift-off loads were graphed similarly to the "sentinel" anchors (Figure 7). Extrapolation of the data indicates that the anchor load available at the end of the 100-year service life will be between 99% and 112% Design Load.

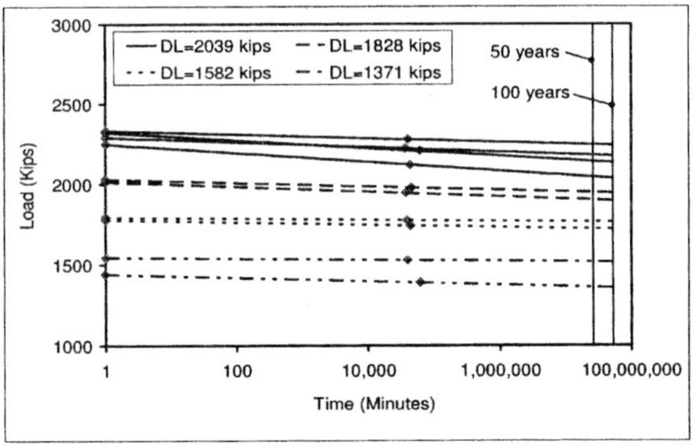

Figure 7. Results of initial and extended lift-off testing of 10 production anchors.

Conclusions

The fast-track interim improvement of Gilboa Dam provided several opportunities to "think outside the box" during rock anchor design and construction. Long-term performance of these anchors was crucial and was verified using design conservatism, extensive testing during construction, and instrumentation and monitoring. The success of the Gilboa Dam anchor project can be attributed to the following factors:

- Design conservatism was beneficial during construction. Since anchors were grouped during design, several strands breaking during stressing allowed an anchor to be restressed at a lesser load while maintaining the minimum factor of safety for stability.
- Pre-production anchors were installed in the same manner as the production anchors to provide insight into the construction procedure, load carrying capacity of the rock, and tendon creep.
- Long-term monitoring of instrumented "sentinel" anchors was utilized to estimate the anchor load for the service life of the dam.
- Corrosion protection of the tendons and anchor heads was assured using thorough testing procedures and inspection at each construction step.
- Performance Testing of all production anchors provided a very high quality of data on anchor behavior. This was especially important since the amount of subsurface information, including rock type at each anchor location and strengths, was limited.
- Extended lift-off tests performed on a small percentage of the production anchors were used to measure the anchor load several weeks after construction and to estimate the available load for the service life of the dam.

Acknowledgements

A key factor of successfully completing this project was clear communication among the owner, engineers, contractors, and inspection team. The authors would like to express their appreciation to all the contributors of this rock anchoring project who helped make it a success. In particular, thanks to the New York City Dept. of Environmental Protection; New York State Dept. of Environmental Conservation – Dam Safety; Hazen and Sawyer, P.C.; Nicholson Construction Co.; and Dvirka & Bartilucci Consulting Engineers.

References

Chas T. Main of New York, Inc., Prattsville Project, F.E.R.C. Project No. 2719 Boring Logs, F. Existing Gilboa Dam and Spillway, Volume 4, prepared for the Power Authority of the State of New York, December 1977.

GZA GeoEnvironmental of New York, Final Report for the Gilboa Dam and Schoharie Reservoir, Contract CAT-146, prepared for The City of New York Department of Environmental Protection, September 2003.

JV (Gannett Fleming/Hazen and Sawyer A Joint Venture), Technical Memorandum on the Final Design Geotechnical Analyses for Interim Stability Improvements, prepared for New York City Department of Environmental Protection, May 16, 2006.

JV (Gannett Fleming/Hazen and Sawyer A Joint Venture), Technical Memorandum on the Final Design for Post-Tensioned Anchors for Interim Stability Improvements, prepared for New York City Department of Environmental Protection, May 30, 2006.

JV (Gannett Fleming/Hazen and Sawyer A Joint Venture), Technical Memorandum on the Stability Analyses Supporting the Rock Anchor Design, December 19, 2005.

Kline, R.K. and D. Cordell, Fast-Tracking Stability Improvements at Gilboa Dam, Dam Safety 07, Annual National Conference, Austin, TX, September 2007.

Littlejohn, G.S. and D.A. Bruce, Rock Anchors – State of the Art, Foundation Publications LTD, 1977

NYCDEP (New York City Department of Environmental Protection), Gilboa Dam Original Construction Drawings, Contract 203, Drawing Acc No. RS-1091, February 5, 1927

PTI (Post-Tensioning Institute), Recommendations for Prestressed Rock and Soil Anchors, 2004.

Conversion Table

1 foot = 0.3048 meter	1 mile = 1.61 km	1 inch = 2.54 cm
1 gallon = 3.785 liter	1 psi = 6.894 kPa	1 kip = 4.448 kN

The Performance of Pacoima Dam in Two Major Earthquakes, With and Without Post-tensioned Anchors

Mark Schultz, Shawn Jones, and Jeff Howard, Division of Safety of Dams
California Department of Water Resources, Sacramento, CA, USA

Abstract

Pacoima Dam is a 110-meter high arch dam located in a mountain canyon north of Los Angeles, California. It is owned and operated by the Los Angeles County Department of Public Works, and regulated by the California Division of Safety of Dams. After sustaining significant cracking and rock block movement during the 1971 San Fernando Earthquake, the upper portion of the left abutment was anchored across a downhill-sloping discontinuity in the foundation rock with 35 multi-strand anchors. Following the Northridge Earthquake in 1994, eight additional high-capacity anchors were installed and have been well monitored, resulting in 8 years of load cell data and two sets of lift-off tests. This paper discusses the seismic performance of the dam and left abutment, the post-tensioned anchors, as well as the long-term performance of the anchors based on 8 years of high-quality data.

Figure 1: Aerial View of Dam and Left Abutment
(Los Angeles County Department of Public Works)

Introduction

Pacoima Dam is a 110-m high variable-radius arch dam located north of Los Angeles, California, with a crest length of 195 m at elevation 614 m. Constructed in the 1920's, the dam is used primarily for flood control. The dam is owned and operated by the Los Angeles County Department of Public Works and regulated by the State of California Department of Water Resources, Division of Safety of Dams. The dam was subjected to intense shaking during the 1971 San Fernando earthquake, and even stronger shaking 23 years later during the Northridge earthquake in 1994. Accelerations recorded at the left abutment during the San Fernando earthquake were some of the largest ever measured at the time, over 1.2g.

Geologic Setting

Pacoima Dam overlooks the northern San Fernando Valley from a narrow canyon in the San Gabriel Mountains. The dam is in a seismically active area and has been almost directly above two large, nearby earthquakes. The 1971 M6.6 San Fernando earthquake ruptured along the north-dipping Sierra Madre-Santa Susana reverse fault system about 5 km below the dam, and the deeper 1994 M6.7 Northridge earthquake ruptured a south-dipping blind fault below the Sierra Madre fault to within 10 km of the dam.

The dam is founded on highly fractured crystalline bedrock. The dam abuts the left side of the canyon at a spur ridge, as shown in Figure 1. This alignment may have minimized the required length of the arch, but the orientation is less than optimal for resisting arch thrust forces. During excavation, the left abutment encountered "badly shattered" rock in the upper portion. As a result, the dam's design height was lowered 3 meters, the left-side spillway was relocated as a tunnel through the spur ridge, and a concrete gravity thrust block was built on the left abutment to help resist arch thrust loads. Eventually, the entire abutment was covered by shotcrete.

In a 1927 construction inspection memo, a seam in the left abutment was noted in the abutment rock. The importance of this feature was not recognized at the time. The discontinuity is an adverse-dipping bedrock fracture set underlying the rock mass upon which the thrust block sits. A 1939 grouting program reported very large takes in this vicinity. More than 400 sacks (43 kg/sack) in total were pumped into two holes at 20 m and 30 m depths, with decreasing takes below this depth. In 1968, three years before the San Fernando Earthquake, a Consulting Board recommended additional grouting, drainage, and rock bolts to improve the stability of identified sliding blocks in the left abutment.

During the 1971 earthquake, thrust block and rock block movements occurred throughout the upper left abutment, and especially on two distinct block masses, shown below in Figure 2 and Figure 3 as Rock Mass A and Rock Mass B. Subsequent studies identified a northwest-striking, southwest-dipping failure plane representing this fracture set. Additional thrust block and rock block movements occurred during the 1994 earthquake, which involved this failure plane (References 5 and 11). Rock Mass B was anchored after the 1971 San Fernando earthquake, prior to the 1994 Northridge earthquake.

Both the 1971 and 1994 earthquakes caused amplified shaking at the site. In general, spectral acceleration was stronger at long periods in 1971 and stronger at short-periods in 1994 (Reference 5). Peak horizontal acceleration (PGA), as recorded just above the thrust block, was about 1.25g in the San Fernando Earthquake of 1971. In the 1994 Northridge Earthquake, the PGA was between 1.5g and 1.7g. Conversely, peak ground velocity was larger in 1971 (~120 cm/s) than 1994 (~105 cm/s).

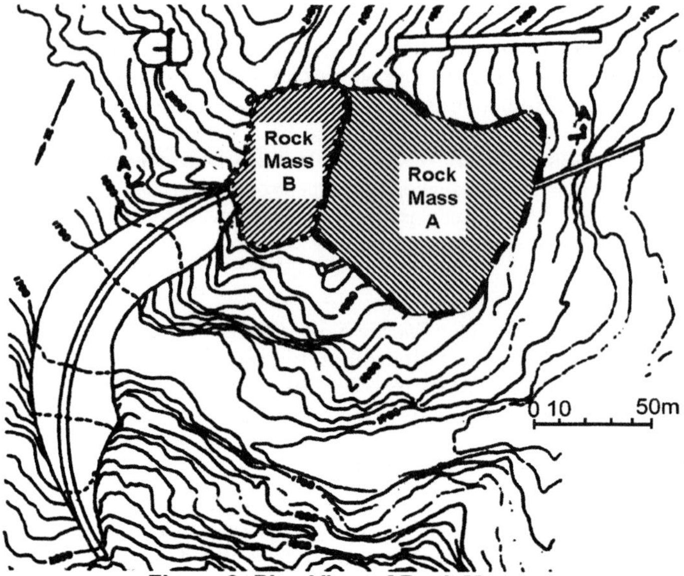

Figure 2: Plan View of Rock Masses

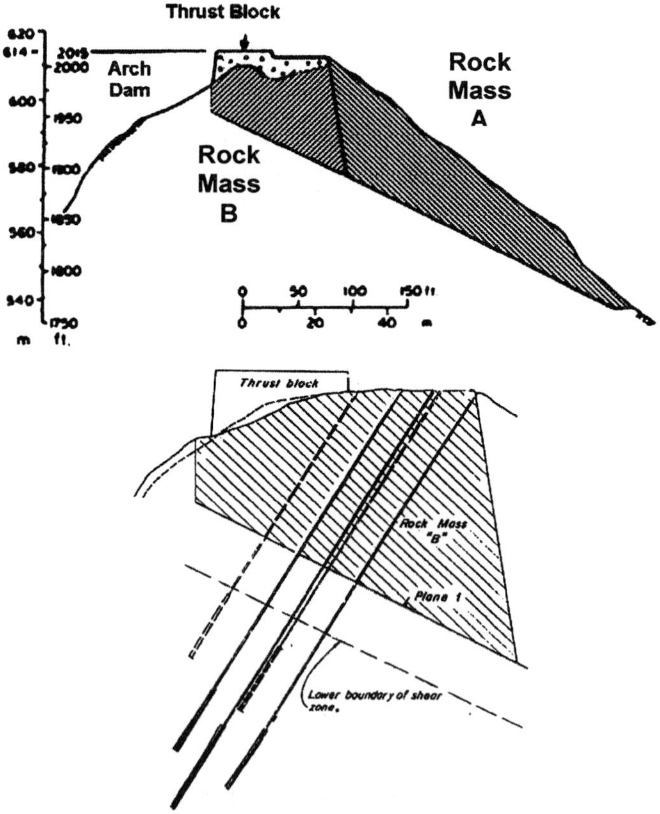

Figure 3: Section Views of Rock Masses & Anchored Mass B

Performance Without Anchorage – San Fernando Earthquake

Surveys of the dam site following the 1971 San Fernando earthquake showed 2.4 cm of shortening between the canyon abutments, and a slight increase in arch curvature, indicating net movement of the entire dam – abutment structure. The contraction joint between the dam and the left abutment thrust block showed a permanent opening of 1 cm, indicating minor net movement of the left abutment thrust block relative to the arch. Major displacements occurred elsewhere in the left abutment.

A seismic remediation program was initiated as a result of the damage during the San Fernando Earthquake. A total of thirty-five ER5-28 Rock Anchors, manufactured by VSL Corporation, were installed in 1976, located as shown in Figure 5. Each of the anchors consists of twenty-eight 0.5" (1.27 cm) diameter strands with 12-m of bonded length, as shown below in Figure 4. The design specifications are listed in Table 1. Eight of the tendons were equipped with load cells. Two of the load cells failed prior to the next earthquake – the 1994 event, and three failed during the shaking.

Table 1: Summary of ER5-28 Anchor Specs

No of anchors	35
Ultimate stress of Anchors f_s'	1860 MPa
Ultimate Strength of each Anchor	5140 kN
Max Tendon Stress during Installation	80% f_s'
Tendon Lock off Load	70% f_s'
Design Load	60% f_s'

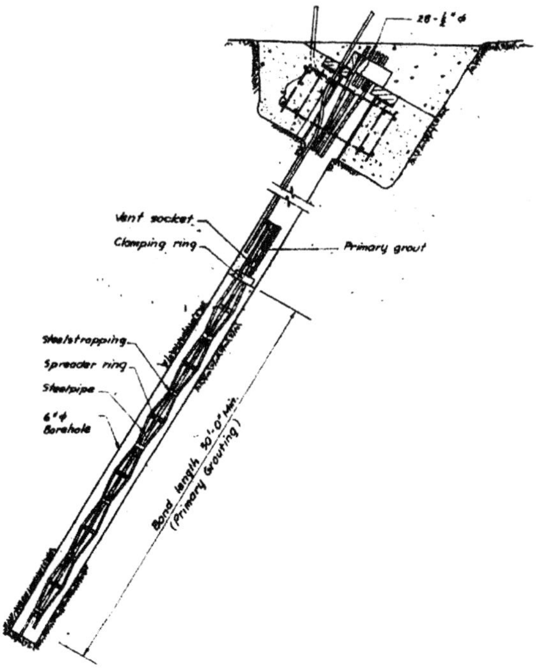

Figure 4: VSL ER5-28 Anchor Design

Two large bedrock masses, one underlying the thrust block (Rock Mass B) and the other west (downstream) of the thrust block (Rock Mass A) were defined by shotcrete cracks across an 8,000-m^2 area. Surveys of Rock Mass A showed displacements of up to 25 cm horizontal, and 20 cm of vertical movement. The perimeter of cracking indicated a failure surface oriented at N26°W, 27.5°SW, defined as Plane 1 in Figure 3. Southern (lateral) boundaries were identified as discontinuities oriented at N39.5°E, 77°NW, and N72°E, 75°N, respectively. Mass A separated from Rock Mass B along a near vertical, north-striking plane (N11°W, 82°W).

Post-1971 earthquake repairs also included consolidation grouting *above* Plane 1 in Rock Mass B, installing drains below Plane 1, and grouting the thrust block contraction joint. A controversial recommendation of a Consulting Board member resulting in grouting mostly areas *above* Plane 1 in order to avoid blockage of natural drainage beneath the sliding plane. Plane 1 was described as a "*rough, irregular zone of failure*" and through back-calculation using assumed psuedostatic loading of 0.3g H and 0.15g V, estimated a φ-angle as high as 47°. Subsequent anchor design assumed a strength of φ = 40-45° and c = 10-20 psi, representing a grouted condition.

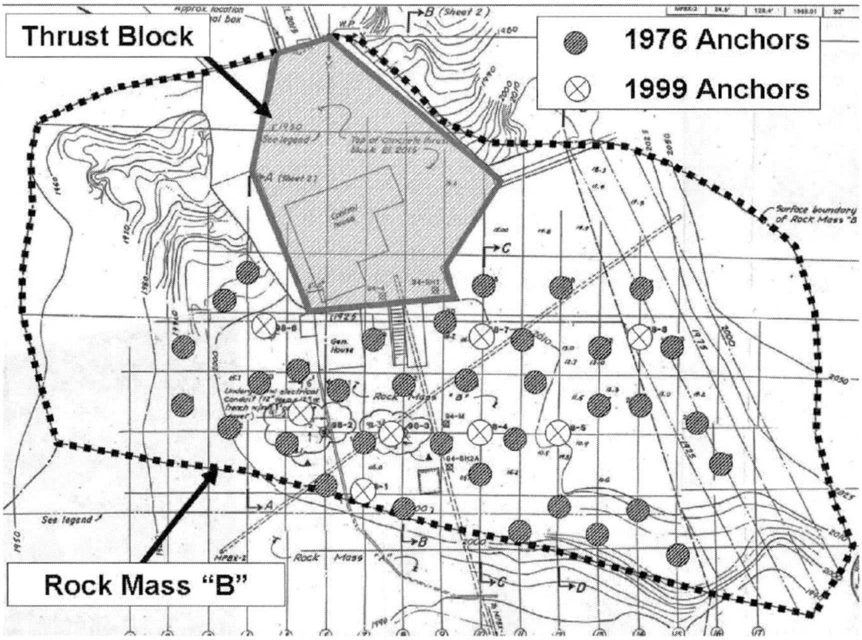

Figure 5: Anchor Locations

Performance With Anchorage – Northridge Earthquake

As a result of the 1994 earthquake, the left abutment sustained rock shattering, numerous cracks, and shotcrete failures. The free-standing concrete spillway chute walls collapsed and massive rock slides continued for days. Based on instrumentation records, visible damage, and survey measurements, the Northridge event seems to have shaken the dam significantly more than the San Fernando Earthquake.

The most significant dam safety effect of the Northridge Earthquake was the re-opening of the left-most vertical contraction joint at the dam-thrust block contact. This same joint opened 1 cm in the 1971 San Fernando Earthquake without the benefit of anchors. The 1994 Northridge Earthquake caused this joint to open 5 cm at the crest of the dam with the thrust block anchored.

The reservoir elevation is controlled passively by a spillway invert more than 20 meters below the crest, which keeps the reservoir low except for brief periods during large flood events. The reservoir was below the invert at the time of both earthquakes. Had the reservoir been high, the 5 cm joint opening would have released a portion of reservoir, and may have opened wider due to larger thrust forces on the abutments. Following the Northridge earthquake, the spillway was greatly enlarged to pass larger storms and minimize the duration of high reservoir elevation.

The dam monoliths were mostly undamaged, but there was evidence of opening and closing of all the keyed contraction joints in the dam, and at some horizontal lift lines. Laser plumb line measurements showed a relative upstream displacement of the dam of approximately 5 cm. A net upstream movement of the crest is not surprising under these conditions.

The 5 cm measured joint opening at the crest can be a misleading indicator of the performance of the anchors. It is important to note that net sliding was limited to approximately 1 cm at the base of the anchored block. The 5 cm measured joint opening occurred at the top of a 20-meter vertical distance and tapered to 1 cm at the base. This indicates either a relative rotation of the block, or cumulative movement along multiple horizontal planes, or a combination of effects.

A second important consideration is the relative movement of the unanchored Rock Mass A, which is downslope of the left Thrust Block/Rock Mass B and shares the same sliding plane. Rock Mass A was not anchored for either earthquake, which provides an indication of what might have happened to Rock Mass B without anchors. The 1994 Northridge Earthquake caused unanchored Rock Mass A to slide a net 58 cm downslope, compared to 31 cm from the 1971 San Fernando Earthquake.

It seems clear that the 35 anchors were effective in reducing the seismic displacement of the thrust block and improving the seismic performance of the dam. Unfortunately, it is difficult to accurately quantify the specific contribution of the anchors alone due to a lack of information on the sliding plane, the effect of multiple grouting programs, and uncertainty of the actual thrust loads generated by the two earthquakes. The authors are modeling the dam and abutment using nonlinear finite element techniques to compare computational methods and to advance dam safety analyses, as shown below in Figure 6.

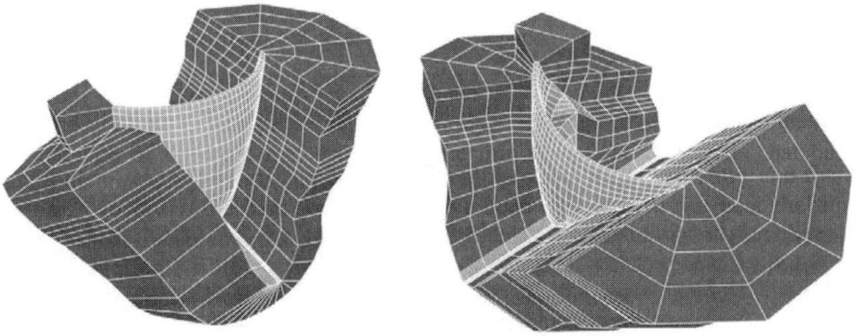

Figure 6: Nonlinear FEM of Dam and Anchored Rock Mass B

Three load cells were temporarily operational following the earthquake, but all load cells failed within the next few days. The load cells were retrofitted in 1996, as shown in Figure 7. A summary of load cell test data is shown in Table 2. Unfortunately, these anchors are no longer monitored because the measurements are labor intensive to acquire. Measurement of the remaining stress in the anchors led to the installation of eight additional anchors to compensate for the reduced effective working force.

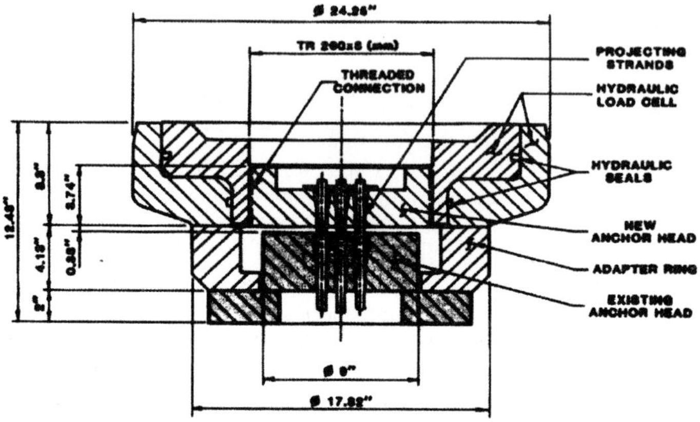

Figure 7: Retrofit Load Cell (Following 1994 EQ)

Table 2: Summary of ER5-28 Anchor Load Cell Data

Anchor No.	% GUTS (5140 kN)			
	1977	1990	1994 Post EQ	1996 retrofit load cells #Strands
3	60	62	64	52 (28)
8	70	58	-	69 (24*)
17	63	63	83	66 (28)
18	67	72	-	59 (28)
20	62	69	-	56 (28)
29	68	71	-	69 (28)
32	67	48	-	45 (27*)
35	68	69	85	61 (3**)

*One strand found broken for anchor 32 and four strands found broken for anchor 8.
**Only three strands could be tested in anchor 35 due to strands being cut too short to grip.

Monitoring Data of 8 New Anchors

A second seismic remediation program was completed following the Northridge Earthquake that included another consolidation grouting program and the installation of eight additional, essentially identical (but epoxy-coated) post-tensioned anchors. The locations of the newest anchors are shown in Figure 5. Each anchor is monitored by vibrating wire load cells comprised of six vibrating wire strain gauges around the perimeter of the load cell. The load cell force over time is plotted in
Figure 8 and
Figure 9 for the eight new anchors. These data provide an accurate estimate of the total prestress losses over a period of several years. The data were cleaned up to correct obvious typos, and anomalous strain gauge readings were removed. The periodic variation of the load cell data is assumed to be due to annual temperature fluctuation. In addition to the load cell

record, lift-off test loads are shown in Table 3.

Table 3: Lift-off Test Data (8 New Anchors)

Anchor No.	Initial Lock-off Load (kN)	Lift-off Test (kN)	
		January 2001	December 2006
98-1	3290	3520	3605
98-2	3290	3195	3115
98-3	3290	3325	3115
98-4	3290	3285	3380
98-5	3290	3285	3600
98-6	3290	3285	3290
98-7	3290	3455	3650
98-8	3290	3285	3380

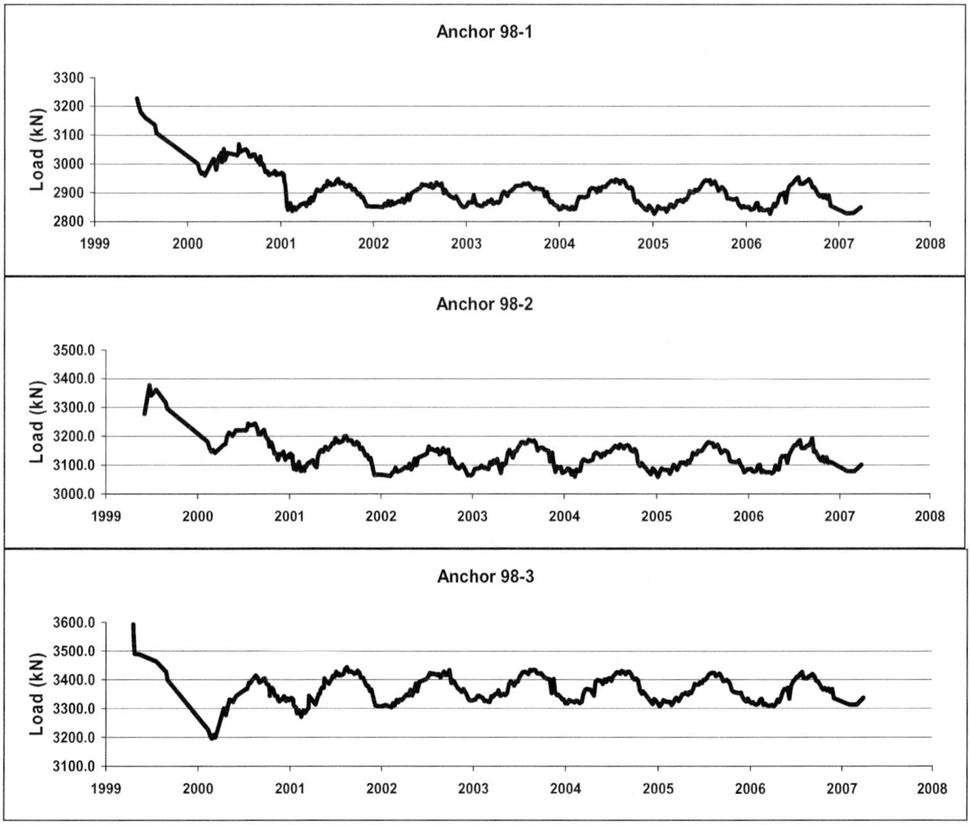

Figure 8: Load Cell Force vs Time

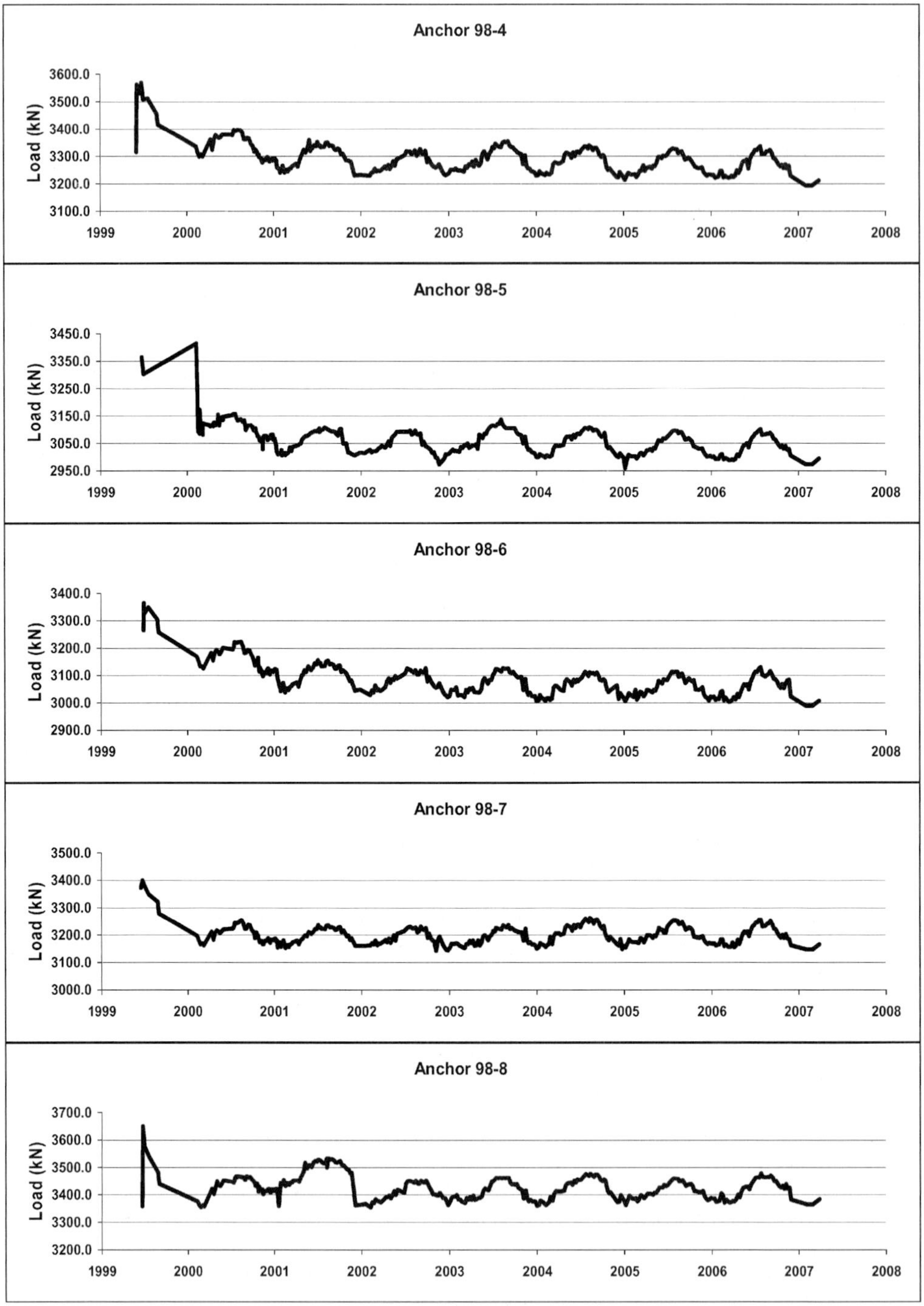

Figure 9: Load Cell Force vs Time

Conclusion

The 35 post-tensioned anchors at the left abutment thrust block appear to have improved the performance of Pacoima Dam during the 1994 Northridge Earthquake. The joint opening at the crest of dam-thrust block contact increased from 1 cm during the San Fernando Earthquake, to 5 cm during the Northridge event. However, the 35 anchors helped limit the magnitude of sliding to approximately 1 cm along the sliding plane. By contrast, an adjacent unanchored rock mass sharing the same sliding plane moved downslope 58 cm. The load cells at the anchor heads failed, but the anchors remained operational following the event. The readings are shown in Table 2.

Eight years of load cell force data, and two lift-off tests are provided for eight anchors installed in 1999. This data shows an average prestress loss of approximately 6% for seven of the anchors. According to the dam owner who performed the load testing, slightly excessive force was used to be certain that liftoff occurred, which explains the net gain in reported prestress. No shims were added for any anchor in either test.

Opinions expressed by the authors are not necessarily those of the California Department of Resources or the Los Angeles County Department of Public Works.

References

1. Pacoima Dam, Effects of the January 17, 1994 Earthquake – Additional Investigation and Structural Review – Phase IIB, Report on Task 2 – Data Evaluation, Morrison Knudsen Corporation, July 1996.
2. Pacoima Dam, Instrumentation Data Review and Evaluation, Christensen Associates Incorporated, July 2001.
3. Personal Transmittal, Pacoima Dam Anchorage lift off data, Michele Chimienti, Los Angeles Department of Public Works, July 19, 2007.
4. Personal Transmittal, Pacoima Dam Anchorage data, Jack Husted, Los Angeles Department of Public Works, May 3, 2007.
5. EERI, Northridge Earthquake Reconnaissance Report, Vol.1, April 1995.
6. Shakal, A., et al., Processing of the Upper Left Abutment Record from Pacoima Dam for the Northridge Earthquake, OSMS Report 94-13, 1994
7. Borcherdt, R., "Empirical evidence for acceleration-dependent amplification factors"; Bulletin of the Seismological Society of America, v.92, p. 761-782, March 2002.
8. Reimer, R.B., Deconvolution of Seismic Response for Linear Systems; EERC 73-10, 1973.
9. S.W. Alves and J.F. Hall, Generation of spatially non-uniform ground motion for non-linear analysis of a concrete arch dam; Earthquake Engineering and Structural Dynamics, v.35, p.1339-1357, April 2006
10. International Engineering Co. (IECO), Geologic Investigations and Stability Analyses, Vol. I & II, July 1968.
11. IECO, Investigation and Evaluation of Effects of San Fernando Earthquake, V.1, August 1972.
12. Morrison Knudsen, Effects of the January 17, 1994 Earthquake – Additional Investigations and Structural Analyses- Phase IIB, Report on Task 2 – Data Evaluation, July 1996.
13. Morrison Knudsen, Effects of the January 17, 1994 Earthquake ...Phase IIB, Report on Model Verification and Sensitivity Analyses, October 1996.

Session four:

Rock bolts in tunnels and mines

A review of Glass Fibre Reinforced Polymer (GFRP) tendon for rock bolting in tunnels.

Alison Littlejohn Ramboll Whitbybird, London, England

Abstract

Historically steel bolts form the most frequently specified material for permanent rock strata support. However, in corrosive or space constrained environments, glass fibre reinforced polymer (GFRP) has been introduced as a viable alternative. GFRP is more frequently specified in soil nailing techniques or as temporary or passive support. Whilst manufacturers are advancing GFRP systems, less is widely known regarding the relative merits of GFRP for permanent tensioned rock bolt support, particularly in tunnels where they are easily cut through by TBM's or road headers.

This paper provides a review of the material and mechanical behaviour characteristics of GFRP and raises sensitivities associated with installation. A comparison of costs between GFRP and steel is also attempted. The paper aims to engage the interest of clients, contractors, consultants and manufacturers and promote discussion on the specification of GFRP with respect to steel in current practice.

Introduction

The paper focuses on the application of tensioned rock bolts for new and existing tunnel support in the UK, however other environments and applications utilising GFRP tendons are examined where relevant e.g. strata control in mining, soil nailing and concrete reinforcement. Rock bolts for tunnelling applications are currently designed in accordance with publications such as CIRIA (1983), BS8081 (1989) and EN1537 (2000) but there are no guidelines for GFRP material. As a consequence, the paper seeks to determine useful references and data for the specification and use of GFRP as a material for rock bolts.

Understanding GFRP

Glass reinforced polymer (GFRP) is a composite material of glass fibre filaments in a resin matrix. The resin matrix bonds the fibres together to allow load transfer, prevent shear between individual fibres, protect the fibres and maintain dimensional stability. Commonly the resin matrix comprises polyester or vinylester resin, but other resins such as epoxy are also used. GFRP is sometimes referred to as resinglass, glass fibre, fibreglass, glass reinforced polymer or in some circumstances simply polyester. These terms may represent alternative composites of resin and glass fibre with differing material and behaviour properties. Such terms should be clarified with manufacturers prior to specification.

More generally, fibre reinforced polymers can contain a number of fibre composites which extend and enhance the behaviour of the resin. Manufacturers use various admixtures, catalysts and accelerators to improve the required performance of the tendon. Commonly

manufactured fibre reinforced polymers (FRP) include glass FRP, carbon FRP and aramid FRP. This paper focuses on glass fibre reinforced polymers.
The development of glass fibre and plastic as independent materials has a relatively long history and glass fibre was first used as a composite with polyester resin in 1942 (Leggatt, 1984). GFRP has a number of physical material properties which make its specification advantageous for a variety of building and industrial processes, namely: high tensile strength, resistant to chloride ions and low pH chemical attack, lightweight, flexible, anti-static and thermally non-conductive.

The last 10 to 15 years has seen FRP more widely introduced for rock bolts and as reinforcement in concrete (Benmokrane et al, 1996). Typically for rock bolting applications the key driver for the use of GFRP is the inherent corrosion resistance of the material and the flexible and lightweight properties which mean it can be utilised in space and access constrained environments, such as mine roadways. These material advantages can therefore lead to an improvement in the durability of rock bolts and facilitate transport, handling and installation. However, some aspects of the behaviour of GFRP such as load-carrying capacity, bond strength, long-term strength and durability are less widely understood.

Existing specifications and guidance

UK anchorage standards and rock bolting guides provide limited information on composite materials but alternative publications on soil nailing and coal mine strata reinforcement provide useful practical guidance on GFRP (Table 1). In addition, use of GFRP bars for reinforcement in concrete structures, e.g. ACI 440 (2001) and Chinese standard CECS 146 (Yuan, 2006), is well established and although not directly related to rock bolting, some data are transferable when investigating material properties and behaviour.

Table 1: Summary of UK standards and guides on GFRP for rock bolts

Standard	Guidance available for GFRP composites
CIRIA (1983) A guide to the use of rock reinforcement in underground excavations	No guidance on GFRP. Adequate protection for steel rock bolts is normally achieved by full embedment in grout. Additional protection includes plastic sheathing for the free length and galvanising for the bond length.
BS 8081 (1989) Code of practice for Ground anchorages	BS8081 states non-metallic fibres may be used for tendons, subject to investigation of their effective life in stressed condition when exposed to potentially aggressive environments that may differ from those aggressive to steel. No further guidance is provided
EN 1537 (2000) Execution of special geotechnical works- Ground anchors	No guidance on GFRP but standard makes a series of recommendations for single and double corrosion protection of steel tendons.
CIRIA 637 (2005) Soil nailing best practice guide	Evaluates ground aggressivity and corrosion protection. Outlines comparisons of different FRP composites and comments on effects on durability.
BS 7861-1 (1996) Strata reinforcement support system components used in coal mines: Part 1 Specifications for rock bolting	Dimensional, material and performance requirements for rock bolting support components used in coal mines. Includes steel rock bolt, glass reinforced plastics rock bolts, resins, conical seats and domed washer plates. The standard does not consider assembly or installation of these components on site to form a rock bolting system.

Based on a specific review of BS 7861 (1996) for rock support systems in a mining environment, comparisons can also be made for steel and GFRP in order to make some preliminary assumptions about differences in behaviour (Table 2).

Table 2: Review of recommendations for steel and GFRP bars, 22mm diameter (after BS 7861-1,1996)

Parameter	Steel	GFRP
Tensile strength	Threaded portion for rockbolt or thread of assembled nut shall not fail at less than 295kN.	Rock bolt and threaded end must have minimum failure load of 300kN and 60kN, respectively.
Shear strength (across rock bolt/resin system)	Minimum of 250kN	Minimum of 120kN
Bond strength (rock bolt/resin interface)	200kN for a bond length of 125mm Minimum system stiffness shall be 70kN/mm measured between loads of 50kN and 150KN	>245kN for a bond length of 450mm and 22mm diameter bar. Bond strength shall be determined at the point where system stiffness falls to 20kN/mm.
Nut torque	95-175N.m roof rock bolts 40-125N.m rib rock bolts 125-185N.m high-torque roof rockbolts	40-125N.m
Note. Values to be achieved in accordance with associated test procedure in BS7861		

The tensile strengths required by BS7861 for GFRP and steel bars are comparable. However, GFRP is required to have only a minimum of 20% of the total tendon tensile strength at the threaded section, whereas steel must confirm 100% of the tensile strength at the threaded nut end. In addition, steel bolts can withstand higher torques during installation. The test for interfacial bond strength requires steel to achieve higher strength over a bond length of 125mm for a ribbed steel bar, as opposed to a 450mm length for GFRP. When testing using the specified double embedment arrangement for bond strength in BS7861, a steel tube provides radial restraint. The results derived will therefore only be comparable with those found in competent rock providing a similar radial restraint.

Properties of GFRP

Glass fibres

The principal advantages of glass fibre are low cost, high tensile strength and high insulating properties. Types of glass available include E-glass, S-glass, C-glass and AR-glass, although E-glass and S-glass are most common. The relative advantages are outlined in Table 3.

Table 3: Types of glass fibre (adapted from Benmokrane et al, 1996)

Glass type	Relative advantages
E-glass	Lowest cost type commercially available, used typically where strength, electrical resistance and acid resistance are important.
S-glass	Higher strength, stiffness and ultimate strain than E-glass but more susceptible to degradation in alkaline environments.
C-glass	Greater chemical stability in acidic environments, c.f. E-glass.
AR-glass	Alkali-resistant to minimise weight and strength loss in alkaline environments.

Resins

Polyester is the most commonly used resin due to a combination of low cost, excellent material properties for FRP and ease of processing. Depending on the mix of ingredients, the properties of resins may vary widely. The relative advantages of the three most common resin types are outlined in Table 4.

Table 4: Material advantages and disadvantages for FRP (adapted from CIRIA 637)

Resin Type	Advantages	Disadvantages
Polyester	Resistant to fire (upper use temperature of 120°C) Resistant to moisture Low viscosity Fast cure time Dimensional stability Chemical resistance to acids and alkalis Moderate material cost	Degradable by chlorinated solvents (avoided for use in concrete) High volumetric shrinkage of unfilled polyesters during processing
Epoxy resin	Improved resistance to alkalis (chloride salts and chlorine), acids (sulphuric, hydrochloric, hydrofluoric, phosphotic, nitiric) and solvents c.f. polyesters	Less weathering resistant Two to five times cost of polyester. Curing process requires more control
Vinylester	Chemical similarities to epoxy resin with curing advantages of polyester Often specified for highly corrosive environments	Cost and performance lie between polyester and epoxy resin

Manufacture of GFRP

GFRP is manufactured by the process of pultrusion where fibre reinforcement is pulled through a resin impregnation area to coat the reinforcement with resin, through preformed plates to shape the fibre/resin bundle, and through a heated die to cure the resin. Numerous process variables, such as pull speed, die temperature, fibre/resin wet-out (i.e. how well fibres are covered by resin) and fibre volume, affect the quality of pultruded composites. GFRP bars are available in nominal diameters typically comparable with those of steel.

Benmokrane et al (1996) highlight two key distinctions between steel and GFRP tendon:
(i) The pultrusion process for the manufacture of GFRP allows high-fibre content products to be obtained, typically 60-80% by volume with homogeneous distribution over the cross-section of the bar. As the tendon is produced at a fairly constant cross section with a smooth surface, indentations are produced on the bar by wrapping a fibreglass string around the bar before the resin sets. These indentations are different from the deformation in steel bars and generally shallower, compared with ribbed bars.
(ii) The stress-strain relationship of GFRP bars is linear up to failure. The elastic modulus of GFRP bars is approximately 20-25% that of steel. The ultimate tensile stress varies from 500 to 1600MPa, depending on glass content, type of fibre and resin, and manufacturing process.

Properties of GFRP composite

The specification of the appropriate composite will depend on the design criteria and environment. Table 5 outlines typical ranges for the mechanical properties of composites and their constituents. Steel tendon is also referenced for comparison and a GFRP composite is included as an example of a known resin matrix and glass fibre content (Weldgrip Co. UK).

Table 5: Typical physical and mechanical properties of commercial glass fibres (adapted from Benmokrane et al, 1996, Yuan at al, 2006 and ACI 440.2R-02)

Parameter	Unit	Glass Fibres				Resins		Tendon		
		E	S	C	AR	Polyester	Vinylester	Steel	GFRP	Gripforce
Ultimate tensile stress	GPa	3.45	4.3	3.0	2.5	0.02 - 0.1	0.095	0.483-0.85	0.483-1.6	1.04
Modulus of elasticity	GPa	72.4	86.9	69	70	2.1 – 4.1	-	200-210	39-50	47
Ultimate strain	%	4.8	5	4.8	3.6	1-6	6.1	-	-	2.3
Poisson's ratio	-	0.2	0.2	-	-	-	-	-	-	-
Density	g/cm^3	2.54	2.49	2.49	2.78	1.0-0.45	-	7.8	1.2 -2.1	
Diameter	μm	10	10	4.5	-	-	-	-	-	-
Longitudinal CTE	10^{-6}/ °C	5	2.9	7.2	-	55-100	-	6-10	6 -11.7	8
Dielectric constant		6.3	5.1	7.2	-	-	-	-	-	-
Cure Shrinkage	%	-	-	-	-	5-12	-	-	-	-

Notes. CTE is the Coefficient of Thermal Expansion
Values for Gripforce relate to 25mm diameter bar, hybrid Vinylester resin matrix, 82% E-glass fibre

Durability of GFRP

Factors influencing durability

CIRIA (2005) and Malvar (1998) identify the primary factors affecting the durability of fibre composites as (i) sustained long term stress leading to the onset of creep rupture, (ii) pH of environment, particularly highly alkaline environments such as cement, (iii) presence of water, (iv) ultra violet radiation and (v) temperature.

The resin matrix, the fibres and the interface between them are all susceptible to degradation by a combination of these factors. In addition, the durability of the GFRP bar is also governed by choice of resin matrix, type of fibre reinforcement and the quality of the manufacture that includes features such as resin wet-out, cracks, voids, degree of cure of resin, and shear strength of the fibre/matrix interface.

Malvar (1998) includes a literature review of the durability of composites in reinforced concrete that suggests using significant restrictions in allowable working stress for GFRP in this application. Reinforcement in concrete places GFRP in an alkaline environment that is simulated for rock bolt installed in cementitious grout. However, for rock bolts installed in an inert resin such reductions in allowable working stress may be considered conservative.

Time-dependent behaviour

Creep

Performance of GFRP can be affected by creep, primarily occurring as a resin matrix deformation but CIRIA (2005) suggests that creep is unlikely to be a problem for FRP where the fibres are aligned in the direction of the applied stress and produced by pultrusion. CIRIA suggests for general guidance that FRP materials should be limited to axial strain levels less than 0.2%. CIRIA addresses passive soil nailing, whereas rock bolts are tensioned to form active support. Malvar (1998) has reviewed a number of long-term static strength tests for tensioned GFRP of varying composites and the key results are outlined in Table 6 below.

Table 6: Results of long-term static strength tests for varying Glass fibre composites (after Malvar, 1998)

Composite	Time period (years)	Percentage of short-term strength at onset of creep rupture
E-glass fibres in a polyester resin	1	70%
Glass fibre in Epoxy Resin	7	50% (some samples as low as 33%)
Hand laid-up fibreglass	10-15	40%
Filament wound glass fibre composites	10-15	55%

Although the above table provides a general guide it should be noted the nature of the composites, i.e. the type and percentage of glass fibre and resin matrix, are not defined.

Creep tests undertaken in Germany (Bundelmann & Rostasy, 1993) on GFRP bars indicate that if sustained stresses are limited to less than 60% of short term strength, failure by creep rupture does not occur. As a consequence, ACI 440 (2001) recommends that the sustained tensile stress should not exceed 20% of the minimum ultimate tensile stress and reports residual strengths of GFRP bars after 50 years to be 30% of the original ultimate tensile strength.

For permanent rock bolts, the designer should establish likely rates of creep and account for these in any performance specification. Malvar suggests in a review of Canadian, Japanese and European recommendations that a proposed allowable ultimate stress ratio for GFRP should be 0.25 (i.e. 25% ultimate tensile stress) or 172MPa, whichever is lower. This provides a level of safety accounting for the measured values in service of 32% of tensile strength.

Effects of adverse environments

Corrosion in aggressive environments

As a minimum CIRIA (2005) suggests the following chemical investigation should be undertaken to establish risk of degradation as a result of the surrounding ground and ground water: pH, chloride ion content, sulphate content and tests recommended from associated standards for cement or resin grout, where appropriate.

The loss of initial strength can be accelerated in adverse environments i.e. the presence of water, acid and alkalis. In a cementitious, i.e. high alkaline environment, Malvar emphasises that the composite must provide a two-fold protection through high resin toughness to prevent development of matrix micro-cracking and diffusion through the matrix must be minimal. Tests undertaken on E-glass fibre show improvements in micro-cracking can be achieved by specifying vinylester over polyester. In relation to chlorides, C-glass and AR-glass fibres have improved resistance.

With reference to the degradation of GFRP under acidic conditions, Mizoguchi et al (2004) immersed a number of GFRP composites in acid and observed behaviour. E-glass fibres were seen to corrode and strongly affect the mechanical properties of the GFRP. The C-glass fibres examined did not corrode but the interfacial properties were observed to degrade leading to a reduction in the interfacial bond shear strength. In order to design the composite materials with higher corrosion resistance, a high cross bonding density and interfacial adhesion should be generated. This is a matter for the GFRP manufacturers who should be consulted as to the optimal material solution when the aggressivity of the environment has been quantified.

Malvar (1998) reported that glass fibres are also affected by moisture, typically losses of 10% in tensile and flexural strength can be encountered after a few months of exposure, although some studies indicate losses may be negligible. The integrity of the resin surrounding the fibres is crucial in minimising the exposure of the glass fibres to moisture. Malvar states the resins used must provide a two-fold protection, where the matrix toughness must be high enough to prevent development of matrix micro-cracks, and diffusion through the matrix must be minimal. Micro-cracking or poor adhesion of the resin will expose the fibres and ultimately lead to reductions in tensile strength. Control in the development of micro-cracking can be improved by specifying more resistant resins at the manufacturing stage. In addition, to protect the tendon from development of cracks during installation, or service, the integrity of the surrounding grouted (resin or cement) annulus is also crucial in minimising moisture exposure.

Temperature effects on strength and stiffness

Fire resistance is a design consideration when specifying rock bolts for tunnelling and limited guidance is provided for FRP reinforcing bars in ACI 440 (2001) and Canadian CSA-S806 (2002). For concrete reinforcement, the bars are protected by increased concrete cover which may not be the case for rock bolts in tunnelling depending, for example, on the use of shotcrete or concrete linings, and bolt head cover design. Fire resistance can be determined through testing or calculation and BS 7861 (1996) provides some indication of testing to determine fire resistance. GFRP manufacturers should provide guidelines on these properties for the tunnelling market.

Wang and Kodur (2005) investigate the impact of temperature on both strength and stiffness of 9.5mm and 12.7mm diameter GFRP reinforcing bars. The bars retained about 90% of their elastic modulii up to approximately 400°C, although in one GFRP sample some resin in the 12.7mm GFRP bar was observed to ignite at 500°C. In contrast, steel reinforcement bars experience a steady decline in their elastic modulus above 200°C. Typically, GFRP composites were observed to lose approximately 50% of their strength at 325°C, the strength variation up to this point being almost linear. The strength of GFRP bars was reduced to zero at 550°C. A 50% strength loss can be related to a 'critical temperature' concept, often taken to represent failure from a fire resistance perspective.

Bond behaviour of GFRP

Benmokrane et al (1996) evaluate bond properties at the grout/bolt interface and pull-out behaviour of cement grouted GFRP bolts, determined from laboratory (concrete) and field pull-out tests on four types of GFRP bolts of varying surface properties and two types of steel bolt. The indentation or surface deformation produced on the tendon affects the bond strength as well as load-slip behaviour and the results illustrate that the deformed GFRP bar anchor bolts have a failure mechanism similar to that of steel anchor bolts due to the crushing and shearing at the grout/bolt interface. Smoother GFRP bolts typically fail due to the shear and delamination of the loose fibres on their surface and are not considered suitable for rock bolts.

The maximum pull-out load capacity increases as the bond length increases but with an excessive bond length the tendon will fail in tension rather than by being pulled out, assuming no partial de-bonding. The concept of critical bond length is used and defined by the minimum embedment length necessary to bring the pull-out load up to the ultimate tensile strength of the bolt. The critical bond length is therefore a function of the interfacial bond strength and the ultimate tensile strength of the bar. The critical bond length of the bars varies considerably and GFRP bars have a longer critical bond length than steel ribbed bars due primarily to the greater surface roughness of ribbed steel bars. A range of critical bond length for the four types of GFRP bolt and the two steel bolts are shown in Table 7.

Table 7: Critical length ranges for GFRP bolt and steel bolt installed in concrete and rock mass (after Benmokrane et al, 1996)

	Concrete		Greywacke rock mass (RMR & Q indicate good quality)	
Tendon type	L_c (mm)	L_c / d_b	L_c (mm)	L_c / d_b
Steel	195-505	7.6-19.9	790	31
GFRP	365-1305	14.3-51.3	470-2905	18.6-114.4
Note. L_c is critical bond length and d_b is bolt diameter				

Yuan et al (2006) also examine bond behaviour of mortar grouted GFRP rock bolts (mortar UCS = 56MPa). Pull-out tests were undertaken on three alternative diameters of bar (10, 13 and 16mm) over increasing bond lengths (65 to 390mm). In this case the rock mass was modelled by a concrete block (UCS = 57MPa). Three failure mechanisms were identified that confirm the development of critical bond length:

1. Failure at concrete/mortar interface and bolt/mortar interface at bond lengths of 0 to 10 times diameter. In this case, the cracking forms a conical zone near the open end of hole.
2. Failure at the bolt/mortar interface with surface cracking near the open end of the hole in the concrete at bond lengths 10 to 20 times bolt diameter.
3. Where bond length is equal to or greater than 20 times bolt diameter, the GFRP bolt was observed to fail in tension.

Clearly, the bond length must exceed the critical bond length in order to make full use of the tensile capacity of the bar. The concept of critical bond length is also reflected in the tests recommended in BS7861, and explains the reason for the longer bond length required for the double embedment test in GFRP in order to develop the ultimate tensile strength of the bar.

Installation of GFRP rock bolts

It is generally acknowledged that GFRP materials are weak in transverse shear and strong in axial tension. However, the reduction in tensile strength of GRFP at the threaded end of the bar and in the capacity of the locking nut is a limiting factor on performance of the bolt in terms of inducing load transfer. It is suggested that a reduction in end bearing capacity of the tendon can be compensated by the axial load transmitted through the bond length and typically manufacturers recommend a fully or 90% bonded bar. However, for rock bolts, as opposed to dowels, a known tensile load is locked off after installation and thereafter, it is good practice to monitor load in a selection of bolts during service.

The tensile capacity of the head detail is typically 25-50% of the bar's tensile strength. In addition, the practicalities of loading can be problematic in terms of crushing of the composite when attempting to grip the bar in the wedge to transfer tensile load into the bar.

The literature reviewed within this paper describes GFRP installed using mortar grouted bolts but in rock bolting practice, resin capsules are commonly used for ease and speed of installation, in addition to potential advantages of chemical resistance and limiting grout loss around the tendon due to an absence of bleed. With GFRP, the torsional effects of spinning the bar into the resin may lead to installed length limitations where high speed resins (short setting times) are required. Typically, where a nominal tensile load of 100kN is applied this may limit the installed length to 2.4m. Such issues can be overcome by using hybrid systems of slow and super slow speed resins.

Designers should be aware of these practical limitations as a number of contractors have described difficulties in not only achieving the load transfer but also slip or twisting of GFRP bars when using resin capsules. The development of appropriate guidelines in collaboration with the manufacturers of resin capsules and GFRP bolts and site installation contractors is required to ensure that feasible installation methods and appropriate individual rock bolt designs are specified. In addition, QA/QC systems should be implemented to ensure high quality practice on site and provide records confirming that the performance specification for installation had been attained, as well as providing feedback to assist in further optimisation of both the design and construction of GFRP rock bolts.

Case studies and applications

GFRP is widely used in mining applications and is usually specified via an observational process which is not always transferable to civil engineering works that require measurement of performance. In addition, mining applications can often be considered temporary where long term support systems are allowed to fail. Nevertheless, much of the experience can be used in terms of initial design parameters for predicting short term behaviour, such as the recommendations in BS7861. Variation in the design parameters for different types of glass fibre, resin and grout type typically rely on the recommendations of the manufacturers, where comprehensive research into the performance of rock bolts of varying material properties, installation and geology has not been published. Practical project case studies are less readily documented and whilst rock bolting with GFRP is known to be undertaken in the UK case studies are typically of a commercial nature and long term monitoring, design parameters or installed pull-out tests are unavailable.

Cost comparison

The total cost per bolt installed is difficult to evaluate particularly in a competitive commercial market. Given the design life of any permanent structure, it is also necessary to evaluate the whole life cost of any remediation, the two contrasting options being:

- a higher initial capital cost associated with minimum maintenance costs, guaranteed durability and minimum risk
- a lower initial capital cost with ongoing durability assessment and maintenance costs necessary to reduce the risk of failure.

The material cost may represent only a small proportion of the total mobilisation and installation cost anticipated. Evaluation of the cost benefits of steel or GFRP tendon for rock bolting can sometimes equate to a comparison of the implementation of varying levels of corrosion protection that can ensure design life and minimise whole life costs. Where the specification of GFRP is driven from the perspective of handling within a space or access constrained site, the influence of material cost may also be reduced in terms of increased installation time for steel. In considering any specification of materials the relevant testing and monitoring requirements for rock bolting should also be established and related to cost.

In order to provide a crude comparison, typical UK material costs (excl. VAT) for high yield Gewi steel, stainless steel and GFRP bar are provided in Table 8.

Table 8: Comparison of UK material costs for Steel, stainless steel and GFRP tendon (July 2007)

	Material	Ultimate tensile strength (kN)	Cost per 6m length (£)
Steel Grades	Dywidag Gewi Steel Threadbar, including Nut 20mm	188	10.54
	Dywidag Gewi Steel Threadbar, including Nut 25mm	295	16.17
	Dywidag Gewi Steel Threadbar, including Nut 32mm	482	28.24
	Stainless UK Gripbar 20mm (Grade 304)	196	76.20
	Stainless UK Gripbar 30mm (Grade 316)	466	238.20
	Stainless steel accessories grade 304: 20mm bar hemispherical load nut £7.90, 20mm gripbar locking nut £2.92, 125mm x 125mm x 10mm dished £7.50. *Staineless steel accessories grade 316: 30mm hemispherical load nut £16.66, 30mm grip bar locking nut £7.75, 200mmx200mmx12mm face plate dished £29.34*		
GFRP Composites	Gripforce 22mm with standard high load nut (Vinylester)	470	35 .00 *(7.50 for 1m; 5.50/m thereafter)*
	Gripforce 31mm with standard high load nut (Vinylester)	730	63.00 *(15.50 for 1m; 9.50/m thereafter)*
	FT-500 (82% E-glass fibre & polyester resin) 24mm	470	24.00
	GFRP Accessories *Standard nut £1, 150mm diameter face plate £1, 250mm diameter face plate £1.75, 250mm diameter high load face plate £2.75*		

CIRIA Report 637 (2005) also provides unit costs of various grades of steel and composite materials that permit useful comparisons.

Final remarks

The lightweight nature of GFRP makes handling and installation easier and can be preferred in terms of health and safety. The flexible properties of the tendon permits the use of relatively long lengths that can be installed in space constrained environments without the need for couplers. The inherent corrosion resistant properties of GFRP in environments aggressive to steel are also attractive.

Where the installation requires low or no induced load (dowels), GFRP is much easier to install e.g. tunnel face support prior to advancing the face. However, for permanent tensioned rock bolt design, clearer guidelines and the introduction of improved interfaces at the bolt head are required to overcome current issues of tensile load capacity and load transfer at lock-off. Such improvements can be evolved but for some contractors these issues are currently seen as limiting factors regarding the use of GFRP.

Looking ahead, there is a need for industry guidelines on the physical properties and geometries of GFRP bars considered appropriate for rock bolt systems in civil engineering, design recommendations, e.g. grout/GFRP bar interfacial bond values for individual GFRP bolts and durability in different aggressive environments, bolt installation techniques and QA/QC procedures to ensure satisfactory performance. In addition, field case histories should be published covering the design, construction and service behaviour of GFRP rock bolts, in order to increase confidence in the use of composite materials.

References

ACI 440.1R-40 (2001) "Guide for the design and construction of concrete reinforced with FRP bars", American Concrete Institute, Farmington Hills, MI, USA.
BENMOKRANE, B., XU, H. & BELLAVANCE, E. (1996) "Bond strength of cement grouted glass fibre reinforced plastic (GFRP) Anchor bolts", International Journal of Rock Mechanics, Abstr. Vol. 33, No.5 pp. 455-465.
BS 7861-1 (1996) "Strata reinforcement support system components used in coal mines: Part 1 Specifications for rock bolting"
BS 8081 (1989) "British Standard Code of practice for Ground anchorages"
BS EN 1537 (2000) "Execution of special geotechnical works-Ground anchors"
BUDELMANN, H. & ROSTASY, F.S. (1993) "Creep rupture behaviour of FRP elements for prestressed concrete – Phenomenon, Results and forecast models", ACI Special publications, vol. 138, pp.87.
CAN/CSA-S806 (2002) "Design and construction of building components with fibre reinforced polymers", Canadian Standards Association, Ottawa, Ont., Canada.
CIRIA (1983) "A guide to the use of rock reinforcement in underground excavations", Report 101.
CIRIA (2005) "Soil Nailing – Best practice guidance", Report 637.
LEGGATT, A. (1984) "GFRP and Buildings – A design guide for architects and engineers", Butterworths, London.
MALVAR, L.J. (1998) "Durability of composites in reinforced concrete", CDCC First International Conference on Durability of Composites for construction, Sherbrooke, Quebec, Canada.
MATSUMOTO, D.S. (1983) "Fatigue initiation in a short glass fibre composite", Journal of Material Science letters 2, 7-11.
MIZOGUCHI, M., HAMADA, H. XU, L. & DRZAL, L.T. (2004) "Interphase degradation in GFRP under acidic condition", Technical Paper - Society of Manufacturing Engineers, ICCM-14, 2004, pp 167-171, ICCM-14, Jul 14-18 2003, San Diego, CA, USA.
Stainless-UK www.stainlessuk.com
WANG, Y.C. & KODUR, V. (2005) "Variation of strength and stiffness of fibre reinforced polymer reinforcing bars with temperature", Cement and Concrete Composites, 27, 9-10, pp. 864-874.
WELDGRIP GEOTECHNICAL (2006) www.weldgrip.co.uk (last accessed June 2007).
YUAN, Y., JIA, X. & LEE, C.F. (2006) "Bond behaviour of mortar grouted GFRP rock bolt", Proceedings of the Soil and Rock Behaviour and Modelling Conference, Shanghai China. ASCE Geotechnical Special Publication 150, pp 228-234.

Design and construction of large rock caverns supported by ground anchorages

Kenji Aoki, Kyoto University, Kyoto, Japan

Abstract

This paper describes the experiences of Japan in the development and utilization of underground openings for the energy storage facilities and principal technological accomplishments in the design and construction of large rock caverns supported by ground anchorages. A fundamental study was carried out on the estimation of the interaction of ground anchorage and rock mass behavior. New technologies for the future development are introduced, by using the AE behavior and DEM with Bonded Particles Model, for the evaluation of the behavior of large rock cavern.

Introduction

Large rock caverns have been used mainly for energy-related facilities. Full-scale utilization of underground spaces are closely related to the energy policy of the Japanese government after the war. In late 1950's, energy demand started to rapidly increase along with steady restoration of its economy. In late 1960's, when the demand of electric power rapidly grew and the output from nuclear electric power generation steadily increased, construction of large-scale underground pumped storage power plants started to cover temporary power shortage during peak loads. After the oil crisis, reservation of petroleum started in 1975 to ensure stable supply of petroleum. Oil storage caverns were noticed as a new method for storing petroleum, and the first test oil cavern in Japan was constructed in 1982. In 1982, the construction of full-scale oil storage caverns started in three different site and the facilities started to operate in 1994. Theses power plants and oil storage caverns are most important and marked examples of technological contributions to the utilization of rock caverns and rock engineering. In this paper the actual behaviors of large rock caverns supported by ground anchorages for pumped storage power plant and their performance are described.

Design and construction of large rock caverns

Underground pumped storage power plants

In Japan, approximately 50 underground power generation plants have been constructed, including almost 20 hydraulic plants and about 30 underground pumped storage power plants. Many large-scale hydraulic power plants have been built underground, which enables large pressure head to be formed and plants of large outputs to be built with small costs. Underground pumped storage power plants are systems for storing electric power by pumping water from lower reservoirs to upper reservoirs using nighttime surplus power from nuclear and other base power plants, and generating electricity during daytime peaks of power demand. The first underground hydraulic power generation plant in Japan is Uryu Power Plant (Hokkaido, completed in 1943). The first underground pumped storage power plant is Shiroyama Power Plant (Kanagawa, completed in 1965). After the construction of Shin Takase-gawa Power Plant that started in 1971, various plants, which have large cross section area of 200 to 300 thousand cubic meters are constructed by solving various technological difficulties at each site, such as high earth pressure and highly fractured rocks. Recently, Kazuno-gawa, (Figure 1) and Kanna-gawa pumped storage plant have been completed and also in 1998, a test pumped storage power plant was completed in Okinawa, which uses seawater, and started to operate. An underground pumped storage power plant is also planned, which has its power generation part and the lower reservoir underground to obtain a large pressure head. The feasibility of other new types of power storage facilities is also being investigated.

Figure 1. Kazunogawa pumped-storage power plant

Design of the caverns supported by ground anchorages

Power plant caverns are designed based on the results of surveys and tests and according to the procedure shown in Figure 2 and 3. In practice, finite element analysis is performed to analyze the excavation of caverns, as well as slip line analysis, key block analysis that assumes discontinuities of rock, and circular slip analysis. In basic designing, the shapes and directions of caverns are first determined. Supports of power plant caverns are generally shotcrete, rock bolts, and prestressing anchors, which apply loads by tensioning strands (Figure 4). The specifications are determined for each of these support members. The tensioning load of a prestressing anchor is usually 40 tf to 120 tf, and prestressing anchors are the most important support member.

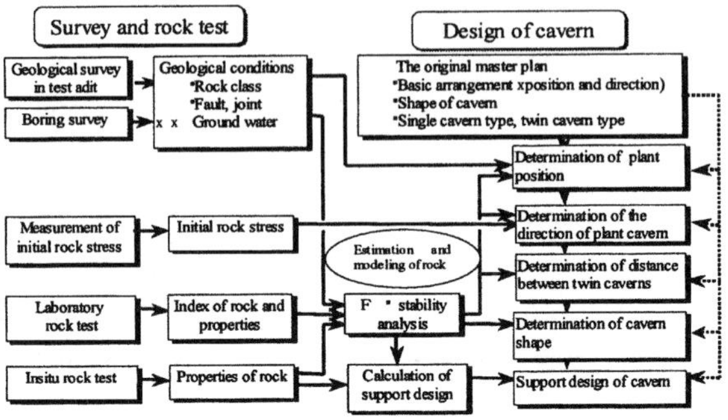

Figure 2. Cavern design, survey and testing

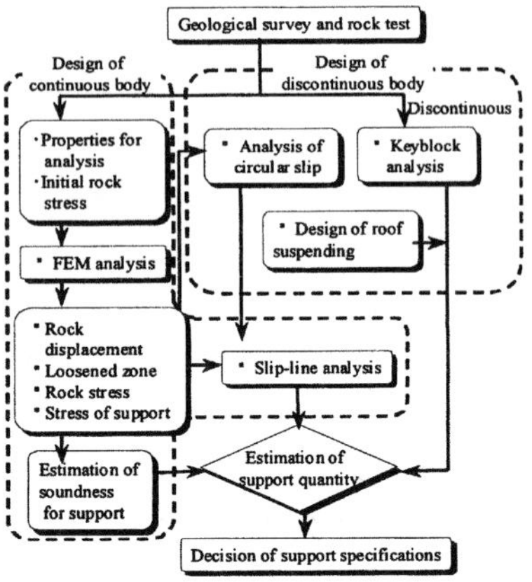

Figure 3. Support design

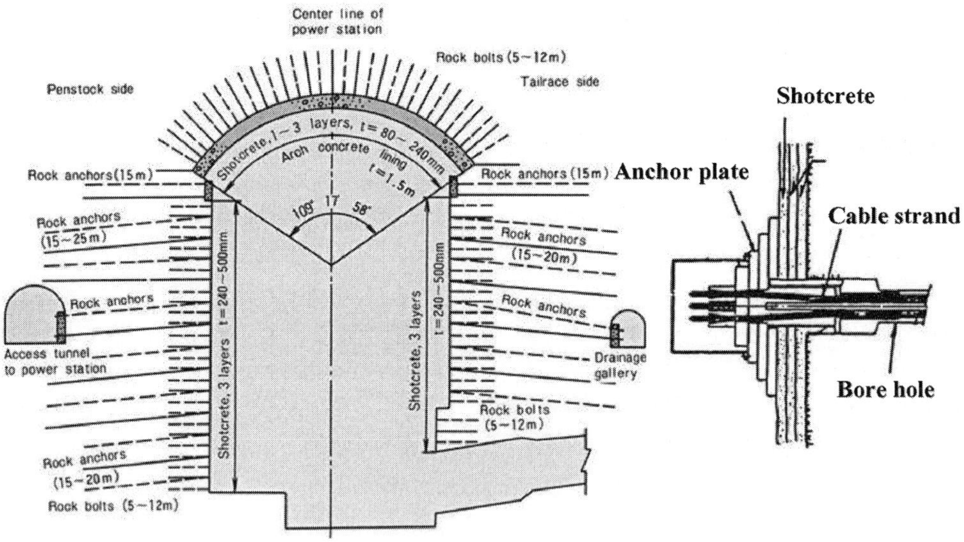

Figure 4. Typical rock support of powerhouse cavern by ground anchorages

Design concept of the support by rock anchors

The maximum principal stress in the stress concentrating zones increases as the excavation advances. Change in stress around the cavern by the installation of prestressed rock anchors, which input stress, are schematically shown in Figure 5. An input to minimum principal stress by the installation of rock anchors (confining pressure of ΔP)

increases the maximum supportable stress from σ_1 to $\sigma_1 + \Delta\sigma_1$. This shows that the effect of rock anchors restrict the progress of the loosened zones and stabilize the cavern.

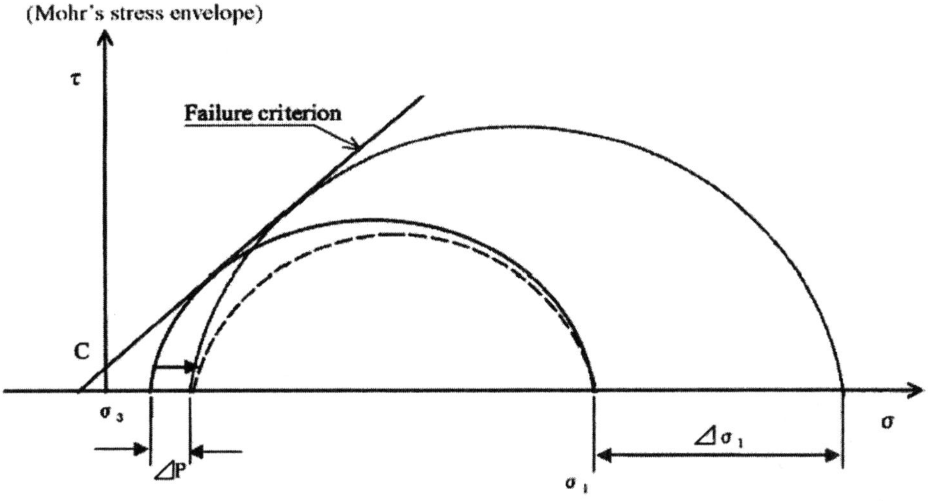

Figure 5. The mechanism for effect of input stress by rock anchor.

Actual behavior of the caverns

The Kannagawa power plant is a pumped storage power station of maximum output of 2820 MW and is being constructed by the Tokyo Electric Power Co. Inc. 150 km north-west of Tokyo at the boundary of the Gunma and Nagano Prefectures. The underground cavern of the power station has dimensions of: length 216 m, width 33 m, height 52 m, approximately 220 000 m^3 in total excavation volume, and maximum cross-sectional area of 1400 m^2. Rock strain measurements and acoustic emission (AE) measurements were conducted as well as conventional strain and rock displacement measurements to directly evaluate the loosened zones. Prediction analyses were performed by conducting precise preliminary geological surveys and considering the results of inverse analyses to improve the precision of prediction

Geology

The geology of the Kannagawa site mainly consists of Paleozoic–Mesozoic sedimentary rocks. As shown in Figure 6, the strata around the powerhouse cavern are classified into six subgroups with different rock properties. The initial rock stresses are also shown in Figure 6.

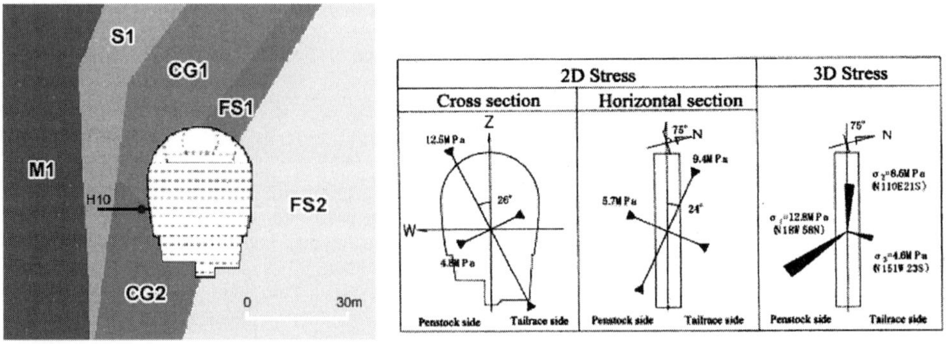

M1 = mudstone, S1 = medium grained sandstone, CG1 = conglomerate, CG2 = conglomerate and coarse grained sandstone,

FS1 = fine grained sandstone with lens shaped layers, and FS2 = fine grained sandstone.

Figure 6. Geological section of the cavern for research and initial in-situ stress.

Design of the cavern

The Analysis for predicting the behaviors of rocks during excavation were conducted using the numerical model that reflects the geological structures. Since highly-stiff triaxial compression tests of boring cores showed that the rocks had strain softening characteristics, the models were analyzed by non-linear multi-step excavation FEM analysis that considered strain softening. In the actual design, rock anchors were more densely installed in the stress concentrating zones than in the other zones to actively restrict the progress of loosening near the cavern. The support arrangement in the power plant cavern is shown in Figure. 7. Supports were cross-sectionally non-uniformly arranged in the cavern to make rational arrangement of the supports.

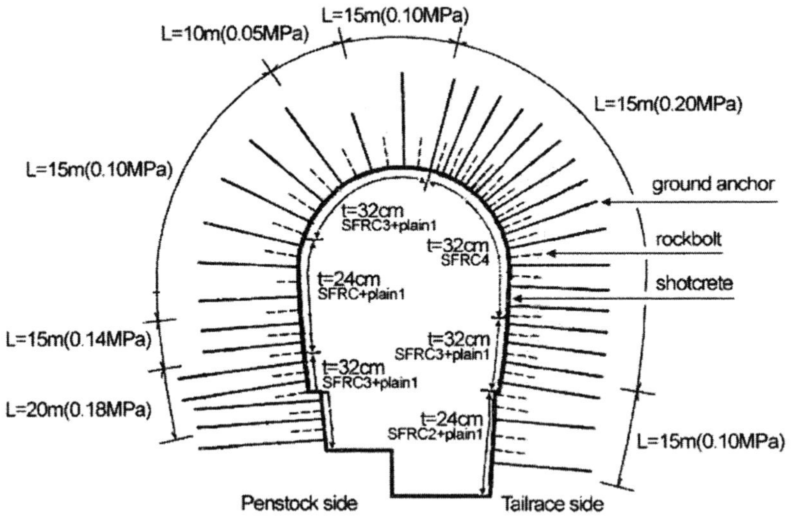

Figure 7. Final support arrangement for the cavern

Behavior of the cavern during excavation

The typical displacement of the wall and the sectional strain distribution are shown in Figure 8 at the completion of excavation. The displacement was measured as time elapsed as shown in Figure 9. The strain was discontinuously distributed around the cavern especially at the tailrace side. The sections of large strains almost agreed with the sections where the highly consecutive geological discontinuities existed.

As shown in Figures 8 and 9, the support of the cavern by rock anchorages have been evaluated as to show sufficient performance to accomplish the mechanical stability of the cavern.

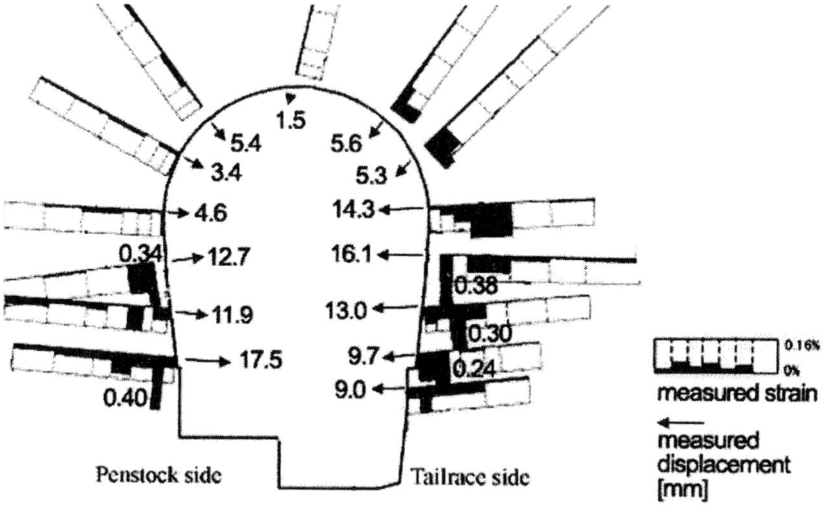

Figure 8. Distribution of rock displacement and strain

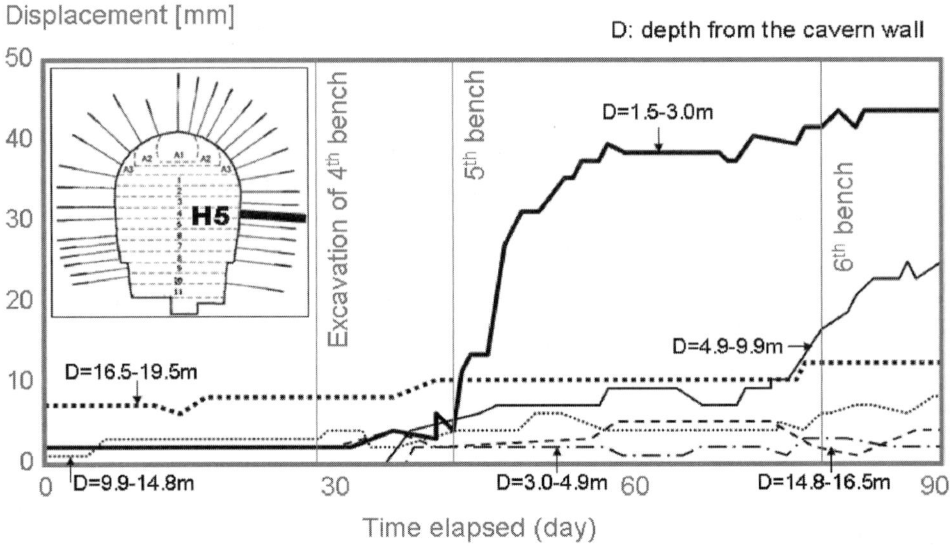

Figure 9. An Illustration of Local displacement (H5-line at the 4th bench)

New technologies for the future development

The ground anchorage is the major support member for the large rock caverns such as pumped storage power plant. An observational design and construction system actively utilizes various kinds of information, such as field measurements, geological conditions, and construction procedures, and it helps engineers to make decisions concerning measurements, analyses and designing, construction of supports (i.e. ground anchorages, rockbolts), and optimization of designing and construction. In this context, the numerical modelling technologies by using AE and DEM analysis are expected to be utilized in the design and construction for nuclear waste repository and storage caverns (Figure 10) with great depth.

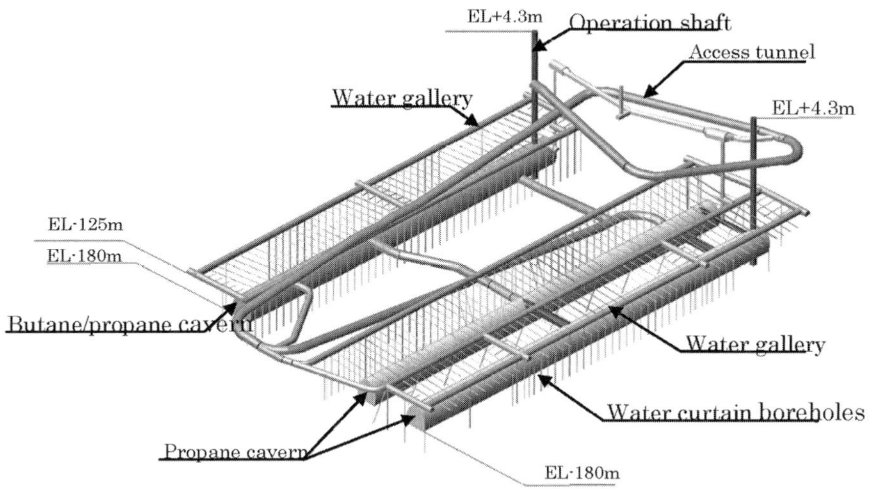

Figure 10. LPG underground storage facility

Relationship between stress, strain and AE parameters

Figure 11 shows the examples of time series data of change in stress, strain, AE counts, predominant AE frequency, and m-value at depth of 3m from the cavern wall. The following relationship between stress/strain and behaviors of AE parameters can be inferred based on the conclusion of the laboratory test.

(1) AE event was first observed when the bench No.1 was excavated. A little change in stress was ob-served although strain could not be perceived at this time. This fact shows that AE can capture a premonition of EDZ with high accuracy.

(2) When excavating the benches No.2 and No.3, AE counts and predominant AE frequency increased and m-value decreased, as stress level increased. Those behaviors of AE parameters consistently show the fact that stress concentration progressed and stress induced cracks increased at this time.

(3) When excavating the benches No.4 and No.5, AE counts showed maximum, predominant AE frequency decreased, and m-value became constant as stress showed its maximum level and strain increased. These behaviors of AE parameters consistently show the fact that the stress had already reached the peak strength of the rock at this time.

(4) When excavating the benches No.6, AE counts became close to 0 and m-value suddenly increased and became its initial level, as stress and strain level became constant. These behaviors of AE parameters consistently show the fact that the stress had already reached the residual strength of the rock and the EDZ had completely been formed at this time.

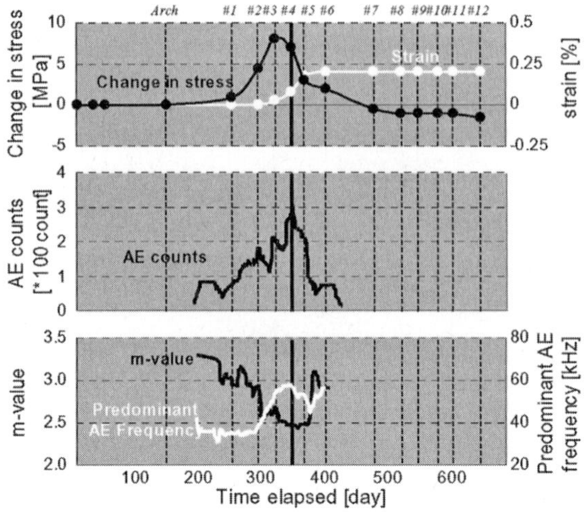

Figure 11. An illustration of the measurement result (at depth of 3m from the cavern wall)

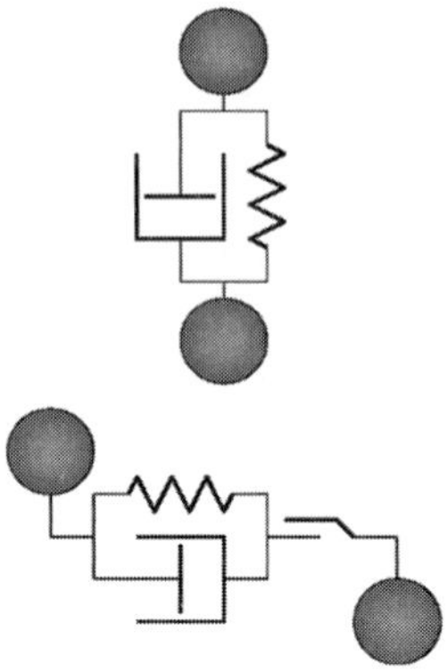

Figure 12. Concept of contact bonded particles

DEM simulation with Bonded Particles Model

The contact Bonded Particles Model (Itasca, 1999) as shown in Figure 12 was used for the DEM simulation of the excavation of the rock cavern to grasp the detailed distributions of stress and initiated cracks in rock mass, as the bench cuts progresses (Aoki et al., 2004). The proposed model considering the effect of input stress by rock anchor is applied as the failure criteria of the contact bond that connects the particles.

The values of micro-properties of the Bonded Particles Model are specified such that the macroscopic behavior of rock mass is corresponded to the result of a series of in-situ tests on rock deformability and rock strength. A set of numerical biaxial tests is implemented to calibrate the micro-properties for all types of rocks. Then the inclined maximum and minimum principal stress, which is predetermined by initial rock stress measurement, is applied to the model and the DEM simulation is implemented at each step of excavation (top headings and bench cuts)..

Figure 13 shows the computed stress distribution (differential stress) and initiated crack distribution as the bench cut excavation progresses The typical stress redistribution caused by the excavation under the stress condition with inclined stress axes is observed. The cracks mainly initiate at the regions of stress concentration. Tensile cracks are dominant in the beginning of the excavation, and then shear cracks increasingly occur. Such fracturing process is analogue of the rock failure observed in the laboratory tests. Cracks mainly propagate in the stress concentrating parts and in the rocks with poor properties (FS1 and S1). Figure 14 shows the comparison between the measured strain and the computed stress distribution. It is found that the large strain is mainly observed in the stress relaxation part computed by DEM.

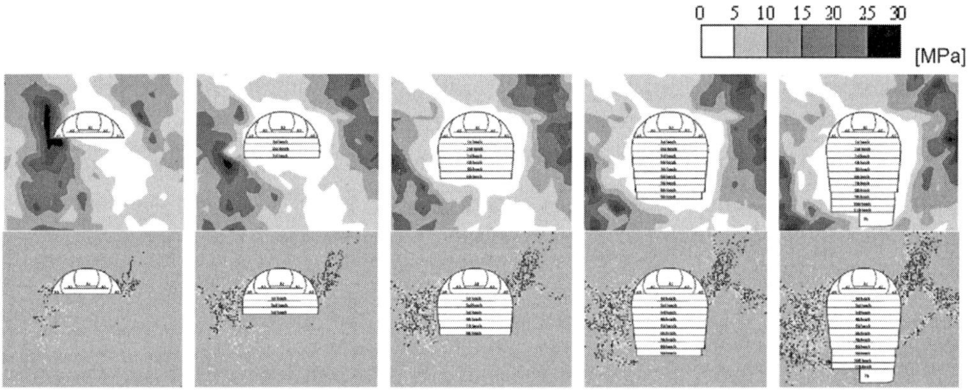

(1) top heading (2) the 3rd bench cut (3) the 6th bench cut (4) the 9th bench cut (5) completion

Figure 13. The results of DEM simulation: the computed stress distribution (top), and crack distribution (bottom)

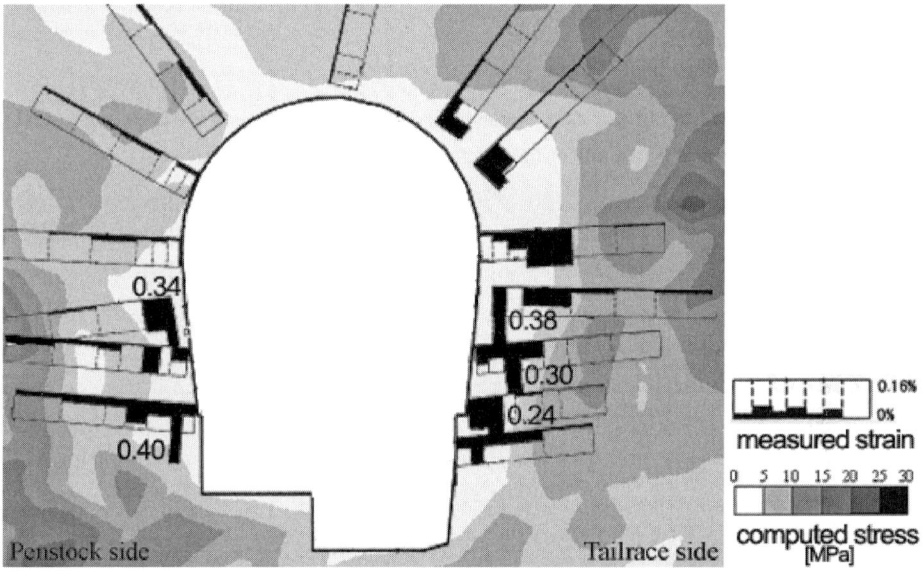

Figure 14. The comparison between the measured strain and the computed stress

Conclusion

An effective design and construction system for underground rock caverns and the actual cavern behavior supported by ground anchorages were introduced. AE (Acoustic Emission) has been thought as useful parameter of stress change in rock mass around underground caverns to evaluate the stability of the large cavern supported by ground anchorage. By using the AE behavior and DEM with Bonded Particles Model, the effect of ground anchorage for large rock cavern can be accurately evaluated. The relationship between in-situ AE behavior and stress paths during a cavern excavation were established and a fundamental study was carried out on the estimation of the interaction of ground anchorage and rock mass behavior.

Reference

Aoki, K., Mito, Y., Mori, T. and Maejima T. (2004) "Numerical modelling in Micromechanics via Particle Method" Numerical modelling in Micromechanics via Particle Method, pp.73-83.

Aoki, K., Mito, Y., Matsuoka, T. and Kondoh D. (2004) "Design of gas storage rock cavern by the hydro-mechanical coupled discrete model" Numerical modelling in Micromechanics via Particle Method, pp.289-330.

Aoki, K., Mito, Y., Mori, T., Morioka, H. and Maejima T. (2004) "Evaluation of behavior of EDZ around rock cavern by AE measurements and PFC simulation" 3rd ARMS, Kyoto, Japan.

Aoki, K., Mito, Y., Mori, T. and Maejima T. (2005) "Evaluation of EDZ around highly stressed rock cavern by AE measurements" EUROCK2005, Brno Czech Republic.

Aoki, K., Mito, Y., Chang, C. S. and Maejima T. (2006) "Numerical analysis of rock fracturing process by DEM using bonded particles model" 4th Asian Rock Mechanics Symposium.

Aoki, K., Mito, Y., Chang, C. S., Tasaka, Y. and Maejima T. (2007) "Hydromechanical coupled discrete modelling for the assessment of airtightness of unlined large rock cavern" 11th International Congress on Rock Mechanics, Lisbon, Portugal, 2007.

Itasca Consulting Group, Inc.,"Particle Flow Code in 2 Dimensions Theory and Background", 111p. (1999) Itasca.

Mito, Y., Chang, C. S., Aoki, K., Matsui, H. Niunoya, S., and Minami M. (2007) "Numerical analysis of rock fracturing process by DEM using bonded particles model" 11th International Congress on Rock Mechanics, Lisbon, Portugal.

Monitoring and testing of mine anchor systems: case studies and applications to the civil industry

Peter Altounyan, David Bigby and Lorraine Kent, Rock Mechanics Technology Ltd., Burton-on-Trent, UK

Abstract

Anchor systems, mainly in the form of rockbolts, but also as cables and other long tendons, are the main support systems used in mining worldwide and many millions are installed each year.[1]

In response to demand for improved ground control safety in mines, new procedures for support design, monitoring and testing of these anchor systems have been developed. Some of these procedures developed for mining, are directly applicable to civil engineering, and are finding increasing use in this application. This includes instrumentation to measure in situ anchor performance, for example new extensometer types, and load measuring rockbolts, together with new laboratory test methods and standards, safety monitoring based on ground movement detection and NDT techniques for detecting broken rockbolts and tendons.

These instruments and techniques and their applications are described in the paper.

Introduction

The application of instrumentation and measurement in mining tunnelling has developed considerably in recent years, primarily in response to demand for improved ground control safety.

The adoption of rockbolting in coal mines in the UK in particular was the catalyst for significant innovation in this area. Instrumentation and techniques were required to confirm rockbolt patterns and ensure safety. The large extent of typical mine workings means that a correspondingly large number of instruments were needed. Consequently these had to be inexpensive, easy to install and simple to read and maintain. This is in contrast to civil applications such as NATM where instrumentation (rod extensometry and total station surveying for example) tends to be relatively specialist and expensive. Following the introduction of rockbolting, additional techniques for measuring rock deformation, detecting broken rockbolts and for evaluating anchor systems in the laboratory have been developed and applied.

Rock Mechanics Technology is a UK based specialist consultancy whose staff have played a major role in the development of these systems from their inception and subsequently in applying them to anchor support design in mines in the UK and worldwide. Much of this work has been part funded by the UK HSE or the European Community. This paper describes these developments and some of their recent application outside the mining industry.

Monitoring and observational design of rockbolt systems

Anchor systems, mainly in the form of rockbolts, but also grouted cables and other long tendons, are the main support systems used in mining worldwide and many millions are installed each year. The predominant system consists of steel bolts spun through polyester resin cartridges into predrilled holes to provide fast installation and early support.

The typical mining support system is not designed in any formal way, rather rockbolt patterns have evolved locally by trial and error, perhaps from an initial design or experiment. The pattern may subsequently be varied in response to conditions encountered, based on the judgement of mine engineers. This practice is effectively a crude form of observational design in which the level of support is varied depending on simple observation of resulting ground deformation and support response. In addition mining regulations sometimes include minimum bolt lengths and maximum spacings which effectively determine the pattern used. In the USA for example the maximum legal spacing of roofbolts in coal mines is five feet (1.5m) and most mines install four bolts per row to give around 1.2m bolt spacing. Local standards or codes of practice may also impose limits on the design.

A crude observational approach worked reasonably well in the past in conjunction with steel arch mine tunnel support because this type of support maintained some capacity to high levels of deformation, at which the need for additional support became visually obvious. However with rockbolts critical levels of roof deformation at which the bolts fail, or the anchorage becomes ineffective, can be as small as 50mm or less. Bolt system failure at these levels of deformation is not always evident from within the tunnel, and consequently although rockbolts are the favoured support worldwide due to the associated production and cost advantages, their use in the past has not been associated with the highest levels of ground control safety.

This was the difficulty faced by British Coal in the late 1980s when the decision to adopt rockbolt support was made as part of a drive to modernise coal production methods and prepare the industry for privatisation. It was essential that the highest levels of safety were achieved, and to do this it was decided that a very high performance rockbolt system and a more sophisticated observational design approach should be adopted, based on Australian practice at the time.[2]

The success of this system, further developed by British Coal and now incorporated in UK coal industry guidance[3], has been well documented.[4] The successful adoption of rockbolt support resulted in large increases in mining productivity and at the same time ground control safety improved significantly. The system developed by British Coal is still in use in the five large coal mines currently in production in the UK and elements of the system have subsequently been adopted in many mines worldwide including coal mines in the USA, Canada, India, South Africa, Germany, Poland, Japan, Russia and China, platinum mines in South Africa and evaporite and stone mines in the UK.

The observational design system makes use of data from borehole extensometers and strain gauged rockbolts to confirm support effectiveness. In addition simple telltale extensometers are used at regular intervals to give visual warning of significant levels of roof movement. Additional support is installed if roof movement exceeds predefined action levels. These instruments, and subsequent developments made from them, are described in more detail below.

The rockbolt support system used by British Coal, Figure 1, was purpose developed to a specification which was subsequently incorporated in BS7861.[5] The standard also includes a specification for birdcaged cablebolts.[5] Associated laboratory testing methods for anchor systems were incorporated in the standards. Further developments of both bolting systems and test methods have subsequently been made, including non destructive testing methods to detect broken or rockbolts, and these are also described below.

Figure 1. UK Coal Mine Rockbolt System

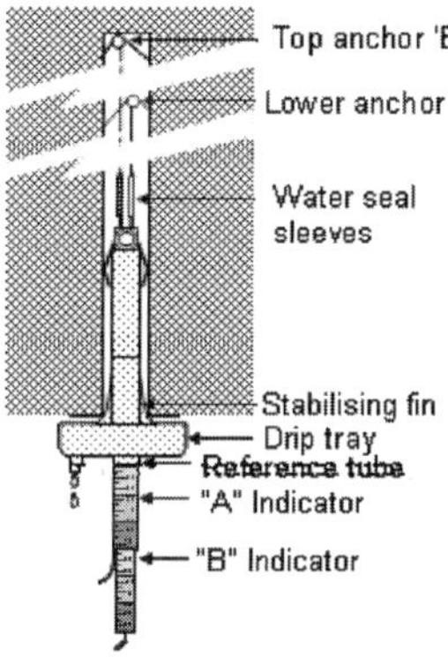

Figure 2. Dual Height Telltale

Extensometer systems

Extensometers are used to determine the magnitude, position and rate of movement of soil or rock surrounding an excavation. They are widely used in the mining industry to obtain support design information and as the basis of safety monitoring systems. Extensometers are installed into boreholes and, in mining, the smaller the diameter the better to minimise drilling costs. The essential features of an extensometer installation are a stable reference anchor position at the far end of the borehole, a hole mouth anchor at the tunnel wall and a means of indicating or measuring change in distance between them. There may be up to 20 intermediate anchor positions to determine the pattern of strain along the hole.

The simplest form of extensometer makes use of a stainless steel spring reference anchor with a tube indicator attached to it by stainless steel wire and visible at the hole mouth. Movement is indicated by coloured reflective bands on the indicator which are progressively covered as movement develops. In mining a simple extensometer such as this is known as a 'telltale' because it gives a visual indication of roof movement. A single point rotary version is used in South African coal mines. Here roof movement is converted to rotation of a pointer around a dial. This has the advantage that small roof movements can be easily read even when the tunnel height approaches five metres.

The most common form of telltale is the dual height version which is in use in coal mines in the UK and elsewhere, Figure 2. This was developed and patented by British Coal in 1992 as a safety device for coal mine tunnels where rockbolts were being introduced as support. The device is installed at the same time as the rockbolts into a 5m long roof hole of 27 to 35mm diameter. The dual height telltale uses two concentric indicator tubes attached to spring anchors by stainless steel wire, positioned so that the upper "A" indicator shows movement within the rockbolted height and the lower "B" indicator shows movement above the rockbolts. Dual height telltales are installed into the roof at 20m intervals in all rockbolted tunnels in UK coal mines and the use of these devices as the basis of a safety monitoring system is generally considered to have been a major factor in the successful conversion of the coal industry to rockbolt support. [6]

One problem associated with use of wire extensometers in mines is that shear movements within the hole length can trap the wires, resulting in false readings. Where this is a problem, it has been overcome by a simple modification. The wires are attached to the upper anchors using small springs. This allows confirmation that the wire is free in the hole by gently tugging on the indicator, which should return to the current reading position when released.

Wire extensometers have limited accuracy and consequently are unsuitable for gathering data for use in geotechnical design. This role is typically undertaken by the sonic probe extensometer, which is used in mines and tunnels worldwide. This extensometer uses up to 20 anchors containing ring magnets which are installed at intervals in the borehole around a central plastic guide tube. By inserting a flexible sonic probe into the guide tube, the position of these anchors can be read to a resolution of 0.1mm. The sonic probe makes use of the Wiedemann effect by which a mechanical pulse is generated at each ring magnet position when current is passed through the ferromagnetic probe material. One of the advantages of the sonic probe extensometer is that hole furnishings are low cost, making it particularly suitable where a number of installations are made at one site, as these can be read by one probe. However the system is limited by probe length (7.5m), and manual reading, requiring continuing access.

A new electronic extensometer system, again originally developed for mining, combines high accuracy with low cost and local or remote reading options. The basic measurement principle used is that the inductance of a coil varies depending upon the position of a ferrite rod within the coil. This has the advantage over potentiometric devices of being contact-less, requiring very low power and not being susceptible to moisture. The signal conditioning electronics are well suited to an analogue, frequency based interrogation system. Numerous forms of this electronic extensometer are available based on a single basic transponder design with a range up to 75mm at 0.1mm resolution. Transponders can be interrogated remotely or locally with a portable readout unit which also provides the power source. They can be read separately, or connected together by a single cable to a central monitoring point.

The first use of this measurement concept in tunnelling was a remote reading dual height telltale system. Up to 4 sets of 100 dual height electronic telltales can be remotely monitored, via a telephone connection, from a PC, Figure 3. The first full-scale application of the system, employing 67 telltales, was in a 1300m longwall access tunnel in a German coal mine in 2002.[7]

A closure/convergence transducer, comprising a spring loaded retractable wire in a cylindrical, plastic housing and based on the same transponder components, has been applied in stope closure monitoring projects in South African gold mines over the last four years. A "blast-proof" multi point extensometer has also been developed. This is installed entirely within the borehole, allowing it to be placed close to the face in "drill and fire" operations. A two-point version is currently in use in South African platinum mines, and was used in the Joskin Tunnel project, a bolted granite tunnel in the English Midlands, Figure 4.

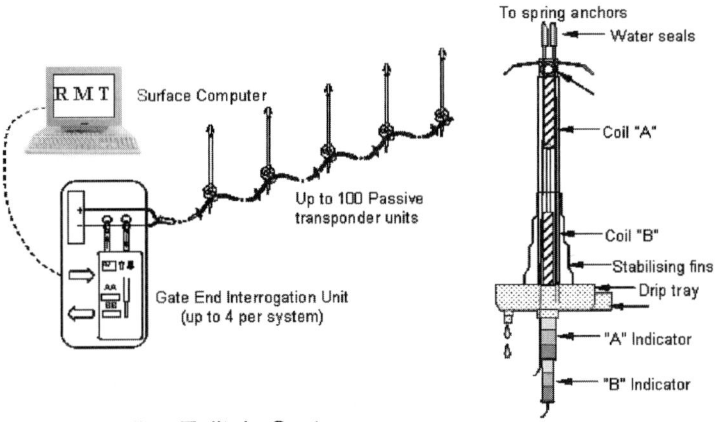

Figure 3. Remote Reading Telltale System

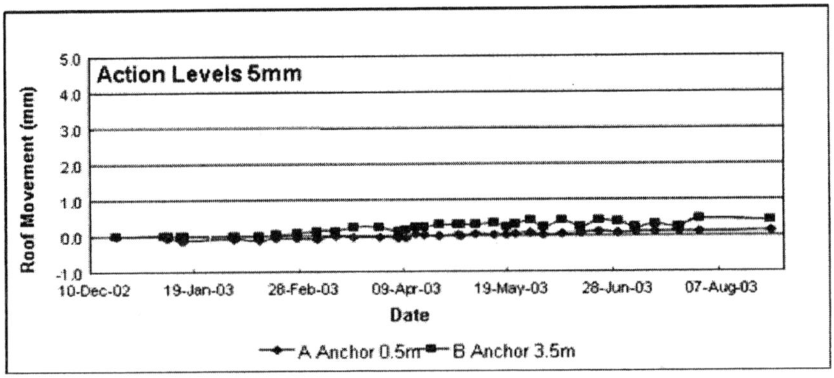

Figure 4. Results from a Blast Proof Multi-point Extensometer

A fully groutable extensometer, deploying the same measurement mechanisms within individual sealed plastic tubes grouted along the full length of the borehole, has been installed in a UK rail tunnel, Figure 5. It is envisaged that this groutable extensometer will be of particular value in civil tunnelling where full hole sealing is usually necessary and other, more fragile and expensive extensometers are currently used. Measurements are made of the extension of each tube within the hole. There is therefore no limit on the number of bays which can be installed in a single hole, the only limitation being imposed by the minimum bay length of 325mm and the hole diameter.

The standard measurement range for all the transponders is 0 - 75mm per "bay", a "bay" being the section of borehole between two adjacent anchors. This provides a measurement resolution of 0.04 mm. Transponders with a higher resolution and smaller range (0 - 40mm) can also be manufactured.

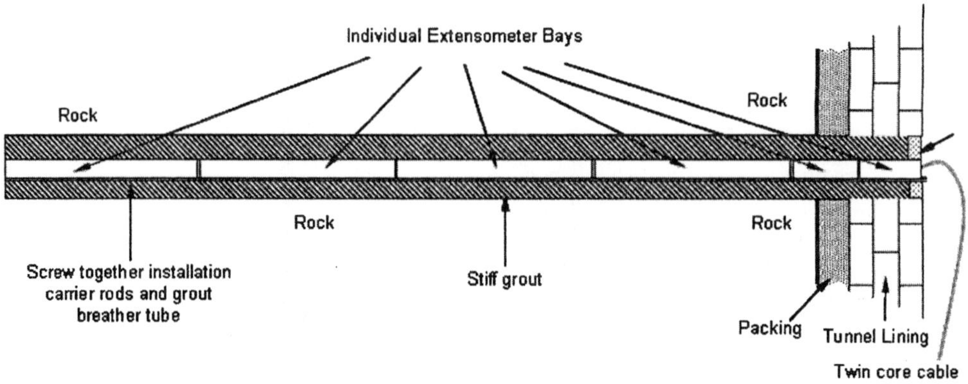

Figure 5. Remote Reading Multi-Bay Fully Groutable Inductive Extensometer

Figure 6 shows the output from the grouted extensometer in the rail tunnel over a two year period indicating total deformation of approximately 1.5mm with indication of a seasonal variation in reading also present.

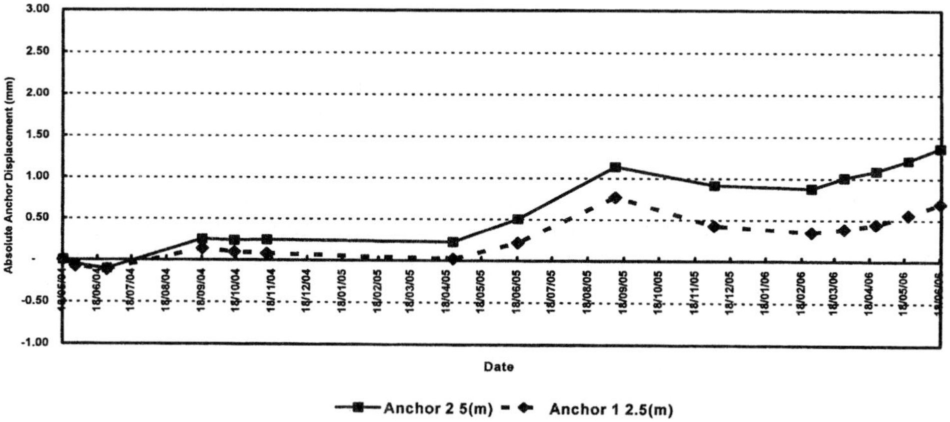

Figure 6. Results From a Remote Reading Multi-Bay Fully Groutable Inductive Extensometer installed in a Rail Tunnel

Anchor in-service load measurements and NDT

The load distribution in rockbolts and cablebolts is an important support design parameter, but one which is difficult to measure. Load cells mounted against the end plate of an encapsulated anchor, for example, provide no information on the load profile along it. To date RMT have manufactured around 4500 strain gauged rockbolts supplied to mine and tunnel projects in 9 countries. They typically have 9 pairs of diametrically opposed resistance strain gauges, allowing calculation and display of mean and bending strains. The technology has recently been extended to include flexible bolts which are encapsulated multi-wire steel strands. Figure 7 shows typical output from a strain gauged rockbolt indicating the maximum load over the central portion of the bolt, significant bending loads and a low end plate load.

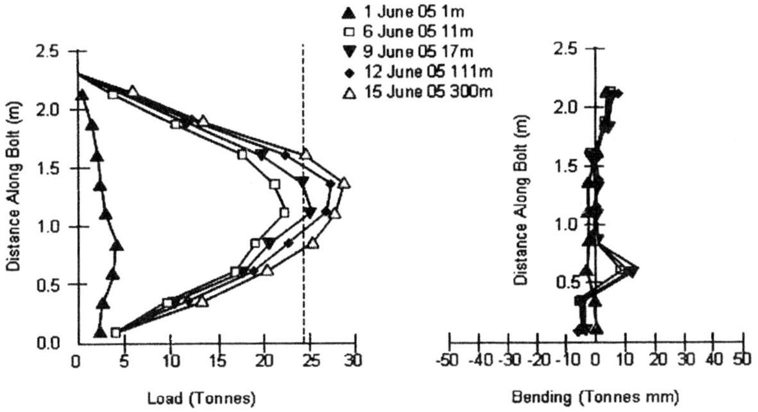

Figure 7. Typical Results from a Strain Gauged Rockbolt

Bolt and anchor loads in mining applications are typically high, deformation beyond the elastic limit being a frequent occurrence. In extreme cases this can lead to bolt breakage within the roof resulting in risk of a roof fall. Methods of detecting bolt breakage have therefore been actively researched.

Ultrasonic techniques for confirming installed bolt length and detecting bent and broken bolts have been successfully developed and used to measure bolt lengths in coal mines and in a large storage cavern. The method was able to confirm that rockbolts in a mine tunnel being recovered after 7 years submergence were still intact for example.[8] At its present state of development, the ultrasonic method however requires end preparation of bolts to allow the signal to be transmitted and the maximum length of fully encapsulated bolt giving a return signal from the far bolt end using this system is about 3m, depending upon the surrounding rock type and bolt properties.

An alternative, radio frequency technique for measuring installed tendon length, applicable to long tendons as well as rockbolts, has been developed and patented by RMT.[9] When a radio frequency signal is applied to a pair of parallel support elements, the system behaves as an antenna, with a fundamental radiating frequency depending on the mean tendon length, Figure 8. Support pairs containing short or broken elements will have a higher resonant frequency than intact ones. The technique was developed for UK coal mines to determine if rockbolts were broken insitu but has since been successfully applied to rockbolts and cablebolts in evaporite mines, platinum mines and limestone mines. It has also been used to determine rockbolt length in a large UK hydro-electric storage cavern and for tendons installed in a road cutting where some bolts had failed by hydrogen embrittlement near their outer ends.

The method requires that the support element pair are electrically isolated from one another and so in some cases it is necessary to remove steel mesh elements in contact with them. However there is no limit on the length of tendon and no need for preparation work beyond ensuring good electrical contact to each tendon and in some cases installing additional tendons of known lengths for calibration purposes.

An alternative means of detecting bolt breakage is by building in a mechanism which indicates when the bolt loses integrity. 'Sentinel' rockbolts in which an electrical circuit is

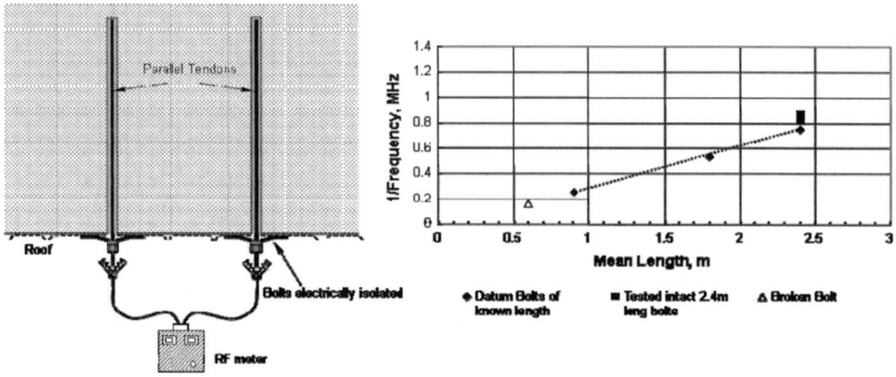

Figure 8. Parallel Tendon Method and Results from a Coal Mine

formed between the bolt and an insulated wire embedded in a groove along its length have been used for a number of years. However application has been limited by the need to manually check the circuit resistance and the relative costs of adding the circuit. In collaboration with UK Coal, RMT recently commenced work on development of an optical fibre based integrity monitoring rock bolt. The new work is based on the principle that a broken optical fibre will no longer transmit light. A special bolt which utilises this principle successfully would indicate its integrity visually without any requirement for special interrogation equipment. Two sets of prototypes have been manufactured, one based upon the transmission of light from a cap lamp and one based upon using an integral radio isotope light source. The light transmission version is shown in Figure 9.

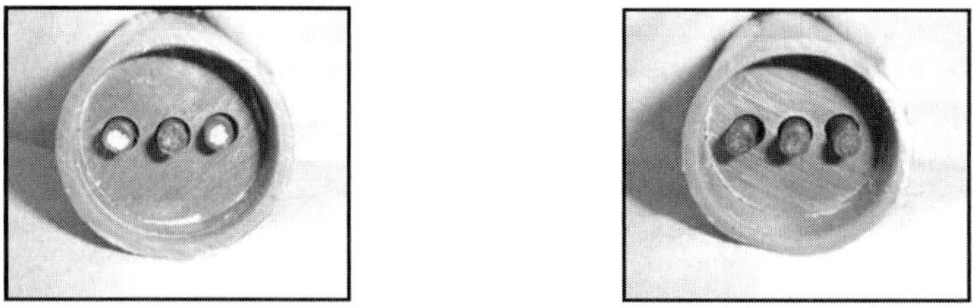

Optobolt "intact" Optobolt "failed"

Figure 9. Optical Integrity Rockbolt

Anchor performance testing and standards

The current British Standard[5] on rock reinforcement components used in coal mining was formulated as the direct result of British Coal's experience in implementing rockbolt and cablebolt reinforcement over the period 1988-94. The standard employs the axial double embedment tensile (DET) test (commonly known as a 'gun barrel' test) to determine the main system load transfer performance criteria, in terms of bond strength and stiffness. This test involves measurement of load/displacement characteristics of a tendon fully grouted into internally threaded thick-walled steel tubes. Part 1 of the Standard covers rock bolting systems and Part 2 covers birdcaged cable bolting systems. Both parts are highly prescriptive, defining a specific design of bar (AT) and cable bolt (birdcaged).

In 2000, it was agreed by the UK Rockbolt Research Liaison Committee that Part 1 of the Standard should be revised to allow further industry innovation of bolts and resins. Amongst other aspects, it was recognised that the double embedment test was limited by its inability to examine load transfer between the bolt and the host rock and the UK Health and Safety Executive commissioned an RMT research project to develop a new test to replace the DET test.

The new test, the Laboratory Short Encapsulation Pull Test, LSEPT, test measures the bond strength and stiffness of the tendon when encapsulated in a confined rock core, Figure 10.

A series of tests using the LSEPT has shown the following[10]:

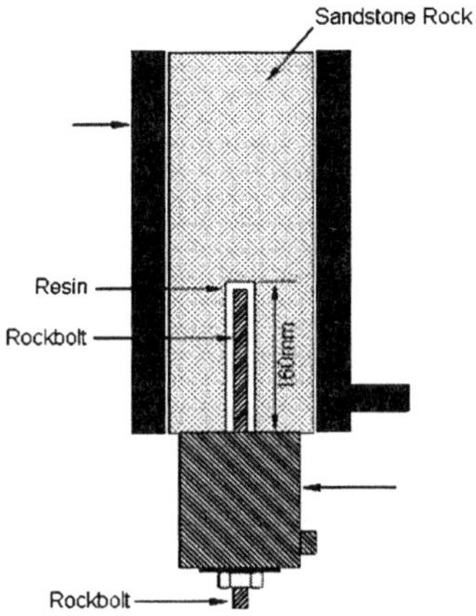

Figure 10. Schematic of LSEPT

• The LSEPT, which incorporates the interaction of the reinforcement tendon, encapsulant and hole drilled in rock, can show significant differences in system load transfer performance to a test such as the double embedment tensile test (DET) which ignores interaction with the hole wall.

• The design of a rockbolt rib profile is not only important for obtaining a good bond with the encapsulant, but also for ensuring the generation of radial forces against the side of the hole. For this reason, two designs of bolt which may behave in a similar manner in a DET can behave very differently in an LSEPT. In particular a concave profile is likely to be less efficient than a convex profile.

• Reinforcement systems which display high bond stiffness characteristics when tested using the DET may display much lower bond stiffness when tested with the LSEPT in rock.

The new LSEPT has now been incorporated into a draft revision of BS7861 Part1 which has been issued for industry wide consultation.

Part 2 of the standard is currently being revised to encompass the many types of long tendon reinforcement which are being used in the UK coal mining industry. These include flexible bolts, deformed strands and tensionable systems encapsulated with resin, cementitious grout and combinations of the two. The LSEPT is being adapted for use with long tendon systems as part of this work.

The use of the LSEPT method gives fundamental data on anchor system performance under controlled laboratory conditions which would otherwise be very difficult to obtain. The method is therefore also being used to compare the performance of a wide range of anchor systems.

Application to the Civil Engineering industry

The role of geotechnical instrumentation in both mining and civil engineering tunnelling for design, design verification, and safety purposes will continue to expand, and insitu measurement will play an increasing role in optimising support design and ensuring safety.

The devices and techniques described here were developed primarily for the mining industry to perform specific safety and design related roles associated with the use of rockbolts and other anchor systems as tunnel support. The circumstances in which they are used generally dictates that mining instruments should be low cost, quick to install and capable of use by non specialist personnel. However demand from outside the mining industry demonstrates the wider applicability of these new or enhanced measurement techniques and devices. Already the electronic extensometer, strain gauged rockbolts and the RF bolt inspection techniques have been used successfully in non mining applications and it is anticipated that this trend will continue where the new techniques have significant advantages over existing methods.

Conclusion

The role of geotechnical instrumentation in tunnelling for design, design verification, and safety purposes will continue to expand, and insitu measurement will play a major role in optimising support design and ensuring safety. A range of innovative instruments, developed initially for mining use, is being adapted for wider use. Early civil tunnelling applications have demonstrated the applicability of these new or enhanced measurement techniques, combined with easier installation and use, at lower cost.

References

1. Minova International Ltd. *The Minova Guide to Resin Grouted Rockbolts*, 2006.
2. Hurt, K. Bigby, D. and Altounyan, P., Mine Tunnel Reinforcement Design Using the Observational Method, *Int. Conf. on Tunnelling in Difficult Conditions*, Basel, Switzerland, 27-29 Oct. 1997.
3. HSE 1996, Deep Mines Coal Industry Advisory Committee, *Guidance On The Use Of Rockbolts to Support Roadways in Coal Mines,* HSE Books.
4. Hindmarsh, W.E. The Management of Rockbolting within British Coal, *2nd International Symposium Roofbolting in Mining*, Aachen, March 1995.
5. British Standard 7861, *Strata Reinforcement Support System Components Used in Coal Mines: Part 1: 1996, Specification for rockbolting and Part 2: 1997, Specification for birdcaged cablebolting*, British Standards Institution, 389 Chiswick High Road, London, W4 4AL.
6. Bigby D.N., Arthur J., Improved Rockbolting Safety Worldwide through Recent UK Instrumentation and Testing, *International Symposium on Roofbolting in Mining,* RWTH, Aachen, Germany 2001, pp463-480.
7. Bigby D., DeMarco, M., Development of the remote reading dual-height telltale system for monitoring mine roof deformation, In: *20th Int. Conf. on Ground Control in Mining, Morgantown,* WV. August 2001, pp 163-172.
8. Clifford, B. Bloor, A. Bigby, D. and Bowler, J. The use of NDT Methods to Determine the Condition of Rockbolt Support During the Recovery of a Flooded Coal Mine Roadway at Thoresby Colliery. *21st Int. Conf. on Ground Control in Mining*, Morgantown, WV. August 2002.
9. UK Patent No: 2 304 417, *Parallel Tendon Integrity Monitoring Method.*
10. Bigby D. Reynolds C. Development of a laboratory Short Encapsulation Pull Test for a Revised British Standard on Rock Reinforcement Systems used in Coal Mining. In: *24th International Conference on Ground Control in Mining*. Morgantown, USA, August 2005.

Session five:

Non-destructive integrity testing

Common anchorage issues addressed by numerical modelling

Ana Ivanović, Richard Neilson, Andrew Starkey and Albert Rodger, University of Aberdeen, School of Engineering, Aberdeen, UK

Abstract

There are a large number of anchorages that are never examined or tested throughout their lifetimes. This has prompted the development of a non-destructive anchorage integrity testing system, GRANIT (GRound ANchorage Integrity Testing). More recently a dynamic model has been developed allowing the different parts of the anchorage to be examined to identify their influence on the overall dynamic response, as measured at the top of the anchor bar. The model complements the GRANIT system and allows better understanding of the dynamic behaviour of anchorages. The model has been used to tackle several common issues that occur in modern anchorage practice including load estimation, debonding of the fixed anchor length, estimation of anchor free length, the influence of the anchor head influence and more recently anchorages placed in soil. This paper will review the use of the model in tackling these issues.

Introduction

The non-destructive testing system, GRANIT has been developed with the aim of assessing the integrity of anchorages through their vibrational response [1]. In the early days of GRANIT the main features of the system were the impact device, attached at the end of the protruding length, by which an impulse of a small amplitude is applied to the anchorage, a laptop, into which all the data are transferred and stored, and an Artificial Neural Network (ANN), in which, in the majority of cases, the relationships between the input and acceleration response are established. A schematic diagram of the original system is shown in Figure 1 (without the dynamic model shown with dotted lines).

The vibration response, captured by an accelerometer, from the anchorages after installation or at any other point thereafter indicates changes in the anchorage but initially the cause of these changes could not be identified. This highlighted the need for a model in which the main features of the system could be examined separately. To this end, a numerical model of ground anchorages was developed and this has been in use as part of GRANIT since 2001.

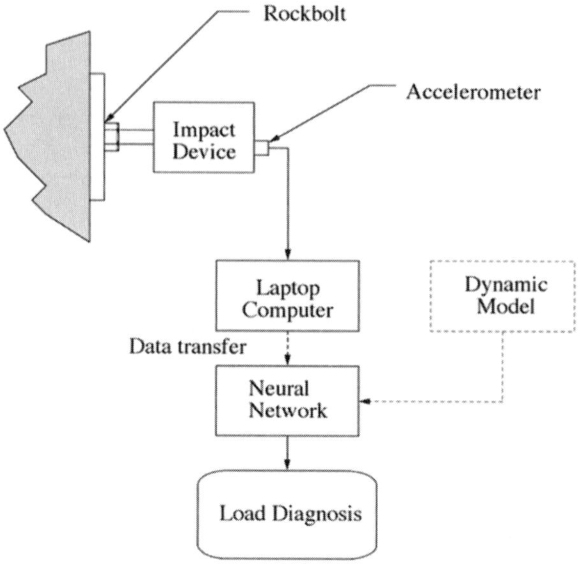

Figure 1 Schematic diagram of the GRANIT system

The usual procedure for the neural network in the GRANIT system to be trained is to obtain data for a variety of load levels from site anchorages that have been destressed and stressed again or from newly installed anchorages of the same geometry as the ones to be tested. This procedure is usually time consuming and in some cases clients are unwilling to allow existing anchorages to be destressed for safety reasons. With the development of the model this can be avoided if the geometry of the anchorage, grout and rock properties are known. The neural network can be effectively trained on data from the model and the system that operates in accordance with Figure 1 with the dynamic model shown with dotted lines.

Model

A schematic diagram of the lumped parameter model developed is presented in Figure 2. The bar and rock mass are represented with masses, springs and dashpots while the anchor head and the bonding between the bar and grout are represented with non-linear springs and dashpots. During the development stage the material properties, required for the model to allow validation with experimental data, were obtained from a series of laboratory experiments undertaken at the University of Aberdeen. [2]

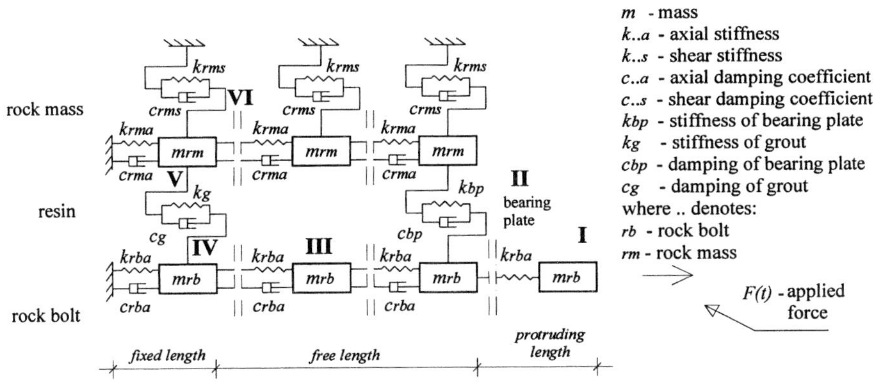

Figure 2 Lumped parameter model [2]

The main features of the model, have been reported elsewhere [3,4]. The model has been used to investigate several common issues that occur in modern anchorage practice including load estimation, debonding of the fixed anchor length, estimation of anchor free length, influence of the anchor head, and more recently the response of soil anchorages. The influence of the model related to the issues listed above will be covered in more detail in the rest of this paper.

Load detection

Estimation of the load carried in bolts and strands is currently undertaken by means of pull out and lift off tests respectively, methods proven to be expensive and potentially harmful to the anchor head and corrosion protection measures. The GRANIT system, as a non-destructive technique provides load detection while producing no damage to the anchorage itself.

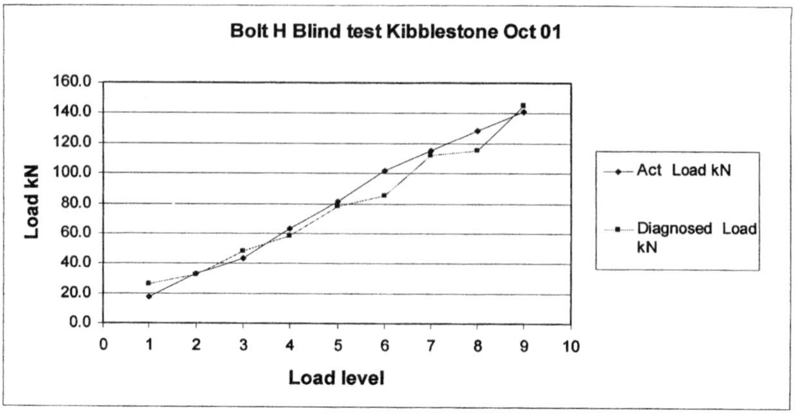

Figure 3. Comparison between the actual load and the load diagnosed using the GRANIT system trained on data produced by the lumped parameter model

The GRANIT system originated when it was noticed during blasting of tunnels at Penmaenbach and Pen y Clip in Wales that the dynamic response of the instrumented rock bolts installed in the tunnel depended not only on their geometry/construction but also on the load they were carrying [5]. In order to understand the cause for this variation in the frequency response at different load levels a series of experiments were undertaken on a specially constructed anchor head rigs with different anchor bore holes and bearing plate dimensions. All the experiments showed a non-linear relationship between the loads applied and the displacements obtained. When this feature was introduced into the model the change in frequency response with a change in load coincides with the results obtained from the experimental and field trials [4]. This feature of the model has now been included in the GRANIT system.

Two test and development sites in England were chosen by AMEC in order to assess the suitability of GRANIT to check the integrity of mining support bolts. The first site used was the disused concrete bridge deck in Swynnerton where 12 2.4 m long M24 mining type resin bonded rock bolts were installed under controlled conditions. Usually 10 samples were taken at each post tension level, the signals of which were shown to be consistent. Typical diagnosis of load shows a good agreement between the actual and trained data. Another set of test of anchorages of the same dimensions was then undertaken at Kibblestone but this time the bolts were placed in a natural weathered sandstone face. Neural networks trained on data from mathematical models of the datum stratabolts successfully diagnosed loads from testing of the stratabolts with unknown load at Kibblestone, (Figure 3) thus confirming and validating the mathematical approach.

Following the successes at Swynnerton and Kibblestone, tests were undertaken in the UK Coal's Thoresby Colliery where 264 rock bolts (strata bolts) were tested in total in two different location in mines, determined by the age of the bolt (30 and 2 months). The results obtained from the datum tests were deemed a success with the same trend between the frequency and load being determined. Thereafter, once an estimate of the distribution of free lengths of the rock bolts was obtained for the site, the model was used to simulate a number of scenarios with different free lengths and loads. Using the frequencies obtained from these, the statistical likelihood of a particular load being present in any bolt could be assessed. This data was presented to the client as a means to assess the effectiveness of the bolting system and to highlight any anomalies or exceptions. Analysis of the data showed a marked difference in installation or loading between areas of the mine as well as identifying a number of bolts designated as "exceptions". The model was used to estimate the unbonded length of these bolts and all three were found to deviate drastically from the installation specification. Subsequent exhumation of the bolts showed that all had either failed or been poorly installed.

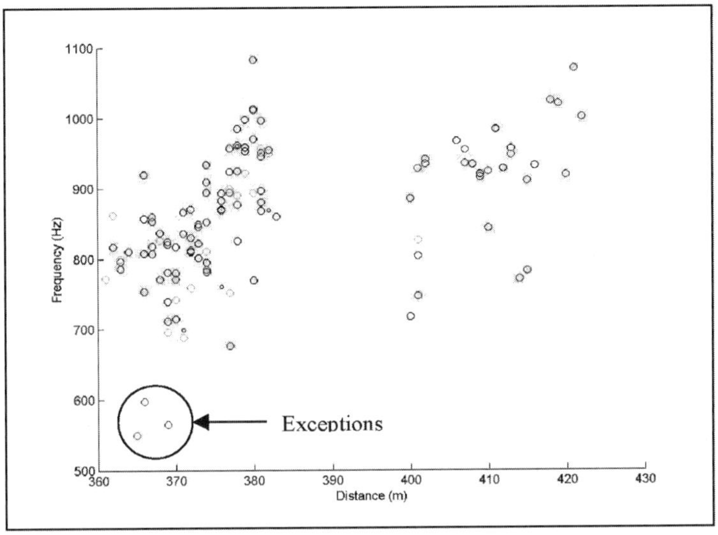

Figure 4 - Frequencies of bolts tested in Thoresby Colliery.

Anchor head

With further investigation using the model it was found that the geometry and configuration of the anchor head assembly (i.e. bearing plate and nut or wedges and barrel for rock bolts and strands respectively) determines the capability of the GRANIT system to detect the load.

The initial investigative field tests were conducted on two strand anchorages installed at Stone in 2001. The geometry of both anchorages was nominally the same with a fixed, free and protruding length of 6, 4 and 1m, respectively. During the first set of tests the stressing jack and chair, used to prestress the anchorages (Fig 5a) was left in position while the second set of tests was undertaken without the stressing jack and chair (Fig. 5b). The results of the first set of tests were encouraging showing an increase in the natural frequencies with an increase in load. In contrast the second set of tests showed only one peak in the frequency range, which did not appear to increase appreciably with increasing load level. This causes difficulties, since using frequency analysis for the successful diagnosis of load, requires the presence of a change in frequency response with increasing load level. The reasons for the result were believed to be due to the thickness of the plate and the same head configuration was introduced into the laboratory anchorage for investigation. In this case the GRANIT system again did not detect any change even when the strand was loaded up to 59 kN.

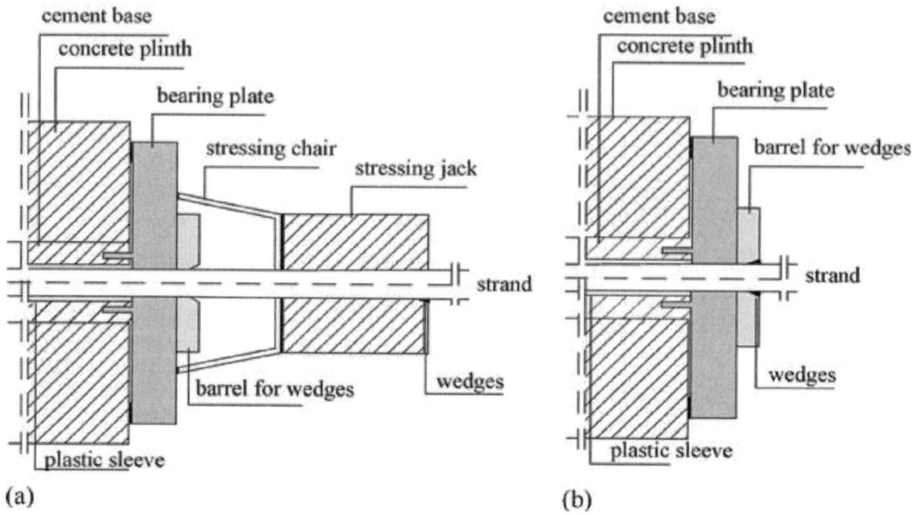

Figure 5. Schematic diagram showing anchor head configurations (a) with the stressing chair and jack in place (b) when stressing chair and jack were removed. [2].

The results obtained suggest that the difference in frequency responses between the anchorages tested with the stressing jack and without the stressing jack in place during field and laboratory tests is primarily due to different anchor head configurations which in turn gives different stiffness characteristics. From Fig 5a it can be seen that the anchor head assembly with the jack in place includes more interfaces between components, hence it has higher flexibility. The anchor head assembly without the jack in place is basically a thick bearing plate fixed on the concrete surface and is therefore much stiffer.

An investigation was carried out on the relative stiffnesses associated with two anchor head configurations and it was concluded that stiffness of the anchor head configuration shown in Fig 5b was ten times higher than that shown in Fig 5a. The effect of such stiffness changes was modelled and Fig 6 shows the relationship between the frequency change and stiffness ratio obtained. When the stiffness ratios for the two different head configurations were compared with Fig 6 it was found that the stiffness ratios for the flexible anchor head falls in the range where a significant frequency change is present while the stiff anchor head falls in the stiffness ratio range where frequency changes are not appreciable.

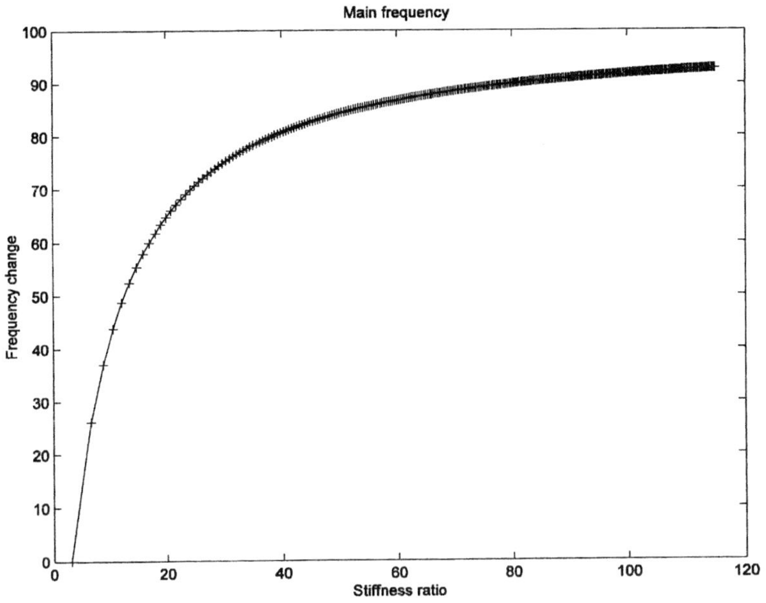

Figure 6. Frequency change vs. stiffness ratio relationship obtained from the model

These model results showed that if the anchor head configuration is too stiff and therefore does not allow much deformation of the bearing plate, load detection is difficult to achieve. On the other hand if nonlinear flexibility of the anchor head is present, load detection is achievable in most cases. This important finding determines whether the load can be detected using GRANIT technique or not. This led to the classification of Class 1 and Class 2 anchorages determined by the configuration of the anchor head. Class 1 heads are those amenable to diagnosis by GRANIT [6]. Currently many multi-strand anchorages are Class 2. However, recent investigations indicate that with a nonlinear elastic element introduced into the anchor head assembly the problems inherent in Class 2 heads may be overcome by providing the flexibility required into the system. Work is ongoing to ascertain the optimal form of this modification.

Debonding of the fixed anchor length

Once anchorages are installed little is known about the condition of their fixed anchor length - whether the bond is as designed or whether any cracks have occurred along the fixed anchor length. All these features can alter the designed fixed length and therefore the load capacity and free length which anchorage practitioners are concerned about. With the analysis undertaken by the model it has been shown that if the bond characteristics of the grout at the top of the fixed anchor length (local debonding) or along the entire fixed length (poor overall installation of the anchorage) are not as specified in the design, changes in the vibrational response are very likely to occur. Newly installed anchorages are particularly prone to initial debonding of their proximal end, a phenomenon which is confirmed by field experiments [2]. Initially this debonding was modelled linearly and the debonding length was defined by changing the stiffness properties of number of masses determining the length. This required changing manually the properties of the grout for all the masses concerned in the code of the model. This inconvenience however has been overcome more recently with the inclusion of a non-linear model for the grout where an appropriate interface model can be chosen for the particular geometry and material properties of the bolt and resin.

Field experiments undertaken on a newly installed bolt suggest that with increased load debonding of the top of the fixed anchor length is very likely to occur. [2] The model has shown that with an increase of load debonding of the top end of the fixed anchor length is possible if the properties of grout are non-linear. Figure 7 shows the effect of debonding on the dynamic response of an anchorage. The figure shows the natural frequency before stressing, at 136kN and after destressing back to zero. The results show a clear shift in frequency after destressing, caused by the increase in free length due to debonding. The figure also shows the frequency that would be achieved if the grout were perfect. It can be seen clearly that this is higher than is shown in the debonding model.

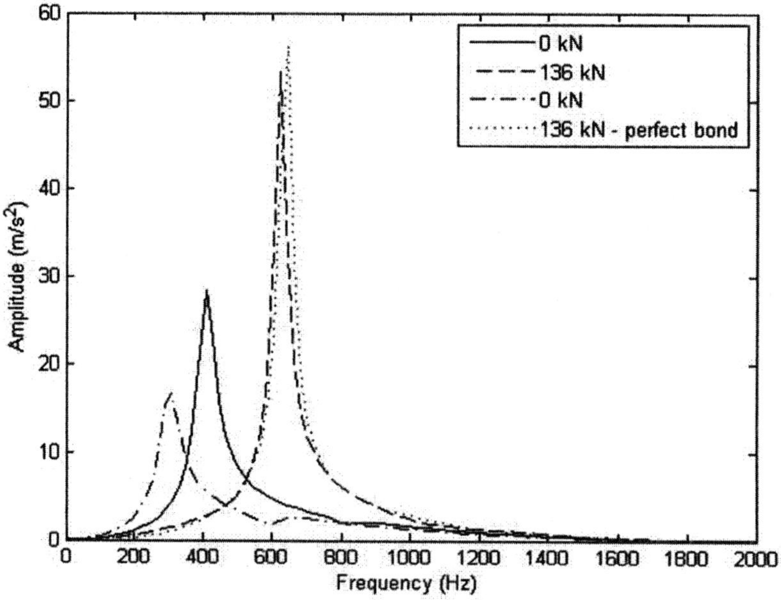

Figure 7. FFT of the free vibrations of the first mass after the impulse for different load levels

Anchor free length

Debonding is closely linked to the actual free/fixed anchor length, an issue of concern to a number of practitioners. It is even more critical when anchorages are to be assessed by the GRANIT system, which relies on an accurate geometry before being able to give any load diagnosis.

This was highlighted when the GRANIT system was used to diagnose the data obtained from cable bolts in the North Parkes hard rock mine in Australia during September 2003 as part of a project concerned with the construction of a crusher chamber required to extract deeper reserves down to 900m below the surface.

According to the North Park Management construction specification 10 m long cable bolts with a 3m de-bonded length were installed to support the crusher chamber roof. In addition 6m long cable bolts with a 2 m de-bonded length were installed in the chamber walls. These bolt geometries were given to the GRANIT team and were then incorporated in the model and the GRANIT system. The results obtained suggested that the free length of the 10 m long bolts tested was substantially different from the geometry supplied initially by the mine practitioners/contractors. An investigation using the model showed that these inconsistencies disappeared when an unbonded length of 2m was applied to all the bolts tested.

After further examination of the "as installed" design documentation it was found that the geometry found by the model i.e all free lengths found to be 2m, was correct. Subsequent re-analysis of the data using the revised geometry in the model provided the mine with useful data for assessing the integrity of the chamber roof. [7]

This shows the importance of the geometry information for as accurate as possible load diagnosis.

Further development

The application of the GRANIT system for rock anchorages is being expanded to develop a technique for detecting the integrity of soil anchorages. In order to understand such systems further development of the existing model and geotechnical centrifuge modelling were used. In order to replicate the GRANIT system within the centrifuge, an in-flight centrifuge robotic manipulator was designed to prestress the anchorages and to apply an impact to the anchor heads. (Figure 8)

A preliminary study, using anchorages in dry non-cohesive material shows that there is a change in anchorage response with an increase of load, which is encouraging since this is the initial requirement for deployment of the GRANIT system. A numerical model for soil anchorages, based on the model used for rock anchorages, has been developed and the results obtained so far suggest that there is a similar trend regarding the load behaviour as found in anchorages installed in rock. Changes in the natural frequencies with load are however of a much lower range than found in rock bolts. The aim of future work is to use the model to optimise the response to provide as much information as possible from the anchorages.

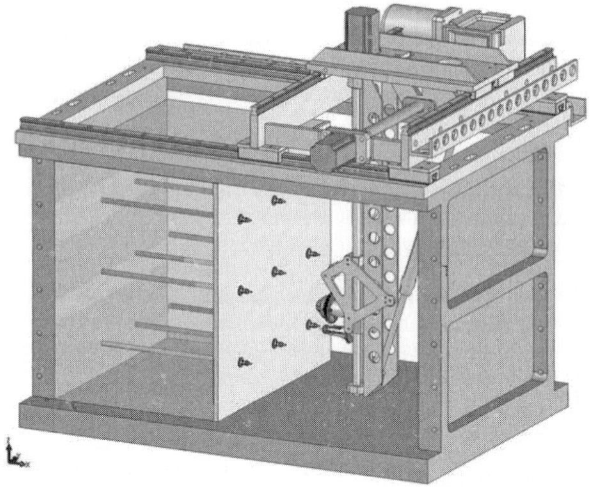

Figure 8. In flight centrifuge robotic manipulator [8]

Conclusion

The development of the dynamic model provides a better understanding of how changes in parameters of materials and geometry affect the dynamic response of ground anchorages.

Several common issues that occur in modern anchorage practice have been addressed using the dynamic model and the following conclusions can be drawn:

- Load estimation can be achieved if the anchor head properties and free length are known
- Inclusion of the debonding features in the model shows the change in frequency found in the field tests when anchorages are initially stressed.
- Estimation of the anchor free length can be used to verify the geometry and data provided by the clients
- Presence of high anchor head stiffness may make load estimation difficult. Different elastic elements are currently being considered to overcome this issue.

The main advantage of the introduction of the model into the GRANIT system has been using the data provided by the model to train the neural network, rather than going on site and getting training data. This in turn saves both time and expense.

Acknowledgements

The authors wish to thank Amec, Rio Tinto and UK Coal for supporting aspects of this work and permitting the use of the data presented.

References

1. EP1324031 *"Ground Anchorage Testing System"*, 2003, granted to the University of Aberdeen and University of Bradford, Inventors: A. A. Rodger G. S. Littlejohn, R. D. Neilson and J Penman
2. Ivanović A, 2001 *The dynamic response of ground anchorage systems,* PhD Thesis, University of Aberdeen.
3. A. Ivanović, R.D. Neilson, and A.A. Rodger "*Lumped parameter modelling of single tendon ground anchorage systems*", Geotechnical Engineering 149 (2), 103-113, 2001.
4. A Ivanović, R.D. Neilson, and A.A. Rodger, "*Influence of Prestress on the Dynamic Response of Ground Anchorages*", Journal of Geotechnical and Geoenvironmental Engineering, 128(3), 237-249, 2002.
5. Rodger et al. The behaviour of resin bonded rock bolts and other anchorages, subjected to close proximity blasting. *Int Cong. On Rock Mechanics*, Tokyo, Japan, 1996, pp665-670.
6. A Starkey, A Ivanovic R.D.Neilson, A.A.Rodger "Condition Monitoring of Ground Anchorages by Dynamic Impulses: GRANIT system" Meccanica 38:131-142, 2003.
7. Neilson, R. D., Rodger A.A., Ivanović, A, Starkey, A. J., Chappell M., Jackson B. H. and Ritson A., "Rock/Cable Bolt Integrity Testing using GRANIT – Recent Field Applications", 5th International Colloquium on Roofbolting in Mining, Aachen, 2-3 June 2004, p523-528.
8. Hao J, Neilson R.D., Ivanović A, Rodger A.A., Starkey A.J., Li J., Davies M.C.R. 2006 "Model Testing to assess a novel technique for condition monitoring of soil anchorages" International Conference on Physical Modelling in Geotechnics, Hong Kong, 4-6 August 2006.

Non-destructive testing of deep mine anchorages

Andrew Starkey, Richard Neilson, Ana Ivanović and Albert Rodger, University of Aberdeen, School of Engineering, Aberdeen, UK

Abstract

Anchorages are used throughout the coal industry for supporting tunnels leading to and from the active coal face. A large scale non-destructive testing programme using technology developed at the University of Aberdeen was undertaken at an underground coal mine on 264 anchorages in total. The paper discusses the effective free length results and the subsequent detailed statistical analysis of the diagnosis. A number of different scenarios for the anchorages were developed based on their specific attributes, with each scenario comprising a newly developed mathematical model of the anchorage system. This analysis indicated that a number of anchorages had failed. Subsequent exhumation of the anchorages identified by this process by the mine operator confirmed this conclusion. In addition, analysis of the effective free length diagnosis shows that at the time of testing some of the anchorages did not fit the design specification at installation. This has important implications for similar anchorage installation in other environments and a number of explanations for this difference are given.

Introduction

It is widely recognised that non-destructive testing methods for ground anchorages need to be developed as a high priority [1], with only between 1-5% of anchorages currently being monitored in service [2]. Anchorages are also used throughout the coal industry, the tunnelling industry and the civil engineering industry as a whole as means of support and a similar need and similar numbers of anchorages tested apply. Previous work [3-12] conducted by the University of Aberdeen had shown the potential for a non-destructive testing method for anchorages, based on the dynamic response of an anchorage to an applied tensile load. A large scale testing programme using this technology was undertaken at an underground coal mine on 264 anchorages in total. The main goal of the study was to examine the condition of anchorages in different parts of the mine, and of different ages. This paper studies the results of the tests that were undertaken, and shows that a number of anchorages were easily identifiable as being different from the rest. A mathematical model of the anchorage system was developed which could be configured for different characteristics of the anchorage system. This allowed different scenarios to be investigated and to examine their probable effect on the frequency response. Finally, a study of the effective free length (efl) of a number of anchorages was undertaken, where the effective free length is defined as being the length of the anchorage at the proximal end that is free to vibrate to an applied impulse and is not fully secured to the surrounding rock mass The results from this study showed that the effective free length was outside the expected value, as indicated by the original design specification of the anchorages. The effective free length may therefore be different to the measured free length if, for example, the resin does not fully encapsulate the tendon. A discussion of these results will be undertaken, with the implications of these results for other types of similar anchorages.

The bolts used in the mine at the time of the tests are known as AT (or "Australian Type") bolts and are used throughout Australia and in the UK. They are typically supplied in various lengths with M24 thread, and are anchored with a two-speed resin bonding system enabling a pre-load to be locked in during the standardised installation process. This process is commonly undertaken using an automated continuous bolting machine, enabling maximum production of coal extraction from the coal face. The bolts installed at the mine were all 2.4m fully encapsulated AT bolts, with a domed Exchem anchor plate resting against a strap that reached horizontally across the roadway. The strap and road way are a maximum of 6m wide (common width of 4.8m) and typically contained six to eight anchorages (depending on the width of the roadway), with the straps positioned a maximum of 1m apart.

Data capture

Technology developed at the University of Aberdeen was used to undertake non-destructive testing of the anchorages in the coal mine. The process involves attaching a specially designed hand held impact device to the end of the anchorage using an inbuilt clamping mechanism that ensures good contact with the anchorage, regardless of the state of corrosion on the surface of the anchorage. A schematic of the system can be seen in a sister paper [13]. This impact device uses compressed air (freely available in the coalmine) to produce a repeatable, controllable impact. The impact device is controlled by a laptop PC, which fires a solenoid valve that controls the flow of air to the impact device. The laptop also records the subsequent acceleration response as measured by an accelerometer that is mounted on the impact device itself. This equipment has been tested above ground on many thousands of tests and has proven itself to be reliable and capable of repeatable impact tests. Time traces of typical tests from the same anchorage could be plotted on top of each other with very little difference seen between tests, showing extremely consistent responses in the time domain. Proving trials of the system on anchorages of the same type as installed in the mine were undertaken above ground at a specially designed test site and accurate diagnosis of these test anchorages was achieved. A typical test run on a single anchorage involved the laptop system recording 20 consecutive impacts of the impact device, which typically could be undertaken in about one minute. The total number of anchorages that can be tested in a single session is therefore largely determined by the accessibility of the anchorages to be tested. Due to the consistency of response of the system under normal conditions, this number has been found to be sufficient to validate the consistency of the data in terms of its time and frequency content.

Mathematical model

In parallel to the testing program, a mathematical model of the anchorage system was developed. This has been reported on more fully elsewhere [9] and a sister paper [13] describes the model in more detail, and for more information regarding the mathematical model the reader is directed to this sister paper.

The mathematical model had been fully tested prior to the tests undertaken at the mine and extremely accurate results in terms of the frequency response of the anchorage system under a range of loads, and for a range of effective free lengths had been obtained. The use of the mathematical model allows certain characteristics of the anchorage system to be altered (such as fixed/free length, diameter, borehole diameter, anchorage head type) and for the subsequent change in the frequency response of the anchorage to an applied impulse to be determined. This allows a profile of the expected frequency response of the anchorage system to be built up before any tests are even undertaken, on the assumption that the details of the anchorage system are correct. This may seem an obvious assumption but previous trials at other sites have shown that records of anchorage installation are not always kept or different anchorage types (typically a different length) can be installed in place of the recommended design. The mathematical model also allows any differences in expected frequency response to actual frequency response to be investigated for any individual anchorage, by modifying characteristics as described above for the anchorage to see if these changes result in a frequency response more closely matching that obtained in the field. Furthermore, differences in frequency response between anchorages can be examined, and this aspect is particularly relevant for the anchorages tested here.

It should be noted that the mathematical model has been able to identify the dominant characteristics that affect the frequency response of an anchorage to an applied impact – the efl, the load level and the anchorage head characteristics. Unfortunately the efl and load level affect the same frequency range, and so cannot be calculated independently of each other. However, if one is known, or can be estimated, then the other can be determined. Current and future research work undertaken by the University of Aberdeen is looking to address the independent measurement of load and/or efl. In addition previous work has shown that if the anchorage head configuration does not give a non linear stiffness response curve in relation to the applied load then there will be little or no change in frequency with an increasing load and therefore the load in the anchorage cannot be determined. Fortunately, the anchorage head characteristics for the majority of cases in mines is suitable for load diagnosis.

A large number of tests were undertaken on anchorages installed above ground in a similar methodology to that used underground. These tests were used to calibrate and fully test the mathematical model. Previous studies in concrete and sandstone have shown that the same anchorage in these different types of rock mass will still give broadly the same frequency profile and that the diagnosis of the load of the anchorage is still possible. Therefore, although the anchorages above ground were effectively installed in large concrete blocks and not similar to the potentially varied rock strata in the mine, the results from these tests could be used for comparison with the tests undertaken below ground.

Once the calibration of the model has been completed, the main characteristics that will be investigated are the load level of the anchorage and the effective free length of the anchorage. It is known that the anchorages will possess a load level immediately after installation due to the preload installed during the installation process, and also that the load of the anchorage will change due to movements in the rock mass underground, increasing the loading in the anchorages. However it was expected that the effective free length of the anchorage would be consistent across all anchorages in the mine since best practices employed by the mine ensure that exactly the same procedure for installing the anchorages is used throughout the mine. This standardised installation process uses the same amount of resin (normally two resin filled cartridges that are mixed when the bar is spun into the borehole that has just been drilled) and as mentioned earlier a bolting machine that should result in extremely consistent bolting conditions.

Results

Effective free length distribution

The first step in diagnosing the condition of the anchorages in the mine was to ensure that the effective free length for all anchorages was consistent, and within a narrow range. This is to ensure that the groups of anchorages being tested are broadly similar to each other and also to establish the consistency of the installation procedure. The frequency response of the anchorage system can change considerably for different effective free lengths, particularly if the effective free lengths are short (for example 200mm). This can therefore make the diagnosis of load difficult, particularly since these two characteristics affect the same frequency range of the dynamic response of an anchorage. However, if the effective free lengths are broadly similar then a comparison between different groups of anchorages can be made using standard statistical techniques.

Tests were conducted on newly installed bolts with the locking nut of the anchorage loosened so that there was no load active on the bolt from the plated end. The nut had to be loosened as the anchorages are installed with a preload during the installation process. With the load at zero, the dominant characteristic governing the anchorage's response to an applied impulse is its effective free length, which can therefore be estimated directly from the impulse tests.

The tests were divided into three sections, which correspond to the three distinct chainages of the roadway that were used for these tests. These sections are described in Table 1.

Table 1: Sections for effective free length tests

Section	Chainage (m)	Number of tests
1	310-320	8
2	360-369	12
3	370-381	10

Analysis of the spread of frequencies for the anchorages in these three sections of the roadway can be seen in Figure 1.

Figure 1 shows the analysis of the comparison of medians of the effective free length frequency analysis for the three sections. The tops and bottoms of each "box" are the 25th and 75th percentiles of the samples, respectively. The comparison of medians in this way can be thought of as a visual hypothesis test, and it can clearly be seen that the first two sections are similar to each other but different from the third section.

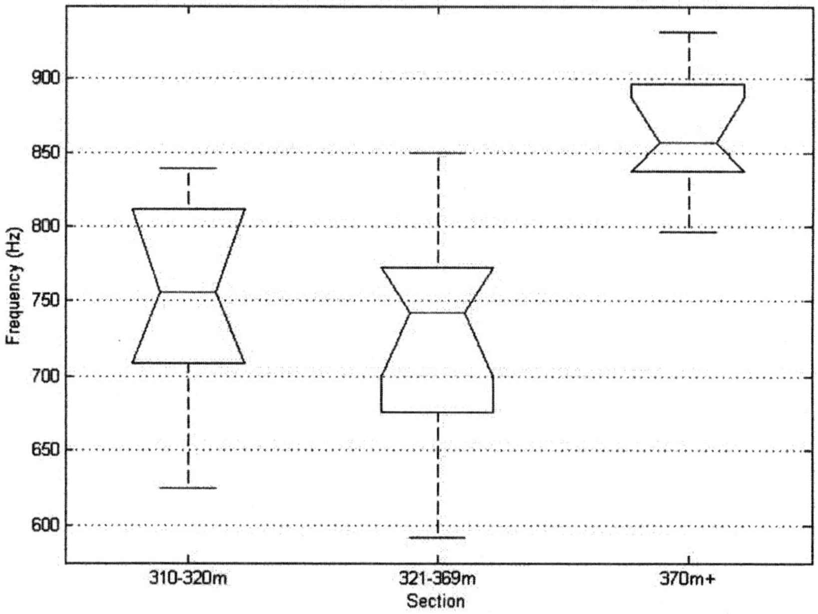

Figure 1: Median and range analysis of three sections in mine showing median, with the box showing the 25th and 75th percentiles of the samples respectively. The "whiskers" show the range of the remaining values.

A multiple comparison of means can also be undertaken at a 95% confidence level which allows the simultaneous analysis of all results to identify which sections are different and how significant this difference is. This analysis confirms that statistically Sections 1 and 2 can be thought of as being drawn from the same distribution. However the results clearly show that Section 3 is different from Section 1 and Section 2, with the means of these two sections when compared with Section 3 calculated to be significantly different at a confidence level of 95%.

This hypothesis can be tested further using the Kolmogorov-Smirnov goodness-of-fit test which compares two distributions and returns a value whether or not the two distributions have been sampled from the same underlying distribution. This test accepts the samples from Section 1 and 2 as being from the same underlying distribution with a significance level of 0.05, but rejects the hypothesis of Section 3 being from the same underlying distribution as Section 1 or 2 using the same significance level.

This is strong evidence that the tests undertaken in Section 3 exhibit a different frequency response than in the other two sections of the mine. This would indicate that the effective free lengths for Section 3 are different than in the other two sections of the roadway, since this is the dominant factor in the frequency response of anchorages under no loading.

This is an important result, since it shows that the anchorages in adjacent sections of the roadway are significantly different in their effective free length. The analysis shows that the average effective free lengths for Sections 1 and 2 are longer than Section 3. This is despite the fact that the same methodology is used throughout the mine, and emphasis is placed on the anchorages once installed having the same characteristics. This process is also to ensure that the design specification for the anchorages in the mine is met consistently, to ensure continued support of the mine roadways and for the Health & Safety of the workers within the mine itself.

This difference in effective free length is perhaps indicative of a change in installation procedure for the anchorages (for example, a different drilling machine), a change in ground conditions (for example, more fractured rock at the proximal end of the bolt, resulting in a change in free length), longer hole diameter or longer drill rod.

The analysis also shows that Sections 1 and 2 have means of frequency analysis that are statistically consistent with each other. This indicates that their effective free lengths are also similar and the average frequency relates to an efl of 330mm for these two sections of the roadway. It should perhaps be noted that Section 3, with an average efl of 197mm, is closest to the design specification of the anchorages, since the specification requires that they are installed to be fully encapsulated with minimal effective free length. Therefore Sections 1 and 2 are actually furthest from the design specification for these anchorages.

The analysis of the spread of data in all three sections of the roadway test areas also indicates that there is a relatively large variation in the effective free length of the anchorages, notwithstanding apparent full encapsulation. The actual effective free length of the anchorages tested was calculated using the mathematical model, and this can be used to estimate the spread of effective free lengths for the anchorages in these two sections. This analysis can be seen in the tables below, which give an indication of the spread of effective free lengths and the percentage of anchorages that possess these effective free length ranges. In other words, these tables give the percentage likelihood of any randomly chosen anchorage in each section possessing any particular effective free length. In view of these likelihoods, it is appropriate to recommend testing of all bolts routinely after installation, as a quality assurance measure. In the calculation of these figures Sections 1 and 2 have been integrated together since the earlier statistical tests have shown that they are equivalent.

The differences in probability and frequency ranges between Sections 1 and 2 (Table 2) and Section 3 (Table 3) are due to the different spread of data for the frequencies for these two sections of the roadway.

It should be noted that it is likely that these ranges and likelihoods will change when more data (in terms of more tests for effective free length) becomes available. For example, the free length distribution for Section 3 has been based on only 9 bolt tests, and may not representative.

Notwithstanding the less than ideal quantum of effective free length data, it can still be seen that there is a significant difference between Sections 1, 2 and Section 3 in terms of their effective free lengths, and that the anchorages in these two adjacent parts of the roadway are not consistent (in terms of their effective free length).

Table 2: EFL scenario analysis for sections 1 and 2 (310-320m and 360-369m)

Scenario	Range (Hz)	Mean (Hz)	Mean efl (mm)	Likelihood (%)
1	0-657	600	508	11.3
2	657-707	682	390	23.0
3	707-757	732	330	31.3
4	757-807	782	282	23.0
5	807+	850	226	11.3

Table 3: EFL scenario analysis for Section 3 (370m+)

Scenario	Range (Hz)	Mean (Hz)	Mean efl (mm)	Likelihood (%)
1	0-807	782	274	2.7
2	807-857	832	231	23.3
3	857-907	882	197	47.9
4	907-957	932	171	23.3
5	957+	982	146	2.7

Frequency analysis for tested anchorages

The results of the effective free length diagnosis show that the anchorages do not possess a consistent effective free length. In addition the results show that the range of effective free length values is too large for each section in the mine to allow a single average effective free length to be used for diagnosis purposes. For the purposes of diagnosis of load or exception testing (where anchorages significantly different from the rest are identified) a number of different scenarios are required, as outlined in Table 2 and Table 3.

To begin with, exception testing of tested anchorages is undertaken, to identify anchorages that have a significantly different frequency response than those around it.

Figure 2 shows the analysis of all of the anchorages tested. As can be seen in the figure, the anchorages were tested in two main areas, relating to Section 2 and beyond Section 3 in the earlier tests, and which appear to show similar characteristics. The figure is shown with a dotted line representing two standard deviations from the mean of the frequencies in these two areas, and can be used as a guideline for identifying outliers in the data. Where the frequency is very low (below the dotted line around 600Hz), this indicates an extremely long efl (up to the half of the length of the anchorage itself) and low loading of the anchorage.

There are therefore a number of exceptions, or outliers, in the data that can be seen below the dotted line representing two standard deviations from the mean, with three between 365m and 370m along the roadway. There is another anchorage at 420m that is also close to this dotted line and that could also be labelled as an exception. To analyse the possible efl and load levels for these anchorages, the scenarios outlined in Table 2 will be used. This gives five different possible scenarios (or efls) for each anchorage, with their expected likelihood. The diagnosis of load based on these scenarios can be seen in Table 4.

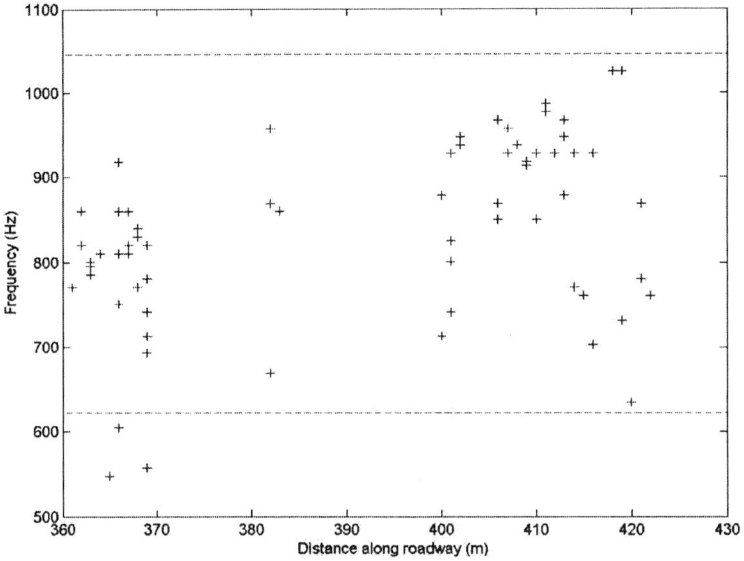

Figure 2: Frequency analysis of individual anchorage tests showing two standard deviations from the mean

The results show that all of the scenarios give a negative load diagnosis, which suggests that none of the scenarios given in the table have an effective free length that is long enough to account for the low frequencies exhibited by these anchorages. This suggests that the effective free lengths for these anchorages are even longer than shown in Scenario 1. This would result in an effective free length of over 500mm, which is over one fifth of the anchorage length.

It was evidently very important to verify these results and the mining operator undertook to overcore the identified anchorages in order to examine them in more detail. An overcoring process was undertaken on a small number of anchorages that were identified as outliers by this process. This showed conclusively that the anchorages had failed and that the effective free length was indeed extremely long in each case. The reason for failure for these anchorages was the catastrophic failure of the resin grout, in most cases due to failure of the resin to bond during the installation process.

Table 4: Load diagnosis for anchorages with low frequency analysis

Effective free length range	Likelihood of occurrence	Predicted mean efl (mm)	Stratabolt distance – strap position (kN)		
			365m- 6	366m- 2	369m- 2
Scenario 1	11.3%	508	-11	-3	-11
Scenario 2	23.0%	390	-36	-19	-36
Scenario 3	31.3%	330	-86	-51	-86
Scenario 4	23.0%	282	-194	-124	-194
Scenario 5	11.3%	226	-387	-263	-387

Conclusions

The analysis in this paper shows that the anchorages tested at this site did not fit within the initial design specification. Further, it would seem that the majority of anchorages had a substantial effective free length. If the intention of the design specification was to have fully grouted anchorages with no effective free length then this was not achieved. The results are also able to show for the first time the actual installation efficacy and give geotechnical engineers feedback on what actually has been installed at their site.

In addition, clear differences between different sections of the site have been identified and this may be due to a difference in installation procedure / equipment, changes in rock mass characteristics or to damage that has occurred at a specific section of the mine.

The anchorages that have been overcored show that the diagnosis made by this process for anchorages that have failed was entirely correct. This demonstrates clearly that the process described in this paper should be extremely useful for geotechnical engineers in identifying satisfactory performance or potential problem areas at their sites, and allowing them to take immediate and effective remedial action, as appropriate. This is an important step in improving the safety of the site and ensuring its long term economic viability.

It is clear that it would be possible to continue testing these anchorages to further monitor their behaviour over time, and identify any anchorages that show signs of a marked change in frequency response – denoting possible failure or impending failure of the anchorage. This type of approach could also be used to identify 'hotspots' in a site, where ground movements are more likely or stresses in the rock mass are increasing to a dangerous level.

The results from the mathematical model show that the diagnosis of load is very difficult if the effective free length is not consistent, and for long term condition monitoring of any type of anchorage it is extremely important that this characteristic of the anchorage is controlled. If the effective free length can be specified to an accurate level then long term monitoring of the condition of the anchorage using its frequency response to an applied anchorage can be undertaken with ease, using the process as described in this paper. The mathematical model shows that this aspect is a fundamental physical property of the anchorage system, and that by removing this variable from the anchorage design then load diagnosis and exception testing will be much more effective. In the opinion of the authors this is an important consideration for the future design of any sites that utilise ground anchorages so that the long term condition and health monitoring of the anchorages and therefore the site can be undertaken.

In addition, as mentioned earlier, the anchorage head characteristics are extremely important and are one of the dominant characteristics when determining the frequency profile of the anchorage to an applied impulse. Therefore it is extremely important that these two aspects – the effective free length and the anchorage head – are controlled during the construction of the anchorage in such a way as to ensure long term condition monitoring of the anchorages in question is possible. If either of these two aspects is not controlled or is defined outwith the recommendations as predicted by the mathematical model then long term non-destructive condition monitoring of these anchorages will not be possible and will increase the cost for future risk assessments of the site in question.

Acknowledgement

The authors would like to thank AMEC plc for their assistance in the taking of data for analysis in this paper.

References

1. The Institution of Civil Engineers., 1992, *A National Agenda for Long-term and Fundamental Research for Civil Engineering in the United Kingdom.* Thomas Telford Ltd, London.

2. British Standards Institution, 1989, *British Standard Code of Practice for Ground Anchorages (8081).*

3. Holland, D.C., 1989, *The influence of close proximity blasting on the performance of resin bonded rock bolts,* MSc. Thesis, University of Aberdeen

4. Holland, D.C., 1993, *The behaviour of resin bonded rock bolts and other anchorages, subjected to close proximity blasting,* PhD Thesis, University of Aberdeen

5. Rodger, A.A., Holland, D.C., Littlejohn, G.S. and Mothersille, D.K.V., 1989, *Instrumentation used to monitor the influence of blasting on the performance of rock bolts at Penmaenbach tunnel,* Instrumentation in Geotechnical Engineering, Nottingham, April 1989, TTL, pp267-279.

6. Littlejohn, G.S., Rodger, A.A., Mothersille, D.K.V. and Holland, D.C., 1989, *Dynamic response of rock bolt systems,* Proc. Conf. On Foundations and Tunnels, London, pp57-64.

7. Mothersille, D.K.V., 1989, *The influence of close proximity blasting on the performance of resin bonded bolts,* PhD Thesis, University of Bradford.

8. Rodger, A.A., Littlejohn, G.S. and Holland, D.C., 1993, *Dynamic response of rock bolt systems at Pen y Clip in North Wales,* Options for Tunnelling, 1993, Developments in

Geotechnical Engineering, 74:719-727.

9. Ivanović, R.D. Neilson, and A.A. Rodger, 2001, "*Lumped parameter modelling of single tendon ground anchorage systems*", Geotechnical Engineering 149 (2), 103-113.

10. A Ivanović, R.D. Neilson, and A.A. Rodger, 2002, "*Influence of Prestress on the Dynamic Response of Ground Anchorages*", Journal of Geotechnical and Geoenvironmental Engineering, 128(3), 237-249.

11. A Starkey, A Ivanovic R.D.Neilson, A.A.Rodger, 2003, "Condition Monitoring of Ground Anchorages by Dynamic Impulses: GRANIT system" Meccanica 38:131-142.

12. Neilson, R. D., Rodger A.A., Ivanović, A, Starkey, A. J., Chappell M., Jackson B. H. and Ritson A., 2004, "Rock/Cable Bolt Integrity Testing using GRANIT – Recent Field Applications", 5th International Colloquium on Roofbolting in Mining, Aachen, p523-528.

13. Ivanovic, A, RD Neilson, A Starkey, AA Rodger, 2007, *"Common anchorage issues addressed by numerical modelling"*, International Conference on Ground Anchorages and Anchored Structures in Service 2007

Ultrasonic Technique for Rock Bolt and Rock Mass Bond Quality Control

Takayuki MORI, Kajima Corporation, Tokyo, JAPAN
Makoto NAKAJIMA, Kajima Corporation, Tokyo, JAPAN
Tohru TSUTSUI, Tsutsui Denshi Co., Ltd., Tokyo, JAPAN

Abstract: A rock bolt is an essential component to stabilize rock cavern and slope etc.. The annulus around the rock bolt in the hole is usually fully filled up with filler such as mortar grout. In order to keep the support strength and the long-term stability of the excavated rock mass, it is necessary to check the degree of grouting after construction. An experimental study has been started to develop a non-destructive inspection technique for quality control of the grouting. In this article, several cases of laboratory and in-situ experiments using a prototype of the inspection equipment are described. The prototype comprises an ultrasonic wave generator and its receiver, which can detect the minute reflective signals that are generated at the boundaries between rock bolt and filler. As a consequence of these experiments, a basic idea for diagnosis of the grouting state is proposed.

Introduction

A rock bolt is an essential component of the NATM (New Austrian Tunneling Method) construction method for stabilizing rock cavern. In the NATM the lining is fastened to the rock mass through a radially straightened hole wherein the rock bolt is settled. The annulus around the rock bolt is usually fully filled up with filler such as mortar.

However, the fully bonded type installation sometimes performed poorly due to the affinity between the filler and the rock mass or the existence of cracks contained in the rock mass. Imperfect installation can result in a lack of support strength and lead to corrosion of the rock bolt, and it can reduce the long-term stability of the excavated rock mass. Thus, development of an inspection technique to estimate the filling state after construction is proposed to contribute to the improvement of the quality of support.

Already some inspection methods to assess the filling quality have been developed: measurement technique of the insert depth of the bolt using ultrasonic transducer enables to detect the damage of the bolt, and studies on the minute reflective signals that have been generated at the boundaries between rock bolt and filler are proceeding.

In this article, firstly, fundamental principle on the ultrasonic wave propagation is described and numerical simulations on the wave propagation in the rock bolt are performed. Then, some laboratory experiments and practical in-situ experiments were performed. As a consequence of these experiments, a basic idea for diagnosis of the filling states is proposed.

Theoretical Verification

Fundamental Principle of Wave Propagation

In this section, fundamental principle concerning sound wave propagation is described. Understanding how sound wave will spread out and reduce is supposed to support the theoretical verifications of the properties of the experiment results.

Supposed that you watch a surface of a probe (dS in Figure 1) that is vibrating repeatedly in the Z-axis and is generating sinusoidal wave to half space. Then, the velocity potential ($d\dot{\Phi}$) at P ($|r|$ far from the wave source) is described as follows:

$$d\dot{\Phi} = \frac{\dot{V}e^{-jkr}}{2\pi r} dS \quad (1)$$

where k is wave length constant and $\dot{V}$ is particle velocity vector of the probe. While plane probe S vibrates, velocity potential of radiation waves ($\dot{\Phi}$) is described as follows:

$$\dot{\Phi} = \frac{\dot{V}}{2\pi} \iint_S \frac{e^{-jkr}}{r} dS \quad (2)$$

The exact solution of this formula cannot be generally obtained by means of non-linear matter. But assuming that the shape of the probe surface is a circle, it can be obtained along central axis of the circle. Because an approximate velocity potential can be obtained in the outer field where the distance from the sound source is more than the distance to the bound of the near field (L), the velocity field of outer field is often discussed. The value of L is calculated as follows:

$$L = \frac{D^2}{\lambda} - \frac{\lambda}{16} \quad (3)$$

where D is the diameter of the bolt and λ is the wave length. In the near field where the distance from the sound source is less than L, the velocity potential shows very complicate form. On the other hand, in the outer field it can be described relatively simply, and it declines in inverse proportion to the distance from the sound source.

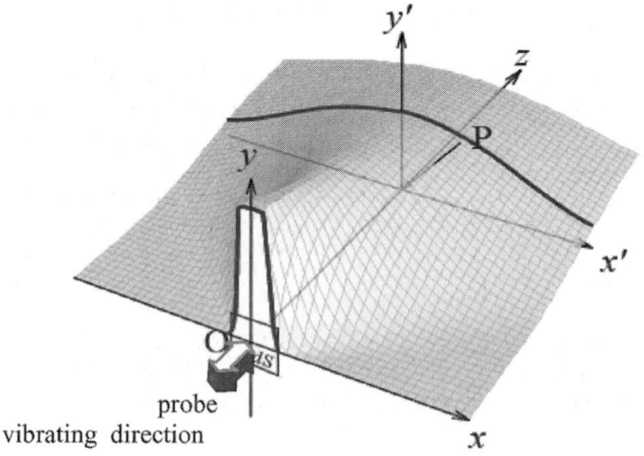

Figure 1. Conceptual model of the wave propagation

Numerical Calculation of the Wave Propagation in a Rock Bolt

According to the previous theory, numerical calculations on the wave propagation in a rock bolt were performed. In all calculations, the probe was defined as a rectangular of 20mm by 20mm, and the Formula (2) was solved using numerical integration by Simpson method. The medium area of sound wave propagation is 3 dimensional infinity, did not use diameter of bolt.

In order to verify the effect of the wave frequency to the sound wave field, several cases of calculations were performed with frequencies of 3.66MHz, 4.5MHz, and 5.0MHz. These values are clock frequencies of the manufactured prototype device.

The results of the calculations are shown in Figure 2. Each vertical axis of the figure indicates the difference of the sound pressure from the central axis, and the absolute value of sound pressure declines in inverse proportion to the distance from the probe as above. Fig.2 a'),b'),c') shows zoom up vertical axis of each fig.2 a),b),c) graph. The results are summarized as follows:

- The distribution of the sound pressure in the bolt becomes uniform as the distance from the probe becomes larger, and the difference of the sound pressure between the central and the edge is 0.4dB at the distance of 2m from the probe in the case of 5.0MHz.
- The sound pressure concentrates around the central axis of the rock bolt as its frequency is higher. On the other hand, it tends to diffuse as its frequency becomes lower.

As the result, the higher the frequency of the input wave is, the larger the directivity (the directional characteristic of wave source) of the wave becomes. And the distribution of the sound pressure in the bolt becomes uniform as the distance from the probe increases. Therefore, it is supposed that the input wave with low frequency is better than with high frequency for filling state inspection because wave with low frequency easily spreads out and the propagation of the wave with low frequency is more strongly influenced by the state of the boundary condition between bolt and rock mass. Further, it is necessary to choose the frequency of the input wave in accordance with the length of the bolt because sound pressure decreases as the distance from the probe becomes larger.

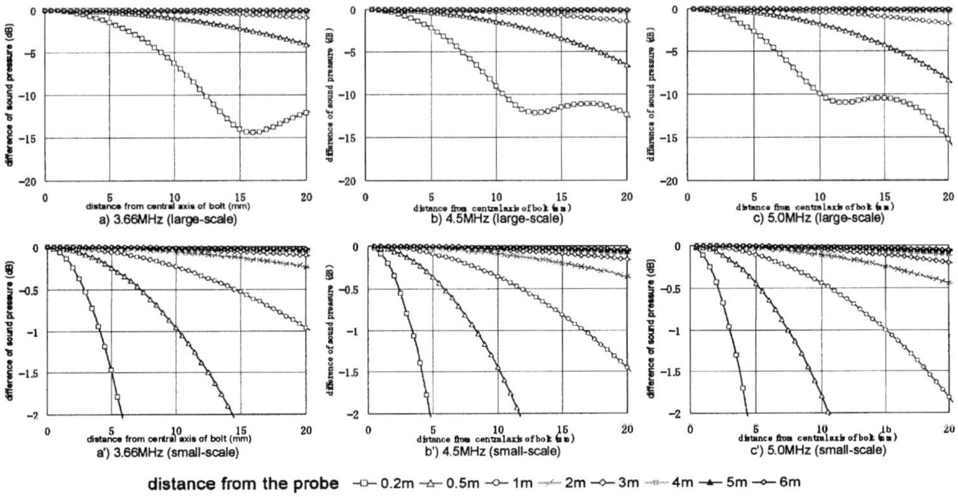

Figure 2. Result of the numerical calculation of the sound wave propagation

Experimental Verification

Development of Inspection Equipment

A prototype of the inspection equipment is shown in Photo 1, and its composition is shown in Figure 3. It is mainly composed of an ultrasonic wave generator and its receiver. The generator can send a pulse wave with high intensity and pure frequency into a rock bolt, and the receiver can detect the minute reflective signals that are generated at the boundaries between the rock bolt and the filler. The reflective wave forms are displayed by the logarithmic amplifier, and filling state inspection is performed by the analysis of the minute reflective signals contained in the waveforms.

Laboratory Experiment

In this section, the differences of the minute reflective signals due to filling states of the installed bolt were verified by laboratory experiments.

Methods

At first, two types of the test models shown in Photo 2 have been prepared. One is a bared SN 2.1m long twist bolt (TD24) in imitation of the imperfectly bonded type (Photo 2a), and the other is a wrapped bolt that is packed by the plaster around the TD24 in imitation of the perfectly bonded type (Photo 2b).

With these two types of test models, the minute reflective signals were measured. The measurements were performed with some different input waves: 3.66MHz, 4.5MHz and 5.0MHz.

Further, to estimate the minute reflective signals due to the filling states qualitatively, measurements of the minute reflective signals were repeated just after the filler was installed until it fully hardened. In this experiment, mortar was used as the filler and the bolt length is 1.0m. The time series changes observation were done while filling state was changing in the hardening process.

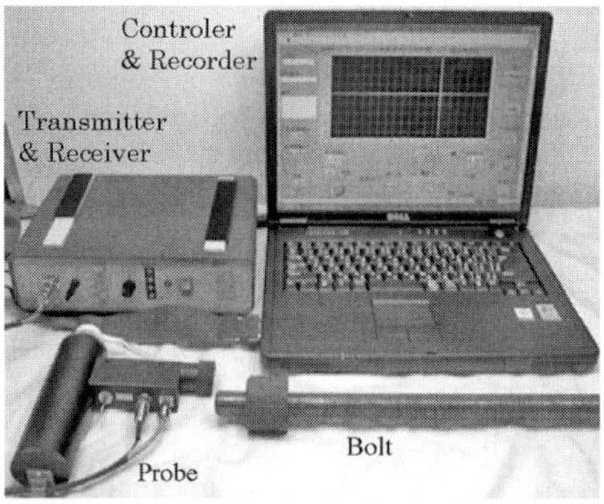

Photo 1. Prototype of the inspection equipment

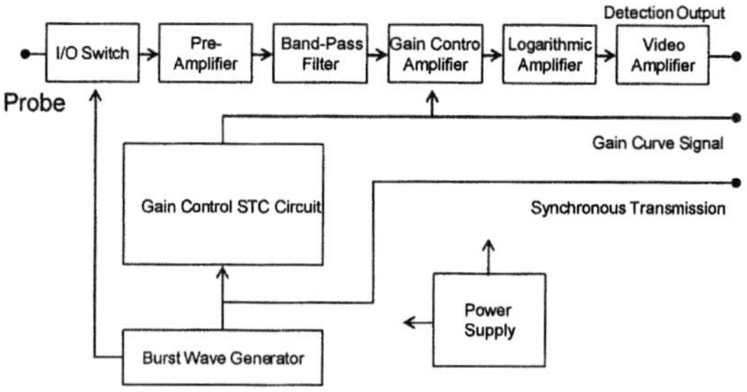

Figure 3. Composition of the inspection equipment

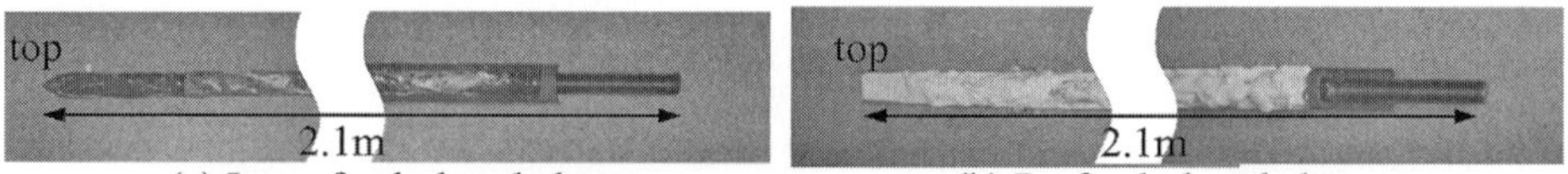

(a) Imperfectly bonded type (b) Perfectly bonded type

Photo 2 - Test Models for the Laboratory Experiment

Results

The results of the minute reflective waveform generated by the input waves with different frequencies are shown in Figure 4 (2.1m long bolt). Where horizontal axis indicates the distance from the probe along the bolt (calculated as the product of the sound wave velocity and the reflection time), and the vertical axis indicates the amplitude level measured by the logarithmic amplifier. In the case of imperfectly bonded type, the minute reflective waves were constant and there was no difference between the cases of the different frequencies. On the other hand, there were differences in the case of perfectly bonded type, that is, there is a tendency that the amplitude of the minute reflective wave becomes larger as the frequency of the input wave becomes larger. It is proposed that an input wave of low frequency easily reflects at the surface of the bolt before it reaches at the top of the bolt, because the directional characteristic is small and it spreads out easily. This tendency is corresponding with the calculations.

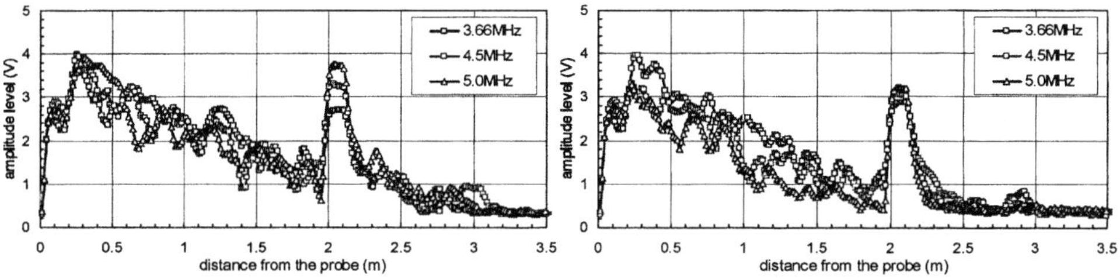

(a) Imperfectly bonded Type (b) Perfectly bonded type

Figure 4. Differences of the reflective waveform due to the frequency of the input wave

The results of the minute reflective waveforms of the imperfectly bonded type and of the perfectly bonded type were shown in Figure 5 (2.1m long bolt). Both waveforms are obtained as the result of the case of 5.0MHz. It can be found that the amplitude of the minute reflective wave is less in the case of the perfectly bonded type than in the case of the imperfectly one. It is supposed that the waves mainly transmit in the case of the perfectly bonded type, and they mainly reflect in the case of the imperfectly case.

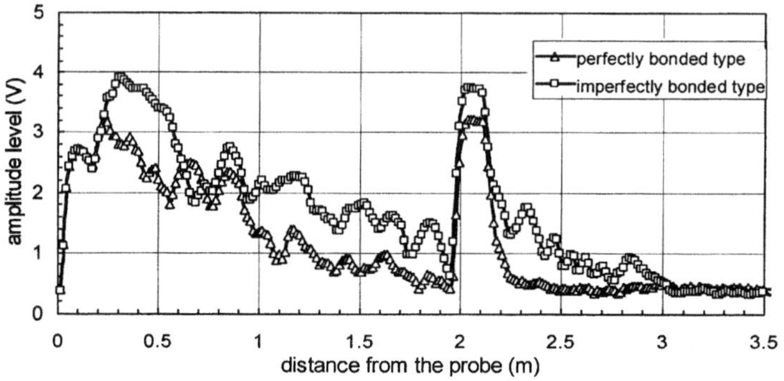

Figure 5. Differences of the reflective waveform due to the filling state

The time series changes of the minute reflective waveform while filling state is changing in the hardening process are shown in Figure 6 (1m long bolt). There is a tendency that the minute reflective waves decreased as time passed. It indicates the wave propagation relies on the filling states, especially the stiffness of the filler. It is supposed that the sound waves begin to transmit through the boundary between the bolt and filler as the hardening process proceeds.

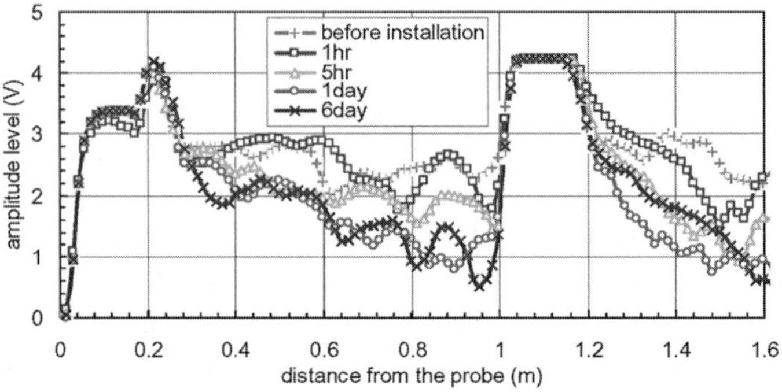

Figure 6. Differences of the reflective waveform due to the hardening process of filler

In-situ Experiment

As an in-situ experiment, the minute reflective waves of installed rock bolt (6m long) were measured in the tunnel, and examined whether the imperfectly bonded bolt could be found using this system. The typical result is shown in Figure 7, and an example that seems to contain somewhat matter in the filling states is shown in Figure 8. It can be found that the minute reflective wave of the latter case is quite larger than of the typical case. Though the filling states weren't confirmed actually, it is supposed that there is a difference in the filler

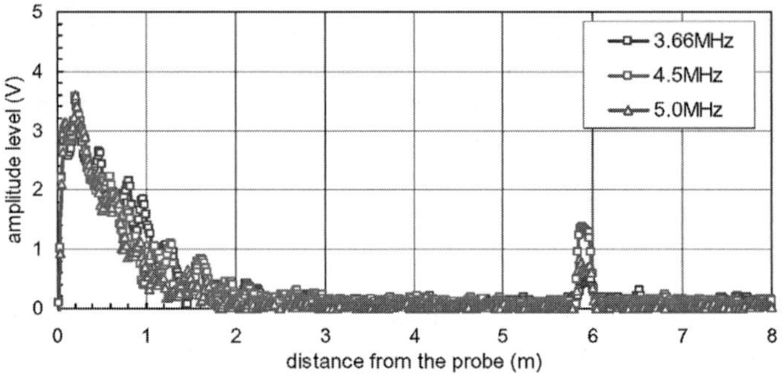

Figure 7. Examples of good grout quality condition

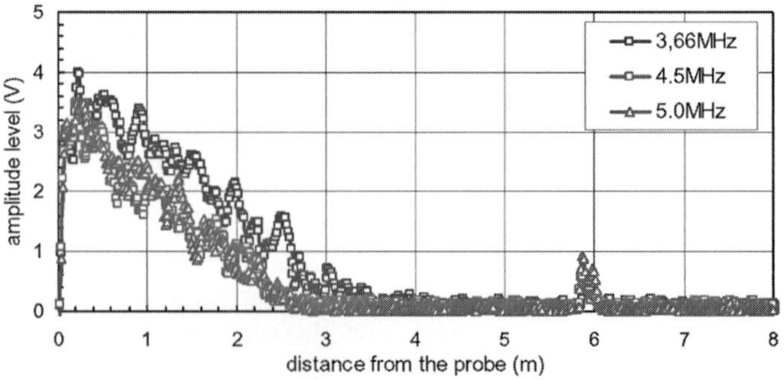

Figure 8. Examples of poor grout quality condition

states.

Estimation of the Filling State

The conceptual model of the criterion to diagnose the filling states using the minute reflective waves is shown in Figure 9. The amplitude of the reflective wave becomes lower as the filling state is better, because the input waves are supposed to transmit to the rock mass if the spaces between the rock bolt and the rock mass is fully filled up with filler. On the other hand, if any spaces remain around the rock bolt, reflective waves generated at the border of the rock bolt are supposed to increase and the amplitude of the minute reflective waves are supposed to become higher. This conceptual model of the criterion to diagnose the filling states is expected to aid the development of a non-destructive inspection for bond quality control. This inspection method can estimate relative quality.

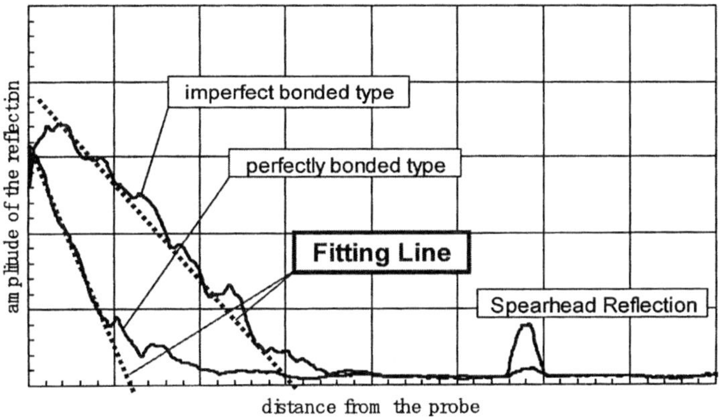

Figure 9. Conceptual model of the criterion to diagnose the filling states

Conclusion

For the purpose of developing the inspection technique to check the grouting condition around the rock bolts, fundamental experiments on the minute reflective wave were performed using a prototype of the inspection equipment. As the result, it seems to be possible to diagnose the filling states by recognizing the difference of the amplitude of the minute reflective wave between in the case of perfectly bonded type and in the case of imperfectly bonded type. However, the reflective wave amplitude will rely on the type of the bolt, rock mass and the filler, so it is necessary to add other case studies to apply this technique to the actual structures.

The development of the method of detecting the anchor tensile force to the impact elastic wave

Kuniyasu Tanaka, Civil & Architectural Engineering Dept, Shikoku Electric Power Co.
Hidenobu Okada, Ikata Civil & Architectural Engineering Joint Venture,
Masayasu Izakura, Hydraulic Power Dept, Shikoku Electric Power Co, Japan

Abstract

One of issues triggered by events such as the failure of fixings to concrete panels or linings in tunnels or to bridges is the need for efficient soundness diagnosis technology to check the safety of infrastructure. Considering the potential safety risk that aged structures pose to the users of infrastructure the requirement for efficient and uninterrupted transportation network, the need for soundness diagnosis technology is heightened even further. This paper, focusing on the nut fixing-type anchor which is frequently provided at the cut slope of power generation facilities, reports a non-destructive inspection method, which is capable of detecting the tensile force of anchors with a light and small device by impacting elastic waves to the anchor. This method is an alternative to conventional tensile tests which are used as the soundness diagnosis method for anchors of several years of age since installation and need a large amount of labor.

Introduction

In Japan, the majority of the land is occupied by steep mountain areas, and slope disasters such as landslides occur frequently due to attacks by excessive precipitation or earthquake. Meanwhile, in power generation facilities, ground anchors are installed at many locations as slope countermeasure work to premises and approach roads in electric power plants and transformer stations. Whether or not these anchors maintain the required tensile force is one of the essential inspection items.

Anchors fall into wedge fixing type, nut fixing type, etc. depending on the method of fixing of the head. This time, rationalization of the inspection method is attempted focusing on the nut fixing type anchors.

Conventionally, inspection of tensile force of the nut fixing type anchor was made by constant measurement by load cells or by tensile test using a hydraulic jack. However, although tensile force can be confirmed as a real value with these methods, for the load cell, cost of installation per cell is high. For the tensile test, works using a large size crane are involved for movement of the hydraulic jack, and installation of sturdy temporary scaffolds, pressurizing devices such as hydraulic jack, pump, etc. and various measuring instruments including load meter are necessary. Furthermore, expertise and skill are required for manipulation of the measurement equipment, and the labor and costs involved are substantial.

Given this situation, the authors paid attention to changes in the dominant frequency by

Ground anchorages and anchored structures in service, Thomas Telford, London

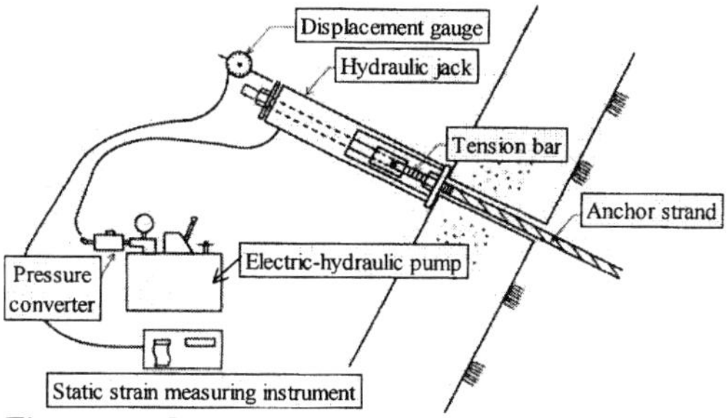

Figure 1. Conventional tensile test

Figure 2. Field set-up of tensile test

impact elastic waves at the anchor head part and developed a method for detecting the anchor tensile force. With this method, the device used is light and small, and measurements can be completed in a short time. Therefore, this method, compared to conventional methods, is extremely simple and it is expected that the efficiency of testing works will be greatly improved and a remarkable cost reduction will be accomplished. This report deals with the currently developed method of detecting anchor tensile force by impact elastic wave.

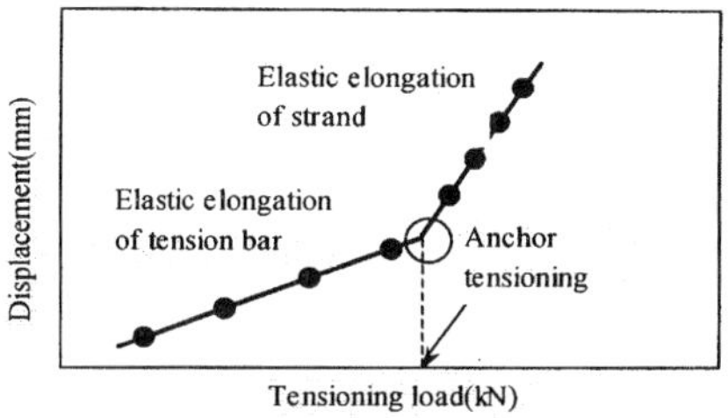

Figure 3. Load vs. displacement curve by tensile test

Conventional method of inspection

Conventionally, tensile test by hydraulic jack is used in many cases for inspection of anchor tensile force. The testing apparatus is shown in Figure 1 and testing status is shown in Figure 2.

For the tensile test, a hydraulic jack is mounted onto the anchor head part for loading, and while the load is increased displacement and load are measured by dial type displacement gauge pressure converter, respectively.

Figure 3 shows the load vs. displacement graph using the tensile test. Elongation of the tension bar alone at the anchor head part appears as the displacement up to the value of tensile force locked in the anchor. Above this tensile force, elongation of the strand is added and thereby the displacement sharply increases. A point of transition from elastic elongation of the tension bar to elastic elongation of the strand is measured as the anchor tensile force in service. It normally takes 40 ~ 50 minutes for this tensile test, although the amount of time required depends on the magnitude of the tensile force.

This method needs a large amount of labor and costs for movement of hydraulic jack weighing 200 ~ 300 kg, erection and dismantling of sturdy temporary scaffolds, fitting of connections and adjustments of the pressurizing and measuring devices.

Proposed inspection method by impact elastic wave

The inspection method developed currently detects anchor tensile force by impact elastic wave. The measuring device is shown in Figure 4 and measuring set-up is shown in Figure 5.

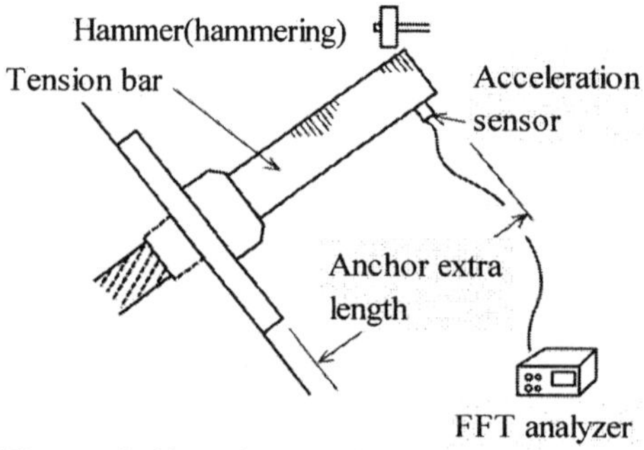

Figure 4. Non-destructive inspection and measuring device

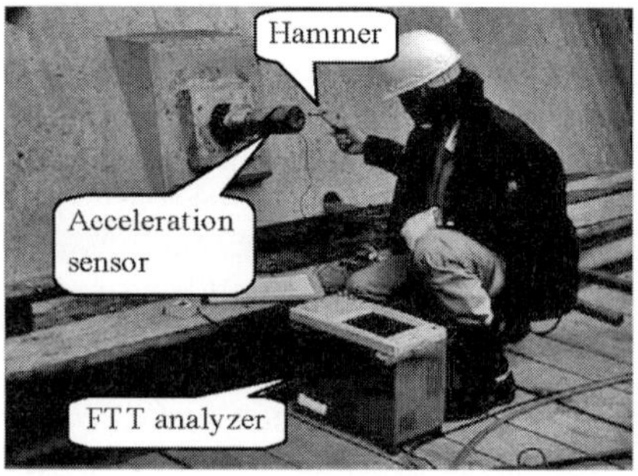

Figure 5. Field set-up of non-destructive inspection

The measuring equipment is composed of an acceleration sensor (23 g), FFT analyzer (portable type is also available), and a small hammer (170 g) only. The equipment is smaller and lighter than that used in the conventional method, thereby ensuring that the inspection is carried out in a shorter time and with less cost.

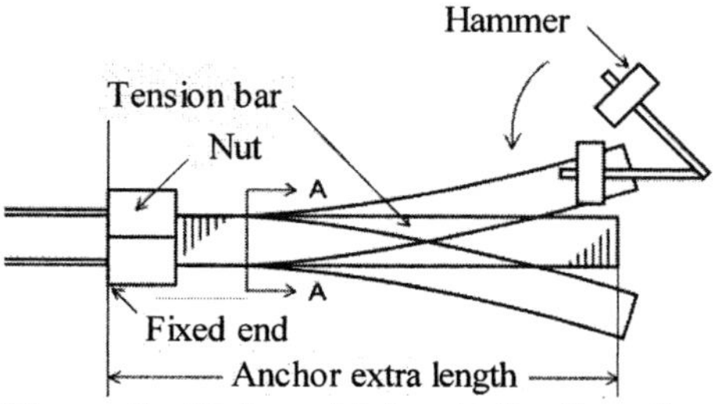

Figure 6. Status of lateral vibration (free vibration) generation at anchor head part

Principle of anchor tensile force detection

The anchor extra length part (protruding length of bar) is considered to be cantilever (nut one side fixed), and when an impact elastic wave is given by light hitting, lateral vibration (free vibration) is generated (Figure 6). This lateral vibration is analyzed by the analyzer and the dominant frequency is calculated.

For the case where the anchor is post-tensioned to the design tensile force, the degree of fixity of the nut part is high and the dominant frequency is higher. On the hand, if the anchor tensile force reduces, the degree of fixity of the nut part is reduced and the dominant frequency is lower level than the vibration frequency at the design tensile force. By paying attention to these characteristics, the tensile force of the anchor may be estimated from the relationship between tensile force and dominant frequency.

Generalized solution of free vibration

On the assumption that fixity of the nut part is a perfect rigid connection integrated with the plate, the generalized solution of free vibration at the anchor head part is expressed by the vibration equation (1) in the cantilever having uniform cross section and is governed by rigidity of anchor head part and overhanging length of bar (anchor extra length). In other words, for a constant rigidity case, the longer the anchor extra length, the lower is the vibration frequency; and the shorter the length, the higher is the vibration frequency.

$$f = \frac{\lambda^2}{2\pi L^2}\sqrt{\frac{EIg}{\gamma A}} \qquad (1)$$

where, f = Vibration frequency (Hz), L = Length of bar (m), E = Elastic coefficient (N/m^2), I = Geometrical moment of inertia (m^4), g = Acceleration of gravity (m/s^2), γ = density (kg/m^3), A = cross-section area(m^2), λ = Dimensionless coefficient determined by boundary conditions and oscillatory type (Primary case = 1.875)

Figure 7 shows sectional profile of a medium type 7-strand anchor of tensile strength 1700 kN class which is the target of the present study, and Table 1 shows the principal particulars. This anchor is consisting of 7 strands, and the strand is composed of 7 PC steel core wires, whose diameter is 5.2mm. Figure 8 shows the relationships which are calculated using the generalized solution of free vibration, between anchor extra length and vibration frequency. In Figure 8, two free vibration curves are shown: curve (1) in which fixity of the nut by Equation (1) is neglected and curve (2) in which fixity of the nut is taken into consideration.

Since rigidity is higher in case (2) where fixity of the nut is considered, case (2) shows a higher vibration frequency than case (1) where the nut is not taken into account. This tendency becomes more marked as the anchor extra length is shorter. This is attributed to the fact that the shorter the anchor extra length, the more significant will become the effect of the nut in terms of anchor extra length.

Meanwhile, in relation to degree of fixity, it is apparent that an actual anchor is weaker than case (2), that is ideal and has perfect fixity, and the vibration frequency measured is lower. It is further predicted that as tensile force is reduced, vibration frequency will become lower (Figure 9).

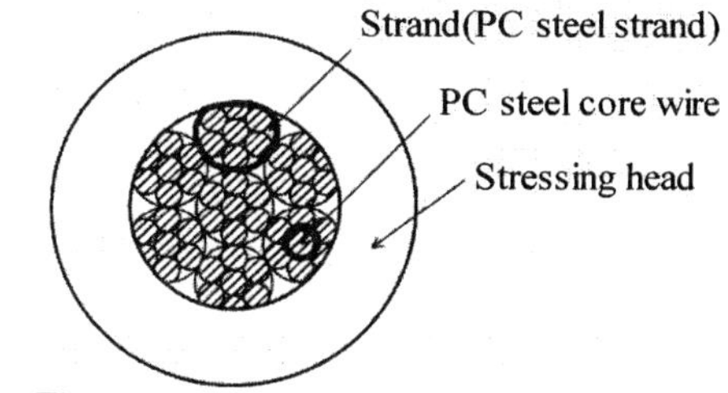

Figure 7. Sectional profile of medium type 7-strand anchor (A-A)

Table 1. Particulars of anchor (Class 1700kN)

	Strand	Stressing head	Nut
Tension strength Pu(kN)	1700		
Sectional area A(m^2)	9.71×10^{-4}	2.71×10^{-3}	2.71×10^{-3}
Geometrical moment of inertia I(m^4)	1.11×10^{-7}	1.05×10^{-6}	7.48×10^{-6}
Elastic coefficient	1.90×10^{11}	1.85×10^{11}	2.10×10^{11}

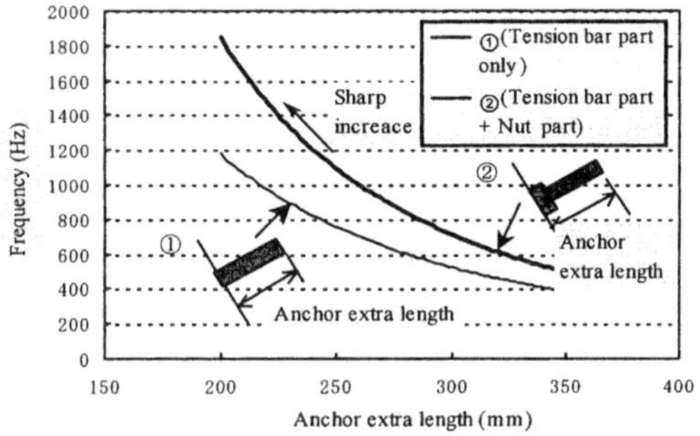

Figure 8. Relationship between anchor extra length and vibration frequency under free vibration

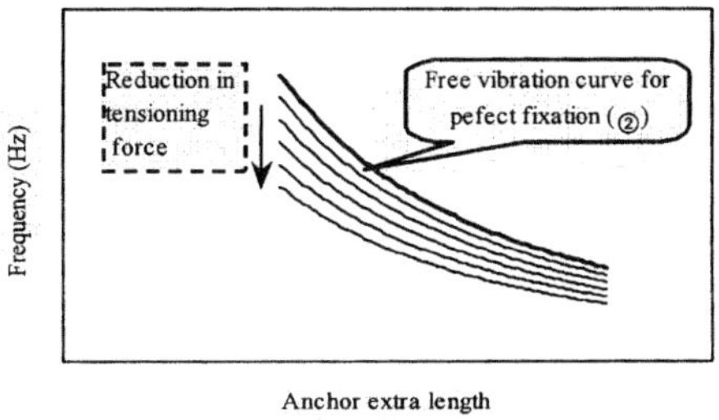

Figure 9 Relationship between anchor extra length, vibration frequency, and tensile force

Confirmation test by actual anchor

Outline of test

In order to check changes in the dominant frequency depending on anchor extra length and magnitude of tensile force, tests were carried out using an actual anchor. The anchor subjected to the test is of tensile strength class 1700 kN. Figure 10 shows the installed test anchor and Table 2 shows the specifications for the measuring device.

Tests were carried out for three cases (I to III) each having a different anchor extra

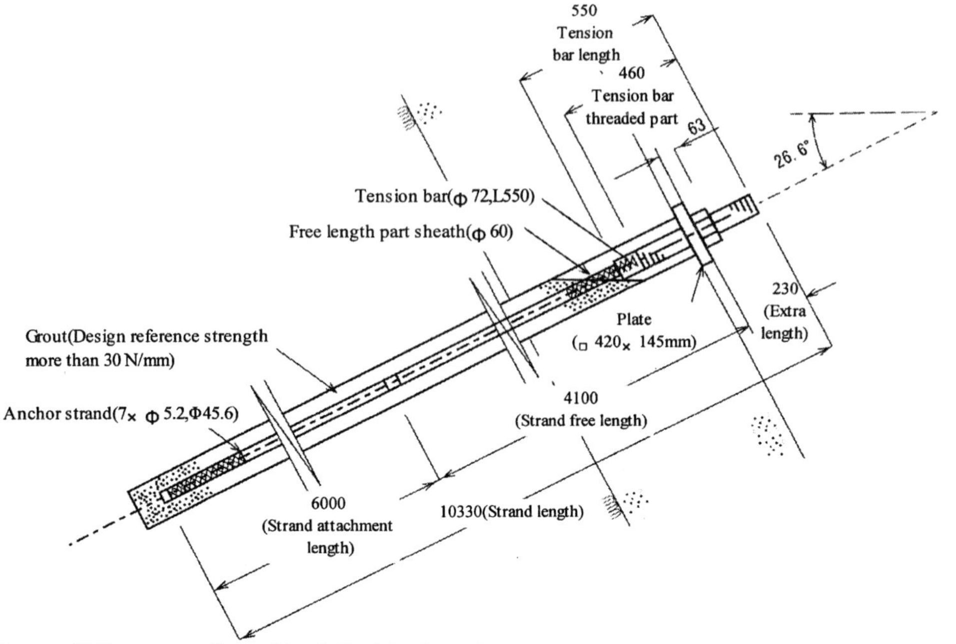

Figure 10 Cross section of installed test anchor

Table 2 Measuring devices

Measuring device	Measuring item	Specifications
Acceleration sensor	Vibration waveform	AE175L manufactured by AFT
FTT analyzer	Dominant frequency	CF-5220 manufactured by Ono Sokki

length as shown in Figure 11. The tensile load was changed from the allowable tensile load 0.6 Pu(2) (Pu = ultimate tensile strength) down to 0.2 Pu in 0.1 Pu decrements. Each correct load, which was fixed with nut, was checked by a load lift-off check using a jack.

After generating free vibration by hitting the end of the protruding bar with the inspection hammer, anchor vibration frequency was measured with an acceleration sensor fixed at the Stressing head front edge using elastic tape. The dominant anchor vibration frequency is obtained by analyzing the results of measurements with a FFT analyzer (see Figure 5).

Results of test and discussion

As a result of the test, first, it could be confirmed that the longer the anchor extra length, the more the dominant frequency was reduced (Figure 12).

Secondly, the dominant frequencies due to changes in anchor extra length and tensile force obtained in tests I to III were plotted in Figure 9 mentioned previously. It was then found that since the degree of fixity of the actual anchor is weaker than the ideal perfect fixity, the free vibration curve (1) is substantially lower than curve (2) (Figure 12). Meanwhile, the anchor extra length is reduced gradually for the plotted values of tests I to III, and this is attributed to the fact that elongation of the anchor free length becomes small as the tensile force is reduced from 0.6 Pu to 0.2 Pu. Anchor extra length is shortened by about 5 mm every 0.1 Pu.

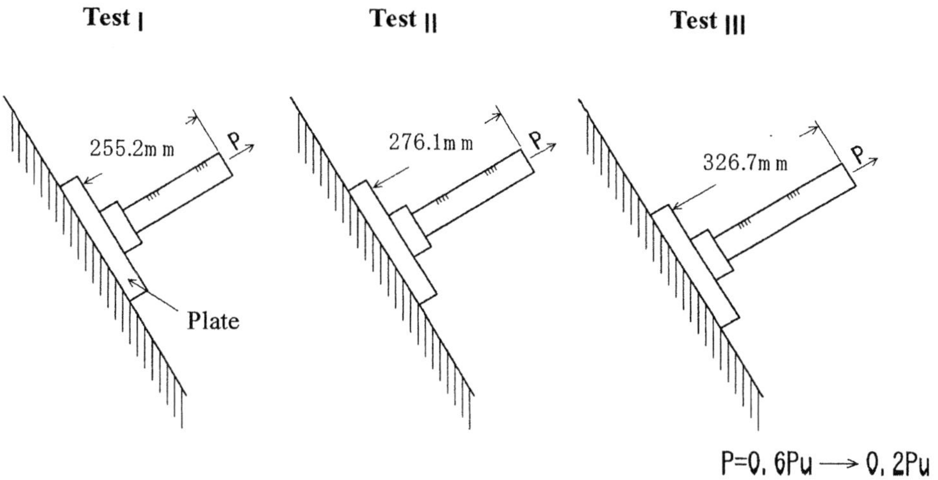

Figure 11 Status of head part of test anchor (P=0.6 Pu)

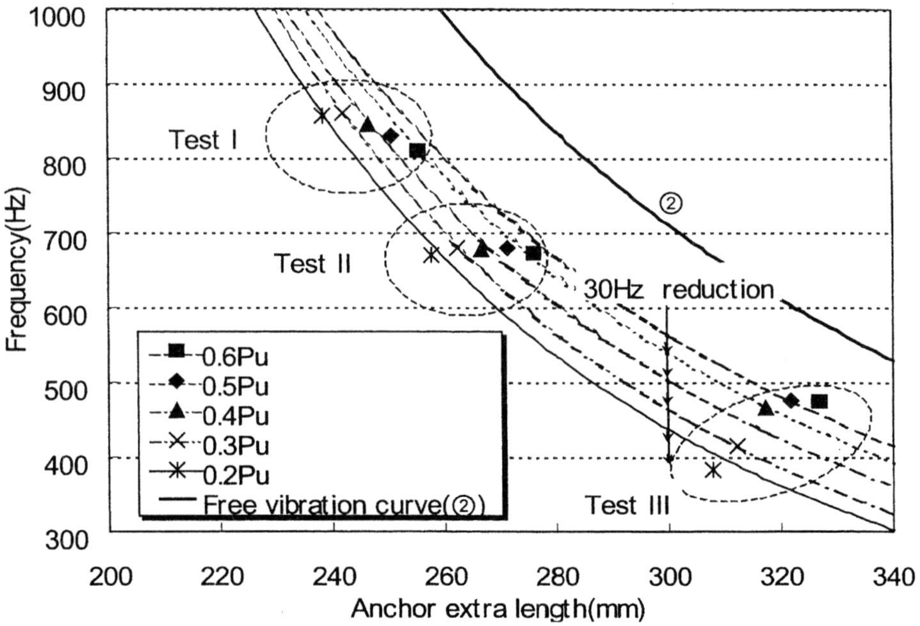

Figure 12 Relationship between anchor extra length, tensile force vibration frequency

Approximate curves are drawn for the plotted values of the same tensile force in tests I to III. It is shown that these curves exhibit near similarity with the free vibration curve (2), and the gradient becomes gentle in the range where the anchor extra length is longer and its effects on the dominant frequency becomes less significant. Conversely, when the anchor extra length is shorter, the gradient becomes steep thereby increasing the effect of anchor extra length on vibration frequency.

Further, it is noticed from separation of approximate curves for each tensile force, that when the anchor extra length is the same, the dominant frequency is reduced by about 30 Hz for every 0.1 Pu reduction in the tensile force. This suggests that in the actual anchor with a constant anchor extra length, if a variation of the dominant frequency of about 30 Hz is found, an increase or decrease of around 10% of the tensile strength Pu can be predicted. If measurement of the initial value of the dominant frequency is taken for the design tensile force following anchor installation, the tensile force in service can be assumed from changes in the dominant frequency with good accuracy.

When the tensile force is changed from 0.6 Pu to 0.2 Pu, plotted values in test I show upward sloping to the left and conversely, those in test III show downward sloping to the left. The following reasons are considered to be responsible for this:

- In test I, in the range where anchor extra length is shorter, the degree of influence of the nut that has high fixity is high. Therefore, an increase in vibration frequency (about 60 Hz per 0.1 Pu) due to reduction in extra length accompanied by reduction in the tensile force is greater than a reduction in the vibration frequency (about 30 Hz per 0.1 Pu) due to the reduced degree of fixity, and changes in the vibration frequency exhibit upward sloping to the left.
- In test III, in the range where anchor extra length is longer, since the degree of influence of fixity of the nut is small, an increase in the vibration frequency (about 20 Hz per 0.1 Pu) due to a reduction in extra length is smaller than the reduction in the vibration frequency (about 30 Hz per 0.1 Pu) due to the reduced degree of fixity, and therefore, changes in the vibration frequency exhibit downward sloping to the left.
- In test II, since influences of reduced extra length and reduced degree of fixity are nearly identical, the said changes remain at the same level.

Analytical examination of factors for changes in anchor vibration frequency

So far, factors for changes in anchor vibration frequency due to the magnitude of the tensile force are discussed using a one-dimensional concept of degree fixity of the nut. However, if details of the anchor structure are observed minutely, there is an air gap (pocket) at the rear of the anchor plate, and it is considered that vibration characteristic of this air gap (factor A) in the anchor body has an influence on vibration of the anchor extra length part. It is further considered that the tensile force of the anchor is conveyed also to the anchor extra length part in the form of stress transfer, and geometrical rigidity of the extra length part* (factor B) is elevated and thereby the vibration frequency is changed. FEM was done to see stress transfer.

To identify factors that change the vibration frequency of the anchor extra length part due to the magnitude of the tensile force, dynamic vibration analysis using an analytical model including the air gap in the anchor hole was attempted.

* A phenomenon that even for a material without rigidity such as a string, an apparent rigidity is caused by applying a tensile force.

Analytical conditions

The analytical model is shown in Figure 13 and the physical properties in Table 1 are used. In the hole at the rear of the plate, the air gap part is perfectly free and the end part filled with grout is supported by rollers. A joint element is inserted between the nut and the plate.

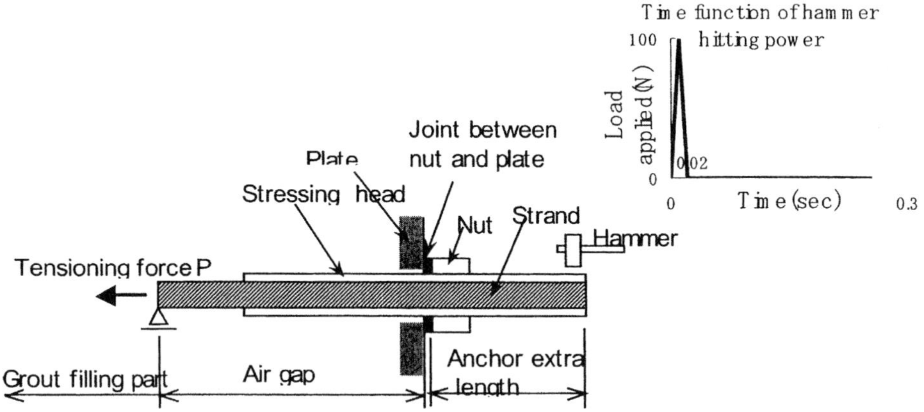

(a) Concept of model

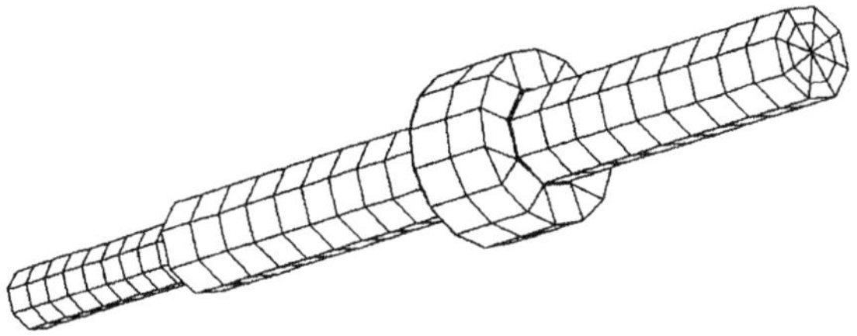

(b) Mesh illustration

Figure 13 Analytical model

Table 3 Analytical conditions

Analysis method	Three-dimensional dynamic vibration analysis
Upper limit vibration frequency f_{max}	10,000 Hz
Time interval Δt	0.05 msec
Duration time T	0.1 sec
Anchor tensile force	0.2 Pu ~ 0.6 Pu, every 0.1 Pu
Anchor extra length	200 mm ~ 350 mm

$f_{max} = 1/(2\Delta t)$ where, f_{max}: Upper limit vibration frequency (Hz), Δt: Time interval (sec)

Table 3 shows the analytical conditions. For hitting power to the anchor extra length, a load of 100N was dynamically applied in 0.02 sec to the front edge of the anchor extra length to simulate hitting with the inspection hammer.

Influence of vibration characteristics at air gap in the hole (Factor A)

Influence of changes in length of air gap in the hole

Since the air gap in the hole is perfectly free, free vibration is generated due to hitting of the anchor extra length, and it is considered that its vibration frequency depends on length of air

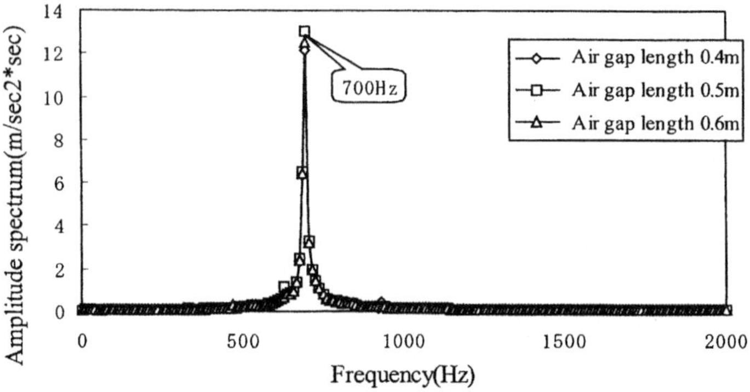

Figure 14 Changes in dominant frequency due to changes in air gap length

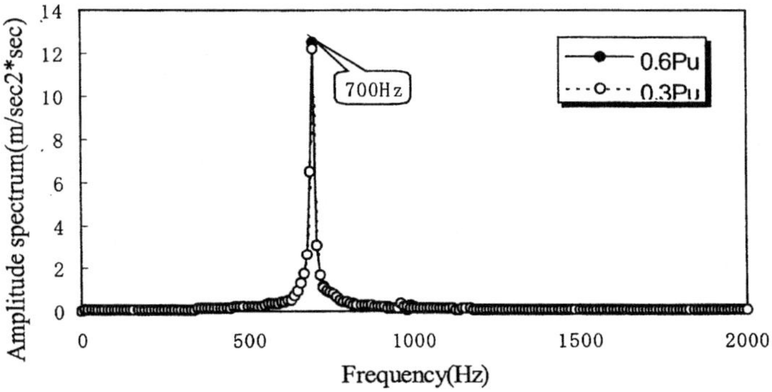

Figure 15 Changes in dominant frequency due to changes in tendon tensile force

gap. Therefore, vibration analysis was carried out on the condition that in the analytical model is shown in Figure 13. Length of the air gap part was changed to 0.4 m, 05 m, and 0.6 m to reflect different depths of grout filling in practice while the tensile force was not applied.

Results of the analysis are shown in Figure 14. It is revealed from this that the dominant frequency of the anchor extra length part is nearly 700 Hz in the air gap lengths investigated and that changes in the length of air gap in the hole do not affect vibrations of the anchor extra length part.

Influence of changes in tendon tensile force of air gap in the hole

It is also considered that vibration frequency of the extra length part will vary by changing vibration characteristics of the tendon with an air gap in the hole due to magnitude of the tensile force. Static analysis was carried out by applying a tensile force to the tendon with an air gap in the hole and subsequently dynamic vibration analysis was carried out.

Figure 15 shows the results of vibration analysis for an anchor extra length of 300 mm and tensile forces of 0.6 Pu and 0.3 Pu.

According to this analysis, it is confirmed that the dominant frequency of extra length part is

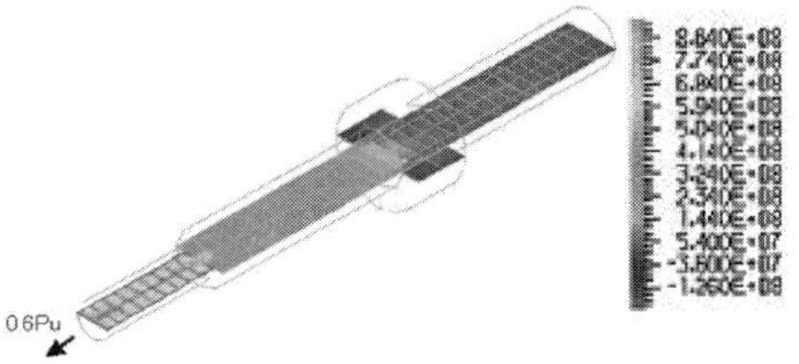

Figure 16 Stress distribution in anchor

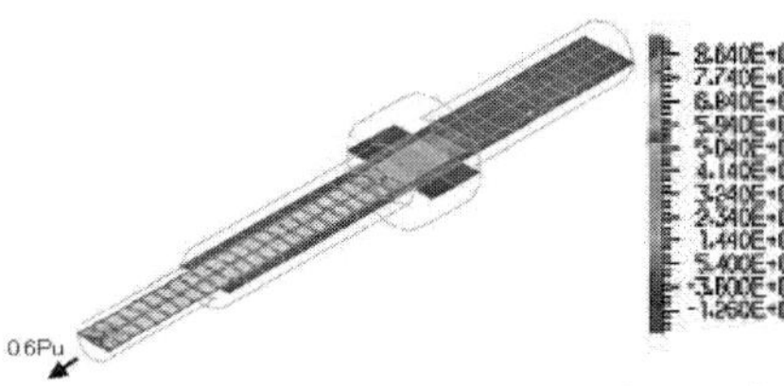

Figure 17 Stress distribution in anchor (Rigidity of Stressing head: 1/100)

700 Hz for both tensile force 0.6 Pu and 0.3 Pu and magnitude of strand tensile force with an air gap in the anchor hole does not affect vibration frequency of the anchor extra length part. This is considered to be attributable to the fact that since rigidity of anchor body is sufficiently large with or without the tensile force, the influence of magnitude of the tensile force on vibration characteristics is small.

Influences of changes in geometrical rigidity of anchor extra length part (Factor B)

It is also considered that when an anchor tensile force is introduced, increase in stress of the anchor body extends not only to strand with an air gap in the anchor hole, but also to the anchor extra length part, and vibration frequency will be increased by enhancement of geometrical rigidity of the extra length part. Especially for the anchor investigated in the present study, the elastic coefficient of the strand tendon is slightly greater than that of the Stressing head (Table 1), so stress transmission to the anchor extra length part then becomes marked due to tension of the tendon, and influences of geometrical rigidity becomes significant.

However, if reference is made to stress distribution at the anchor extra length part with tensile force of 0.6 Pu (Figure 16), the tensile force does not affect the anchor extra length part, while only a slight stress increase is noticed with the part of strand fastened by nut.

Even when the elastic coefficient of the Stressing head is considered to be 1/100, the tensile force has no impact on the anchor extra length part and the dominant frequency is not affected (Figure 17). It has been confirmed from this that the influence of changes in geometrical rigidity of the anchor extra length part on the dominant frequency is a little.

Influences by changes in degree of fixity between nut and plate

Based on the results of investigation mentioned above, factors which are responsible for changes in the anchor vibration frequency due to magnitude of the tensile force, were narrowed down to degree of fixity of the nut part and investigated in detail.

For analysis, rigidity (EI) of joints (minute elements) between the nut and plate was changed and dynamic vibration analysis was performed considering varying degrees of fixity. From the fact that for rigidity of this joint, the degree of fixity is increased with greater tensile force and that the longer the anchor extra length, the weaker the influence of degree of fixity on the vibration frequency, dimensional analysis was carried out using tensile force P, anchor extra length L and area A of Stressing head as the parameters, and the following equation was obtained. Constant c was set based on the test results of the actual anchor.

$$E \cdot I = \left(\frac{c \cdot A \cdot P}{L'} \right) \cdot I \qquad [c : \text{Constant}] \tag{2}$$

where, E = Elastic coefficient of fixed part (N/m^2), I = Geometrical moment of inertia of fixed

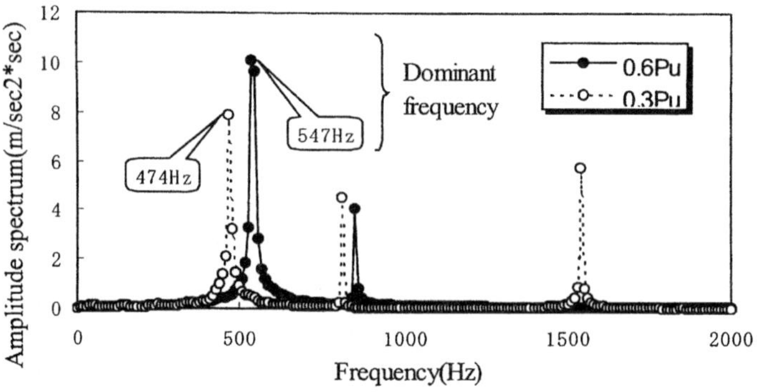

Figure 18 Changes in dominant frequency due to difference of tensile force

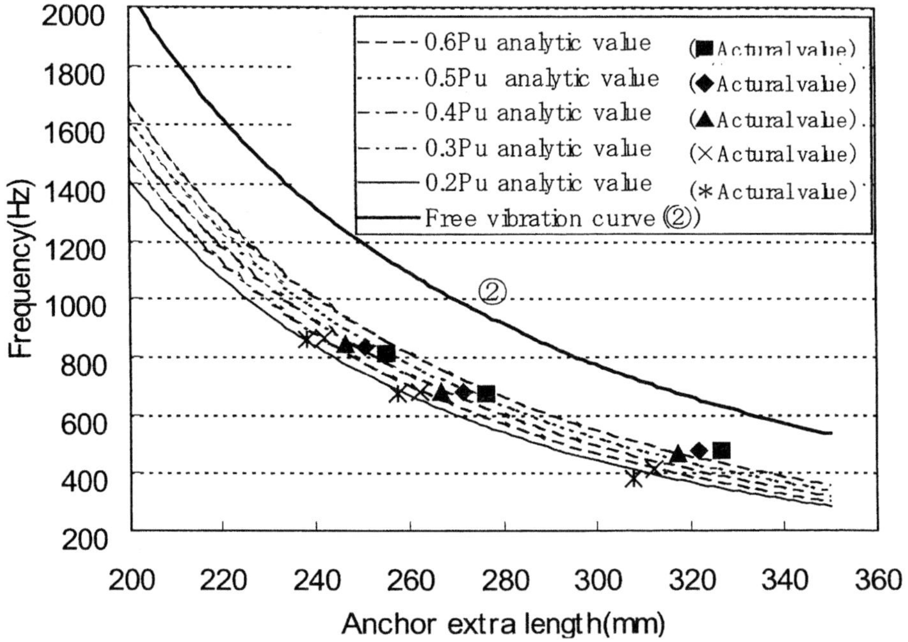

Figure 19 Analytical curves and plotting of actual measurements

part (m^4), A = Sectional area of Stressing head (m^2), P = Tensile force of anchor (N), L = Anchor extra length (m)

Results of vibration analysis for a case where anchor extra length is 300 mm and tensile force is 0.6 Pu and 0.3 Pu are shown in Figure 18. It is noticed from this that the dominant frequency for tensile force of 0.3 Pu is 474 Hz while the same for tensile force of 0.6 Pu is 547 Hz, which means that a reduction as much as 73 Hz (about 25 Hz per 0.1 Pu) is attained.

The analytical curve and the plotted actual measurements, while tensile force P and anchor extra length L are changed, are shown in Figure 19. As a result, the analytical curve can reproduce virtually a reduction in the vibration frequency due to a reduction in the tensile force. Rate of reduction in the analytical curve was 25 ~ 35 Hz per 0.1 Pu, which approximated to the results of actual measurement and presented a similar profile to the free vibration curve.

Conclusions

Effectiveness of the method for estimating tensile force of the anchor from changes in dominant frequency by impact elastic wave has been verified by site confirmation tests.

From the results of analysis, it has been found that degree of fixity between the nut and the plate has a big influence on factors for changes in the dominant frequency depending on the magnitude of tensile force of the anchor. The authors consider that in order to apply this method to small-size or large-size anchors in the future, it is necessary to improve general versatility by assessing rigidity EI paying attention to the degree of fixity and promoting verification from accurate testing.

According to the unprecedented method for detecting tensile force of the anchor, as proposed currently, the tensile force of the anchor can be detected in terms of changes in lateral vibration frequency of the anchor head part, and it is expected that this method will contribute greatly to the efficiency of inspection jobs and reduce cost.

The authors intend to widen the range of application to small-size or large-size anchors and to evolve this method towards establishment of a new soundness diagnosis technology. This method is currently patent pending (Japanese Patent Application No. 2001-316014).

In closing, the authors wish to express thanks to those who collaborated with us in developing this method.

References:

1) Japan Society of Mechanical Engineers: Mechanical Engineers' Handbook, Revised Ver. 5, 1968, p.3-48 ~ 3-50

2) The Japanese Geotechnical Society: Ground Anchor Design and Construction Standards, Explanatory Notes, 2000, p95~128

DEVELOPMENT OF LOADING APPARATUS AND TEST PROCEDURE OF RAPID ROAD TEST FOR GROUND ANCHORS

Tatsuro SUEYOSHI, Raito Kogyo Co., Ltd, Tokyo, Japan
Kazuo TANI, Yokohama National University, Yokohama, Japan

Abstract

A rapid load testing method has been proposed by authors as an effective measuring method to evaluate the pull-out resistance of ground anchors. This testing method has several advantages such that testing time is very short compared to static load tests, and that analysis is simpler compared to dynamic load tests, and damage to a ground anchor is minimal. In our previous studies, the applicability of the rapid load testing was examined in field tests. A rapid loading device using disk springs showed satisfactory performances in laboratory model tests and obtained results were shown analyzable using a single mass model to be used for ground anchor evaluation. In this study, a modified rapid loading device, which overcomes shortcomings of the previous models, is developed and examined, and the testing procedure and the measurement method are established through the field tests using this device.

Introduction

The structure of the loading device using disk springs used for laboratory model tests in our previous study is shown in Fig.1. This loading device is designed to exploit the elastic potential energy stored in the disk springs. When applying a load to a ground anchor, the slide ring slide down to release steel ball stoppers and the disk springs bounce back to apply the load. However, it became clear that this device had the following shortcomings and needed further developments.

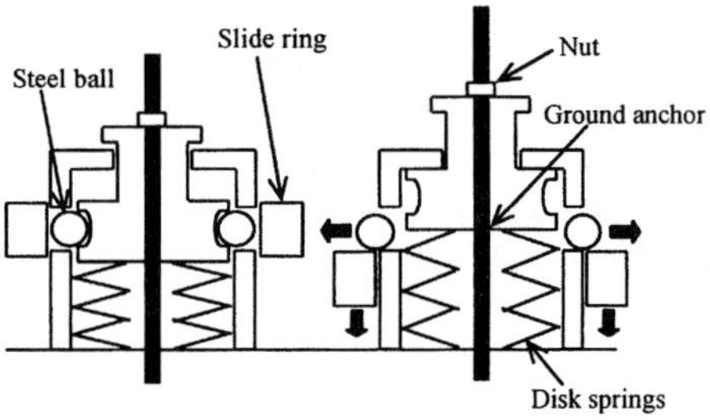

Fig.1 Loading device developed for laboratory tests

Development of loading device

i) Loading velocity and unloading mechanism

This device was not able to control loading speed and was not equipped with unloading mechanism. According to the Japanese Geotechnical Society's standard (JGS 1815-2002), the duration of the rapid loading is in the range of $5 \leq T_r \leq 500$, where T_r is a relative loading duration. The relative loading duration T_r is expressed in Eq.1 as the function of load time t_L, a bond length l_a, and bar wave velocity c.

$$T_r = \frac{t_L}{2l_a / c} \qquad \text{Eq.1}$$

Although this loading device complied with the standard, its T_r value was close to the lowest boundary of the range. Another problem was that the loading speed that this device provided

was too high to control. Another shortcoming was that this device was not equipped with a mechanism to remove the load and return to the initial loading condition.
To overcome these shortcomings, the structure of the device needed to be redesigned.
ii) Load and displacement
In order to adjust the amount of load and displacement, it is necessary to change the rigidity and the number of disk springs as well as the way to stack disks. The load in ground anchors in construction is diverse. The amount of displacement varies depending on test load, free length, section area, and material of an anchor.
To obtain useful results by testing ground anchors in a target range of capacity, the configuration of disk springs needs to be studied.

Method of rapid load testing for ground anchor

A new rapid loading device was developed and tested in the field. To establish the testing method, the followings were to be investigated.
i) Test procedure
The structure of rapid loading devices is unique. Therefore it is necessary to establish a testing procedure unique to the loading device. A major issue is the calibration of the device to be able to initiate unloading at an intended unloading value.
ii) Measurement method
It is necessary to confirm that load transducers and displacement transducers provide accurate measurements under rapid loading and unloading conditions. It is also important in dynamic measurement to determine the measuring method and the measuring intervals. Whether or not the arrangement of instrumentation affects accuracy should also be investigated.
iii) Applicability of the device
The type of ground anchors that the new rapid loading device can deal with should be identified by determining the range of load and displacement that the device can provide.

Performance required for the new loading device

Structure of the new loading device

The loading device requires the capability of controlling the loading speed. The loading speed needs to satisfy the T_r of Eq.1. Therefore when bond length is l_a=3~10m[5] and elastic wave velocity is c=4000m/s[4], the loading time will be 0.025~0.75 second for l_a=3.0m, and 0.75~2.5 second for l_a =10.0m. The device is also required to have unloading mechanism to release the load and return to the initial loading condition.

Specification of the disk spring

The disk springs to be used for the loading device needs to satisfy the load and displacement requirements in testing the ground anchors of a target range. Due to cost constraints, it is not realistic to make a large loading device applicable to all types of ground anchors. Therefore, the applicability of a loading device needs to be narrowed. Based on the track record of anchor installation by Raito Kogyo Co., Ltd., in 2004, the test load for more than 50% of the ground anchors installed is less than T_t=300kN and the anchor length is less than l_A=20.0m. Therefore, it seems reasonable to set our target at T_t=300kN and l_A=20.0m.
The displacement of a ground anchor δ_e is related to T_t, anchor free length l_f, the sectional area of the tendon A_s, and the elastic modulus of the tendon E_s as expressed in Eq.2.

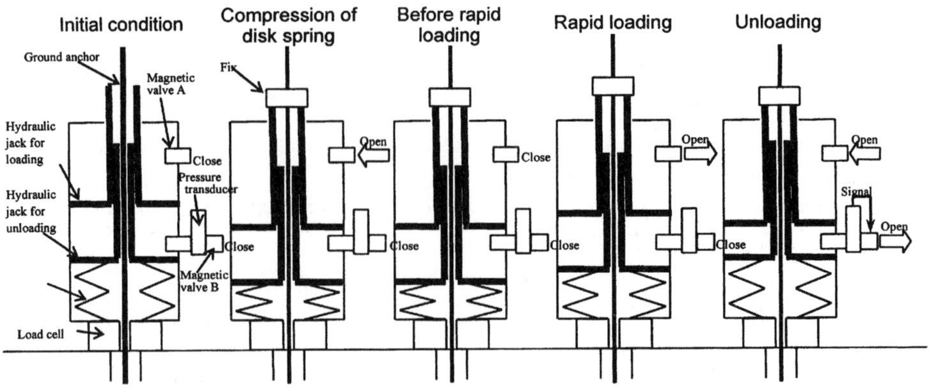

Fig.2 Procedure of rapid loading for field tests

$$\delta_e = T_t \cdot l_f / (E_t \cdot A_s) \quad \text{Eq.2}$$

If the test load is T_t=300kN and, an average bond length l_a=5.0m, an average free length l_f=15.0m, sectional area of the tendon A_s=296.13mm^2(3 strand, d_t=12.7mm) and, elastic modulus of the tendon E_s=193kN/mm^2, the amount of displacement is 79mm. Then 10% of this amount should be allowed for creeping of the ground and pull out of the tendon. Furthermore, about 2mm is expected for the ground subsidence. Therefore, the disk springs needs to be capable of providing $\delta_{e,max}$=90mm at $T_{t,max}$=300kN.

Development of the new loading device

The outline of the loading mechanism and the motion of each cylinder are shown in Fig.2. The double cylinder system was adopted for the purpose of downsizing the device. One cylinder performs compression and release of the disk springs, and the other cylinder introduces the initial load and releases the whole load.

Loading velocity and unloading mechanism

The device needs to be equipped with the mechanism which can adjust the force that the disk springs release in order to adjust the loading speed. For this purpose, the method using the oil pressure to compress disk springs was proposed. As the trigger which releases the compressed disk springs, a magnetic valve is suitable because it can control the loading speed by adjusting the amount of oil release.

The unloading mechanism using oil pressure was also proposed. The unloading mechanism detects the target load with a pressure transducer, and releases the load automatically by opening a magnetic valve.

When compressing the disk springs, the Valve B is closed, and oil is sent through the magnetic Valve A. The elastic energy is held by closing the magnetic Valve A.

Rapid loading tests start by opening the magnetic Valve A and releasing the elastic energy stored in the disk springs. The applied load is measured by a pressure transducer. When the pressure reaches a predefined value, the magnetic Valve B opens automatically and initiates unloading.

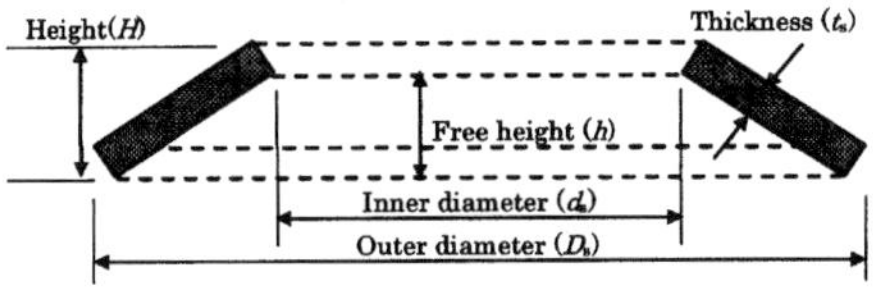

Fig.3 Cross-section of disk spring

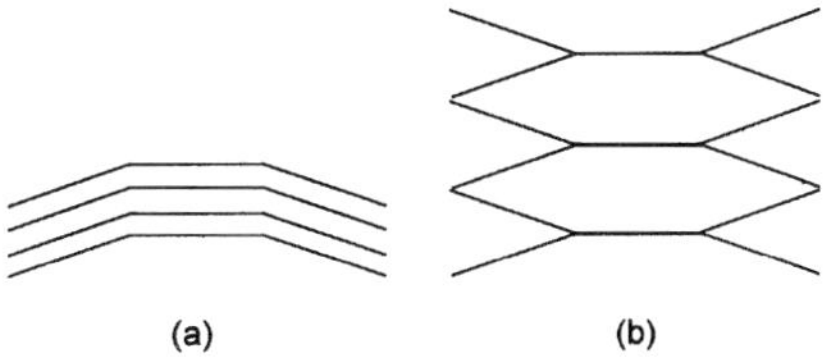

Fig.4 Superposition of disk springs (a) Parallel (b) Series

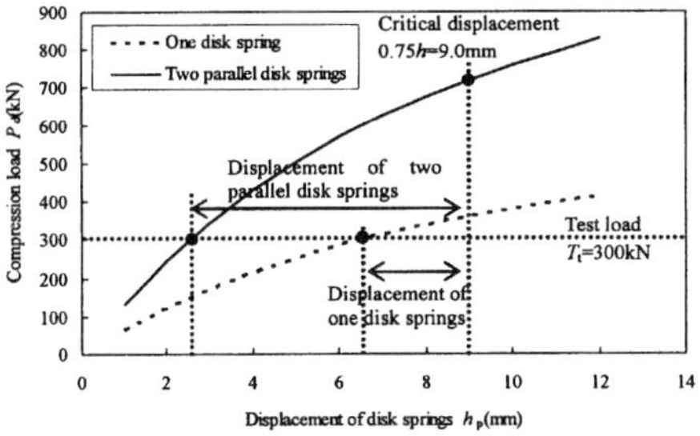

Fig.5 Relationship between displacement and compression load of disk spring

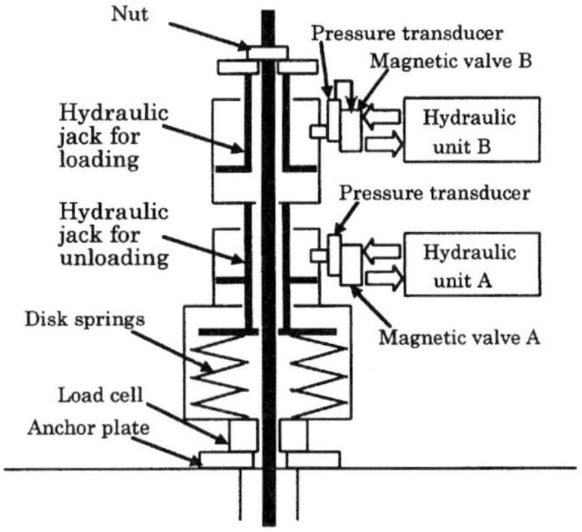

Fig.6 Loading device for field tests

Design of disk spring

The structure of a disk spring is shown in Fig.3. How to pile up disk springs is shown in Fig.4. The rigidity of the disk spring is influenced by its shape (D_s, d_s, t_s), and the way of superposition. In order not to enlarge a loading device, the size of a disk spring is constrained by the size of the anchor plate, and the outer diameter of the tendon. The disk spring thickness of 12 mm is regarded as the maximum due to degradation after repetitive use. Free height h is influenced by thickness t_s and generally is designed as h/t_s=1.0. The shape of a disk spring is determined in consideration of the form of an anchor plate and maximum outer diameter of tendon. The dimension of the disk spring is determined as follows: D_s=240mm, d_s=100mm, t_s=12mm, and h=12mm.

Relationship between displacement of disk spring h_p and compression load T_p is shown in Fig.5. A disk spring is compressible to 75% of its free height. The difference between the free height at the test load and the free height at the maximum compression is available for use in load testing. However, this difference with a single disk is too narrow. Therefore, two disk springs are superposed in parallel to provide a 6.4mm

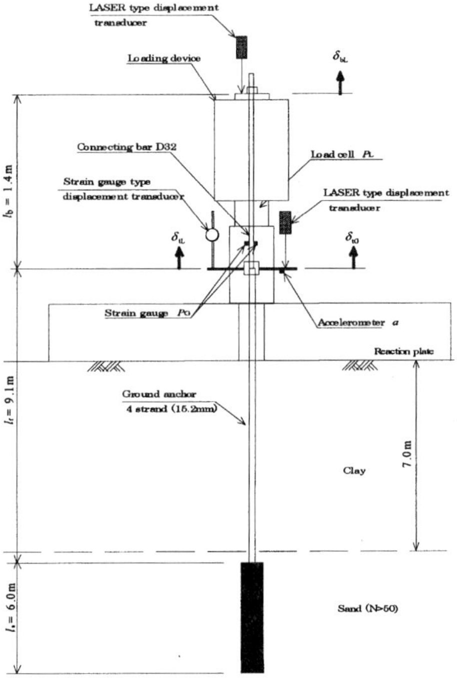

Fig.7 Ground condition and arrangement of test equipment

displacement margin. In order to obtain the maximum load $T_{t,max}$=300kN and the target displacement $\delta_{e,max}$=90mm, the disk springs consisting of 14 layers of the double disks (total 28 disk springs), stacked as in Fig.4 (b), are needed.

Table 1 Testing cases

Case	Initial load T_0(kN)	Maximam test load $T_{t,max}$(kN)	Trigger load for unloading T_s(kN)	Interval of masurement (s)
1	30	-	200	1/5000
2	30	-	150	1/5000
3	30	-	100	1/5000
4	30	-	50	1/5000
5	30	-	200	1/2000
6	30	-	150	1/2000
7	30	-	100	1/2000
8	30	-	50	1/2000
9	30	-	200	1/1000
10	30	-	150	1/1000
11	30	-	100	1/1000
12	30	-	50	1/1000
13	10	200	185	1/5000
14	20	200	185	1/5000
15	30	200	185	1/5000
16	40	200	185	1/5000
17	50	200	185	1/5000

Field experiment of rapid load test

Field experiment were carried out in a testing site in Tochigi, Japan.

Loading device

The schematic diagram of the newly developed loading device is shown in Fig.6. To reduce the manufacturing cost, the device utilizes 2 hydraulic jacks available on market instead of manufacturing special double cylinder structure. Disk springs which can conform the requirements are also procured from the market.

Ground condition

Ground conditions and arrangement of the test equipment are shown in Fig.7. The top layer consists of clay underlain by sand. The relation between maximum pull-out load T_{ug}, bond length l_a, borehole diameter d_a, and ultimate bond stress τ is shown in Eq.3. From the past test data, the ultimate bond stress of 0.35-0.45MN/m^2 has been obtained in sand layer. The ground anchor will not be pulled out during rapid load tests with $T_{t,max}$=200kN.

$$T_{ug} = l_a \cdot \pi \cdot \tau \qquad \text{Eq.3}$$

Experimental case

The combination of parameters to be tested is shown Table 1. Experimental cases are developed based on different initial loads, trigger loads for unloading, sensor types and sensor positions, and interval of measurement. A total of 17 cases were tested with two tests for each case in this experiment.

Ground anchor

The ground anchor used for this experiment consists of 4 PC strands with an outer diameter of d_t=15.2mm, a bond length of l_a=6.0m, a free length of l_f=9.1m, and borehole diameter of d_a=140mm. The loading time t_L should conform 0.015-1.5 seconds from Eq.1.

Method of rapid load test

Testing procedure and control of load

i) Procedure of the work

The tests were carried out with the same work sequence as that of static load tests in order to confirm this sequence was free from problems in terms of workability, overall work time, and smooth measurement. In this regard, when to apply compressive load to the disk springs, before or after installation of the device needed to be investigated.

ii) Control of load

The unloading mechanism detects the load having reached an unloading trigger load T_s with a pressure transducer, and opens the magnetic Valve B. Due to time difference between the sensor reading and oil pressure release by valve operation, the maximum test load T_t to the ground anchor becomes greater than T_S. In this experiment, the relation between T_t and T_s was investigated at T_s =50, 100, 150, and 200kN.

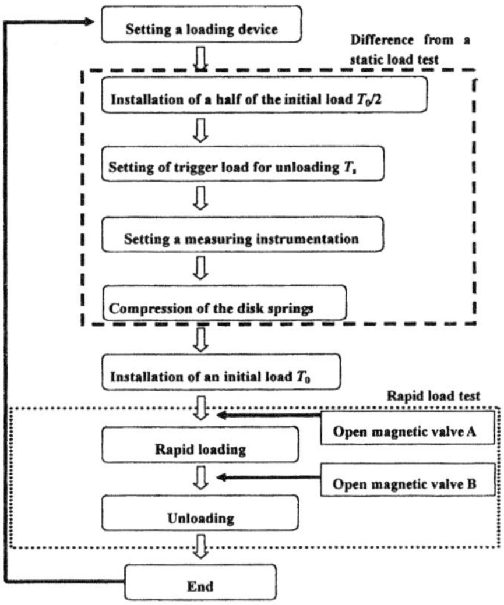

Fig.8 Procedure of work for rapid load

Measurement method

i) Measurement of load

In static load tests, oil pressure measurement is normally used. The load cell measurement is not believed quite accurate due to eccentricity when it applies to a slanting ground anchor. In the rapid load testing, the load needs to be measured by the load cell only because the response of oil pressure is too slow. Therefore the accuracy of load cell measurement needs to be verified by comparing results using the load cell with results obtained from strain gauges attached to the connecting bar.

ii) Measurement of displacement

The strain gauge type displacement transducer is normally used for the measurement of ground anchor displacement. This type of displacement transducer may not follow quick movement of the ground anchor during rapid load tests, and the contact position of the displacement transducer may move by rotation of the tendon. Therefore the accuracy of strain gauge type displacement transducer needs to be investigated by comparing results from the strain gauge type with that of a LASER non contact type displacement transducer.

iii) Measurement of acceleration and velocity

Velocity is calculable by differentiation of displacement or integration of acceleration with respect to time. Acceleration is calculable by two times differentiation of displacement, or measured by an accelerometer. The calculated results are compared with measured results to verify accuracy of the results calculated from the displacement measurement.

iv) Time interval of measurement

The time interval used in rapid loading tests for foundation piles is 1/1000 second or less. In our previous experiments, 1/5000 second was used because the loading speed was very fast. Because the new loading device can control velocity, the measurement for 1/5000 seconds may be unnecessary. Measurements are taken and compared in accuracy at 1/1000, 1/2000, and 1/5000.

v) Arrangement of instrumentation

For some types of ground anchors, it is often required to use a connecting-bar. In this case, displacement measured at the tendon end and the connecting-bar end may differ. The difference in the displacement measurement at the tendon end and that at the connecting-bar end needs to be investigated by comparing results measured at these points.

Applicability of loading device

The applicable range of the loading device needs to be examined from the maximum loading capability of the device and the maximum displacement allowable.

Test results and discussion

Testing procedure and control of load

Procedure of the work

The work sequence to perform rapid load testing is shown in Fig. 8. When the disk springs are compressed before installing the device, the work sequence is similar to that of static load tests. However, the installation works required about 30 to 40 minutes in this case because pressurized oil hoses dangling around the device hindered the installation work. Moreover, 5% to 10% decline in compression force introduced to the disk springs was observed. When the disk springs were compressed after installing the loading device, the time between the compressive load introduction to the load start became 3-5 minutes, and the fall of compressive load did not occur. The time required between the installation start to the load start was about 20 minutes.

Control of load

The relationship between the unloading trigger load T_s and the maximum test load $T_{t,max}$ is shown in Fig. 9. Because of the time lag caused by the magnetic valve operation, $T_{t,max}$ becomes larger than T_s. This difference becomes smaller if T_s is closer to the maximum. The relation can be approximated by Eq.4.

$$T_s = 0.002T_t^2 - 0.090T_t + 148.14 \qquad \text{Eq.4}$$

In the experimental case 13-17, the unloading trigger load was set to T_s=185kN based on the Eq.4 in order to obtain $T_{t,max}$ =200kN. Unloadings occured at $T_{t,max}$=200kN with an accuracy of ±3%.

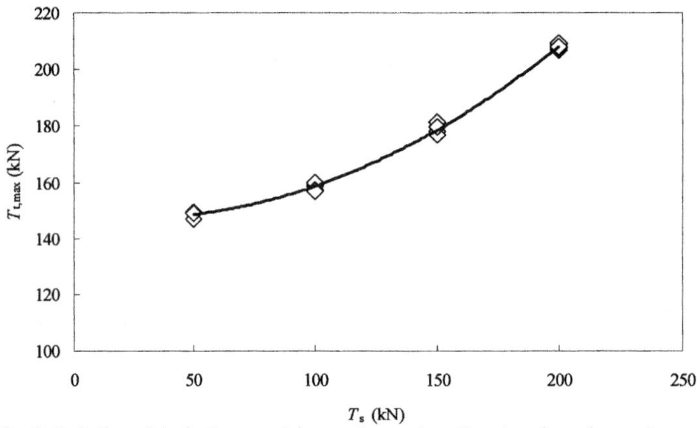

Fig.9 Relationship between trigger for unloading load and maximum test load

Measurement method

Measurement of load

The relationship between the load measured by the load cell P_L and by the strain gauge P_G and time is shown in Fig.10. Although similar results are obtained from the initial load T_0 to the maximum test load $P_{G,max}$, P_L becomes smaller than P_G starting from P_G =50kN during unloading. The ratio of P_L to P_G is plotted over P_G in Fig.11. This seems because the contact between the loading device and the reaction plate becomes somewhat loose. Based on the results, $T_0 \geq 50$kN is necessary to secure a good contact between the device and the reaction plate. By securing contact in this way, load cells can be used to obtain accurate measurements.

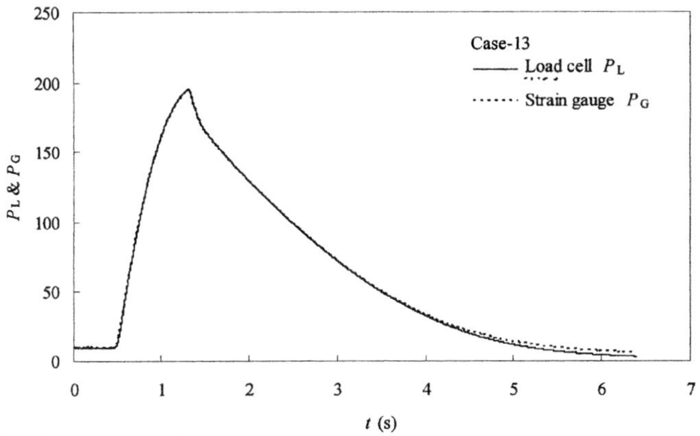

Fig.10 Relationship between test load and time

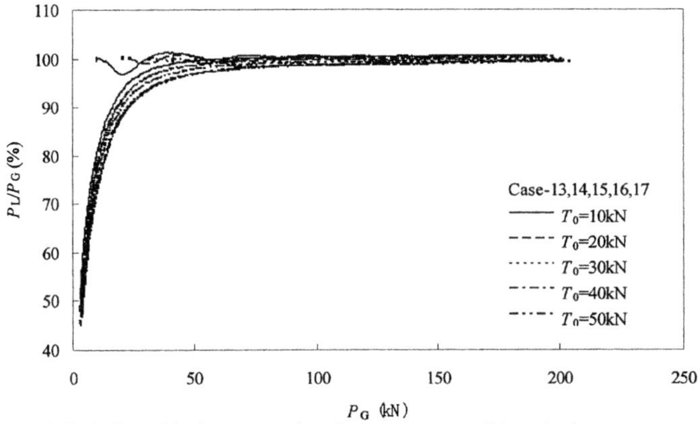

Fig.11 Relationship between load measurement by strain gauge P_G and load rates P_L/P_G

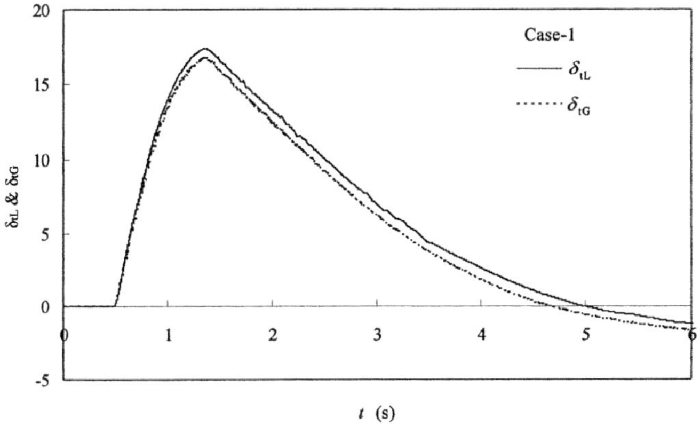

Fig.12 Relationship between displacement and time

Measurement of displacement

The relationship between displacement δ_{tL} measured with the LASER non contact type displacement transducer and displacement δ_{tG} measured with the strain-gauge contact type displacement transducer and time t is shown in Fig. 12. Although the time when displacement starts, and the time when the maximum load reaches are completely the same with δ_{tL} and δ_{tG}, δ_{tG} always shows smaller values than δ_{tL}. This seems caused by rotation of the measuring position. Tendons using PC strands which tend to untwist under load may cause rotation of the anchor head. The rapidly moving anchor head may cause a drag of the strain-gauge contact type displacement transducer, resulting in smaller displacement readings. Based on the results, the LASER non contact type displacement transducer is more suitable for this type of rapid load testing device.

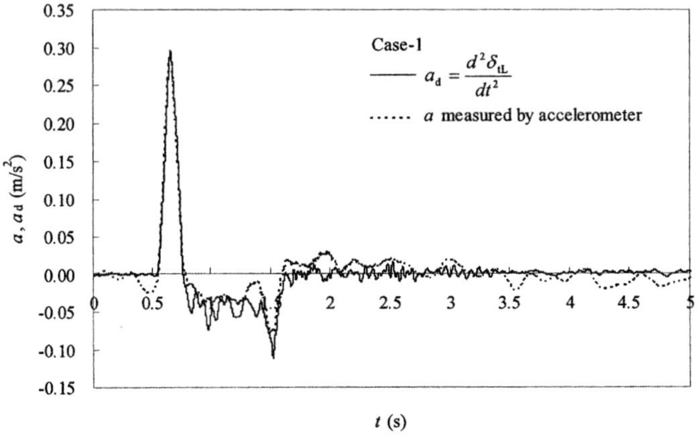

Fig.13 Relationship between acceleration and time

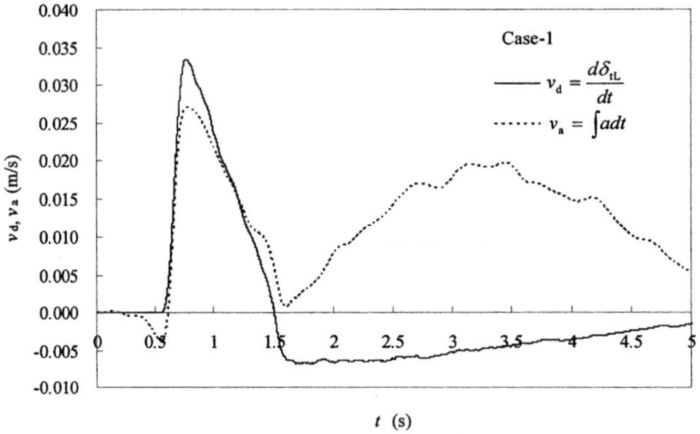

Fig.14 Relationship between velocity and time

Measurement of acceleration and velocity

The relationship between the acceleration measured by accelerometer A and the acceleration calculated by 2 time differentiation of displacement A_d and time is shown in Fig.13. The relation between the velocity obtained by integration of acceleration V_a and the velocity obtained by differentiation of displacement V_d is shown in Fig.14. In acceleration, two results mostly approximate each other and the peaks are practically identical (0.29m/s^2). However in velocity, the calculation results do not approximate to measured results. This is because the accelerometer results seem to be affected by vibrations and the size of the accelerometer may be too large for the range of measurements. Therefore using calculation from the displacement results is more suitable to determine velocity and acceleration because the use of calculation is less influenced by measurement conditions and the number of measurements can be reduced.

Arrangement of instrumentation

The relationship between the displacement measured at the tendon end δ_{tL}, the displacement measured at the connecting-bar end δ_{bL}, the calculated tendon end displacement δ_c, and time t is shown in Fig.15. δ_c is obtained from Eq.6, where l_b is the length of the connecting-bar (l_b =1400mm), A_b is the sectional area of the bar (A_b=415.5 mm^2), and E_b is the elastic modulus of the bar (E_b =200N/mm^2). The difference between δ_c and δ_{tL} stays within ±1.0% range. Therefore when the specifications of a connecting bar are given, the displacement measured at the end of the connecting-bar can be converted into the displacement at the tendon end.

$$\delta_c = \delta_{bL} - T_t \, l_b / (E_b A_b) \qquad \text{Eq.6}$$

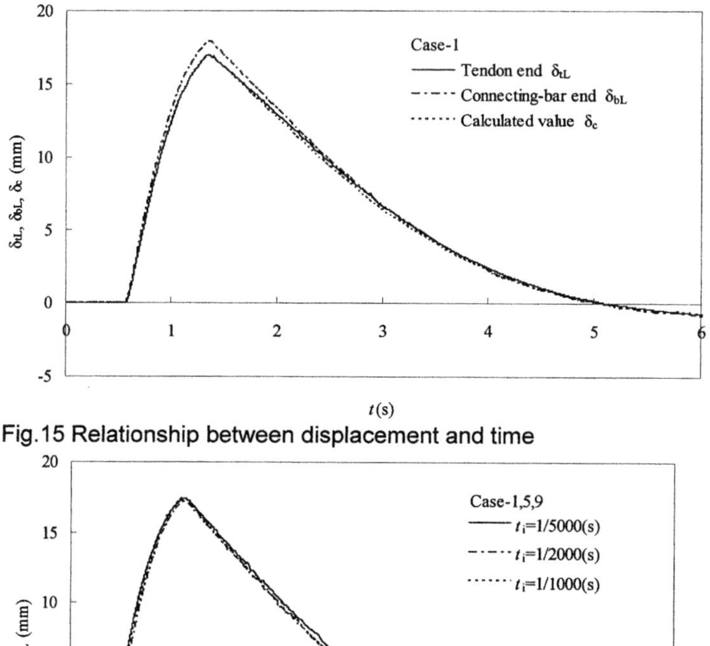

Fig.15 Relationship between displacement and time

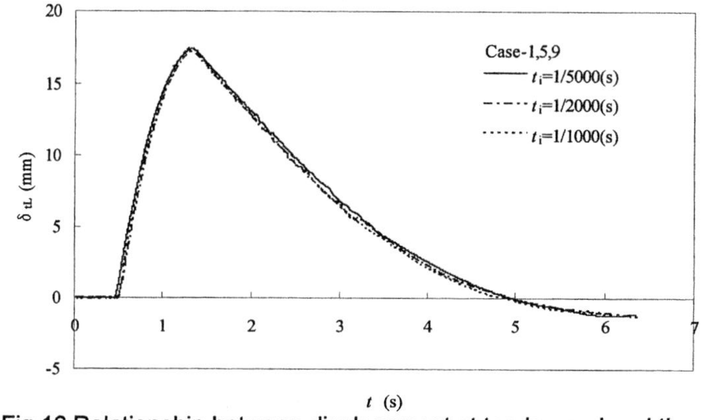

Fig.16 Relationship between displacement at tendon end and time

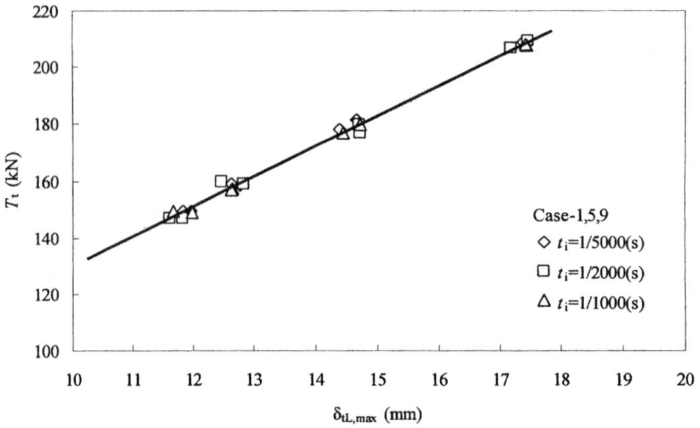

Fig.17 Relationship between test load and maximum displacement

Time interval of measurement
The relationship between displacement δ_{tL} and time t, at different measurement intervals of 1/5000, 1/2000, and 1/1000 second, is shown in Fig.16. The same results are obtained regardless of measurement intervals. The relationship between test load T_t and the maximum displacement $\delta_{tL,max}$ is shown in Fig.17. T_t and $\delta_{tL,max}$ are in linear relationship. Therefore 1/1000 second interval is enough for the measurement.

Applicability of loading device

The loading device is shown to be able to provide the maximum load of $T_{t,max}$=207kN and the maximum displacement of $\delta_{tL,max}$=17mm. The anchor types corresponding to $T_{t,max}$ are D23 (T_{as}=231kN) or smaller in the case of PC-bar tendons, and 3 strands (d_t=12.7mm, T_{as}=280kN) or smaller in the case of PC-strand tendons. The relation between the amount of displacement δ_t of the corresponding anchors and the free length l_f is shown in Fig.18. Based on the maximum displacement provided by the loading device, the applicability range is $l_f \leq 10$m for PC-bar tendons (D23) and $l_f \leq 5$m for PC-strand tendons (3 strand, d_t=12.7mm).

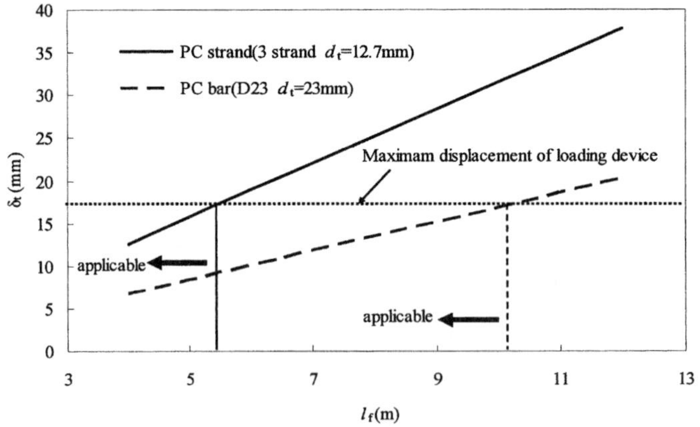

Fig.18 Applicability of the loading device

Conclusions

The following conclusions were obtained through the development of the rapid load testing device and the field experiment.

1) Development of a loading device

i) Loading speed and unloading

The loading device which can control loading velocity and initiate unloading at a target load is devised utilizing disk springs and oil pressure.

ii) The design of the disk springs

The disk springs capable of $T_{t,max}$=300kN with a displacement of $\delta_{t,max}$=90mm was designed and the disk spring structure was determined.

2) Method of rapid load testing for ground anchor

The method of the rapid load testing for ground anchors was established.

i) Test procedure

An optimum test sequence is identified, where compressive force is introduced to the disk springs just before loading starts.

Unloading is successfully carried out at the target test load by calibrating the unloading mechanism.

ii) Measurement method

Load measurement is carried out accurately with a load cell instead of oil pressure measurement as widely used for static load tests. The initial load should be more than 50kN.

A laser displacement transducer should be used instead of strain-gauge type displacement transducer to eliminate noises caused by rotation of the measurement point.

The velocity and acceleration should be obtained by calculation from the displacement at the tendon end.

If the specifications of a connecting-bar are given, the displacement can be measured either at the end of the connecting-bar or at the end of the tendon.

The interval between measurements of 1/1000 second is determined adequate.

iii) Applicability of the loading device

The loading device is applicable to the anchors with the maximum load of 200kN and the free length of $l_f \leq$10m for a PC-bar tendon (D23) and $l_f \leq$5m for a PC-strand tendons (3 strand, d_t=12.7mm).

For further research, data accumulation is needed by applying this device at many sites in order to verify the analytical method of rapid load testing of ground anchors proposed in this paper.

REFERENCES

1) Sueyoshi T. and Tani K. : Field tests to investigate applicability of rapid load test for ground anchor, Proc. 48th JGS Sym. on Geotechnical, pp.351-358, JGS, 2003.

2) Sueyoshi T. and Tani K. : Laboratory tests to investigate applicability of rapid load test for ground anchor, Proc. 34th JSCE Sym. on Rock Mechanics, pp.333-338, 2005.

3) Sueyoshi T. and Tani K.: Loading apparatus of rapid load test for ground anchor, Proc. 35th JSCE Sym. on Rock Mechanics, pp121-126., 2006

4) JGS : Method for Rapid Load Test of Single Piles, JGS 1815-2002, 2002

5) JGS : Design and Construction Standard of Ground Anchor, JGS1401-2000,

Session six:

Corrosion monitoring and corrosion protection systems

Durability of buried and encased ground anchorages and structural components; requirements for corrosivity assessment and protection to ensure service life

Gareth John, CAPCIS Ltd, Manchester, UK

Stuart Littlejohn, University of Bradford, Bradford, UK

Abstract

Corrosion of steel components in ground is often given insufficient detailed attention during the initial design and subsequent service period of anchorages whether rock bolts or tie bars using reinforcement steel or multi-unit tendons using high yield steel. The corrosion process is dependent on many factors such as ground type, and ground and ground water factors including resistivity, chemistry and microbiology. Numerous procedures relating to the assessment of soil aggressiveness/corrosivity are available but most are aimed at general metal loss and not localised corrosion in terms of rate of attack or consequence, where pitting of pre-stressed components can be a precursor to environmentally assisted cracking and ultimately failure. The paper identifies the requirements for assessment of corrosivity and discusses the provision of durability for anchorage components in aggressive environments.

Introduction

Corrosion of buried steel components can range from negligible (typically < 0.001 mm/yr) for virtually non-aggressive conditions to very high (up to 1 mm/yr localised attack) in strongly aggressive conditions. Several interrelated factors affect the overall aggressiveness of the soil and ground water, for example,

- Soil type (clay, sand, gravel, loam, ash etc.)
- Soil condition (disturbed or undisturbed)
- Soil resistivity
- Soil chemistry (pH, water content, chloride/sulfate content, buffer capacity)
- Ground water chemistry (pH, chloride/sulfate content, buffer capacity)
- External factors such as differential aeration/concentration cells, microbiological activity, stray currents (dc or ac).

For temporary anchorages (design life ≤ 2 years), normally only strongly aggressive conditions give concern; for permanent systems (design life ≥ 50 years) the cumulative damage from soils and ground water considered weakly aggressive can be a problem.

Ground anchorages and anchored structures in service, Thomas Telford, London

Assessment of aggressiveness/corrosivity.

The most extensive investigation of corrosion of buried metals/alloys was undertaken in the US in the 1940s and 1950s by Romanoff [1]. This study was mainly aimed at disturbed soils, e.g. where the soil is removed and then replaced, as for many pipelines. A separate study was carried out on corrosion rates for piles in disturbed and undisturbed soils[2].

Methodologies were then developed to provide an assessment of soil corrosivity. All assign ratings to the various influencing factors and the sum of the ratings is used to assess aggressiveness. Schemes include Eyre & Lewis system[3] and DIN 50 929 Part 3[4]. DIN 50 929 Part 3 provides the most comprehensive methodology by far and has been validated on many different projects, following comparison of predicted metal loss to actual metal loss.

In DIN 50 929 Part 3 the aggressiveness of soil is determined by consideration of soil type, extent of contamination, resistivity, pH, moisture content, soil condition (i.e. disturbed/undisturbed), total alkalinity/acidity (buffer capacity), combined chloride/sulfate content (aqueous extract), sulfate content (acid extract), sulfide content, presence of ground water, vertical/horizontal homogeneity and external stray current. DIN 52 929 Part 3 also assesses ground water in a similar manner by considering water mobility, position (fully submerged, at water/air interface or in splash zone), chloride/sulfate content, buffer capacity, calcium content, pH and external stray current.

The individual items are assigned values based on defined criteria ranging from **+4** (virtually non-aggressive) to **-12** (strongly aggressive). The sum of the various items can range from a best case of **+4** to a worst case of < **-30**, and from this the overall corrosivity/aggressivity is assessed. This also allows an estimate of the expected corrosion rate and the practical range (essentially a factor of ×2 on the median corrosion rate), as shown in Tables 1 and 2.

Table 1. Assignment of aggressiveness and expected corrosion rates for soils (after DIN 50 929 Part 3, 1985)

Sum of ratings		Aggressiveness	General corrosion rate (mm/yr)	Localised attack (mm/yr)
≥ 0	**Ia**	Virtually non-aggressive	0.005 (range 0.0025 to 0.01)	0.03 (range 0.015 to 0.06)
-1 to -4	**Ib**	Weakly aggressive	0.01 (range 0.005 to 0.02)	0.05 (range 0.025 to 0.1)
-5 to -10	**II**	Aggressive	0.02 (range 0.01 to 0.04)	0.2 (range 0.1 to 0.4)
< -10	**III**	Strongly Aggressive	0.06 (range 0.03 to 0.12)	0.4 (range 0.2 to 0.8)

Table 2. Assignment of aggressiveness and expected corrosion rates for ground water (after DIN 50 929 Part 3, 1985)

Score		Aggressiveness	General corrosion rate (mm/yr)	Localised attack (mm/yr)
≥ 0	**Ia**	Virtually non-aggressive	0.01 (range 0.005 to 0.02)	0.05 (range 0.025 to 0.1)
-1 to -4	**Ib**	Weakly aggressive	0.02 (range 0.01 to 0.04)	0.1 (range 0.05 to 0.2)
-5 to -8	**II**	Aggressive	0.05 (range 0.025 to 0.1)	0.3 (range 0.15 to 0.6)
< -8	**III**	Strongly Aggressive	0.2 (range 0.1 to 0.4)	1.0 (range 0.5 to 2)

The main items in this method of assessment for soils are listed below.

- All soils contaminated with coal, coke, industrial waste etc. are strongly aggressive.
- Very low resistivity soils (< 30 Ω m) are normally associated with strongly aggressive conditions.
- Whilst pH can range from < 4.5 to > 9.5, its effect can be less important than the buffer capacity of the soil. Soils which are strongly alkaline buffered (as measured as alkalinity $K_{S\,4.3}$ of > 1000 mmol/kg or carbonate content > 1500 mmol/kg) have sufficient residual alkalinity which is beneficial. Soils which are strongly acidic buffered (as measured by acidity $K_{B\,7.0}$ > 20 mmol/kg) are very corrosive, whilst a low buffering capacity (e.g. $K_{B\,7.0}$ < 5 mmol/kg) has little impact on overall aggressiveness.
- Sulfides reflect whether the soil contains active sulfate reducing bacteria (SRB) which can lead to microbiologically influenced corrosion (MIC). Other assessment methods use an in situ measure of the redox potential* of the soil to determine this risk. In general a redox potential < 0 mV SHE indicates strongly anaerobic conditions which can result in severe pitting; whilst redox potential > 400 mV SHE indicates strongly aerobic conditions which would generally indicate a lower corrosion rate and less tendency for pitting.

  * Redox potential is the potential difference measured between a platinum electrode and a suitable reference electrode in contact with the soil, normalised to standard hydrogen potential (SHE) and a pH 7[5].

- Where vertical/inclined anchorages pass through different soils, the main concern relates to the formation of differential aeration and/or concentration cells, where corrosion occurs generally in the lower oxygenated soils, at or just below the interfacial layer.
- Stray current sources (from cathodic protection systems, dc welding, dc railway systems) are assessed by the change in potential on site and hence is only directly applicable to existing facilities. Other methods are available to assess stray current risks at the design stage based on computer modelling[6]. Stray current corrosion can result in severe metal loss and is typically characterised as localised (pitting) corrosion. For example dc stray current discharging from a steel component, will result in a total metal loss of approx 9 kg per amp per year. If the current discharge is spread over a wide area then it may not be significant, but if it is concentrated over a small area it can lead to rapid perforation/dissolution of metal (for example a current density of 0.1 mA/cm^2 (1 A/m^2) gives an approx metal loss rate of 1.1 mm/yr). Whilst ac stray current can also occur, it is normally only a problem on well coated items which insulate the steel from the ground and where there is significant parallelism between ac source and the component (e.g. well coated pipelines running adjacent to high voltage overhead power lines) and hence is not usually an issue for ground anchorages.

Whilst DIN 50 929 Part 3 states that it covers both disturbed and undisturbed soil conditions, the corrosion of components in undisturbed soil is significantly lower than that shown in Table 1. Romanoff [2] concluded that for driven piles the portion in the undisturbed soil does not suffer significant attack. This concept is adopted in Eurocode 3 "Design of steel structures" Part 5[7], where projected corrosion rates for a particular soil show essentially a doubling of corrosion rate from the compacted/driven to non-compacted conditions.

The aggressiveness of soils (disturbed or undisturbed) is directly relevant for anchorage systems where the steel component is in direct contact with the soil (e.g. where the tendon has lost its cement grout cover) or the protective system fails (e.g. tie bars with damaged protective plastic tape wrapping). For permanent anchorages where the components are protected by corrosion protection systems, in accordance with most national standards, the steel tendon is not in direct contact with the ground, hence it is the aggressiveness of the ground water that is more relevant for both soils and rocks, since it is the ingress of ground water through any breaks in the protective system that is of concern.

Ground water ingress into anchorages can occur early in the life of an anchorage, typically due to poor installation practice (Figure 1). This type of problem can lead to early corrosion of strands and wedges, resulting in strand slippage and loss of load (Figure 2).

(a) (b)

Figure 1. (a) Ground water seepage at stressing head and (b) through sheathed strand following installation

(a) (b)

Figure 2. Corrosion of wedge serrations due to ingress of aggressive ground water (after 6 months of exposure)

In addition to general corrosion/metal loss, there is also the risk of brittle failure due to hydrogen embrittlement/stress corrosion cracking (SCC) in high strength strands exposed to aggressive ground water. Here monatomic hydrogen, evolved during the corrosion process, diffuses into the tendon causing dislocation of the metal matrix. When associated with a surface defect/stress concentrator, this results in brittle failure (Figure 3).

(a)

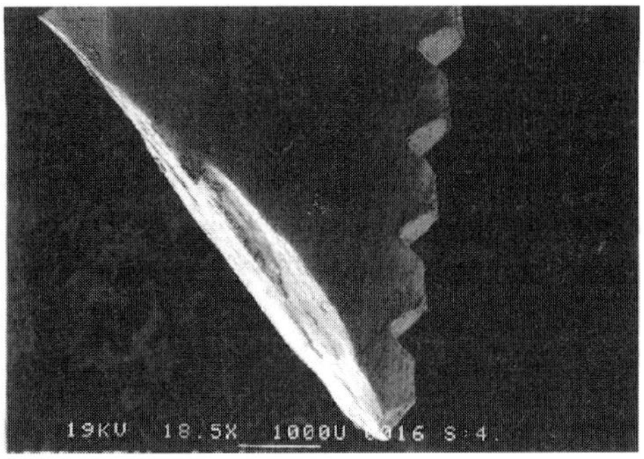

(b)

Figure 3. Failed wire in multi-strand tendon due to stress corrosion cracking

The surface defect where the crack is initiated may be due to a manufacturing defect, installation defect such as wedge serration indentation within the stressed length or a pit induced by localised corrosion. The latter can be a particular problem if the anchorage is subject to contamination or water ingress resulting in sulfate reducing bacteria (SRB) leading to microbiological influenced corrosion (MIC), e.g. failure after 5 years[8]. Depending on the quality of steel and applied stress levels defects/pits greater than 0.4 mm can be critical and lead to failure.

Depending on the local conditions, the threat of corrosion can vary significantly from site to site and also for the different components of the anchorage. Therefore, in order to assess the integrity of existing anchorages and to provide a sound engineering basis for the design of new anchorages, a rigorous assessment of the aggressiveness of the soil and ground water that will impact on anchorages is required on a case-by-case basis. Based on the authors' experience, such a rigorous approach is rare and explains the current dearth of case histories where both the condition of anchorage components and the aggressiveness of the environment are quantified accurately.

Corrosion of steel tendon within grouted length

For many anchorage systems, a significant proportion of the anchorage tendon is embedded in cementitious materials, i.e. grout. Dense cement grout provides a high degree of protection to embedded steel due to the high alkalinity provided by the cement (typically pH > 13), which forms a highly stable thin, protective iron oxide film on the surface, a condition referred to as passivity. The cement grout also provides a uniform, low porosity/low diffusivity cover, limiting water and oxygen migration to the steel surface. Whilst these conditions persist the corrosion rate is effectively nil (< 0.1 μm/yr). However, protective conditions can be lost due to one of more of the following factors.

- Loss of alkalinity due to reaction with acidic ground water/carbon dioxide waters. For grout pH < 10 the protective film is no longer stable and corrosion can occur.
- Ingress of chlorides that disrupts the protective passive film allowing corrosion
- If cracking extends to the steel, the conditions at the base of the crack are different to the section still embedded and hence a differential concentration cell can be formed.

Corrosion only occurs once the grout protection is lost, and the time for this is dependent on the quality of the grout, width of any crack, aggressiveness and mobility of the ground water.

Loss of alkalinity can be due to attack by acidic groundwater, either due to industrial contamination or from CO_2 content. Generally, ground water conditions are assessed in accordance with EN 206-1[9] e.g. aggressive for pH < 5.5 or CO_2 content > 40 mg/L, and strongly aggressive for pH < 4.5 or CO_2 content > 100 mg/L. In the UK, BRE SD-1[10] is used for assessment for aggressive conditions to cementitious materials but this is mainly aimed at sulfate attack.

The chloride level required to disrupt the passive film is pH dependent; generally corrosion can occur where the chloride/hydroxide molar ratio in the grout pore water exceeds 0.6. In practice this is difficult to measure and chloride levels within grout >0.4% m/m cement content are normally considered sufficient to allow corrosion to proceed. BS 8081[11] stipulates that the total chloride derived from all sources should not exceed 0.1% m/m cement content for stressed tendon bonding. However, the actual corrosion rate will still be dependent on the oxygen availability at the tendon surface. In fully saturated conditions this can be very low, since the oxygen has to diffuse through the water in the grout pores to get to the tendon. The main concern relates to cases where, for example, there is a differential oxygen cell with part of the tendon exposed to free flowing aerated ground water and other parts exposed to static conditions resulting in the oxygen content reducing with time. In these cases the corrosion is concentrated in the low oxygen zone and driven by the oxygen reduction elsewhere on the tendon.

The effect of cracks is to increase the likelihood of differential concentration cells being formed and hence allowing corrosion to occur. Recent developments in predicting durability of reinforced concrete (ACI 365.1R[12] and Concrete Society TR61[13]) allow an assessment of the time to first damage for concrete structures to be determined. This method of assessment, based on a combination of determining the concrete quality from the mix design and then estimating the rate of attack dependent on exposure conditions, is increasingly being used in current practice. Whilst these overall principles apply they have not yet (to the authors' knowledge) been applied to ground anchorages. A methodology for modelling the degradation of concrete from acids is currently being developed[14] and is expected to allow the rate of deterioration from exposure to acidic conditions to be assessed and hence provide a predictive tool for the onset of corrosive conditions for steel components embedded in cement grout

Principal methods of protection against corrosion

Most national anchorage standards recommend corrosion protection of all permanent steel components at time of construction. For example, BS 8081[11] requires either one or two impermeable protective barriers ("single" and "double" protection) and provides guidance on system selection that is the responsibility of the designer. The barriers normally comprise grouted plastic ducts over the tendon bond length and grease and plastic sheaths and/or ducts over the free length. In "double protection", dense cement grout can be considered as one of the two permanent protection barriers provided (i) the grout is injected under factory or equivalent controlled conditions with a cover between the tendon and the outer barrier of at least 5 mm and (ii) the crack width during serviceability loading has been proven to be less than 0.1 mm. This traditional crack width limit is considered appropriate for high yield pre-stressing steel with high carbon content (i.e. with specified yield stress > 1200 MPa; specified tensile stress > 1500 MPa or hardness > 450 Hv_{10}), but for standard and high strength reinforcement steels (i.e. with specified yield stress < 500 MPa or hardness < 210 Hv_{10}), the crack width limit can be relaxed to 0.2 mm, in accordance with the recommendations of Eurocode 2 "Design of concrete structures" (EN 1992)[15]

In principle, a protective system should aim to exclude a moist gaseous atmosphere by totally enclosing the tendon and anchor head with an impervious covering. The following properties are pertinent[11].

- have an effective life at least equal to that required of the anchorages
- not adversely affect the environment nor the efficiency of the protected anchorage
- not restrict movement of the free tendon length (at least until the anchorage has been tensioned and locked–off, where re-stressability is not required)
- comprise materials that are mutually compatible with respect to deformability, permanence and avoidance of corrosive circumstances
- allow for once only treatment because, with few exceptions, protective systems within the free length and tendon bond length can neither be replaced nor maintained
- not fail during stressing to proof load, especially at junctions between components of the protective system
- be tough enough to withstand handling during manufacture, transport and installation
- permit inspection prior to installation, where practicable.

The European Standard EN1537 [16] requires that *"The minimum corrosion protection surrounding the tendon(s) of the anchor shall be a single continuous layer of corrosion preventive material which does not degrade during the design life of the anchor."* EN 1537 also recognises the problems with corrosion of buried components and states "*There is no certain way of identifying corrosive circumstances with sufficient precision to predict corrosion rates of steel in the ground. All steel components which are stressed shall be protected against corrosion for their design life*". For permanent anchorages, EN 1537 requires as a minimum the tendon to be surrounded by a single continuous layer of corrosion preventive material which does not degrade during the design life of the anchorage. A variety of systems is described, e.g. two protective barriers without testing in situ and a single barrier where the integrity of the single barrier must be proven by testing each anchorage in situ.

Testing single protective barrier systems during construction to demonstrate integrity is an important quality control but whether the initial demonstration is a reliable indicator of the long-term performance of the protective system is questionable[17, 18]. The assessment of the integrity of existing anchorages, many of which were installed with corrosion protection measures now considered as temporary, is also an issue that should be addressed.

Typically, such integrity tests comprise electrical resistance measurements and the insulation is deemed to be acceptable provided the measured resistance to earth is > 0.1 MΩ. Such a methodology can, theoretically, identify the presence of defects in any non-metallic sheath, however the size of defect that can be detected is proportional to the square of the soil resistivity (e.g. for soil resistivity of 500 Ω m typical minimum detectable defect size would be approx 10 mm^2, whilst for soil resistivity of 50 Ω m, minimum detectable size would be approx 0.1 mm^2). This is the reason why systems that "*pass*" one day, may be found to have "*failed*" if re-tested after period of heavy rain fall. Also in the discussion at the ICE Ground Anchorage Conference in 1997it became apparent that where ≤ 10% of the anchorages failed the integrity test, they might still be accepted into the works[18, 19]. This undermines confidence in the sole use of this test when assessing the integrity of a single protective barrier once installed and there is a strong argument for a supplementary in situ water test to provide reassurance.

The entire anchor has to be protected from corrosion and importantly the anchor head. EN1537 states "*Where the environment is aggressive, early protection of the anchor head shall be applied to both temporary and permanent anchors*". A major area of concern is the void space behind the anchor head (Figure 4a), where lack of protection permits exposure to ground water. The void can be filled with a suitable protective material. Cement grout has been used but this brittle material can suffer cracking and undermines the facility to carry out load lift-off tests in service, so a flexible system is preferred. High viscosity industrial greases are used but proprietary systems that include selected corrosion inhibitors should be preferred[20].

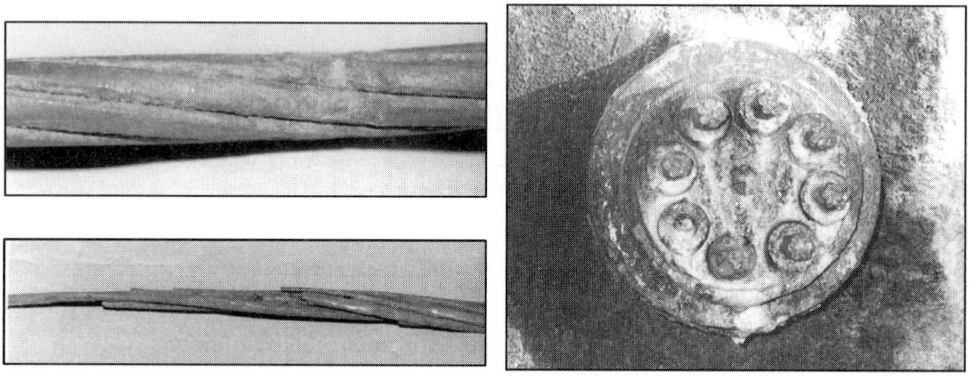

(a) (b)

Figure 4. Corrosion and failure of tendon in void space behind the stressing head due to inadequate protection (a) and to stressing head itself (b).

For the outer head, some systems have used cement grout or mortar to provide external protection (Figure 5). Aside from preventing load lift-off tests, long term protection, may be lost, especially in marine environments where chloride ingress would be expected. In addition, cement that gains access to the serrated wedge interface within the stressing head can reduce grip efficiency leading to strand slippage, bearing in mind that the wedge/tapered hole interfaces should be oiled or greased to lubricate the gripping mechanism and provide a seal to prevent corrosion and the ingress of deleterious materials[21].

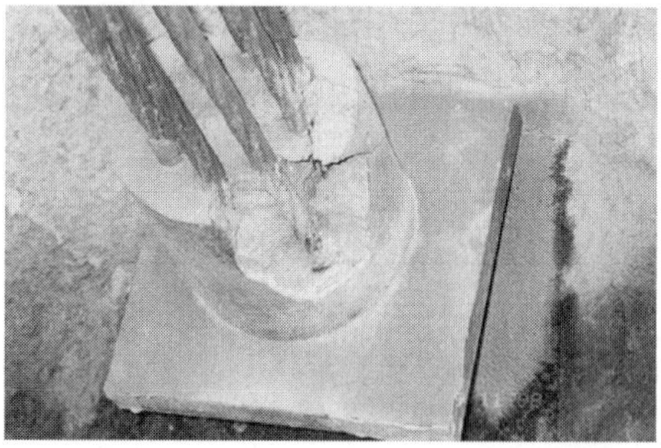

Figure 5. Inappropriate cement grout protection of anchor head

Figure 6. Protective wrap to bar anchor.

Other forms of protection for the anchor head include grease filled "top hat"[21]. For bar tendons, the anchor head area can be protected using a combination of corrosion inhibiting grease and petrolatum tape (Figure 6).

Aside from the prudent specification of regular inspections, maintenance and service behaviour monitoring, additional measures may be taken at the design stage to improve long-term integrity of the system, e.g. site drainage. For existing structures the options are more limited and include injection of protective compound behind the anchor head, maintenance and remedial protection of the exposed components at the anchor head.

In some cases the application of cathodic protection where current flows from the surrounding ground to the tendon may be applicable, but only where the free and/or fixed lengths are not covered by a metallic or non-metallic protective sheath (e.g. two-stage fully bonded tendon) and where cathodic polarisation will not result in hydrogen damage to the tendon. Any such system must be installed to ensure protection to the entire non-sheathed portion of the anchorage and requires careful design to ensure that over-protection does not occur, since there is the chance of generating hydrogen on the steel surface if the applied current density is too high. For most normal steels (i.e. with tensile stress < 1100 MPa / hardness < 350 Hv_{10}) this may only cause a problem if the resulting potential on the steel surface exceeds -1200 to -1500 mV (as measured against copper/copper sulfate reference electrode, CSE). However, anchorage tendons fabricated from very high strength steels (with specified yield stress > 1200 MPa / specified tensile stress > 1500 MPa / hardness values of > Hv_{10} 450) and such steels are particularly sensitive to hydrogen. Sufficient hydrogen to induce damage in these very high strength pre-stressing grade steels can be generated at potentials in the range -900 to -1100 mV CSE, which is within the normal protection range for cathodic protection systems. Hence, if cathodic protection is considered for existing anchorages, it is vital that the polarisation limit for the particular strength/alloy material is determined under simulated operating conditions[22] and the final system is subject to detailed design and close operational monitoring to ensure that excessive polarisation is prevented.

Conclusions

The aggressiveness/corrosivity of ground and ground water should be quantified accurately using a well proven method such as included in DIN 50 929 Part 3 for each proposed anchorage site, in order to predict the rates of corrosion and determine the most appropriate type of corrosion protection required for the design life of the ground anchorages. Even for weakly aggressive conditions, permanent anchorages with a design life ≥ 50 years should be provided with a corrosion protection system.

Dense cement grout cracks, particularly within the tendon bond length under tensile stress, and differential concentration cells can be formed. Cement grout is also susceptible to a loss of alkalinity in acidic ground water/carbon dioxide waters and when exposed to the ingress of chlorides. As a consequence, it is considered appropriate that one and two physical barrier protective systems have been developed for permanent anchorages, particularly in Europe, where the barriers, e.g. impermeable plastic ducts/sheaths, comply with EN1537 [16].

However, cement grout can be considered as one of the physical protective barriers when the grout is placed around the tendon within a robust duct under factory or equivalent conditions with a specified cover (≥5 mm) and where the crack width during serviceability loading has been proven. Currently, the appropriate crack width limit is 0.1 mm for high yield steel tendons (i.e. yield strength > 1200 MPa) and this crack width can be relaxed to 0.2 mm for standard reinforcing steels (i.e. yield strength < 500 MPa), in accordance with EN1992 [15].

Bearing in mind the potential condition of old anchorages with inadequate corrosion protection, there is a growing need to monitor regularly their condition where the aggressiveness of the environment is also known, in order to provide relevant information to predict the remaining service life that in turn dictates if maintenance or remedial measures are necessary to maintain anchorage performance and stability of the anchored structure.

References

1 Romanoff, M (1957) *Underground Corrosion.* National Bureau of Standards circular 579 (republished by NACE International, Houston, TX, 1989, ISBN 0-915567-47-4).

2 Romanoff, M (1972) *Underground corrosion of steel piling,* National Bureau of Standards Monograph 127 (and paper no 82 *Performance of steel pilings in soils,* presented at NACE Corrosion '69, Houston, TX, 1969)

3 Eyre, D & Lewis, D A (1987) *Soil corrosivity assessment.* Report No. CR54, Transport Road and Research Laboratory, Crowthorne, England

4 DIN 50 929 Part 3 (1985) *Probability of corrosion of metallic materials when subject to corrosion from the outside – Buried and underwater pipelines and structural components.*

5 BS 1377-9 (1991) *Methods of test for soils for civil engineering purposes — Part 3: Chemical and electro-chemical test*

6 Aylott, P, Cotton, I & Hassanein, A (2003) *The application of modelling systems at the design stage to the mitigation of stray current interference.* Paper No. 03706, presented at NACE Corrosion 2003, San Diego, CA. USA

7 EN 1993-5 (2007) *Eurocode 3 – Design of Steel Structures. Piling*

8 Littlejohn, G S and Mothersille, D (2007) *Case histories of detailed investigation and testing of permanent ground anchorages in service,* Geotechnical Engineering (October Issue)

9 EN 206-1 (2000) *Concrete - Specification, performance, production and conformity*

10 BRE (2005) *Concrete in aggressive ground,* SD-1 (3rd ed). Building Research Establishment, Garston, England.

11 BS 8081 (1989) *British Standard Code of practice for Ground anchorages*

12 ACI 365.1R (2000) *Service-life prediction – state-of-the-art report*

13 Concrete Society (2004) *Enhancing reinforced concrete durability – guidance on selecting measures for minimising the risk of corrosion of reinforcement in concrete.* Report TR61

14 Beddoe, R E and Dorner, H W (2005) *Modelling acid attack on concrete: Part 1 – The essential mechanisms.* Cement and Concrete Research, 35, 2333-2339.

15 EN 1992-1-1 (2004) *Eurocode 2 - Design of concrete structures*

16 EN 1537 (2000) *Execution of special geotechnical work- Ground anchors*

17 Klemenc, I and Kuhar, V (2007) *Some examples of the poor installation of electrically isolated prestressed ground anchors in Slovenia.* Proc. Int. Conf. "Ground anchorages and anchored structures in service", Institution of Civil Engineers, London

18 Kuo, C and Liao, H-J (2007) *A case study on corrosion protection effect of using field-mixed resin as tie-back anchors' grout material for retaining wall and its inspection method.* Proc. Int. Conf. "Ground anchorages and anchored structures in service", Institution of Civil Engineers, London.

19 ICE (1997) *Long term performance: discussion Session 6.* Proc. Int. Conf. "Ground anchorages and anchored structures", Institution of Civil Engineers, London.

20 Anon (1999) *Denso Void Filler protects naval dockyard anchors,* Journal: Anti-Corrosion Methods and Materials, Vol. 46, Issue 6 (December).

21 Littlejohn, G S and Mothersille, D (2008) *Maintenance testing and service behaviour monitoring of permanent ground anchorages,* Geotechnical Engineering (April Issue)

22 Billingham, M and John, D G (2008) *Determining compatibility of high strength steels to cathodic protection.* Paper to be presented at NACE Corrosion 2008, New Orleans, March 2008.

Application of Linear Polarisation Resistance Measurements and System Modelling to Corrosion Assessment of High Strength Rock Anchorages

Gareth John and Petra Ernst, CAPCIS Ltd, Manchester, UK

Abstract

As part of an integrity assessment of prestressed ground anchorages installed in a gravity dam, a trial on-site corrosion monitoring investigation was carried out, in which three 122 mm diameter cores, up to 70 m deep were drilled into the dam adjacent to three of the anchorages. A novel probe arrangement was then lowered into the core (which was filled with water) and electrochemical potential, corrosion rate (LPR) and system resistance measurements made at 2 m intervals down the core.

The objective of the trial was to attempt to establish if corrosion of the tendon was active, and if that was the case to provide an estimate of the rate of corrosion of the tendon.

In order to determine the area of anchorage being polarised in the LPR measurement, a finite element analysis model of the system was developed and used in conjunction with the resistance data obtained. The rates were also compared to laboratory test results carried out on actual ground water samples taken from within the dam.

Introduction

An on-site corrosion investigation was carried out on a dam which was completed in 1962 and is about 50 m high and 260 m wide. The dam was designed as a pre-stressed concrete dam utilising the Coyne system. The foundation of the dam is located entirely in dolerite, which was shown to be sound and above average hardness. Further, the thickness of the dolerite is well above 200 ft (66 m) with no major changes in its composition and properties. The trial corrosion monitoring system was applied to three of the pre-stressed ground anchorages installed on the dam.

The ground anchorages in the dam, which run from the top ridge beam and up to 54 ft (16 m) down into the dolerite, comprise 102 no pre-stressed high strength quenched and tempered parallel wires, each of 0.2 inch (5 mm) diameter. Each wire was separated from its neighbouring wires using a template, to allow the grout to flow evenly between the wires. The lower end of the anchor is grouted into a sheet steel shoe and then the fixed anchor is stressed to a force of approximately 220 tons (2192 kN). Thereafter, the free length of the tendon in the borehole is grouted in place to provide a fully bonded anchorage. Finally, the anchor head was encased and overlain with reinforced concrete making it inaccessible.

The overall concrete structure consists of several blocks. The anchorages investigated were situated in two blocks on the right abutment (identified here as #A and #C) and one block on the left abutment (identified here as #B). For anchorages #A and #B the nominal depth of the concrete dam (i.e. start of the dolerite) was 32 m, whilst for anchorage #C the dolerite started at a depth of about 40 m.

The specification for the wires required a guaranteed ultimate tensile stress (GUTS) of 224 ksi (1544 MPa), equivalent to a hardness value of approximately 470 H_V10. Such high strength components raise the possibility of the wires being susceptible to hydrogen damage if any corrosion takes place.

Within the dam, two rows of anchorages exist, approximately 640 mm apart with the individual anchorages spaced by approximately 560 mm. The anchorages were installed during the early 1960s and whilst they are nominally protected by the neat cement grout (with a water-cement ratio of 0.36 and ca 0.7 by weight over the fixed and free lengths, respectively) injected into the anchorage column they are not in compliance with more modern codes (such as BS 8081 [1]) which considered this type of anchorages construction as "temporary" (i.e. a maximum design life of 2 years); BS 8081 requires *single* or *double protection* for permanent anchorages.

For corrosion to take place on the anchorages in question two separate conditions are required, viz

- Loss of protection from the grout and
- Sufficient oxygen available to drive the corrosion process.

Loss of protection could occur either due to cracks or bleed pockets formed in the grout column or due to the grout being exposed to acidic ground waters. Provided the grout surrounding the tendons was chloride free the tendons would have been protected initially by the formation of a stable iron oxide passive film. However, it is understood that the grout column is probably cracked in places with additional bleed pockets, which could allow differential concentration cells to be formed. Here the portion of the tendon that spans the crack would be exposed to different chemical conditions to the majority of the tendon that is still embedded in the grout. Whilst initially the tendon would also have an adherent cementitious film, with time the passive film and hence protection could be lost. The extent of this would be dependent on the chemistry of the local ground water and the rate of flow through the crack.

As with reinforced concrete structures, if the cracks in the grout column are sufficiently small (normally considered as being <0.1 mm wide) then it is likely that the cracks would close due to *autogenous healing* [2]. However, if the cracks are of sufficient width (normally considered as being >0.3 mm wide) then it is unlikely that any significant healing would take place.

The trial corrosion investigation reported here was part of an overall major engineering review of the dam by its operators, which also included structural monitoring and structural assessment and previous laboratory studies on the corrosivity of the seepage water extracted from the dam.

Investigation

Corrosion Probe

For this trial corrosion investigation a prototype probe was designed, comprising a 300 mm long, 50 mm diameter, stainless steel cylinder, 2 no silver/silver chloride/potassium chloride

(Ag/AgCl/KCl) reference electrodes, a temperature sensor, 70 m of 5-core wire, and a central support wire to hold the probe's weight, see Figure 1. Foam was wrapped around the probe to prevent the probe from spinning in the borehole. The probe was installed into vertical boreholes (122 mm diameter) which had been previously cored adjacent to the three test anchorages at a separation distance (edge-to-edge) of 600 mm. A separate electrical connection was made to the anchorages utilising the existing anchorage resistance monitoring panels. The borehole depths ranged from 45 m to 65 m and were filled with reservoir water to provide a conductive media.

The probe was lowered into the boreholes and the following measurements were taken at 2 m intervals

- Corrosion potential of the anchorage
- Temperature of the probe
- Corrosion rate of the anchorage (using Linear Polarisation Resistance Method)
- Electrical Resistance between the probe and the anchorage (at 135 Hz).

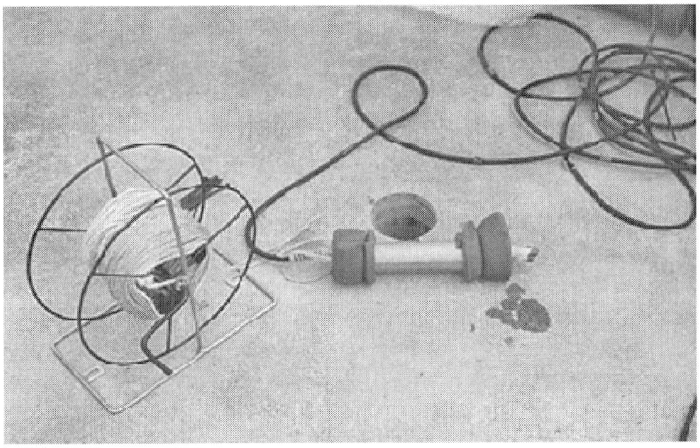

Figure 1. Corrosion sensor as used on site.

Corrosion Potential Measurements

The corrosion potential measurements where taken between the tendon and the reference electrodes at each measurement depth down the borehole. Potentials were recorded against both references in the probe to provide a check reading. In all cases any variation in potential from the two electrodes was negligible (< ±2 mV).

Linear Polarisation Resistance

Electrochemical measurements were carried out using a portable ACM Instruments Field Machine, Grange-over-Sands, Cumbria, UK, to measure the linear polarisation resistance (R_p); firstly measurement of the free corrosion potential (E_{corr}), then a potential equivalent to E_{corr} minus 20 mV was applied for 30 seconds after which the potential was raised by 40 mV (i.e. to E_{corr} plus 20 mV) for a further 30 seconds. The current response was recorded throughout. The linear polarisation resistance was then calculated as shown in eqn 1.

$$R_p = \frac{\Delta E}{\Delta I} \qquad (1)$$

where R_p = Polarisation Resistance (Ω),

ΔE = the applied potential difference (in this case ΔE = 40 mV), and

ΔI = the measured current difference (mA)

The polarisation resistance is then used to estimate the corrosion rate using the Stern Geary equation [3] (eqn 2):

$$corrosion\_rate = \frac{k \cdot B}{A \cdot R_p} \qquad (2)$$

where B = Stern Geary Constant (for carbon steel in natural waters B ≈ 24 mV, for steel in concrete / grout B ≈ 26 mV when active and B ≈ 52 mV when passive) [4],

A = Steel tendon surface area (cm^2),

k = constant to convert from current density to corrosion rate (for B in mV, A in cm^2, R_p in Ω and corrosion rate in mm/yr then k = 11.5)

It should be noted that typical corrosion rates for steel embedded in grout that is fully passive will be <0.01 μm/yr. Corrosion rates for steel that is active (i.e. has lost passivity) can range from ~1 μm/yr (which would be consistent with oxygen starved conditions) to >30 μm/yr (which would be considered as very high and only found in well aerated structures).

For high resistivity environments the measured polarisation resistance includes an ohmic error (referred to as an IR drop) which also has to be taken into consideration. This IR drop affects the potential difference actually applied to the anchorage. This was compensated for by measuring the system resistance (see below).

Resistance measurements

The electrical resistance measurement was to be used to provide both a correction for the Linear Polarisation Resistance Method (LPR) and to provide an estimate for the length of the anchorage (and hence surface area) polarised in the tests.

The resistance between the probe and the anchorage was measured by establishing the corrosion (rest) potential at the position and applying an AC signal with 10 mV rms at a frequency of 135 Hz. Ten measurements were made and the average calculated. The resistance measurement is used for a) IR drop compensation and b) estimation of the polarised area using the following formula.

$$A = \frac{\rho \cdot L}{R_M} \qquad (3)$$

where A is the area of the polarised tendon (m^2)

ρ is the resistivity of concrete and dolerite, respectively (Ω m),

L is the distance between the probe and the anchorage (0.64 m), and

R_M is the measured resistance (Ω)

The above analysis is considered justified as a reasonable first approximation, assuming that (a) the resistivity of the media is consistent and (b) the effective area of the tendon in the resistance measurement is the same as during the corrosion rate (linear polarisation resistance) measurement. The resistivity of the concrete and dolerite was determined from 20 mm diameter core samples using an Insulation Tester (UNILAP ISO X NORMA). The resistivity for wet concrete was determined as 370 Ω m and for wet dolerite 19,000 Ω m (i.e. some 50x higher).

Computer modelling

When carrying out polarisation resistance measurements in the laboratory the surface area of the electrode under test is known and controlled. Unfortunately for on-site monitoring of "real" structures, a non-uniform current distribution can exist between the auxiliary electrode and the working electrode (in this case the anchorage tendon) which determines the actual length of steel tendon (and hence the surface area of the tendon) that the probe polarises.

The length (or area) is dependent upon the distance between the probe and anchorage and the resistivity of the environment. Due to the high resistivity of the dolerite and in more general terms the different resistivities of the materials involved i.e. water, concrete etc, the current distribution is difficult to determine (Figure 2). In order to aid the interpretation of the resistance data a computer model was developed.

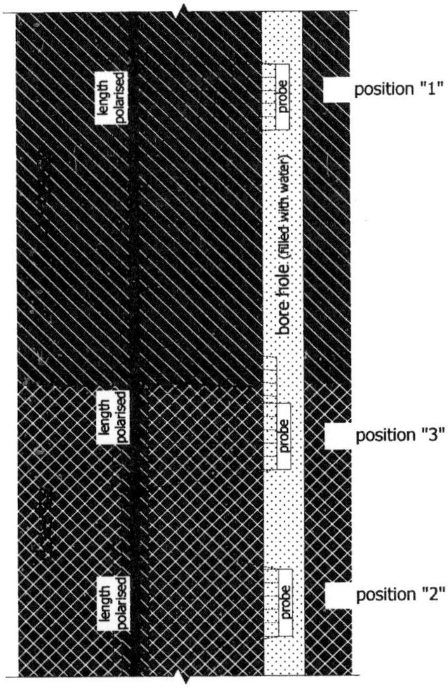

Explanatory notes (Figure 2):

The probe is lowered down the borehole and, measurements are taken at 2 m intervals.

The anchorages in the upper part of the dam are surrounded by concrete, hence at position "1" the nominal current path is expected to follow simple geometric pattern as shown.

At lower depth the anchorages are surrounded by bedrock (dolerite) with significantly higher resistance. Hence there are two possible current paths that could be followed.

- *At position "2" a similar geometric pattern occurs.*
- *However, at position "3" some current could also travel up the water column and then pass trough the lower resistivity concrete to the tendon. In this case the length of tendon polarised could be significantly larger.*

Figure 2. Impact of the surrounding media upon current distribution and path

The computer model was constructed within the MALZ module of the CDEGS [Current Distribution, Electromagnetic Fields, Grounding and Soil Structure analysis] software, from Safe Engineering Services & Technologies, Montreal, Canada. The module allows currents to be injected and collected at various points in a conductor network. It computes the flow of these currents through each individual conductor within the network. This computer model represents all major parts of the system investigated. The following physical constants were used in the model:

Borehole length	50 m	Resistivity of borehole water	30 Ω m
Borehole diameter	122 mm	Resistivity of concrete	370 Ω m
Steel tendon circumference	345 mm	Resistivity of dolerite	19 kΩ m
Distance between borehole and probe	640 mm	Concrete depth	30 m
Probe length	300 mm	Injected current	1 A

Results

On site Corrosion measurement

Corrosion potentials

ASTM C876 [5] gives guidance on the interpretation of potentials for steel embedded in concrete / grout. Typically measured potentials more positive than –100 mV Ag/AgCl/KCl are generally indicative of passive (i.e. non-corroding conditions) and potentials more negative –300 mV are generally indicative of active corrosion.

Figure 3 illustrates the measured potentials for the individual anchorages at different depths down the borehole. The start of the bedrock for each anchorage is also shown.

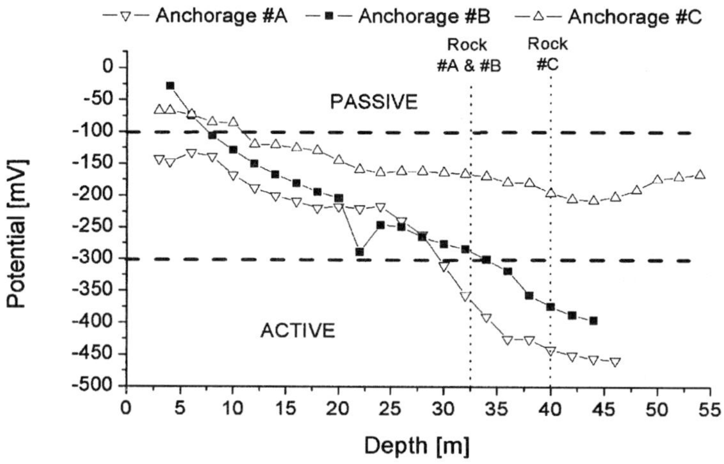

Figure 3. Comparison of corrosion potentials for all three anchorages. (The start of the dolerite in each bore hole is also indicated)

As can be seen the corrosion potentials generally decrease (become more negative) with depth. The corrosion potentials for anchorage #A and #B are similar, and in both cases are more negative than those obtained on anchorage #C. This is indicative that #A & #B have a higher risk to corrosion compared to #C, all other factors being equal. The change in slope of the corrosion potential for anchorage #A between 25 m and 35 m to about –450 mV might be a sign for natural water having direct access to the anchorage through cracks in both the concrete and the bedrock, reducing the protection from the grout to the tendon. In all cases, only the potentials in the top 6 m of the anchorage give values indicative of passive conditions.

Resistance Measurements

The measured resistances are shown in Figure 4. As can be seen the system resistance with the probe in the concrete area is reasonably consistent for all three anchorages, between 30 to 50 Ω. The interface between concrete and rock can be clearly seen by the sharp increase in the resistance measurements, to >1000 Ω in all three cases and also generally increases with increasing depth. For anchorages #A and #B the rate of increase with depth is similar, whilst for anchorage #C there is a decrease of the resistance at around 50 m, which is probably due to either changes in the rock itself or cracks which are filled with water.

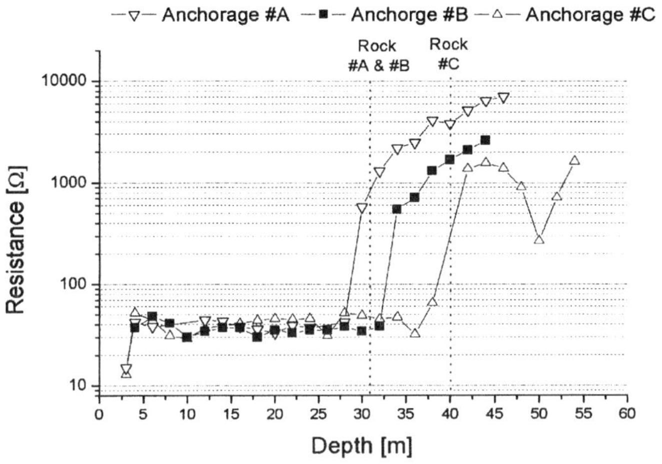

Figure 4. Comparison of resistance measurements for all three anchorages. (The start of the dolerite in each bore hole is also indicated)

Area calculation

As noted above where the probe was located in the concrete portion of the dam, the measured average resistance was 40 Ω, whilst the corresponding resistance values for the dolerite were not constant and rose with depth. Using the resistivity values previously determined the polarised area in wet concrete and dolerite (assuming an average resistance value of 1600 Ω) can be calculated and is approximately 5.9 m^2 and 7.6 m^2 respectively. For dolerite, this calculation is very approximate.

Corrosion Rates

Figure 5 shows the estimated corrosion rates for the individual anchorages. Generally, the corrosion rates within the concrete are fairly low decreasing even further within the dolerite. The low corrosion rates within the dolerite are thought to be due to the extremely high resistance of the dolerite.

For anchorage #A the estimated corrosion rate within the concrete dam is low at around 1 to 3 μm/yr (with peaks of 8 μm/yr and 15 μm/yr at 6 and 24 m depth) dropping to < 0.02 μm/yr once in the dolomite. For anchorage #B, the corrosion rate in the concrete is slightly higher ranging from 1 to 5 μm/yr (with peaks of 15 μm/yr at 18 m and 50 μm/yr at 24 m depth). Again the corrosion rate drops once in the dolerite from 0.1 to 0.01 μm/yr. For anchorage

#C, the estimated corrosion rate is generally higher between 0.8 to 9 μm/yr in the concrete (with no obvious peaks). The corrosion rate in the dolerite ranged from 0.015 to 0.8 μm/yr.

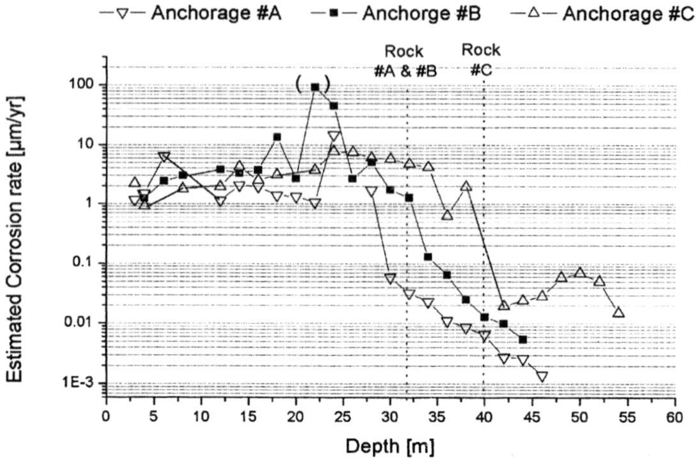

Figure 5. Comparison of the estimated corrosion rates for the individual anchorages. (The start of the dolerite in the different bore holes is also indicated)

Computer model

The results of the computer model (see Figure 6) confirmed that the current distribution is not only focused at the location where the probe is but is distributed along the length of the tendon and is dependent upon the probe location.

As already assumed in Figure 2, the current is not just travelling into the concrete or dolerite or both but also up and down the borehole. If the probe is located within the concrete, the current is only entering the tendon over the depth of the concrete. Closer analysis of the general current distribution revealed that only 50% of the current is entering the tendon within a length of 3 m either end of the probe. The remaining 50% of the current is distributed along the remaining tendon encased by the concrete.

At the interface between the concrete and dolerite, the majority of the current is still confined within the concrete. Once the probe is within the dolerite, the current travels into the dolerite but also up the borehole to enter the tendon along the concrete. Here the analysis revealed that 50% of the current is entering the tendon via the dolerite and the remaining 50% is entering the tendon via the concrete. This makes the corrosion analysis uncertain within the dolerite, since the area polarised is not just within the dolerite but also in the concrete.

The computer model was also able to calculate resistance values and more importantly changes of the resistance versus depth (Figure 7). Within the concrete, the resistance values are very constant. At the interface, the system (measured) resistance increases sharply and once in the dolerite, the system resistance still increases. This is due to the current travelling up the water column. The overall response is very similar to the actual measured resistances from the site work (Figure 4), and hence confirms that the basic assumptions in the model are sound and consequently provide a reasonable description of the overall system.

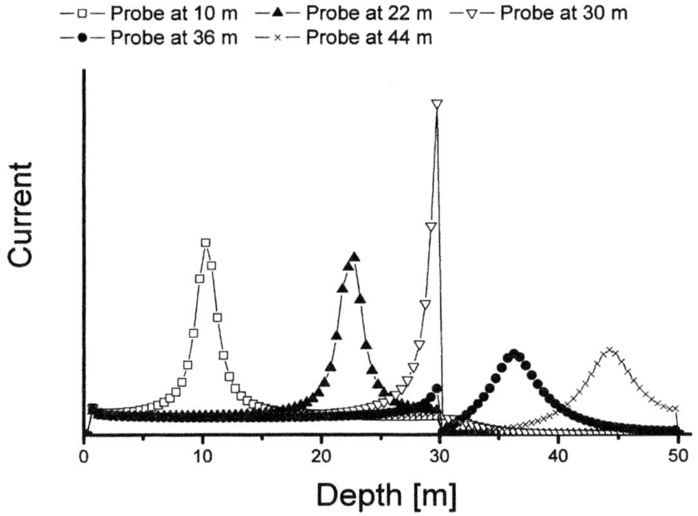

Figure 6. Computer model of current distribution due to probe location along the anchorage

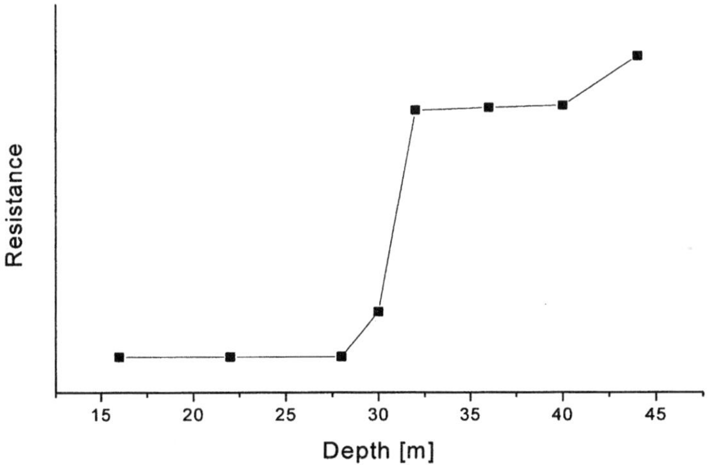

Figure 7. Computer model of the relative changes in resistance versus depth

Pressure testing

As a direct result of the corrosion measurements, the drilled borehole #B was pressure tested in order to locate any possible paths where water could escape from the borehole. This pressure test confirmed that for anchorage #B leakage of water occurred at 10 m above the bedrock which coincides with an active spot found during the potential survey and the estimated results for the corrosion rate. It is not clear at this moment of time whether any crack or joint is present at this elevation.

Conclusions

- On-site corrosion investigation of prestressed ground anchorages, where the anchorages are not surrounded by a duct, has been shown to be practical.
- In the trial described herein it is seen that in all cases there is a clear distinction in the conditions of the anchorages within the concrete dam, compared to the dolerite bedrock.
- Higher than expected resistivity of the surrounding concrete and bedrock, has meant that the spatial resolution of the system was not as refined as had been hoped, but computer modelling has confirmed the overall probable sensitivity.
- The measured system resistance values are consistent for all three anchorages and clearly follow the change from concrete to dolerite. The potential distribution of the anchorages implies significantly different conditions for anchorage #C compared to anchorages #A and #B. Estimated corrosion rates for all three anchorages were similar in the concrete (typically 1 to 9 μm/yr), dropping away to negligible values in the dolerite (all < 0.1 μm/yr).
- The site data are comparable with the previous laboratory assessment of ground water corrosivity.
- Based on the experience gained from the trial investigation it was concluded that modifications to the probe arrangement and measurement procedures and obtaining measurements at closer intervals with a smaller probe (possibly including a guard ring system to allow for closer control of the areas of the tendon being polarised), would provide more detailed assessment and hence a higher confidence in the survey results.

References

1 BS 8081 : 1989 *Code of Practice Ground Anchorages*, (section 8.2 and table 11).

2 A. Neville, "*Autogenous healing – a concrete miracle?*", Concrete International, Nov-2002, pp 76-82

3 M. Stern and A. L. Geary "*Electrochemical Polarization: I. A Theoretical Analysis of the Shape of Polarization Curves*", J. Electrochem Soc, Vol. 104, No. 1, 56-63, 1957

4 Andrade C. and Gonzales J. A., "*Quantitative measurements of corrosion rate of reinforcing steels embedded in concrete using polarization resistance measurements*", Werkstoffe und Korrosion, Vol 29,1978, pp515-519.

5 ASTM C876 "*Half Cell Potentials for Steel in Concrete*"

Corrosion Protection Options for Permanent Ground Anchorages

Chris Irvin, Dywidag-Systems International Ltd., Warwickshire, UK
Devon Mothersille, Geoserve Global Ltd., London, UK

Abstract

The subject of appropriate corrosion protection for permanent ground anchorages can appear daunting to an engineer faced with a new anchorage design for the first time. There are at least ten different methods of corrosion protection available, not all of them applicable to permanent anchorages.

Defining the appropriate materials or methods which constitute an acceptable physical barrier for corrosion protection, in accordance with the codes of practice, is one of the most important subjects within ground anchorage practice. This paper aims to clarify the subject by briefly describing the current methods available, redefining terms that have been confused or misused and highlighting some practical issues in implementation.

Introduction

The six most common methods considered for corrosion protection are: Double Corrosion Protection, Single Barrier Protection, Epoxy Coating, Galvanizing, Sacrificial Corrosion Allowance and Borehole Grout. A number of these are not suitable for permanent anchorages and one has practical limitations during site handling.

Further confusion as to which method to apply, has arisen due to three reasons:

a) The belief that a lower level of corrosion protection applicable to different installations (i.e. mini piles or soil nails) may be suitable for ground anchorages.
b) Incorrect reclassification of the anchorage as a tension pile, in order that a lower level of protection may be applied.
c) The desire to utilise a self-drilling hollow bar where ground conditions are variable or where cased installation is problematical.

The importance of corrosion protection for permanent anchorages is illustrated below in figures 1 and 2, where the durability of the anchorage is relied upon as a dependable element of the permanent works.

Figure 1 A472 Pontymoile by-pass, showing anchorage installation.

Figure 2 Completed works, anchorages concealed behind a masonry facing.

Classification of Installation (Ground Anchorage, Mini Pile, Soil Nail)

In order that the appropriate level of corrosion protection may be selected, the installation should be classified according to load, mode of loading (active or passive, tension or compression), consequences of failure and required design life. Table 1 summarises typical installations together with their requisite levels of protection.

Table 1

Installation / Component	**Load Classification** (indicative)	**Mode of Loading**	**Consequence of Failure due to Corrosion**	**Corrosion Protection, Technical Standard**
Ground Anchorages (soil anchorages, rock anchorages)	Often heavily loaded 500-20,000 kN	*Active* (debonded, prestressed free length)	Risk of overload to adjacent anchorages	Double corrosion protection to BS8081 or Single barrier protection, Tube-a- Manchette / sheathing combinations to EN1537
Rock Bolts for tunnels or caverns (*internal*)	Loads: 150-1,100 kN	Fully Bonded (with or without prestress)	Risk dependent on loading and application	Dependent on load and aggressivity. Double corrosion protection or epoxy coating. (BS8081)
Mini Piles / Micro Piles (Tension)	Up to 3,500 kN (single bar reinforcement)	Passive (Fully Bonded)	Risk dependent on loading and application	Double corrosion protection (BS8081) or lower level of protection depending on risk or aggressivity levels (EN14199)
Mini Piles / Micro Piles (Compression)	Up to 2,000 kN (single bar reinforcement)	Passive (Fully Bonded)	Settlement (Lower corrosion risk due to compression loading)	Sacrificial corrosion allowance, galvanizing, grout cover (EN14199)
Soil Nails	Lightly loaded 30-150 kN	Passive (Fully Bonded)	Low risk (built-in system redundancy)	Sacrificial corrosion allowance or galvanizing (CIRIA 637, EN14490)
Rock Bolts / Rock Dowels (*external*)	Lightly loaded 30-100 kN	Passive (Fully Bonded)	Low risk (built-in system redundancy)	Sacrificial corrosion allowance, galvanizing or stainless steel

Common Corrosion Protection Methods

The six most commonly considered methods for corrosion protection, together with their constraints for use on permanent anchorages, are outlined below.

Double Corrosion Protection

Double corrosion protection (i.e. double barrier protection) is the provision of two impermeable barriers between the tendon and the external environment, such that the outer barrier ensures the integrity of the inner barrier should the potential for damage arise during handling or installation.

Protection measures that constitute acceptable barriers in accordance with BS8081:1989 and EN1537:2000 are often misrepresented, leading to confusion. Borehole grout, galvanizing or sacrificial corrosion allowance are all not accepted by the above standards. Acceptable barriers must ensure:

a) durability over the lifespan of the anchorage (often up to 120 years),
b) integrity, i.e. an impermeable barrier over the full length of the tendon,
c) remain intact during handling, installation and stressing.

Double corrosion protection for bar anchorages and strand anchorages are summarised below.

Bar Anchorages

Permanent bar anchorages comprise a threadbar, of either high yield or prestressing steel, surrounded by an internal cement grout encapsulated within a corrugated plastic duct (see figure 3). The internal grout should be applied under factory conditions and is not to be confused with borehole grout, which does not constitute a fully protective barrier (see separate section, borehole grout).

The internal grout of bar anchorages may be considered as an acceptable barrier to corrosion on the basis that crack widths generated under load at each threadform, can be limited to less than 0.1mm. Each crack width is uniform in thickness, due to the regularity of the thread pitch in conjunction with the linear extension of the bar. Crack width control is reliant upon the confinement from the corrugated plastic duct in order that the cracks propagate at each threadform as opposed to intermittently (see figure 4).

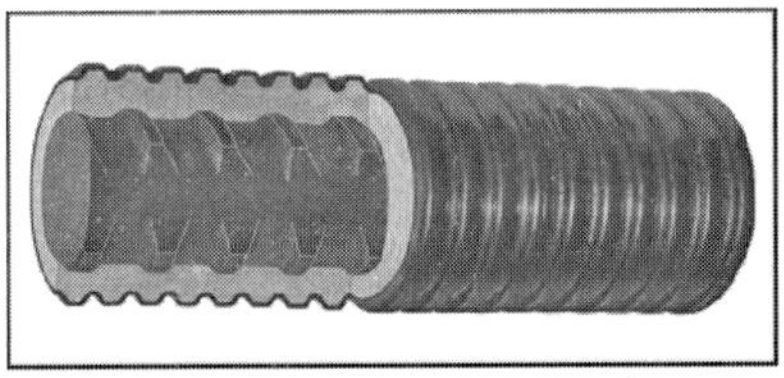

Figure 3. Section of permanent bar anchorage, showing duct encapsulation of pregrouted bar

Figure 4. Crack width control on pregrouted section

Strand Anchorages

Corrosion protection for strand anchors is provided by two plastic sheaths as the internal grout is unable to act as a protective barrier. Crack width control of the grout on strand anchors is not possible due to:

a) the lack of ribs or threads on the strand,
b) the increased strain of the strand (strand is worked much harder than bar).

Triple Corrosion Protection

This is somewhat of a misnomer and usually arises from either, the aspiration to add a further protective layer on the assumption that it is a simple process, or the misinterpretation of borehole grout or galvanizing as a protective barrier. Neither is credible. Double corrosion protection remains one of the most practical methods of protecting permanent anchorages and provides the highest level of protection in comparison to other systems.

Single Barrier Protection

This method has become popular for high capacity strand anchorages employed in the stabilisation of dams. It comprises one physical barrier, the integrity of which shall be proven in-situ by either a falling head water test, or an electrical integrity test (see electrically isolated anchorages).

Corrugated Plastic Ducting - Wall Thicknesses

The corrugated plastic duct is an integral part of corrosion protection systems for both bar and strand anchorages.

For double corrosion protection, current developments are incorporating increased wall thicknesses of 1.0-1.5mm for the corrugated ducts (previously 0.8mm taking account of manufacturing tolerances), on the basis that there are two barriers.

For single barrier protection, guidelines laid down by EN1537:2000 (Clause 6.10.1) have incorporated even greater wall thicknesses for the single corrugated duct (see table 2). Both systems now provide better abrasion resistance, should the potential for damage arise during handling or placement of the anchorage.

Table 2

External Corrugated Duct (i.d.)	Wall Thickness
$\leq$ 80 mm	1.0 mm
> 80 mm but $\leq$ 120 mm	1.5 mm
> 120 mm	2.0 mm

For high capacity strand anchorages used in dam stabilization, even larger corrugated ducts of diameters up to 254mm with a 3mm wall, are now being used.

Limitations of Borehole Grout as a Protective Barrier

Borehole grout (i.e. grout injected during installation or subsequently) is often misquoted as a recognised barrier for protection. Whilst it will provide a degree of protection, its cover and integrity cannot be verified in-situ, hence its use for permanent anchorages is not recognised by either BS8081:1989 or EN1537:2000.

Where borehole grout is used with self-drilled hollow bar systems, installed using the simultaneous drill and grout technique, its effectiveness as a protective barrier is problematical due to:

a) Reduced grout cover at coupler locations or at points along the full length of the drilled bar, which often takes the form of a lazy spiral. Spacers are disruptive, restricting grout circulation / flush and causing disturbance to the borehole wall.
b) Crack width control of the grout body, irrespective of threadform, is inconsistent at coupler locations or in areas of poor confinement.
c) Grout placement is workmanship sensitive. Simultaneous drill and grout can produce grout bulbing through the assistance of reaming, but it is irregular over the full length of the tendon. Exhumed tendons have shown wide variations in grout cover.
d) Contamination of the grout from the drill spoil or ground water.

BS8081:1989 Clause 8.2.3 summarises the limitations of borehole grout as "*grout injected in situ to bond the tendon to the ground does not constitute a part of a protective system because the grout quality and integrity cannot be assured*".

In recent years, self-drilled systems have become popular for soil nailing, as the concept of a self-drilled system in loose overburden has attractive cost benefits over conventional drilled and cased systems. However, their use as permanent anchorages is universally not accepted.

Epoxy Coatings

Epoxy Coated Bars

Epoxy coatings are occasionally applied to rock bolts or anchors in hard rock tunnelling applications, where slenderness of the borehole is a key cost requirement. The coating has the added benefit in that its application process or chemical composition is not detrimental to prestressing grade steel threadbars, which are sometimes used in areas of high loading in tunnels.

Damage during handling remains the limiting factor for epoxy coated threadbars. The baked coating is brittle, with the highpoint of each thread particularly susceptible to nicks or chipping during transportation and handling. Each bar requires wrapping individually in a disposable protective sleeve, which is removed prior to installation. On site repairs can be time consuming as multiple nicks or chips all require repairing (see figure 5).

Figure 5 Repairs to chipped epoxy coatings.

Figure 6 Damage to Duplex coated hollow bar, caused by hydraulic jaws used for rod release.

Duplex Coatings - Self-Drilled Hollow Bar

Duplex coating is a term used to describe two protective coatings, typically comprising a galvanized coating overlaid with a layer of epoxy or polyester. Its application has led to confusion as it does not offer double corrosion protection. It is rarely used on solid bars and has real limitations when used on self-drilled systems as the coating becomes compromised through:

a) Abrasion damage from rotary installation.

b) Perforation of the coating by hydraulic jaws or stilsons, used to grip the bar for release from the drill head (see figure 6). Soft jaws have been trialed, but are ineffective.

Epoxy Coated Strand

The epoxy coating of strand for permanent anchorages is acknowledged as a popular process in the USA and Japan, but its application to strand anchorages in Europe is rare. The principle areas where attention is required are:

a) Creep under load of the bonded strand within the grout. Creep values need quantifying.

b) Adhesion of the epoxy to the strand. Adhesion is heavily dependant on the strand preparation process (i.e. shot blasting or pickling in acid).

c) Abrasion damage during handling. Sand added to the epoxy coating to enhance its bond in the grout is extremely abrasive, requiring special care on multi-strand anchorages.

d) Coating thickness and uniformity. Both require careful checking.

e) Grip of the wedges. Long teeth are required to bite through the epoxy coating, in order that the steel can be gripped.

Sacrificial Corrosion Allowance

This method has evolved from sheet piling and buried corrugated steel structures. It is now also regularly applied to soil nails and rock bolts (both passive installations), as these are generally classified as low risk installations, on the basis of an element of redundancy. This is not routinely the case for ground anchorages.

In addition, the application of sacrificial corrosion allowance to anchor tendons is not acceptable as corrosion rates are rarely uniform and not predictable, tending to extend more rapidly at localized pits or surface irregularities.

Both BS8081:1989 and EN1537:2000 preclude the use of sacrificial corrosion allowance for the permanent protection of anchorages, with EN1537:2000 (Clause 6.9.1) stating "*There is no certain way of identifying corrosive circumstances with sufficient precision to predict corrosion rates of steel in the ground. All steel components which are stressed shall be protected against corrosion for their design life*".

Galvanizing

Galvanizing is universally recognised as a method of protecting exposed steel surfaces. Its use on low risk installations such as rock bolts (figure 7) or soil nails (figure 8) is popular as it provides a robust coating, but its use as a protective coating for anchor tendons is not accepted.

The use of galvanizing as a means of increasing lifespans for buried installations, should be considered on the following basis:

a) Lifespans are variable and are dependent upon the aggressivity levels of the environment, typically 5-15 years for buried installations.

b) Coating thicknesses vary according to the application process. Hot Dipped Galvanizing (in accordance with EN1461:1999) provides a minimum zinc cover of 85 μm (microns), Electro Plated Galvanizing only offers 5-12 μm.

c) It is not suitable for prestressing grade threadbars as the preparation process (pickling in acid) can induce the potential for hydrogen embrittlement.

d) Both BS8081:1989 and EN1537:2000 preclude the use of galvanizing as permanent protection, on the basis that it is a sacrificial coating.

Figure 7. Galvanized Rock Bolts

Figure 8. Galvanized Soil Nails

Alternative Protection for Permanent Anchorages

EN1537:2000, The European Standard for Ground Anchors, includes a range of protection options:

a) Double Corrosion Protection, as detailed above
b) Tube-a-Manchette and Compression Tube Anchorages. Each system features different mechanisms for load transfer in the bond length, but both anchorages include for two impermeable barriers as corrosion protection.
c) Electrically Isolated Anchorages, as detailed below.

Electrically Isolated Anchorages

These anchorages feature a single protective barrier (typically a corrugated plastic duct), which is subjected to an electrical resistance test. The principle of the system is based upon proving the integrity of the protective barrier by measuring the resistance between the anchorage and the surrounding soil. The concept is attractive to designers, as it enables the integrity of the protective barrier to be checked, following installation and stressing.

There are however two considerations where a detailed understanding of the system is required:

a) Contractual implications of the system are quite complicated, as the integrity of the plastic sheathing should be tested at the time of delivery to site, where ownership of the anchor is transferred to the installer. Conductivity tests prior to installation require a water bath.
b) Moisture is a problem, as resistivity readings can be unreliable and difficult to interpret.

Other Corrosion Protection Options

Stainless Steel and Cathodic Protection are occasionally considered for corrosion protection of permanent anchorages. The use of stainless steel is rare on permanent anchorage applications and there is little guidance on its use without protective barriers in either BS8081:1989 or

EN1537:2000. Its application for external rock bolting or dowelling is more common. Durability of stainless steel is dependent on the grade, also the cost of higher grades can be prohibitive.

Cathodic Protection, whilst popular in other applications, has practical limitations for ground anchorages in the following respects:

a) the determination of the correct electrical current sufficient to provide protection,
b) the need for continual monitoring and maintenance of an impressed current system.

Ground Anchorage or Tension Pile?

The key distinction between ground anchorages and tension piles, is that ground anchorages are *active* installations (featuring a prestressed free length), as opposed to tension piles which are *passive* installations (fully bonded, initially unstressed). A debonded tension pile requiring stressing is a misnomer and should be classified as a ground anchorage.

The extension characteristics of tension piles (figure 9) under service loadings differ from those of ground anchorages (figure 10). Tension piles will extend under load as restraint is progressively mobilised in the ground, whereas ground anchorages do not, as they feature a free length which enables the extension to be stressed out at the time of installation.

For temporary works the increased extension of a tension pile is often not critical, but for permanent works the increased extension often requires careful consideration in the context of the overall structure, where additional movement could lead to cracking in the concrete faces. Typically a design incorporating tension piles will have to allow for movement on a global basis.

Figure 9 Tension piles for uplift resistance

Figure 10 Ground Anchorages for wall tieback

Critical Areas of Anchorages Requiring Protection

The location at the underside of the bearing plate is typically the area of highest risk. At this point both oxygen and moisture are present, which when combined will promote corrosion if the tendon is unprotected. Access to this area for inspection, even with the use of an endoscope, is notoriously difficult (see figure 11). Repairs are usually only possible by de-stressing the anchorage and removing the full head assembly.

Protective end caps, used on the face of the bearing plate to enclose the anchor head, do not contribute to protection below the plate. It is the section of anchorage directly below the bearing plate which requires comprehensive protection.

Figure 11. A typical application for anchorages. Access to the underside of the bearing plate is notoriously difficult.

Practical Issues on Site

Flushing Out of Grout at the Anchor Head

Flushing out of the grout (wash-back) over the top 1.0m of the anchor tendon is required in order to avoid any strut effect during stressing and also to facilitate the placement of the sealing tube (at the underside of the bearing plate) over the anchor tendon, see figure 12.

This is of particular importance for bar anchorages, where the grease filled sealing tube is an integral part of the bearing plate. Failure to flush out can lead to the sealing tube not being properly placed around the anchor tendon, resulting in the potential for corrosion.

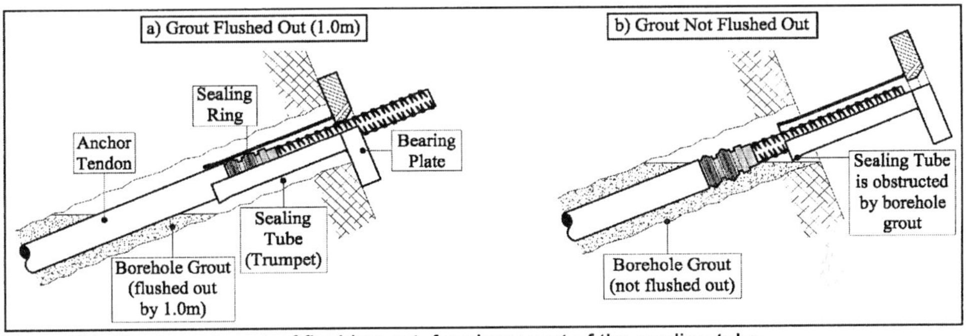

Figure 12. The importance of flushing out, for placement of the sealing tube.

Prevention of Ingress of Ground Water in Anchor Heads

In areas of high ground water pressure, additional sealing measures may be required. This is a complicated detail that needs to be carefully thought through, as practical limitations during installation require a specific order of assembly.

The use of additional sealing components or gaskets as well as the injection of specialist resins should be considered where high water pressures are anticipated.

Conclusion

Ground anchorages remain a popular choice with engineers for a range of applications. Their durability in permanent works is of utmost importance, requiring careful attention in respect of corrosion protection. The re-classification of the installation in order that a lower level of protection may be applied, or the application of unsuitable barriers, is to ignore the fundamental issues of durability. It is the area of least protection that defines the effectiveness of all corrosion protection measures.

In practice, the corrosion protection principles outlined in both BS8081:1989 and EN1537:2000 tend to complement each other, in that either double corrosion protection or single barrier protection (coupled with an in-situ integrity test) will ensure permanent protection for the anchorage.

Acknowledgements:
The use of photographs from the following contractors are acknowledged: Ritchies, CAN, Systems Geotechnique and Cementation.

References:
1. British Standards Institute. BS 8081:1989 British Standard Code of Practice for Ground Anchorages.
2. European Standard. EN 1537:2000 Execution of Special Geotechnical Work - Ground Anchors.
4. European Standard. EN 14199:2005 Execution of Special Geotechnical Work - Micropiles.
5. European Standard. EN 14490: (2004) Execution of Special Geotechnical Work - Soil Nails.
6. CIRIA. C637 Soil Nailing. Best practice guidance.
7. British Standards Institute. BS EN ISO 1461:1999 Hot dip galvanized coatings on fabricated iron and steel articles - Specifications and test methods.

Replenishment of Underhead Protection (Below the Bearing Plate)

Kouichi SUGA, Akira Yonemura,Yoshio Iwaida
Technical Committee, Japan Anchor Association, Tokyo, Japan

Abstract

In order to maintain the performance of existing anchors it is important that the replenishment of underhead protection (below the bearing plate) should be considered. In this paper, the basic concept of anchor repair and an example of repair work are given.

The Public Works Research Institute and Japan Anchor Association have started joint research in order to publish the "Manual of inspection, integrity investigation and repairing of anchors" described in this paper.

Basic concept of anchor repair and types

Normal anchors having no need of repair work retain sufficient performance with respect to allowable limited level at both the initial stage of service and throughout the service life (Figure 1) even though the performance of the anchor deteriorates year by year. On the other hand, anchors that have deteriorated to below the level required for service at the time of investigation (A in Figure 2) or to the allowable limited level in some cases (B in Figure 2) require repair work. Some anchors provide an acceptable level of performance at the time of investigation but show signs that they will deteriorate further, and fall below the level required for service, before reaching the design service life (C in Figure 2). For these anchors, considered to be abnormal, repair work should also be provided.

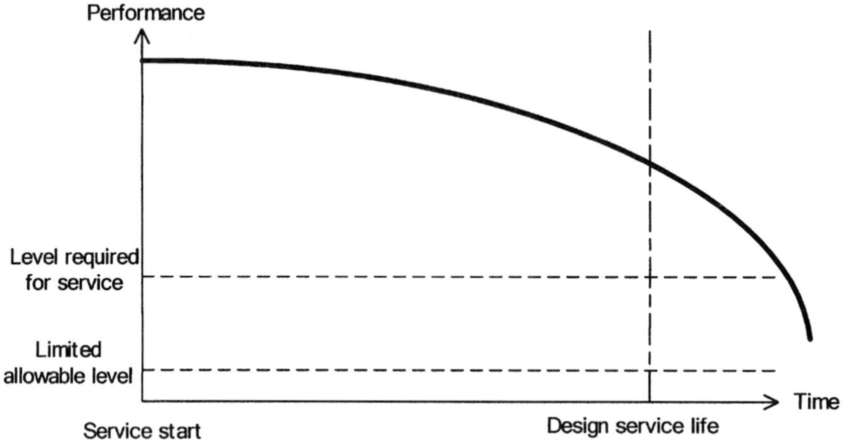

Figure 1. Illustration of normal anchor performance

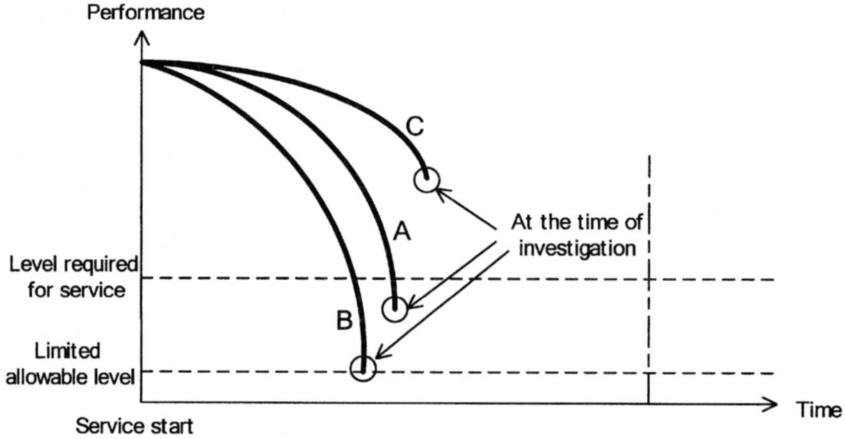

Figure 2. Illustration of anchors that cannot be considered normal

There are several types of repair that can be applied such as (i) durability enhancement, (ii) repair enhancement, (iii) renewal, (iv) emergency countermeasures and (v) life extension countermeasures

The basic flow chart for repair type selection is shown in Figure 3. Types and outline of repair works are also stipulated below.

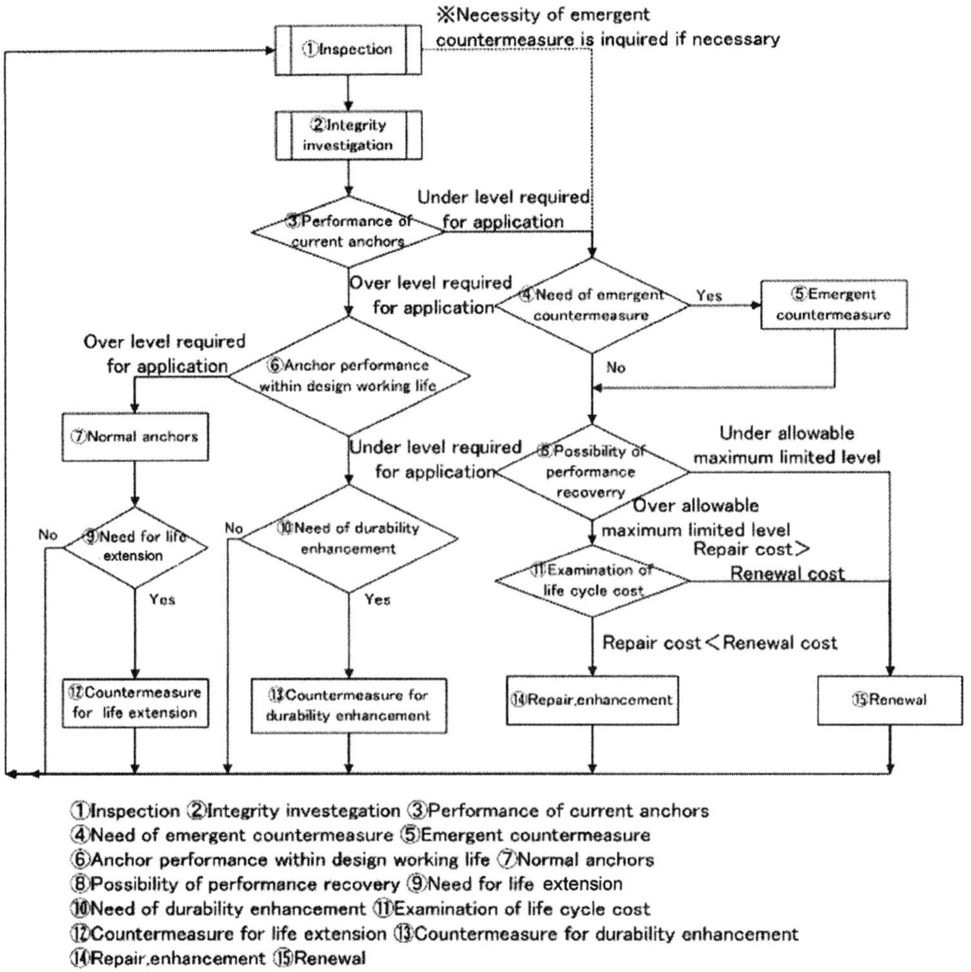

Figure 3. Flow chart for selection of repair work

Durability enhancement

Based on the condition of the anchor at the time of investigation, Figure 4 indicates that the anchor is still above the level required for service, but the performance deteriorates drastically after the service start and is expected to fall below the required level for service within the design service life. In such case, a countermeasure, or durability enhancement, to maintain the performance of the anchor is necessary so that the required level of service is maintained for the design service life. The durability enhancement can extend the life of the anchor without improving performance, shown as ① in Figure 4, or with improving performance, shown as ② in Figure 4.

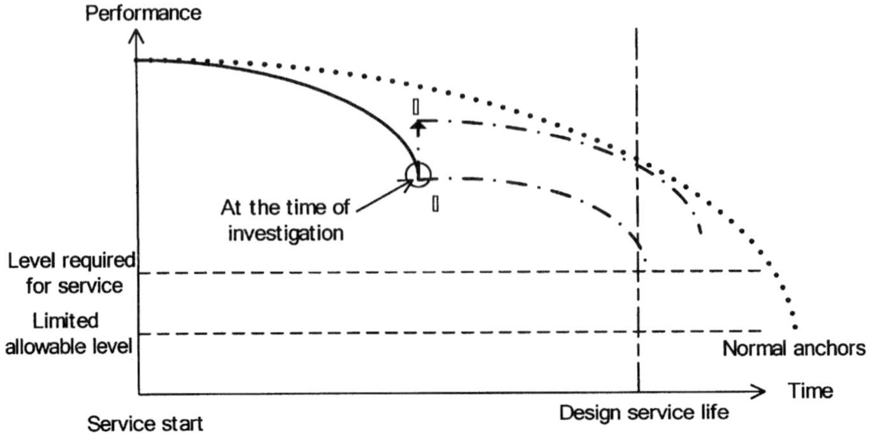

Figure 4 Illustration of durability enhancement

Repair enhancement

For anchors shown in Figure 5, performance deteriorates drastically after service start and is below the required level for service at the time of investigation. In such a case, performance improvement to the required level for service is implemented.

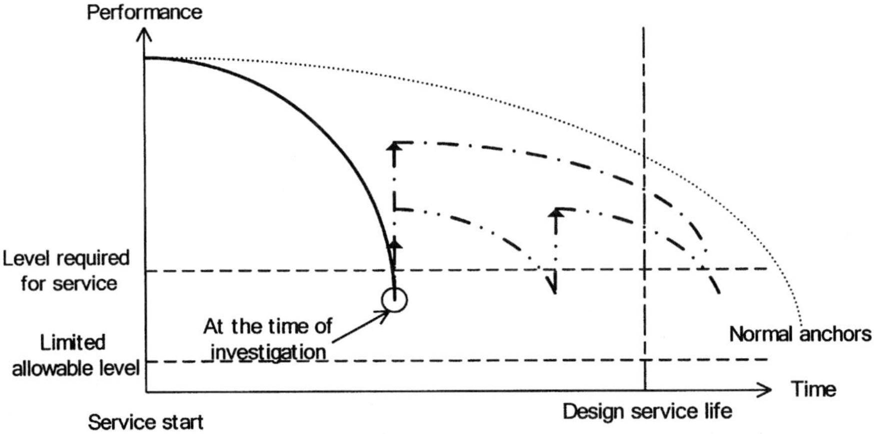

Figure 5 Illustration of repair enhancement

Renewal

New anchors are required in the case where the installed anchors are below the level required for service at the time of investigation and it is difficult to recover performance by repair enhancement due to technical constraints or would be disadvantageous in terms of cost or construction time, (Figure 6).

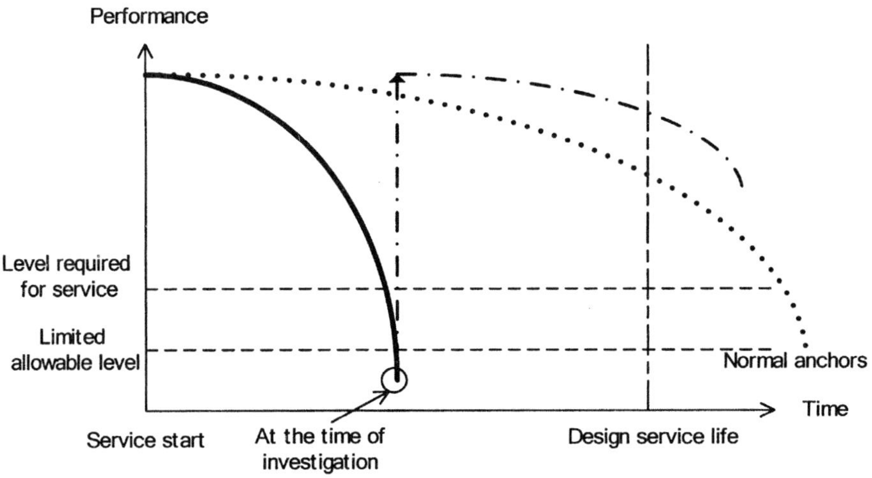

Figure 6 Illustration of renewal

Emergency countermeasure

Where anchors, anchored structures or slopes reach the allowable maximum limited level at the time of investigation, an emergency countermeasure should be taken in order to prevent third party damage before full-scale countermeasures are taken, (Figure 7).

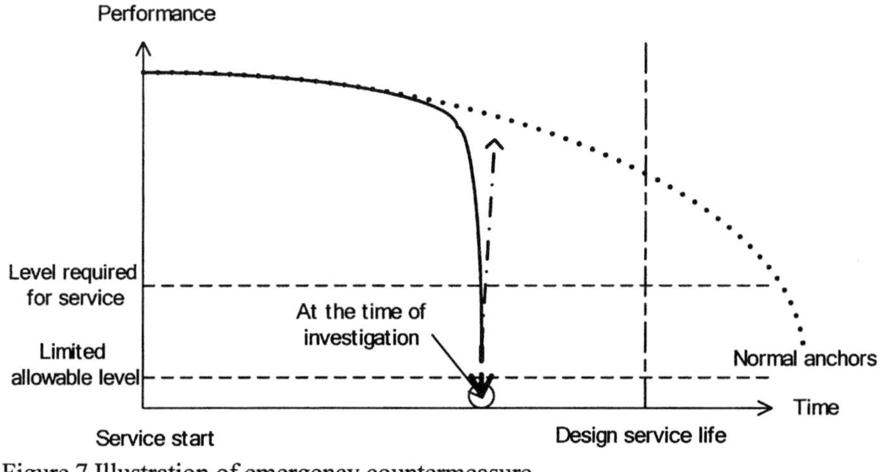

Figure 7 Illustration of emergency countermeasure

Life extension countermeasure

Normal anchors are expected to remain above the level required for service during their design service life. For such anchors, as they approach their design service life it is possible to keep them in service and thereby extend their renewal time by undertaking life extension countermeasures (Figure 8).

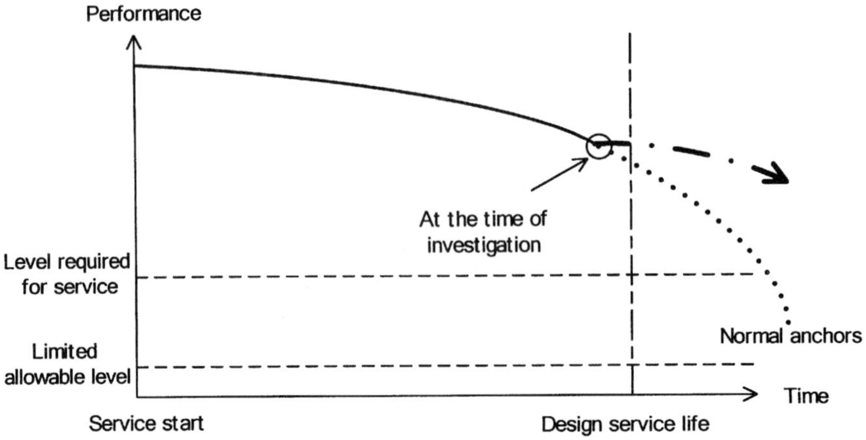

Figure 8. Illustration of life extension

Cause of and countermeasure for corrosion of anchor head and underhead (below the bearing plate)

The causes of corrosion of the anchor head and underhead (below the bearing plate) are: -

① Insufficient corrosion inhibiting measures

② Deficient construction

③ Deterioration of the corrosion inhibiting compound

The structure of the old type of anchor is shown in Figure 9. In old type anchors a discontinuous part existed below the bearing plate i.e. between the anchor head and the free anchor length sheath. Over this part the tendon easily corrodes when surrounded by air and water. For some anchor head parts, anti-corrosion treatment was not considered at all at time of construction, some of them were untreated by any corrosion inhibiting compound and some of them were provided with only a simple cap. Often where anti-corrosion treatment was considered such treatment was insufficient. Some of them were simply covered with concrete or a corrosion-inhibiting compound that leaked easily when the cover became loose. Such insufficient corrosion-inhibiting treatment is considered as the most common cause of corrosion for both the anchor head and the underhead part (below the bearing plate).

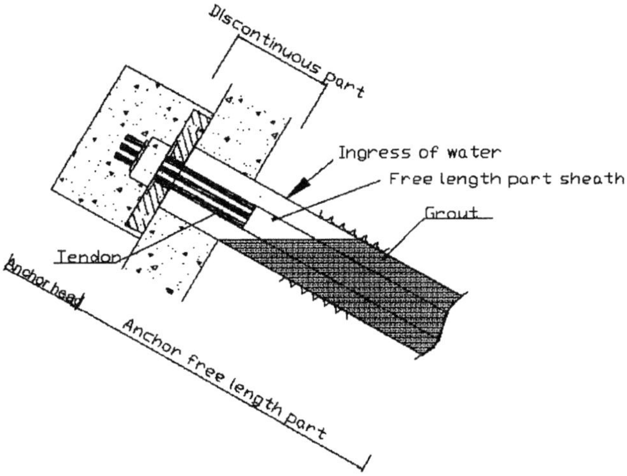

Figure 9. Constitution of old type anchor

The inside of the anchor free length part should be filled completely with corrosion inhibiting compound at the time of construction. However, during integrity investigation, it is often found that corrosion inhabiting compound for the free anchor length part is either insufficiently filled with, or completely empty of, corrosion inhibiting compound due to deficient construction or through damage to the free anchor length sheath. When the corrosion inhabiting compound does not completely fill the space between the grout that surrounds free anchor length and underside of the bearing plate, the ingress of water and air will cause corrosion to occur. Since the underhead part (below the bearing plate part) of the old type anchor includes a discontinuous structure, insufficient filling of corrosion inhibiting compound will increase the possibility of corrosion.

Corrosion inhibiting compound does not easily deteriorate in the absence of water and air. However, corrosion inhibiting compound in the anchor head cover will be deteriorated by air when it is insufficiently filled. Even if filling is complete at the construction stage, it will deteriorate when some of it leaks out and is replaced by water and air as a result of deterioration of the anchor cover seal or the head cover it is damaged.

An example of a typical repair and enhancement of the anchor head and underhead part (below the bearing plate) of old type anchors considering the causes of corrosion is shown in Figure 10.

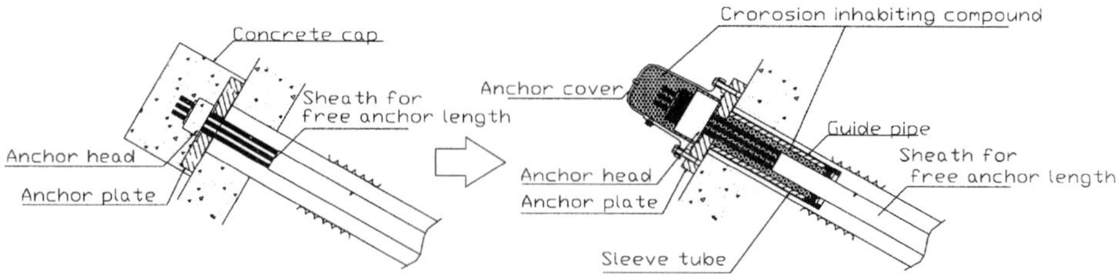

Figure 10. Repair and enhancement for the anchor head and underhead part (Below the bearing plate)

For the discontinuous part of the underhead part (below the bearing plate), joining is achieved by installing a sleeve pipe and guide pipe. Inside the sleeve pipe, the water stoppage rubber seal is set in advance. The new water stoppage rubber seal is operated through the anchor plate and guide pipe at the time tensile load is added to the anchor then worked as a watertight seal (Figure 11). Then, the anchor cover is attached and pressurized corrosion inhibiting compound is grouted via a grease nipple. By pressurized filling, corrosion inhibiting compound fills the inside of the anchor cover, inside of the guide pipe (under head part-below the bearing plate) and inside the free length sheath (free anchor length). During the installation (using visual observation equipment) the corrosion-inhibiting compound was seen to push out and replace water and air as it filled. As for the management procedure, the filling pressure and amount of corrosion inhibiting compound are measured and recorded. Filling pressure should be confirmed to be over 0.4 MPa at the time of measurement (Figure 12.).

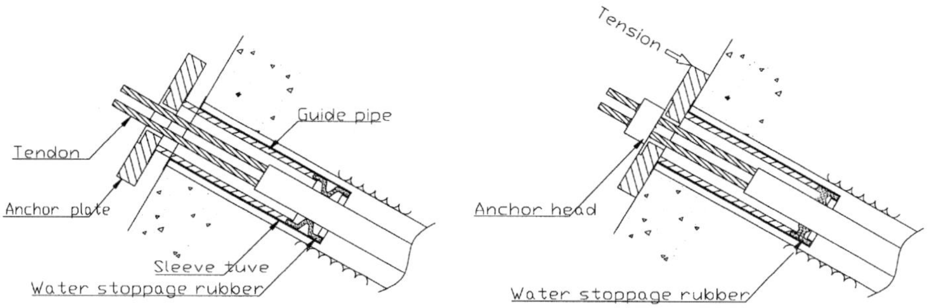

Figure 11. Example of water stoppage rubber

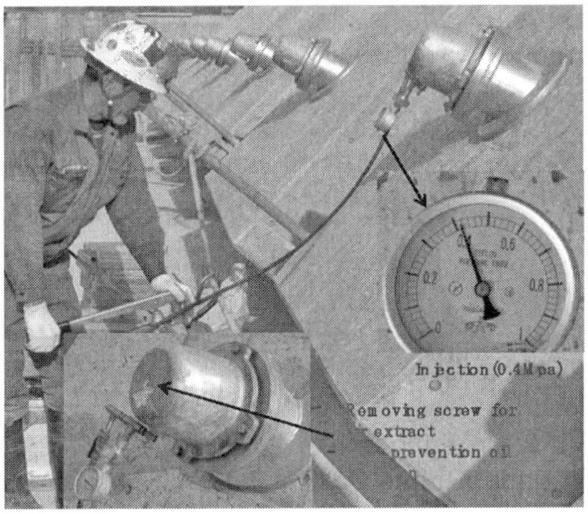

Figure 12. Pressure measurement

Final remarks

According to the FIP report published in 1986, 45% of anchor failures resulted from damage below the bearing plate. Although an anchor maintenance system and development of associated equipment are in progress, until now no documented guidelines have been produced. This paper will serve to provide guidance. The condition of existing anchors differs in each case depending on circumstances and for this reason maintenance and construction information should be collected to form a database for future reference.

Reference

FIP (1986) "Corrosion and corrosion protection of prestressed ground anchors" State of the art report, Institution of Structural Engineers, London

Durability of Ground Anchors and other Geotechnical Stabilisation Applications in Concrete Reinforcement Steel Quality, Subject to New European Regulations

Dieter Jungwirth, TU Munich, Germany
Susanne Adler, Stahlwerk Annahütte Max Aicher GmbH & Co. KG, Ainring, Germany

Abstract

Based on current European standards such as EC1, EC2 and EC7 as well as application standards for anchors, piles, nails and bolts, new corrosion protection systems for geotechnical stabilisation systems in concrete reinforcement steel quality, e.g. SAS 670/800 threadbars, will be discussed. Grout cover, permissible crack widths, zinc and corrosion rates will be defined and lifetimes of up to 120 years for different soil aggressiveness specified. The data for integrated systems have already been included in various approvals and approval drafts. Integrated systems mean that consistent durability for pre-stressed permanent anchors up to robust reinforcement steel, e.g. rock bolts, for different aggressive soils under tensile load as well as compression load can be covered. Main criterion gives EN 1537, e.g. chapter 6.9.

Introduction

Current European Eurocode design standards [1], material standards [2] as well as standards for ground anchors and other geotechnical application standards [3-5] make it possible to develop and approve new geotechnical stabilisation systems in concrete reinforcement steel quality. New findings in structural engineering, EC1 and EC2, provide some interesting additional information. The corrosion protection scenario ranges from bare steel and its corrosion rates (sacrificial corrosion) to permanent protection using cement grout (Fig. 1).

In geotechnics, in addition to geotechnical concerns, the system reserves must be taken into consideration as well. Anchors, piles and nails are usually installed in large numbers. In the case of premature failure of a load-bearing element it must be verified that the load is redistributed to the adjacent elements. A potential premature failure is therefore negligible with regard to the overall safety (EC1).

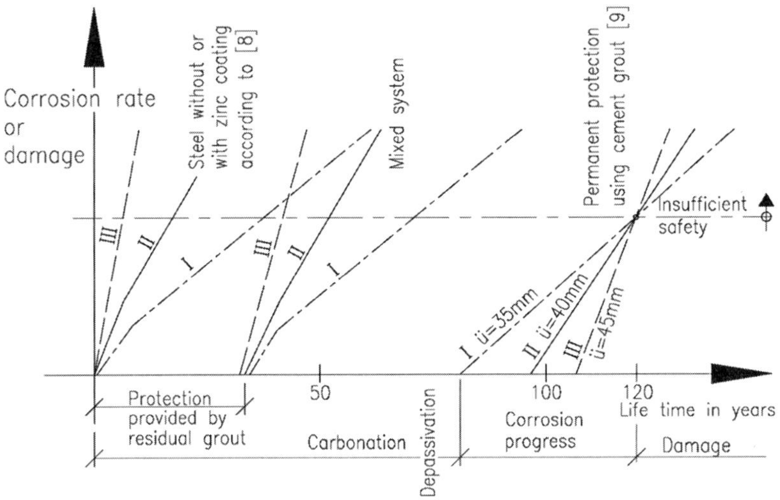

Figure 1. Scenario from bare steel to permanent protection using cement grout according to [6 and 7]: soil class I to III according to Table 1: ü = grout cover according to Table 2

Solution approaches

Steel

First it must be noted that the steel used, SAS 670/800, is an insensitive threadbar comparable to concrete reinforcement steel. The Annahütte steel works have developed and manufacture SAS 670/800 steel. According to Nürnberger's expertise [10], stress corrosion cracking does not occur. The steel meets the requirements for concrete reinforcement steel according to prEN 10080 and has been tested according to EN ISO 15630-1. The minimum guaranteed yield stress is 670 N/mm^2, the characteristic tensile strength 800 N/mm^2 and the overall elongation under ultimate load is 5%. Diameters range from d_s = 18–75 mm, failure loads from 200 to 3535 kN.

Corrosion rates

Another expertise by Nürnberger on corrosion rates [8] specifies corrosion rates (sacrificial corrosion) for bare and galvanized threadbars according to [11] for service lives of $\leq$ 50 years (Fig. 2) without taking the favourable influence of cement grout into consideration.

A large number of corrosion rates, assigned to the soil classes listed in Table 1, were evaluated and the approximately 90% fractile of the corrosion rates was applied. Thus, these values have been confirmed by testing as well as in standards and practical applications. The values specified in literature [12] or in the respective standards, e.g. [4] and [13], [14] are somewhat lower and cannot be clearly assigned to soil classes. To avoid the formation of macro elements (large cathode in conjunction with small anode) in moist and aggressive soils, there must be no conducting connection between the pile/nail made of SAS 670/800 and the reinforcement of the construction.

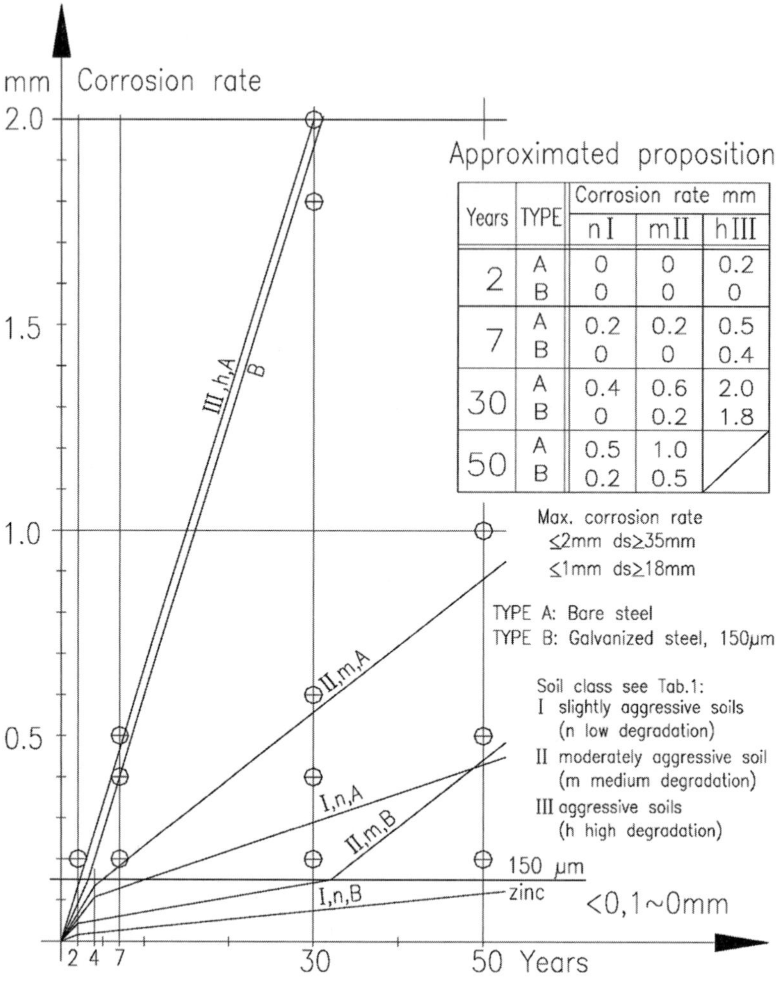

Figure 2. Corrosion rates (sacrificial corrosion) according to Nürnberger [8], without the positive influence of cement grout

The soil classes (listed in Table 1) correspond to EN 12501-1 and 2 [15]. For classification, the principle of the evaluation value sum according to [13, 14] may be of assistance.
Fields of application are the free length of temporary anchors and rock anchors/bolts as well as self-drilling anchors without grout cover. To verify the stability, the corrosion rates must be taken into consideration. The corrosion rates can be reduced by using galvanizing, plastic coatings or increased cement grout thickness.

Table 1: Soil classes
With pH-values > 8 and < 5, soil class I advances to soil class II and soil class II to soil class III. For class III with pH-values > 8 and < 5, special corrosion protection measures are required (e.g. sheathing).

Slightly aggressive soils I ~ XC1 according to [16]
• moderately well to very well aerated soils
• predominantly sandy to gravelly (coarse to medium-textured), friable rock
• low water content (drainable), dry
• low neutral salt content
• pH-values 5 to 8
• specific soil resistance > 7000 Ω cm
Moderately aggressive soils II ~ XC2, XC3, XS2, XD2 according to [16]
• poorly to moderately well aerated soils
• high silt and fine sand content (medium to fine-textured)
• in general medium water content, moist
• increased neutral salt content
• pH-values 5 to 8
• specific soil resistance 1000 – 7000 Ω cm
Aggressive soils III ~ XC4, XD1, XS1, XS3, XD3 according to [16]
• very poorly to poorly aerated soils
• high clay content (fine-textured)
• in general high water content or alternating water level
• high neutral salt content possible, de-icing salt, industrial waste
• pH-values 5 to 8
• specific soil resistance < 1000 Ω cm

Cement grout cover, permanent and with a limited protection period
The alkaline cement grout forms a protective layer on the steel surface. When depassivation sets in due to carbonation resulting from the CO_2 in the air and if oxygen and moisture are present, iron will dissolve = corrode upon electrical/electrolytic contact.
Pitting corrosion mechanisms induced by de-icing salts (chlorides) follow similar and other parameters. These however will not be discussed in this paper since this kind of corrosive action in soil requires double corrosion protection.
For certain chemical actions, e.g. SO_4, special cements must be used, see [1]. The first sign of corrosion is spalling, followed by extensive corrosion degradation and eventual damage, i.e. insufficient stability (see Fig. 1). A zinc layer, known to be compatible with cement grout, additionally accelerates these processes.
If geotechnical standards and regulations do not provide specifications on permissible crack widths or grout cover, these can be inferred analogously from the structural concrete and design standards, EC2 and EN 206. Fig. 3 compares the corner reinforcement of a foundation beam with a cement-grouted geotechnical element. The grout usually used according to EN DIN 445 – 447 [17] is comparatively safer in terms of corrosion protection, since it has a denser and higher alkaline content (cement) than concrete. The concrete covers specified in EC2 dependent on the exposition class correspond to the values listed in Table 2. The crack width of 0.1 mm specified in EN 1537 for sensitive pre-stressing steel could be raised to 0.2

for concrete reinforcement steel following EC2 (EC2 lists crack widths of up to 0.4mm, e.g. very low risk of corrosion).

A group of experts [9] specifies requirements for threadbar anchors, piles and nails/bolts sheathed using cement grout for service lives of up to 120 years, based on geotechnical and structural standards. Under tensile stresses, the crack widths must be limited to ≤ 0.2mm. The relevant European standards [3-5] are thereby complied with. The grout cover is specified in Table 2 dependent on the soil aggressiveness.

Table 2: Grout cover in mm dependent on soil type/class for service lives of up to approximately 120 years (Grout cover is ensured by centralisers.)

Soil class	Class	Compression load	Tensile load (Crack widths $w \leq 0.2$ mm)
Slightly aggressive	I	25	35
Moderately aggressive	II	30	40
Aggressive *	III	35	45

* Double corrosion protection with corrugated sheathing should be provided (see also Table 1)

The soil classes are specified in Table 1 according to [15]. Thereby, the exposure classes are assigned according to Gehlen [16] following EC2 and EN206, dependent on the corrosion probability.

It is only logical that between the corrosion rate of bare steel and permanent cement grout protection, a mixed system of reduced grout cover and corrosion rate should be feasible. The grout cover increases the period until depassivation due to carbonation sets in and reduces the following corrosion rates (Fig. 1). In [16 and 18] the influence of carbonation was examined for a reduced grout cover of ~ 10 to 15mm.

In dry environments, carbonation proceeds rapidly, the resulting corrosion is however negligible and can be disregarded. Conversely, in moist environments carbonation proceeds slowly, the corrosion probability is however considerably higher and therefore the resultant damage as well. Carbonation and corrosion probability thus proceed inversely to each other. Based on these assumptions, the time axis according to Fig. 2 shifts for soil class III by 10 years, for soil class II by 12 and for soil class I by 15 years to the right.

Scatter in the corrosion pit depth should not be considerably higher than those that have already been taken into consideration by Nürnberger with regard to the corrosion degradation of bare steel in soil [8].

Fields of application are self-drilling anchors with uncertain cement grout covering as well as rock bolts and nails with insufficient grout covering, larger crack widths or partial destruction of grout layers induced by soil shearing (see Fig. 4). Due to the large relative rib areas of SAS 670/800 ($f_R \sim 0.08$), there is still sufficient bond between threadbar and grout/soil even in the case of local corrosion spalling.

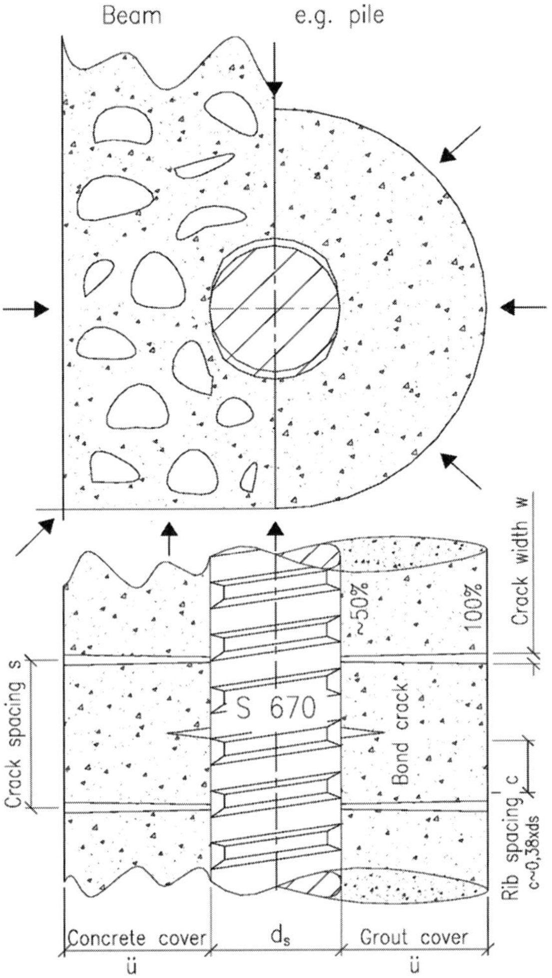

Figure 3. Comparison of a concrete cover in structural engineering according to EC2 with a grout cover in geotechnics (S 670, relative rib area f_R ~ 0.08)

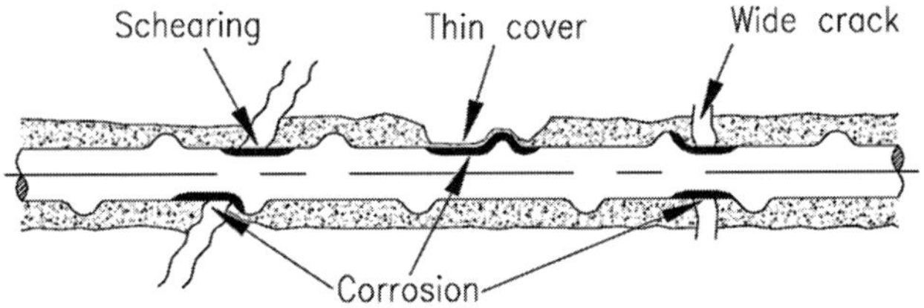

Figure 4. Rock bolts/Nails with insufficient grout cover

Permissible crack width, crack width limitation

Cracking in grout or concrete cannot be avoided. In concrete reinforcement steel crack widths must be limited to 0.2 mm so as not to reduce the active corrosion protection provided by cement grout. At this point, the crack laws for geotechnical tensile elements shall be briefly referred to. The 18 tensile tests carried out by Wichter [19] on various SAS 670/800 diameters and differing grout covers (test length ~ 1,5m) revealed low scatter so that the simple law for crack spacing shown in Fig. 5 could be determined:

$$s_{medium} = 1.10 \cdot (2 \cdot c + ü) \quad (1)$$

$$= 0.85 \cdot (d_s + 4/3 \cdot ü) \quad (2)$$

Dimensions in mm
ü grout cover
c rib diameter = $0.38 \cdot d_s$
d_s bar diameter

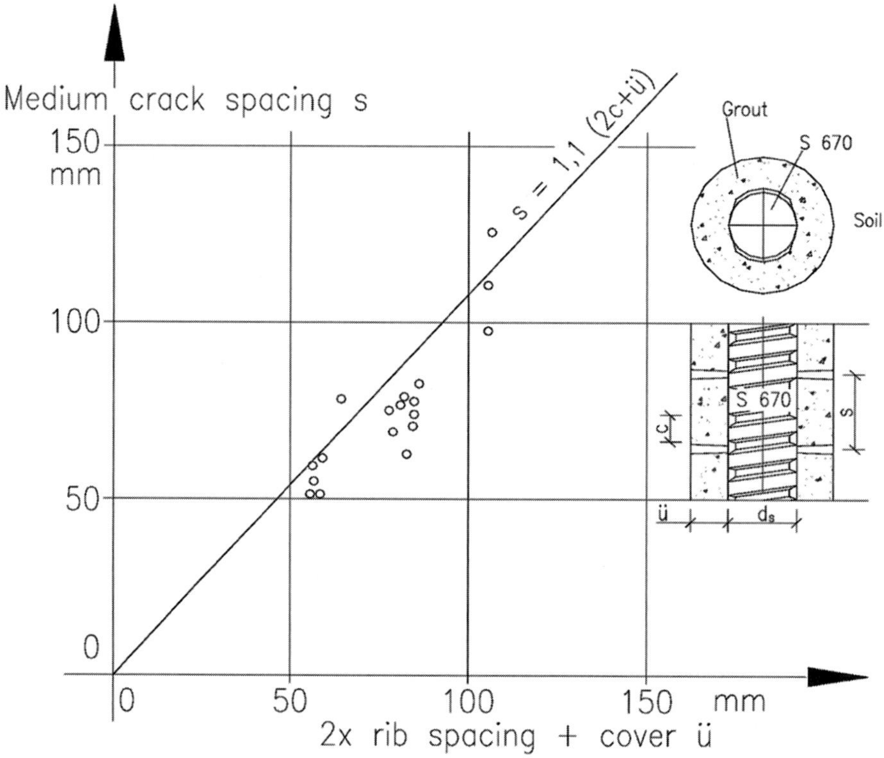

Figure 5. Assessment of crack spacings in the tests by Wichter [19] according to EN 1537, Test A (unconfined conditions)

"Freezing" of the cracks using resin injection under load (yield stress/1,75 = 383 N/mm²) revealed that the relevant crack width on the steel surface was approximately half as wide as on the grout surface, due to small undetectable intermediate bond cracks (Fig. 3). The crack width on the steel surface is therefore (approximately 95% fractile):

$$W_{95\%,steel} = 0.6 \cdot (\sigma/205{,}000) \cdot (d_s + 4/3 \cdot ü) \qquad (3)$$

Dimensions in mm
σ steel stress in N/mm^2
E-modulus = 205,000 N/mm^2

The equation includes a factor of 0.95 for the load sharing action of the grout in tension and a multiplication factor of 1.5 for uneven crack distribution [7]. The calculated values show excellent correspondence with the measured values.
A bar with d_s = 63.5mm and ü = 50mm would only show crack widths of 0.18 mm under a maximum tensile steel stress of 480 N/mm^2. The requirement mentioned above of 0.2 mm is thus met.
With compression piles, additional cracks can occur in the load introduction zone of the structure. These must be limited by appropriate tests and reinforcement, e.g. tests according to ETAG 013 [20]. Bursting forces that may occur at the pile head and cause grout spalling must be controlled e.g. using loop reinforcement. Due to the differing transverse strain behaviour of steel/grout, longitudinal cracking may occur in the free length of the pile. This can be avoided by a grout cover of $> 0.8 \cdot d_s$.

Corrosion protection according to EN 1537

Double corrosion protection according to EN 1537 with cement grout and plastic duct, whose high quality is ensured by prefabrication, enables permanent control of the most extreme conditions. The tensile tests to verify the crack widths according to EN 1537, Annex B, must of course be performed.

Recommendations

The corrosion protection systems described above that are recommended for ground anchors and other stabilisation applications (piles, nails) using concrete reinforcement steel SAS 670/800 are summarized in Table 3, dependent on the service life (temporary, semi-permanent and permanent) and soil aggressiveness.

Table 3. Corrosion protection systems

Chemical aggressiveness of the soil (see table 1)	Service life in years					
	2	5	7	30	50	< 120
low	A	A	A	A	A	-
	-	-	B	B	B	B
	-	-	-	C	C	C
medium	A	A	A	A	A	-
	-	B	B	B	B	B
	-	-	C	C	C	C
high	A	A	A	-	-	-
	B	B	B	B	-	-
	-	C	C	C	C	C

Type A: Corrosion rates [8] (bare steel, galvanized, cement grout cover 10 mm

Type B: Cement grout cover
Compression 25, 30, 35 mm
Tension 35, 40, 45 mm
(crack widths < 0.2 mm)

Type C: according to EN 1537

Outlook

For all systems of geotechnical stabilisation applications durability models have been proposed following European standards and regulations. Thus, SAS 670/800 threadbars can be used as anchors, tension or compression piles and nails/bolts:

- for a limited period dependent on the corrosion rates as bare or galvanized bar (up to 50 years)
- carbonation with a guaranteed minimum grout cover (semi-permanent)
- or with sufficient grout cover as permanent stabilisation application or according to EN 1537

The anchorage components and couplings were tested successfully following new European approval guidelines for pre-stressing concrete construction [20]. The installation of stabilisation systems is supervised by qualified and quality-assured specialist companies according to [21]. Design engineers can thus choose a variety of application options within the European market.

References

[1] Eurocode 1, 2,7 and EN 206
[2] Material standards: EN 10138 pre-stressing steel, EN10080 Steel for the Reinforcement of concrete, EN ISO 15630-1 Test methods
[3] prEN 14490 Soil Nailing
[4] EN 14199 Micropiles, 2005
[5] EN 1537 Ground anchor, 2001
[6] CEB: Durable Concrete Structures N^0 182, 1989 and Durability Design for Carbonation Induced Corrosion N^0 238, 1997
[7] D. Jungwirth: Application of crack width limitation in concrete engineering, Concrete Day 1985; Beton- und Stahlbetonbau 7, 1985, Verlag Ernst und Sohn, Berlin
[8] U. Nürnberger, expertise: Corrosion rates of S670 anchors, nails and piles in ground; 27 April 2004, TU Stuttgart, Stuttgart-Hofen, Stirnenweg 7
[9] CUAP-expert group for European approval of GEWI piles, anchors and nails made of S670 at the OIB Vienna; 2004 to 2007, as yet unpublished
[10] U. Nürnberger, expertise: Assessment of S670 threadbar with regard to its susceptibility to corrosion and stress corrosion cracking; 15 July 2004, TU Stuttgart, Stuttgart-Hofen, Stirnenweg 7
[11] ISO/DIN 14657: Galvanized concrete reinforcement steel, prEN ECO 19023-2005

EN ISO 1461; 2000: Hot-dip galvanized coatings on fabricated iron and steel articles, Specifications and test methods
[12] Recommendations Clouterre 1991 as well as Federal Highway Administration Soil Nail Recommendations
[13] ÖNORM B5013: Surface protection using organic protection materials in municipal hydraulic engineering: Corrosion probability
[14] DIN 50929: Probability of corrosion of metallic materials when subject to corrosion from the outside: Assessment criteria
[15] EN 12501-1 and 2: Protection of metallic materials against corrosion: General, Corrosion protection probability
[16] C. Gehlen: Probabilistic lifetime calculation of reinforced concrete structures; D.A.f.Stb. Number 510, 2000, Beuth-Verlag. Burggrafenstraße 6
[17] EN 445 to 447: Grout for pre-stressing tendons, 1996
[18] Jungwirth, Jeltsch: Pragmatic considerations regarding the corrosion protection of geotechnical stabilisation applications according to new European regulations, ÖIAZ; Number 2-3/2005, Ingenieur- und Architektenverein Wien, Eschenbachgasse 9
[19] L. Wichter: Investigation into the grout body's crack behaviour under tensile stress: FMPA Test report TU Cottbus, March 2004, Cottbus, Universitätsplatz 3-4
[20] ETAG 013: Guideline for European technical approval of post-tensioning kits for pre-stressing of structures, 2002
[21] EN ISO: 2000: Quality management systems – Fundamentals and CEN Workshop Agreement, 2003; Requirements for installation of post-tensioning systems and qualification of the companies

A Case Study on Corrosion Protection Effect of Using Field-Mixed Resin as Tie-Back Anchors' Grout Material for Retaining Wall and Its Inspection Method

Chihping Kuo and **Hung-Jiun Liao**, Department of Construction Engineering, National Taiwan University of Science and Technology, Taipei City, Taiwan, R.O.C.

Abstract

The long-term behavior of working ground anchors can be measured from the long-term stability of the lock-off anchor load and the effectiveness of anti-corrosion measures for the anchor components. The paper describes measures that have been taken and tested in the field including the in situ resin grouting method for the tendon bond length and the measurement of electrical resistance between anchor components and the surrounding ground. After a period of one and a half years, the test anchor showed a minor decrease in lock-off load, but the electrical resistance dropped significantly from 155.9 MΩ to almost 0 MΩ. This indicated that the resin grout might have cracked over that time period and permitted groundwater to seep into the corrosion protection measures and contact the anchor components. It is considered that the process of in situ resin grouting may be feasible during anchor installation but the durability of field mixed resin needs to be further improved to ensure its long-term performance for corrosion protection.

Introduction

There are many kinds of landslide in Taiwan with different mechanisms to induce slope failures. To deal with deep and large landslides, structural retaining measures are adopted and tie-back anchors are additionally installed to provide reinforcement and improve stability. These ground anchors have been widely used nationally for more than half a century. However, a serious tie-back retaining wall failure took place in the Lincoln residential community (Xijih town, Taipei County) in 1997 and induced a major disaster. The investigated reasons for the failure pointed to inadequate durability of the anchor materials.

Generally, the geology where anchors are adopted in Taiwan is saturated with groundwater or is located under the groundwater table and many anchors encounter rock joints, weathering rocks and the interface between soil and rock. The mobile and coagulating behavior of grouting anchors is complicated to understand and control. How to maintain the durability of anchors becomes an important issue for engineers. Recently, permanent anchor development has started to tackle the corrosion issue and resin is one of grouting materials to replace cement in Taiwan. Although different from overseas, in Taiwan the resin adopted is grouted in the field and not in a factory. The advantage of manufacturing resin anchors in a factory is that it permits quality control and examination. However, due to transport convenience the size of pre-manufactured resin anchors has to be limited. A large anchor size is required in Taiwan due to the complex geological environment and to ease shipping, the use of in situ grouted resin anchors is preferable.

Two anchors, grouted with cement and resin (traditional and permanent), were installed in this study to be compared in relation to their durability. This study followed Fischli's method (1997) to test the durability of the anti-corrosion protection, i.e. the electric resistance measurement method (ERM) was used to check the integrity of each installed corrosion protection system.

Design and performance of permanent anchors

In permanent anchors provided to tie back retaining walls, the corrosion protection should cover the entire anchor, including the grouted section, free section, and anchor head. Where unfavorable environmental conditions are considered to have caused the anchors to suffer, examination of the corrosion protection has to be performed before and after the anchor is stressed.

Corrosion protection of anchors

The permanent anchors in this test complied with two important points (1) and (2).
(1) Protection for the steel tendon bond length. This was provided by wrapping with a wave-shaped plastic tube and then grouting the annulus with cement grout (water/cement ratio by weight = 0.45). Based on experience in Taiwan, the quality preparation and construction of the cement grout is hard to control, and shrinkage and cracking are often induced inside the grout annulus. This results in reduced protection and even loss of protection due to water intrusion. Due to these known drawbacks, specially made macromolecule resin (polyethylene resin with hardener) was adopted to replace cement for the second anchor.
(2) Protection for the steel tendon free length. The strands of the steel tendon in this part were each coated with oil and wrapped with PE sheathing. However, for cost saving and convenience, the latter step is often ignored in Taiwan. This may cause evaporation or degradation of the oil. Therefore, the two different free length protection conditions were also compared in this test, namely, oil and PE sheathing for the resin grouted anchor and simply oil treatment for the cement grouted anchor.
(3) Protection for anchor head. In routine practice, this part of the protection includes tendon grippers, clips and the space between protective cover and the anchor head.

Test method of corrosion protection

Generally, the effect of corrosion protection is difficult to detect in the short term. The authors used the electrical resistance measurement ERM method (Fischli 1997) in this study. The protection effect can be evaluated by measuring the electric resistance between the tendon and the surrounding earth. The method measures the electrical resistance between anchor head and surrounding earth in two stages:
(1) Before the anchor is stressed. When applying a voltage of 40V DC, the measured electrical resistance should be between 100~200KΩ.
(2) After the anchor is stressed. When applying a voltage of 500V DC, the measured electrical resistance should be between 10~100KΩ.
If the measured resistance is high, the encapsulation is presumed to be intact. A low measured resistance indicates a problem. The operation of this method is shown in Figures 1 and 2.

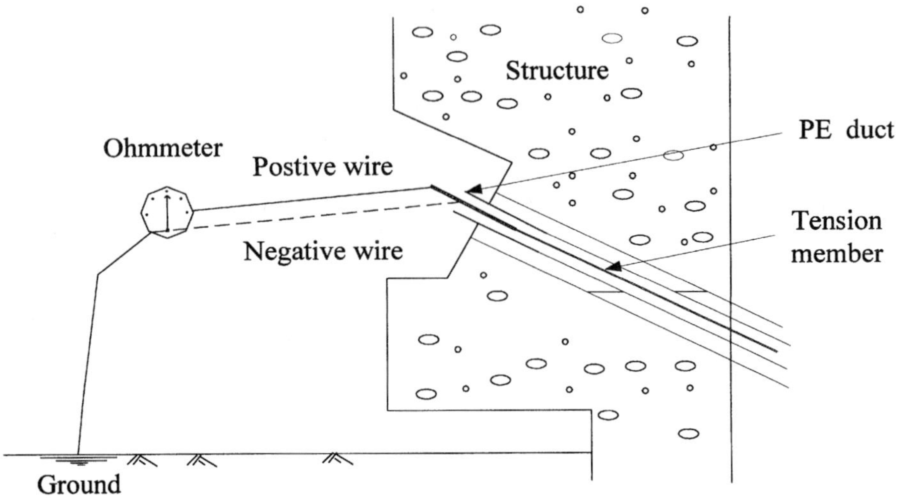

Figure 1. Schematic diagram of measurement of electrical resistance by DC current before being stressed (after Fischli, 1997)

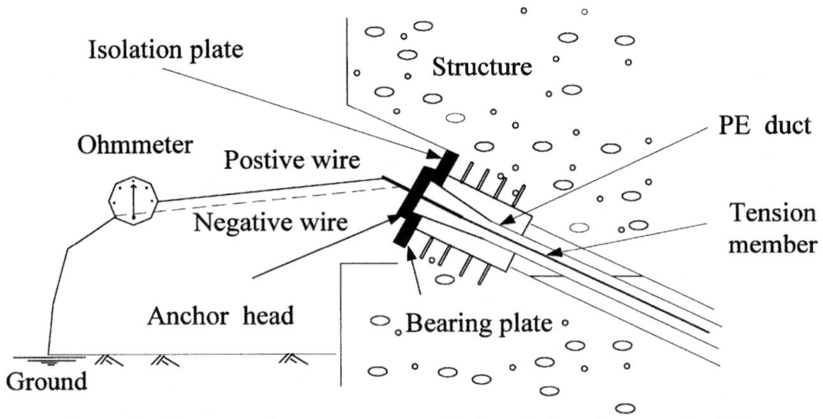

Figure 2. Schematic diagram of measurement of electrical resistance by DC current after being stressed (after Fischli, 1997)

In situ test

Arrangement on site

The test was performed in a slope located in Keelung. A landslide area of height 80m and length 150m was chosen to install and test the experimental anchors. The slope supported by a grid-beam retaining structure was built in 1980 and failed due to a huge typhoon in 1995. The primary geological structure in the area is a cataclinal slope with a dip of 20° and a strike of N40°E. The ground profile from the surface contains an overburden layer and a fallen-deposited rock layer overlying a sandstone layer that in turn is underlain by a shale/sandstone layer and underlying sandstone layer. The test anchors were placed through the first three layers with the tendon bond length installed in the first sandstone layer. The geologic profile of the test site and anchors' location are shown as Figure 3.

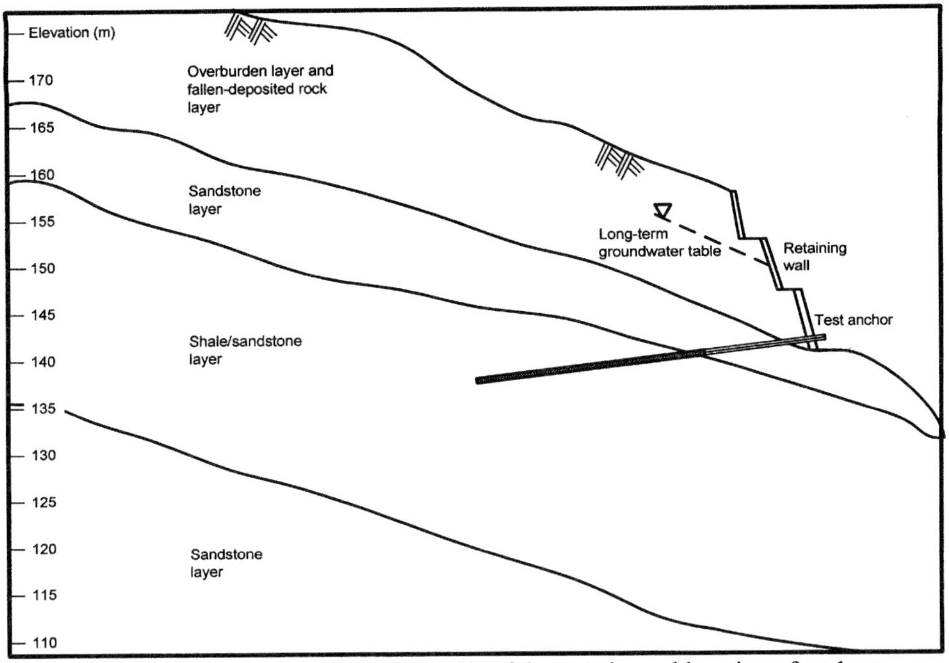

Figure 3. Schematic diagram of geologic profile of the test site and location of anchors

The test was performed on an existing retaining wall whose elevation was located between 141.0m~146.9m, and the grid size was 2m×2m. The long-term groundwater table elevation is 150.3m. The ground water seeps from wall surface through rock joints as shown in Figure 4. This reveals that the drainage was not working well and that the ground water table was high. By observation during the test period, the ground water kept seeping and increased with rainfall. The rainfall log during test is shown as Figure 5.

Figure 4. Groundwater seeping out from the slope at test site (2005-07-19)

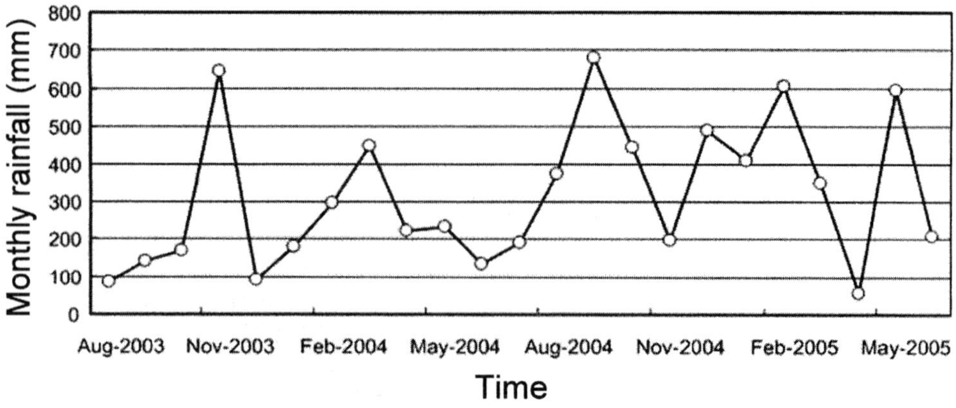

Figure 5. Rainfall data between 2004-03-01 to 2004-07-12 at test area

Set-up of anchors

The test began in August, 2003. According to the geological investigation, the potential slide surface might be the interface of the fallen-deposited rock layers and underlying sandstone layer. In order to have the anchors penetrate the potential slide surface, an overall anchor length of 12m was adopted, i.e. 4m long tendon bond length and 8m long free length. The diameter was 75mm and the tendon was composed of four 7-wire strands. Both tendons were fabricated in the factory, the main difference being the in situ grouting materials. The free length of one tendon was oiled and wrapped in PE sheathing while the second free length was simply oiled. A stainless steel rod was welded to the tendon and extended outside the anchor head to use as one electrode for ERM. The other electrode for the surrounding earth was also a stainless rod welded with steel to the RC retaining structure.

Drilling began in the middle of August, 2003 and adopted a rotary drilling method. Two 15m long boreholes were made for the anchor installation. The anchors were placed at the end of August, 2003. The grouting works were performed from early to middle September, 2003. The RC beam structure for bearing the anchor stress was constructed from middle to end September, 2003. Thereafter, all of the works were allowed to cure.

Prior to anchor stressing, phase 1 of the ERM was performed in early February, 2004. After that, the two anchors were stressed and locked off. The lock-off loading of each anchor in this test was 40t. In order to establish unfavorable environmental conditions, drains were not installed in the slope. Figure 5 shows the daily rainfall log (CWB 2005) during test performance to June, 2005. That could be taken as field weather condition.

After anchor stressing, phase 2 of the ERM was performed during July, 2005.

Installation of corrosion protection

The grouting material is macromolecule resin. The resin had good mobility after mixing and coagulated after 20 minutes, so grout injection should be finished within 20 minutes. 24 hours after coagulation the one-dimensional compressive strength of in-situ grouted resin sample is about 450kg/cm^2. Compared with cement grout, the resin provides higher compressive and tensile strength and lower shrinkage.

The resin grouting was applied between tendon strands and the wave-shape tube, and the cemented grouting was applied outside wave-tube. Low strength cement grout (water/cement ratio = 0.5) was injected along the free section.

To clarify the difference between the protection in the two anchors tendons in this test, the permanent anchor had low strength cement grout for free section of the tendons and resin grout for bonded section of the tendons. The free section of the tendon was also oiled and wrapped with PE sheathing. The cement anchor has only cement grout for anchored section tendons but no grout for the free section of tendon that was only oiled.

Oil was applied in the anchor head set-up. Figure 6 shows the set-up of the anchor head before covering. From left to right are the bearing plate, insulation plate, load cell and bearing plate. One strand of the tendon was extended outside to perform ERM.

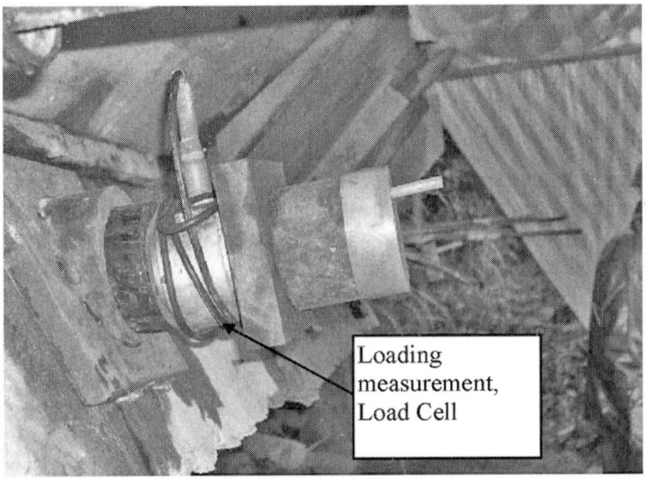

Figure 6. Anchor head and load cell after lock-off (2004-02-07)

Corrosion protection test results

The final test period of anchor performance was carried out from 3 to 7 February, 2004 during a period of heavy rainfall that represented a corrosion protection test in the worst environment. The ERM was also performed on both anchors in this period. The test result of the traditional cement grouted anchor before being stressed showed the electrical resistance stayed at zero, that meant the DC current loop was short. This indicated that the protection had failed and it was conjectured that the shrinkage of the cement grout was so serious that it had permitted water intrusion. The test result of resin grouted anchor before being stressed showed the protection was intact (the electrical resistance was around 200MΩ, far larger than 0.1MΩ). After stressing, the test result of the resin grouted anchor showed that the protection was also intact (the electrical resistance was around 155.9MΩ, far larger than 0.1MΩ (Figure 7). These results indicated that the resin grouted anchor performed well after installation and initial stressing.

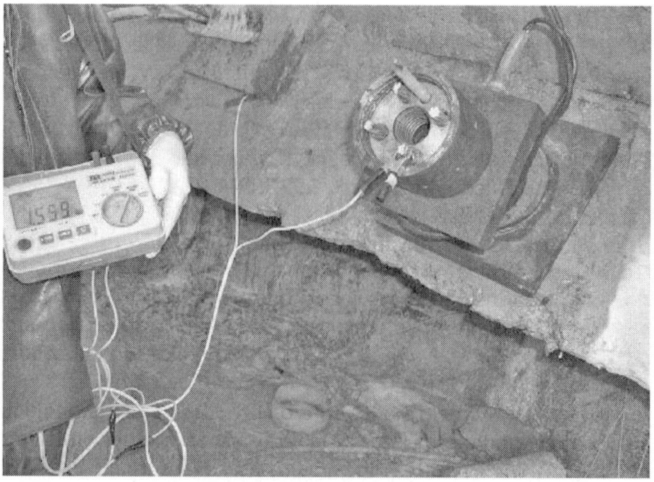

Figure 7. Electric resistance measurement before being stressed = 155.9MΩ (2004-02-06)

After one and half years, there was no rainfall according to the monitoring log from 1 to 14 July, 2005. The water supply should be only ground water stored in rock joints without any surface seepage or run-off. Therefore, the surface of the tie-back retaining structure was supposed to be dry. The ERM was performed again to check the protection durability on 14 July, 2005. As shown in Figure 8, the measured electrical resistance was 0.369MΩ. Even though this value was much lower than previously recorded, it was still acceptable, i.e. larger than 0.1 MΩ. On 18 July, 2005, the typhoon HAITANG passed Taiwan Island and brought 186mm rainfall in one day at Keelung. It still rained on 19 July, 2005 and the ERM was performed again. The test area was then entirely wet. As shown in Figure 9, the test result showed the measured electrical resistance had lowered to 0.027MΩ, i.e. much lower than four days previously and the code recommendation. However, the locked-off load had not decreased according to the load cell.

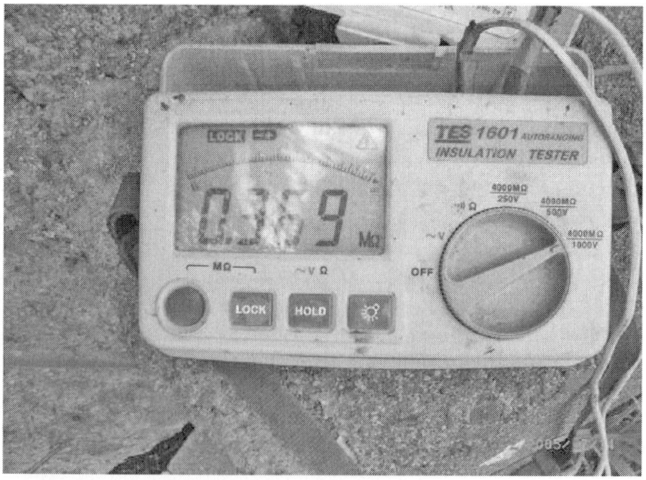

Figure 8. Electric resistance measurement after about 1.5yrs = 0.369MΩ in dry conditions (2005-07-14)

Figure 9. Electric resistance measurement after about 1.5 yrs = 0.027MΩ in wet conditions (2005-07-19)

Remarks

In reviewing the reasons for the poorer results relating to the corrosion protection, there are several possible explanations.

(1) The macromolecule resin has a higher one-dimensional compressive strength after coagulating but it also has a lower ductility. Therefore the resin grout might have cracked due to a small strain and permitted the moisture to permeate into the steel strands of the tendon;

(2) The moisture might have spread along the extended stainless steel rod to permeate into strands;

(3) The moisture might have permeated from the interface of RC beam and anchor head cover.

In spite of careful preparation and installation of the resin grouted anchor, it still failed to comply with the standard.

Conclusions

A corrosion protection integrity test was carried out on a resin grouted permanent anchor in unfavorable geological, weather and ground water conditions. The results showed that prior to and after initial stressing, the electrical resistance was acceptable.

After one and half a years, although the service loading on the anchorage did not decrease, the electrical resistance of the corrosion protection effect reduced. Although the corrosion protection passed the ERM test in dry conditions, it failed a few days later in wet conditions. This result shows that the anti-corrosive resin grout is incapable of complying with the durability requirements of a permanent anchor.

According to a comparison of the results from both anchors, the corrosion protection of the resin grouted anchor was better than the cement grouted anchor, both before and after stressing.

Acknowledgements

The authors would like to thank the National Science Council for financial support (NSC 92-2622-E-011-002-CC3), the co-operating factory (Poly engineering) for the anchor fabrication and installation, and the National Taiwan University of Science and Technology for providing the test site.

References

Central Weather Bureau (2005), "daily rainfall log", http://www.cwb.gov.tw/V4/index.htm.

Fischli, F. (1997). "Electrically Isolated Anchorages," Proc. Int. Conf. "Ground Anchorages and Anchored Structures", Edited by G.S. Littlejohn, Thomas Telford, London, England, pp. 335-342.

Advancements in Ground Anchors: Carbon Fibre Reinforced Polymer (CFRP) Strands

Matthew Sentry, Department of Civil Engineering, Monash University, Australia
Abdelmalek Bouazza, Department of Civil Engineering, Monash University, Australia
Riadh Al-Mahaidi, Department of Civil Engineering, Monash University, Australia
Darren Loidl, Geotech Pty Ltd., Australia
Chris Bluff, Geotech Pty Ltd., Australia
Len Carrigan, Geotech Pty Ltd., Australia

ABSTRACT
Steel tendon ground anchors are an integral construction technique for numerous civil engineering applications ranging from deep excavation support to resistance of structural uplift and overturning of superstructures. Corrosion and human error generally causes failure of steel tendon ground anchors. Several methods of minimising anchor system corrosion have been adopted over time to minimise ingress of corrosive substances. Anchors are still failing due to corrosion. Advancement in the development of corrosion resistant materials has been at the forefront of materials research. Research and development of FRP materials has enabled the progress of providing the industry with a more potentially robust anchor system aimed at eliminating current limitations encountered with steel strand ground anchors. This paper investigates current developments in FRP materials for ground anchor applications as an alternative to conventional steel tendon ground anchors.

Introduction
Technological advancements of fibre reinforced polymer (FRP) products for applications in civil construction has allowed new products such as glass fibre (GFRP), aramid fibre (AFRP) and carbon fibre (CFRP) to pave the way for research into further improvements to the currently favoured steel tendon ground anchor system. This paper investigates current developments in FRP materials for ground anchor applications as an alternative to conventional steel tendons.

Corrosion limitations to steel tendon ground anchors
Corrosion in steel tendon ground anchors occurs as a consequence of in-homogeneities or impurities in the steel tendon or grout, or by the existence of salts, sulphates and other dissolved solids present in grout mixtures, soils or groundwater (Littlejohn and Bruce, 1977).

International standards including EN1537:2000, AS5100.1-2004, BS8081:1989 require some type of corrosion protection for permanent ground anchor applications. Permanent steel tendon ground anchors require several areas to be corrosion protected, including tendon bond length, tendon free length, transition between anchor head and free length and the anchorage head. This paper will focus primarily on permanent ground anchor systems.

Littlejohn (1987) reported on 35 known cases of corrosion related failures to steel tendon ground anchors, concluding that free length corrosion comprised 60% of failures, 34% of failures occurred in at the anchorage head and only 6% of failures occurred in the fixed anchor length. Catastrophic structural failures can occur as a direct result of ground anchor failure due to corrosion (Figure 1). Littlejohn emphasised that the small percentage of failures within the fixed anchor length was a result of inadequate grout placement.

Steel tendon corrosion occurs locally where the tendon intersects a crack in the surrounding grout, or as a result of damage to the corrosion protective sheaths (Weerasinghe and Adams, 1997). Grout cracks occur as a result of either shrinkage strains during curing or tensile loading of the anchor. Due to grout micro cracking, majority of standards do not permit grout to be considered as one of the two means for corrosion protection. Weerasinghe and Anson (1997) and Mothersille (2006) provided project examples where corrosion occurred above and below the anchor head, resulting in anchor strength reduction (Figure 1).

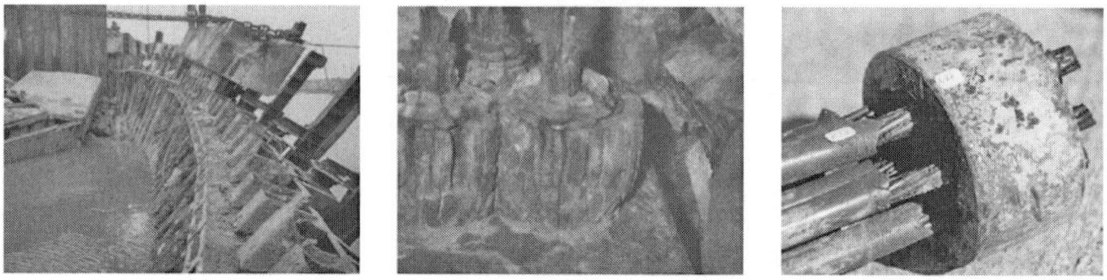

(a) (b) (c)

Figure 1: Failure of anchored quay wall (b) corroded barrels and wedges in anchor head (c) lack of corrosion protection behind anchor block (Mothersille, 2006)

Tendon material choice is limiting ground anchor development. Anchor corrosion and its need for protection, performs an integral element of a permanent ground anchor.

Measures to prevent corrosion failure of permanent ground anchors has developed the double corrosion protection system (EN1537:2000) (Figure 2). Double corrosion protection is currently applied over the tendon bond length, tendon free length, the anchor head and the transition between the anchor head and the tendon free length.

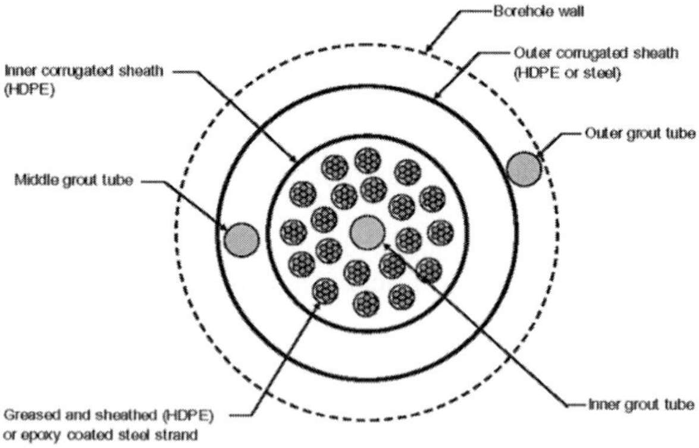

Figure 2: Schematic of double protection system for permanent ground anchor

Numerous methods are adopted to provide double corrosion protection to permanent ground anchors. Corrosion protection systems are project driven, but standard protection requirements as specified in EN1537 (Europe), BS8081 (UK), AS5100 (Australia) should be adhered to.

Common systems used to meet standards requirements

Epoxy coated steel strand tendons are commonly used throughout Northern America as a means to provide one of the two corrosion protection systems throughout the anchors free length (Figure 3). Weerasinghe and Adams (1997) concluded that epoxy coated strand tendons have been observed (Upper Occoquan Dam project) to creep by approximately one order of magnitude compared to bare wire strand. Damage to epoxy coating during transport or installation can also void protection (Figure 3).

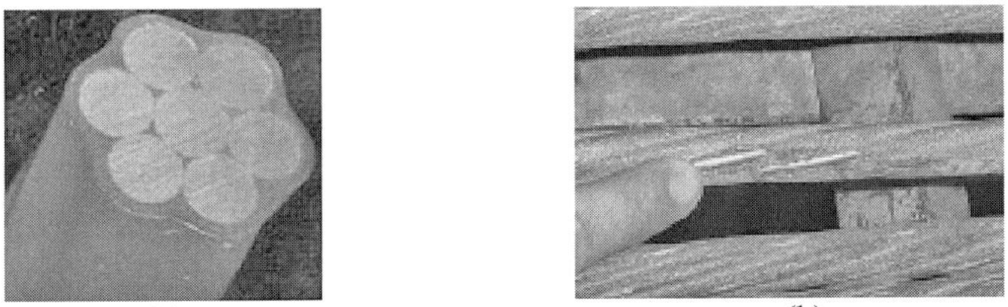

(a) (b)

Figure 3: (a) Epoxy coated strand; (b) damaged epoxy coated steel strand (Mothersille, 2006).

Greased steel tendons and smoothed walled HDPE have successfully been used to minimise corrosion while allowing each individual steel tendon the mobility required throughout the tendon free length (Figure 4). Practical experience shows that the greasing and sheathing is a messy and time consuming process. Should the greased tendons require de-greasing, considerable effort is required to ensure successful removal of grease from the steel tendons.

(a) (b)

Figure 4: (a) section of a greased and sheathed steel strand tendon; (b) greasing process of the individual steel strand (7 wire) ground anchor tendon.

Corrugated ducting is a critical protective barrier preventing corrosion of steel tendons in ground anchors. Double layered ducting is an approach to eliminate potential system corrosion failure (Figure 5). Internal corrugated ducting is required to be impermeable, continuous and semi-rigid, HDPE is a common product used (Barley, 1997). Damage to the internal system can be fatal to the validity of the ground anchor. External corrugated duct is used to ensure the inner corrugated duct is in sound condition during transport and once the ground anchor is homed and stressed. The outer corrugated duct can be made from HDPE type continuous system. The need for a double ducting system increases the diameter of the borehole and overall weight of the anchor itself, increasing fabrication and installation times.

Protection for the ground anchor head includes both the inner and outer anchor heads. The inner anchor head is protected by using an HDPE trumpet sleeve welded to the corrugated sheath (Weerasinghe and Adams, 1997). Grease or polyester resin is used to protect the inner head. The outer anchor head is protected by grease with a steel or plastic cap. Appropriate interlocking between the head cap and bearing plate must ensure no leaking of the grease/resin or ingress of water/air. Bearing plates can be painted with bitumastic paint (or similar) for corrosion protection or newer plastics and resins can also be used.

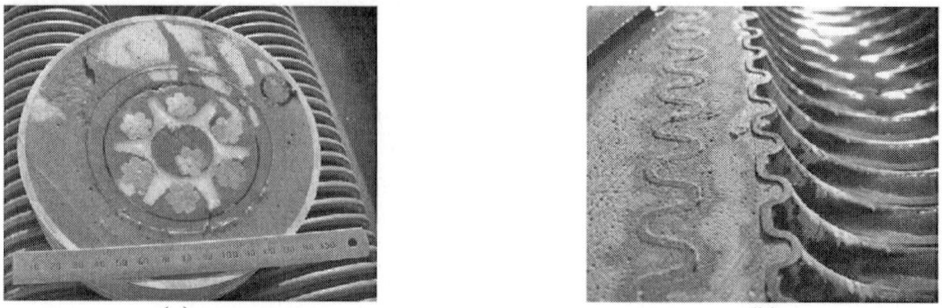

(a) (b)

Figure 5: Details of double corrosion protection through the bond length of a ground anchor (a) cross section (b) longitudinal section (Mothersille, 2006)

Although Portland cement grouts are not considered a means of corrosion protection, EN1537:1996 allows for a dense cement grout injected under controlled conditions to be considered as one of the two layers of corrosion protection required, providing it can be proven that crack widths during service loading do not exceed 0.1mm. Additionally EN1537:1996 allows for a HDPE corrugated duct to act as a permanent barrier on the provisions that a 20mm grout cover injected under a minimum pressure of 500KPa with a crack width during service loading not exceeding 0.2mm. Experts have questioned the ability to prove these magnitudes of crack widths in practice (Weerasinghe and Adams, 1997).

Advancements in corrosion resistant composite materials – FRP

Current complexities of double corrosion protection has shifted research into investigating the use of alternative materials to assist in minimising or eliminating risk of anchor failure due to corrosion, including investigations into the use of fibre reinforced polymers (FRP).

Fibre reinforced polymer (FRP) materials

FRP consists of a group of materials containing aligned continuous organic (aramid or carbon fibre) or inorganic (glass fibre) filaments embedded in a polymer matrix and shaped into a product for use. Organic filaments are the primary choice for ground anchors (Zhang and Benmokrane, 2004). North America is continuously developing and refining the use of inorganic filaments as a substitute for steel in concrete reinforced structures (Tardif et al., 2007, Van der Wal and Boulfiza, 2007).

FRP filaments consist of various kinds of load bearing fibres of high tensile strength and high modulus of elasticity. The ultimate tensile strengths of CFRP filaments range from 290-758GPa, AFRP ranges from 70-179GPa compared to prestressed steels ultimate tensile strength of 195GPa (Zhang and Benmokrane, 2004).

Isotropic-based (pitch) and polycrylonitrile-based (PAN) fibres are two types of carbon fibre filaments used in CFRP composites. Carbon fibre filaments are also classified based on their modulus; low modulus ranging between 230-250GPa, medium modulus ranging between 290-300GPa, high modulus ranging between 350-380GPa and ultrahigh modulus ranging between 480-760GPa. Low modulus carbon fibre filaments have a lower density, higher tensile and compressive strengths, higher tensile strain to failure than high modulus carbon fibre filament products and are considered less expensive than other carbon fibres. PAN based carbon fibres compared to Pitch based carbon fibres have a higher ultimate strain, but a lower modulus of elasticity. PAN based carbon fibres are more expensive than Pitch based carbon fibres. The main advantages of carbon fibres over other FRP products are high strength to weight ratio, high modulus to weight ratio, high fatigue strength, low coefficient of thermal expansion, and excellent moisture and chemical resistance (Mallick, 1993).

Europe, Asia and North America manufacture aramid fibre filament products. Kevlar (modulus 82-179GPa), Twaron (modulus 70-130GPa) and Technora fibres (modulus 74GPa) are used depending on manufacture (Mallick, 1993). Aramid fibres benefits include excellent impact resistance, high strength, high stiffness and thermal stability.

The achilles heel of aramid fibre filaments is that fibre strength decreases as water is absorbed into the fibres (Zhang and Benmokrane, 2005). Additionally, aramid fibres are sensitive to ultraviolet light and care must be taken during storage and transportation.

For all FRP materials, a polymetric matrix is used to bond the fibres, protect the fibres against environmental effects and assist in the equalization of fibre forces and load transfers in the transverse direction. Thermoplastic and thermoset polymers can be applied with FRP fibre filaments to form an FRP composite material.

Thermoplastic polymers including PVC, polyethylene and polypropylene are viscous polymers capable of being modified under temperature and more environmentally friendly than other products, but are difficult to combine with continuous fibres in a production operation (Zhang and Benmokrane, 2005). Thermoset polymers are classified as low viscous, low molecular-weight liquid with thermal stability and chemical resistivity and are said to exhibit reduced creep and stress relaxation when compared to thermoplastic polymers. Thermoset polymers including epoxy, polyester and vinyl ester are preferred resins for FRP material in civil engineering applications (Zhang and Benmokrane, 2005).

Properties of FRP Composites

Composite products currently able to provide the industry with an environmentally resistant alternative product to steel tendons for ground anchor applications include glass fibre (GFRP), aramid fibre (AFRP) and carbon fibre (CFRP) reinforced polymers. FRP properties can vary depending on quality, manufacturer and manufacturing process (Table 1).

Table 1: Properties of FRP composites (after Zhang and Benmokrane, 2005)

Property	GFRP	AFRP	CFRP	Prestressing steel
Minimum fibre volume ratio	0.55	0.6	0.63	-
Density (g/cm^3)	2.1	1.38	1.58	7.85
Longitudinal tensile strength (MPa)	1080	1280	2280	1865
Transverse tensile strength (MPa)	39	30	57	1860
Longnitudinal modulus (GPa)	39	78	142	190
Transverse modulus (GPa)	8.6	5.5	10.3	190
In-plane shear strength (GPa)	89	49	71	
In-plane shear modulus (GPa)	3.8	2.2	7.2	73.1
Major Poisson's ratio	0.28	0.34	0.27	0.3
Maximum longitudinal strain (%)	2.8	1.5	1.5	4
Maximum transverse strain (%)	0.5	0.5	0.6	4
Relaxation ratio (%)			2-3	8

FRP composites are known to have a brittle mode of failure compared to yielding failure of steel (Figure 6). Designers need to consider the brittle failure nature of FRP composites when designing.

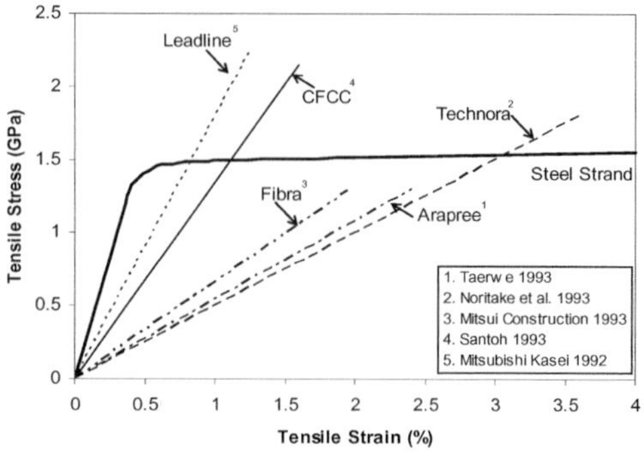

Figure 6: Typical tensile stress-strain behaviour of various FRP products compared to steel strand (after Benmokrane et al., 1997)

Tensile strength is dependant on length of tendon, anchorage system being used and the loading rate. CFRP ranges between 13-18% higher tensile strength than steel (Zhang and Benmokrane, 2005). Brittle nature of FRP composites result in the modulus of elasticity for CFRP composites to be in the order of 30% lower and AFRP composites to be 65% lower than prestressing steel (Zhang and Benmokrane, 2005). Tensile strain of CFRP ranges between 1.3-1.6% and AFRP ranges between 2.0-3.7%, compared to prestressing steel which is at 6%.

FRP fibres are weak in transverse loading. FRP shear strength is a result of the matrix used to construct the composite, which is considerably lower than its tensile strength. Santoh (1993) compared AFRP, CFRP and prestressing steel showing lateral shear strength of 11% (CFRP), 15% (AFRP) and 47% (prestressing steel) of the materials tensile strength. FRP (CFRP and AFRP in particular) density is approximately 15-20% that of prestressing steel (Benmokrane et al., 1997).

FRP creep rupture may be more prominent than in steel strand due to brittle failure of FRP tendons over time under sustained loading or under adverse conditions (Benmokrane et al., 1997, Zhang and Benmokrane, 2005). Under initial stress of 65%f_{pu}, creep strain of CFRP tendons is approximately 0.007% after 100 years (Benmokrane et al., 1997). Zhang et al. (2005, , 1997) reported that the relaxation for AFRP was 12% after 1000 hours (under load of 0.6f_{pu}), while CFRP ranged between 0.48% and 0.96% (0.5f_{pu} – 0.8f_{pu}) when compared to prestressing steel relaxation rate of 1.02% to 7.35% (0.5f_{pu} – 0.8f_{pu}).

Advantages and Limitations of FRP composites for ground anchor application

FRP composite tendon fatigue strength exceeds that of prestressing steel, while all FRP composite tendons are generally more resilient to all kinds of aggressive environments compared to steel. Little research has been conducted to verify manufactures recommendations for environmental durability.

Europe and Asia use FRP composites over steel where durability has been a serious issue. Numerous Japanese projects use FRP ground anchors for slope stabilisation in costal and snowy environments. CFRP and AFRP have been adopted for permanent ground anchorage use while GFRP has been limited to temporary soil nails and anchors. The advantages and of

CFRP include high tensile strength and modulus of elasticity, excellent fatigue and corrosion resistance, low thermal expansion, low stress relaxation and excellent moisture resistance. CFRP is the most expensive FRP tendon on market. The advantages to AFRP include excellent relaxation and corrosion resistance, high fatigue strength, moderate cost. AFRP limitations are its lowest strength and modulus of elasticity of all FRP tendons, strength loss whilst moist and at temperatures in excess of 150°C. GFRP is the cheapest FRP on the market; however, their limitations include lowest tensile strength, most sensitive to stress fatigue damage and creep rupture, most vulnerable to alkaline environment, may lose strength when exposed to moisture and increased temperature. All FRP materials are brittle in nature.

FRP composite ground anchorage design

FRP tendon ground anchors must consider the same design criterion and procedures as with conventional steel tendon ground anchors. Several quality standards are available to assist designing with FRP tendons. In addition to standards, government organisations have established rigorous guidelines complementing local and international standards. Adjustments to current standards and guidelines are required when replacing steel tendons with FRP tendons. Current documentation relevant to ground anchorage construction and for application of FRP tendons in ground anchors include (but not limited to) are EN 1537:2002–Execution of special geotechnical works–Ground Anchors; BS 8081:1989–Ground Anchorages; ACI 440.4R-04-Prestressing Concrete Structures with FRP Tendons; CAN/CSA S806–Design and Construction of Building Components with Reinforced Polymers; ISIS Canada-Educational Modules No. 1– 8; Post Tensioning Institute (USA)-Recommendations for Prestressed Rock and Soil Anchors and RTA QA Specification B114 Ed 2/Rev 3 1999.

Design working loads

Maximum working load of $0.6f_{pu}$ is recommended when designing high capacity ground anchorage systems with FRP (Benmokrane et al., 1997). Various design working loads have been trialed and adopted for a range of project specific FRP ground anchor field applications ranging between $0.22f_{pu}$ to $0.66f_{pu}$ (Table 2).

Table 2: CFRP ground anchors field application in Japan (Benmokrane et al., 1997)

Project	Year	Anchor number	Rock type	Tendon Size (mm)	Ultimate Load, Pu (kN)	Design Load (%Pu)	Total length (m)	Bond Length (m)
Slope stabilisation, Niigata	1993	42	Granite	6ϕ12.5	852	0.6	10.5	6.5
Slope stabilisation, Ito, Shizuoka	1995	20	Soft Rock	4ϕ12.5	568	0.63	9-11.5	-
Slope stabilisation, Nazusa, Gifu	1995	30	Soft Rock	3ϕ12.5	426	0.43	7.6-10.6	-
Slope stabilisation, Ishikawa	1995	40	Soft Rock	6ϕ12.5	852	0.57	10-11.5	-
Slope stabilisation, Yamanashi	1996	10	Soft Rock	3ϕ12.5	426	0.53	7.3	-
Slope stabilisation, Kyoto	1996	10	Soft Rock	2ϕ12.5	284	0.66	7.3-17.3	-

Fixed Anchor Design

Fixed anchor length calculations for FRP anchors assumes the design is for a straight shafted ground anchor into rock based on uniform bond stress distribution at the ground and grout interface. From these assumptions, Benmokrane (1997) developed a relationship between fixed anchor length and anchor design working loads (Equation 1) and at the grout-tendon interface (Equation 2).

$$L = \frac{T_w S_f}{\pi D \tau_{ult-1}} \tag{1}$$

$$L = \frac{T_w S_f}{n \pi d \tau_{ult-2}} \tag{2}$$

Where: L=fixed anchor length, T_w=working load, and τ_{ult-x}=ult shear strength, D,d=diameter.

Bond values at the interface for grout and ground used for FRP tendons are coherent with what is currently being established for conventional prestressed steel tendon ground anchors (Littlejohn and Bruce, 1977). The wide variety of surface textures available in FRP tendons still requires considerable amounts of research to establish various bond values between tendon and grout for all usable FRP tendons (Zhang and Benmokrane, 2002). Some results published by Benmokrane et al. (1997) show the relationship of bond values between FRP tendons and cement grout under field conditions (Table 3) on selective FRP materials.

Table 3: Working bond values back-calculated from field applications of FRP ground anchors (after Benmokrane et al., 1997)

Tendon Type	Tendon	Design load (kN)	Bond length (mm)	Working bond strength (MPa)
CFCC	6ϕ12.5	490.0	7.5	0.28
	6ϕ12.5	510.0	6.5	0.33
Technora	9ϕ7.4	400.0	6.5	0.44
	9ϕ7.4	400.0	3.0	0.96

FRP Anchorage Systems

Tensile forces for FRP anchorage systems are transferred to the FRP anchor head system by anchorage in the same manner as steel tendon anchor systems. This load transfer requires lateral pressure and shear stresses to act on the FRP tendon surface in conjunction with high axial stresses applied to the anchor (Benmokrane et al., 1997). Sensitivity to lateral stresses and any notch or defect can limit its performance during stressing.

Standard steel tendon based wedge type anchorage systems are currently inefficient and cause premature failure of the FRP tendon at the wedge interface. To accommodate for the transverse material property limitations of FRP at the anchor block research into different methods for anchorage has taken place over the past ten years (Zhang and Benmokrane, 2004, Campbell et al., 2000). There are no single method of anchorage to suit a wide variety of FRP materials and their unique profiles, but several anchorage methods have been developed including soft metal wedge and soft metal swage anchorage system, plug-in-cone anchorage system, bond-type anchorage system and soft cone anchorage system (Campbell et al., 2000, Reda and Shrive, 2003, Tardif et al., 2007).

Soft metal wedge anchorage system is dependant on the gripping action of the wedges and/or teeth around the tendon perimeter accompanied by compression exerted by the wedges onto the FRP tendon (Reda and Shrive, 2003). There is great potential for this system as it is easy to assemble and use on site, but localised FRP tendons rupture at the wedge/tendon interface by excessive shear stress has restricted the potential (Tardif et al., 2007). Trials using a soft swage system, such as aluminium or copper minimises the excessive shear stress between the wedge and tendon have showed promise (Reda and Shrive, 2003, Tardif et al., 2007)(Figure 7(a)). Non metallic wedges are currently being researched as an alternative to soft metal wedges (Sireg, 2007) (Figure 7(c)).

Plug-in-cone anchorage system accommodates multiple tendons substituting the wedges with a conical solid cone. Anchorage is dependant on the effects from the inner solid cone pressing the FRP tendon against the teeth of the outer conical socket. Common failure is as result of shear rupture of the tendon at the base of the anchorage system. Limited success has been had using AFRP material (Reda and Shrive, 2003)

Bond type anchorage systems have had large repetitive success (Figure 7(b)). Zhang et al. (2004) developed a bond type anchor using an internally deformed cylindrical metallic shell to host the number of tendons in the anchor. High strength, expansive grouts are commonly used, but resins (epoxy) can also be successful. Resins with low modulus of elasticity can potentially limit the peak shear stress during load transfers, while high modulus resins are more efficient at long-term creep control (Benmokrane et al., 1997). The anchorage system can only be locked off at the anchor block using an externally threaded nut. Bond type anchorages are quite large and complex systems, restricting the clearance and tendon extension at the anchor head and may impact on space constraints of the anchor head and minimise additional re-stressing. The authors believe that further research into the anchorage methods at the anchor block is needed for FRP tendons to be incorporated into future ground anchor standards.

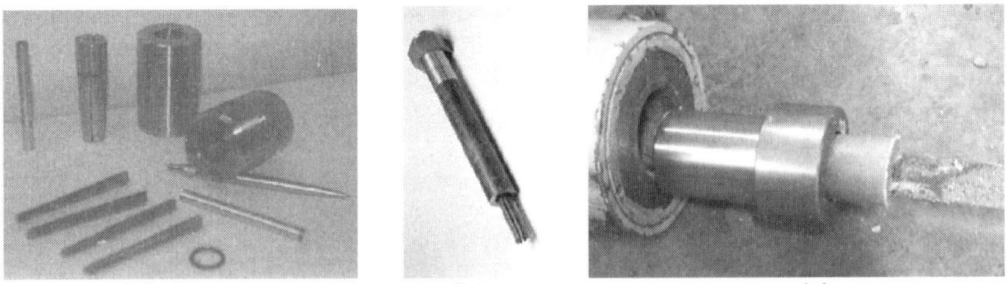

(a) (b) (c)

Figure 7: FRP anchorage systems (a) soft metal swage wedge anchorage system (Campbell et al., 2000) (b) bond type anchorage system with external lock-off nut (Benmokrane et al., 1997) (c) non-metallic wedge anchorage system for flat CFRP tendons (Sireg, 2007)

Research potential for FRP ground anchors

Current limitations to steel tendon ground anchorage systems highlight the need to research and develop new methods to improve the durability, reliability and overall performance of ground anchors. With newly developing material technology, FRP is a viable alternative to the corrosion prone steel tendon. Unfortunately limited knowledge in the effectiveness of FRP materials exposed to extreme aggressive conditions, has restricted the development of FRP ground anchors. The authors believe further research into the effective sustained durability of various FRP materials exposed to extreme aggressive environments under sustained loading is essential to establish the feasibility of FRP materials as a viable alternative. Additionally, knowledge into the behaviour of FRP materials at the anchor head may assist in developing a general anchorage device with continuous success as with the current wedge system for steel tendons.

Conclusion

Recent developments have seen not only the improvements to steel tendon ground anchors regarding corrosion protection but the increased interest in the developments of using new age materials. Fibre reinforced polymer (FRP) composites have been used in the aeronautical and motor industry for decades, it is only now that these advanced composite materials are becoming available for research into use for civil construction. FRP composites have the potential to minimise or eliminate current corrosion problems faced by designers and construction companies fabricating and installing steel tendon ground anchors. Research is still required prior to adopting FRP tendons as an alternative to steel strand into standards. Over the next few years the development of FRP composites will see more and more standards incorporate the new advanced composite materials for common project use.

Acknowledgements

The authors would like to acknowledge Monash University, project sponsor Geotech Pty Ltd and Devon Mothersille.

REFERENCES

ACI (2005) ACI 440.3R-04 Guide Test Methods for Reinforcing or Strengthening Concrete Structures. American Concrete Institution (ACI).

BARLEY, A. D. (1997) Ground Anchor Tendon Protected Against Corrosion and Damage by a Double Plastic Layer. *ICE Conference on Ground Anchorages and Anchored Structures.* London.

BENMOKRANE, B., XU, H. & NISHIZAKI, I. (1997) Aramid and carbon fibre-reinforced plastic prestressed ground anchors and their field applications. *Canadian Journal of Civil Engineering,* 24, 968.

CAMPBELL, T. I., SHRIVE, N. G., SOUDKI, K., AL-MAYAH, A., KEATLEY, J. P. & REDA, M. M. (2000) Design and evaluation of wedge-type anchor for fibre reinforced polymer tendons. *Canadian journal of civil engineering,* 27, 985.

LITTLEJOHN, G. S. (1987) Ground Anchorages: corrosion performance. *Proceedings - Institute of Civil Engineers,* Part 1, 645-662.

LITTLEJOHN, G. S. & BRUCE, D. A. (1977) *Rock Anchors - State of the Art,* Essex, England, Foundation Publications Ltd.

MALLICK, D. K. (1993) *Fibre-reinforced composites: materials, manufacturing, and design,* New York, Marcel Deckker, Inc.

MOTHERSILLE, D. (2006) ANCOLD Conference Presentation - Recent Development in Permanent Ground Anchor Technology. Geoserve Global Ltd & Geotech Pty Ltd.

REDA, M. M. & SHRIVE, N. G. (2003) New Concrete Anchors for Carbon Fibre-reinforced Polymer Post-Tensioning tendons - Part 1: State-of-the-Art Review/Design. *Aci Structural Journal.*

TARDIF, D., NEALE, K. W., DEMERS, M., ZAKI, A. R. & TADROS, G. (2007) Potential use of CFRP and GFRP in precast prestressed reinforced bridge deck system with extended service life. *Durability and field applications of fibre reinforced polymer (FRP) compoisites for construction.* Canada.

VAN DER WAL, J. S. & BOULFIZA, M. (2007) A step forward in characterizing thhe performance of GFRP rods in concrete. *Durability and field applications of fibre reinforeced polymer (FRP) composites for construction.* Canada.

WEERASINGHE, D. R. B. & ADAMS, M. D. (1997) A Technical Review of Rock Anchorage Practice 1976-1996. *Ground Anchorages and Anchored Structures.* London, Thomas Telford.

WEERASINGHE, D. R. B. & ANSON, R. W. W. (1997) Investigation of long term performance and future behaviour of existing ground anchorages. *International Conference on Ground Anchorages and Anchored Structures.* London, Thomas Telford publications.

ZHANG, B. & BENMOKRANE, B. (2002) Pullout Bond Properties of Fiber-Reinforced Polymer Tendons to Grout. *Journal of Materials in Civil Engineering,* 14, 399-408.

ZHANG, B. & BENMOKRANE, B. (2004) Design and evaluation of a new bond-type anchorage system for fibre reinforced polymer tendons. *Canadian Journal of Civil Engineering,* 31, 14-26.

ZHANG, B. & BENMOKRANE, B. (2005) Large model test of prestressed carbon fibre reinforced polymer ground anchors. *Canadian Journal of Civil Engineering,* 32, 1064.

GROUND ANCHOR TENDONS IN CARBON FIBRE-REINFORCED POLYMERS

Tony Barley, Single Bore Multiple Anchor Ltd, Harrogate, UK
John Hartley, Exel Composites Ltd, Runcorn, UK

Abstract
Carbon fibre plates and flat bar have been used extensively for structural repairs for over a decade. Carbon fibre is inert in the majority of natural environmental conditions and does not deteriorate in the alkaline environment of cementitious materials. As a tensile member it has numerous advantages to offer as a replacement of high grade steels which themselves are highly vulnerable to corrosion, and over a time period loss of section.

This research investigates the performance and behaviour of CFRP flat bar with regard to ultimate tensile strength, elastic tensile modulus, bond capacity with cement grout and development of bar gripping systems. The results are indicative that carbon fibre anchor tendons can comply with the requirements of anchor codes and be installed as individual or as a multiple of bars each with a load capacity of between 400 and 500 kN. The use of such lightweight components may allow fabrication at site level and avoid factory application of corrosion resistant barriers as demanded in the use of steel tendons

Introduction
The following introduction published in "Guide Test Methods for Fibre-Reinforced Polymers (FRPs) for reinforcing or strengthening Concrete Structures" – Ref ACI 2004 provides a fitting introduction to the challenges in the testing of FRP for use in cement grouted ground anchors and the potential benefits of FRP usage:

Conventional concrete structures (or the geotechnical equivalents in anchors and soil nails) *are reinforced with nonprestressed steel, prestressed steel, or both. Recently, composite materials made of fibres embedded in a polymeric resin, also known as fibre-reinforced polymers (FRPs), have become alternatives to steel reinforcement. Because FRP materials are non-metallic and noncorrosive, the problems of steel corrosion are avoided with FRP reinforcement. Additionally, FRP materials exhibit several properties, such as high tensile strength, that make them suitable for use as structural reinforcement. FRP materials are supplied as bars for reinforced and prestressing applications* (or for ground anchors) *and in flat sheets or laminates for use as repair materials for concrete structures.*

Ground anchorages and anchored structures in service, Thomas Telford, London

The mechanical behaviour of FRP differs from the behaviour of steel reinforcement, FRP materials are anisotropic due to the fibre orientation in the bars and laminates and are characterised by high tensile strength only in the direction of the reinforcing fibres. This anisotropic behaviour affects the shear strength and dowel action of FRP bars and the bond performance of FRP bars to concrete.
FRPs are available with a wide range of mechanical properties (tensile strengths, bond strengths, and elastic moduli). Generally, FRP reinforcements are not covered by national material standards, as few such standards exist. Instead, manufacturers of FRP provide test data and recommend design values based on these test data. Unfortunately, also due to the lack of material standards, few standard test methods exist for FRP concrete reinforcements. Therefore, it is difficult to compare test results between product manufacturers. In addition, research has considered the durability of FRP concrete reinforcements in environments containing moisture, high and low temperatures, and alkaline environments. Test methods that allow for the comparison of mechanical property retention in a wide range of standard environments are needed so that durable FRP reinforced concrete structures can be ensured."
This lack of guidance in the testing of CFRP materials relating to their use in structures also exists in their use in anchors. For this reason the authors have chosen to establish test methods that can be, where possible, related precisely to the performance of CFRP tendons at full scale in simulated ground anchors.

Potential CFRP Products Available for Ground Anchors

These are generally available from specialist suppliers in 6 to 40mm diameter circular bar form or in flat bar form in a wide variety of widths and thicknesses. The general range suitable for anchor tendon usage in a conventional size borehole is 3 to 8mm in thickness and 25 to 76mm in width. Members may have either ultra smooth faces or surface roughened faces. Clearly in order to enable bond with the cement grout in the fixed anchor length then roughened or deformed surfaces are required over these bond lengths. Conversely smooth surfaces or sheathed surfaces are essential in the free tendon length in order to reduce or eliminate cement bond or the existence of unacceptable frictional forces.

Notable advantages available from CFRP products are: -

- High absolute stiffness (130GPa for flat bar)
- High absolute strength (1200 to 1800MPa for flat bar)
- Very lightweight (350gms/metre)
- High stiffness to strength ratios
- Coefficient of thermal expansion approaching zero (-0.6×10^{-6} per degree C)

Typical CFRP composites have 65% fibre content in a variety of available resins, a specific gravity of 1.5 to 1.6 and an elastic tensile modulus of 150 GPa. Undoubtedly the high quality of this alternative material (conforming to the requirements of IS09002) for use in anchor tendons, in lieu of steel, has been strongly influenced by the universal adoption of this product in the Aerospace industry. Furthermore the tensile modulus is relatively close to the range presented by steel bar or prestressing strand and thus problems due to large extension during CFRP prestressing are not expected

Development of a Gripping System for Trial Work

Extensive knowledge from the steel prestressing industry combined with that from other material gripping systems (even ropes) allowed the development of an appropriate flat bar gripper at the commencement for the investigation period (Figures I and 2) .

This system allows the combination of:

- A materials transitional interface
- A rough steel interface
- A taper wedge system
- A tensile bolt pre stress system

The bolt system allows transitional application of stress along the loaded interface prior to and during execution of the test.
The use of this transitional load transfer mechanism has allowed the application of prestress up to 520 kN prior to bar failure (Figure 3). Over the 5% to 65% UTS loading range the tested bar exhibited an elastic tensile modulus of 163 GPa.

Figure 1 Gripping arrangement set up for early tests with preloaded tensile bars allowing grip increase on bar prior to and during loading

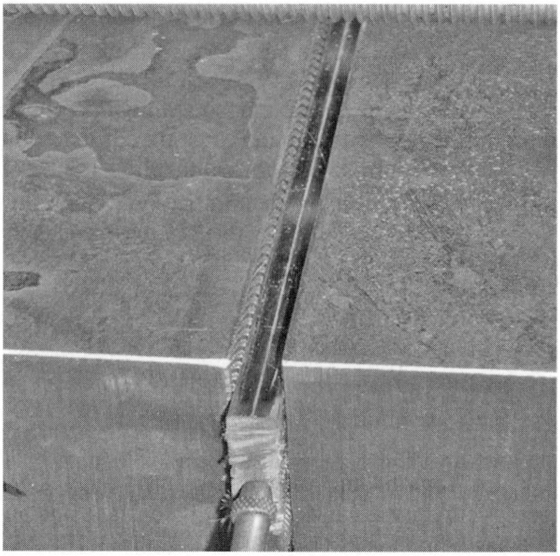

Figure 2 End view of Carbon Fibre flat bar being gripped and movement monitored during tensile tests.

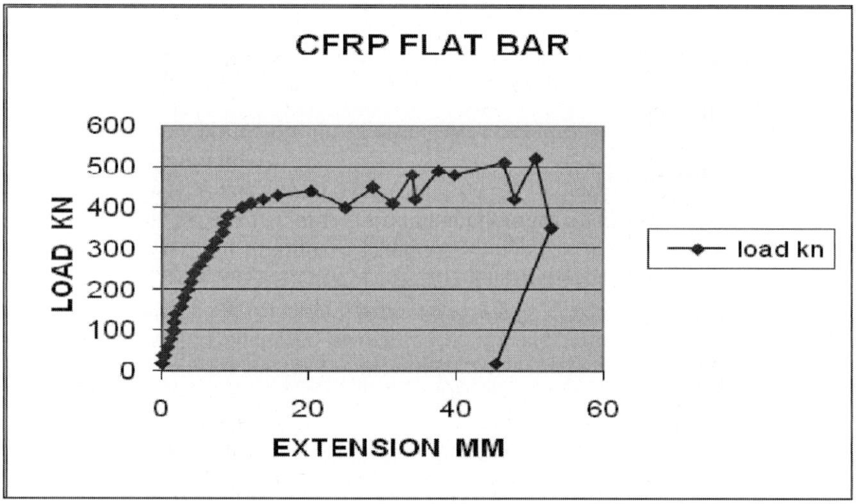

Figure 3 Carbon Fibre Reinforced flat bar(38 x 8mm) achieved 520 kN at failure

CFRP Bond with Cement Grout

"Gunbarrel testing"

Tensile members encapsulated within a grouted steel tube, then subjected to pull out, have been carried out regularly over a 35 year period (Barley 1978, Littlejohn and Weerasinghe1997, Barley 1997). Little modification of the gunbarrel test procedure or the grouting techniques was required to adapt the system to establish the pull out capacity of CFRP flat bar.

Initial tests demonstrated that the rough faced exposed surface on the CFRP flat bar (Figure 4) was inadequate to provide the satisfactory resistance to pull out over the two tested bond lengths (0.8 and 1.6m). Loads of 150, 138, 210 and 300 kN were achieved exhibiting average failure bond stresses ranging from 1.97 to 2.13N/mm². Although these bond stress values are not inconsistent with the pull out capacity of non deformed strand anchors over similar lengths, they were considered inadequate to comply with the severe demand requirements of the CFRP multiple unit anchor system under development. The low capacity demanded a review of the method of surface roughening utilised by the manufacturer in his application to standard products.

After considerable research the 6mm flat bar thickness was increased to 8mm and a greater surface roughness was achieved (Figure 5).

In order to eliminate any effect of mechanical gripper efficiency in the second set of trials, two identical 2m long gun barrels were grouted a distance of approximately 1m apart. The metre gap contained a 60 ton capacity hollow ram jack which allowed complete equal and uniform loading of the opposing gun barrels. (Figure 6)

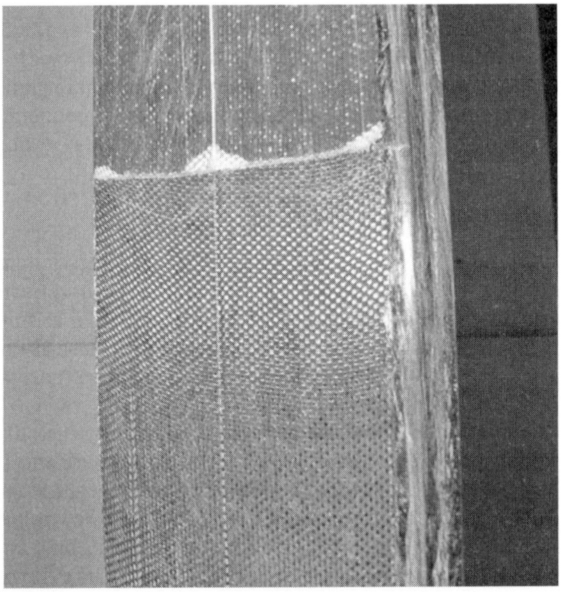

Figure 4 Surface roughness of CFRP bars (6mm) used in initial tests were inadequate to provide high enough bond capacity with cement grout to comply with the severe test requirements.

Figure 5 Increase surface roughness has been achieved in development of a slightly thicker CFRP Bar (8mm)

Load cycles were built up whilst monitoring ram extension and base movement of each gun barrel and movement of the contained grout and bar at the base. This extensive monitoring allowed the accurate assessment of the relative movements of the components and result analysis allowed prediction of the final failure mechanism even prior to the exhumation of the tested bars.

Figure 6 Double Ended Gunbarrel Test with hydraulic jack contained between the two gunbarrels to ensure equal and uniform loading.

Figure 7 Longitudinal bar splitting at the junction with the grout

Observation at Cement/Bar Failure and Non Failure Interfaces

The dissection of the steel gunbarrels allowed access to the grouted body and the subsequent exposure of each grout to bar interface. Visual inspection then allowed the determination of the failure mechanism over the full length of the exposed interfaces in both gun barrels.

Progressive debonding takes place in all load transfer involving longitudinally bonded elastic members such as steel or FRP bonded in cement grout. The occurrence of such a mechanism is apparent and dominant in these trials.

Figure 7 illustrates the longitudinal splitting of the CFRP bar at the entrance to the gunbarrel, whilst Figure 8 illustrates the pull out mode (progressive debonding) from zero to 900mm depth. This involves the shearing of the white cement grout nominally outside the roughened bar surface.

Between 900 and 1100 mm into the gunbarrel (Figure 9) some transition in the failure mechanism begins to occur. The whole failure interface becomes grey which suggests that the failure interface is integral with both the black resin protrusions and the white grout both in shear. In Test A, beyond 1100mm the CFRP bar remains fully bonded within the cement grout (Figure 9) and the test demonstrates an overall average bond stress of 3.4 N/mm² without failure.

However, with due consideration of the exhibited progressive debonding and the likely hood that the sliding friction was low over the top metre it may be realistic to consider that an adhesive bond as high as 10 N/mm² was exhibited over a bonded 0.5 m length.

However, in Test B failure took place at approximately 80% of the capacity of the 520 kN bar and an average bond stress at failure of 2.6 kN/mm² was exhibited. Despite failure this is considerably better than that exhibited in the preliminary trials on the smoother facing.

The test results generally support the benefits of utilizing a multiple of short bond length anchor tendons associated with the internationally patented system (UK Patent –GB2,340,144) as compared with the severe problems encountered with debonding in long individual members

Figure 8 Exposed CFRP bar near the top of Test A illustrates the shear of cement grout

Figure 9 Progressive debonding along Test A to a fully bonded non slip area

Figure 10 No debonding of bar from grout exists at 1200mm depth in Test A

Figure 11 Test B gunbarrel exhibited complete pull-out of the bar from the cement grout at the base (2m bond length)

Development of Permanent Gripping System for use in CFRP Ground Anchors

Development of permanent fixed end grippers has followed a number of possible systems involving CFRP to resin bond, a combination of CFRP to resin bond incorporating wedge grippers, and the jointing of CFRP with steel members.

This development is well advanced after considerable liaison and co-operation with specialists in the Industry. Owing to the confidentiality demanded by the parties more precise details cannot yet be published.

Conclusion

The development of CFRP permanent anchor tendons and associated gripper and bonding systems to provide the industry with a non-corrodible, non-degradable tensile member within a cementitious environment is well advanced. Existing Codes of Practice for Ground Anchorages generally consider steel tendons and demand pre-application of corrosion protection under factory conditions. This inevitably demands expensive double handling of heavy steel tendons.

CFRP bar typically 38mm wide, 4 to 8mm in thickness can be delivered to site in coils and simply assembled into multiple bar anchor tendons. Consistent with current multiple anchor practice unit anchor bond lengths would range from 2.5 to 5m and the tendons could be fabricated on site to lengths to suit the encountered ground conditions.

CFRP multiple anchor type tendons with a working capacity 1000 kN of typical anchor length of 35m and a weight of only 75 kg is a realistic option for the future.

References

American Concrete Institute (2004) "Guide Test Method for Fibre-Reinforced Polymers (FRP) for reinforcing or strengthening Concrete Structures" ACI : 440.3R-04

Barley A.D. (1978) "The study and investigation of under-reamed anchors and associate Load transfer mechanism". – Post Graduate Thesis, at Marischal College, Aberdeen.

Littlejohn G.S. and Weerasinghe R.B. (1997) "Instrumental full-scale re-usable laboratory apparatus for rock anchorage testing" – Proc. Int. Conf. on Ground Anchors and anchored structures ICE London, UK, on 20-21 March 1997- Pages 34 - 44

Barley A.D. (1997) "Ground Anchor Tendon Protected against Corrosion and damage by a Double Plastic Layer"- Proc Int. Conf on Ground Anchors and anchored structures = Pages 371-383 ICE London UK on 20-21 March 1997

Ground Anchor Made of New Materials Applicable for Highly Corrosive Environments

Tsuyoshi Enomoto, Tokyo Rope Mfg. Co., Ltd., Tokyo, JAPAN
Toshio Sugisaki, Sekisui Chemical Co., Ltd., Tokyo, JAPAN
Kouichi Suga, Nittoc. Construction Co., Ltd., Tokyo, JAPAN
Tatsuro Sueyoshi, Raito Kogyo Co., Ltd., Tokyo, JAPAN

Abstract

Conventional ground anchor systems designed for slope stabilization/landslide prevention have been degraded under highly corrosive environments, e.g. in a hot spring area, in Japan. In 1993, an R&D project was started with the aim of developing special anchors applying carbon fiber composite cables (CFCCs), pressure plate made of continuous-glass-fiber reinforced plastic and a resin grout suitable for use in such highly corrosive environments. 5 years of monitoring in field have been carried at a hot spring area under a pH range of 3-4 and at temperatures ranging from 60^0C to 100^0C. Conventional steel strands and cement grout deteriorated severely. In contrast, the carbon fiber strands (CFCCs) do not show any degradation in strength or visible defects, resin grout keeps its required compressive strength (24 N/mm^2) for design even with little change in the weight and long fiberglass reinforced plastic has stable properties. New material ground anchor systems now can be deployed even in such highly corrosive environments.

Introduction

Ground anchor systems that are designed for slope stabilization/landslide prevention have been used in Japan for more than 40 years. Since the publication of JSF standard D1-88 "Ground Anchor Design and Installation Standard" in 1988 which required that permanent anchors have redundant corrosion protection capabilities, corrosion protection-minded design philosophies have been spreading and ground anchors have been becoming more reliable, which has contributed to the wider use of ground anchor works.

However, ground anchor systems have not been considered applicable for highly corrosive (high temperature and/or highly acid) environments, e.g. in a hot spring area, because, even if the bearing plate has been fabricated of corrosion resistance materials, the tendons, cement paste and concrete can be degraded by such a severe condition. Therefore, where such a corrosive environment exists, alternative slope stabilizations such as earth removal works, counterweight fill and/or groundwater drainage have been applied rather than ground anchors.

In 1993, an R&D project was started with the aim of developing special anchors that would be suitable for use in such highly corrosive environments. This work gave rise to a new tendon made of acid and heat resistant carbon fiber composite cables (CFCCs) and a new pressure plate made of continuous-glass-fiber reinforced plastic. Further, with the introduction of resin grout, ground anchor systems now can be deployed even in such highly corrosive environments.

To demonstrate the life of the newly developed material under such highly corrosive environments, a five year long term exposure test has been performed since 2001.

Overview of NM ground anchor system

New Material (NM) ground anchor system has been jointly developed by the Tokyo Rope Mfg. Co. Ltd., Zenitaka Corp., Kowa Corp. and Sekisui Chemical Co. Ltd. in 1993 in Japan. The NM ground anchor system consists of Carbon Fiber composite cable (CFCC) tendons, Fiber Reinforced Foamed Urethane (FFU) passive pressure plates and epoxy resin grout for the highly corrosive environment fields as illustrated in Figure 1.

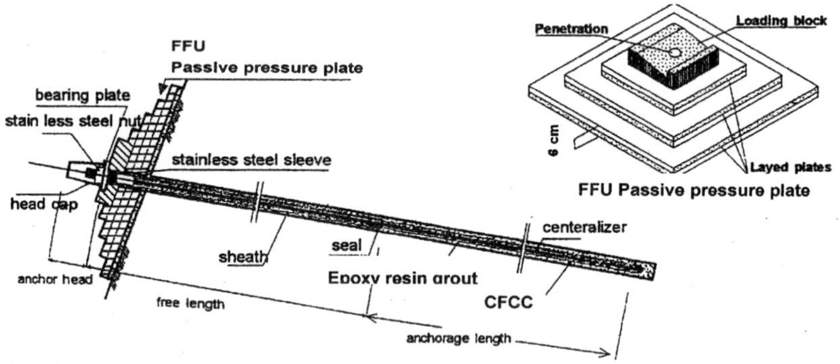

Figure 1. NM Ground Anchor System for the highly corrosive environment fields

Exposure test field

Exposure tests were conducted in the Myoban Area that was affected by past hydrothermal landslides. The Myoban Area is located at the foot of Mt. Tsurumidake on the northwest of Beppu City Oita Prefecture in Japan. Its soil consists of altered and weathered andesites covered with a clastic rock layer and a cohesive soil layer composed of solfataric clays. Therefore, from the topographic point of view it presents a typical terrain that involves hydrothermal landslides.

The Myoban Area has suffered hydrothermal landslides twice in the past: the first in September 1966, and the second in September 1997. The landslide in 1966 was caused by a localized torrential downpour induced by Typhoon No. 24, with its magnitude extending over an area of 120 meters wide and 80 meters long. Earth removing and counterweight fill were executed after the landslide as preventive measures against future recurrence of landslides. However, the area was attacked once again by landslide in 1997 in a similar situation, i.e. it was caused by local heavy rainfall brought by a typhoon. This time the area suffered two landslides: one measuring 50 meters wide by 110 meters long, and the other, 60 meters wide by 130meters long. One of the countermeasures taken after these landslides in 1997 was the ground anchor system using new materials (Figure 2).

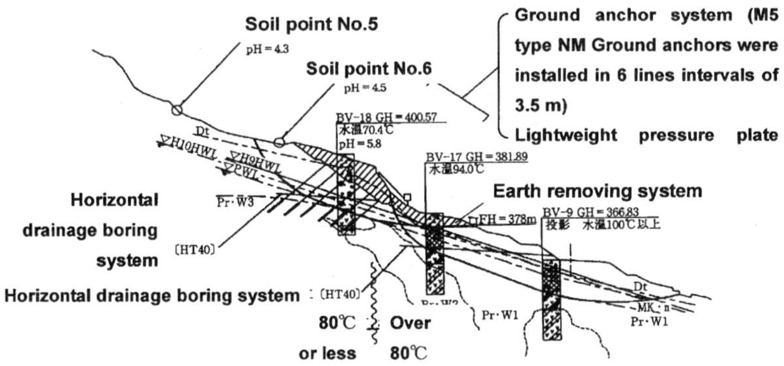

Figure 2. Landslide control construction systems at the Myoban Area

Figure 3 shows the test points and temperatures as well as pH distribution of the soil in the area. All these points are in a high corrosion environment: according to previous survey, the test points showed pH values 3 or 4 and in-tunnel temperatures ranging from 60^0C to 100^0C. Table 1 shows the analysis results of the hydrothermal water existing in the Myoban Area. The concentration of hydrogen sulfide (H_2S) in the soil of the area remains at levels of 0.1 to 0.6 ppm.

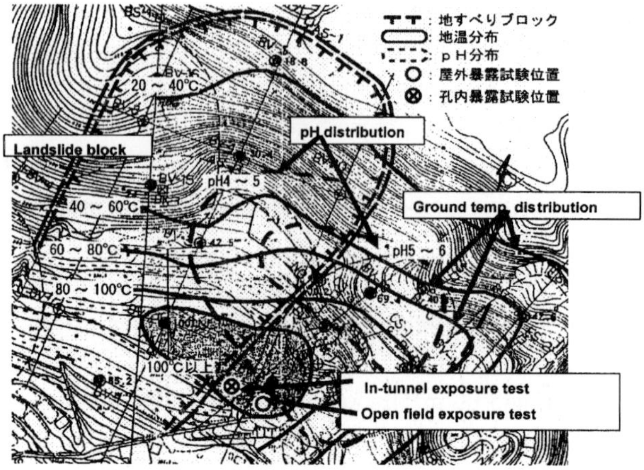

Figure 3. Test points, temperatures and pH distribution

Table 1. Hydrothermal water analysis results

Temp.	pH	Na_+	K^+	Ca^{2+}	Mg^{2+}	Fe^{2+}	Al^{3+}	Cl^-	SO_4^{2-}	HCO_3^-	H_2S	Water quality
82℃	2.25	11.6	5.4	10.1	4.0	5.8	23.2	1.07	550	0	6.05	Acid hydrogen Sulfide water

[The unit for concentration of dissolved ion is ppm]

Exposure test

The test used two different types of corrosive environment: open field and an underground vertical tunnel in a hot spring area. The tunnel had a very corrosive atmosphere showing a pH value of 3 to 4, temperature of 60^0C to 100^0 C and hydrogen sulfide (H_2S) concentration of 0.1 to 0.6 ppm. Photograph 1 shows an overview of the open field exposure test. Figure 4 shows an overview of the in-tunnel exposure test. The samples were placed in stainless steel baskets located at depths below grade of GL 10, 11, 12, 13, and 14 m. The components of the pressure plate were placed in the basket at GL 12 m. In the other baskets (excluding GL 12 m), the sample grouts (cement paste and epoxy resin) and tendons (PC steel strand and CFCC) were placed.

The following items were measured for the specimens sampled from the respective test points:

Grouts: Compressive strength (JIS A 1108)

Tendons: Tensile strength (JIS Z 2241, JSCE E 531)

Pressure receiving plates: bending strength, Young's modulus for bending, vertical compressive strength, horizontal compressive strength, partial compression proportional limit strength, shear strength, and shear strength of adhesive bonds (JIS Z 2101 and JIS K 6852)

Photograph 1. Overview of the open field exposure test

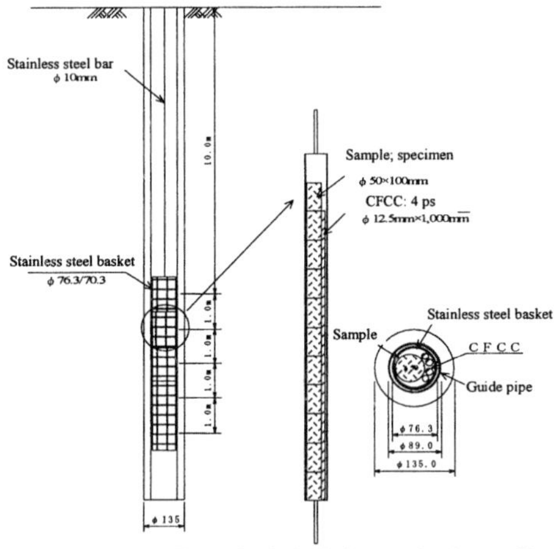

Figure 4. Overview of Exposure Test in Well Sample Installation Method

The grout specimens, 50 mm in diameter and 100 mm in length, were cured for a period of 7 days in the laboratory before placing on the test site. Each tendon specimen was cut in a length of 1,000 mm.

The specimens were sampled after 1, 2, 3, 4, and 5 years of placement on the test site.

Results and discussion

Condition of test environment

Figure 5 shows the in-tunnel temperature fluctuations observed when the specimens were sampled. It may be observed that the deeper positions had higher temperatures at the beginning of the test. These temperature differences according to the depth decreased over the course of years. Overall temperatures were on the rise as time went by. This might have been caused by earth and sand that entered the test tunnel blocking up the upper part of them and causing heat sources to migrate.

No significant variation was observed in pH value throughout the test period: it was pH 4.4 according to the H_2O method and pH 2.4 according to the H_2O_2 method in Year 5.

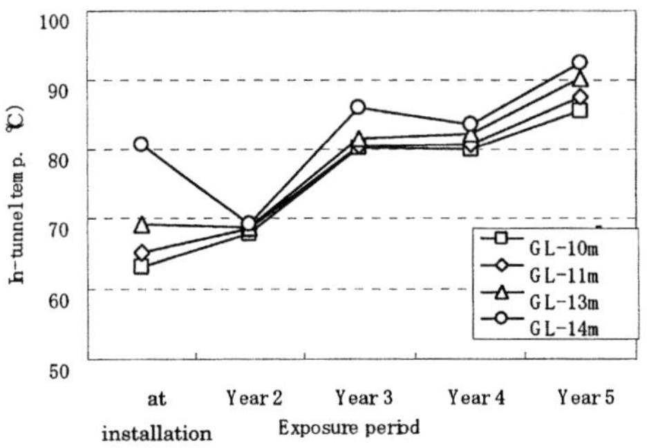

Figure 5. In-tunnel temperature fluctuations

Property of grouts

The deterioration of the property of grouts was considered by a decrease of compressive strength. The compressive strength test results of the epoxy resin grout and the cement grout are shown in Figure 6. While the resin grout strength shows a decrease from the initial value in Year 1, there is no significant change in the subsequent years. The measured values of its compressive strength also meet the required design standard strength (24 N/mm^2). Cement grout, on the other hand, showed values below the required design standard strength in Year 1, at the depth of GL-13 m and deeper levels. It did not meet the design standard strength at any level of depth from Year 2 on, and from Year 4 on, its deterioration was such that strength measurement was impossible.

Figure 6 (cement grout) shows the decrease rate of compressive strength according to the depth. Difference according to the depth is on the decrease from Year 3 on for the resin grout, and from Year 2 on for the cement grout. While it is estimated that the difference in compressive strength according to the depth depends on the temperature to which each material is exposed and also on the pH level, actually there was little fluctuation in pH value from the initial period of test. Therefore, the cause is probably attributable to the reduced temperature difference among depth levels (Figure 4).

The mass change of each specimen over a period of three years also is shown in Figure 7 while there was little change in the weight of resin grout, the cement grout decreased about 30 % in weight at the depth of GL-13 m and GL-14 m. photograph 2 shows the sampled specimens. As can be seen in photograph 2, the surface of the cement grout specimen melted down while the resin grout merely presents discoloration. This result proves that in no way can skin friction between ground and grout be expected from cement grout.

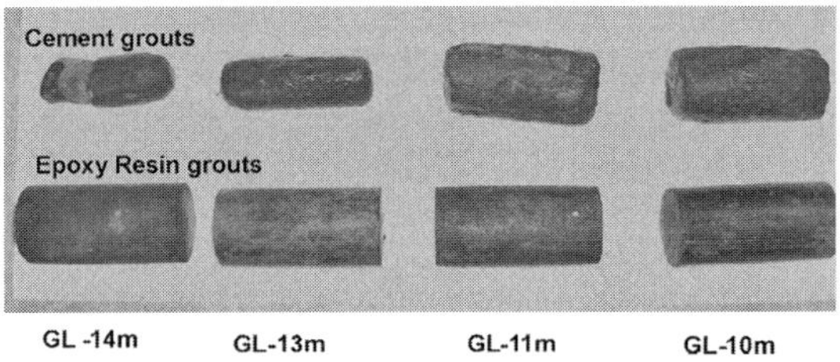

Photograph 2. Deterioration of sampled specimens

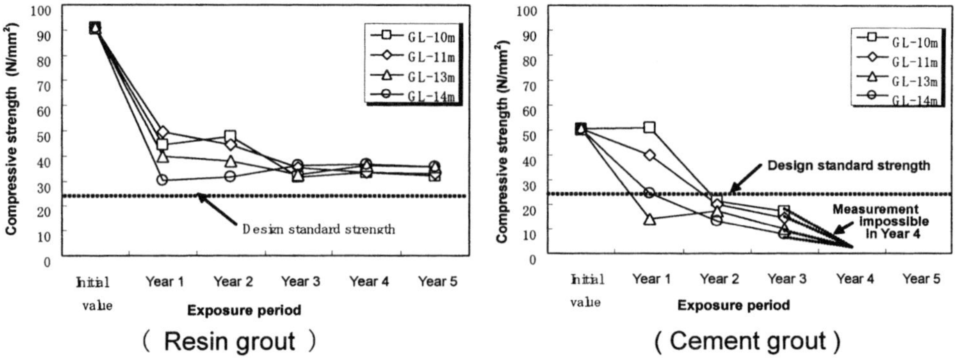

Figure 6. Compressive strength test results of grout

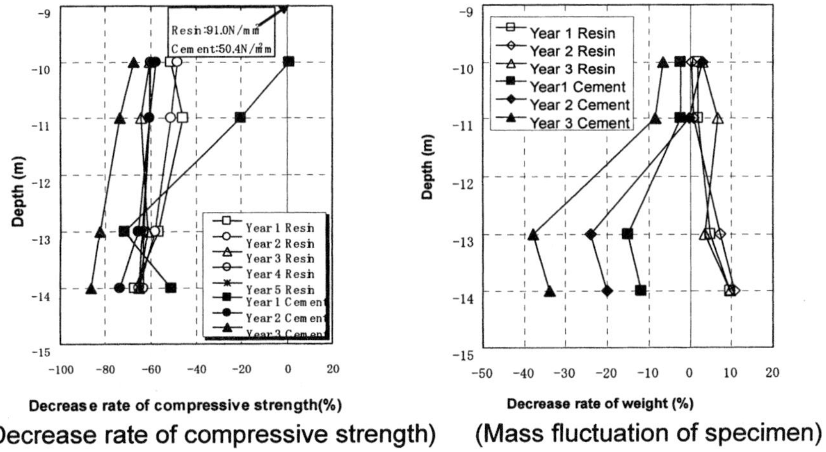

Figure 7. Physical properties test results of grout

Property of tendons

The deterioration of the property of tendons was considered by a decrease of tensile strength. The tensile strength test results of tendons obtained after in-tunnel exposure conditions and open field exposure conditions are shown in Figure 8.

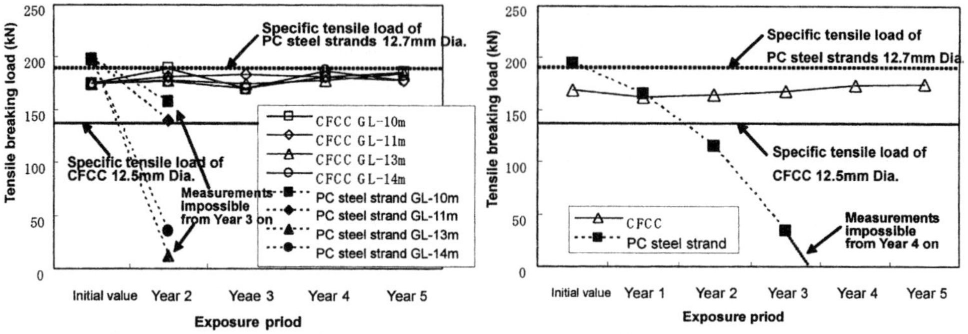

(In-tunnel exposure) (Open field exposure)

Figure 8. Tensile strength test results of tendons

Carbon fiber stranded wire exhibited a tensile strength far over the specified tensile load even after five years. Whilst PC steel strands was seriously deteriorated, showing strength far below the specified tensile load from Year 1 on, in all the exposure depth levels and environments. Its deterioration was so serious that measurements were impossible from Year 3 for the in-tunnel exposure test and from Year 4 for the open field exposure test.

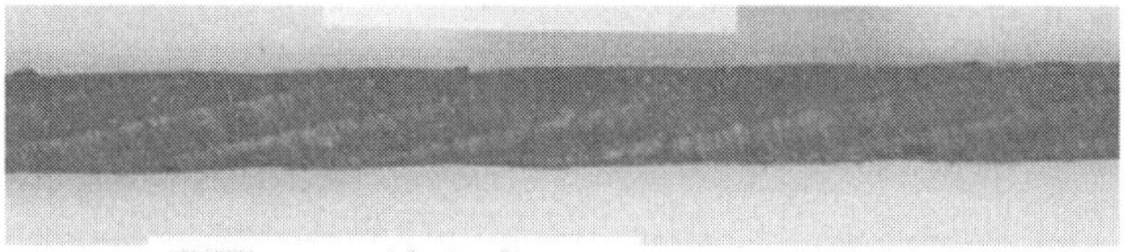

CFCC was no deterioration.

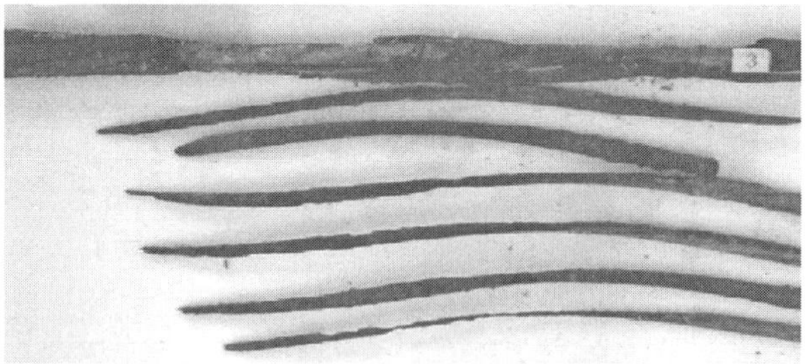

PC steel strand was crumbled since with rust.

Photograph 3. Deterioration of sampled specimens (Tendons)

Physical properties of pressure receiving plate

The physical properties test results for the pressure receiving plate are shown in Figures 9. These tests were conducted according to the methods specified in the following JIS specifications: "Methods of test for woods" (JIS Z 2101) and "Testing methods for shear strength of adhesive bonds by compression loading" (JIS K 6852). The test results verified that the physical properties of pressure receiving plate undergo little change over a span of five years in all exposure test conditions. It has therefore been verified that pressure receiving plates of long fiberglass reinforced plastic provide steady physical properties even in aggressive environments where they are exposed to high temperatures and strong acids.

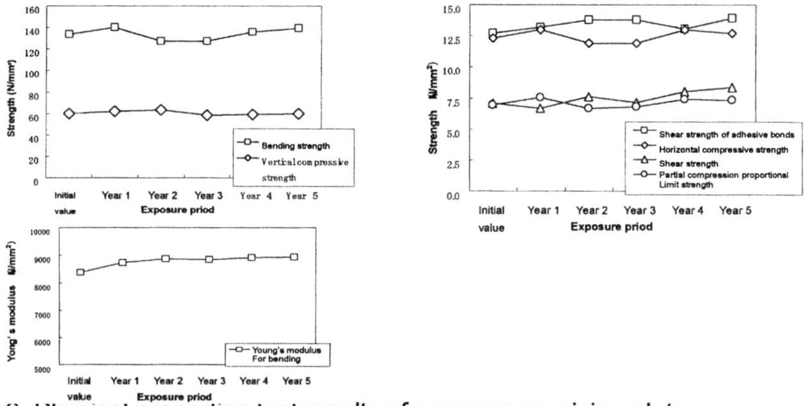

Figure 9. Physical properties test results of pressure receiving plate

Conclusion

As a result of the five-year severe corrosive environment exposure test, the following conclusions were obtained for the new materials (CFCC, epoxy resin, and continuous-glass-fiber reinforced plastic) of the highly corrosive environment ground anchor system.

(1) The carbon fiber strands (CFCCs) do not show any degradation in strength or visible defects after exposure to such highly corrosive environment. They exhibit stable properties as required for the tendons for permanent anchors.

(2) PC steel strands without corrosion protective treatment were degraded within several months in these highly corrosive environments, which results in the decrease in tensile strength to lower than the acceptable range.

(3) The epoxy resin grout shows a rapid decrease in strength at the beginning of use, though it does not show further decrease in strength in the second and later years. It can, therefore, be considered a long term stable material.

(4) The cement grout shows a significant decrease in strength and surface deterioration under high temperature and highly acid soil environment. It cannot, therefore, be considered effective as a grout material of the anchor.
(5) The pressure plate made of continuous-glass-fiber reinforced plastic shows stable properties under high temperature and highly acid soil environment.

The ground anchor components consisting of the new materials (carbon fiber tendons, epoxy resin and continuous-glass-fiber reinforced plastic) has proved stable over a 5 year test program and thus may be considered for use in corrosive environments represented by a pH value of 3 to 4 and ground temperature of around 80^0C.

To gather additional data, further tests are planned to be performed by using different environmental conditions.

Acknowledgment

We would like to express our gratitude to all those who gave us helpful assistance and cooperation. We would like to thank the Beppu Civil Engineering Bureau of Oita Prefecture for providing the test site, and Meidai Kogyo Co., Ltd. for cooperation given in installation and sampling of specimens.

References

1. Suga, K. et al.,: "the material strength change of ground anchor in high corrosion environment," *J. of the Jpn. Landslide Soc.*, Vol.43, No.4 215 (2006), pp. 31-35.
2. Aoyagi, K., et al.,: "NM (New Material) Ground Anchor System,", *Proceedings of the 2nd International Conference on Advanced Composite Materials in Bridge and Structures,* Montreal Quebec, Canada, August 1996, pp. 727-734.
3. JSCE,: "Recommendation for Design and Construction of Concrete Structures using Continuous fiber reinforcing Materials," Concrete Engineering Series 23, October 1997
4. http://www.tokyorope.co.jp/cfcc/, "CFCC"
5. Tokyo Rope, "Technical Data on CFCC," Tokyo Rope Mfg. Co. Ltd., Japan, 2003.
6. JIS A 1108 "Method of test for compressive strength of concrete"
   JIS Z 2241 "Method of tensile test for metallic materials"
   JSCE E 531 "Test method for tensile properties of continuous fiber reinforcing materials"
   JIS Z 2101 "Methods of test for woods"
   JIS K 6852 "Testing methods for shear strength of adhesive bonds by compression loading"

New Development Of Low Relaxation Epoxy Coated Strand

Toshiro "Tiger" Kido, Takayuki Yamagiwa, Toshihiko Niki, Yoshihiko Touda

Sumitomo (SEI) Steel Wire Corp, 1-3-12 Moto-Akasaka, Minato-ku, Tokyo, 107-8468 Japan

Abstract

Epoxy Coated Strand (ECS) was originally developed in the USA in the early 1980s. Since then ECS has been widely applied to various applications in Japan, USA, China, and Spain. Currently ASTM and ISO specify that ECS shall have relaxation losses of not more than 6.5% at 1000hours after being stressed. Although this level of relaxation property has been accepted so far, it seems that the market is eager for it to be improved to as low as that for low relaxation bare strand. Based on this background, development of the epoxy coating technology was started in order to improve the relaxation property of the ECS, and at last optimum manufacturing conditions to be able to manufacture the low relaxation ECS have been identified.

Introduction

When ECS was invented, the primary type was coated only on the outside of the strand and not in the interstices of the seven wires. For this product, there was always a concern about the possibility of migration of corrosive media in the interstices of the seven wires. In the early 1990s, filled ECS was developed to eliminate this concern. The filled ECS has epoxy resin filling in the interstices of the seven wires as well as the outside of the strand. Since then the filled ECS has been widely applied to dam retrofit, slope stabilization, land slide prevention, concrete pile driver, pretensioned bridge girder and slab, external cable for bridge, extradosed bridge (a cross between a girder bridge and a cabled stayed bridge), cable stayed bridge, and others in Japan, USA, China, and Spain.

When manufacturing the filled ECS product, the entire strand is heated up in order to apply the epoxy resin to the strand. This heating process causes the steel to lose some of the dislocations in the steel's atomic structure that were created originally during the thermo-mechanical stabilization process utilized to make the bare low relaxation strand. Because of that, currently ASTM and ISO specify that ECS shall have relaxation losses of not more than 6.5% instead of not more than 2.5% after 1000hours when initially loaded to 70% of the specified minimum breaking strength of the strand. Although this level of relaxation property has been accepted so far, the market in the future seems to require the low relaxation ECS.

There were two key factors to explore to develop the low relaxation ECS. One of them was to use special raw materials when making the bare strand that resists the negative effects by heating the strand during the epoxy coating process. It is known that silicon doped steel gives a high level of heat-resistance. After several trials for producing ECS using the bare strand that contains 0.6-0.8% of silicone doped steel, it was confirmed that the low relaxation property (<2.5% after 1000hours with an initial load of 70% of the specified minimum breaking strength of the strand) for ECS could be achieved. The other factor was to alter the manufacturing technique to reduce the plastic deformation caused by unravelling the strand during the powder coating process. Several studies were performed to identify the impact on the strand relaxation caused by both pre-heating of the strand and unravelling of the strand. Eventually the optimum conditions to achieve the low relaxation ECS were identified.

Background of ECS

To enhance the strand's corrosion resistance, ECS is manufactured by applying the high-quality epoxy powder coating to the strand as shown on Figure 1. It is utilized when high corrosion resistance is required, such as for bridges in salt exposed areas as well as for non-grouted cables and ground anchors as shown on Figure 2

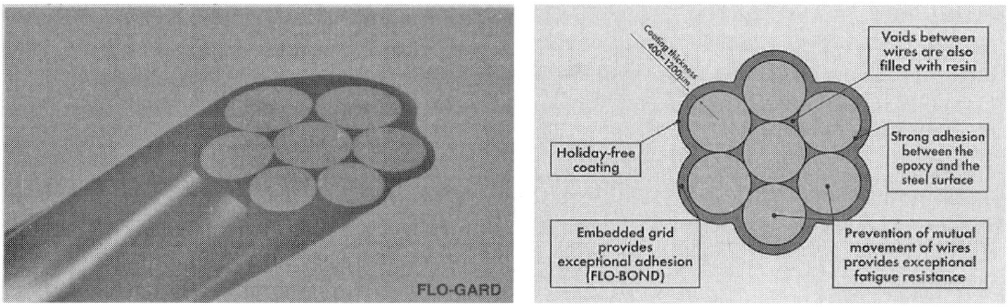

Figure 1: Cross Section of ECS

Figure 2: Applications of ECS

Relaxation properties of ECS

Even though ECS shows high corrosion resistance, the relaxation losses after 1000 hours when initially loaded to 70% of the specified minimum breaking strength of the strand are as high as 4.0 %, as shown in Figure 3.

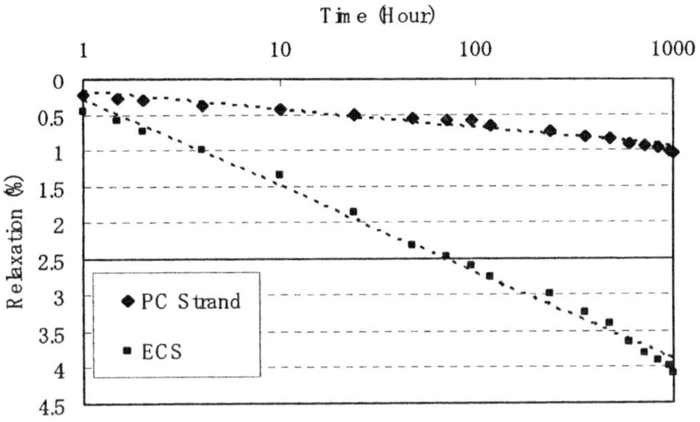

Figure 3: Relaxation losses of ECS and low relaxation bare strand

The relaxation losses of low relaxation bare strand before epoxy coating are typically 1.0-1.4% and so it is clear that the relaxation property is changed by the epoxy coating process. Initially, the factors that could affect the relaxation property had to be investigated.

Powder coating

The strand is coated by a "powder coating". This coating technique is a painting method in which powder resin is applied instead of liquid paint as shown on Figure 4. This method is commonly used for foundation coating of automobile bodies and household appliances such as refrigerators and microwave ovens.

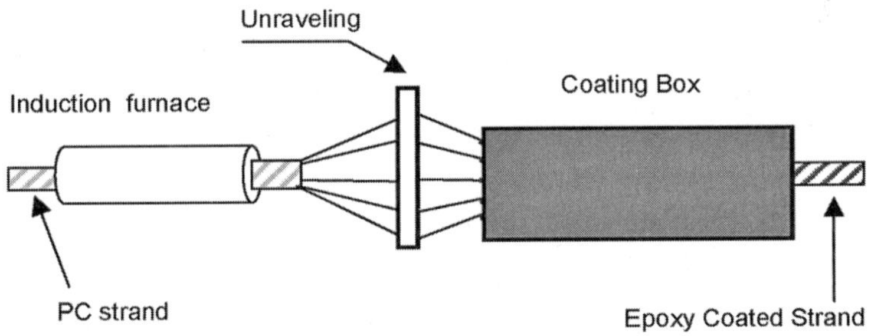

Fig.4 Schematic diagram of the epoxy coating process

In the powder coating process the target is first heated up to 150°C to 250°C using an induction furnace as shown on Figure 5. The heat energy is then transferred to the powder from the target when the powder contacts the target. As a result, the powder melts down and spreads out on the surface of the target. After the resin is cured, the coating layer is formed.

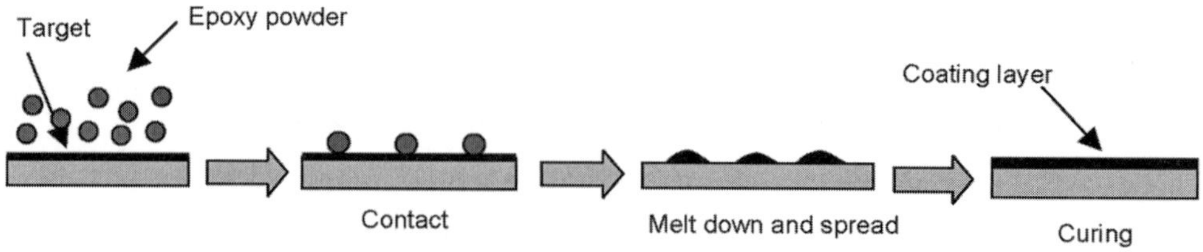

Figure 5: Mechanism of the powder coating

Two processes during powder coating were focused upon: -

1. Pre-heating by the induction furnace
2. Unravelling

The detailed mechanisms involved in each operation and the corresponding effects on the strand relaxation properties are considered in the following sections.

Pre-heating by induction furnace

In the "hot stretching process" for manufacturing the low relaxation bare strand, the strand is stressed at a certain level, simultaneously heated up to 300 °C to 400 °C and then is rapidly cooled down with water. In view of the microstructure, the dislocations that affect the plastic deformation are fixed by the rapid cooling process. As a result, the fixed dislocations prevent micro-plastic deformation under the continuous stressing of the strand. That is why the strand can achieve the relaxation property, which is not more than than 2.5%.

In the process of pre-heating by induction furnace, the strand is again heated. The heating temperature is about 200 °C. Compared with the temperature of the "hot stretching process" utilized for making the low relaxation bare strand, it is rather low. Therefore, the mechanical properties, such as tensile strength, yield strength or elongation, are not affected by this process. However, the relaxation property changes after the epoxy coating process, which brought up a hypothesis that the pre-heating temperature affected the steel microstructure. The pre-heating process might cause the fixed dislocations to become loose and be able to move.

Unravelling

Filled ECS has epoxy resin placed in the interstices of the seven wires as well as the outside of the strand to prevent the migration of corrosive media in the interstices of the seven wires. In addition, the strong adhesion between the epoxy resin and steel prevents each wire from moving and therefore the filled ECS has an anti-fretting corrosion protection.

However, the strand has to be unravelled during the application of powder coating to fill the epoxy coating in the interstices of the seven wires. Moreover this unravelling process is carried out while the strand is in the pre-heated condition. In this process, the bending force load to the outer wires and the synergistic effect with the pre-heating and the unravelling process may alter the relaxation property of the ECS. The next section describes how these two items were overcome.

Approach to achieve the low relaxation ECS

Special material

It is known that doping silicon as an additive in steel increases its strength and provides increased heat resistance. It is also known to affect the relaxation property when it is used in the prestressing materials. This is also applied to galvanized PC wire and strand.

ECS was manufactured using different steel chemistries containing 0.9-1.5% of silicon and the corresponding relaxation property was recorded and compared with the ECS manufactured using conventional 0.25% silicon high carbon wire rod. Figure 6 shows that the relaxation property is influenced by the quantitiy of silicone, and that >1.0% of silicon can achieve the low relaxation ECS without any other process changes.

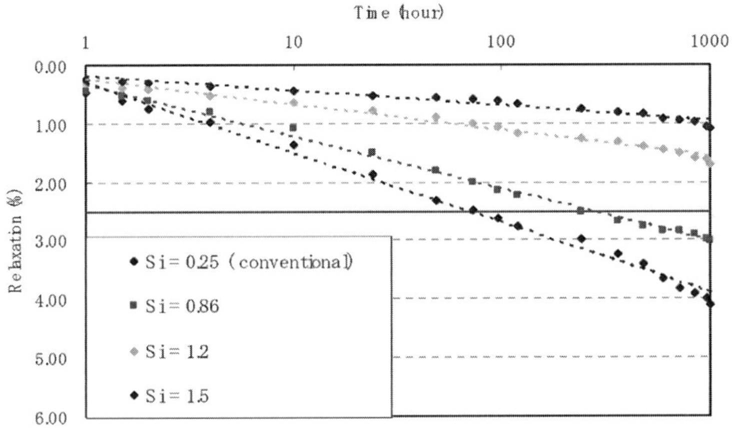

Figure 6: Relationship between quantity of silicon and relaxation property of ECS

Optimization of manufacturing conditions

The low relaxation ECS was achieved using the silicon doped wire rod, however this type of steel doped with more than 1.0% silicon may not be recommended for mass production because high silicon wire rod increases the tensile strength of the wire rod and requires reduced wire drawing speeds. In addition the high silicon wire rod can be much more expensive than the conventional high carbon wire rod.

As a result, the cost of the final product becomes higher and it was therefore attempted to achieve the low relaxation property by using a silicon percentage that was higher than the conventional wire rod, but below the 1.0% level, and by modifying the other process conditions.

The chemical compositions of the selected material is shown on Table 1

Table 1 Chemical composition of the material (wt%)

C	Si	Mn	P	S	Cr
0.83	0.86	0.83	0.09	0.04	0.02

At first, the pre-heating temperature was reduced down to 200°C to 220°C versus the conventional condition of 220°C to 240°C.

The tension between the bare strand payoff equipment and the unravelling equipment was minimized. This kind of production condition made it difficult to control the thickness of the epoxy coating layer. However, an improved production condition to achieve the low relaxation ECS was eventually found as shown in Figure 7.

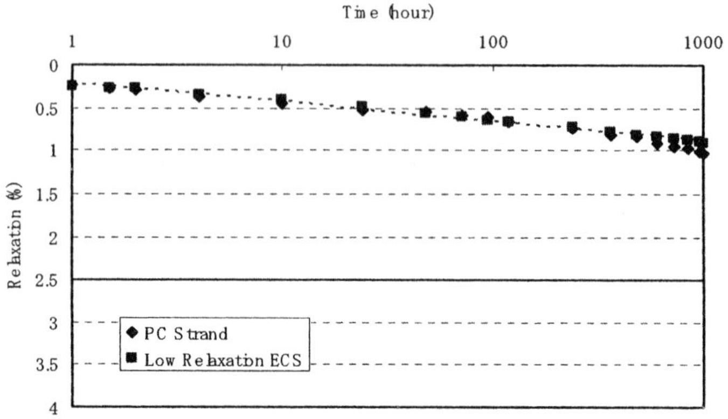

Figuer 7: Relaxation property of the developed ECS

Conclusions

By increasing the silicon concentration in the steel and reducing the tension of the strand during the pre-heating and the unravelling process, it was possible to develop the low relaxation ECS. In addition:

(1) The pre-heating and the unravelling in the epoxy coating process affect the relaxation property.

(2) Silicon doped wire rod can improve the relaxation property. More than 1.0% of silicon doped wire rod can achieve the low relaxation ECS with standard coating condition.

(3) Even though not more than 1.0% of silicon is utilized, it is possible to achieve the low relaxation property with the optimum coating conditions.

The ECS is expected to expand its range of applications as a result of the success of this fundamental investigation.

References

1. Victor Dorsten, Frederick F. Hunt. "Epoxy Coated Seven-Wire Strand For Prestressed Concrete". *Journal of Prestressed Concrete Institute*, Vol 29, No.4, July-August 1984.
2. Shigeki Hada, Yasuharu Mikami, Toshihiko Niki, "Epoxy Coated Prestressing Strand". *FIP Symposium '93 Kyoto, Japan, Oct 17-20,* (1685-1691), 1993
3. Prof.Dr.-Ing. D.Jungwirth and Dr.Techn. Andor Windish, "Coating of tendons", *FIP '94 Congress May 29 -June 2, Washington D.C J23-J27*
4. Yasuhiro Hoshino, Masato Yamada, Motonobu Nishino, "External cable using epoxy coated strand", *The first fib congress 2002 Concrete Structures in 21st Century Volume2 (65-66)", 2002*

Session seven:

Case histories of satisfactory performance, shortcomings and failures in service

Assessment of cement grout cover as sole corrosion protection of permanent tensioned steel tendons

Stuart Littlejohn, University of Bradford, England

Abstract

In ground anchorage practice a few exhumed cement grouted tendons have been found in pristine condition after several years of service, although the environmental conditions are seldom quantified. As a consequence, some engineers believe that cement grout cover alone is sufficient for the protection of permanent tensioned steel tendons. A review of field case histories on tendon corrosion of post-tensioned concrete structures and ground anchorages illustrates the shortcomings of duct and borehole grouting in situ because it is workmanship sensitive and grout quality and integrity cannot be assured. This factual information supports the view advocated in standards since the 1980s that cement grout cover alone does not provide adequate corrosion protection for permanent tensioned steel tendons when installed in an aggressive environment.

Influence of cracks in cement grout surrounding steel tendon

Where non-contaminated cement based grout surrounds a clean steel tendon, the steel is protected due to the high pH environment, e.g. pH = 9.5 – 13.5. Where the grout is free of aggressive ions, it reacts with the steel to form a passive oxide film and thereby provides both a chemical and a physical protection. In such circumstances, satisfactory performance has been observed over the long term, where an alkaline environment has been maintained and is the only protection against corrosion. However, the intrusion of chloride and/or other aggressive ions, either directly through the grout or via cracks, can undermine these protective qualities and lead to steel corrosion if oxygen and moisture are also present.

The time taken for a depassivating front to penetrate through cement grout to the steel tendon depends on grout quality and cover, whether or not the grout is cracked, crack width and nature of the environment. Cracked grout will afford little protection in a marine environment. A further issue is that cracks become preferential anodes and the remaining (non-cracked) areas become preferential cathodes that induce localised accelerated corrosion at the cracks. Both axial and radial cracks can develop as the tendon bond length is post-tensioned. In addition, when steel corrodes it forms rust that occupies a volume greater than the original steel volume. This expansion exerts lateral stresses and can lead to grout cracking in ground of low lateral constraint, because cement grout has a low tensile strength.

In the short term up to two years and based on observations of 0.05-1.5mm wide cracks, there is a significant influence of crack width on the amount of corrosion found at the crack. In the long term, the effect of differences in these crack widths reduces dramatically (Beeby,1978). For small apertures, the smaller the crack width the lower is the corrosion risk and current practice accepts a crack width of 0.1mm as an upper limit for a cementitious protective barrier around prestressing steel (BS8081,1989; FIP,1996; EN1537,2000). Nevertheless, reports of crack widths greater than 0.1mm and up to 1mm or more in cement grout around tensioned steel tendons include Meyer (1977), Graber (1981), Natau & Wullschlager (1983), Littlejohn & Weerasinghe (1997) and Adams & Littlejohn (1997).

Influence of tendon steel grade on corrosion and crack widths

The corrosion of prestressing steel is fundamentally the same as for reinforcing bars, but it is recognised that high yield prestressing steel is marginally more reactive than low yield reinforcing grade steel (RILEM,1976; Manning,1988). Stress corrosion and hydrogen embrittlement are more likely to be encountered in highly stressed high yield strength steels and corrosion damage when it does occur can have more serious consequences in prestressed structures. As a result, permissible crack widths are relaxed to 0.2mm for reinforcing grade steel (EN 1992, 2004), c.f. 0.1mm for prestressing steel.

Irrespective of steel strength and composition, these permissible crack widths do not eliminate the need for impermeable corrosion protection barriers in permanent tensioned steel tendons because the grouts routinely applied in practice (w/c = 0.4-0.5 by weight) also bleed during the stiffening phase and the resulting voids formed can expose the tendon.

Sacrificial corrosion allowance for steel tendon

Some engineers believe that the longevity of a tensioned bar tendon in a ground anchorage can be extended by simply increasing the bar diameter, since it is common practice to make allowances for rate of corrosion in buried steel structures and sheet steel piling where general corrosion rates are applied, depending on the degree of corrossivity. This concept is inappropriate where the environment is highly aggressive and FIP (1986) states *"The use of thicker metal sections for the tendon, with a sacrificial area in place of physical barriers, gives no effective protection, as corrosion is rarely uniform and extends most rapidly and preferentially at localised pits or surface irregularities* Nevertheless, it is common practice in permanent rock bolting and soil nailing to incorporate an additional thickness or a sacrificial bonded metal coating. For aggressive conditions, this approach should only be considered if the design incorporates additional bolts or nails as a contingency against random failures.

Durability of cement grout in soil

Although it is often assumed that cement grout in soil will act as a permanent encapsulating medium for anchorage tendons, there are occasions, e.g. in the presence of carbon dioxide in saturated or partially saturated soils, where the surface of cement grout can suffer degradation due to attack by carbonic acid (H_2CO_3). Manns (1997) describes the storage of neat OPC cement grout cylinders (160mm long x 40mm diameter) made from w/c ratios of 0.40, 0.45 and 0.50 in running water with CO_2 contents of 44, 66 and 96 mg/litre for up to 600 days. The cement grout surface became soft and slippery in wet conditions and weak and highly erodible in dry conditions with no influence of w/c ratio on the extent or rate of degradation. After 500 days of storage, the thickness of the degraded surface layer was

- 0.5 – 0.7 mm at a CO_2 content of 44 mg/litre
- 0.7 – 1.2 mm at a CO_2 content of 66 mg/litre
- 1.5 – 2.0 mm at a CO_2 content of 96 mg/litre.

Subsequently, 400mm long x 110mm diameter cylinders of neat OPC grout (w/c = 0.44) were embedded in compact sand (D_{max} = 2mm; k = 2-5 x 10^{-4} m/s) and subjected to (i) drinking water, (ii) carbonic acid 20mg CO_2/litre and (iii) carbonic acid 100mg CO_2/litre. The ultimate load holding capacity was measured at 3, 6, 9 and 12 months (Figure 1).

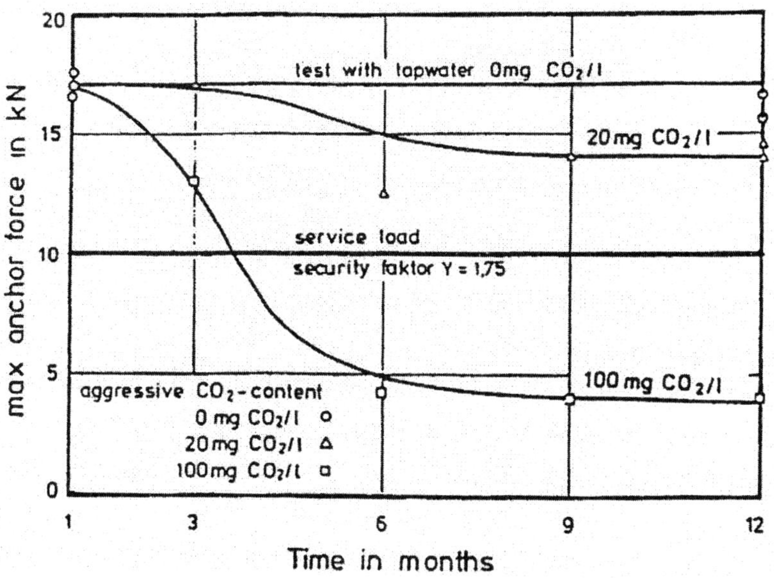

Figure 1. Development of maximum anchorage force due to long-term influence of aggressive CO_2 content in water (after Manns, 1997)

After 12 months a reduction in anchorage force of about 20% and 75% is observed for a CO_2 content of 20 mg/l and 100 mg/l, respectively. Given the degraded layer thickness of about 2mm, the soil/grout bond may suffer significant deterioration where the soil is fine grained and the construction technique (e.g. grout injection during steel casing withdrawal) forms a relatively smooth cement grout surface (Figure 2). Accordingly, CO_2 content of soil should be measured as part of routine ground investigation. Levels of corrosivity for ground water can be related to pH and CO_2, as shown in Table 1.

Figure 2. Low pressure cement grouted fixed anchor in uniform fine to medium sand

Table 1. Levels of corrosivity of ground water in relation to CO_2 attack of concrete (after DIN 4030, 1969)

Level of corrosivity of ground water in relation to CO_2 attack on concrete	pH	CO_2 mg/litre (ppm)
Low	5.5 – 6.5	15 – 30
Medium	4.5 – 5.5	30 – 60
High	< 4.5	> 60

Aside from CO_2 content, other aggressive agents in the ground or ground water such as acids, sulfate salts or pyrite (FeS_2) should be measured, together with water mobility and pH, as the results determine the appropriate type of cement for the encapsulating grout (BRE, 2005).

Case histories of loss of cover and corrosion of steel tendon protected by cement grout in post-tensioned concrete structures

There is a significant amount of published data on exhumation of tendons from post-tensioned concrete structures around the world (Clarke, 1992 reviews over 100 publications) and incomplete cement grouting of ducts is among the main factors contributing to corrosion damage. Tendon corrosion has often resulted from inadequately grouted ducts linked to excessive bleed from the grout and subsequent ingress of chlorides. Bleed water is generated in long ducts, even when neat grouts with w/c of 0.4-0.45 are employed. The following historical field examples are included simply to illustrate typical deficiencies in duct grouting that have led to unacceptable tendon corrosion or complete structural failure.

Manning (1981) reports on a bridge in Ontario where the tendons were examined as the structure was replaced after 16 years. The quality of the grout was highly variable and corrosion was present in a number of tendons. Where the ducts were not fully grouted there was a significant reduction in cross-sectional area of the tendon units. Woodward (1981) in an investigation of twelve bridges in the UK found voids in ducts in ten of these structures illustrating that even above ground level, the efficiency of duct filling with cement grout is workmanship sensitive and efficient void filling cannot be guaranteed.

In 1985 in Wales the Ynys-y-Gwas single span bridge collapsed after 22 years (Woodward & Williams, 1988). The cause of failure was severe corrosion of the tendons at both longitudinal and transverse joints, due primarily to ingress of chlorides. The 25mm thick mortar transverse joints between segments had permeabilities 100 to 1000 times that of concrete. The transverse ducts varied from well grouted to almost empty. The transverse tendons were corroded with up to a 90% loss of section over a distance of a few mm even in otherwise well-grouted ducts.

In spite of the recorded high incidence of deficiencies in duct grouting, the resulting voids are surrounded by low permeability concrete in compression and lead to a relatively low incidence of severe tendon corrosion in bridges world-wide (FIP, 1996a). Nevertheless, cement grouts for improved corrosion protection of bonded post-tensioned structures have been developed over the past decade, based on low w/c ratios of 0.33 to 0.35 (by weight) with the aid of superplasticisers (e.g. Schokker et al, 1999)

Case histories of loss of cover and corrosion of tensioned steel tendon protected by cement grout in ground anchorages

In comparison with post-tensioned concrete structures, there are few published examples on the in situ condition of anchorage steel tendons due to the difficulty and expense of exhuming the tendons. Nevertheless, several trials were instigated in the early 1960s, in order to investigate the integrity of cement grout cover as the sole corrosion protection for tendons, bearing in mind the number of dams known to incorporate 2-stage grouted post-tensioned tendons in Africa, Australia, Europe, India and North and South America at this time,.

In 1964 at an Australian dam site, a multi-wire tendon (102No. 5mm diameter wires, spacers at 1.5m centres) was stressed to a nominal 50t in a 15.2m deep, 102mm diameter vertical drill hole in sandstone. Over the free length, neat OPC grout (w/c = 0.40-0.45), mixed in a paddle mixer, was placed around the tendon by gravity displacement, injection continuing until the quality of grout exiting was the same as that being injected, after which the tremie pipe was withdrawn at a rate of 0.3m/min, whilst injection continued. In spite of this carefully controlled and supervised procedure, gaps of 302mm and 203mm were found at depths of 4.9m and 12.2m. At each location, numerous wires were exposed and the cavities extended to the centre of the tendon (Figure 3).

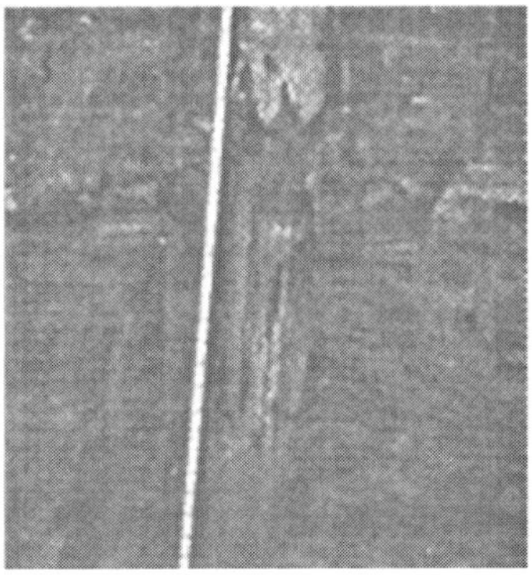

Figure 3. Loss of cement grout cover at a depth of 4.9m within a vertical 102mm diameter drill hole in sandstone

Figure 4. Loss of cement grout cover at a depth of 7.0m within a vertical 118mm ID asbestos cement duct

In a simulation test using a vertical 15.2m x 118mm ID asbestos cement duct, an unstressed tendon comprising 72No. 7mm diameter galvanised wires with spacers at 0.76m centres was grouted carefully as before. 305mm, 152mm and 203mm long gaps in the grout column were found at depths of 0.9m, 7.0m and 14.3m, respectively, although they did not penetrate right through the tendon (Figure 4). The defects in both cases were attributed to a combination of bleed and arching of the grout between the tendon and the side of the hole. It was judged that such gaps could occur typically at 5.2m intervals in a vertical hole and whilst wires smeared with cement would still provide some protection, this could be removed by water ingress, if the gaps in the grout cover coincided with permeable joints in the rock mass or concrete.

Similar simulation tests were carried out in the UK, based on vertical 15.2m grout columns formed around unstressed multi-wire tendons inside 114mm ID asbestos cement pipes, from which it was established that even with circulation of high shear and paddle-mixed neat OPC cement grout (w/c = 0.485 by weight) over 40-55 minutes for the top and bottom of the tendon, respectively, bleed voids still developed within the boundaries of the flow paths after circulation ceased. Voids were narrow (25-75mm) of varying heights, random in location around the periphery of the tendon, distributed fairly uniformly throughout the length of the tendon, and located mainly in the annular space between the tendon and the wall of the hole. No problems were encountered for tests in 2.4m to 3.7m long ducts (Cementation, 1965).

These early trial investigations led to the use of finer cements and admixtures such as methyl cellulose to reduce bleed settlement and aluminium powder to overcome shrinkage. Although, initially popular, such grouts had practical disadvantages such as increased viscosity and shorter stiffening times and were ultimately abandoned in routine practice.

In 1980, grouted free length sections of eight multi-strand tendons (24No. 15.2mm diameter strands post-tensioned to 250t and encapsulated in neat high early strength cement grout (w/c = 0.48 by weight with aluminium powder) were exposed adjacent to a gallery at a depth of 29m in Hume Dam, Australia. After 19 years the strands were in pristine condition in the benign environment but the grout contained a variety of defects, namely a large void in one tendon (Figure 5), small voids with entrapped bleed water in several tendons and near vertical/near horizontal crack patterns around three tendons. Concern was expressed that as the environment was likely to more aggressive lower in the dam where H_2S had been detected, pH was reducing and cracks were known to exist through the concrete and underlying rock. Houlsby (1980) concluded that the use of sheathed tendons would have overcome all the uncertainties prompted by the deficiencies in cement grout cover observed

Figure 5. Loss of cement grout cover at a depth of 29m within a 127mm diameter hole at Hume Dam, Australia (after Houlsby, 1980)

To eliminate bleed and improve durability (water content determines the porosity of the grout), it is necessary to reduce the w/c ratio, a fact recognised by Houlsby (1988) in his trials investigating grouting improvements for large tendons for dams. For hole depths of 90m, the value of API Oilwell cement Type G grouts (w/c = 0.36) and superplasticisers such as sulfonated melamine formaldehyde condensates was highlighted at the time.

In the early 1980s, a world-wide review by FIP of steel tendon failures due to unacceptable corrosion or fracture, included the following case histories relating specifically to deficiencies where reliance was placed solely on cement grout cover were included (FIP, 1986).

In an underground power station crane beam in Sweden, a 26mm diameter bar anchorage installed in 1955 failed after 16 years, within the cement grouted length but where there was virtually no trace of grout at the location of failure. Groundwater seepage was recorded (pH = 7-8) and overall, the reduction in cross-sectional area was found to be 6.8%. Significant pitting was observed, together with corrosion cracks. Fracture was triggered by a 5mm deep crack originating from the bottom of a 0.8mm deep pit. As more than 50% of the bars examined by boltmeter and probe testing were grouted unsatisfactorily, all 118 anchorages were replaced (FIP Case 4).

Three multi-wire tendons installed in 1963 to tie back a retaining wall in Germany failed after a few years. The tendons were encased in cement based grout but pitting was observed where there was insufficient cover of grout. The groundwater contained a high chloride ion content (FIP Case 7). After an unknown period of service (but less than 16 years after installation in 1968), corrosion failure of steel strand tendons tying back a retaining wall in USA failed where the bentonite/cement grout cover had dropped 1-1.2m below the anchor heads in an aggressive landfill environment with a high organic content and brackish water (FIP Case 16). In 1969 within a few weeks, six multi-ribbed wire tendons tying back a retaining wall in France failed due to incomplete filling of the hole with protective grout beneath the anchor head. Brittle failure of the tendons was observed and the water contained chlorides and sulfates from sewer leakage (FIP Case 18).

Within a year of going into service (installation prior to 1971), five multi-wire tendons tying back a retaining wall in Germany failed where each tendon was encased in cement based grout for its protection. Pitting and general corrosion was observed in the free length at locations where there was no adherent cement grout. No corrosion was observed where the tendon was still bonded to grout. Analysis of corrosion products exhibited 0.63% sulfates but no chlorides or sulfides (FIP Case 22). Two years after installation in 1972, unacceptable corrosion in the form of pitting was observed in the grouted fixed anchor zone of one of nine anchorages providing restraint for a cantilevered grandstand in South Africa. When the fixed anchor was exhumed, the grout cover ranged from nil to 6mm and pitting up to 1mm in depth was measured. All nine anchorages constructed in fill were condemned (FIP Case 25).

A pipeline bridge constructed in 1976 in Switzerland collapsed after 5 years in service, due to failure of three rock anchorages tying back one of the bridge abutments. Although permanent, the multi-strand tendons were designed to be encased only in cement based grout over the fixed length, but this section was only partially grouted in practice due to loss of grout through gravels. This exposed the tendon to aggressive groundwater containing chlorides and sulfides and resulted in severe corrosion and ultimately failure of the strands (FIP Case 31). In Hong Kong, ten bar anchorages installed for rock stabilisation in 1980 failed after 1 to 3 years within the free length but always adjacent to a coupling joint, where no cement grout or grease was in contact with the bar. The atmosphere was humid and slightly saline. Traces of chloride salts were present on the bar and the metallurgical examination showed pitting corrosion and hydrogen embrittlement (FIP Case 35).

It is considered that these summarised cases from various countries provide ample field evidence to support the view expressed by BS DD81 (1982), FIP (1986), BS8081 (1989) and FIP (1996b) that *"grout injected in situ to bond the tendon to the ground does not constitute a part of a protection system because the grout quality and integrity cannot be assured"*.

Corrosion protection strategy for ground anchorages

Ground anchorages employed to address the construction requirements of a project, typically have a service life of 2 years or less. Consequently, a 2-year service life is considered to be a reasonable limiting period for temporary anchorages where no special protection other than cement grout is required, unless the conditions are strongly aggressive. Generally, a minimum grout cover of 5mm or 10mm is specified in practice for encapsulation of the tendon. FIP (1986) and BS8081 (1989) illustrate a typical unprotected bar anchorage where the tendon bond length is encased in cement grout only (Figure 6).

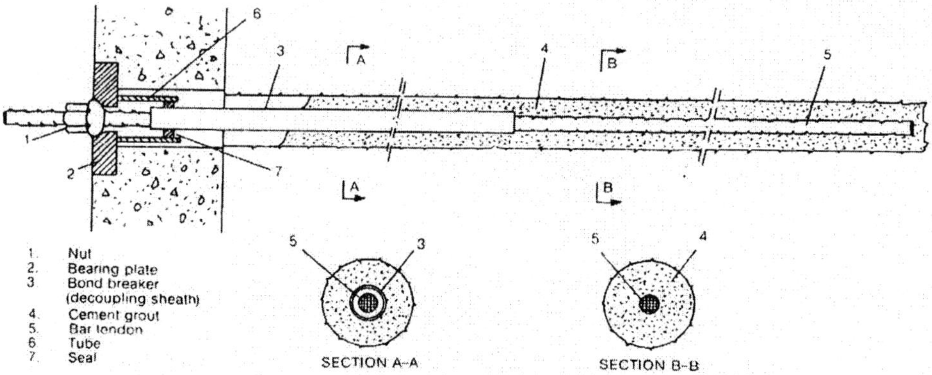

Figure 6. Unprotected bar anchorage (after FIP, 1986)

For strongly aggressive conditions, FIP (1986) includes case histories where 32mm and 35mm diameter steel bars installed in 1971 as temporary anchorages with unprotected free lengths failed after 4-6 weeks due to stress corrosion cracking in the free length. As a consequence, FIP (1996) warns that *"...in highly aggressive conditions, unprotected anchorages may only last a few weeks."*

EN1537 (2000) states without ambiguity *"There is no certain way of identifying corrosive circumstances with sufficient precision to predict corrosion rates of steel in the ground. All steel components of permanent anchorages shall be protected against corrosion for their design life."* This means that irrespective of the general environment or ground conditions, the steel components of permanent anchorages must be protected. By reference to EN1537, most steel tendons in new permanent anchorages in Europe are protected by either (i) two protective barriers, such that if the outer barrier is damaged during installation, the second inner barrier remains intact, or (ii) a single protective barrier, the integrity of which can be proven by water testing the barrier in situ. In common jargon the solution ranges from single to double protection and the protective barriers are usually plastic sheaths or ducts.

The resulting minimum recommendation of a sheath encapsulation for protection of the full length of tendon, i.e. free stressing length and tendon bond length, is included in FIP (1986),

BS8081 (1989), FIP (1996) and EN1537 (2000) and specified in Hong Kong (GCO, 1989) and New South Wales, Australia (RTA, 1997 and draft RTA, 2007).

Typically, cement grout is only considered as a reliable physical barrier within a protection system for prestressing steel tendon, if (i) the grout cover is at least 5mm between the tendon and a surrounding plastic duct and (ii) crack widths are limited to a maximum of 0.1mm. This limiting crack width criterion is only applicable in practice where field and/or laboratory evidence has confirmed grout crack widths not exceeding 0.1mm for the anchorage system proposed, e.g. ribbed bar with a cement grout surround inside a duct.

In the United States, PTI (2004) addresses the requirements of permanent anchors and states *"A Class II Protection system encases the stressing steel over the free length and relies on the cement grout to protect the prestressing steel along the bond length."* The accompanying commentary indicates that *"Class II Protection is often referred to as a grout protected tendon or a single corrosion protected tendon."* This interpretation of "single corrosion protection" contrasts with the definition developed in Europe, where the single corrosion barrier is a sheath or duct that must comply with minimum thickness criteria and be *"continuous, impermeable to water, resistant to age brittleness and resistant to ultraviolet radiation damage during storage, transportation and installation."*

PTI (2004) states further *"Cement grout can be used to protect the tendon bond length in non-aggressive ground when the installation methods ensure that the grout will remain fully around the tendon."* No guidance is provided on installation methods that ensure that the cement grout will remain fully around the tendon and PTI is silent on the influence of cracking in the tendon bond length that impairs the ability of the cement grout cover to provide protection. Importantly, aggressivity of ground is seldom quantified accurately in current practice to substantiate the non-aggressive assumption over the anchorage service life.

Even for low capacity grouted nails, the French National Research Project Clouterre (FNRPC, 1991) acknowledges that it is impossible to assess with any degree of accuracy the width of grout cracks and therefore, it cannot be proved absolutely that the cement grout will act as an impermeable barrier between the steel bar tendon and the surrounding soil. Where cracks occur, the steel can come into contact with the electrolyte carried by the moisture in the soil and FNRPC considers that corrosion will be generated as if the steel was in direct contact with the ground. Based on this hypothesis, any steel tendon not covered by a designed protection system is judged by FNRPC to be in direct contact with the soil.

Conclusions

There are sufficient case histories of loss of grout cover and tendon corrosion in the field of post-tensioned concrete structures and ground anchorages to illustrate the shortcomings of duct or borehole grouting in situ and the serious deficiencies that can occur in practice. This factual evidence confirms that reliance should not be placed on cement grout cover alone for the corrosion protection of steel tendons in permanent ground anchorages, particularly where the environment is aggressive.

Reasons include (i) a grout placement process that is workmanship sensitive, (ii) a lack of assurance as to the quality and integrity of the grout in situ, (iii) localised voids due to bleed and/or arching, and (iv) the occurrence of cracks, particularly where ductile tendon transfers tensile load via bond to brittle cement grout.

Whilst the design of grouts with low w/c ratios (e.g. 0.35) can eliminate bleed and improve durability, it is considered essential to incorporate at least one physical protective barrier against corrosion of permanent tensioned steel tendons in aggressive ground conditions, or where the consequence of failure could lead to loss of life and/or catastrophic damage, in accordance with published standards of practice that include FIP (1986), BS8081 (1989), SAICE (1989), GCO (1989), FIP (1996), RTA (1997), EN1537 (2000), JGS 4101 (2000) and PTI (2004).

Where cement grout surrounding the tendon is considered as part of the permanent protection, there is an onus on the designer to provide assurance that there will be an adequate grout encapsulation of the tendon after installation and that crack widths over the tendon bond length will not exceed the specified limits for the type of steel tendon under service loading and in ground conditions pertinent to the project.

Although not all published standards on the corrosion protection of anchorage tendons are strictly compatible, there should be an awareness of the need to design reliable and durable corrosion protection systems for permanent ground anchorages, if a legacy of reliable anchored structures is to be left secure for future generations.

References

Adams, D A & Littlejohn, G S (1997) "Anchorage load transfer studies using an instrumented full-scale re-usable laboratory apparatus", Proc. Int. Conf. on Ground Anchorages and Anchored Structures, Thomas Telford, London pp 411-421.

Beeby, A W (1978) "Corrosion of reinforcing steel in concrete and its relation to cracking", The Structural Engineer, Vol. 56A (3), 77-81.

BRE (2005) "Concrete in aggressive ground", Special Digest SD1, Building Research Establishment, UK

BS8081 (1989) "British Standard code of practice for ground anchorages".

Clark, L A (1992) "Performance in service of post-tensioned concrete bridges", Report for the British Cement Association, 59p

Cementation (1965) "Cable grouting – Carshalton Tests" (unpublished report by P I Parker)

DIN 4030 (1969) "Assessment of waters, soils and gases aggressive to concrete".

EN1537 (2000) "Execution of special geotechnical work – ground anchors".

EN1992 (2004) "Eurocode 2: Design of concrete structures" Part 1-1.

FIP (1986) "Corrosion and corrosion protection of prestressed ground anchorages", Fédération de la Précontrainte, Thomas Telford, London.

FIP (1996a) Proc. FIP Symposium "Post-tensioned concrete structures", London

FIP (1996b) "Design and construction of prestressed ground anchorages", Fédération de la Précontrainte, c/o Institution of Structural Engineers, SETO, London.

FNRPC (1991) "Soil nailing recommendations", French National Research Project Clouterre. English translation published by Federal Highway Administration, Washington, D.C, USA.

GCO (1989) "Model specification for prestressed ground anchors", Geospec 1, Geotechnical Control Office, Hong Kong.

Graber, F (1981) "Excavation of a VSL Rock Anchor at Tarbela", VSL Silver Jubilee Symposium, Losinger Ltd., Berne, Switzerland.

Houlsby, A C (1980) "Hume Dam Post-tensioning : Examination of stripped cables – October 1980". Private correspondence with G S Littlejohn – 28.11.80 (also Houlsby, A C (1981) "Grouting and sheathing of post tensioning tendons". Bull. Australian National Committee on Large Dams, No. 58, 33-35).

Houlsby, A C (1988) "Improvements in grouting of large ground anchors". ASCE Journal of Geotechnical Engineering, Vol.114 (4), 448-468.

JGS 4101 (2000) "Design and construction of ground anchorages and commentary", Japanese Geotechnical Society, 4-38-2, Sengoku, Bunkyo-ku, Tokyo, 112-0011, Japan.

Littlejohn G S & Weerasinghe, R B (1997) "Instrumented full-scale re-usable laboratory apparatus for rock anchorage testing", Proc. Int. Conf. on Ground Anchorages and Anchored Structures, Thomas Telford, London, pp 422-431.

Manning, D G (1981) "Corrosion resistant design of concrete structures", Proc. Can. Struct. Conc. Conf, Toronto, pp 199-233.

Manning, D G (1988) "Durability of prestressed concrete highway structures", Transportation Research Board, National Co-operative Highway Research Program Synthesis Highway Practice Report 140, pp 65.

Manns, W (1997). "Long-term influence of aggressive carbonic acid upon bearing capacity of ground anchors", Otto-Graf Journal, Vol. 8, 63-74 (also Proc. Int. Conf. On Ground Anchorages and Anchored Structures, Thomas Telford, London, pp 393-397).

Meyer, A (1977) "Report on Discussion to Session VI" by J Mitchell, A Review of Diaphragm Walls, Instn. Civ. Engrs., London, p111.

Natau, O P & Wallschlager, D H (1983) "Theoretical and experimental studies in bearing behaviour and corrosion protection of rock anchors up to a load limit of about 4900kN", Proc. 5th Cong. ISRM, Melbourne.

PTI (2004) "Recommendations for prestressed rock and soil anchors", Post-Tensioning Institute, Arizona, USA

RILEM (1976) "Corrosion of reinforcement and prestressing tendons. A state of the art report", Materiaux et Constructions, Vol. 9 (51), 187-206.

RTA (1997) "Permanent rock anchors", Roads and Traffic Authority, New South Wales, QA Specification DCM B114 (also draft RTA, 2007).

SAICE (1989) "Lateral support in surface excavations", Code of practice, Geotechnical Division, South African Institution of Civil Engineers.

Schokker, A J, Koester, B D , Breen, J E & Kreger, M E (1999) "Development of high performance grouts for bonded post-tensioned structures". Research Report 1405-2, Centre for Transportation Research, University of Texas, Austin, USA.

Woodward, R J (1981) "Conditions within ducts in post-tensioned concrete bridges", TRRL Laboratory Report 980, Transport and Road Research Laboratory, Crowthorne, England.

Woodward, R J & Williams, F W (1988) "Collapse of Ynys-y-Gwas bridge, West Glamorgan", Proc. Instn. Civ. Engrs. Part 1, Vol. 84, 635-669.

The Failure of Anchorage B160 at Devonport Royal Dockyard, Plymouth; Reasons for Failure and Lessons Learnt.

Michael Whitworth Devonport Royal Dockyard Ltd, Plymouth, United Kingdom
Steve Parrish Jacobs, Exeter, United Kingdom

Abstract

In 2004 as part of the routine maintenance of ground anchorages within 15 Dock, part of the Nuclear Facilities within Devonport Dockyard an anchorage known as B160 failed. The maintenance comprised of check lifts and visual inspections, in total a further 70 anchorages underwent check lifts without failure.

Due to the nature of the facility a significant investigation was undertaken in an attempt to identify the cause and evaluate implications for anchorages throughout Devonport Dockyard. The investigation comprised extraction of the anchorage enabling visual inspection of the fixed and free length, integrity of couplers and double corrosion protection, CCTV survey, coring of the base of the hole and laboratory analysis of grout.

Due to anchorage B160 being extracted with the couplers intact and with no break in the bar or failure of the encapsulation (factory) grout, the possible remaining factors contributing to the failure of the anchor are:

1. Grout to rock bond failure
2. Outer encapsulation to grout bond failure
3. Contamination of Grout
4. Incomplete Grouting

Due to the information obtained from the exhumation and subsequent laboratory testing, it is considered most likely that the failure occurred at the encapsulation / primary grout interface.

Background

Devonport Royal Dockyard, located in Plymouth, SW England, currently carries out the maintenance of the UK's fleet of Nuclear Submarines. To allow this to take place, between 1995 and 2002 the existing docks and associated structures were upgraded to modern nuclear standards of seismic qualification. Part of this upgrade required the installation of 247 multistrand anchorages with a working load of 3150kN, and 531 bar anchorages with a working load of 2040kN.

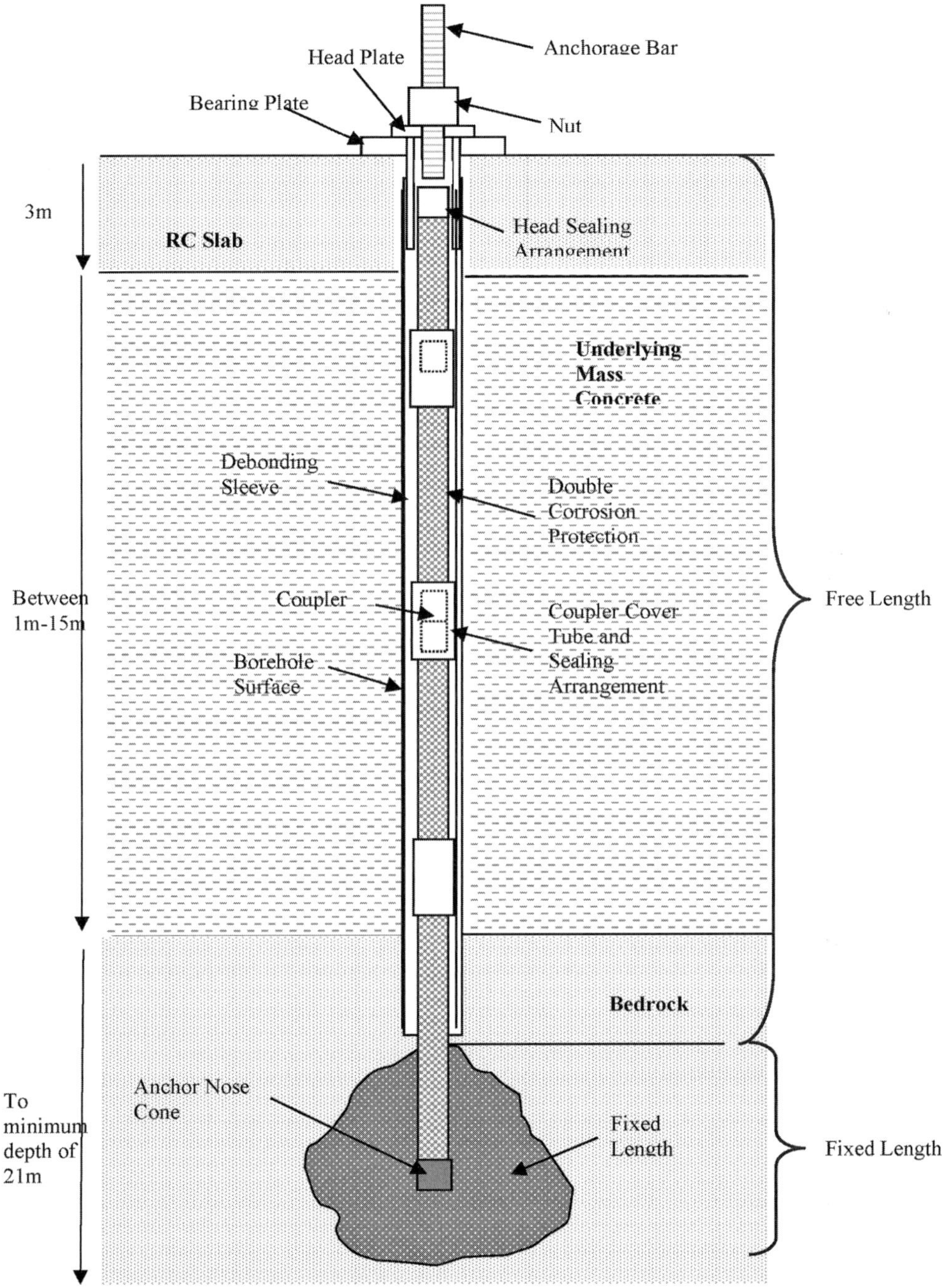

Figure 1. Bar anchorage Scematic

During annual maintenance of bar anchorages within one of the dock facilities, a failure occurred to one of the bar anchorages during routine check lifts. The failed anchorage known as B160 was subsequently exhumed and examined and an investigation into the likely causes of failure occurred. This paper discusses the potential cause of the failure of B160, the repercussion of a failure on a high risk structure and the subsequent lessons learnt to prevent future failures.

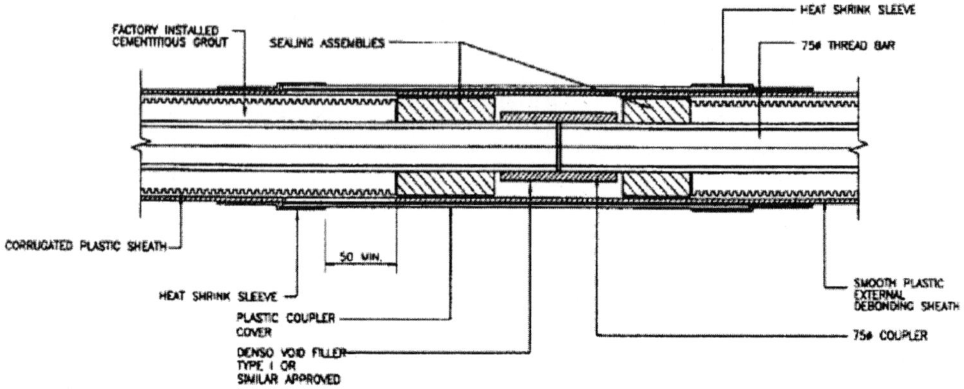

Figure 2 Cross Section through a bar anchorage showing the coupler arrangement

Details of Failure

During the period of maintenance the demand on the anchorage under a seismic event was such that the anchorages required testing to150%Tw. The anchorage failed at 148%Tw, characterised by a large "jump" in displacement, and a reduction in load shown on the jack. For information; approximately 150 other anchorages visited during this period of maintenance, maintained load at 150%Tw. Subsequent analysis on the demand placed upon the anchorages is now on average 120%Tw and the requirement to test to 150%Tw has been removed.

Extraction Process

Shortly following the failure of anchorage B160 during routine maintenance, a trial jacking of the anchorage succeeded in pulling out the anchorage by about 320mm.

The jacking operation of anchorage B160 was carried out over a 3 day period, the operation is detailed below:

1. A 4000kN stressing jack was placed over the anchorage and a 1.2m long draw bar attached to the anchorage with a stressing nut on top. Load increments to 50, 200, 700,

1200, 2200kN with corresponding displacements were then recorded. Also any sudden load loss was recorded and displacement measured. Following the extraction of a further 350mm of anchorage, the 1.2m drawbar was removed.

2. A 2.1m draw bar was then attached to the top of the anchorage and the 2.0m sleeve was placed over the top. This enabled the anchorage to be jacked up through the sleeve for a distance of approximately 1.9m.

3. The same load increments were applied and re-applied until the maximum extension (250mm) of the jack ram was reached. The jack was then retracted and the nut wound back down and the operation re-commenced.

Initially the anchorage took the load up 1621kN before suddenly slipping about 41mm. For the next 400mm, the anchorage would increase resistance to a load of between approximately 1400kN and just over 1700kN before slipping about 40mm and the load dropping to about 300kN.

The displacement in mm was recorded at each load increment and load loss. This sudden loss in load is known as "saw tooth" behaviour due to the relationship between load and displacement.

For the next 3.5m of extraction, the load required for jacking stayed steady at between 1300 and 1800kN before dropping down to under 700kN. Following the extraction of a total of 7.0m of anchorage, the load dropped off to below 350kN. Following the extraction of a total of 15m of bar, the load dropped to below 50kN. This allowed the remaining anchorage (including all of the fixed length and bottom coupler) to be pulled out by the dockside crane intact and in one operation.

Figure 3 is a photograph of the extraction process. Figure 4 shows the 2m sections, with figure 5 showing the graph of load vs displcacement during the extraction process. Following the removal of the anchorage, a CCTV survey was carried out to inspect the condition of the hole. Also a rock core of the base of the hole between 23.3m and 24.8m was taken.

Figure 3. The extraction of the failed anchorage. Due to the length of the anchorage and equipment available, it had to be chopped into sections.

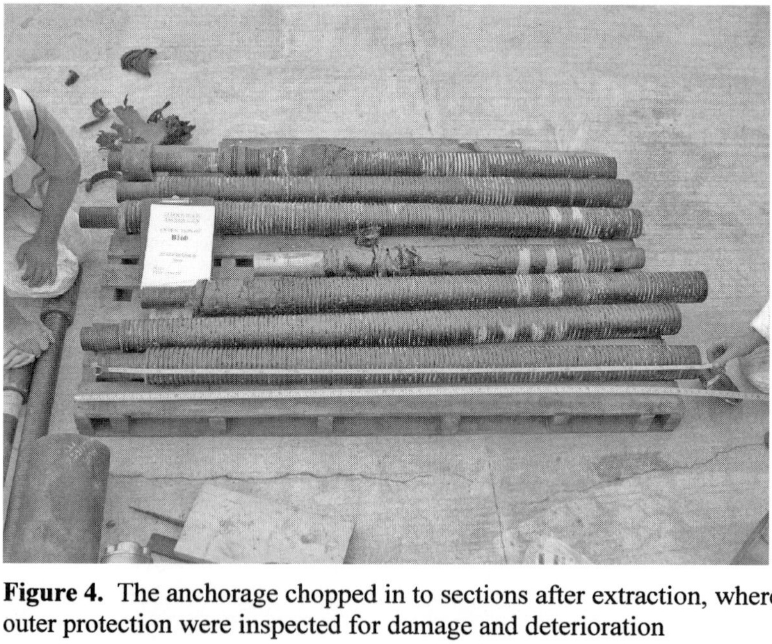

Figure 4. The anchorage chopped in to sections after extraction, where the couplers and outer protection were inspected for damage and deterioration

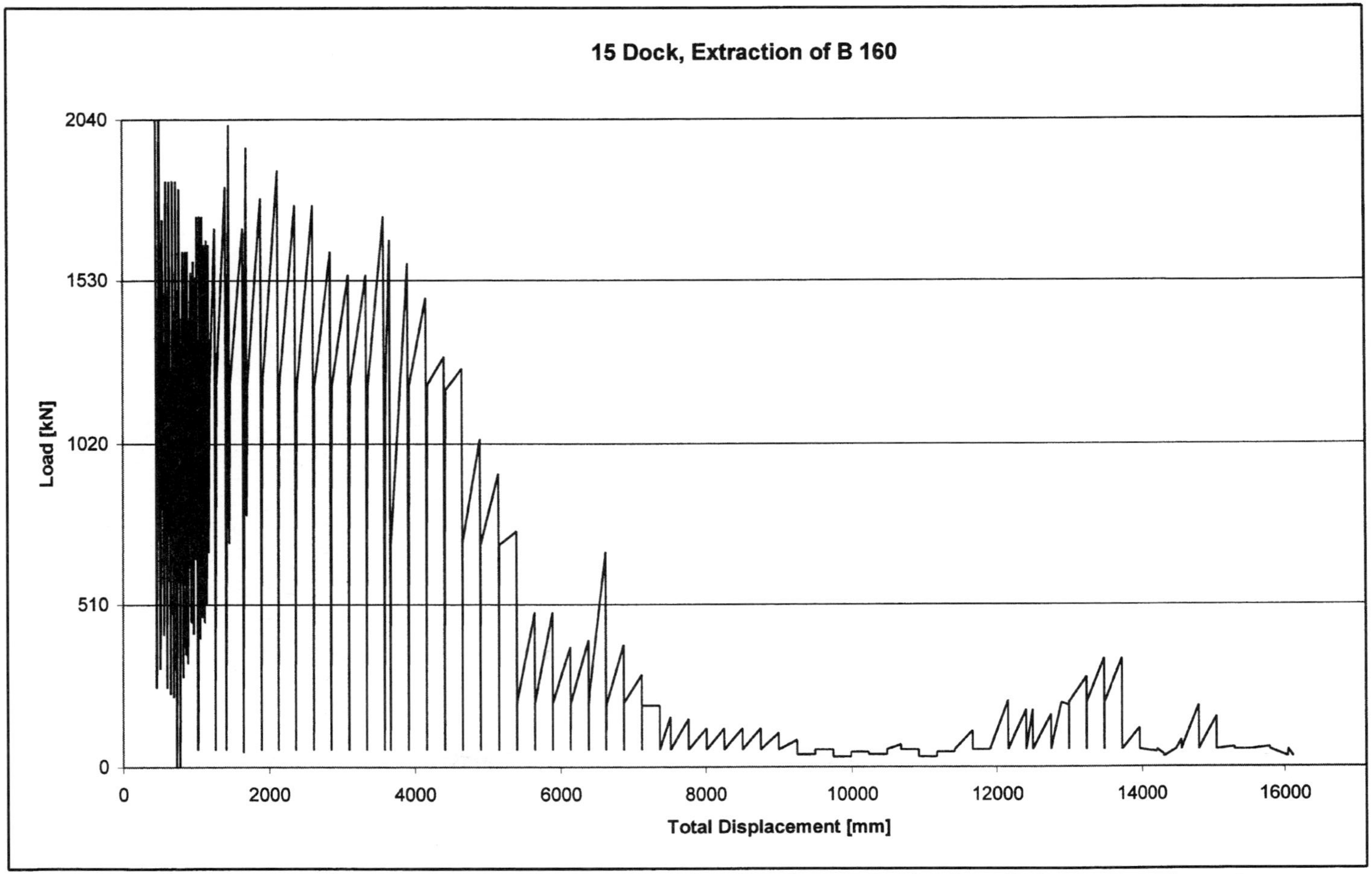

Figure 5. Graph of the extraction process of anchorage B160 showing the varying load applied to extract the anchor

Examination of Exhumed Anchorage

Double Corrosion Protection

The removed sections all showed the corrugated double corrosion protection outer sheathing to be intact although some distortion (scuffing) of the corrugations had occurred. This was probably due to the anchorage catching on the side of the borehole during extraction.
The de-bond sheath, which covers the free length of the anchorage between the top of the fixed length and the trumpet tube near the top of the anchorage was not recovered and is still contained by the grout within the borehole.

Condition of Coupler, 75mm Bar and Encapsulation (Factory) Grout

The two couplers were removed intact. A close inspection of the coupler assemblies show the top and bottom pair of 'O'-rings to be in place. The condition of the coupler and the bar were both found to be in excellent condition. No signs of corrosion were noted hence confirming that the double corrosion protection for the coupler assemblies (twin cover tubes plus two 'O'-rings each end) were working.

The 75mm bar was saw cut in half and showed to be in excellent condition.

The factory grout between the bar and the outer sheathing was also inspected and found to be intact except for some minor, >1mm radial cracks in the grout. These cracks may have been caused by the cutting operation during removal of the anchorage.

Fixed Length

The 6m fixed length was removed in one section together with 4.80m of free length and bottom coupler by the dockside crane

.

The fixed length was found to be intact except for minor scuffs on the corrugations, which in places have exposed the factory grout. The scuffs are a result of extracting the anchorage out of the hole.

A cohesive or cementitious material was noted between some of the corrugations on the outer sheathing of the fixed length. Samples of this material were removed for laboratory analysis and results detailed under the discussion section.

The CCTV survey of the fixed length noted that the rock-grout bond was largely intact, confirming that the failure occurred at the encapsulation-grout interface.

Discussion of Factors Leading to Failure

Due to anchorage B160 being extracted with the couplers intact and with no break in the bar or failure of the encapsulation (factory) grout, the possible remaining factors contributing to the failure of the anchorage are discussed in detail below: -

1. Grout to rock bond failure
2. Outer encapsulation to grout bond failure
3. Contamination of Grout
4. Incomplete Grouting

Grout to Rock Failure

The rock quality of B160, was assessed by a data logger attached to the drilling rig which by means of a number of drilling associated parameters (for example, drilling rate) gives the quality of the rock during drilling operations.

Prior to any production drilling works being carried out, a trial was carried out whereby a rotary percussion hole (diameter 250mm) (as undertaken in the production works) was drilled adjacent to a previously drilled site investigation cored borehole in the dock floor of 15 Dock. From a correlation of the parameters recorded from the data logger of the percussion hole and from the core log of the cored borehole, an assessment of rock quality can be made. For example, a drilling rate (DR) of less than 25 metres per hour was shown to indicate good quality rock.

The data logger print out for anchorage B160 is given in Figure 6. This shows a high drill rate (DR) through the plastic guide tube for the top 3 metres of hole and also between 11.5 and 14m below pit level (possible shear zone). Below 14 metres and including the fixed length, the drill rate decreases to around 12 metres an hour indicating strong competent bedrock. This rate is consistent for the drilling of the majority of fixed lengths on 15 Dock.

The "Ground Anchorage Installation Record" sheet shows the bedrock to be intact grey slate with no mention of soft rock. The suitability of the bedrock to provide a good bond was also assessed by carrying out a water packer test. This test is used to test the permeability of the bedrock to ensure that no grout loss occurs. The results of this test show that the water take was negligible hence indicating intact and tight bedrock.

The specification required that 1 in 8 boreholes be surveyed by CCTV. The reason for the survey is to visually assess the quality of the bedrock and confirm the parameters recorded from the data logger during drilling operations. Also the Dock's supervising engineer can request a CCTV survey on any borehole if he has any doubt on the adequacy of the bedrock over the fixed length. For anchorage B160, no CCTV survey was carried out which indicates that the slate bedrock over the fixed length was considered suitable for the installation of anchorage B160.

The CCTV survey carried out in January 2005 showed grout over the fixed length. The upper portion of the fixed length showed grout shearing and removal during exhumation of the anchorage. In the lower portion of the fixed length, faint impressions of the corrugation in the fixed length are evident.

From the above, it seems unlikely that the failure occurred at the grout to rock bond interface. Furthermore, the consistent pattern of scouring along the corrugated sheath on the encapsulation indicates that the shear failure occurred over this interface rather than the grout generally breaking up due to grout to rock failure.

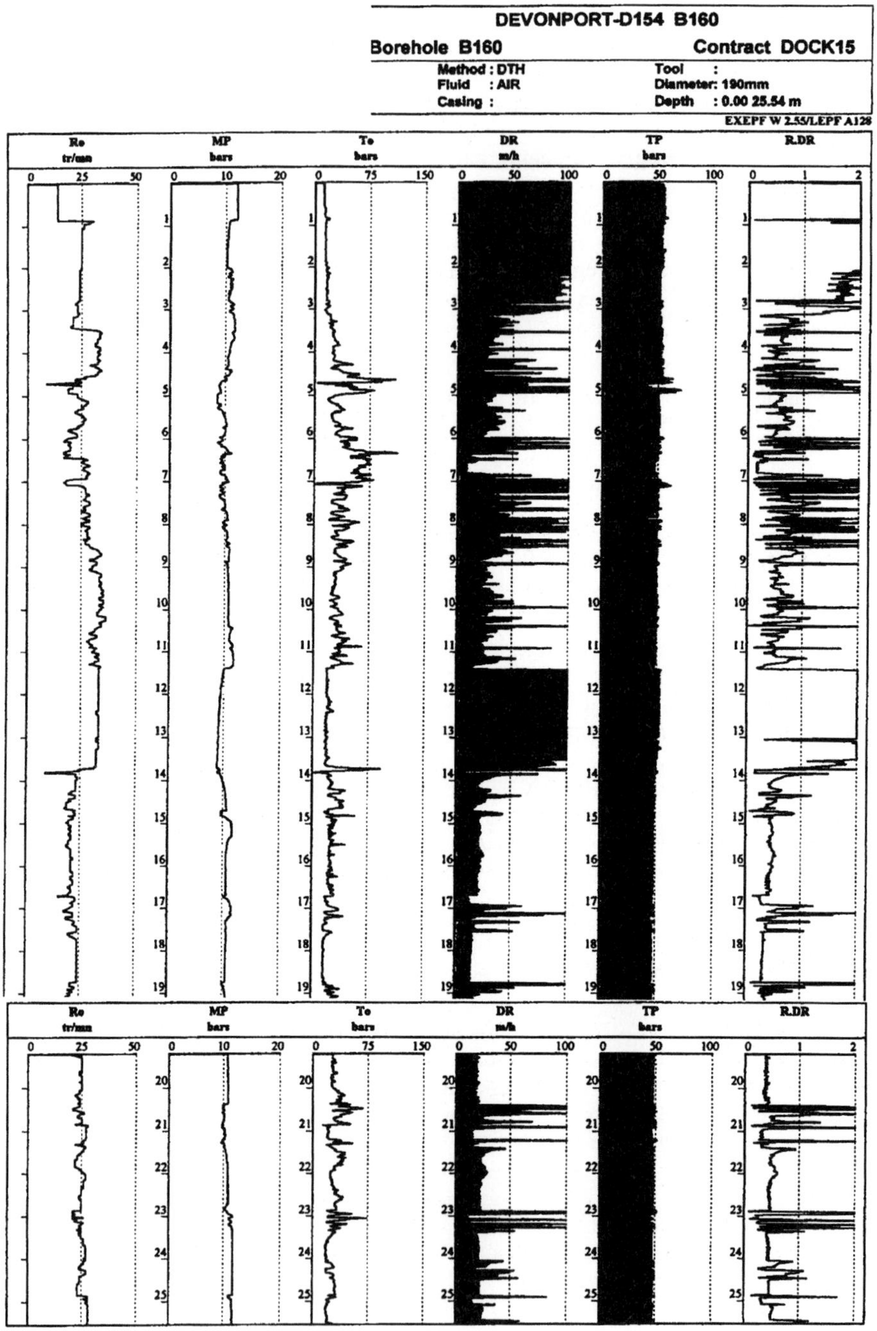

Figure 6. Drilling record for anchor B160

Outer Encapsulation to Grout Failure

The installation methodology requires that the ground anchorage is installed with a sacrificial grout tremmie tube attached to the base of the anchorage. Following installation the hole is grouted to the surface. This procedure ensures that the void surrounding the fixed length is flushed by upwards flowing grout, creating a fully grouted bond.

However, a number of situations where grout may not have fully surrounded the fixed length are discussed below:

Tremmie tube becoming unattached or ripped during installation.

A detachment of the tremmie tube could allow for the grout not to completely fill the area around the fixed length as the grout would not be pushing any slurry/detritus from the base of the anchorage.

From a review of the grout logs during installation, the amount of cement recorded for anchorage B160 is consistent with other similar length 15 Dock anchorages).

During grouting operations, the grouting is stopped once it has displaced all the water and grout consistency material is noted as emitting at the surface. Grout design was neat cement grout of grade 32.5 with a water to cement ratio of 0.45. Grout cubes are taken from the grout plant during the mixing and pumping operation. Grout cube strengths at 7 days for B160 were 36.5 and 38N/mm^2 which are above the specification requirement of 30N/mm^2 at 7 days.

Results from the CCTV survey following exhumation showed the tremmie tube to be present to at least a depth of 21m. This was the maximum depth the CCTV could be taken and hence is 2 to 3m above the base of the anchorage.

Rock collapse causing a blockage around the fixed length.

It is possible that during installation the anchorage could dislodge any loose rock especially in the weaker zones around the concrete/rock interface or at any shear zone. This rock could be caught up around the fixed length. Also following installation any loose rock could collapse into the borehole. This could occur due to adjacent drilling operations disturbing any loose ground prior to grouting. Three drilling rigs were in operation 24 hours a day, with a minimum separation of approximately 8m.

The data logger plot from B160 shows a weak zone over a length of 2 metres at the concrete to rock interface and also a shear zone between 12 and 14m below pit level. It also confirms the presence of competent rock over the design fixed length.

Records for anchorage B160 show that grouting was carried out the day after installation, which could have allowed for spalling (local collapse) of the rock. Also adjacent drilling operations may have "interconnected" with this shear zone, hence dislodging material down the hole. However due to the small annulus (20mm) between the de-bond sleeve (150mm) and borehole (190mm), this seems unlikely.

Contamination of the fixed length prior to grouting.

During drilling operations, the action of the percussion hammer cutting through the slate produces cuttings, which are fine grained and cohesive in nature. These cuttings when mixed with water form a cohesive paste which can cause smearing of the fixed length and borehole sidewalls.

In order to prevent this occurring, the specification requires that the hole is jetwashed prior to installation.

However, with adjacent drilling works in operation (including night drilling), an overspill of drilling derived material may have occurred. Also during the tendon installation stage, the fixed length may have accidentally been covered in thick slurry from the dock floor or adjacent drilling operations possibly covering the fixed length.

From an examination of the installation records, it is noted that the anchorage was grouted the day after installation; hence drilling derived material from adjacent drilling operations could have entered the borehole in the intervening period and following jetwashing. General practice is for any open hole with an un-grouted anchorage to be sandbagged around its perimeter to prevent this happening.

Furthermore sounding of the borehole following extraction and from analysis of the core sample showed a minimal sump, >1m. Hence without any sump, any detritus falling into the hole would have contaminated the grout. This could have the effect of not fully allowing the bond length to be grouted.

Low Strength Grout

A grout below the specified limit of 30N/mm^2 would reduce the bond at the interfaces. The presence of contaminants can affect the strength of the grout but this can also be affected by

- Cement used
- Aggregate
- Admixtures
- Miscellaneous properties including water / cement ratio.

Control of the above is by quality procedures and audits, together with cube tests to confirm the strength. An assessment of the grout installation records do not indicate that there is anything untoward with the grout mixture.

Conclusions

Following failure of anchorage B160, the anchorage was removed intact from the anchorage hole. Due to the information obtained from the exhumation and subsequent laboratory testing, it is considered most likely that the failure has occurred at the encapsulation / primary grout interface. The following is a list of potential reasons for the conjectured failure mechanism:

- Contamination of the fixed length or borehole sides by drilling induced "rock flour/cohesive paste". This could have occurred due to drill induced slurry entering the hole after jet washing or from the fixed length being covered in slurry prior to installation or a combination of both.
- Dilution of the grout due to either water entry over the bond length, from the surface or from the grout tremmie tube becoming detached at the lowest un-surveyed portion of the anchorage or damaged during installation. However the level of water entering the hole would probably be insufficient to dilute the grout enough.
- Contaminated grout surrounding the fixed length caused by rock debris from either the borehole sides or from the sump at the base of the anchorage hole.

- A grout mix outside the specifications of $30N/mm^2$ within 7 days which was not sampled via cube tests as only 1 in 4 batches are tested.
- A shallow or infilled sump, allowing deleterious materials that would have otherwise been removed from the area of the fixed length into this sump to mix with the grout preventing full encapsulation of the fixed length and/or dilution of the grout.

Suggestions for Future Rock Anchorage Installation Contracts

Notwithstanding the conclusions raised and the comments above, the following recommendations should be included as an addition to any future specification for the installation of ground anchorages:

- In order to prevent any debris or contamination to the anchorage jet washing must be carried out immediately prior to installation of the anchorage and grouting of interspace should be carried out on the same day. Ideally, grouting should be carried out immediately after installation.

- To prevent any drilling arisings / detritus entering the borehole from adjacent drilling operations, the drilled hole must be adequately sandbagged and / or blocked, ideally in two concentric circles around the perimeter of the borehole before and following installation, using sandbags and a bung.

- Randomly selected samples of the grout mixtures from the installed anchors should be tested to demonstrate correctness of cementitious mixtures.

- To prevent the grout tremmie tube becoming unattached during installation, the grout tube should be securely fixed at least every 0.5m over the fixed length. Alternatively consideration could be given to filling the hole with grout prior to installation.

- Possibly increase the specified frequency of CCTV inspection surveys to monitor any anomalies in the rock strata or at the concrete/rock interface, and check the degree of cleanliness of the holes prior to installation.

- Ensure that the grout emitting from the borehole is of the correct consistency, the grout should be tested and sampled as it emerges at the surface as well as from the mixing plant. Also the batch of grout which will cover the fixed length should be tested.

Service Performance of Multiple Anchors in difficult ground conditions for both Permanent Anchors and Temporary Removable Anchor usage

Tony Barley, Single Bore Multiple Anchor Ltd, Harrogate, UK
Mark Edwards, Keller Ground Engineering, Wetherby, UK

Abstract

At the site location in Ashford, Kent, the underlying Hythe Beds typically consist of alternating layers of hard sandy limestone and weakly cemented sandstone interspersed by beds of silts or clays of variable thickness. The irregularity of the strata made preliminary anchor design difficult since it was not possible to predict that inclined wall retention anchors distributed around the site would encounter either rock bands or entirely clay within the proposed fixed length. The multiple anchor design offers flexibility which is particularly beneficial in such varied ground conditions.

Both permanent anchors and temporary anchors were required to support sides of the secant pile wall cofferdam. Furthermore the full length tendons were required to be removed from the temporary anchor borehole when the adjacent structural works were complete. The satisfactory performance of each and every individual anchor (typically 4 "units" per anchor) in compliance with the acceptance criteria for both load/extension behaviour and creep characteristic is described. Temporary anchors were load checked during the service period and prior to successful removal of the anchor tendons.

Introduction

Permanent anchors were required to support one side of a secant pile wall and temporary fully removable ground anchors were required to provide temporary support to a second length of secant pile wall on the opposite side of the site. On completion of usage of the temporary anchor some anchors were lift-off tested to confirm their load performance in service, whilst the contractor confirmed that movement of the retained wall was within acceptable limits. Finally the entire steel strand tendons were completely removed from the borehole of the temporary anchors.

Ground Conditions Fixed Length Design and Supplementary Safeguards

Typically the anchor contractor was provided with the load per metre run of wall, level of the single row of anchors and a complete package of drawings, plans and site investigation data.

The geological sections showed "made ground", "drift", "Hythe beds", "Artherfield clay" and "Weald clay" with the lower three beds being appropriate for anchor founding strata. The "Weald clay" was dismissed owing to its depth below ground level and the uneconomic length of anchors. Consideration was given to the "Artherfield clay" consisting of predominantly very stiff sandy clay which would require either undereaming or post grouting to provide anchors of adequate capacity.

However, the mixed soil conditions described as the "Hythe beds", occurring at between 5 and 15m below the existing ground level were selected as being the most suitable stratum for founding the multiple anchor system. These comprised of medium dense to dense clayey sand interbedded with Limestone bands, or firm to very stiff sandy clays interbedded with either limestone or some weak sandstone bands (See Figure 1).

Multiple anchors consist of a number of short unit anchors founded at staggered depth in a single anchor borehole (Barley 1997). Typical unit anchor fixed lengths are varied from 3.0m to 4.0m depending on the envisaged strength of the ground and the bond capacity anticipated in that ground at that unit anchor founding depth. The use of a multiple of short fixed lengths reduces the effect of progressive debonding encountered in long conventional anchors and allows more efficient load transfer and hence mobilisation of higher anchor loads.

Owing to the variable nature of the Hythe beds and the possibility that a unit anchor may be located solely in a clay bed and possibly fail, and then more unit anchors were installed than the basic design required. For example in certain areas anchors with 3 units each with a 4m unit fixed length were adequate from a tendon performance viewpoint. However, in order to ensure the complete anchor fully performed in any expected ground condition then 4 unit anchors were installed and tested. In addition to this safeguard the actual depth and length of each unit anchor installed could be varied, based on the ground conditions encountered during the anchor drilling. These precautionary measures proved effective throughout the entire anchor works involving 29 permanent anchors with 356 kN working load and 27 temporary removable anchors with working load of 590kN. During cyclic testing to the required test loads in actual fact no individual unit anchors failed. Despite the extreme ground variation no redrilling or testing of supplementary anchors was required.

Anchors were installed nominally 1.0 to 2.0m below the wall crest and inclined at alternate 37.5 and 40° angles to allow utilisation of the appropriate level in the Hythe Beds whilst maintaining a safe distance between the total fixed length of individual anchors. Anchor free lengths were between 8 and 10m and a total of the unit fixed lengths summated to 13m.

Tendon Systems

All permanent anchor tendons were fully protected complying with the corrosion protection requirement of BS 8081thropughout the fixed length, free length and the capped anchor head. Each 2.5m long unit anchor encapsulation contained two concentric pregrouted corrugated plastic ducts and each free length strand was surrounded by grease, a 1mm thick yellow plastic sheath and a 1mm thick black plastic sheath.

Research work on totally removable anchors was started in the mid eighties by the investigation of the capacity of a looped strand around a small diameter saddle (Figure 2). Inspection of a loop after testing did indicate the expected damage to the sheathing but the sheath in the fixed anchor constitutes a barrier to prevent tendon bond to the cement grout and not a protective medium (Figure 3).

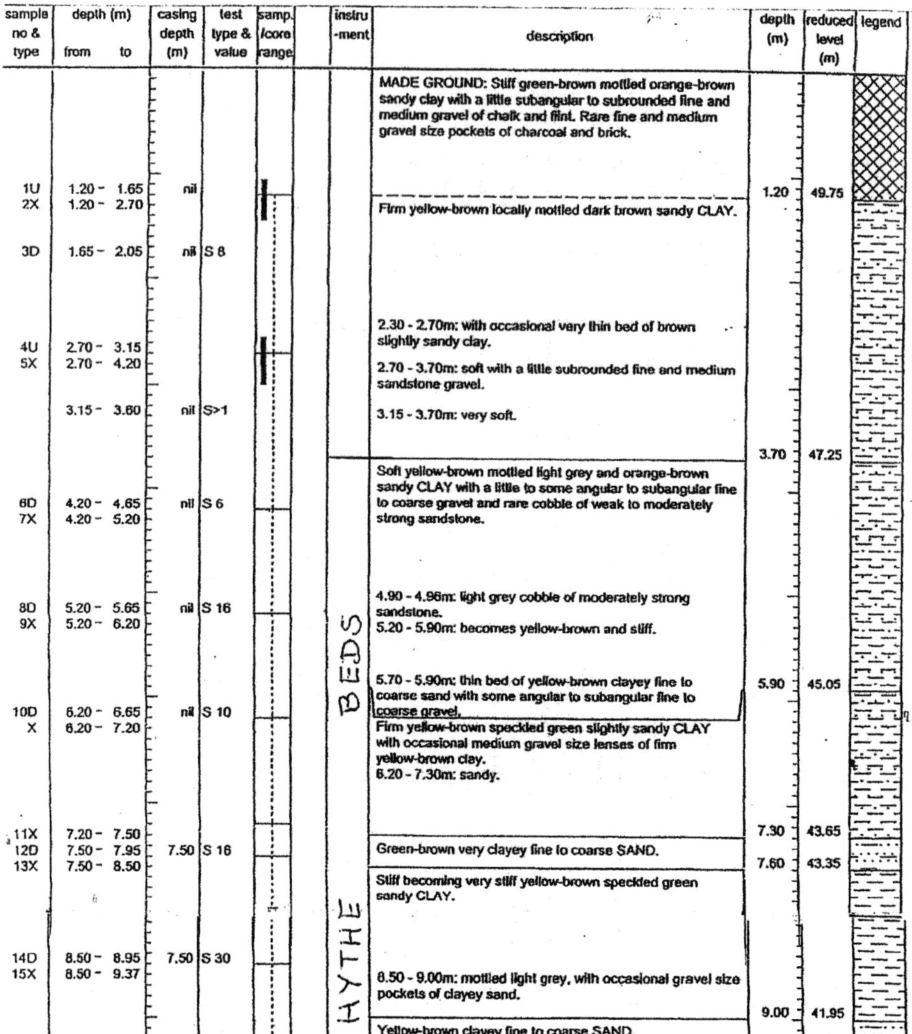

Figure 1. Illustration of the varied and in some cases the weak nature of the Hythe Beds in which the 20° inclined fixed anchor lengths were founded

Figure 2 Research work on strand
Loading around a steel saddle (Barley 1986)

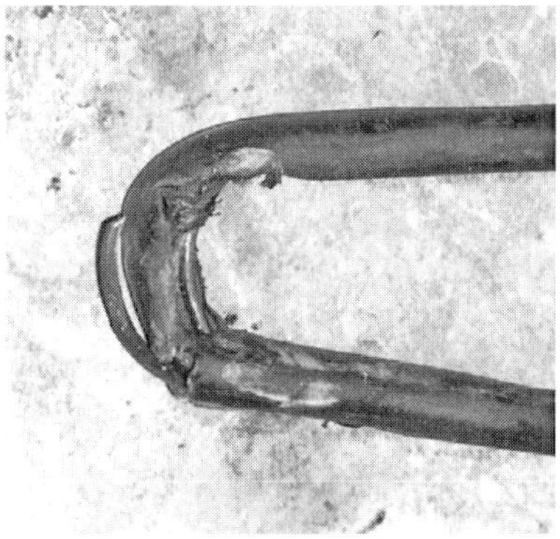

Figure 3 Example of a looped strand
after removal from test arrangement

Figure 4 Tendon removal by Winch Method

After successful trials temporary removable anchor tendons consisted of a number of 180° looped strands each fully greased and sheathed over its entire length. (Trummer et al 1995). Within the head of each loop was located a steel saddle ("first size") and a short glass reinforced plastic compression bar. On completion of usage of the temporary anchor the lubricated unbonded strands can be easily withdrawn from the grouted boreholes by the use of a 10T winch (Fig 4) or a crane or where access space is limited by the use of a long stroke stressing jack

Performance in Service

Permanent anchors were test loaded in accordance with on Site Suitability Test and On Acceptance Test requirements of BS 8081. This involved application of load cycles up to 1.5 x working load and resulted in compliance with apparent tendon free length requirements and load hold capacities.

The looped strand temporary anchors were also subjected to the two test methods requiring load cycles of up to 1.25 x working load and compliance with the same acceptance criteria (Fig 5).

During the works wall movement was monitored by the Contractor and no movement was detected

On completion of the satisfactory service performance of the temporary anchors 25 out of 100 unit anchor strands were subjected to lift of tests to investigate the potential load loss in this looped strand system founded on the variable Hythe Beds. A typical example of the establishment of initial anchor lock-off load (627 kN) followed by a load check at 24 hours (617 kN),and a lift-off check prior to anchor removal (625 kN) almost a year later is present in Figure 6.

In other load checked anchors load loss varied from zero to 5% of the original lock-off load. Hence no unit anchors indicated a loss of load greater than the 10% normally accommodated in anchorages complying with the acceptable load change requirements of BS 8081.

Strands were inspected after tendon withdrawal and exhibited no indication of corrosion or corrosive products or indeed any damage to the seven individual wires making up each of the removed looped strands.

Summary and Recommendations

Subject to the provision of adequate lengths of unit anchor tendons that can be adjusted on the anchor site, then the anchor lengths can be controlled during anchor drilling to utilise the stronger grout to ground bond capacity of stratum encountered in such variable materials as on the Hythe Beds.

Records of movement of the retained wall indicated that no movement took place during the service life of the temporary fully removable anchors. After 4 and 8 months when the temporary anchors were surplus to requirements, lift-off checks established that the load loss ranged from zero to 5% of lock-off load. On completion of lift-off checks the entire length of the fully lubricated prestressing strand were easily removed from the anchor boreholes. This completed the special service performance required of the temporary removable anchor system.

The use of removable tendons in temporary work anchorage is becoming more common owing to the modern advancement and development of ground anchorage technology. In many cities throughout the world including major conurbations such as Hong Kong and Singapore it is considered that the abandonment of used prestressing tendons within the ground is unnecessary and an unacceptable long term contamination of the subsoil.

National and International Codes of Practice for Anchorages should fully acknowledge the extensive development of such available systems (Herbst 1997) and incorporate an acceptance criteria for tendon removal from temporary work anchorages.

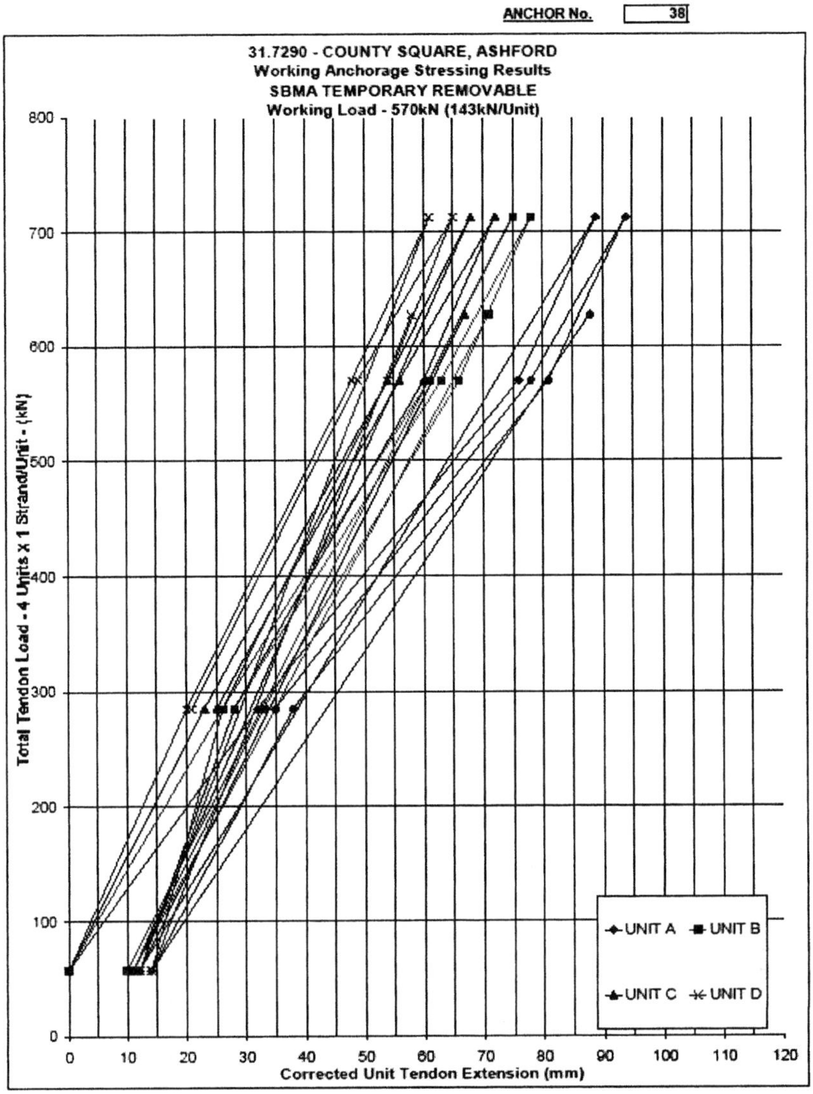

Figure 5 Proof loading of a temporary removable anchor to Test Load of 720 kN involving four simultaneously stressed unit anchors.

31.7290 CONTRACT - COUNTY SQUARE, ASHFORD
MONO-JACK LOCK OFF - ANCHOR 36

ANCHOR 36 WAS LOCKED OFF ON THE 24 AUGUST 2006 & 24 HOUR LOAD CHECKED ON THE 25 AUGUST 2006

% Tw	HOLD PERIOD	ANCHOR LOAD (kN)	UNIT LOAD (kN)	GAUGE PRESSURE/ STRAND (psi)	UNIT RAM MEAS.	UNIT TENDON EXTENSION (mm)	ANCHORHEAD MOVEMENT Measured (mm)	ANCHORHEAD MOVEMENT Actual (mm)	CORR. UNIT EXTENSION (mm)	COMMENTS 4 Unit SBMA with 2 Strand/Unit LOCK-OFF
110	0	627	78	2351	880	86		0	86	
			78	2351	881	84		0	84	
			78	2351	867	74		0	74	
			78	2351	872	75		0	75	
			78	2351	867	63		0	63	
			78	2351	871	61		0	61	
			78	2351	858	49		0	49	
			78	2351	849	48		0	48	
										ΔP %Po Allowable
108	24 Hrs	617	78	2300						10 1.5 6.0
			76	2300						
			78	2350						
			78	2350						
			76	2300						
			76	2300						
			78	2350						
			78	2350						

ANCHOR 36 WAS LOAD CHECKED & DE STRESSED ON THE 23 JULY 2007

% Tw	HOLD PERIOD	ANCHOR LOAD (kN)	UNIT LOAD (kN)	GAUGE PRESSURE/ STRAND (psi)
108	NA	625	78	2050
			76	2000
			78	2050
			80	2100
			78	2050
			76	2000
			80	2100
			78	2050

Figure 6 Precise load holding capacity of a temporary anchor initially, after 24 hours and after nearly a year

Acknowledgments

The authors thank Laing O Rourke for their pro-environment approach in specifying the removal of steel tendons from the temporary work anchors and for their provision of technical data.

References

British Standard, Code of Practice for Ground Anchorages, BS8081:1989

Trummer F., Stockhammer P., "Der wiederausbaubare Litzinanker, System Keller" Int. Conf on Anchors in Theory & Practice – Salzburg, October 1995

Barley A.D., "The Single Bore Multiple Anchor System" Int. Conf on Ground Anchorages and Anchored Structures, ICE London, November 1997

Herbst T., "Categories of Removable Anchors" Int. Conf on Ground Anchorages and Anchored Structures, ICE London, November 1997

Investigation of the failure of permanent anchorages adjacent to a tidal surge barrier

Tony Barley, Single Bore Multiple Anchor Ltd, England
Devon Mothersille, Geoserve Global Ltd, England
Peter Chamley, Ove Arup, New York, USA

Abstract

To provide protection against flooding of densely populated areas, a tidal barrier and anchored sheet pile walls were constructed as part of overall flood protection scheme across a tidal estuary in England. The piled structures, completed in 1979, were supported by thirty four, 48m long permanent ground anchorages installed below the capping beam at an angle of 40 degrees to the horizontal.

In September 1991 a 25m length of the pile wall was subject to a lateral deflection of up to 450mm at the pile head. The paper describes the visual examination of the wall, exhumation behind the wall to anchorage level and inspection of the exposed anchorage components. These included failure of the anchor tendon and pull out of the tendon from the anchor head. In addition there were indications of the use of inappropriate construction techniques employed at the time of installation.

Given anchored structures of similar type and age exist at many UK locations, there is concern that similar standards of construction and design specification may have been adopted elsewhere. By highlighting the features faced at this site, the paper gives an insight into the issues that could potentially be presented in other structures and provides recommendations on forward strategies that could be adopted to prevent failures in the future.

Introduction

The vulnerability of the east coast of England and its estuaries to flooding has been an age old concern. During the mid to late 1970s tidal barriers were constructed on the Thames and on estuaries to the north of England. Downstream of these severe tide controlled dams were the river bank raising schemes, many of which used ground anchored steel piles as the new barriers to prevent tidal overtopping. The barrier in question was completed in 1979 and involved some 34 ground anchorages retaining new sheet pile wing walls adjacent to the barrier and for 50m downstream (Figure 1).

Figure 1 Elevation of anchored sheetpile wall located adjacent to tidal surge barrier

Figure 2 South-east pile wall showing deflection immediately after failure

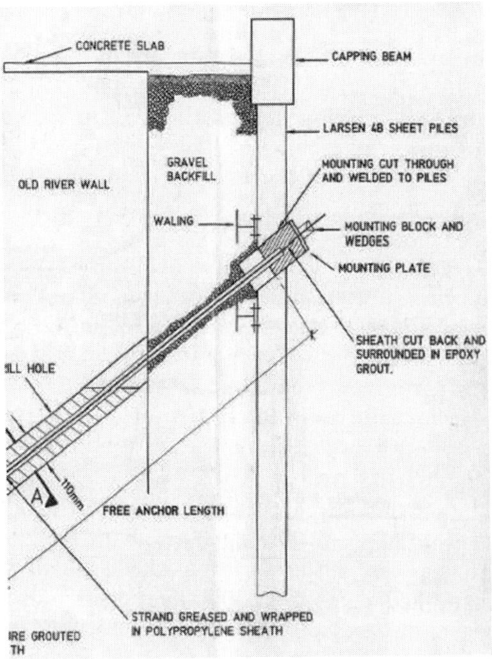

Figure 3 Cross-section through South-east pile wall

Structure and movement

In September 1991 it was noted that an approximately 25m length of the south-eastern sheet pile wall had undergone a sudden lateral deflection of up to 450mm (Figure2). The cross-section through the south east wall is shown in Figure 3, comprising a line of Larssen 4B sheet piles with a concrete capping beam. The piles were generally located some 2m in front of the former river wall which comprises a mass concrete or masonry structure with timber fenders. The gap between the old and new walls was stage backfilled with gravel. To support the sheet piles, permanent ground anchorages were installed, comprising 48m long anchorages installed at an angle of 40° to the horizontal. Each anchorage was formed of two Dyform prestressing strands (15.2mm diameter, 300kN ultimate tensile capacity) fixed to the river side of the sheet piles by an anchor head and gusset plates and the fixed length founded by grouting into a dense sand stratum.

Initial Inspection

Excavation of the retained soil behind the wall relieved the wall loading and exposed the ground anchor tendons (Figure 4). In the area of the failure a small number of the anchor tendons could be seen to have broken whereas the majority appeared to have pulled through the restraining anchor heads located on the river side of the sheet piles.

Figure 4 Exposed anchor tendons after excavation to relieve load on piled wall

Construction and Construction Techniques

Sheet Piling

The south-eastern sheet pile wall was installed a short distance (1-2m) in front of the existing timber and concrete river wall. Due to a number of problems which would appear to have included the presence of the old wall, a setting out error and over correction for the error, the resulting line of the sheet piles was not straight. The magnitude of the mis-alignment is not known but it may be assumed to have been significant.

Anchorage installation, backfilling and stressing

The sequence of events after the anchorages were installed is of particular importance to the subsequent failure. Following anchorage installation, the anchorages were stressed and locked off at low loads of approximately 1 tonne per strand. This would have been done to ensure that the tendons were tight and straight and providing some support to the wall during backfilling. In an effort to improve the alignment of the sheet piles at the northern end of the wall a number of anchors were loaded to 20 tonnes. This apparently gave a noticeable improvement in the wall alignment. Backfilling was then carried out up to the level of the waling.

The backfilling appeared to have been undertaken by simply end tipping from trucks or by dropping the fill in with a crane mounted grab. This operation caused a significant amount of damage to the tendon sheaths where they bridged the gap between the old and the new walls. Repairs to the sheaths were made by wrapping either the damaged sheaths or the exposed strand itself in grease impregnated tape. There was evidence of the presence of strand couplers fitted between the sheet pile and the rear wall indicating either strand damage or installation of strands too short to allow stressing. These were also protected only by grease impregnated tape.

When approaching the completion of back filling it was decided that further adjustments would be made to the wall alignment by altering some of the anchorage loads to 10 tonnes per anchorage compared to the design load of 34 tonnes.

Corrosion protection provided and standards at time of construction

At the time the ground anchorages at this site were installed there was no British Standard relating to their design, construction or testing. BS8081:1989 was eventually published in 1989 which was first issued as a draft for development DD81 in 1982. More recently the Euro Norm EN 1537:2000 was published as an execution code to provide guidance on aspects relating to the construction of anchorages. There was, however, a significant amount of published information available at the time (notably the GLC Specification for anchor protection on the Thames Bank Raising Scheme) which recommended quite clearly types and extent of corrosion protection to be provided although this was not embodied in this Contract Specification.

Although the specification called for ground anchorages to be 'sufficiently protected against corrosion to provide a working life of not less that 100 years, no reference was made in the specification to concepts such as single or double protection. The tender drawings showed bar anchorages with no specific corrosion protection measures to the anchor head.

Prior to the anchorage installation it was known that the anchorage contractor offered additional corrosion protection to the Client. This comprised a steel, bitumen coated, grease filled cap to cover the anchor wedges and exposed strand (Figure 6). This offer was not taken up, primarily on the grounds that it did not appear necessary to provide a greater degree of protection to that given to the ends of the anchor tie bars. It is unfortunate that the implications of this decision were not realised. This type of protection was subsequently developed into glass reinforced plastic caps as shown in Figure 7.

Figure 6 Grease filled bitumen coated steel caps at Greenock, 1981.

Figure 7 Grease filled glass reinforced plastic caps at Lowestoft, 1987.

Overall examination

Examination of the failure comprised a visual inspection of the failed anchorages and strands, as well as the remaining intact anchorages, together with a more detailed examination of the recovered portions of some of the failed anchorages.

It was clear from the outset that the failure of the sheet pile wall was a result of a failure of the ground anchorages by one of three mechanisms:

- Failure of the anchor strands
- Pull out of the strand from the anchor head
- Pull out of strand from an in-situ free length coupler

The wedges and strand at the anchor head were completely exposed to the environment, with no covering or protective cap as shown in Figure 4. The corrosion protection applied to the anchor heads at the time of construction, in the form of bitumen coating, had long since cracked and peeled off.

The examination showed that the majority of tendon strands had pulled through the locking wedges (Figure 8). The number of strands that had actually failed in tension was limited with only four, of a total of 22 that had been affected.

Figure 8 Unprotected barrel and wedges exposed to the environment

Of the anchorages that had pulled through the locking wedges, a limited number of locking wedges and barrels were recovered, including two retrieved from the mud below the wall. The recovered wedges showed corrosion of the outer surface underneath the remains of the bitumen paint. The inner serrated surfaces of the wedges were also corroded. The wedges were generally corrosion "welded" into the barrel and virtually immovable. The strands that had pulled out were particularly heavily corroded on the short section that had been exposed outside the wedges. The degree of corrosion and the volume of corrosion products was such that these exposed sections of the wires, above the wedges, were forced apart as individual wires (Figure 9). Where the locking wedges had gripped the strand could be determined by an area of lesser corrosion, but still exhibiting a clearly reduced section of the strand.

Below the wedges the protective polypropylene sheath had been cut back over a short length and leaving some strands unnecessarily exposed. This had been done to allow a considerable amount of strand extension during stressing. Such "extension" would accommodate true free length extension and also pull-in of some out-of-line piles. This length of tendon was generally slightly corroded, but to a degree much less than the length exposed outside the wedges probably a result of some of the initial grease remaining as a coating.

There were also two anchorage strands whose condition and location within the backfill suggested that they had been exposed for a considerable length of time. One strand was permanently bent into a curve over its exposed length. The section of strand that had been exposed by stripping back the sheath was relatively long, about 0.5m, and was uniformly more corroded that most other strands inspected.

The second strand had pulled back in behind the sheet piles and had pushed downwards so that it was resting against the rear of the sheet pile pointing vertically upwards. The length over which the strand had been cut back was very short, only about 150mm, and the exposed section was quite heavily corroded. The strand must have pulled back and pushed downwards by the backfill. It is believed that these strands were remains of the original sacrificial test anchorage.

An examination of the exposed anchor heads and tendons on the anchorages on the north-west wing wall showed them to be more heavily corroded than those on the south-east wall where failure had occurred. The individual wire segments were in general almost completely corroded through and could easily be broken off by hand above the wedges (Figure 9). The greater degree of corrosion was probably due to these anchorages being located at a slightly lower level within the splash zone. Below the anchor head plate could be seen the underhead protection in the form of an epoxy resin. Of the strand examined most showed that the polypropylene sheath had extended up into this resin. However, a number of the sheaths ended below the resin.

Detailed Examination

A number of the affected anchor tendons and components were taken for detailed laboratory examination. One of the most important aspects shown by the detailed examination was the lack of wedge serration markings on the strand that had pulled out of the wedges. Strand subjected to high anchor forces had quite distinct serration marks where the teeth had gripped the tendon. Conversely those that had pulled out exhibited shallow and indistinct wedge serration markings. These strands also showed a greater degree of corrosion at the location of the wedges.

One particular strand showed a series of teeth marks over an approximately 250mm length. At the lower end these were quite deep and distinct. Moving up the strand there were further teeth marks but becoming progressively less distinct and shallower. The location that the wedge block had finally taken up was shown by the corrosion thinning immediately above and below. Within the area that the wedge block had been located, for what was obviously a long period of time, the wedge marks were almost non-existent. These same features were noted on several of the recovered strands which had pulled out of the anchor wedges.

Figure 9 Heavily corroded north-west wing anchorage where exposed strands could be broken off by hand

Clearly the anchorages had been stressed to proof load before being locked off at lower loads. The application of the proof load resulted in deep teeth marks, the lower lock off resulting in virtually indistinct teeth marks, particularly after loss of section.

It is known from the construction records that although the anchorages had a design load of 34 tonnes, many were locked off at 10 tonnes load. The calculated elastic extension of the strand between these two loads almost exactly matched the distances between proof wedge marks and final wedge lock off wedge marks.

Summary and Recommendations

The construction techniques were extremely poor when considering the detrimental effect they had on the anchorages and the anchor corrosion protection. Realignment of the sheet pile utilising a variety of anchor loads resulted in misalignment of the anchor tendon from the original bore line and its direct contact with the sheet pile sheet and the rear wall. This was damaging to the strand sheathing.

The pouring of backfill behind the piles directly onto the strand caused extensive damage to the sheath, and where observed, necessitated repairs. The sheath repairs were poorly executed (grease impregnated tape only) and yet even afterwards direct backfilling over unprotected strands proceeded. It should be born in mind that the coupler would move extensively during testing and such movement, without coupler protection damage, must be accommodated.

The vulnerability of a single sheath around each individual strand free length is now generally acknowledged and double plastic sheathing is now frequently recommended. However even this system would not tolerate direct tipping of fill without damage.

Evidence suggests that on some anchorages the under head resin grout protection system was incomplete: the free length of strand did not extend fully into the resin thereby leaving a very vulnerable short length of strand below the resin. When applying under head protection and loading/unloading and strand tendons, the fact that nominal frictional forces may move the sheath up and down, must be considered. Furthermore the original cut-off length of sheath must be correct. In such circumstances the under head protection on a number of anchorages should be removed after setting, and broken out to confirm its suitability for its intended long term purpose.

The provision of bitumen coatings as permanent overhead protection is unsuitable for ground anchorages in any environment. Its effective lifespan on the anchor heads in the marine environment was probably a matter of months.

An example of the successful use of the proposed grease-filled capping system when applied in a similar environment is illustrated in Figure 7.

The principle of the taper wedge in the taper hole gripping the prestressing strand is that the wedge serrations indent in to the strand steel and the wedges slide further into the tapered hole. The sliding movement in turn increases the lateral force on the strands and the serrations indent deeper. However, the system only works efficiently if there is a lubricant between the wedge and hole interface.

Observation of the state of the wedges, the taper holes and the strand were consistent with those reported by Barley (1997) whilst analysing contributory factors to a major anchored wall collapse on the Thames in 1990. In exposed conditions the wedges become "corrosion bonded" within the taper hole; the strand corrodes within the wedges resulting in a reduced section. If the wedges could move freely they would advance into the taper hole and grip the reduced strand section.

However when wedge movement is not possible then the reduced section strands pull through the wedges leaving an empty hole (Figure 4). The anchorage then fails shedding increased load onto adjacent strands and then anchorages. Thus failure of unprotected anchor heads and the damage of strands behind the sheet pile occurred progressively in such a manner that total structural failure was avoided. In addition to the omission of outer protection of the anchorages, the poor construction techniques and failure to appreciate the necessary precautionary measures to prevent tendon damage, contributed to the short twelve year lifespan of anchorages.

All the in situ anchorages were subsequently replaced by new double corrosion protected anchorages complying with BS 8081:1989 requirements.

With regard to the future mechanism associated with anchorages on this contract, three matters should always be addressed:

- Any structural movement that occurs during anchorage stressing should be measured in terms of the effect of change of alignment of the drilled anchorages and potential damage to the corrosion protection.
- Complete anchor head protection is difficult to achieve in a marine environment yet without its guaranteed integrity the lifespan of anchorages may be relatively short.
- Anchor head protection performance should be checked periodically (Littlejohn and Mothersille, 2007).

References

BARLEY, A.D. 1992 The use of ground anchorages in waterfront structures. International Shipping and Marine Technology Market with Congress, Hamburg, 1992.

BARLEY, A.D. (1997) "The failure of a 20 year old anchored sheet pile quay wall on the Thames". Ground Engineering, March issue, 42-45.

BS8081 (1989) "Code of practice for ground anchorages". British Standards Institution, London.

EN1537 (2000) "Execution of special geotechnical works – Ground anchors". European Committee for Standardisation, Central Secretariat, rue de Stassart, 36, B-1050 Brussels, Belgium.

LITTLEJOHN, G.S. and MOTHERSILLE, D.K.V. (2008) "Maintenance Testing and Service Behaviour Monitoring of Permanent Ground Anchorages". Geotechnical Engineering, (to be published).

MOTHERSILLE, D.K.V and BARLEY, A.D (2007) "Inspection anchorages supporting waterside structures" Proc. Int. Conf. "Ground anchorages and anchored structures in service", Institution of Civil Engineers, London.

MOTHERSILLE, D.K.V., JACKMAN, S and FERRIER, J (2007) "Performance and condition assessment of 30 year old anchorages, River Clyde, Glasgow" Proc. Int. Conf. "Ground anchorages and anchored structures in service", Institution of Civil Engineers, London.

Reliability-based Failure Probability and Resistance Factor for Flush-Drilled Anchors in Taipei Basin

Hung-Jiun Liao, Jianye Ching and Chia-Wei Sue, National Taiwan University of Science and Technology, Taiwan

Abstract

The goal of this research is to calibrate the reliability-based resistance factor (= reciprocal of safety factor) of flush drilled soil anchors for their ultimate pullout capacities based on in-situ anchor pullout test data in the alluvial soils of Taipei basin. Efforts are taken to quantify the uncertainties with a full probabilistic analysis approach. The resistance factor is calibrated using the in-situ test results of 46 anchors with a rigorous theoretical approach which constructs the relationship between the resistance factor and failure probability. With this relationship, the reliability corresponding to the code regulation can be verified. From the results of the analysis, it is found that the borehole enlargement due to the flush drilling is quite significant. In other words, the actual diameter of the fixed anchor end may be much larger than the nominal diameter of the drilling casing. Consequently, the safety factor of three which is recommended by most anchor codes for the working anchor pullout capacity is found to be too conservative. The results of this study should be of value for the reliability-based design of flush drilled soil anchors in alluvial soils.

Introduction

Traditionally, soil anchor designs are usually following the safety-factor approach which accounts for the uncertainties in soil anchors (e.g.: uncertain soil parameters, uncertain mechanisms, uncertainties in the laboratory and in-situ tests, etc.) with an empirical basis. The safety-factor approach has been working reasonably well: as long as a large safety factor is adopted, the anchor system can usually be made quite reliable.

However, the safety factor is not a consistent measure of uncertainties: it does not tell us how reliable the designed anchor system is. Mathematically, the safety factor is not a consistent indicator of reliability: the safety factor of the same limit state may change depending on the mathematical expression of the limit state function (Phoon 1995). As a consequence, geotechnical engineers tend to be conservative in choosing safety factors, and sometimes they can be overly conservative in order to ensure reliable anchor designs. This over design does not fit the current need in terms of economic and ecological considerations. In contrast, since safety factor does not directly provide information about reliability, it is possible that designers may choose a safety factor corresponding to unsatisfactory reliability.

Therefore, it is essential to verify if the safety factor adopted in the design code is appropriate: What is the corresponding reliability for the safety factor adopted in codes? Does the corresponding reliability meet the design requirements? Is the adopted safety factor too conservative? One way of answering those questions is through a probabilistic analysis that constructs the relationship between the factor of safety and reliability (= one minus failure probability). In this paper, the term "resistance factor", simply the reciprocal of the safety factor, will be employed instead of the factor of safety.

More specifically, the afore-mentioned verification will be only conducted for soil anchor designs in the Taipei basin through a reliability-based theoretical design approach. Reliability-based analysis and design methods have been proposed for deep foundations (Zhang et al. 2001; McVay et al. 2003; Paikowsky et al. 2004), for shallow foundations (Phoon et al. 2003), for retaining walls (Chalermyanont and Benson 2004), and for slope stability (Christian et al. 1994; Low et al. 1998) to name a few. For soil anchors, a systematic study of the possible implementation of a reliability-based design method is still lacking.

In this study, a database containing the in-situ test results of 46 soil anchors in the Taipei basin is compiled and analyzed to construct the relationship between the resistance factor and target reliability. This relationship will be used to verify the anchor design code in Taipei. Since only a local database of Taipei is used for the analysis, the conclusion may be only applicable to anchor designs in Taipei.

Database

The database contains the in-situ pullout test results of 46 soil anchors in the Taipei basin. For each anchor, the following information is available: (a) the pullout force v.s. anchor head movement of the test anchors; (b) the soil profile nearby the anchor location, including soil classification, soil layer thickness, SPT-N value, unit weights, location of water table, and, in some cases, the undrained shear strength of the clayey layers; (c) the configuration of the anchor, including the free length and fixed length, the diameter of the drilling casing, the number of steel tendons, etc. The actual pullout capacity of each anchor is determined from the curve of in-situ pullout force v.s. anchor head movement using the method recommended by FIP (1982). Table 1 lists the data required by the following analysis for the 46 anchors. The pullout capacity of each anchor is determined from the curve of in-situ pullout force v.s. anchor head movement using the method recommended by FIP (1982).

Table 1 Basic parameters for the 46 test anchors in the database

Anchor Index	Total length (m)	Fixed length (m)	Free length (m)	# of steel tendons	Inclination angle (°)	Observed pullout capacity (kN)	Dia. of drilling casing (mm)	Depth of anchor head (m)
F1	37	32	5	12	Vertical	1250	114	13
F2	37	32	5	12	Vertical	1200	114	13
F3	42	37	5	12	Vertical	1570	114	10
F4	37	32	5	12	Vertical	1570	114	10
F5	37	32	5	12	Vertical	1290	114	9
F6	37	32	5	12	Vertical	930	114	9
TS1	37	32	5	9	Vertical	1400	152	10
TS2	37	32	5	9	Vertical	1500	152	10
TS3	37	32	5	9	Vertical	1350	152	10
TS4	37	32	5	9	Vertical	1250	152	10
TS5	37	32	5	6	Vertical	950	114	10
TS6	37	32	5	6	Vertical	900	114	10
TS7	37	32	5	9	Vertical	1300	152	10

TS8	37	32	5	9	Vertical	1500	152	10
TS9	37	32	5	6	Vertical	1350	152	10
TS10	37	32	5	6	Vertical	1450	152	10
A	27	22	5	5	Vertical	640	102	6.6
B	33	28	5	6	Vertical	860	102	6.6
P1-17	26	16	10	5	35	490	91	2.2
P1-19	30	17	13	5	20	520	91	2.2
P1-20	20	10	10	5	35	410	91	2.2
P1-22	22	12	10	5	35	380	91	2.2
P1-29	20	10	10	5	35	410	91	2.2
P1-30	22	12	10	5	35	520	91	2.2
P1-38	22	12	10	5	35	520	114	2.2
P2-32	18	10	8	5	35	330	114	5.5
P2-33	16	9	7	5	35	510	114	5.5
P2-35	16	9	7	5	35	510	114	5.5
P2-38	16	9	7	5	35	530	114	5.5
P2-39	18	10	8	5	35	430	114	5.5
P2-46	14	10	4	5	35	340	114	5.5
TS-11	26	15	11	10	Vertical	900	152	11
TS-12	29	18	11	10	Vertical	1360	152	11
TS-13	32	21	11	10	Vertical	1330	152	11
TS-14	32	21	11	10	Vertical	990	152	11
TS-15	36	18	18	10	Vertical	1290	152	11
TS-16	24	13	11	6	Vertical	760	152	18
TS-17	31	13	18	8	Vertical	1010	152	18
A1	22	5	17	3	35	150	100	7.4
A2	36	10	26	3	35	150	100	7.4
A3	47	5	42	6	35	200	100	7.4
A4	61	5	56	6	35	390	100	7.4
NO1	24	10	14	6	60	160	100	7.9
NO2	49	10	39	4	60	470	100	7.9
NO3	45	7	38	6	60	470	100	7.9
NO4	27	13	14	4	60	460	100	7.9

The boreholes of all test anchors were drilled with rotary casing and the debris was flushed out with water from outside of the casing. Since the amount of alluvial soils being flushed out is difficult to control, the process might result in a borehole with a larger diameter than that of the drilling casing and an irregular borehole surface as well. So the exact borehole diameter could not be determined and the diameter of the drilling casing was usually assumed as the nominal anchor diameter in the practice. After drilling was completed, the assembled tendons were installed into the casing and cement grout with a water/cement ratio 0.45-0.5 was injected into the borehole under a pressure of 200kPa. During grouting, the drilling casing was withdrawn gradually. After the casing for the fixed anchor end has been removed, a pressure of about 1000kPa was used to pressure-grout the anchor. In general, the grout takes for anchors were about double the volume calculated from the nominal dimension of anchors. Due to the high fines content, cement grout cannot permeate into the voids of the alluvial soils. The main purpose of pressure-grouting is to recompact the soils loosened during borehole drilling. The further details of anchor installation can be referred to Liao et al. (1996).

Models for pullout capacity

For the probabilistic analysis, a non-probabilistic predictive model for the anchor capacity must be provided. The following model is often adopted to compute the nominal pullout capacity C^n of an anchor:

$$C^n = d\pi\left[\sum_{i=1}^{N_c} s_u^{(i)} l_c^{(i)} + \sum_{i=1}^{N_s} K^{(i)}\sigma^{(i)} \tan\left(\phi^{(i)}\right) l_s^{(i)}\right] \quad (1)$$

where d is the nominal diameter of the anchor fixed end, often taken as the diameter of the drilling casing; N_c and N_s is the total number of clayey and sandy soil layers within the fixed end; $l_c^{(i)}$ (or $l_s^{(i)}$) is the fixed length within the i-th clayey (or sandy) layer, while $S_u^{(i)}$ (or $\phi^{(i)}$) is the undrained shear strength (or effective friction angle) of the i-th clayey (or sandy) layer; $\sigma^{(i)}$ is the average vertical effective stress in the i-th sandy layer; $K^{(i)}$ is the earth pressure coefficient in the i-th sandy layer, i.e. the effective normal stress on the anchor shaft divided by $\sigma^{(i)}$.

A preliminary study is taken to compare the actual pullout capacities of the 46 anchors (i.e. the in-situ test results) with their predicted values. In order to compute the predicted capacities, all variables in the model are fixed at their nominal values. The computed nominal capacities of the 46 anchors and their actual capacities are plotted in Figure 1 for comparison. The following observations are evident: the actual capacities are significantly larger than the predicted ones. This may be mainly due to the enlargement effect of borehole drilling and grouting: after the anchor hole is drilled and the steel tendons are placed inside the hole, grouting under a pressure of about 1000 kPa is taken to fill the irregular borehole space and to some extent recompact the soils surrounding the borehole. The pressured grouting may also intrude into the surrounding soils, so the "effective" diameter of the fixed end will be much larger than the diameter of the drilling casing.

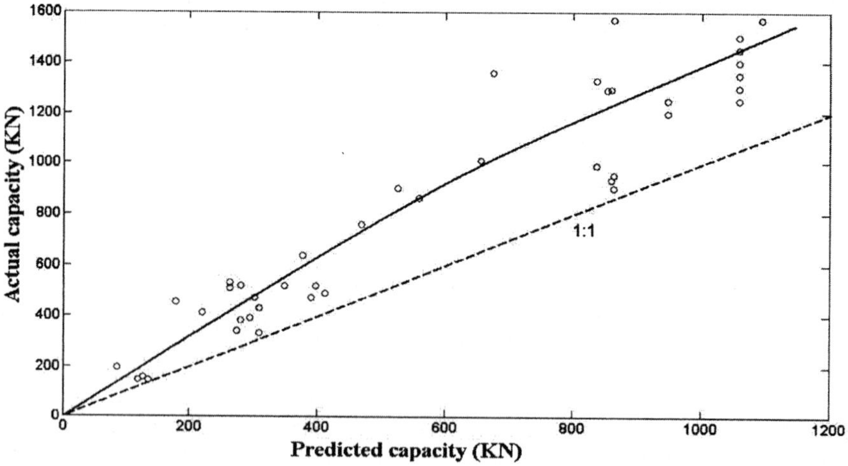

Figure 1 The actual v.s. nominal pullout capacities for the anchors in the database.

Equation (1) assumes that the developed shear resistance along the fixed end is uniform within the same soil layer. However, it is reported by Hawkes and Evans (1951) that the distribution of developed shear resistance along the fixed anchor length is not uniform even in a homogeneous soil layer: the developed shear resistance is fully developed at the top of the fixed end but is only partially mobilized away from the top. This non-uniformity is more pronounced when the surrounding soils are stiff. Hawkes and Evans (1951) proposed the following equation to account for such decay: $\tau_x/\tau_0 = e^{-\alpha x/d}$, where τ_0 is the shear resistance developed at the top of the fixed end; τ_x is the shear resistance developed at a distance of x from the top of the fixed end; α is called the decay parameter, whose value usually depends on the stiffness of the surrounding soils.

In reality, the α parameters of different soil layers in the Taipei basin should be different, but in this paper, the α parameters of different soil layers are assumed to be identical since the soil stiffnesses within the fixed end do not drastically change. Under such a simplifying assumption, a more sophisticated model for pullout capacity C^α can be derived as follows:

$$C^{\alpha} = d\pi\left[\sum_{i=1}^{N_c}\left(\int_{L_c^{(i)}}^{U_c^{(i)}} s_u^{(i)}(x)\cdot e^{-\alpha x/d}dx\right)+\sum_{i=1}^{N_s}\left(K^{(i)}\tan\left(\phi^{(i)}\right)\int_{L_s^{(i)}}^{U_s^{(i)}}\sigma(x)\cdot e^{-\alpha x/d}dx\right)\right] \quad (2)$$

where $U_c^{(i)}$ and $L_c^{(i)}$ are the distances (along the fixed anchor length) to the top of the fixed end from the upper and lower interface of the i-th clayey layer, and similar for $U_s^{(i)}$ and $L_s^{(i)}$; $\sigma(x)$ is the effective vertical stress at a distance of x from the top of the fixed end, $S_u^{(i)}(x)$ is proportional to $\sigma(x)$, and d is the diameter of the drilling casing.

Modeling of uncertainties

The basic uncertainties include the unit weight, friction angle, ratio of undrained shear strength to vertical effective stress of each soil layer as well as the effective diameter D and the decay parameter α. Uncertainties like $\sigma(x)$, $K^{(i)}$, and $S_u^{(i)}(x)$ are not basic since they are functions of the basic ones. To quantify the uncertainties, it is desirable to first specify the joint prior probability density function (PDF) (Ang and Tang 1975) of all basic uncertainties. The term "prior" PDF denotes the PDF before new data, i.e. the actual capacities of the 46 anchors, is available. Later we will mention the "posterior" PDF (Ang and Tang 1975) of the basic uncertainties, which denotes the PDF after the new data is available. Among them are the uncertain soil parameters, including friction angles and unit weights of all sandy layers and undrained shear strengths and unit weights of all clayey layers. The prior PDF of these uncertainties are modeled as lognormal with mean values equal to their field-measured values or published results for the Taipei basin (e.g.: Moh and Associates 1987) with coefficients of variation (c.o.v.= standard deviation divided by mean) appropriately chosen according to Phoon (1995).

An obvious limitation is that α must be positive. Moreover, according to Hawkes and Evans (1951), $\alpha \fallingdotseq 0.1$ for rock anchors, which implies α for soil anchors should be less than 0.1. Other than these, there is not much prior information about the decay parameter. Therefore, the prior PDF of α is taken to be uniform over [0, 0.1]. The effective diameters of the j-th anchor D_j in the database is assumed to be equal to the diameter of the drilling casing, denoted by d_j, multiplied by an enlargement factor ρ_j, and it is assumed that the enlargement factors of all anchors are independent but identically distributed. It is reasonable to assume the enlargement factors to be independent because the enlargement effect of one anchor does not seem to provide information about the effect of another anchor. They are assumed to be identically distributed because the drilling and grouting procedure for all the anchors is roughly the same. The enlargement factors are assumed to be lognormal with mean value μ_p and c.o.v. δ_p, where μ_p and δ_p are two uncertain parameters. The prior PDF of μ_p is taken to be uniform constrained by $\mu_p>1$, while that δ_p is taken to be uniform constrained by $\delta_p>0$. Perhaps, the most important uncertainty in the problem is the uncertain pullout capacity of an anchor. In most literature, the uncertain (actual) capacity of the j-th anchor C_j is usually taken to be the product of the predicted capacity with a "model factor", which characterizes the model uncertainty. However, for our model, the ρ_j parameter plays a similar role to the model factor. It seems reasonable not to use an extra factor but treat ρ_j as a combined parameter characterizing both the borehole enlargement effect and the model uncertainty. If this approach is taken, it is clear the uncertain capacity of the j-th anchor C_j can be expressed as

$$C_j = \rho_j C^\alpha(Y_j)$$
$$= \rho_j d_j \pi \left[\sum_{i=1}^{N_j^c} \left(\int_{L_j^{c(i)}}^{U_j^{c(i)}} \beta_j^{(i)}(x) \cdot \sigma_j(x) \cdot e^{-\alpha x/d_j} dx \right) + \sum_{i=1}^{N_j^s} \left(K_j^{(i)} \tan\left(\phi_j^{(i)}\right) \int_{L_j^{c(i)}}^{U_j^{c(i)}} \sigma_j(x) \cdot e^{-\alpha x/d_j} dx \right) \right] \quad (3)$$

where Y_j denotes the collection of the uncertain soil parameters for the j-th anchor.

Calibration of resistance factor

Given the in-situ measured pullout capacities of the 46 anchors, it is possible to calibrate the reliability-based resistance factor for anchor designs in the Taipei basin. The reliability-based design can be achieved by restricting the failure probability of a future design conditioning on the past data, i.e. consider the following equation:

$$P\left(C_{new} \le C^* \mid C_{1:46}\right) = P_F^* \quad (4)$$

where C_{new} is the uncertain capacity of a new anchor, and C^* is the design load of the anchor; the $C_{new}<C^*$ is the failure event; $C_{1:46}$ contains all capacity data in our database; $P(C_{new}<C^*|C_{1:46})$ is the failure probability conditioning on the capacity data, which, in principle, can be computed with Bayesian analysis; P_F^* is the target failure probability. Equation (4) can be rewritten as

$$P\left(C_{new} \le C^* \mid C_{1:46}\right) = P\left(\rho_{new} C^\alpha\left(Y_{new}\right) \le C^* \mid C_{1:46}\right) = P_F^* \quad (5)$$

where ρ_{new} is the model factor of the new anchor, which is lognormally distributed with mean value equal to μ_p and c.o.v. equal to δ_p (recall that it is assumed that all model factors are independent identically lognormally distributed with mean equal to μ_p and c.o.v. equal to δ_p); Y_{new} contains all uncertain soil parameters of the new anchor site. Note that $C^\alpha(Y_{new})$ is uncertain because Y_{new} is uncertain.

Let us assume that the PDF of the soil properties Y_{new} is known from site investigation and appropriate uncertainty characterization. Therefore, the relationship between the resistance factor and target failure probability can be derived as follows:

$$P\left(\rho_{new} C^\alpha\left(Y_{new}\right)/C^D\left(Y_{new}^n\right) \le C^*/C^D\left(Y_{new}^n\right) \middle| C_{1:46}\right)$$
$$= P\left(\rho_{new} C^\alpha\left(Y_{new}\right)/C^D\left(Y_{new}^n\right) \le \eta \middle| C_{1:46}\right) = P\left(G\left(\rho_{new}, \alpha, Y_{new}\right) \le \eta \middle| C_{1:46}\right) = P_F^* \quad (6)$$

where Y_{new}^n is the nominal value of Y_{new}, e.g. mean value of Y_{new}; $C^D(Y_{new}^n)$ can be any predictive model chosen by the designer; $\eta = C^*/C^D(Y_{new}{}^n)$ is exactly the resistance factor, which is the reciprocal of the safety factor. One can see that the resistance factor can be simply estimated as the $100 \cdot P_F{}^*$ percentile of $G(\rho_{new},\alpha,Y_{new})$ conditioning on the data $C_{1:46}$. If one can obtain stochastic simulation samples of $G(\rho_{new},\alpha,Y_{new})$ (see Ching et al. (2007) for details), denoted by $\{G^1,\ldots,G^N\}$, conditioning on the data $C_{1:46}$, the relationship between resistance factor and target failure probability can be estimated as the following according to the Law of Large Number:

$$P_F^* \approx \frac{1}{N}\sum_{k=1}^{N} 1\left(G^k \le \eta\right) \tag{7}$$

where 1(.) is the indicator function: it is unity if the inside statement is true and is zero otherwise. Equivalently, η can be estimated as the $100 \cdot P_F{}^*$, the largest number in $\{G^1,\ldots,G^N\}$. Once this relationship is obtained, the calibration of the resistance factor is achieved since one can determine the required resistance factor from the target failure probability. Note that the η v.s. $P_F{}^*$ relationship depends on the configuration of the new anchor, the soil profile at the new anchor site, and the adopted capacity model for design C^D.

Results of stochastic simulations

The samples for stochastic simulations contain interesting information about the behaviour of the 46 anchors. For instance, the samples of the α parameter, as shown in Figure 2, are distributed as $f(\alpha|C_{1:46})$, so the histogram of the sample values, also shown in the figure, basically gives the relative degree of plausibility of the α value learned from the data. From Figure 2, it is clear that the decay parameter is mostly in the range of [0.001, 0.004], and the mean value and c.o.v. of the samples of the decay parameter are 0.0025 and 0.253, respectively. Moreover, the possibility of $\alpha = 0$ is close to zero, i.e. compared to the C^α model, the C^n model is basically rejected.

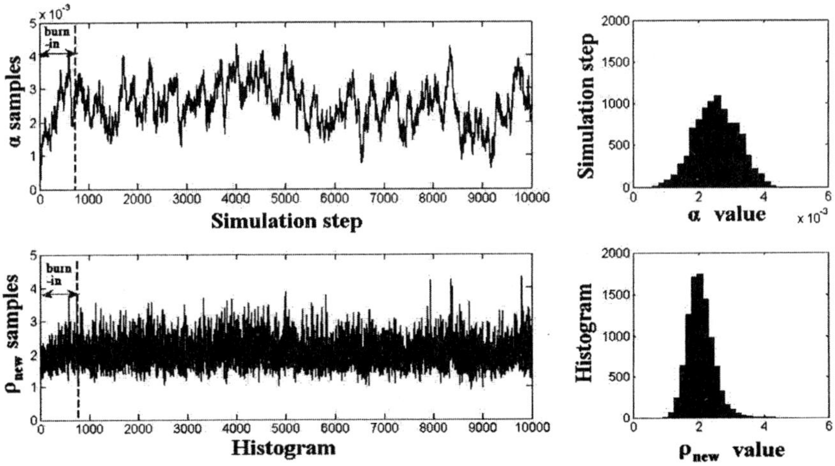

Figure 2 Markov chain samples of α (upper plot) and ρ_{new} (lower plot) and the histograms.

The samples from $f(\rho_{new}|C_{1:46})$, as shown in Figure 2, gives us the possible range of the model factor ρ learned from the data. From the histogram in the figure, it is clear that the model factor is mostly between 1 and 3. This means the actual capacity of an anchor in Taipei is roughly two times larger than the capacity predicted by Eqn. (2) because of the combined effect of the enlargement and model bias. Moreover, the mean value and c.o.v. of the samples of the model factor are 2.049 and 0.186, respectively.

Figure 3 shows the comparison between the actual capacities and the ones predicted by Eqn. (2) multiplied by 2.049, where the uncertain soil parameters are held at their nominal values and the decay parameter is held at 0.0025. It is clear that the predicted capacities unbiasedly reflect the actual ones, indicating that the analysis results are consistent.

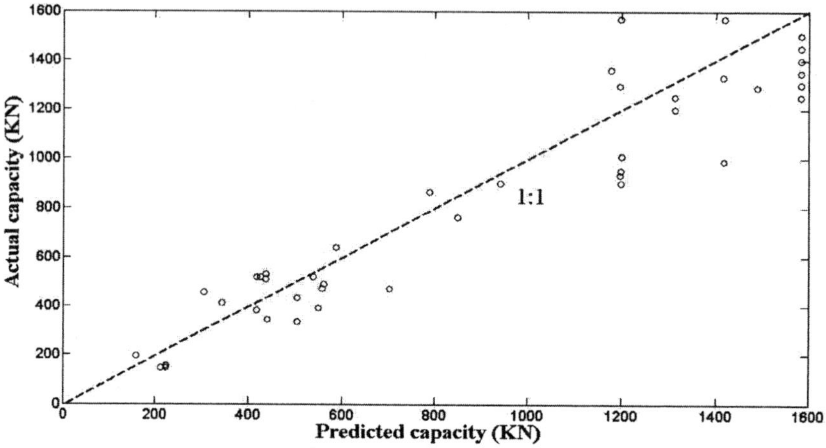

Figure 3 The actual v.s. predicted pullout capacities for the anchors in the database.

Relationship between resistance factor and target failure probability

The procedure introduced in a previous section is employed to estimate the relationship between resistance factor η-P_F^* and target failure probability. The estimated relationship is not unique: it depends on the configuration of the anchor, the soil profiles, and the chosen capacity design model C^D. This non-uniqueness imposes a difficulty, but it can be solved by just selecting a representative set of anchor design scenarios in Taipei basin and estimating the η-P_F^* relationships for all scenarios to obtain the possible range of the relationship. For example, one can first determine the target failure probability P_F^* for anchor designs within Taipei basin, then the range of the required resistance factor can be found.

In this paper, the sites of the 46 anchors in the database are taken as the representative set of soil profiles in the Taipei basin. Numerous scenarios of anchor designs are considered and the corresponding η-P_F^* relationships are computed. It is found that the η-P_F^* relationships strongly depend on the chosen design model C^D, the dominant soil types within the fixed end, and the length of the fixed end.

When the design model C^D is chosen to be C^n in Eq. (1), the η-P_F^* relationships will depend on the dominant soil types within the fixed end as well as the length of the fixed end. It is evident that if the soils within the fixed end are primarily sandy, the calibrated resistance factors are larger, while the resistance factors are smaller if the soils are mostly clayey. The dependency of the η-P_F^* relationship on the soil types is due to the fact that undrained shear strengths of clayey soils are more uncertain than the shear strength of sandy soils.

Moreover, the calibrated resistance factors are smaller for anchors with long fixed ends and are larger for those with short fixed ends. The dependency on the fixed anchor length is due to the fact that the C^n model does not take the decay behavior of developed shear resistance along the fixed end into account. Therefore, the calibrated resistance factor will be smaller for long fixed end than for short fixed end.

When the C^n model is adopted for designs, the ranges of required resistance factors corresponding to various target failure probabilities, fixed anchor lengths, and soil types are compiled in Figure 4. Note that the required safety factors for target failure probability 10^{-2} and 10^{-3} rarely exceed 3, the safety factor required by most anchor codes. For target failure probability 10^{-2}, the required resistance factor is around 0.6 to 1.2 (safety factor is around 0.8 to 1.7), depending on the soil type and fixed-end length. In all sand conditions, the resistance factor ranges from 0.8 to 1.2 and safety factor ranges from 0.8 to 1.25; in all clays condition, the resistance factor ranges from 0.6 to 0.8 and the safety factor ranges from 1.25 to 1.7. It indicates that the borehole enlargement is less significant in clays. So the safety factor is larger when anchor is installed in clays but smaller when in sands.

For target failure probability 10^{-3}, the required resistance factor is around 0.4 to 1 (safety factor 1 to 2.5). The safety factor of 3 recommended by anchor codes corresponds to failure probability less than 10^{-4}, which is somewhat over-conservative for flush drilled anchors in Taipei. Note that this conclusion may not apply to different types of anchors nor to anchors outside Taipei. In addition, one must be careful that the recommended resistance factors and safety factors in Figure 4 for fixed anchor length between 30m-40m are extrapolation results because the longest fixed end in the database is 32m. So, care must be taken when adopting the resistance factors and safety factors if the designed fixed end is longer than 32m.

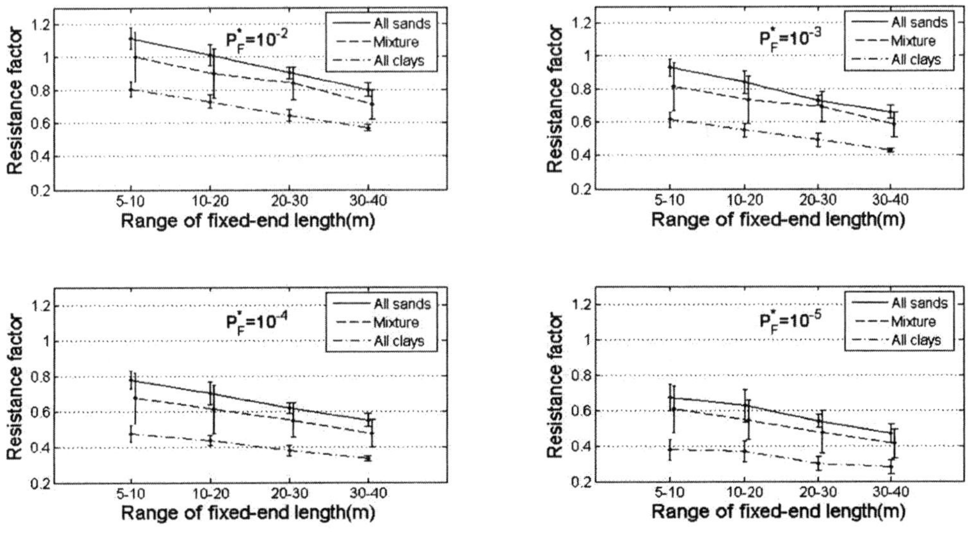

Figure 4 The recommended ranges of the resistance factors for different design scenarios and target failure probabilities based on the Taipei database when C^n is adopted.

Conclusion

A database of 46 anchors in the Taipei basin is analyzed in this study to construct the relationship between the resistance factor and target failure probability of flush drilled anchors. Due to the borehole enlargement resulted by flush drilling, the required safety factors for target failure probability of 10^{-2} and 10^{-3} rarely exceed 3 which is the safety factor required by most anchor codes. When the safety factor is equal to three, it corresponds to a failure probability less than 10^{-4}, which is somewhat over-conservative for flush drilled anchors in Taipei. Although the above resistance and safety factors may be of value for the reliability-based design of flush drilled soil anchors in alluvial soils. Care must be taken when applying it to anchors with fixed length longer than 32m or applied to outside Taipei.

References

Ang, A.H.S., Tang, W.H. (1975) *Probability concepts in engineering planning and design: Volume I – basic principles*, John Wiley and Sons, New York.

Chalermyanont, T., Benson, C.H. (2004) Reliability-based design for internal stability of mechanically stabilized earth walls, *ASCE Journal of Geotechnical and GeoEnvironmental Engineering*, 130(2), 163-173.

Ching, J., Liao, H.J., and Sue, J.W. (2007) Calibration of reliability-based resistance factors for flush drilled soil anchors in Taipei basin, submitted to *ASCE Journal of Geotechnical and GeoEnvironmental Engineering.*

Christian, J.T., Ladd, C.C., and Gregory, B.B. (1994) Reliability Applied to Slope Stability Analysis, *Journal of Geotechnical Engineering*, 120(12), 2180-2207.

FIP Commission on Practical Construction (1982) *Recommendations for the design and construction of prestressed concrete grout anchors*, FIP Wexham Springs.

Hawkes, J.M., Evans, R.H. (1951) Bond stresses in reinforced concrete columns and beams, *Structural Engineer*, 29 (12).

Liao, H.J., Ou, C.D., and Shu, S.C. (1996) Anchorage behavior of shaft anchors in alluvial soil, *ASCE Journal of Geotechnical and GeoEnvironmental Engineering*, 122(7), 526-533.

Low, B.K., Gilbert, R.B., and Wright, S.G. (1998) Slope reliability analysis using general method of slices, *ASCE Journal of Geotechnical and GeoEnvironmental Engineering*, 124(4), 350-362.

McVay, M., Kuo, C., and Guisinger, A.L. (2003) *Calibrating resistance factors for load and resistance factor design for statnamic load testing*, Final Report # 4910-4504-823-12, Florida Department of Transportation, March 2003.

Moh and Associates (1987) *Engineering properties of the soil deposits in the Taipei Basin, Report No. 85043*, Retired Servicemen Engineering Agency and Taipei Public Works Department, Taipei (in Chinese).

Paikowsky, S.G., Birgisson, B., McVay, M., Nguyen, T., Kuo, C., Baecher, G., Ayyub, B., Stenersen, K., O'Malley, K., Chernauskas, L., and O'Neill, M. (2004). Load and Resistance Factor Design (LRFD) for Deep Foundations, NCHRP Report 507, Transportation Research Board, Washington, 2004.

Phoon, K.K., Kulhawy, F.H., and Grigoriu, M.D. (2003) Development of a reliability-based design framework for transmission line structure foundations. *ASCE Journal of Geotechnical and Geoenvironmental Engineering*, 129(9): 798-806.

Phoon, K.K. (1995) *Reliability-based design of foundations for transmission line structures*, Ph.D. Dissertation, Cornell University.

Zhang, L.M., Tang, W.H., and Ng, C.W.W. (2001) Reliability of axially loaded driven pile groups, *ASCE Journal of Geotechnical and GeoEnvironmental Engineering*, 127(12), 1051-1060.

Field Surveys of Anchored Slopes after 2004 Niigataken Chuetsu Earthquake

Jiro Takemura: Tokyo Institute of Technology, JAPAN
Hiroaki Kubo: Japan Anchor Association, Tokyo, JAPAN
Junichi Yamazaki: Sanshin Corporation, Tokyo, JAPAN

Abstract

On October 23, 2004 an intense near-field earthquake of magnitude 6.8, shook mid Niigata Prefecture, one of the famous heavy snow fall and landslide prone areas in Japan. The maximum intensity of 7 on the 7-grade Japanese intensity scale was recorded. Due to such a strong motion and the potential vulnerability of slopes, more than 1600 slope failures occurred. Moments after and then two and half years after this Nigataken Chuetsu earthquake, field surveys were carried out for geotechnical structures. This paper describes a case study of the performance of the anchored slopes observed by the field surveys.

Introduction

On October 23, 2004 an intense near-field earthquake of magnitude 6.8, shook mid Niigata Prefecture, central Japan (Figure 1), one of the famous heavy snow fall and landslide prone areas in Japan. The maximum intensity of 7 on the 7-grade Japanese intensity scale was recorded. The main shock was followed by a series of strong aftershocks, including that of M6.5. The maximum acceleration of 1500 gal was recorded at Ojiya station, nearest site to the hypocenter. The acceleration was much greater than those recorded during 1995 Hyogoken Nanbu (Kobe) earthquake. From the Pseudo Velocity Response Spectra, it is known that the observed motions during Chuetsu earthquake were equivalent to those during Kobe earthquake. Due to such a strong motion and the potential vulnerability of slopes, more than 1600 slope failures occurred (Tatsuoka et al., 2005) and the volume of soil mass in slides was expected to be about seventy millions cubic meter (Kokusho, 2005).

The earth structures and slopes protected by ground anchors have shown good performance against earthquakes. Even in the past deadliest earthquake, like the Kobe earthquake, only small damage was observed in the anchored structures (JSCE, 1995). In the design standard for ground anchors (e.g., JGS, 2000), no specific consideration is given for earthquakes, which probably means that the current design standard implicitly secures the stability of the anchored structure during earthquake. However, because of the very strong motion of the Nigataken Chuetsu earthquake and the vulnerable geological conditions of the shaken area, it was considered that from study of the actual performance of anchored slopes in the area more concrete confidence might be provided on the current design standard or an onset of seismic design of ground anchors, if necessary. This report describes the damage of slope protection systems, including ground anchors, observed by the field surveys conducted just after, and two and half years after the earthquake. The restored slopes and earth structures, which had been heavily damaged, were also studied in the surveys.

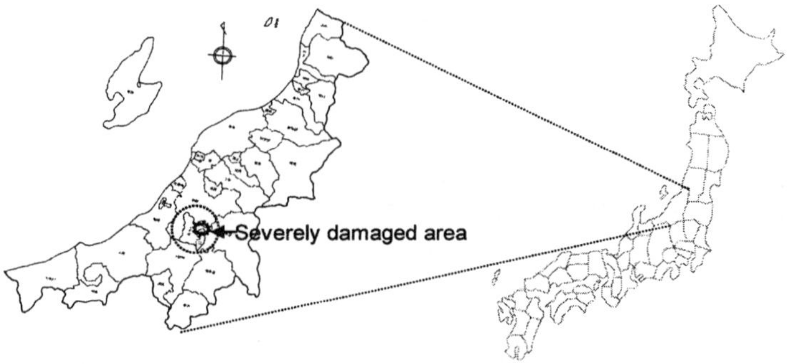

Figure 1. Niigata Prefecture and severely damaged area by 2004 Chuetsu earthquake.

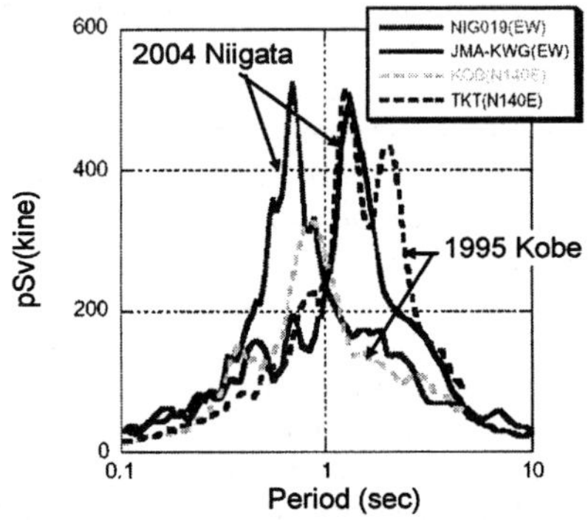

Figure 2. Pseudo Velocity Response Spectra (AFRC IAIST, 2004).

Geological conditions and slope failures

Topological and Geological conditions (Suzuki, 2005)

Slope failures triggered by the earthquake concentrated to Higashiyama hill, especially Yamakoshi village. In addition that this area covered the focal region, geological and topological conditions caused the serious damages in this area.

The damage area represents well known landside-prone areas of green tuff. Active tectonic movement creates folding structures, of which synclines and anticlines are running parallel to north-north east direction together with two active faults. The inclination angles of dip in the folds are typically 10° – 50° with 80° as a maximum. The slopes are composed of soft sedimentary rocks of late Tertiary and Pleistocene, which are alternate layers of sand and mudstones. Solidification of each layer is low with unconfined compression strength of 0.8 – 5.5 MPa. Depth of weathering observed from the exposed rock is from several cm to 1m. On top of the above mentioned geological conditions, fluvial erosion has created a variety of topological features. The slopes of hillside are quite steep ranging from 20° to 70°. Furthermore, a strong rainstorm by Typhoon No.23 hit the area just before the earthquake, which might have affected the damage of the slope failures.

Slope failures by the earthquake (Suzuki, 2005; Kokusho, 2005)

More than 1600 slope failures occurred as a result of the Niigata-ken Chuetsu earthquake and the aftershocks. Expected volume of the failed mass was 70,000,000 m^3, which included 10 large landslides with a sliding mass over 1,000,000m^3. The slope failures can be classified into three types: fall, slip and landslide. The failure types are closely related to the relative inclination between the slope and the dip.

The slopes with dip inclination angle (γ) smaller than 90° are generally called "dip slope" and those with γ greater than 90° "opposite slope" (Figures 3). Among the dip slopes, those with daylighting dip ($\gamma < \theta$, where θ is slope angle) is the most unstable against landslide. While opposite dip is considered to be stable against landslide.

Slips that took place in the opposite slopes and the slip depth were quite shallow, mostly 1-2m, which mainly consisted of vegetated soils and weathered layer removed from the intact layer. On the other hand, large landslides occurred in the dip slope with the sliding plane on the mudstone layer. Some slip planes of old landslides were also reactivated by the earthquake. Kokusho (2005) summarized the relationship between the slope angle, and the number of failed slopes and failure areas as shown in Figure 4. Except for very gentle slopes, it cannot be said that more failures occurred for the steeper slopes. As for the failure area, slopes with middle angles moved more than those with steep angles. These results can be attributed to the difference of failure types according to the slope angles. The dip slopes could hardly exist with steep angles, while the opposite slopes can stand with steep angles. The falls and slips, which mainly occurred at the steep slope, are relatively limited in local area compared to the landslides in the middle angle dip slope with large movement of soils and rocks. The largest landslide occurred on a dip slope with an original slope angle of 16° at Shiotani, Ojiya city. The sliding mass and area of this landslide were more than 10 million m^3 and 550m x 450m respectively. The largest number of failures occurred in slopes facing the south east direction and the least in the north to north east direction. Not only the direction of dominant earthquake motion, but also the other various factors, such as retaining structure, rock properties, and topological conditions are considered to affect the failure of slopes.

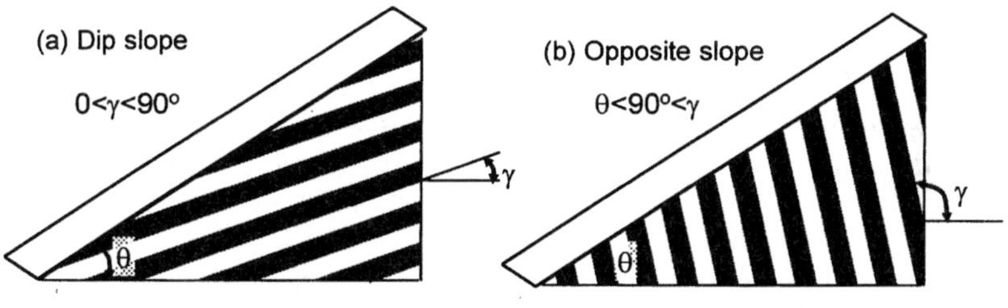

Figure 3. Relative inclination of slope and dip.

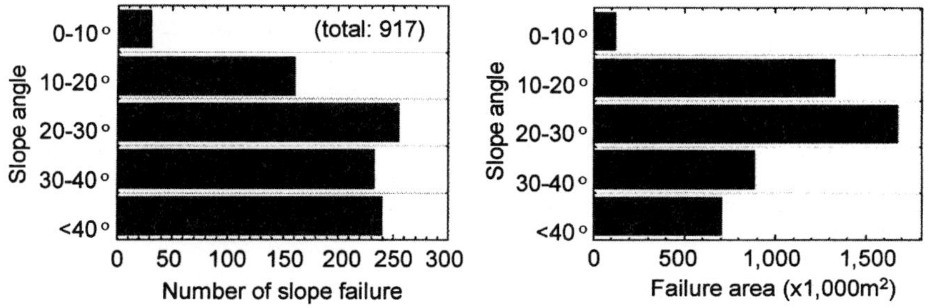

Figure 4. Relationships between slope angle, and number and area of slope failure.

Field observations

The authors conducted the field surveys of the damaged sites three times, moments after the earthquake when the restoration works had not been started and in May 2007, more than two and a half years after the earthquake when most restoration had been completed. Furthermore, just after the 2007 Niigataken Chuetsuoki earthquake, which hit the coastal area of Chuetsu by magnitude 6.8 on July 20, 2007, an additional survey was carried out to investigate the damage of structures restored after the 2004 earthquake.

In the first survey, the damage to slopes protected by various types of anchor and face supporting structures, such as ground anchors, rock bolts, grating cribwork, and pre-cast concrete plates was investigated. Restoration works for the slopes and earth structures, which had been heavily damaged by the earthquake, were studied in the second survey. Although the seismic intensity of the study area in the 2007 earthquake, grade 5+, is much smaller than that in the 2004 earthquake, grade 7, the second earthquake was considered to provide a good opportunity to confirm the performance of the various protection works of the damaged slopes and structures.

Figure 5 shows the locations of the survey sites. According to Kokusho (2005), the slopes within 8km from the hypocenter were intensively damaged in terms of both the number and area. The investigated sites are all close to the hypocenter or in the focal region as shown in the figure. Among the investigated sites, ground anchors had been applied before the earthquake at five sites. These anchors were installed to cut slopes for roads. Many of them seemed to have been done after the construction of the roads, which means that these slopes once became potentially unstable in the static condition. The damage of the slopes protected by the ground anchors was very limited and no visible damage was observed in the anchors, confirming a good seismic performance of the anchored slopes even against the worst conditions explained above. Observed damage of the slope and supporting structures is described and illustrated below.

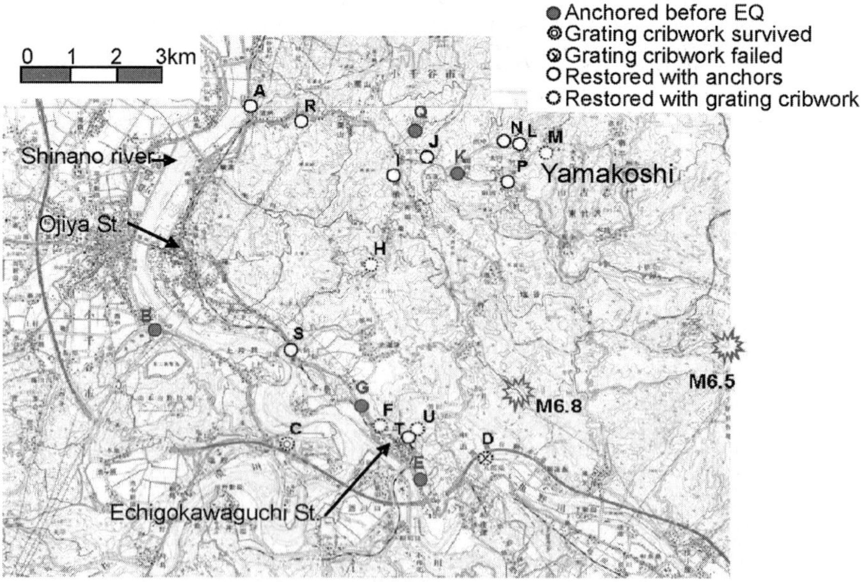

Figure 5. Locations of investigation and conditions of slope.

Performance of slopes with supporting structures

A failure of slope protected by grating cribwork with rock blots

Slip type failure occurred near the hypocenter of the mainshock (Location D). The slid slope was partly protected by grating cribwork with rock bolts just above a newly constructed gallery as shown in Figure 6(a). As the slope facing the south is an opposite slope, the sliding was limited to the shallow depth of the highly weathered soils, although it extended a quite large area. Therefore the snow gallery survived from the sliding with small damages.

Grating cribwork applied to a natural slope (Location C) which is steeper and higher than the failed one survived securely, showing very limited falls of vegetated soils from the slope. The difference in the damage of the two slopes could be explained by the differences in the intensity of ground motion and level of weathering of the rock surface below the cribwork.

Cut slopes protected by ground anchors

At the locations E and G which are west and southwest from the hypocenter respectively, ground anchors were installed to protect the cuts along National Highway 17 before the earthquake. Both cuts were originally supported by concrete block retaining walls as shown in Figures 7(a) and 8(a). At location E the anchors with grating cribwork were installed on the slope above the retaining wall (Figure 7(b)). While at location G the ground anchors were installed on the retaining wall and the slope above it (Figure 8(b)). It is considered that these anchoring works were conducted because of instability of the cuts in the static condition, such as lateral movement.

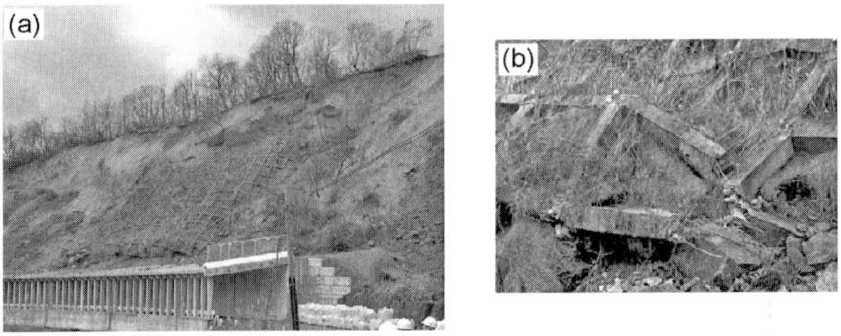

Figure 6. Surface slip failure of slope protected with grating cribwork with rock bolts. (Location D)

Figure 7. Cut slope protected by retaining wall and ground anchors. (Location E)

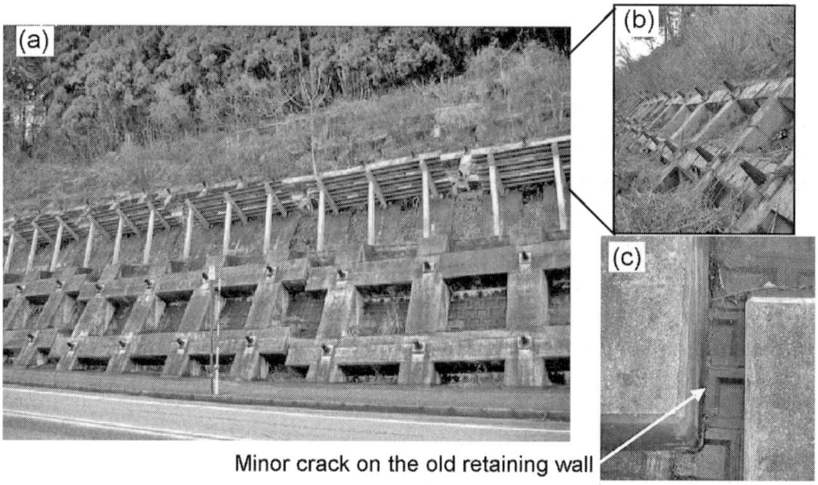

Figure 8. (a) (b) Retaining wall with grating cribwork and ground anchors. (Location G)

From the field observation, no evidence about lateral movement of the slope could be found. Visible damage of the slopes were cracks of the retaining wall behind the cast-in-place RC frame at Location G (Figure 8(c)) and small damage of a U-shape drainage concrete ditch (Figure 7(b)).

Figures 9 show the cut slope protected by ground anchors with grating cribwork at Location B. These anchors are installed to a corner of slope on which a transmission tower rests. The anchors were quite old (Figure 9(b)), which might not satisfy the current standard (JGS, 2000). Furthermore, the slope protected by the ground anchors moved laterally crushing the concrete ditch at its toe (Figure 9(c)). However, the anchors survived with sound conditions as shown in Figure 9(a).

A cut slope supported by ground anchors with PC facing plates (Location K) also survived from the earthquake as shown in Figure 10(a). This cut was originally protected by shotcrete and then the ground anchors were installed for the stability problems. Compression failure of the sidewalk at the toe of the slope is clear evidence that lateral movement of the entire slope took place by the earthquake. Because of the lateral movement, the PC plates were penetrated in the shotcrete. The observed performances at locations B and K imply that the anchored slope can secure the function even if it is subjected to relatively large deformation by earthquakes. This might prevent catastrophic failure of the slopes or retaining structures.

Restoration of damaged structures and slopes

For the restoration of the failed slopes and damaged earth structures, ground anchors with facing works have been adopted as one of the most reliable measures to secure their seismic stability. According to KTB Anchor Niigata Division, the companies in the division had conducted ground anchoring works at 69 sites for various purposes as the restoration works from the damages by the earthquake, such as prevention of landslides and steep slope failures, and reconstruction of failed roads and railways. The followings are the brief explanation at the sites which the authors visited in the second and third field survey.

Figure 9. (a), (b) Cut slope with grating cribwork and ground anchors; (c) crushed concrete ditch by lateral movement of slope. (Location B)

Figure 10. (a) Cut slope covered by shotcrete which was subsequently protected by PC facing plates with ground anchor; (b) compression failure of sidewalk due to lateral movement of the slope; (c) PC plate penetrated into the shotcrete. (Location K)

Restoration from massive rock fall (Location A)

One of the largest disasters caused by 2004 Niigataken Chuetsu earthquake was massive rock falls occurred in a hillside along Shinano river at Myoken, Ojiya city (Figure. 11 (a)). A road running between Shinano river and a steep cut slope fell down to the river with large rocks and debris.

As shown in Figure. 11(a), the route of the road was totally covered by huge rocks and debris, three routes with different structures were considered as the alternatives for the restoration. One is the original route with cut slopes by excavating the soils and rocks plus ground anchors. The other two are the routes avoiding the collapse area using either a tunnel in the hill or a bridge over the Shinano river. The first route was selected because of least cost and construction time. For the first route, the unstable rock mass and transmission towers on the hill required a deep and steep cut. To create the stable conditions under the geological conditions explained above, ground anchors with steel facing plate were adopted for the slope. The anchors were installed to the mudstone underling the potential slip plane (Figure 12). The lengths of the anchors are 7.0 - 46.8m and the total number of anchors is 830, the largest anchoring works among the restoration works. This construction was completed in a year, which is much shorter than the time expected for the other two routes.

Figure 11. (a) Massive rock falls at Myoken; (b) Cut slope supported by ground anchors with steel facing plate. (Location A)

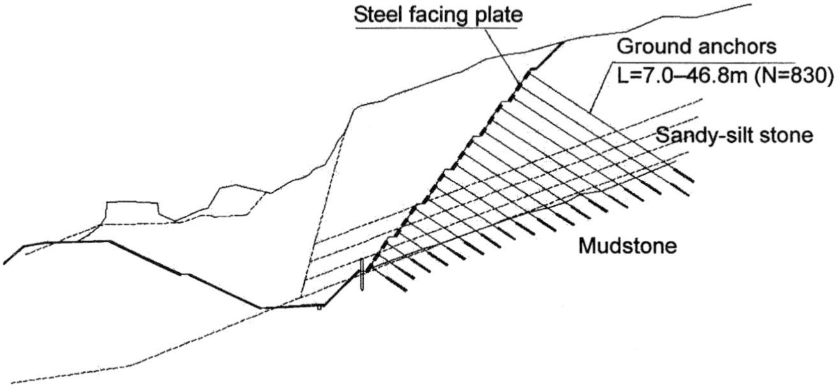

Figure 12. Cross section of anchored slope at Myoken(Location A)

Restoration from a failure of highly weathered mudstone slope (Location L)

A small height slope consisting of highly weathered mudstone collapsed by the earthquake. A new house survived at the shoulder of the collapsed slope as shown in Figure 13(a). The building could avoid the structural failure by a very rigid reinforced concrete made ground floor, which is common in this heavy snow fall area.

Ground anchors with grating cribwork were adopted for restoring the damaged slope. The anchors were installed to less weathered mudstone layer and long fiber reinforced soils were filled behind the cribwork to create an extra ground between the house and the slope as shown in Figure 14.

Protection of retaining wall on potential sliding plane (Location I)

Figure 15 shows a retaining wall made of large concrete blocks. This retaining wall did not collapse by the earthquake but was damaged due to the movement of the slope retained. As an emergency countermeasure against the instability of the walls and the landslide of the large slope containing the retaining wall and the road in front of the wall, ground anchors with PC facing plate were constructed as shown in Figure 15(a). The anchors were installed to soft mudstone below the critical slip plane which passes through the former sliding plane (Figure 16).

Figure 13. (a) Slope failure at Takezawa; (b) Restored slope by grating cribwork with ground anchors. (Location L)

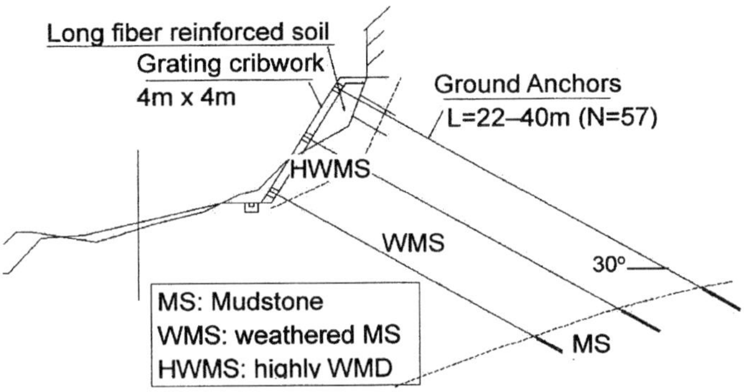

Figure 14. Cross section of anchored slope at Takezawa(Location L)

Figure 15. Restored block concrete retaining wall by ground anchors with PC facing plate at Minaminigoro. (Location I)

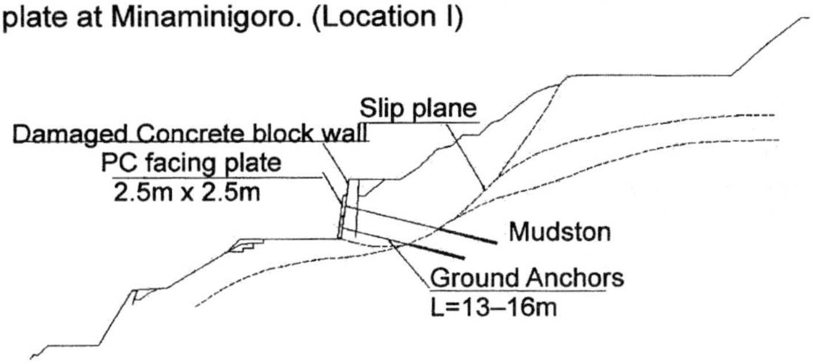

Figure 16. Cross section of restored retaining wall at Minaminigoro(Location)

Investigation of the restored slopes and structures by ground anchors after the 2007 Niigataken Chuetsuoki earthquake

After the second survey of the restoration works from 2004 Niigataken Chuetsu earthquake, the damage area was hit by the 2007 Niigataken Chuetsuoki earthquake. The magnitude of the 2007 earthquake was 6.8, the same as the 2004 earthquake, but the hypocenter was further from the survey sites than the former one. Therefore the seismic intensity of the study area during the 2007 earthquake, grade 5+, is much smaller than that during the 2004 earthquake, grade 7.

In the third survey no visible damage was observed on all structures and slopes newly protected by the ground anchors. The steep slopes protected by grating cribwork with rock bolts also worked well against the 2007 earthquake without showing any damage. Although the intensity is comparably small, it can be confirmed that the ground anchored slopes and retaining structures have well performed against seismic loadings.

Conclusions

Damage of the slope caused by earthquakes is influenced by various factors, such as geometrical conditions, geological conditions, and characteristics of ground motion. However, the types of protection works were found to be a crucial factor for the seismic stability of slopes. From the investigation of slopes and retaining structures in the area severely damaged by the 2004 Niigataken Chuetsu earthquake, it can be confirmed that those protected by the ground anchors worked well against one of the largest earthquake motions in Japan in additional vulnerable conditions, such as high potential of landslide in dip slopes and heavy rainfall just before the shake. The slope surface protection against weathering is probably a rational countermeasure to prevent the fall type failure or the shallow slip of the opposite slopes, which are normally steeper than the dip slope.

References

Active Fault Research Center, National Institute of Advanced Industrial Science (AFRC IAIST)(2004): Quick Report of the Oct. 23, Chuetsu Earthquake.

Japan Geotechnical Society (2000): JGS Standard "Design and Construction of Ground Anchorages and Commentary (JGS4101.2000) (in Japanese).

JSCE(2005) Damage of Steep Slope Control Facilities, Damage of Civil Engineering Structure, Report on the Hanshin-Awaji Earthquake Disaster, pp.534534 (in Japanese).

Kokusho, G.(2005): 3.2 Slope Failure in Chapter 3 on Slope failure, Report on Niigata-ken Chuetsu Earthquake. (in Japanese).

Kokusho, G. Ishizaa, T. and Yoshikawa Y. (2005): Influence factor of the slope failures during Niigata-Chuetsu earthquake, 40th Annual Convention of Japanese Geotechnical Engineering Society, pp.2131-2132 (in Japanese).

KTB Anchor Niigata Division (2007) Photo Archives about the Restoration from the Damages by Niigata-ken Chuetsu Earthquake (in Japanese).

Suzuki, R.(2005): 3.1 Relation of topological and geological conditions to slope failures in Chapter 3 on Slope failure, Report on Niigata-ken Chuetsu Earthquake. (in Japanese).

Tasuoka, F., Konagai, K., Kokusho, T., Koseki, J. & Miyajima, M. (2005): Special session on the 2004 Niigata-ken Chuetsu Earthquake, 16th ICSMGE, Vol.5, pp.3279-3287, Osaka.

Rehabilitation of a Failing Anchored Retaining Wall

Thomas Richards, Nicholson Construction Company, Cuddy, PA USA
Daniel Thome, Nicholson Construction Company, Kalamazoo, MI USA

Abstract
Significant movement to an existing diaphragm (slurry) wall on 'O' Street in Southeast Washington, D.C. was caused by failure of some anchors. Potential causes of failure are presented. Rehabilitation and stabilization for the failed wall was performed by demolishing the failed panels, drilling horizontal weep holes, constructing a new precast lagging and soldier pile wall in front of the existing structure, and improving site drainage system.

Background
In the late 1970s the District of Columbia decided that permanent stabilization of the slope was required due to the continuous slope movements toward 'O' Street. In 1978 an anchored slurry wall was constructed between 'O' Street and Highwood Drive to retain the earth from further movements.

The anchored slurry wall consisted of fifty, 6.7 meter wide reinforced concrete panels designed to be embedded into a hard clay stratum. High strength threaded bar anchors were installed through the slurry wall and designed to be bonded into the hard clay. During construction the bearing plates to the ground anchors were reported to have been visually showing failure after they were stressed. Eventually the initial bearing plates were replaced with larger ones soon after construction. Corrosion protection was a tar like coating applied to the bearing plates and beveled nuts, and anchor heads after they were installed and stressed.

In the winter of 1995 excessive ground movements were noted in approximately 46 meter of the slurry wall. Ground movements were recorded on this particular section of the wall and continued until the spring of 1996 when eventually this section failed. Maximum deflections in the order of 4 meters downslope were recorded. Through time the ground anchors became overloaded and failed causing a global stability failure with a distinct slide plane reaching completely under the toe of the slurry wall.

Photo 1 Overview Distressed Section of Wall

Photo 2 Close-up of Leaning Panels (wall in distance is vertical)

An investigation (Thomas L. Brown Associates, P.C, 1999) of the wall determined that the global failure was caused by three contributing factors:

1. Improper drainage from behind the wall is believed to be the major cause of the stability failure due to the additional hydrostatic pressure acting on the slurry wall
2. Zones of highly plastic slickensided clay below the toe of the wall

3. Lack of corrosion protection on the initial thread bars used as tieback supports.
Conclusions of the study suggested that while each defect in the wall is believed to be a contributing factor in the failure, the combination of all three factors acting simultaneously probably initiated the global failure. Improper drainage from behind the wall increased the earth pressures acting on the wall leading to overload the tiebacks, failing in part due to the lack of sufficient corrosion protection around the ground anchors. Finally, subsurface investigations concluded that the toe of the slurry wall was actually embedded in a stratum of slickensided clay, which created a less resistant path for the failure plane.

Condition of Old Anchors

Bar anchors were installed in 1978. As noted above, the bearing plates and nuts reportedly exhibited problems during testing. By 1999, several of the anchors had failed causing the wall to lean significantly. Detailed condition assessment of the anchors was not a component of our work, since the rehabilitation did not rely on any of the existing anchors.

As part of demolition of leaning wall panels, several anchors were exhumed. The breaks were typically within the wall or just behind the wall as shown in Photos 3 and 4. The pipe was a typical anchor sleeve in the diaphragm wall.

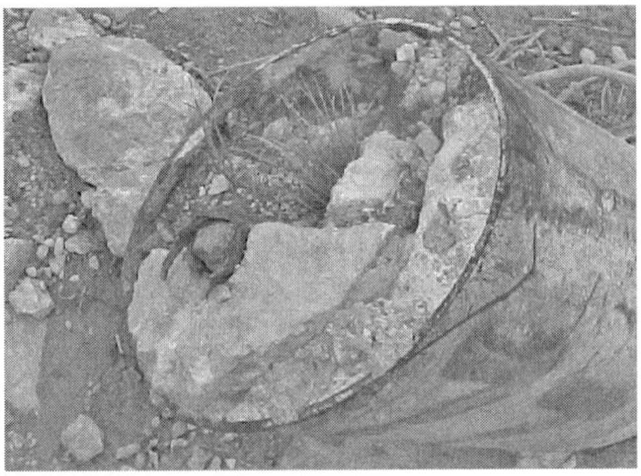

Photo 3 Breaks in Anchors behind Wall

Photo 4 Breaks in Anchors behind Wall

Grouting of the sleeves seemed to have some deficiencies as shown in Photos 3 and 4 and also several of the other anchors which had seepage and grass growing around the bearing plates. It is also unclear how access for secondary grouting was provided.

The broken bar had pitting corrosion and had been protected by plastic sheath as shown in Photo 5. It did not appear that grease had been used or remained inside the plastic sheath.

Water could then get to the bars through the incomplete secondary grouting and then travel down the bar. However, the bars would then have been expected to fail near the top. This suggests that shear loads may have also been present at the back face of wall to ground interface due to possible filling behind the wall or lateral wall movement allowing downward soil movement behind the wall.

Photo 5 Close-up of Failed Bar

Several other anchors were found to have broken and popped out of the wall as shown in Photos 6 to 9.

Photo 6 Popped Anchor note gap between nut and washer and between bearing plates

Photo 7 Missing Anchor and Anchorage

Photo 8 Popped Anchor – note gap between washer and bearing plate and between bearing plates

Photo 9 Popped Anchor

In general, the tar performed well in protecting the bearing plates and nuts, which had less corrosion than the photos above.

Most Aggressive Design Section near Failures
The area of the wall with most of the anchor failures was judged by the authors to be the most aggressive design section. Photos 1 and 10 show that the most distressed section of wall was just beyond a transition from two rows of anchors to one row. The elevation of the ground near houses behind the wall was similar and a significant slope (4.5m high at 2H:1V) existed behind the failed section. This slope and change in top of wall elevation also created a significant water runoff area focused on the failed area with one catch basin inlet. The catch basin inlets were poorly maintained and many were clogged with debris frequently leaves and vines. After failure of anchors in the one row section (right of Photo 10), prying action of the continuous cap beam then would have lead to overstressing of two row section.

Photo 10 Existing Wall Elevation (left of Photo 1)

Probable Cause of Failures
The failure of the anchors is judged by the authors to have been caused by a combination of:

1. Buildup of water pressure due to clogging of weepholes in the existing wall

increasing the load on the anchors

2. Most aggressive design section with a backslope surcharge
3. Movements of the wall in a landslide prone area with questionable toe support
4. Corrosion resulting from 1 and 3

Stabilization of the Existing Wall

The section presents the construction of the new wall.

Demolition of Leaning Panels

For safety and access to build a smooth new wall line, approximately 45 meter length of the failed slurry wall was demolished down to the ground line behind the wall up to 3 meters below the top of the wall. The ground line was below the top of wall due to the considerable lateral wall movement.

Weepholes

Nine weepholes were installed through the downhill face of each panel to relieve water pressures built up behind the existing slurry wall. Each weephole consisted of a 38 mm diameter well screen covered with a filter fabric sock to prevent any loss of ground from occurring behind the wall. These holes were drilled through the existing slurry wall and continued approximately 0.5 meters into the soil in order to promote a proper drainage path. The original weepholes were simply plastic pipe through the wall apparently with filter protection.

Test Anchor Program

The anchors were founded in hard clay below the slickensided zones. A total of eight test anchors each with six strands were installed with differing drilling techniques and casing size. Post grout tubes were also installed along the length of the additional anchors with rubber sleeves covering holes with in the bond zone. We found that the clay was severely softened using water or significant amounts for dust suppression. We chose air as a flush fluid to minimize mess in a residential sloping site and minimal water for dust suppression.

New Wall Construction

A new steel soldier pile and precast lagging wall was constructed in front of the slurry wall. Precast wales were used to connect the anchors to the soldier piles.

The detail section is shown Figure 1.

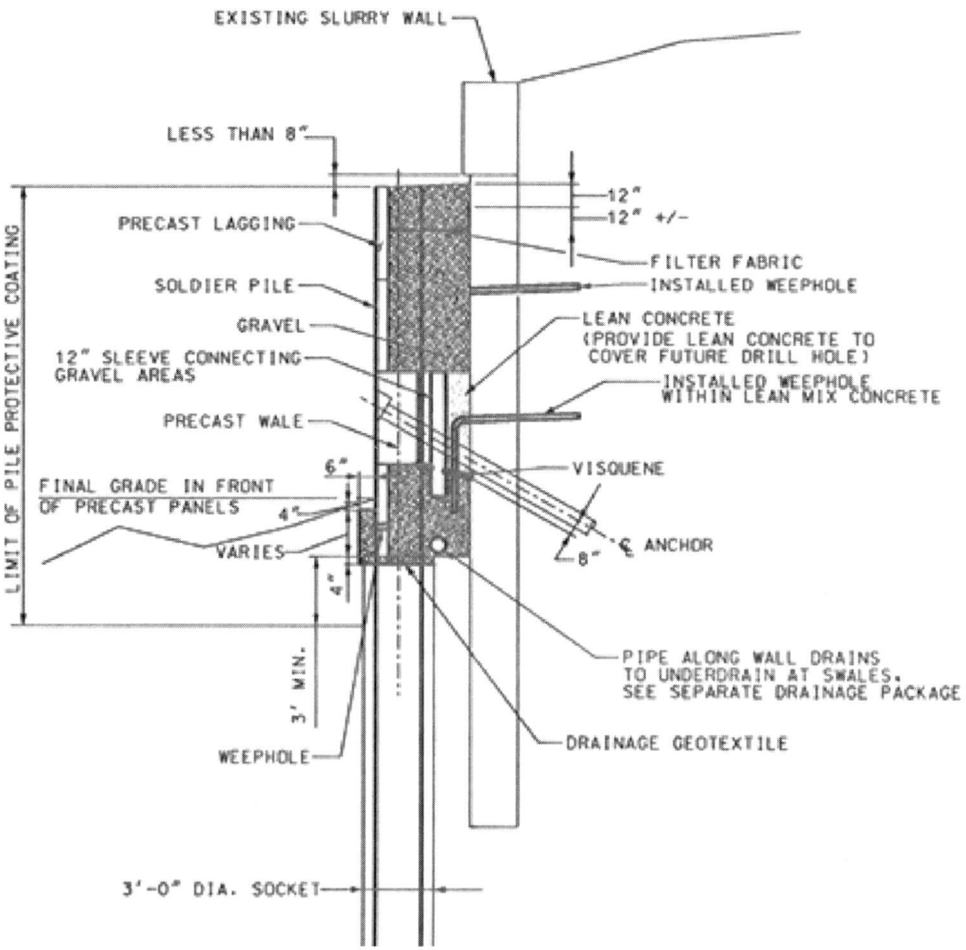

Figure 1 Rehabilitation Typical Design Section

Soldier piles were placed in 0.9 meter predrilled holes. To reduce risk of any slurry wall movement some shafts were skipped and then drilled on another shift. Slurry wall panels were monitored with a total station approximately every ten minutes during drilled shaft installation by shooting prisms mounted to the slurry wall. Readings continued on the wall every half an hour until the soldier piles were placed in the drilled shafts and filled with structural concrete.

Due to the precast lagging and especially the precast wales, accurate soldier beam placement was required. 10 meter long templates were made that would support the beams in the drilled shafts prior to the initial set of the structural concrete.

Prior to setting the concrete waler, existing anchor locations on the slurry wall were compared to proposed anchors through the new wall. If any of the new anchors were expected to hit an existing anchor bearing plate, the waler would be slid to one side if clearance of the bearing plate could be accomplished. As a last resort a portion of the bearing plate would be burned off to clear the way for the path of the new ground anchor. Nonshrink grout was poured in the space between the soldier pile and waler.

Concrete was placed between the precast wale and the slurry wall to:

- reduce deflection in the new wall during anchor stressing operations by transferring some of the load in the new wall to the existing slurry wall and
- provide a conduit so that drill cutting would not contaminate the gravel drainage layer behind the wall.

PVC sleeves were placed prior to concrete to connect the two gravel layers above and below the waler concrete.

Ground Anchor Drilling and Installation

Photo 11 Drilling of New Anchors

A total of 262 anchors were installed with the following details:

- Three to six strands per tendon
- Strand – ASTM A416 15.2mm diameter 261 kN ultimate strength
- Full length corrugated HDPE sheath - 90 mm OD, 76 mm ID, 1.5mm thick
- Free lengths of 16.8 m with grease and extruded sheath on individual strands

Number of Strands	Working Load (kN)	Bond Length (m)
3	605	10.7
4	818	12.2
6	1063	15.2

Duplex drilling with air was used for the majority of the taller end and through the failed section of the wall. For the remainder of the project, duplex drilling was performed for the upper 10 to 16 meters through the softer saturated clay and sand lenses and then the hard clay was drilled open hole. Surface casing was longer on the eastern side of the site because of the deeper saturated sandy silt found below the wall.

For anchors installed in an open hole, great care was taken to clean out the hole prior to anchor installation. PVC centralizers, 127 mm in diameter, were used to keep the corrugated corrosion protection sheath from dragging through the soil and to concentrate the anchor in the center of the hole when installed. These centralizers were placed on the corrugated plastic sheath at 1.5 meter spacings throughout the bond zone. Results showed clean holes with good air return when casing went beyond the problem zone.

Anchors (six strand and about half of the four strand) in the fail zone and tall wall sections were post grouted with approximately six bags per tube. Post grouting ceased after operations went out of the failed portion of the wall. Ground anchors out of the failed zone were only post grouted when the stressing criteria were not met.

The grout mix was neat cement grout with Type I (ASTM C150) cement. The water to cement ratio was 0.45 for initial grouting and 0.67 for post grouting.

Each anchor was tested to 133% of design load and accepted according to the Post-Tensioning Institute, "Recommendations for Prestressed Rock and Soil Anchors" (1996).

Secondary Grouting and Closure Panel Installation

Anchor tails from stressing were removed once the anchor data was proved to conform to the stressing criteria. Approximately two bags of grout were then placed in each anchor head to flush any remaining water existing in the hole. This was done to prevent any water from coming in contact with the anchor in the future. If particular anchors were found to be seeping water they were pressure grouted to remove the excess water and seal the anchor head with grout.

A 50 mm thick closure panel, approximately 2 feet high by 5 feet long, was placed over the two anchors in the recessed portion of the waler. The closure panel was temporarily braced from the front while grout was pumped from holes placed on top of the waler. Once the grout cured the closure panel was held in place by a metal loop previously precast into the panel and grouted in during closure panel installation.

Photo 12 Precast Cover Panel Installation and Grouting

Drainage

Water from in between the existing slurry wall and the new precast wall had to be diverted once it went through the gravel backfill. A wall "French" drain was installed approximately 18 inches in front of the existing precast lagging wall. Excavation for the drain occurred directly in front of the wrapped filter fabric set prior to precast panel installation.

Downhill of the wall, surface drainage swales and numerous new french drains were installed to increase the probability of intersecting the groundwater seepage running from underneath the precast lagging retaining wall.

Conclusions

Failure of some permanent bar anchors led to an extensive rehabilitation of the wall. The failure is considered to have been caused by a combination of

1. Buildup of water pressure due to clogging of weepholes in the existing wall increasing the load on the anchors
2. Most aggressive design section with a backslope surcharge
3. Movements of the wall in a landslide prone area with questionable toe support
4. Corrosion resulting from 1 and 3

The finished wall provided much improved stability and more pleasing aesthetics.

Photo 13 Finished Wall

References

Thomas L. Brown Associates, P.C., "Pre-Design Studies 'O' Street Retaining Wall and Soil Movements", Prepared for Department of Public Works District of Columbia, February 1999

Thome, Richards, Wise, " 'O' Street Anchored Precast Lagging and Soldier Pile Retaining Wall" Contemporary Solutions to Land Mass Stabilization, IX Great Lakes Geotechnical Geoenvironmental Conference, Dayton, Ohio, May 11, 2001

Predicted and Observed Performance of an Anchored Retaining Wall in Granite

Abid Adekunte, Parsons Brinckerhoff Limited, Dublin Ireland
Douglas Madsen, Parsons Brinckerhoff Limited, Godalming U.K
Neil Margetson, Parsons Brinckerhoff Limited, Birmingham U.K
Liam Quinn, Quinn Piling Limited, Downpatrick, Northern Ireland

Abstract

The results of suitability & acceptance tests carried out on tie – back ground anchors for a deep excavation in Dublin are presented. Excavation was made through firm to stiff gravelly clay and weathered granite. The excavation has been supported by 750mm diameter hard – soft secant pile wall rock – socketed into intact granite and tied – back with a single level of ground anchors at capping beam level.

Anchor test results are back – analysed to estimate the apparent free lengths of the anchors, these are compared with original design estimates. Monitored wall deflections at various stages in the excavation process are also compared with design predictions.

Measured results are discovered to be lower than design estimates. Existing design approaches are shown to be quite conservative. It is recommended that wherever possible, designers should place greater reliance on data from relevant local case histories in similar ground conditions.

Introduction

The use of tied-back embedded retaining walls has been a widely adopted practice in civil & geotechnical engineering. While design recommendations and guidelines largely abound in the literature, they are somewhat considered to be only general indicative guides. Experience and knowledge of local ground conditions is thought to be the most valuable tool to a safe, effective and economical design.

This paper is centred on the design, construction, testing and monitoring of the deep basement for a commercial development in Dublin, Ireland. Due to the requirement to provide groundwater cut-off in addition to retaining an excavation of 4.5m-7m depth, temporary works for the basement comprised of a secant pile wall rock-socketed into granite bedrock for effective seal against groundwater ingress. Areas of the basement with retained height $H \leq 5$m were designed as cantilevered wall sections, while areas with $H > 5$m were designed to be tied–back at capping beam level with ground anchors inclined at 45^0.

In recent years, many case histories of tied-back excavation supports have been published in the literature e.g. Finno & Roboski (2004), Symons et al. (1987), Clough & Tsui (1984) and Mulligan (1983). However, most of these have concentrated on overall stability & serviceability performance of the walls, as well as soil deformation around the excavations, with little reference to the performance of the ground anchors. Detailed studies on the monitored behaviour of tie-back ground anchors in service have been scarce in the literature in recent years. While suitability and acceptance tests are often carried out on installed anchors on projects, results are seldom shared through publications.

In the current study, greater emphasis has been placed on the monitored behaviour of the tie-back ground anchors. The results of the suitability tests on some of the installed anchors are presented, these are back-analysed and related to design predictions. Construction stages are related with monitored inclinometer readings of wall deflection, while comparisons are made with estimates of the pseudo – finite element model used for the design.

Local Geology

The development is located on the southern end of Dublin, in the foothills of the Dublin mountains. Demolition of the previously existing public house on the site left a legacy of abandoned shallow foundation systems and buried services on the site. While the site terrain is relatively flat, topographical survey shows rapidly rising ground levels to the rear of the site. Site geology chiefly comprises of superficial glacial drift deposits underlain by granite bedrock.

Site stratigraphy includes 1m of made ground overlying a layer of firm to stiff gravelly clay, which extends to 2m depth. Below this stratum is a 2m thick layer of medium dense to dense silty sandy gravel, which is underlain by highly to moderately weathered granite. The highly to moderately weathered granite overlies the granite bedrock, with rock head varying between 7m – 10m across the site. Figure 1 shows the generalised site stratigraphy in relation to the support system. Static groundwater level at the site is 0.5m bgl.

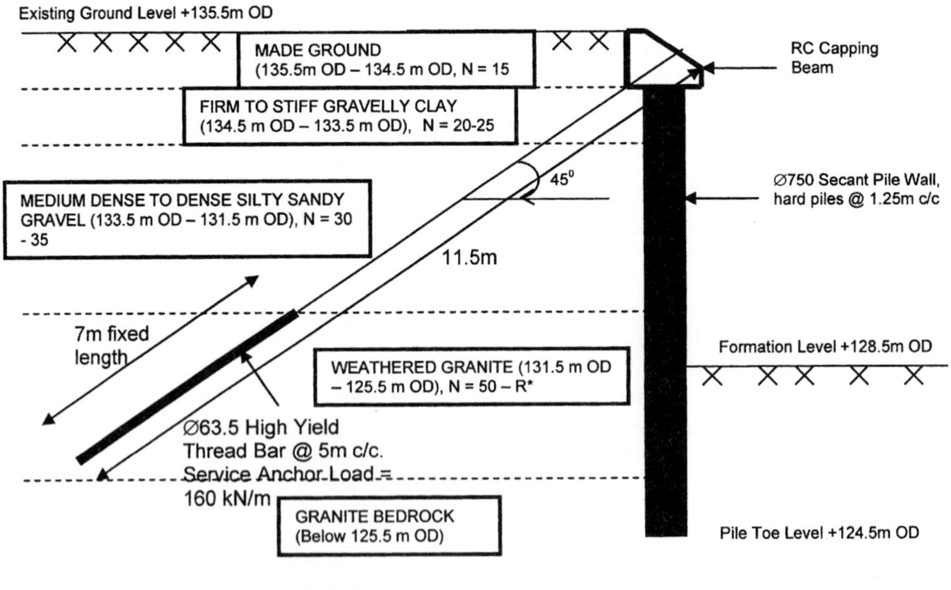

N – Standard Penetration Test N value, R* - Refusal.

Figure 1 – Site stratigraphy in relation to excavation support system.

The firm to stiff gravelly clay is of low to intermediate plasticity with liquid limit ranging between 23% - 43%, while plastic limit varies between 16% - 24%. Clay content in the gravelly clay varies between 5% - 7%. The bedrock is typically strong to moderately strong crystalline fine to medium grained granite. Rotary cores show this to be fresh to locally slightly to moderately weathered, with RQD values of between 48% to 75%. Unconfined compressive strength in the bedrock varies between 40 MPa – 100MPa. Parameters adopted in design are summarised in table 1. It must be noted that coefficients of earth pressure K_o used in design (and shown in table 1) are lower than actual measured values, so as to account for piling – induced stress relief in the ground.

Table 1 – Soil parameters adopted in design.

STRATUM	γ (kN/m^3)	E (kPa)	ϕ' (degrees)	C_u (kPa)	K_o	τ_{ult} (N/mm^2)
Made Ground	18.0	30,000	30.0	-	0.5	-
Firm to stiff gravelly clay	19.0	40,000	30.0*	-	0.5	-
Medium dense to dense silty sandy gravel	20.0	60,000	37.0	-	1.0	-
Weathered Granite	22.0	110,000	41.0	-	1.0	0.7
Intact Granite	25.0	300,000	-	5,000	1.0	-

* - analysis based on drained parameters.

Proposed Development

A large – scale hotel development is being undertaken in South Dublin. The proposed multi – storey development will incorporate the construction of a double–level basement over the entire building foot print for car-parking and storage purposes. The site is bounded on the north & east sides by existing roads and residential buildings. Open grasslands and old stone walls make up the south and west boundaries. Maximum excavation depth for the basement is 7m (6.5m below the static groundwater level on the site), while excavations for the foundations were locally made in the weathered granite to a maximum depth of 1m below basement formation level.

Since excavation was to be made below the existing groundwater level on the site, a secant pile wall was proposed as the most economical and effective soil retention/groundwater cut-off solution for the site. It is anticipated that the lower basement slab will be subject to hydrostatic uplift after construction and hence, it was proposed to install permanent anti-uplift anchors grouted into the underlying intact granite bedrock, to guide against flotation.

The piling and ground anchorage sub-contract for the project was awarded to Quinn Piling Limited on a design & build basis. Parsons Brinckerhoff (Ireland) Limited were appointed by Quinn Piling to provide consultancy services for the design aspect of the piling and ground anchorage sub-contract.

Construction Procedure

The construction of the basement involved 7 major stages as defined in table 2. Similar steps have also been followed in the analysis and design of the soil support system.

Table 2 – Stages in basement design & construction

Stage No.	Description of works
1	Initial conditions; in-situ soil properties and relevant surcharge loads
2	Install guide frames and hard & soft piles
3	Excavate to 1.5m depth and construct RC capping beam
4	In sections where proposed retained height H ≥ 5m, install and prestress tie-back anchors
5	Complete excavation to basement formation level
6	Construct lower basement slab and permanent RC retaining wall
7	Construct upper basement slab and permanent RC retaining wall

Secant Pile Wall Construction

Piles were constructed using a bored cast-in-situ method. Pile length varies from 9m in the cantilevered section to 11m in the anchored section. Prior to piling, guide frames were put in place to ensure the accurate positioning of piles in accordance with proposed basement layout. Drilling in the overburden layers was done with a Ø750 Casagrande B250 rotary rig. A smaller diameter (Ø600) DTH hammer was used for rock open-holing below the overburden layers. With this approach, the required maximum rock-socket of 3m was achieved. Hard piles were spaced at 1.25m c/c. These were reinforced and concreted with grade C35 concrete, while soft piles were concreted with grade C5 concrete without steel reinforcement.

Pile installation followed the following sequence:

- Install every second soft pile (this is to limit disturbance to fresh concrete).
- After 24 hours, install intermediate soft piles and allow a minimum of 24 hours for concrete to set.
- Install every second hard piles to interlock the soft piles
- After 24 hours, install intermediate hard piles.

Excavation & Tie-Back Anchor Installation

Excavation was carried out in two stages. An initial excavation to 1.5m depth was done to allow for the construction of the RC capping beam and the tie-back anchors. Tie-back system comprised of 11.5m length Ø63.5 high yield thread bars installed at 45^0 angles through the capping beam at 5m intervals along the wall. Anchor installation was done with a Casagrande M9 anchoring rig (see figure 2). Figure 3 shows a typical anchorage system after completion, comprising of the grouted tendon, bearing plate and the load bearing threaded nut.

Stages involved in the construction of the anchorage system are summarised below:

- Prebore and install Ø168 steel casing.
- Advance Ø168 borehole to 11.5m length
- Install Ø63.5 high yield thread bar over full length of borehole
- Place OPC grout over full length of borehole (water/cement ratio = 0.4)
- Remove casing and leave grout in place for minimum of 24 hours
- Install 250mm square bearing plate and load bearing nut
- Carry out suitability & acceptance tests and lock off at pre-determined prestress level (50% of anchor safe working load i.e. 400 kN).

Installation and testing of the tie-back anchors was followed by excavation to the required formation level and subsequent construction of the basement slabs & permanent RC walls.

Figure 4 shows the secant pile wall after completion of excavation, while figure 5 shows the construction of the lower basement slab.

Figure 2 – Installation of tie-back anchors

Figure 3 – Completed anchorage

Figure 4 – Wall after complete excavation.

Figure 5– Lower basement construction

Design of Tie-Back Anchors

Prior to the detailed design of the tie-back anchors, overall stability and serviceability analysis of the secant pile wall was carried out. In the overall stability analysis, an ultimate limit state approach was adopted, using factored soil parameters and loads, to estimate the required depth of embedment of the wall for overall stability. Total and effective stress parameters presented in table 1 were reduced with factors of safety of 1.5 and 1.2 respectively, in accordance with CIRIA Report 104 (1984). Analysis has been done with 'Oasys STAWAL' finite difference modelling software. In the serviceability analysis, unfactored soil parameters as presented in table 1 were used to estimate the service anchor loads, bending moments, shear forces and lateral deflection of the wall. This has been done with 'Oasys FREW' pseudo-finite element modelling software.

Anchor Design Methodology

The 'Oasys FREW' pseudo-finite element modelling software estimates the maximum service load on the tie-back anchors to be 160 kN/m run of wall. The piling contractor preferred an anchor spacing of 5.0m c/c, therefore each anchor had to be designed to support a minimum safe working load T of 800 kN. The empirical method of stability analysis shown in figure 6 has been used to determine the location of the uppermost fixed anchor zone, assuming a planar surface.

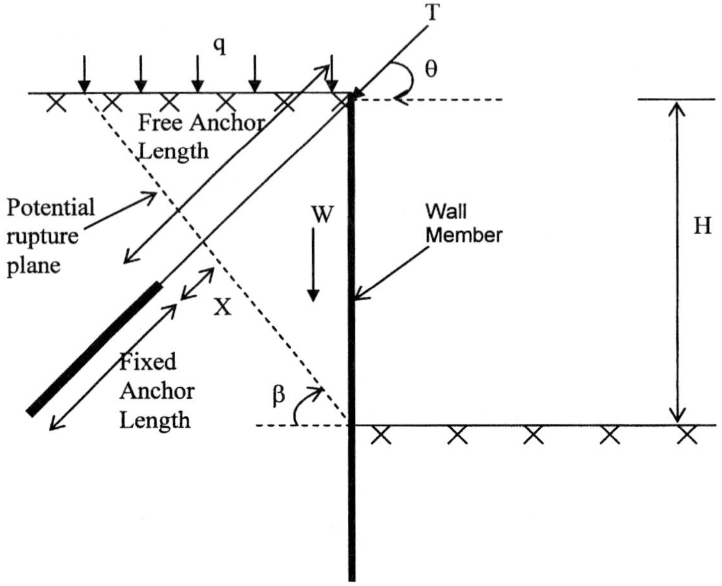

Figure 6 – Wedge method of stability analysis.

In figure 6, β is the angle of inclination of the potential rupture plane, θ is the angle of inclination of the anchorage force, H is the depth of excavation, q is the surcharge and X is the distance between the top end of the fixed anchor and the point where the free anchor intersects the assumed rupture plane. β is estimated from the expression;

$$\beta = 45 - \phi'/2 \qquad (1)$$

Typically, many designers tend to conservatively adopt a value of $\beta = 45^0$ for the rupture plane (e.g. Mansur & Alizadeth (1970), Clough & Tsui (1984) and Booth (1966)). Adopting a similar approach, $\beta = 45^0$ was assumed in design, anchorage angle $\theta = 45^0$, maximum excavation depth H = 7m. Assuming a design fixed length L_2 of 7m in the weathered granite (after Ostermayer (1974), BS 8081 (1989)), analysis based on figure 6 gives a corresponding design free length L_1 = 4.5m, X = 0.7m and a total anchorage length L = 11.5m. A typical cross – section of the anchored retaining wall has been presented in figure 1.

The following design checks were carried out on the anchorage system:

- Tendon-Grout Bond Capacity; for an assumed bar/grout bond stress of 2.0 N/mm^2 and factor of safety F = 2.5 (after BS8081, 1989) on a Ø63.5 tendon, design was found to be satisfactory.
- Grout-Ground Bond Capacity; BS8081 (1989) recommends that ultimate skin friction τ_{ult} could be estimated from the expression;

$\tau_{ult} = 10\% - 30\% \times f_{cu}$ --------------------------- (2)

where f_{cu} is the characteristic compressive strength of rock. Taking f_{cu} in the highly to moderately weathered granite stratum to be 10 MPa, corresponding 10% f_{cu} = 1.00 N/mm^2. Suzuki et al. (1972) recommend τ_{ult} = 1.5 – 2.5 N/mm^2 for weathered granite, while Koch (1972) suggests τ_{ult} = 0.35 – 0.7 N/mm^2 for weak rocks. Accordingly, τ_{ult} = 0.7 N/mm^2 has been adopted in the design of the anchors on the current project. For an assumed τ_{ult} of 0.7 N/mm^2, ∅168 borehole, 7m fixed anchor length in the weathered granite and a factor of safety F = 2.5 (after BS8081, 1989), design was found to be satisfactory.

- Tendon Capacity; based on a factor of safety F = 1.6 for temporary anchors (after BS8081, 1989), the most economical and readily available grade of solid thread bar (∅63.5, 800kN SWL) was recommended for construction.

Anchor Testing

All installed anchors were tested in accordance with clause 11.3 of the BS8081 (1989). The programme of tests comprised of:

- On-site suitability tests; carried out on the first 6 No. anchors installed on site, to verify the suitability of the ground anchors for local conditions on site. Anchors were loaded to 1.5 times their safe working load (SWL) in 3 cycles.
- On-site acceptance tests; conducted in accordance with clause 11.4 of the BS8081 (1989) on all installed anchors. This involved loading the anchors to 1.5 times their safe working load (SWL) in 2 cycles.

Loads were applied with a hydraulic jack at a loading rate of 40 kN/min. For each cycle, load was increased in stages to 10%, 50%, 100% and 150% of the SWL. Subsequent unloading is done in similar stages. Measurements of tendon displacements were made with LVDTs attached to the surface of the tendon. A multi-channel data logger was used to acquire output data from the LVDTs at 1 second intervals. Each stage loading was held for at least 1 minute, while measurements of displacements were recorded at the start and end of each period. However, each peak cycle load was held for 15 minutes, with an intermediate displacement reading taken at 5 minutes.

Experimental Results & Discussion

Anchor Test Results

Measurements of load and displacement from the suitability tests can be illustrated with load-displacement curves. Load displacement curves for the test on anchor No. 13 is presented in figure 7. The figure shows the load-displacement response of the anchor during 3 cycles of loading. After each cycle, the anchor is seen to exhibit stiffer responses on reloading. This may be attributed to cyclic preloading, which causes the consolidation of the surrounding material and mobilizes a longer fixed length. Plastic deformation d_p increases with increasing number of cycles; after the first cycle, d_p = 0.2mm and this increased to 0.32mm and 0.38mm after the 2nd and 3rd cycles respectively. Also, change in plastic deformation Δd_p (which is a measure of the difference between d_p at the end of a given load cycle and d_p recorded at the end of the preceding cycle) is observed to reduce with increasing number of cycles. Δd_p = 0.12mm at the end of the 2nd cycle, this reduced to 0.06mm at the end of the 3rd cycle.

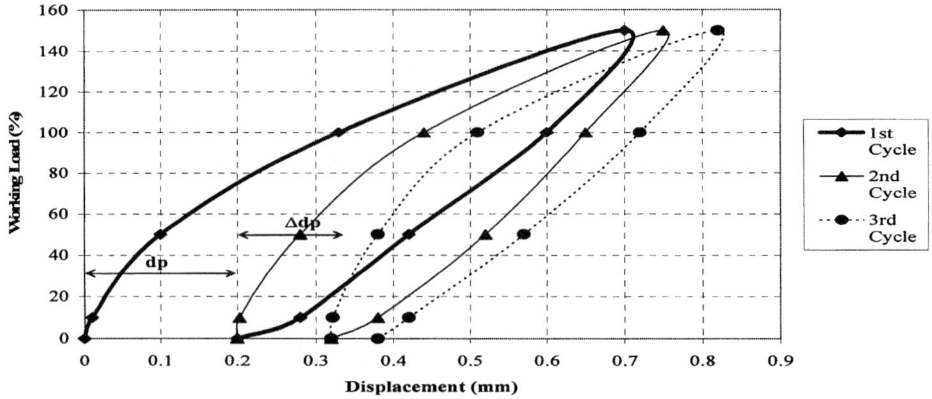

Figure 7 – Load-displacement response under cyclic loading.

In accordance with BS8081 (1989), the apparent free length L_d of the anchor can be deduced from the load-displacement curve over the range 5% f_{pu} – 80% f_{pu}, where f_{pu} is the characteristic tensile strength of the tendon, using the expression;

$$L_d = A_t * E_s * \Delta e / T \qquad (3)$$

Where A_t = tendon cross-sectional area (= 3167mm^2), E_s = tendon elastic modulus as provided by manufacturer (= 200 x 10^6 kN/m^2), Δ_e is tendon elastic displacement given by the difference between the displacement at peak cycle load and displacement at datum load, after allowing for temperature and bedding of the anchor head and T is the difference between the peak cycle load and datum load. Analysis has been based on the unloading curve for the 2nd cycle of each test, results of these for 6 No. anchors are presented in table 3. In table 3, the design free lengths L_1 had been estimated with a 45 degree failure wedge (β = 45^0) as illustrated in figures 1 & 6, while fixed length L_2 = 7m in weathered granite had been assumed in design as described earlier.

Table 3 – Summary of analysis results

Anchor No.	Installed Anchor Length(m)	Design Free Length L_1 (m)	Design Fixed Length L_2 (m)	Maximum Applied Load (kN)	X* (m)	Apparent Free Length L_d (m)	Estimated Redundant Anchorage Length L_r (m)
12	11.5	4.5	7.0	1200.0	0.75	0.3	4.2
13	11.5	4.5	7.0	1200.0	0.75	0.23	4.27
14	11.5	4.5	7.0	1200.0	0.75	0.13	4.37
15	11.5	4.5	7.0	1200.0	0.75	0.19	4.31
19	11.5	4.5	7.0	1200.0	0.75	0.12	4.38
21	11.5	4.5	7.0	1200.0	0.75	0.2	4.3

* - X is the clearance between the top end of the fixed anchor and the point of intersection of the free anchor with the failure plane.

Apparent free lengths L_d have been back-figured from the tensile test results using equation 3. Redundant anchorage length L_r is given by;

$L_r = L_l - L_d$ ---------------------------- (4)

Estimates of L_d vary between 0.12m – 0.3m. The value of L_d gives an indication of the extent of debonding that occurs at the proximal end of the anchor when loaded. The BS8081 (1989) recommends that; $0.9L_l \leq L_d \leq 1.1L_l$.However, test results presented in table 3 show that while estimated L_d values are less than $1.1L_l$ in all cases, L_d generally lies below $0.9L_l$ and consequently points to a significant level of redundancy in design i.e. apparent free lengths L_d are considerably lower than design free lengths L_l. As shown in table 3, corresponding redundant anchorage lengths range between 4.2m – 4.4m.

A number of factors may have been responsible for the wide differences between design values and back-analysed results, these are summarised below:

- Adoption of a 45 degree failure wedge as typically done by many designers (e.g. Mansur & Alizadeth (1970), Clough & Tsui (1984) and Booth (1966)) may have been an overly conservative approach. Considering a representative frictional angle ϕ' of the materials at the back of the wall, such that failure plane angle $\beta = (45 - \phi'/2)$ may have yielded less conservative, yet effective design solutions.
- 7m fixed length (L_2) in weathered granite (as adapted from relevant case histories in the BS8081 (1989)) used in design may have been quite conservative, leading to high factor of safety in design. A safe, effective and more economical solution could have been achieved with a shorter fixed anchor length.
- Ultimate bond stress τ_{ult} of 0.7 N/mm^2 adopted in design may have been quite low, also contributing to a high factor of safety in design. While τ_{ult} & L_2 adopted in design have been deduced from related case histories presented in the BS8081 (1989), rock strength and characteristics vary significantly across different geographical regions and hence, typical parameters and historical data presented in the BS8081 (1989) and other relevant publications in the literature may be considered to be only indicative. Such data should only be used in conjunction with the designer's knowledge of the local geology to achieve a reasonably safe and economical solution.
- Since the entire tendon length has been fully grouted during installation, skin friction mobilised in the boulder clay and the silty sandy gravel may have contributed to the tensile capacity of the anchor, such that actual free length turns out to be considerably lower than the design value.
- Due to slight variations in the depth of intact bedrock across the site, design had been based on the lowest rock head recorded during site investigation. Therefore, at some anchor locations, part of the anchor fixed length may have been located in intact granite and not in the weathered layer as assumed in design, such that actual grout-ground bond stress over a fraction of the anchor fixed length may have been higher than design values.

Overall Wall Performance

The lateral deflection of the anchored secant pile wall has been monitored with inclinometers during and after excavation. Maximum values recorded at each construction stage are compared with original design estimates in table 4. Figure 8 also compares monitored deflections along the pile length after completion of excavation, with design predictions. From table 4 and figure 8, it is clear that measured wall deflections are considerably lower than design estimates (typically 9.2% – 11.5% of predicted values). After the completion of excavation to formation level, a maximum wall deflection of 2.3mm was recorded. The serviceability analysis of the wall had predicted a maximum deflection of 21.5mm. Hence design predictions may be considered to be quite conservative. The service loads on the anchors may have been over-estimated in the wall serviceability analysis, leading to the adoption of higher anchor lengths than necessary. K_o values for the materials had been significantly reduced to allow for piling – induced stress relief in the ground, while the soft piles haven't been considered in the estimation of the overall wall stiffness, although they are believed to influence the wall stiffness to a certain degree.

Table 4 – Comparison of monitored wall deflections with design estimates

Stage No.	Description	Predicted wall deflection (mm)	Measured wall deflection (mm)
2	Install hard & soft piles	0.0	0.0
3	Excavate to 1.5m depth and construct RC capping beam	3.5	0.4
4	Install and prestress tie-back anchors	6.0	0.55
5	Complete excavation to basement formation level	21.5	2.3

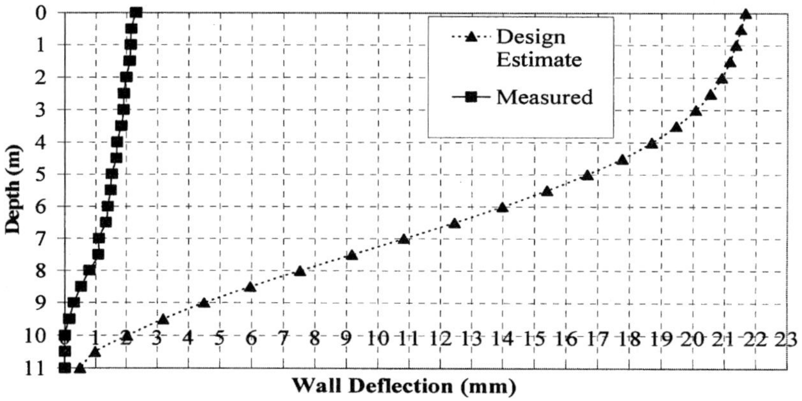

Figure 8 – Comparison of measured wall movement with design predictions.

Also, actual earth pressures acting behind the wall may have been much lower than assumed by the pseudo-finite element modelling software used in the analysis. As suggested by Ivor et al. (1996), Dublin glacial deposits (including the gravels) have the capability to behave as undrained materials and stand unsupported, even when subjected to loading. This has been attributed to negative excess pore pressures, cohesion and cementation of the soil grains. On the contrary, in the current retaining wall analysis, overburden materials were assumed to be drained; an approach taken after the CIRIA C580 (2003), which considers total stress analysis of embedded retaining walls in cohesive materials to be an approach associated with a certain degree of risk, especially in the permanent condition.

Conclusions

The design, construction, testing and monitoring of an anchored retaining wall has been described. The results of the suitability and acceptance tests on some of the installed tie-back anchors have been presented; these have been back-analysed and compared with design predictions. Monitored readings of wall deflection at different stages in construction have also been compared with estimates of the pseudo-finite element model used for the design.

Measured results have been found to be significantly lower than design estimates; likely factors responsible for this have been highlighted. While the majority of anchor design parameters (e.g. bond stresses, fixed anchor lengths) recommended in the literature seem to be sufficient for the safe design of anchors on this project, a more economical design could have been achieved if a sizeable database of published test results in similar ground conditions around the same locality had been available to the designers. Wherever possible, designers should place greater reliance on data from relevant local case histories in related ground conditions.

References

Booth (1966), *"Tie-backs in soil for unobstructed deep excavation"*. Civil engineering, September 1966, pp46-49.

British Standard Institution (1989), *"BS8081 – Ground anchorage"*.

CIRIA (1984), *"Report No. 104 – Design of retaining walls embedded in stiff clay"*. CIRIA, London.

CIRIA (2003), *"Report No. C580 – Embedded retaining walls – guidance for economic design"*. CIRIA, London.

Clough & Tsui (1984), *"Performance of tied-back walls in clay"*. In: Proc. ASCE, 100, (GT12), December 1984, pp1259-1273

Finno & Roboski (2004), *"Three – dimensional responses of a tied-back excavation through clay"*. Journal of geotechnical & geoenvironmental engineering, Manuscript No. GT – 023591.

Ivor et al. (1996), *"The geotechnical aspects of the deep basement for the Jervis Street Shopping Centre"*. Paper presented to a joint meeting of the civil division of the Institution of Engineers of Ireland and the Geotechnical Society of Ireland, 2nd December 1996.

Mansur & Alizadeth (1970), *"Tie-backs in clay to support sheeted excavation"*. In: Proc. ASCE, 96, (SM2), pp459-509.

Mulligan (1983), *"Soil deformations near anchored sheet pile walls"*. Geotechnique, 33, No.1, pp41-55.

Ostermayer (1974), *"Construction, carrying behaviour & creep characteristics of ground anchors"*. In: ICE conference on diaphragm walls and ground anchorages, London, pp141-151.

Suzuki et al. (1972), *"Developments nouveaux dans les foundations de pylons pour lignes de transport THT de Japon"*. In: conference internationale des grande reseaux electriques a haute tension paper 21-01, 13.

Symons et al. (1987), *"Behaviour of a temporary anchored sheet pile wall on A1 (M) at Hatfield"*. Research report 99, TRL, Crowthorne.

The Design, Installation & Monitoring of High Capacity Antiflotation Bar Anchors to Restrain Deep Basements in Dublin

Jim Martin, Byland Engineering Ltd, York, UK
Phillip Daynes, Atkins China Ltd, Hong Kong
Cathal McDonnell, White Young Green, Dublin, Ireland
Martin Pedley, Cementation Foundations Skanska Ltd, London, UK

Abstract

The construction boom in Dublin over the past decade has resulted in a demand for larger and deeper basements. These structures often extend several metres below the local ground water level and may have insufficient dead load to resist flotation.

This paper discusses four recent case histories where passive high capacity bar anchors, often-referred to as Anti-flotation Tension Minipiles (ATMs), provide the necessary restraint against hydrostatic uplift. In particular, it discusses opportunities where close coordination of the ATM designer and the basement slab designer can provide the optimum solution, both from a cost and programme perspective.

Due to the competent founding strata in much of Dublin, either the boulder clays, dense gravels or limestone bedrock, the use of ATMs is commercially attractive and offers several advantages over traditional stressed tendon anchors.

This paper concludes that full understanding of the appropriate design case and close interaction between the structural, geotechnical and ATM engineers is necessary to achieve the optimum basement solution.

Introduction

Ground water is the origin of many geotechnical problems and the understanding of the ground water regime over the life of a structure is of paramount importance. The depth of a basement below ground water level, the dead weight of the new structure and the construction method will dictate the requirement for antiflotation measures. The importance of simple uplift calculations should not be underestimated because hydrostatic forces can be considerable and the consequence of mistakes potentially catastrophic. Plates 1 & 2 show the basement car park of a large hotel in Dublin, constructed during the dry summer of 1996, with no consideration for long-term hydrostatic forces. This relatively light two-storey basement is founded within Dublin black boulder clay, which due to its low permeability appears dry when excavated, masking the actual porewater pressures present. Long-term standpipe readings in this stratum indicated the groundwater level to be 5m above the underside of the basement slab. Following construction, the gradual build up of hydrostatic forces exceeded the capacity of structural elements, resulting in failure of the columns, cracked floor slabs and

Ground anchorages and anchored structures in service, Thomas Telford, London

seepage into the basement. The retrofitting of antiflotation measures is difficult in such circumstances so long-term pumping to reduce uplift pressures is often the most suitable solution, albeit requiring a high ongoing operating and maintenance cost. However, this solution is only viable provided there are no adverse effects on adjacent services and structures, and that there is a suitable discharge location. Normally it is most economic to design in any antiflotation measures at an early stage.

Plate 1: Basement suffering failure due to hydrostatic uplift

Plate 2: Column failure due to hydrostatic uplift of slab

Historically a common solution to providing uplift resistance was to use stressed tendon anchors (Plate 3). This solution applies a pre-stress into the structure, which may be beneficial if there is poor ground below the basement slab, however there are several disadvantages:

- Additional compression piles may be required to provide a reaction force as the anchors only provide tension resistance.
- Time consuming to construct due to the separate installation and stressing phases required.
- Cold joints formed where the anchor passes through the structure, with the risk of subsequent leaks if not sealed correctly.
- Maintenance is required over the life of the structure to ensure durability
- Anchor head pockets are necessary which set a restriction on the minimum slab thickness.
- Highly stressed tendons remain within the basement structure and substructure.
- Corrosion risks in a wet environment.
- Relatively high cost.

Plate 3 –Stressing of Tendon Anchor

Deep basements in Dublin are often within competent strata where the alternative of passive high capacity bar anchors, also known as Antiflotation Tension Minipiles or ATMs, can be suitable (Figure 1). Compared with tendon anchors ATMs exhibit a number of advantages:

- Quicker construction as anchor head pockets are not needed and stressing is not required
- They can act both in compression (temporary) and in tension (permanent)
- The risk of leakage through cold joints does not exist as the anchors are cast integrally within and simultaneously with the basement slab
- No maintenance is required, unlike tendon anchors
- Thinner basement slabs can be employed, especially when integral anchor head plates are used (Plate 4)
- The uplift restraining forces develop passively with only a few millimetres of anchor head movement
- Basement restraint is possible upon the slab gaining its design strength, potentially allowing early decommissioning of construction dewatering systems
- Relatively low supply and installation cost

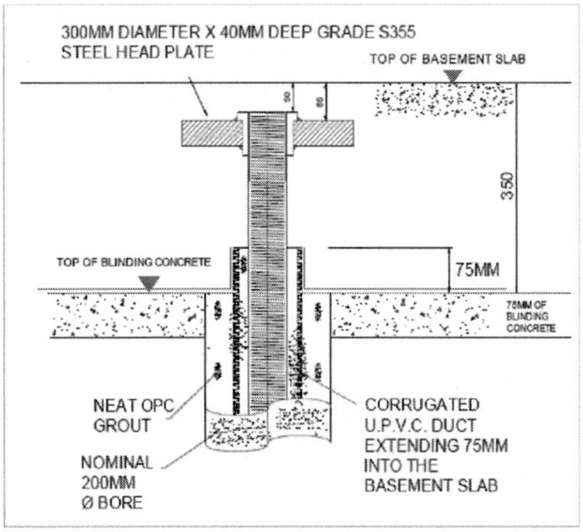

Figure 1 – Example of antiflotation passive bar anchor (ATM) head detail

Plate 4 – ATM head plate prior to casting of basement slab

This paper discusses the design and construction of four deep basement projects in Dublin where ATMs have been used during the past decade. The primary focus of this paper is on the redevelopment of Smithfield Market. This project involved one of the largest basements in Dublin (220m long x 110m wide x 10m deep) where 3No ATMs were instrumented with strain gauges and monitored over a two-year period covering both construction and building commissioning. Also discussed is the successful use of ATMs for deep basements at Spencer Dock (founding in boulder clay), Elm Park and Swords (both founding within strong limestone). Smithfield and Spencer Dock were uniform thickness raft slabs, while Elm Park and Swords were raft slabs with pad thickenings.

Typical Dublin Ground Conditions

The ground conditions in the city of Dublin are particularly suited to passive high capacity antiflotation minipiles (ATMs). Typically, in much of the city centre, made ground overlies glacial tills that overlies strong limestone bedrock. The glacial till consists of weathered upper brown boulder clay overlying a very stiff to hard black boulder clay. Alluvial gravels are present in the down stream port area where they overlie the glacial deposits. A buried preglacial gorge of the River Liffey, filled with dense water bearing gravels, lies to the north of the present course of the river. The Smithfield site discussed later is located on this buried gorge.

Design Philosophy

The design of antiflotation measures cannot be effective or economic unless the long-term ground water regime is fully understood. This can be done through site-specific ground investigation and groundwater monitoring over an appropriate period of time to understand tidal and seasonal effects.

Once this information is available, it is necessary to understand whether the basement is subject to either long term and/or short-term uplift. A simple check comparing the dead weight of the structure relative to the uplift force developed during short and long-term loading cases can be carried out to determine whether the basement requires uplift restraint. In this calculation, it is common practice to apply a partial factor to the dead weight of the structure (typically taking 90% of the building dead weight although this is dependent on the code of practice adopted). The pressure acting on the underside of the basement slab is then calculated from the depth of the underside of the slab relative to the standing groundwater level, or *worst credible* water level likely to occur for the specific loading case.

The total nett 'working' hydrostatic uplift force is therefore:

$$\text{Nett Uplift Force} = \text{Basement Area} \times 9.81\,kN/m^2 \times \text{head of water} - (0.9 \times \text{dead weight})$$

This is the force (if positive) which is to be resisted by the ATMs. The restraining effect of the basement walls is usually conservatively ignored in this calculation. Whilst this check is carried out for the structure as a whole it may also be necessary to check local areas of the structure, particularly where wide column spacing's are employed, as antiflotation measures can lead to savings in member (slab) thicknesses even if only applied locally.

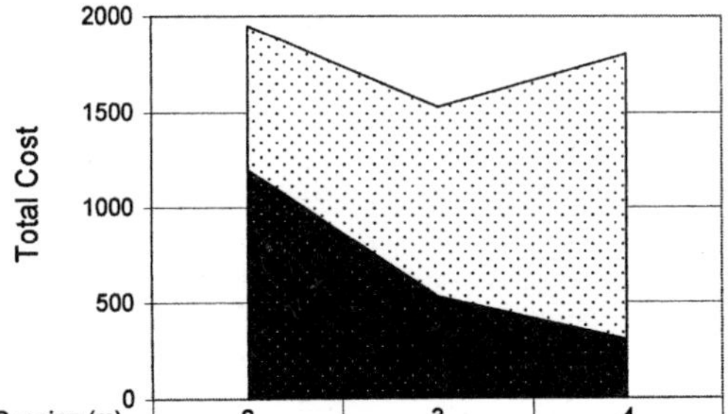

ATM Spacing (m)	2	3	4
■ Slab Thickness (m)	0.5	0.75	1
□ Slab Cost	750	1000	1500
■ ATM Cost	1200	533	300

Figure 2 – Schematic detail of the optimisation process showing variation in total cost with increasing ATM spacing and slab thickness

At this key stage, close cooperation between the ATM and structural engineers is required to obtain the optimum anchor layout and slab thickness. The anchor layout should fit uniformly between the column layouts, because the building dead load transmitted down the columns combines with the ATMs to resist the uplift forces. It is necessary to determine an efficient anchor spacing that reduces the basement slab thickness and reinforcement quantities to produce the optimum solution (Figure 2). Special integral head plates assist with this optimisation by locating the head plate near the top of the slab, which reduces the risk of punching failure. In the case of raft slabs with pad thickenings, this optimisation process can reduce the slab thickness considerably, which can lead to major cost and programme savings via reduced excavation (often in strong rock) and reduced quantities of reinforced concrete.

The calculated antiflotation anchor force from the long-term 'worst case' ground water level is defined as the anchor working load and can be designed for in accordance with the standard methods in the National and European Codes of Practice. In the Dublin area, the 'worst credible' flood condition is when the design water level is at ground level and this defines the 'ultimate' anchor force. This is because the water table cannot rise above the ground level without the basement flooding unless prevented from doing so, although such protection is rare in Dublin. Lower structural load factors (of close to unity) and lower geotechnical factors of safety (approximately 1.5 on the shaft friction) may be permissible with this 'ultimate' design force (which should also be the maximum test load for the project). However, this approach should ensure that a suitable factor of safety still exists under service conditions.

The main reinforcement element of the ATM anchor is typically a fully threaded rebar (GEWI or MAC500) with a yield stress of 500MPa, with diameters in the range 50mm∅ to 75mm∅. This rebar has a coarse thread and can be installed in sections if required, connected together using full strength couplers. The high capacity bar anchors are normally grouted in-situ by pumping a neat colloidally mixed (high shear) Ordinary Portland Cement grout, with a 0.4 water: cement ratio, through a small diameter tremie pipe into the base of the bore until completely full of clean grout. This grout mix typically achieves unconfined compressive strengths of 50MPa to 80MPa at 28 days. As the passive bar anchors generate their restraining forces micro cracks can form within the grout column. Therefore, an unperforated corrugated UPVC duct is installed over the full bar length, extending into the basement slab, to provide continuous corrosion protection to the bar, and protect against possible corrosion at the interface of the ATM and base of slab. As bar anchors operate at typically 30% of the stress levels of tendon anchors they are less susceptible to corrosion because of the large single rebar, rather than the numerous small diameter wires within tendon anchors.

Smithfield, Dublin

The Smithfield project is a 4-acre mixed-use development located in Dublin city centre, surrounded by public roads on all four sides. The development comprises over 350 apartments, a hotel and 139,000 ft^2 of offices, retail and leisure space. It includes a three level basement primarily used for car parking, but which also houses a 4-screen cinema, retail space, gymnasium with pool and plant room areas (Figure 3). The perimeter retaining walls for the basement were constructed using diaphragm walls. The diaphragm walls extended to bedrock, however only limited penetration into the strong limestone was possible and additional shear pins through the base of the walls were required. Penetration into the bedrock would have required costly and undesirable chiselling, resulting in excessive vibration of sensitive services. The structures found on a 900mm thick reinforced concrete raft slab bearing onto dense gravel at a depth of up to 13m.

The buildings at Smithfield are typically eight stories high with a 14-storey tower block. The existing ground water level is approximately 6m above the underside of the basement slab generating approximately 60kN/m^2 hydrostatic water pressure. Over the majority of the site, there was sufficient dead load to resist the uplift pressure; however, in some lightly loaded areas (e.g. below internal streets and atriums), there was a nett hydrostatic uplift force and in these areas ATM anchors were used. In addition, there were also areas where long spans were required in the building superstructure, for example in the cinema space and it was cost effective to utilise ATM anchors to reduce the span of the slab, rather than designing a thicker or more heavily reinforced slab to span the large column spacing.

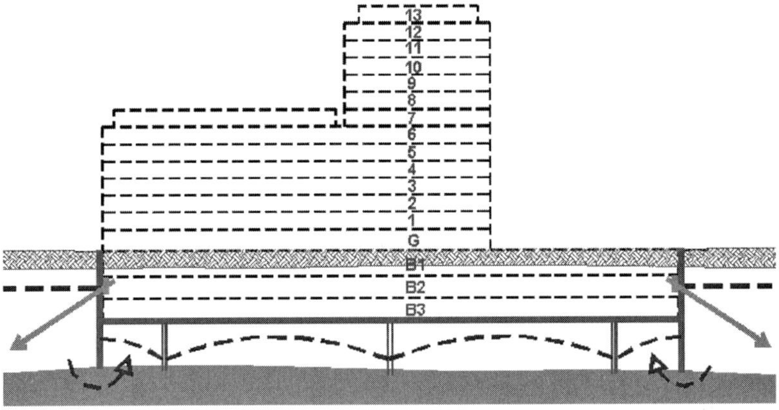

Figure 3: Schematic section of the Smithfield development

Geology & Groundwater

There was a phased site investigation for the Smithfield development, with a total of 20 rotary cable percussion boreholes and 7 rotary-cored boreholes, drilled between 1998 and 2002. These indicated the following general succession of strata:

Made Ground (1.8m to 7.5m thick)
Dense Gravel (9.0m to 13.0m thick)
Limestone Bedrock (encountered 13.8m to 21.0m below ground level)

Isolated sections of site encountered alluvium, medium dense sand and boulder clay. A significant proportion of the Standard Penetration Tests (SPT's) achieved "refusal" (penetration < 0.3m for 50 blows) on cobbles or boulders. All cable percussion boreholes recorded substantial chiselling time (typically 7 hours per location). Bedrock consisted of moderately to very strong, fresh to locally slightly weathered limestone, with Uniaxial Compressive Strength (UCS) values ranging between 11-158MPa (average of 100MPa). Solid Core Recovery (SCR) ranges from 49 to 87% and Rock Quality Designation (RQD) from 0 to 83%. The limestone bedrock in Dublin typically has a karst surface, facilitating flow paths through fissures and joints. The initial site investigation indicated standing groundwater at a depth of 4m to 6m. The second phase of site investigation provided more information on the rock profile as it varied across the site and rising and falling head permeability tests were carried out to estimate the permeability of the gravels. Ground water strikes were recorded in the high permeability gravel at between +0.23mOD and +0.99mOD. Ground level is at +4.5mOD and the underside of the basement slab is at – 6.0mOD.

A pumping test was undertaken within the bedrock to gain a more detailed understanding of the ground water regime and the permeabilities of the overlying gravel and limestone, and to optimise the design of the temporary dewatering system. Standpipes were monitored at varying distances from the pumping test to approximate a drawdown curve applicable to the gravel and limestone water tables.

ATM Design

The ATMs were generally designed in accordance with principles of BS8081:1989. During the design development, early consultation between the structural engineer and the foundation (ATM) contractor determined that a characteristic anchor-working load of 1,100kN was achievable. The engineer used this to determine the optimum anchor layout for the base slab taking into account slab spans. The bar anchor comprised of a 63mm diameter MAC500 (f_y = 500MPa) fully threaded bar with a full-length 160mm diameter semi-rigid unperforated corrugated sheath to provide corrosion protection. Generally, the anchor fixed length was designed upon a minimum 4m socket into strong limestone, but some contribution from the dense gravels was utilised where the bedrock dipped lower and ignoring the gravels was over conservative. An ultimate grout to ground bond stress of 2MPa was adopted for the rock socket design, although pullout tests by the bar manufacturers report ultimate bond stresses of up to 5MPa for ribbed bars of this type.

Cone pullout calculations gave higher factors of safety than an ATM ground to grout failure mechanism meaning the latter was the critical case. The anchors ranged from 10m to 16m below slab invert level. Each ATM was fitted with a 50mm thick by 500mm diameter steel head plate, secured above and below by full strength nuts. This was located within the upper 1/3 of the basement slab, below the top reinforcement mat.

Due to the tight construction programme there were no working ATM tests, however three preliminary tests were undertaken prior to construction of the working anchors. These were installed prior to general excavation due to site and programme constraints. To compensate for additional friction the upper section of the anchor was debonded from the ground. The ATMs were successfully proof tested to 1,552kN or 70% of the guaranteed ultimate tensile capacity of the bar.

ATM Installation

Three hundred and sixteen ATM anchors were installed in two phases spanning 2002 and 2003 (Plates 5 and 6). The bar anchors were installed from the slab blinding concrete cast onto the formation. A 220mm diameter Symmetrix drilling system was used to install temporary casing though the gravel. Once this casing was sealed into bedrock, the drill bit (191mm diameter) was advanced to the design depth. The centralised corrugated sheath and rebar were then installed and grouted insitu as detailed earlier.

Plate 5: Air Flush DTHH Drilling (showing ground water blown out of bore)

Plate 6: Bar anchors with head plates fitted and slab reinforcement being fixed

ATM Instrumentation & Monitoring Results

To examine the long-term performance of the ATMs, three permanent anchors were installed with strain gauges at 2m centres over their full depth and monitored for approximately two years, i.e. during the construction, commissioning and early operational phase of the development. Figure 4 is a typical graph of the forces inferred from the strain gauges for one of the instrumented ATM anchors. From October 2003 to February 2004, the readings are erratic due to the casting of the basement slab and nearby construction activities. Then, from February 2004 to September 2004, the graph shows eight months of gradual build up in compression force as the dead load from the superstructure increases. In September 2004, the dewatering system is decommissioned and hydrostatic uplift forces develop under the basement slab. Tensile forces then develop in the ATMs to resist these flotation forces. It takes approximately seven months for the ATM (hydrostatic) forces to stabilise. The graph shows that the hydrostatic forces are greater than the self-weight of the structure because all strain gauges on the ATM rebar indicate tension forces of between 25 and 75kN. This is considerably less than the 1,100kN design loading and is probably due to inevitable load sharing between the basement structural members.

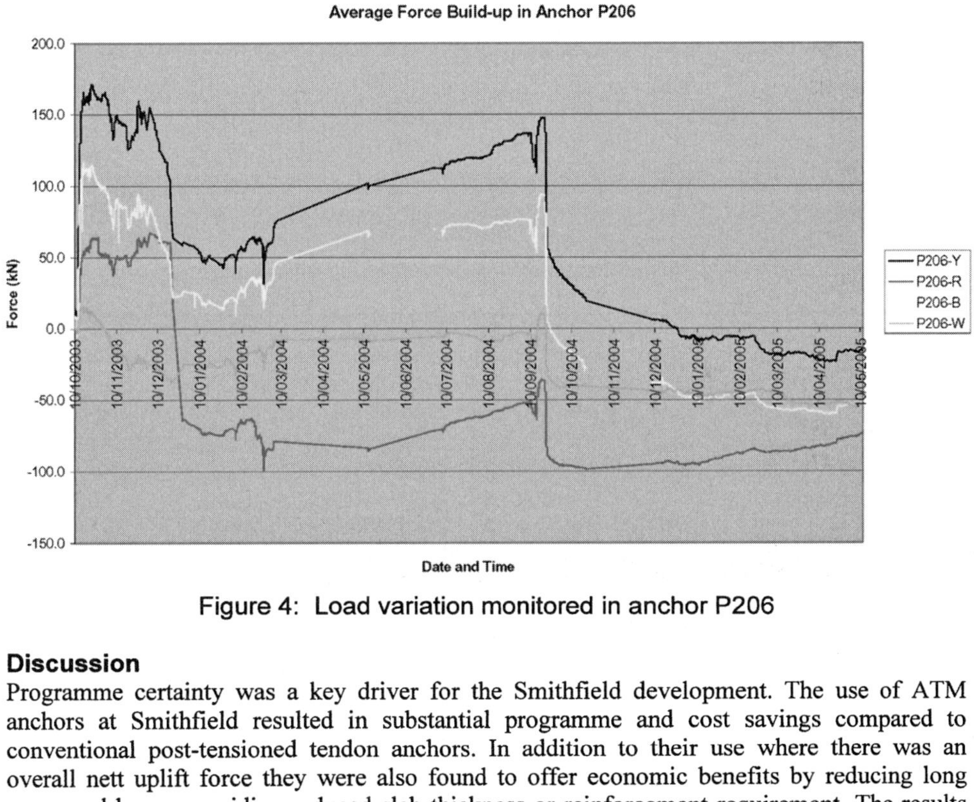

Figure 4: Load variation monitored in anchor P206

Discussion

Programme certainty was a key driver for the Smithfield development. The use of ATM anchors at Smithfield resulted in substantial programme and cost savings compared to conventional post-tensioned tendon anchors. In addition to their use where there was an overall nett uplift force they were also found to offer economic benefits by reducing long spans and hence providing reduced slab thickness or reinforcement requirement. The results of long term monitoring confirmed the build-up of uplift pressure once the temporary dewatering system had been decommissioned. This increase in water pressure occurred rapidly at Smithfield, however, this may not be the case elsewhere in Dublin where lower permeability soils exist below the basement.

Spencer Dock, Dublin

The site is located to the east of Dublin City Centre, immediately north of the River Liffey within the Spencer Dock Redevelopment and comprises a two-level 8m deep basement of approximate plan dimensions 135m x 60m. Ground water level is approximately 3m below street level. The site was for the construction of Price Waterhouse Coopers headquarters building, comprising three tower blocks interspaced with atrium areas and open spaces. Below the atrium and car park areas, a 6m hydrostatic head acts on the underside of the basement slab resulting in significant nett uplift areas. High capacity ATMs founded within the very stiff to hard boulder clay were used to resist these flotation forces.

Geology & Groundwater

The overlying made ground, alluvium and gravels were removed during the basement excavation and the antiflotation bar anchors were installed from the slab blinding concrete level, cast directly on top of the boulder clay. This low plasticity glacial deposit (PI 10-15%) extends down to limestone bedrock, a further 12m below. Cable percussion boreholes normally only penetrate 3 to 4m into the black Dublin boulder clay, even with significant chiselling. SPT 'N' values are between 40 and refusal with mean values of 100 to 200, which classifies the material as very stiff to hard with an undrained shear strength of 500 to 1,000kN/m^2 (Skipper et al., 2005 & Mentiki et al., 2004). Core drilling through this stratum was of limited use, because the water flush washed out the clay matrix, leaving only granular material, meaning good quality undisturbed samples of Dublin boulder clay were difficult to obtain (although the use of triple core barrel sampling has become more prominent in recent times). The groundwater level fluctuates from - 0.7m AOD to +0.3m AOD, relative to a ground level of +3mAOD, and is therefore loosely related to the tidal effects of the River Liffey.

Bar Anchor Design & Installation

Plate 7: Drilling for bar anchors at Spencer Dock using DTHH techniques

Plate 8: Tension test reaction system at Spencer Dock

Six hundred and eighty two x 600kN safe working load ATM anchors were specified by the engineer with a maximum allowable anchor head movement in service of 5mm. Design calculations required a 9.0m penetration into the boulder clay for a 220mm $\varnothing$ anchor, which equates to a working grout to ground bond stress of approximately 100kN/m^2. This 9m penetration is within the 10m maximum bond length limit recommended in clause 6.2.3.4 of BS8081:1989. The reference design, based on anchors founding in the underlying limestone bedrock, would have required significant 'free' lengths, but this was deemed unnecessary due to the competent nature of the overlying Dublin boulder clay. The 5mm allowable anchor head movement at SWL is demanding and therefore a 63mm Ø rebar was used, which ensured a stiff reinforcing element and a high structural load factor (>3.0). The ATMs were drilled using an air flushed down-the-hole hammer (DTHH) technique in order to penetrate the ground quickly and efficiently (Plate 7). The vast majority of ATMs could be drilled using open-hole techniques, but a small number of holes encountered pockets of gravel, which caused the bore to collapse and in such instances a Symmetrix drilling system was used to install temporary 220mm Ø drill casing to the full depth. The ATMs were then grouted in-situ as detailed previously. As the top level of the bar was a critical dimension, the bores were over drilled by 0.5m and the rebar was suspended centrally in the fluid grout at the specified level. A corrugated duct was installed over the full depth of the hole, thus ensuring the rebar had 500mm of grout cover within a corrugated UPVC duct at its lower end. A 300mm diameter x 35mm thick steel head plate, with full strength nuts above and below was provided at the head of each ATM. The column-punching shear onto the slab and the underlying boulder clay dictated the use of a 1m thick basement slab.

Load Test Results & Discussion

Four preliminary and six working (acceptance) tests were undertaken in accordance with Tables 13 and 18 from BS8081:1989. The reaction beams (Plate 8) were located a minimum of 1.5m from the anchor to prevent any influence on the anchor performance. The preliminary tests of up to 1,330kN showed no indication of failure, equating to a proven geotechnical factor of safety of 2.2. The 'working' acceptance test results are summarised in Table 1 and these show average pile head movement of 2.95mm at SWL, comfortably within the 5mm specification.

Table 1 Summary of Static Test Pile Results at Spencer Dock

Minipile Detail	Elastic movement (mm)	Permanent movement (mm)	Total pile head movement (mm)	Apparent free tendon length (mm)	Free tendo n length (mm)	Tendon bond length (mm)	Free length + 50% of tendon bond length mm	Notes
Pile 67	1.88	0.00	1.88	1,360	1,500	9,000	6,000	OK
Pile 16	2.19	0.44	2.63	1,581		9,500	6,000	OK
Pile 76	3.70	0.65	4.36	2,676		9,500	6,250	OK
Pile 84	3.93	0.74	4.67	2,839		9,000	6,000	OK
"OK" BS 8081, section 11.2.12 states that the apparent free tendon length should not be less than 90% of the free intended in the design nor more than the intended free length plus 50% of the tendon bond length intended in the design								

Elm Park & Swords, Dublin

These two projects had several similarities, which allowed full optimisation of the antiflotation schemes. The Elm Park Development is a multi-functional development (residential, commercial, medical and hotel & conference units) located 4km southeast of Dublin City Centre. It covers an area of 75,000m^2 with an 8m deep basement over large areas. The Swords Primark Development, 8km north of Dublin City Centre is a residential and retail development, covering 22,500 m^2 with an 8m deep basement over the majority of the site.

Geology & Groundwater

The basement slabs on both sites found directly onto strong limestone bedrock, with the ground water level approximately 2m below ground level i.e. an uplift pressure of 60kN/m^2. The limestone bedrock is strong to very strong limestone, slightly weathered over the top metre, with UCS values of 100 to 200 MPa, locally referred to as 'Calp' limestone.

Bar Anchor Design & Installation

These most recent projects allowed the full use of the optimisation techniques described above due to the strong limestone at basement slab level and early discussions between structural and geotechnical engineers. ATM working loads of 1,000 to 1,350kN were adopted, in conjunction with integral head plates within the slabs, which allowed reduced slab thicknesses as low as 400mm to be used. These thin slabs were possible because of the strong limestone immediately below preventing the columns punching through the slab. The 6m deep anchor bores were all formed using air flushed open hole DTHH drilling techniques. This quick and efficient drilling system forms holes with good verticality tolerances and a rough bore, resulting in a good grout to ground bond values. The design of the anchor length was governed by global stability and potential interaction, rather than ground to grout bond stresses. These works were carried out on a rolling programme working off blinding concrete using relatively small 12 tonne Casagrande C6 Drill Rigs. This is an efficient method of working with the Main Contractor, using only small rigs in localised areas of the site. The final scheme at Swords used 566No x 1,000kN SWL ATMs on a 4m x 4m grid spacing (Plate 9).

Test Results & Discussion

The anchor test results at both sites were excellent with head movements being primarily elastic extension, moving ~ 2.2mm at 1,700kN (Figure 5).

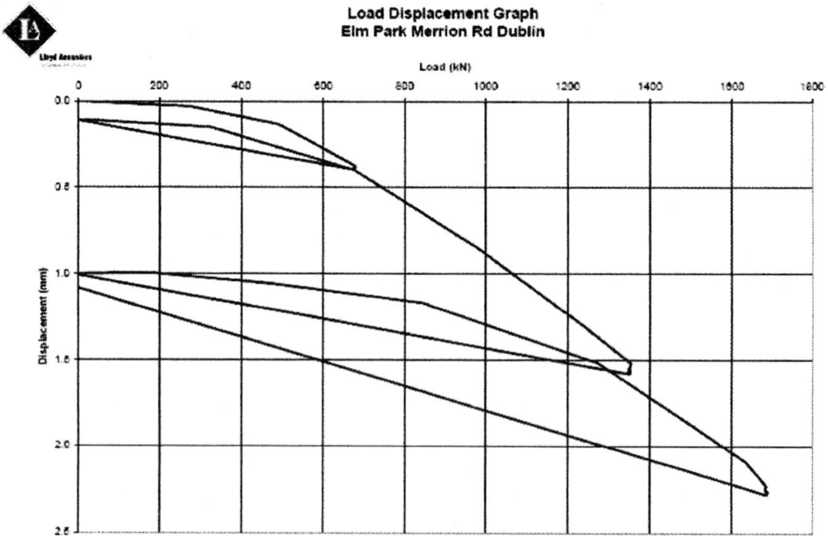

Figure 5: Static load test result at Elm Park

Plate 9: Basement construction at Swords

Conclusions

This paper has presented a number of applications where high capacity bar anchors (ATMs) have been used to resist flotation forces acting on basements in the Dublin area. The ground conditions in the Dublin area make the use of passive high capacity bar anchors particularly appropriate. This type of antiflotation solution offers a number of engineering, cost and construction programme advantages over more traditional forms (such as tendon anchors or permanent dewatering). However, to ensure that optimum benefit is achieved from their use early cooperation and discussion is necessary between the structural engineer and geotechnical designer (specialist contractor). In addition, an understanding of the groundwater conditions during short and long-term load cases is essential.

Acknowledgements

The authors would like to thank Cementation Skanska Ireland Ltd, White Young Green, Frank Motherway Associates, Michael McNamara Ltd, G & T Cramptons Ltd, O'Connor, Sutton & Cronin, Pierse Contracting Ltd, Molony & Millar and Abiola Ojo (Leeds University Engineering Geology Department) for their assistance and support throughout these works and the preparation of this paper.

References

1. B.S.8081:1989 – British Standard Code of Practice for Ground Anchorages
2. MENKITI, C.O., LONG, M., KOVACEVIC, N., EDMONDS, H.E., MILLIGAN, G.W.E. AND POTTS, D.M (2004). Trial excavation for cut and cover tunnel construction in glacial till - a case study from Dublin. Proc. Skempton Memorial Conference Advances in Geotechnical Engineering, Vol 2, pp 1090 – 1104
3. SKIPPER, J., FOLLETT, B., MENKITI, C., LONG, M, CLARKE – HUGHES, J. 2005 - The engineering geology and characterisation of Dublin Boulder Clay, Quarterly Journal of Engineering Geology and Hydrogeology pp 171 – 187

Session eight:

Recommendations and standards for inspection, monitoring and repair

A New Manual for Ground Anchor Inspection, Integrity Investigation and Remedial Measures

Hiroaki Miyatake and Takeshi Oshita, Public Works Research Institute, JPN
Hiroaki Kubo and Mitsunari Takeyama, Japan Anchor Association, JPN

Abstract
In the 50 years since the first ground anchors were used in Japan, the anchors have been used for various purposes in at least 30,000 projects. Some anchors constructed in early times have aged and lost their performance, resulting in slope deformation and exposure of broken anchor heads.

Various integrity investigation techniques and remedial measures have been proposed but have not been systematically organized. It has also been observed that precious data on inspections and measures taken are scattered, and the lack of a system and data have impeded the efficient maintenance of anchor structures.

As a consequence, establishment of an efficient and rational maintenance scheme for existing anchors was deemed necessary. This paper describes the efforts in Japan to improve maintenance of anchors by publication of a new manual for ground anchor inspection, integrity investigation and remedial measures.

Introduction
Ground anchors were first used in Japan in 1957 in a dam construction project. In the 50 years since then, anchors have been used extensively, including the application of permanent anchors in at least 30,000 projects exceeding a total length of 20,000 km, in order to stabilize natural and man-cut slopes for road and dam construction, improve structural stability, control floating of structures caused by underground water, etc.

Structurally, anchors stabilize slopes by introducing initial tensioning force during construction, which is gradually released over time through various mechanisms. Thus, anchors must not only be carefully designed and constructed but also vigilantly maintained, especially by monitoring residual tensioning force and checking integrity.

In the early years of anchor work, execution and rust-prevention technologies were still under development, and some of the anchors constructed during this period have aged and lost their performance, resulting in slope deformation and exposure of broken anchor heads.

Dangerous anchors can easily be found and treated if the construction drawings and specifications and maintenance data are available. However, such references are scarce, and anchors are judged dangerous or safe based only on their type and construction period.

The relevant drawings, specifications, and construction records were not stored because anchors were regarded as an auxiliary method for stabilizing slopes.

Various integrity investigation techniques and remedial measures have been proposed but have not been systematically organized. Precious data on inspections and measures taken are scattered, and the lack of a system and data impede the efficient maintenance of anchor structures. A system for maintaining anchors needs to be constructed including inspection, integrity investigation, and remedial measures. In 2002, the Public Works Research Institute commenced studies on integrity investigation and remedial measures with the co-operation of the Japan Anchor Association (JAA) and conducted a joint study in FY2005. This paper describes resultant technologies for investigating the integrity of ground anchors and remedying damaged anchors.

Background of study

Actual state of existing anchors

In 1977, the Japanese Society for Soil Mechanics and Foundation Engineering (the present Japanese Geotechnical Society) established the "Earth anchor design and construction standard". In 1988, the standard were revised into the "Ground anchor design and construction standard" [(1)], which requires the implementation of enhanced anticorrosion measures including double corrosion protection for permanent anchors. The revision accelerated the development of technologies for improving the corrosion resistance and reliability of each kind of anchor. Anchors that were designed and constructed based on the old standard (hereinafter referred to as the "old-type anchors") are likely to contain insufficiently corrosion-resistant members. The total number of old-type anchors is not known since some data are unavailable, but a survey by the Japan Anchor Association gives an estimate of about 14,000 anchors, most of which were used for highway construction.

To understand the actual state of existing anchors, field surveys were conducted of anchors installed long ago and checked and categorized for major deformation and damage.

Manual for ground anchor inspection, integrity investigation and remedial measures

The objectives of manual preparation were to:

1) Systematically organize the concepts on inspection, integrity investigation and remedial measures needed for the rational and efficient maintenance of anchors;

2) Summarize and state the need for properly designing, constructing and maintaining anchors to retain their performance, in an easy to understand manner to ensure that the owners of projects, especially of public works, understand the need for maintenance; and

3) Emphasize the need for storing information throughout the service life of anchors to promote the development of technologies for investigating, designing, constructing, and maintaining anchors.

To accomplish the objectives, the manual describes the basic properties of anchors and fundamental concepts of maintenance in Chapters 1 and 2, enabling readers to understand the need for maintaining anchors just by reading the first two chapters. As described above, the importance of maintenance is little understood by project owners, leading to failure of systematic budget allotment and loss of data and records. Chapters 1 and 2 were written to raise owner awareness. Chapter 3, which describes inspection, is also for owners, who usually inspect the anchors themselves in the case of highways. Chapters 4 and 5 describe integrity investigation and remedial measures assuming that the readers are owners who have read Chapters 1 to 3 and use the manual as a reference for discussing technical aspects with contractors. The basic concepts for appropriate method selection are described at the beginning of Chapters 4 and 5, followed by a description of integrity investigation methods and remedial measures, respectively, aiming to promote communication between owners and contractors. The general flow in Chapters 4 and 5 is a fundamental point to be understood by owners, and is stated as such in Chapter 2.

In Japan, design standard are established for each kind of anchor, but the technical standard for structures must also be considered since anchors also serve as supplementary members for stabilizing structures. In other words, anchors should be maintained by regarding them as part of the structure that they support. Chapter 2 explains that integrity investigation should be performed not only on individual anchors but also on the entire body consisting of anchors and supporting structure.

The following is an overview of the manual.

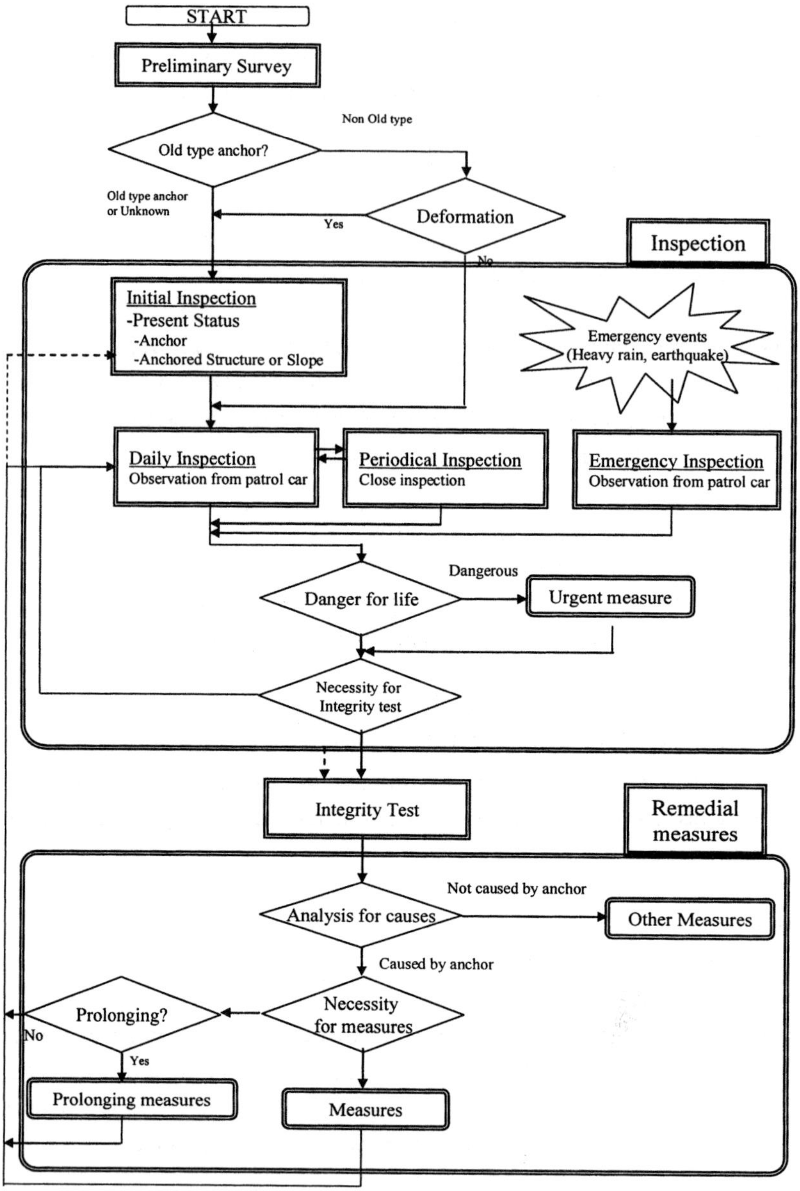

Fig.1 Flowchart for maintenance of anchor

Inspection of anchors

Anchor inspection is classified into periodical inspection to be performed as part of daily maintenance work and emergency inspection after storms, earthquakes and other such events. When any doubt arises as to the integrity of anchors, each anchor is investigated.

Preliminary survey

As a preliminary survey, references are studied before performing in-situ inspection of ground anchors. The objective of the preliminary survey is to collect the necessary data and information for maintaining the anchors, such as drawings and specifications of the anchors, structures, and slopes, maintenance records, and references on the peripheral topography.

When the preliminary survey reveals "old-type anchors", initial inspection is to be conducted to understand the state of the anchors before performing ordinary inspection.

Initial inspection is also conducted on anchors other than "old-type anchors" if any deformation was previously found on the anchors, structure and/or slope. Preliminary survey involves studying references and should be performed for all anchors in principle. When references on the said anchors are not available due to disposal, etc., initial inspection is to be performed.

Initial inspection

During the initial inspection, the specifications of anchors, structure and slope and the state of the heads and pressure plates are inspected. For example, the need for anticorrosion treatment is judged from the year in which the anchor was designed and constructed, the title of the project, use and shape of head caps, etc.

Since initial inspection is conducted to supplement insufficient information and references, all anchors should be inspected, and the results recorded in inspection ledgers and retained for use in subsequent inspections. When the integrity of anchors is in question, an integrity investigation is performed to assess performance.

Ordinary inspection

There are daily, periodical and emergency inspections.

Daily inspection involves visual examination of anchors, anchor heads and pressure plates and other structures for any abnormality, in principle. Daily inspection is performed by the facility manager during patrols. For example, on ordinary national highways, daily inspection is conducted from vehicles and would be better if combined with annual or biannual periodical inspections.

Periodical inspection involves visual examination, in principle, and close inspection of hammering and measuring when necessary. Visual examination should, in principle, be conducted on all anchors, but close inspection can be reduced to 10% of all anchors or three or more extracted anchors. Anchors should be extracted so that different anchors are inspected every year and the state of the entire slope is understood by several periodical inspections.

Emergency inspection is performed upon notification from users as well as immediately after storms, earthquakes and other emergency events. In emergency inspection, safety must be quickly assessed, and all anchors should be visually examined. Anchors at which a slight deformation was found during daily and/or periodical inspection must especially be carefully examined on a priority basis.

When any abnormality that could affect integrity is found during an inspection, the anchors should be investigated for integrity. The results of all inspections should be recorded in ledgers and retained.

Integrity investigation of anchors

Based on the inspection results, judgment is made as to whether or not the anchors need to undergo integrity investigation. Each anchor must be checked and, as described above, the anchors must also be assessed as a group of anchors constituting a structure. The manual describes this point in a manner that is especially easy to understand.

Even when an integrity investigation on individual anchors is not necessary, such as when no abnormality is found in any of the anchors but is found in the entire structure and/or slope, or when an anchor abnormality is so slight that there is no need for integrity investigation but the slight abnormality is concentrated in a certain range and is likely attributable to similar causes or spreads over a large area, integrity investigation is recommended.

The manual was prepared aiming to serve as a reference for assisting communication between owner and contractors, and includes the description of optional investigations and measures as well as obligatory work. For example, the manual recommends integrity investigation for anchors that were installed long ago and have not been examined for integrity and the exchange or repair of parts without integrity investigation when there is obvious damage and/or deterioration.

To better suit practical use, the manual was written to enable effective measures to be taken even under diverse limits, such as cost restrictions. For example, anchors demonstrating slight abnormality hinting of a slight integrity problem, but judged as not warranting integrity investigation, should be specifically monitored during subsequent inspections and quickly treated upon detection of increased abnormality.

Integrity investigations described in the manual include:

Detailed investigation of anchor heads

Detailed investigation of anchor heads involves checking the corrosion state of tension members and anchorages, water seepage from the back, tendon allowance length, and filling and degeneration state of rust-preventive oil as well as rough investigations such as visual examination and hammering (impact acoustics method).

Ultrasonic inspection

Ultrasonic inspection detects damage to tension members by irradiating ultrasonic waves and monitoring reflections, and is used as a preliminary investigation for assessing safety before lift-off tests, capacity tests and re-stressing, which are described below.

Lift-off test

Lift-off test involves measuring of residual load. The resultant load-displacement properties are used to investigate the apparent free length of the anchor and detect abnormalities. The data are also used to judge the need for investigating the back of anchor heads, conducting maintenance tests and re-stressing. Thus, lift-off tests would be better performed on all anchors subjected to integrity investigation.

Investigation of the back of anchor heads

Investigation of the back of anchor heads is conducted on anchors for which the head metal can be removed by releasing tensioning force in order to check for corrosion of tension members, filling and degradation state of corrosion-inhibiting materials, condition of anticorrosion structures, and seepage of underground water.

Performance maintenance test

The performance maintenance test involves releasing the tensioning force of anchors, investigating the locking force of fixed anchors and the tensioning strength and apparent free length of tendons from the load and displacement properties. The data are used as a reference for integrity investigation.

Test of rust-preventive oil

The test of rust-preventive oil involves checking efflux and degeneration of oil by removing the head caps, sampling the oil by releasing the tensioning force and removing the distribution plates when necessary, and conducting chemical tests to examine air and water pollution caused by oil efflux and deterioration by heat and oxidization.

Monitoring of residual load

Monitoring of residual load, which is an index of anchor integrity, is performed through lift-off tests, performance maintenance tests, and sometimes, by installing axial load meters for continuous monitoring.

Supplementary investigations

Supplementary investigations are performed when anchor integrity is vulnerable due to chemical properties of the peripheral ground and underground water. The pH, acidity, and propagation of anaerobic sulfate-reducing bacteria are tested.

Remedial measures

For the various available remedial measures, the manual gives a systematic description of the technologies used so as to serve as a reference for technical communication.

Especially, the manual emphasizes the need for ensuring the performance of anchors throughout their design service life by implementing measures for those that have shown or possibly show a drop in performance to below the required level during inspection. Such measures were previously dismissed since anchors were believed to maintain the required performance level throughout their service life when they were installed.

Remedial measures should be taken by 1) considering anchors as a set when they have two or more factors in common regardless of the deterioration of the anchored structure and the fact that each anchor is investigated for integrity; 2) designing repair and reinforcement measures by assessing the present behaviour, using the latest information and breaking away from the initial design conditions; and 3) thoroughly investigating the need for restoring and using anchors and combination with other appropriate methods.

When a problem is found on the entire slope before taking remedial measures, the causes should be removed first. If this is not possible, the appropriateness of using anchors should be investigated.

Drop in performance, including tensioning force, of anchors and improvement by remedial measures are schematically organized to systematically show the available measures.

Usually, the performance of anchors drops in the course of time, a typical example of which is tensioning force of fixed anchors. Anchors should retain the required performance throughout their service life, and remedial measures should be taken when an inspection identifies anchors that do not satisfy the requirements or ones predicted to show reduced performance within several years. Such anchors are to be treated by one of the following remedial measures based on their performance at the time of inspection and the remaining service life, and are either improved for durability, repaired, reinforced or renewed.

Improving durability

Anchors that demonstrate satisfactory performance at the time of inspection but are predicted to drop below the required level within the design service life are to be improved for durability. There are two methods: 1) decelerating the drop by removing the cause; and 2) improving performance to ensure it for the remaining service life.

Repairing and reinforcing

When the performance drops to a level unsuitable for service, the anchors are repaired and reinforced to improve their performance. Measures that are to be taken several times during the service life are called "repairing", and those to be performed only once are called "reinforcing".

Renewing

When anchorage performance is difficult to ensure by improving durability, repairing or reinforcing, new anchors are installed. This method is called "renewing".

Prolonging life

Ensuring longer service of anchors that are sound at the time of inspection is called "prolonging life". This can be achieved either by decelerating performance deterioration or improving performance as in improving durability.

Temporary measures and urgent measures

Temporary measures are defined as measures taken to retain the immediate performance of anchors when their performance level drops far below the required level and full-scale measures cannot be taken for a while due to monetary restrictions, etc. The manual states that temporary measures, unlike ordinary measures, should be investigated in terms of economy and efficiency. The need for taking full-scale measures afterwards is also emphasized to call owners' attention to the fact that temporary measures are only temporary. Measures taken to ensure the safety of third persons against damage to structures and/or the safety of workers while taking emergency measures are called "urgent measures".

Training engineers involved in maintenance

The Japan Anchor Association issues licenses for ground anchor engineers aiming to cultivate engineers who use advanced technologies for anchor work and maintenance. To date, 3,038 ground anchor engineers have been certified and are involved in maintenance of anchors. The actualization of advanced anchor maintenance requires continuous training of highly capable engineers as well as preparation of manuals.

Conclusions

Methods for maintaining anchors have not been systematically organized, resulting in structures losing their function due to the drop in anchor performance and inappropriate measures causing economic loss.

The safety of structures that use anchors can be ensured and the maintenance cost can be reduced by proposing a flow of maintenance work ranging from daily inspection to integrity investigation and remedial measures. The proposed maintenance methods should be tested at sites and revised into more practical ones.

A standard for the design and construction of anchors has been established by the Japanese Geotechnical Society, but there is little information on maintenance. Since the effects of maintenance methods can be improved by recording data, installing axial force meters, etc. at the time of designing and installing anchors, the precision of the proposed maintenance methods would be enhanced by using them in practice and the results should be reflected in revised standard that cover the entire life cycle of anchors including design, construction and maintenance.

Just as standard on rust prevention of anchor materials promoted the development of rust-prevention technologies, the systematic organization of specific maintenance methods is expected to promote the development of technologies for maintaining existing anchors. There is high demand for structures that have easily exchangeable and repairable parts vulnerable to deterioration and technologies for assessing anchor integrity.

Acknowledgement

The authors wish to thank the members of the Japan Anchor Association for their co-operation in conducting surveys.

References

1. The Japanese Geotechnical Society: Ground anchor design and construction standard and explanations, March 23, 2000

2. Japan Anchor Association: Manual for constructing ground anchors, May 26, 2003

Structural integrity and durability of ground anchors
- general programme of research in Japan -

Junichi Yamazaki, Tatsuro Sueyoshi and Nobuyuki Urakawa, Japan Anchor Association, Tokyo, Japan

Abstract

During the maintenance of ground anchors, anchors requiring detailed further investigation have been evaluated by conducting integrity investigations. The integrity investigations can consist of a preliminary survey, detailed investigation of the anchor head by investigation of the locking unit arrangement, condition of corrosion inhibiting compound after exposing the anchor head, a lift-off test to confirm anchor residual load, investigation below the bearing plate after removal of the locking unit, anchor performance confirmation test, property test of corrosion inhibiting compound and monitoring of axial force in anchors equipped with a load cell.

The Public Works Research Institute and Japan Anchor Association have been jointly compiling the "Manual For Inspection And Integrity Investigation Of Ground Anchors" since 2002. This paper describes anchor integrity investigation in Japan.

Flow of anchor integrity investigation

An anchor integrity investigation will be conducted for anchors whose integrity is brought into question as the result of an inspection. The condition of the anchor will be checked in detail and the integrity evaluated. An integrity investigation will comprise a preliminary survey, detailed investigation of the anchor head, lift-off test, investigation below the bearing plate, anchor performance confirmation test, tests on the corrosion inhibiting compound (CIC), and monitoring. The integrity of the anchor is evaluated from the integrity investigation. However, countermeasures such as repair works or enhancement are considered when any problem is discovered. The flow of the anchor integrity investigation is shown in figure 1.

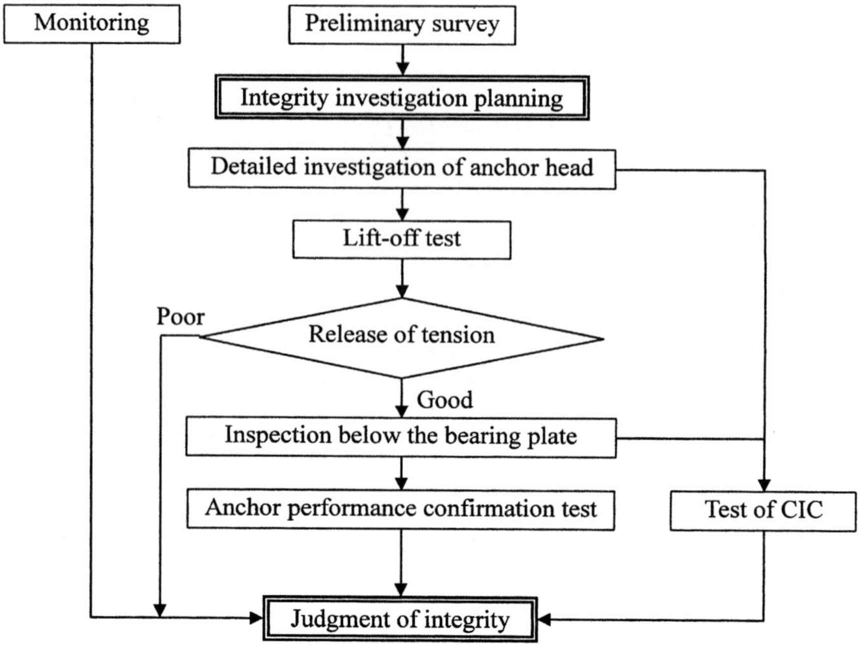

Figure 1. Flow of anchor integrity investigation

Method of integrity investigation

Preliminary survey

A preliminary survey is conducted in order to gather references for deciding upon and planning the type and method of integrity investigation. The preliminary survey includes investigation of published documents, site reconnaissance and primary stage of preliminary survey. The investigation of published documents, such as maintenance, management and investigation records, will include examination of specifications, situation and variation of anchors. During the site reconnaissance, dimensional and positional checks are made against drawings, the site situation is examined and an evaluation made of the requirements for scaffolding, transportation of construction materials, machinery and electric power supply. The preliminary survey also includes a basic visual inspection of the protection to the anchor head and any changes in the anchored structure and variations in the whole structure. Leakage of groundwater is also examined.

Plan of integrity investigation

The type and method of test to be conducted, as well as an investigation into feasibility, are examined from the information collected in the preliminary survey. An integrity investigation planning sheet designating the implementation plan and operation management in detail is then developed. Safety at the site and its vicinity as well as environmental protection are considered in the plan. The planning sheet will include anchor specification data gained from the preliminary survey, the type of integrity investigation to be conducted, the location and identification number of the investigation, the method of test and investigation, the equipment and materials to be utilized, method of operation, management and temporary works planning. The number of tests and investigations differ by grade of importance of the anchors being considered and number of anchors installed. The approximate number of tests is estimated based on Table 1.

Table 1. Estimate of number of test and investigations

Type of test and investigation	Estimate of number of tests and investigation	Reference
Detailed investigation of anchor head (Appearance investigation)	Decided by preliminary survey	Refer to anchor inspection
Detailed investigation of anchor head (Exposure investigation)	Anchor group judged to be required for integrity investigation and 20% from the rest (more than 5 anchors).	
Lift-off test	Anchor group judged to be required for integrity investigation and 10% from the rest (more than 3 anchors).	Confirmation of head part of structure
Investigation below the bearing plate	Anchor group judged to be required for integrity investigation and 5% from the rest (more than 3 anchors).	Possibility of detensioning
Anchor performance confirmation test	All anchors: below the bearing plate investigation	
Corrosion inhibiting compound test	Area where abnormality is found by visual investigation	Anchor head, below the bearing plate
Monitoring	Anchors equipped with load cell	

Note 1) Anchor group means anchors in the vicinity (up and down, left and right)

Note 2) In case an abnormality is found during the integrity investigation, anchors in the vicinity (up and down, left and right) will also be examined.

Detailed investigation of anchor head

The detailed investigation of the anchor head involves a visual inspection and an investigation after exposure of the head part. The visual inspection will consider variations in the concrete (loosening, flaking damage, deterioration), variations in the anchor cap (damage, deformation, deterioration), leakage of corrosion inhibiting compound and presence/source of spring water. In the exposure investigation, the anchor cap is removed and the condition of the locking unit, tendon and filling of corrosion inhibiting compound is investigated. The investigation item and content of the detailed investigation for the anchor head are summarized in Table 2. The result of this investigation is useful for future reference for maintenance and management of the anchor and can be used to diagnose the cause of changes in appearance in the future. The remaining length of the tendon for restressing is then confirmed and implementation of the second examination by lift-off test and possibility of tension load removal are determined.

After conducting the detailed investigation of the head part, protection to the anchor head is replaced to prevent corrosion of the tendon and locking unit.

Table 2. Investigation item and content of detailed investigation for anchor head

Investigation item	Investigation content	Investigation method	Application
Concrete head	Loosening, damage, flaking, cracking, deterioration	Visual inspection	
Anchor cap	Damage, variation, deterioration, fixing, seal deterioration	Visual inspection	
Corrosion inhibiting compound	Oil leakage, amount, change of quality,	Visual inspection	Performance test of corrosion inhibiting compound
Tendon (remaining length part)	Drawing in, corrosion situation, scratch, damage, remaining length for restressing	Visual inspection, measurement	Measurement of section loss rate
Locking unit	Corrosion situation	Visual inspection	
Bearing plate	Loosening, variation, corrosion, water welled out from the below the bearing plate, deterioration of paint	Visual inspection, hammering sound	

Figure 2. Investigation after exposing anchor head

Lift-off test

The residual load in anchors is confirmed by lift-off test and the anchor integrity is evaluated by comparing this load with the design load and lock-off load. In the test, the remaining length of tendon for restressing is first confirmed and, if necessary, a tool for stressing is manufactured. The loading in the test is simple and the load and head displacement are measured from 10 to 20kN. The lift-off load (residual load) is determined from the variation point on the load/displacement curve. The apparent anchor free length and abnormality of anchor can be confirmed from the relationship between load and displacement after lift-off. Anchor integrity is confirmed in case that residual load is more than 80% of the lock-off load and less than the design load. Integrity is evaluated as shown in Table 3.

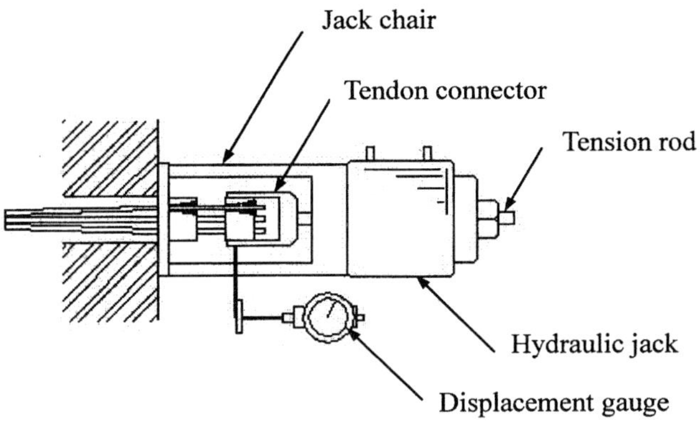

Figure 3. Lift-off test equipment

Figure 4. Lift-off test

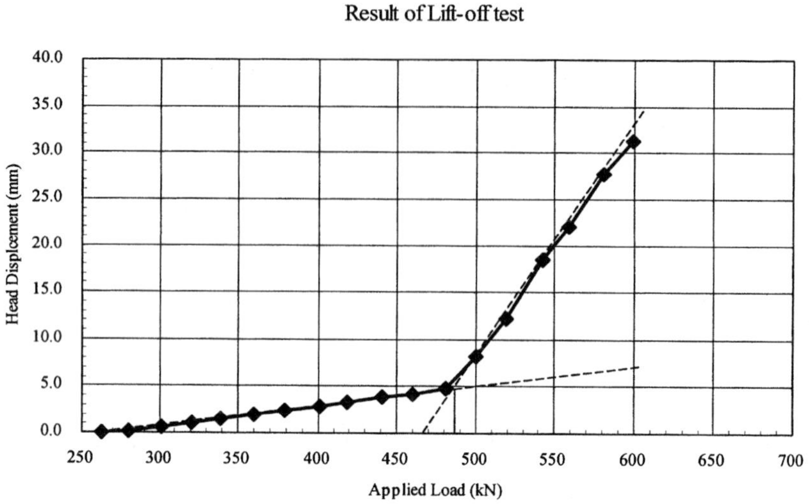

Figure 5. Example of Lift-off test

Table 3. Residual load and anchor integrity

Range of residual load	Integrity	Situation	Countermeasure
0.9Py	E	Possibility to rupture	Emergency countermeasure is necessary.
1.1Ta	D	Possibility of dangerous situation	Countermeasure is necessary.
Safe load of anchorage (Ta)	C	Beyond tolerance	
Design load (Td)	B		Necessity of countermeasure is examined by observing process.
Lock-off load (Pt)	A	Sound	
0.8Pt	A	Sound	
0.5Pt	B		Necessity of countermeasure is examined by observing process.
0.1Pt	C	Function is deteriorated drastically.	Countermeasure is necessary.
	D	Non-Functional	

Below the bearing plate investigation

After releasing the anchor load and removing the locking unit an investigation below the bearing plate can conducted. In cases where anchors are fixed by nuts, the tensile load is easily released and recovered. But when the anchor is fixed by wedges, it is necessary to confirm the remaining length of the tendon is sufficient for restressing and recovery of the load before the investigation.

During the investigation below the bearing plate, the tendons are examined for corrosion, the filling of the corrosion inhibiting compound is checked and any ingress of groundwater or other variations in condition are reported as detailed in Table 4. As with the anchor head investigation, the results of this investigation can be used as reference for future maintenance and management after examining the cause and phenomenon relating to the observed variations.

Table 4. Investigation item and content of below the bearing plate investigation

Investigation item	Investigation content	Investigation method	Application
Below the bearing plate area	Anticorrosion structure, scratch, water stoppage	Visual inspection	
Tendon	Corrosion, scratch, Damage	Visual inspection	
Corrosion inhibiting compound	Oil leakage, amount, changing of quality, deterioration	Visual inspection	Performance test of corrosion inhibiting compound
Infusion of groundwater	Infusion of groundwater, soil	Visual inspection	
Below the bearing plate	Variation, crack, free lime	Visual inspection and measurement	

Figure 6. Below the bearing plate investigation

Anchor performance confirmation test

The anchor performance confirmation test confirms the strength of the tendon, the pull-out load of the anchor and the restraining force provided by the anchor are all above the design load. From the test results the anchor can be judged to be effective for future service. The maximum test load is decided upon the proof test result at the time of construction rather than the specification. However, consideration is also given to the results of the detailed investigation of the anchor head and below the bearing plate investigation and the maximum test load reduced accordingly. In the test, loading is applied in cycles of 5 steps until the maximum test load is reached. The apparent free length of the anchor and the factor of safety over the design load for the anchor is confirmed from the relationship between load and displacement, the relationship between load and elastic movement.

Test of corrosion inhibiting compound (CIC)

When a change in color of the corrosion inhibiting compound is confirmed by visual inspection during the “detailed investigation of the anchor head” or the “below the bearing plate investigation”, a sample is taken and, if necessary, tested for its corrosion inhibiting property. The property test can be omitted when the number of installed anchors is few and replacement of the compound would be more economical than testing. When deterioration of anticorrosion effectiveness is observed from the test, replacement of corrosion inhibiting compound is necessary.

Anchor performance confirmation test result

Anchor No.	A-XX		Date	XXXXX
Anchor specification				
Anchor type	XXXXX		Design load	800.0 kN
Tendon type	φ12.7×8		Lock-off load	560.0 kN
Tendon area	789.7 mm2		Residual load	415.0 kN
Elastic Modulus	192.0 kN/mm2			
Free length	24.0 m		Tendon free length	25.2 m
Fixed length	8.0 m			

Cycle No.	Design load	Datum load	Step 1	Step 2	Step 3	Step 4	Step 5	
Load (kN)	800.0	100.0	240.0	480.0	720.0	960.0	1100.0	
Head displacement (mm)		0.0	21.4	59.8	102.9	148.6	180.2	
Elastic movement (mm)		0.0	19.8	57.0	98.8	142.7	171.6	
Residual movement (mm)		0.0	1.6	2.8	4.1	5.9	8.6	
Theory value (mm)		0.0	23.3	63.2	103.0	142.9	166.2	
Upper bound value (mm)		0.0	25.6	69.5	113.4	157.2	182.8	
Lower bound value (mm)		0.0	20.9	56.8	92.7	128.6	149.6	

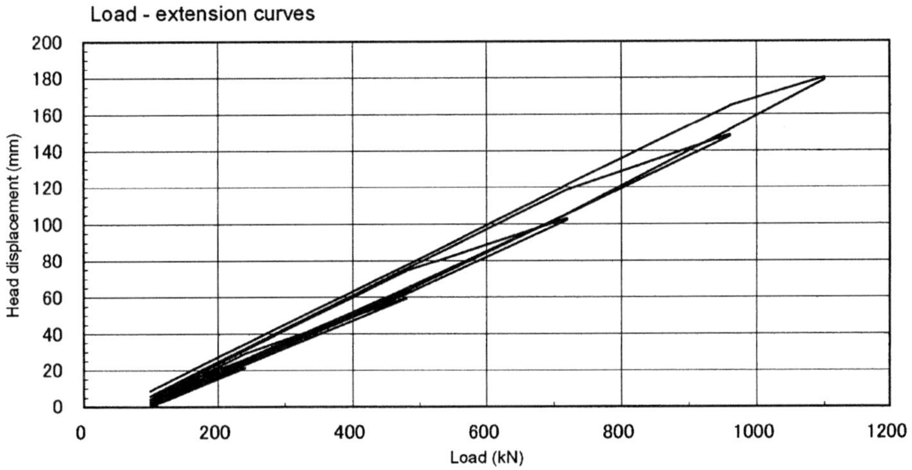

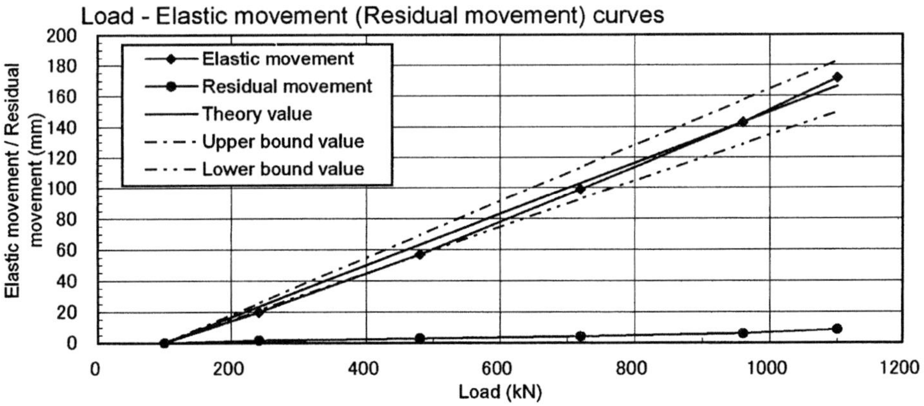

Figure 7. Example of anchor performance confirmation test result

Monitoring

A load cell (axial meter) is installed to monitor changes in anchor residual load. By monitoring the axial load, decreases in load as a result of ground creep, increases of earth pressure, changes in the retained slope or increases in load caused by rising groundwater can be measured regularly. The results of monitoring are useful as a reference for judging anchor integrity and estimation of future trends. However, load cells have some implications for the future as they are expensive and have limited durability. Currently, strain gauge load cells, differential transformer load cells and hydraulic disk type measuring instruments are utilized.

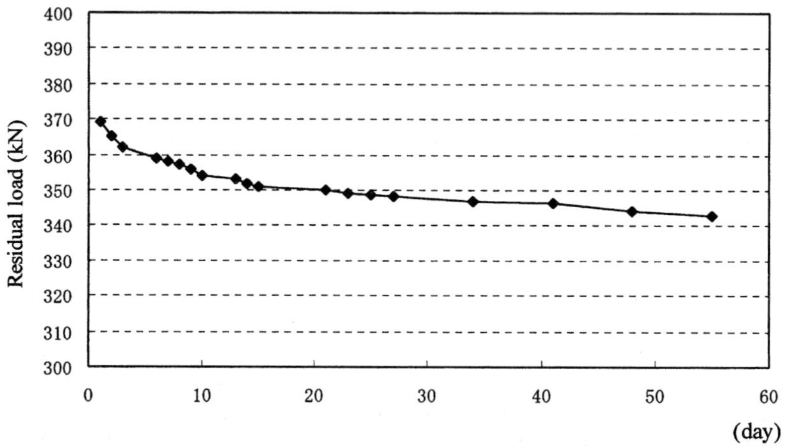

Figure 8. Example of monitoring of residual load

Evaluation of investigation result

The integrity of each anchor is clarified from the results of the anchor integrity investigation. Therefore, the integrity of the structure can be evaluated totally from the investigation result. If the investigation concludes that there is a risk of an accident to a third party, urgent countermeasures should be implemented. Possible countermeasures include repairing, enhancement, improvement to durability and renewal.

Summary

Anchor integrity investigation is important for progressing maintenance management by accurately recognizing the current situation of anchors. In the future, widespread investigation of anchors will be conducted practically in accordance with the new "Manual For Inspection And Integrity Investigation Of Ground Anchors".

Engineering assessment of the grease-filled annulus of sheathed prestressing strand for use in permanent ground anchorages

Tony Barley Single Bore Multiple Anchor Ltd, England
Devon Mothersille Geoserve Global Ltd, England

Abstract

The intent of this paper is to briefly review the processes currently adopted to manufacture greased and sheathed strand. The paper also assesses the success of the greasing operations applied to the free length of strand tendons and reports on tests, carried out to investigate the efficiency of greasing, on samples produced by three different manufacturers.

The test work has revealed an unacceptable variation in product quality. Codes of practice recommend that the annulus between the individual strand and the surrounding sheath should be "completely filled with a corrosion inhibiting compound." Where specialized equipment and factory controlled conditions were employed, the efficiency of grease infilling ranged from 76% to 100%. Investigation of strand greasing from one particular site indicated a range of 0% to 100. A practical acceptance criterion ≥ 80% is recommended and that all sheathed strand suppliers should provide factual data confirming compliance with this criterion that can also be checked routinely in the field.

Introduction

The nature of ground anchorage technology requires the placement of steel elements in ground which are subsequently subjected to tensile forces in order to stabilize a slope or structure. Given these circumstances it is imperative that the steel is protected from the environment to ensure that the anchorage is fit for its intended purpose for the design life of the anchored structure.

In the late 1960's when the relatively novel anchorage technology was introduced into the geotechnical industry, ground anchorage specialists began to realize the potential problem of long term tendon deterioration. By 1970 a prestressing strand protection system comprising both the strand and grease within an extruded plastic sheath had been developed. Since this time concerns have been raised regarding the efficiency of the greasing in products supplied to the industry particularly when grease is pre-applied to the strand which is then advanced into a preformed sheath.

BS 8081 Clause 8.2.3.1 states "to fulfil the recommendations.... protective fluids usually aim to exclude a moist gaseous atmosphere around the metal by totally enclosing it within an impervious covering or sheath". Furthermore, clause 8.2.4.2.3.5 states "The

Ground anchorages and anchored structures in service, Thomas Telford, London

method of filling should be such as to displace all fluids during injection and allow application of pressure to the filler to enable it to occupy all empty volume".

The comparable statement from EN1537:2000 detailed in Table 3 for corrosion protection of the free length of permanent anchors, requires "a plastic sheath to individual tendons filled completely with flexible corrosion protection compound....". It is the industry's adherence or non-adherence of these clauses that forms the basis of the work reported in this paper.

Observation

There is a general assumption throughout the industry that the annulus between the strand and the sheath is filled with grease. However, observation of dissected strands, or longitudinally cut sheaths have over a 30 year inspection period by the authors established that greased and sheathed strands supplied for use in permanent ground anchorages are not always manufactured to the required quality. Visual observations of in-situ anchor heads have revealed that due to inefficient greasing the strand sheathing may in fact form a drainage path for the passage of water (see Figure 1). Alternatively strand sheaths have been observed to contain grout at the head of the anchor which is indicative that the annulus was not originally grease filled (see Figure 2). Such observations raise immediate concern as to whether the corrosion protection prefabricated on strand tendons, are always of appropriate quality to ensure fulfilment of their intended purpose.

Figure 1 Water flowing through annulus between strand and sheathing

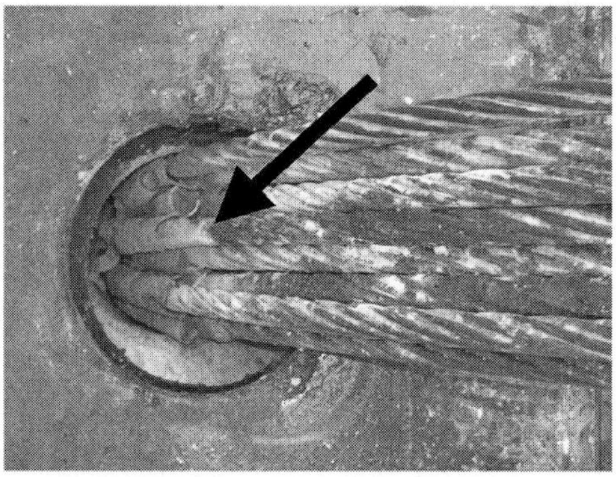

Figure 2 Grouted sheathing at the anchor head

Greased and Sheathed Strand – the process

External and Internal Strand Wire Greasing

Seven wire prestressing strand with a bare steel surface is generally delivered to the factory on a strand reel. The reels are mounted on a frame which allows the rotation and dispensing of the strand through powered rubber faced rollers ("The strand pusher"). Such a system fully aligns the strand onto a horizontal axis that is maintained throughout the application process. The moving strand then enters a rotating wire splitting device which parts the 6 peripheral wires from the king wire and provides typically a gap of 5mm between the wires. The corrosion resistant grease is then poured or injected to within the body of the seven wires such that on wire closure back to its original lay, the interstices between all the wires are grease filled (see Figure 3).

Figure 3 Separation of strand into individual wires for internal greasing (courtesy Geotechnical Engineering)

Pre-formed Sheath

Preformed sheaths typically consist of a semi rigid high density polyethylene with a wall thickness between 0.5 and 1.5mm. Internal sheath diameters are frequently between 0.8 and 1.0mm greater than the external diameter of the strand, resulting in the average annulus thickness of 0.4 to 0.5mm.

Sheath lubrication and Strand Penetration

The entry of a grease coated strand into a sheath of nominally larger diameter does not guarantee that the grease enters the sheath or provides any uniformity in distribution in the annulus, either circumferential or axially. In this respect it is vital that in the grease coating possesses the correct viscosity at the point of entry to the sheath and maintains appropriate viscosity whilst travelling within the sheath. In some instances the advanced internal lubrication of the sheath by the passage of a slug of grease or oil ahead of the strand has been found to be beneficial.

Extruded Sheath Application

This method has been used successfully for application of a continuous grease filled sheathing to long lengths of polystrand since 1970. Immediately after the grease application to the wire core and the external face of the strand the strand is advanced through a heated extrusion process which applies a high density polyethylene sheath of controlled diameter and wall thickness. A water cooling system thus ensures immediate setting on the extruded plastic. Owing to the absence of materials other than strand, grease and extruded polyethylene during the application process no air voids can remain within the annulus. This process is typically applied to the continuous lengths of strand ranging from 300 to 1000m.

Investigatory Work

Sources of Samples

From a delivered reel of sheathed strand or from selected tendons, random samples of one metre lengths of sheathed strand were cut with a disc cutter. On completion of dissection each exposed end of the sample was tightly taped to prevent any discharge of grease from within the sheath during transportation.

Sheathed Strand Type A was sourced from 3 separate coils of full length sheathed strand from one supplier. Three one metre lengths were selected at random during anchor tendon fabrication in a purpose made factory area.

Sheathed Strand Type B was removed from 3 separate anchor strands which were considered as surplus to requirements on a permanent works contract. The strands had been sheathed and greased in typically 25m lengths of free length tendon.

Sheathed Strand Type C were taken from sample strands fabricated by a supplier at a new location specifically for investigation of sheath and greasing performance using a recently assembled fabrication plant.

Sheathed Strand Type D samples were taken from a number of rejected anchor tendons fabricated for permanent works anchors of sheathed strand system C.

Measurements and Presentation

The operations and apparatus used to measure precise displaced and filled volumes at various stages of the grease coating and grease removal are detailed Figure 4. Figures 5 and 6 show the lowering of greased strand into the water filed container and the position of the measuring cylinder, respectively.

Table 1 is prepared from the volumetric measurement data taken throughout the investigation tests.

It records the displacement volume of a one metre length of greased strand withdrawn from the sheath and the measured internal volume of the greased sheath from which the strand was withdrawn. It also reports on the displacement volume of the one metre length of strand after removal of external grease (but not the removal of grease from within the strand wires) and the internal volume of the one metre sheath after removal of all grease from the internal surface. All measured volumes are considered to be to an accuracy of ± 0.5ml based on accuracy of reading of the calibrated measuring cylinder.

Table 2 analyses the volumetric results of Table 1 to establish the difference between those of the removed greased strand and the measured internal volume of the sheath after strand removal. The difference represents the volume of a void within the sheath not filled with grease. The difference between the internal volume of the cleaned sheath and the external volume of the degreased strand represents the measured volume of the annulus (which the Codes require to be grease filled).

The volume of the annulus minus the volume of the void (unfilled with grease) divided by the volume of the annulus provided the percentage “fullness” of the annulus, which can be seen to vary from 0 to 100%. It should be noted that the accuracy of these results diminishes at the extreme ends of the range since the potential error in volume measurement (± 0.5ml) has, where volumetric measurement is small, the greatest influence. Conversely in the middle range of say 30 to 70% the potential cumulative error has a much smaller influence on the presented result (± 2% or less).

Table 3 provides independent data used to estimate volumes based on Vernier readings for diameter of strand and internal diameter of the sheaths. Due to the irregularity and curvature of the empty sheathing, these results present a small degree of error. However they do allow comparison of calculated volumes based on dimension data with volumes established using water displacement methods (Table 1).

Strand Ref.	Volume of withdrawn Strand + Grease (Vgs)	Volume of Strand after degreasing (Vs)	Internal Volume of greased Sheath (Vpgs)	Internal Volume Sheath after internal degreasing (Vps)	Sample Length
A1	188.5	178.0	188.0	193.0	1000
A2	188.5	177.5	188.5	196.0	1000
A3	188.0	177.0	189.5	196.0	1000
B1	194.5	176.0	201.0	218.5	1000
B2	192.5	176.5	200.0	217.5	1000
B3	198.0	179.5	203.0	222.0	1000
C1	200.0	180.0	203.0	220.0	1000
C2	198.0	181.0	198.0	219.5	1000
C3	198.5	180.0	201.0	218.5	1000
D4	173.0	172.0	211.0	218.0	990
D5	184.0	173.0	208.0	220.0	998
D6	191.0	173.0	202.0	221.0	989
D7	174.0	172.0	211.0	219.0	991
D8	188.0	173.0	203.0	220.0	990
D9	187.0	173.0	205.0	218.0	991
D10	184.0	173.0	199.0	219.0	992
D11	183.0	172.0	209.0	221.0	1002
D12	175.0	173.0	216.0	222.0	1002
D13	182.0	174.0	207.0	220.0	1000
D14	180.0	174.0	211.0	221.0	999
D15	189.0	173.0	204.0	220.0	1001
D16	183.0	172.0	210.0	223.0	999
D17	173.0	173.0	208.0	224.0	1001
D18	168.0	173.0	214.0	218.0	1001
D19	191.0	174.0	199.0	221.0	1000
D20	183.0	176.0	209.0	222.0	1000
D21	198.0	173.0	198.0	222.0	1000
D22	181.0	173.0	216.0	221.0	1001
D23	193.0	172.0	199.0	223.0	1000
D24	197.0	173.9	198.0	222.0	1000

Table 1 Measured volumes (ml) and sample lengths (mm)

Strand Ref.	Internal Greased Sheath Volume minus Greased Strand Volume (i.e. void)	Internal Sheath Volume minus Strand after degreasing (Annulus Volume) (Va)	Annulus Vol - Void Volume / Annulus Volume %	Percentage "Fullness" of Annulus (%Gis)
A1	-0.5 (0)	15.0	1.0+	100
A2	0.0	18.5	1.0	100
A3	1.5	19.0	0.92	92
B1	6.5	42.5	0.85	85
B2	7.5	41.0	0.82	82
B3	5.0	42.5	0.88	88
C1	3.0	42.0	0.93	93
C2	0.0	38.5	1.00	100
C3	2.5	38.5	0.93	94
D1	38.0	46.0	0.17	17
D2	24.0	47.0	0.49	49
D3	11.0	48.0	0.77	77
D4	37.0	47.0	0.21	21
D5	15.0	47.0	0.68	68
D6	18.0	45.0	0.60	60
D7	15.0	46.0	0.67	67
D8	26.0	49.0	0.47	47
D9	41.0	49.0	0.16	16
D10	25.0	46.0	0.46	46
D11	31.0	47.0	0.34	34
D12	15.0	47.0	0.68	68
D13	27.0	51.0	0.47	47
D14	35.0	51.0	0.31	31
D15	46.0	45.0	0.00	0
D16	8.0	47.0	0.83	83
D17	26.0	46.0	0.43	43
D18	0.0	49.0	1.00	100
D19	35.0	48.0	0.27	27
D20	6.0	51.0	0.88	88
D21	1.0	49.0	0.98	98

Table 2 Evaluation of the volumetric measurement data to estimate the "Fullness" of the annulus between the strand and the inside of the sheath

Strand Ref	Dimensions; Sheath OD Sheath ID Strand OD	Calculated Sheath Volume per metre	Average of Displacement Volume of Sheath/m	Calculated Strand Volume per metre	Average Displacement Volume of Strand/m
Type A	18.76 16.21 15.28	206.4	195.0	183.3	177.5
Type B	19.2 16.7 15.4	219.1	219.3	186.3	177.3
Type C	18.9 16.6 15.32	216.4	219.3	184.3	180.3

Table 3 Strand and sheath dimensions (mm) and evaluated volumes (ml)

Designated nomenclature and result evaluation method

Volume of greased strand withdrawn from the sheath	**Vgs**
Re-immersion of strand after degreasing with solvent to determine volume	**Vs**
Measurement of volume of plastic sheath, complete with any retained grease by sealing the basal end and measuring volume of water to fill the void	**Vpsg**
Volume of water to fill sheath, having cleaned sheath with solvent	**Vps**
Volume of annulus, Va	**Vps - Vs**
Grease volume around strand, Gs	**Vgs - Vs**
Grease volume remaining within the plastic sheath, Gps	**Vps-Vpsg**
Percentage grease infilling of the annulus, %Gia	**(Gps+ Gs)100/Va**

Discussion

Sheathed Strand Type A utilises a dual extruding process for the application of grease and sheathing. Test results are excellent in establishing almost complete filling of the 0.5mm wide annulus between strand and sheath. Two samples indicated 100% "fullness" and one 92% "fullness"

Sheathed Strand Type B applies thermostatically controlled temperature controlling grease to the strand then advances the grease coated strand into predetermined lengths of open sheathing with an annulus width of 0.6mm. Results indicate "fullness" of 82, 85 and 88%.

Sheathed Strand Type C applies the same principles as System B and achieved "fullness" values of 93, 100 and 94%.

Sheathed Strand Type D applied the same methods as systems B and C but investigation of such bad results confirmed that the control of the grease temperature and hence grease viscosity and general production was subject to poor operation management. As a result the volumetric measurements were indicative of unsatisfactory products with no measurable volume of grease being present in the annulus of one sample. Inspection of that extruded strand showed the existence of only a very thin layer of grease down one side of the strand and the other side totally bare of any grease protection.

Of the remaining tests:

- More than 50% were indicative that the annulus was less than 50% full (17 to 49% range)
- 25% between 50% and 80% full
- Only 20% between 80% and 100% full
- Only two tested strands from Type D, supplied for permanent production anchors, attained a value comparable with the prior investigation tests used to demonstrate the adequacy of the system performance (Type C).

In utilising the developed displacement system it must be acknowledged that the potential cumulative inaccuracy of volumetric measurement (± 0.5ml) is greatest when the annulus is recorded to contain the maximum volume of grease. For example
such cumulative corrections would result in a ± 3.0% deviation on an established fullness value of 80%. This potential deviation must be appreciated when results are considered and recommended acceptance criteria evolved. However such nominal error cannot severely influence conclusions drawn.

Although the data produced using the displacement method cannot positively identify;

- at what percentage fullness the annulus would allow passage of water within the sheath
- whether one side of the annulus is full and the other empty,
- whether the strand is lying eccentrically located within the sheath.

The test can identify when bad practice is applied and when the manufactured products are unacceptable for installation and fail to provide reasonable compliance with international standards.

It may be appropriate to consider that Coated Strand Systems utilised in systems A and C are indicative that good practice does exist and the efficiency of annulus greasing can result in 90 to 100% fullness. In systems A the three test samples exceeded 80%.

The European Standard for Ground Anchors EN 1537:2000 requires that compliance tests are implemented by all Suppliers but the Code does not establish criteria for acceptance.

It is recommended tests of the above type or a type that provide similar data, be carried out to ensure good practice is established by the suppliers. Suppliers should then issue, for all batches, appropriate documentation similar to those recording strand tensile capacity for compliance with the required characteristic strength requirements.

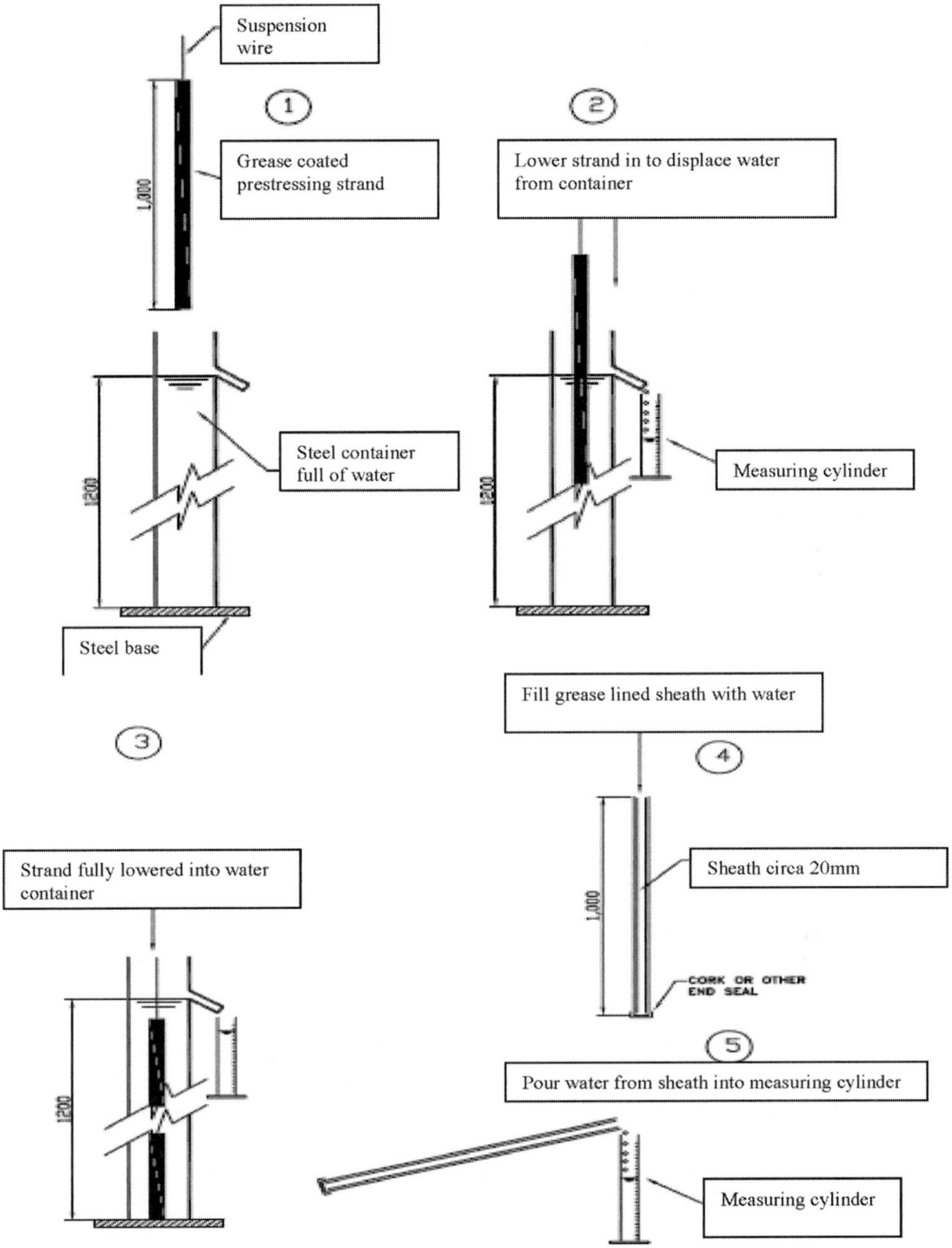

Figure 4 Measurement of volume of grease coated strand after withdrawal from strand sheath and measurement of internal volume of sheath after withdrawal of grease coated strand

Figure 5 greased strand lowered into water filled steel container

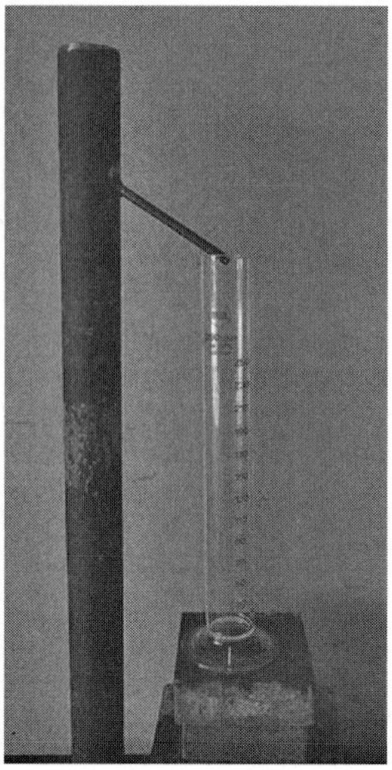

Figure 6 Steel container adjacent to measuring cylinder

Summary and closing remarks

Field observations over a number of years supported by investigation into the quality of the factory applied corrosion protection, encompassing prestressing strand supplied to the ground anchorage industry, has revealed an unacceptable variation in quality. Both in the UK and abroad, codes of practice recommend that the annulus between the individual strand and the surrounding sheath should be "completely filled with a corrosion inhibiting compound." Tests on samples taken from strands supplied for the use in the industry indicate that results that the "fullness" of the annulus can vary from 0 to 100% depending on the material source.

A practical acceptance criterion ≥ 80% fullness is suggested and it is also recommended that all sheathed strand suppliers should provide factual data confirming compliance with this criterion that can also be checked routinely in the field. Results of such tests should be published by the suppliers of the coated strand systems in conjunction with the results of strand load tests currently made available as a quality standard.

References

BS8081 (1989) "Code of practice for ground anchorages". British Standards Institution, London.

EN1537 (2000) "Execution of special geotechnical works – Ground anchors". European Committee for Standardisation, Central Secretariat, rue de Stassart, 36, B-1050 Brussels, Belgium.

Rock Anchors for North American Dams: The Development of the National Recommendations (1974-2004)

Donald Bruce, Geosystems, L.P., Venetia, PA, U.S.A.
John Wolfhope, Freese and Nichols, Inc., Austin, TX, U.S.A.

Abstract

This paper describes the third major goal of Phase 1 of a National Research Program dealing with rock anchors for North American dams. Since 1974, there have been 5 successive versions of the "Recommendations" for ground anchor practice which have each effectively acted as a national standard. An analysis is provided which illustrates the evolution of theory and practice with time, with particular reference to the philosophies and details of corrosion protection.

Introduction

During the period 2005-2006, Phase 1 of the National Research Program into the use of rock anchors for North American dams was undertaken. This work had three goals:

(i) complete a bibliography of all technical papers published on the subject of dam anchoring in North America;
(ii) create a database containing as much information as possible on each dam anchored in North America; and
(iii) conduct a comparative review of each of the five successive versions of the national "recommendations" documents which have been published in the U.S. since 1974.

This paper describes the results of the third research goal, with a particular focus on the evolution of corrosion protection concepts and details. The findings of the first two research goals are presented in the companion paper, "Rock Anchors for North American Dams: The National Research Program Bibliography and Database," also published in these Conference Proceedings. The current authors were the Co-Principal Investigators for this project, funded by a consortium of American and Japanese interests. They relied heavily on the co-operation of specialty contractors and specialist post-tensioning suppliers who provided access to historical records.

General Statement

Current research indicates that the first U.S. dam to be stabilized by high capacity prestressed rock anchors was the John Hollis Bankhead Lock and Dam, Alabama (first 6 test anchors and 16 production anchors installed from 1962). This project was completed for the U.S. Army Corps of Engineers who had sufficient confidence in the technology (and, presumably, a pressing need for it!) that they were the sponsor for most of the half dozen or so similar applications in the six years that immediately followed. The U.S. Bureau of Reclamation began using anchors to stabilize appurtenant structures at dams in 1967. The Montana Power Company was also an early proponent. In those days, the technology was largely driven by the post tensioning specialists, employing the same principles and materials such as used in prestressed/post tensioned structural elements for new buildings and bridges. The "geotechnical" inputs, i.e., the drilling and grouting activities, were typically subcontracted to drilling contractors specializing in site investigation and dam grouting in the west, and to "tieback" contractors in the east.

Recognizing the need for some type of national guidance and uniformity, the Post Tensioning Division of the Prestressed Concrete Institute (PCI) formed an adhoc committee which published, in 1974, a 32-page document entitled "Tentative Recommendations for Prestressed Rock and Soil Anchors." It is interesting to note (Table 1) that half of the document comprised an appendix of annotated project photographs intended to illustrate and presumably promote anchor applications, including dam anchors at Libby Dam, Montana, and Ocoee Dam, Tennessee.

After publication of its document, the Post Tensioning Division of PCI left to form the Post Tensioning Institute (PTI) in 1976. Successive editions of "recommendations" were issued in 1980, 1986, 1996 and 2004. As general perspective to the development of concepts, Table 1 provides an analysis of the relative and absolute sizes of the major sections in each successive edition. It is immediately obvious that the original documents stressed "applications" – in an attempt to promote usage – while the most recent edition provides very detailed guidance (and commentary) on the "big five" in particular (i.e., Materials, Design, Corrosion Protection, Construction, and Stressing/Testing).

Table 1. Number of Pages in Major Sections of Successive U.S. "Recommendations" Documents

ASPECT	1974	1980	1986	1996	2004
Materials	1	2	2	8	10
Site Investigation	0	1	1	1	2
Design	2	6 ½	6 ½	12+ Appendix on grout/strand bond,	14
Corrosion Protection	1	4	5	10	14
Construction	7	9	9	10	15
Stressing and Testing	1	6	8	17	18
Bibliography/References	0	1	1	1 ½	4
Applications	16	18	0	0	0
Recordkeeping	0	1	1	1 ½	1 ½
Specifications	0	1	1 ½	2	2
Epoxy-Coated Strand	0	0	Very minor reference,	Frequent reference but no separate section.	10 Separate sections.
TOTAL PAGES	32	57	41	70	98

Detailed Comparison by Technical Topic

The structure of each successive edition has changed in the same way that the content has, although there are comparatively few structural differences between the 1996 and 2004 versions. The following comparison therefore is based on the structure of the 2004 version.

Scope and Definitions (Chapters 1 and 2)

The scope has remained relatively constant, and focuses on the anchors themselves (as components) as opposed to the analysis and design of the overall anchored system. A total of 72 technical terms are now defined, which represents a major expansion even over the 1996 edition: the first edition has 24 definitions, most of which, incidentally, remain valid and little changed.

Specifications, Responsibilities and Submittals (Chapter 3)

Whereas 1974 provided no insight into specifications and responsibilities, certain records were required to be maintained on the grouting operations. By 1980, however, specifications had been addressed, reflecting the need to tailor procurement processes to "experienced" contractors, "thoroughly experienced" and match the innovation of the technique with alternative procurements methods. It is notable that the three types of specification outlined in 1980 (namely open, performance and closed) have endured, although "closed" is now referred to as "prescriptive." Building on a 1996 innovation, the responsibilities to be discharged during a project — regardless of type of specification — were summarized in 2004 as shown in Table 2. Clear guidance is also provided on the content of preconstruction submittals and as-built records. The former also include the requirement for the contractor to prepare a Construction Quality Plan. Emphasis remains on the need for "specialized equipment, knowledge, techniques and expert workmanship" and for "thoroughly experienced" contractors. The obvious, but often ignored, benefit of "clear communication and close co-operation," especially in the start up phase of a project, is underlined.

Table 2. Tasks and responsibilities to be allocated for anchor works (PTI, 2004)

1.	Site investigation, geotechnical investigation and interpretation, site survey and potential work restrictions.
2.	Decision to use an anchor system, requirements for a pre-contract testing program, type of specification and procurement method, and contractor prequalification.
3.	Obtaining easements, permits, permissions.
4.	Overall scope of the work, design of the anchored structure, and definition of safety factors.
5.	Definition of service life (temporary or permanent) and required degree of corrosion protection.
6.	Anchor spacing and orientation, minimum total anchor length, free anchor length and anchor load.
7.	Anchor components and details.
8.	Determination of bond length.
9.	Details of water pressure testing, consolidation grouting and re-drilling of drill holes
10.	Details of corrosion protection.
11.	Type and number of tests.
12.	Evaluation of test results.
13.	Construction methods.
14.	Requirements for QA/QC Program.
15.	Supervision of the work.
16.	Maintenance and long-term monitoring.

Anchor Materials (Chapter 4)

The 1974 document very briefly refers to wires, strand, and bars, and to protective sheathing. In stark contrast, the current version has built to 10 pages providing definitive detail on materials used in each of the 10 major anchor components, with particular emphasis placed on steel, corrosion-inhibiting compounds, sheathings and grouts (cementitious and polyester).

Strong cross-reference to relevant ASTM standards is provided as a direct guide to specification drafters.

Site Investigation (Included in Chapter 6 – Design)

Not referred to in 1974, recommended first in 1980 and completely revised and expanded in 1996 and 2004, this issue now provides clear guidance on the goals and details of a site investigation program. "Minimum requirements" are recommended. However, this remains an area where the anchor specialist often has less "leverage" to influence since the costs associated with such programs typically exercise strong control over the scope actually permitted by the owner.

Corrosion and Corrosion Protection (Chapter 5)

Given the significance and relevance of this topic to the current conference, this subject is discussed separately below (Section 4).

Design (Chapter 6)

Judging from the relatively short and simplistic coverage of this aspect in 1974, it is fair to say that not much was really then *known* of the subject. Core drilling was considered absolutely necessary and preproduction pullout tests were "strongly recommended." However, two enduring issues were faced:

- The safety factor (on grout-rock bond) "should range from 1.5 to 2.5", with grout/steel bond not normally governing.
- A table of "typical (ultimate) bond stresses" was issued as guidance to designers.

Today even despite superior and often demonstrated knowledge of load transfer mechanisms (i.e., the issue of bond stresses NOT being uniform), the same philosophy prevails:

- The safety factor (reflecting, of course, the criticality of the project, rock variability and installation procedures) is normally 2 or more.
- A table of "average ultimate" bond stresses presented, which is basically identical except for typographic errors, to the 1974 table.

However, the current edition does provide very detailed guidance on critical design aspects, including allowable tendon stresses; minimum free and bond lengths; factors influencing rock/grout bond stress development; anchor spacing; grout cover/strand spacing; and grout mix design.

Construction (Chapter 7)

As noted above, there was a strong bias in the 1974 document towards construction, largely, it may be assumed, because practice far led theory. Furthermore, much of what was described in 1974 remains valid, especially with respect to issues relating to grouts, grouting and tendon placement. Certain features, such as a reliance on core drilling, the use of a "fixed anchorage" (i.e., the use of a plate) at the lower end of multistrand tendons, and specific water take criteria to determine the need for "consolidation grouting" are, however, no longer valid.

The 2004 version expanded upon the 1996 guidance, itself a radical improvement over its two immediate predecessors, and is strongly permeated by an emphasis on quality control and assurance. For example, practical recommendations are provided on the fabrication of tendons (including the pregrouting of encapsulations) and storage handling and insertion. Drilling methods are best "left to the discretion of the contractor, wherever possible," although specifications should clearly spell out what is not acceptable or permissible. In rock, rotary percussion is favored, and the drilling tolerance for deviation of 2° is "routinely achievable," while smaller tolerances may be difficult to achieve or to measure. Holes open for longer than 8 to 12 hours should be recleaned prior to tendon insertion and grouting.

The acceptance criterion for water pressure testing is adjusted to 10.3 liters in 10 minutes at 0.035 MPa for the entire hole. Technical background is provided on the selection of this threshold (based on fissure flow theory). Holes with artesian or flowing water are to be grouted and redrilled prior to water pressure testing. The pregrout (generally WCR = 0.5 to 1.0 by weight) is to be redrilled when it is weaker than the surrounding rock. When corrugated sheathing is preplaced, a water test should be conducted on it also, prior to any grouting of its annulus.

The treatment of grouting is considerably expanded and features a new decision tree to guide in the selection of appropriate levels of QC programs. Holes are to be grouted in a continuous operation not to exceed 1 hour, with grouts batched to within 5% component accuracy. The value of testing grout consistency by use of specific gravity measurements is illustrated. Special care is needed when grouting large corrugated sheaths; multiple stages may be required to avoid flotation or distortion. The cutting of "windows" in the plastic (to equalize pressures) is strictly prohibited.

Stressing, Load Testing and Acceptance (Chapter 8)

Given the professional experience and background of the drafting committee, it is surprising, in retrospect, to note the very simplistic contents of the 1974 document:

- "proof test" every anchor to ≥ 115% "transfer" load (to maximum 80% GUTS),
- hold for up to 15 minutes (but no creep criterion is given),
- lock-off at 50 to 70% GUTS,
- alignment load = 10% of Test Load, with movement only apparently recorded at this Test Load (115 to 150% transfer load). "If measured and calculated elongations disagree by more than 10%, an investigation shall be made to determine the source of the discrepancy,"
- lift-off test may be instructed by the Engineer "as soon as 24 hours after stressing."

Despite significant advances in the 1980 and 1986 documents, reflecting heavily on European practice and experience, significant technical flaws persisted until the completely rewritten 1996 version. The 2004 document was little changed in structure and content, the main highlights being as follows:

- Practical advice is provided on preparatory and set up operations and on equipment and instrumentation including calibration requirements.
- Alignment Load can vary from 5 to 25% of Design Load and 10% is common. This initial, or datum load, is the only preloading permitted prior to testing. On long, multistrand tendons, a monojack is often used to set the Alignment Load, to ensure uniform initial loading of the strands.
- Maximum tendon stress is 80% F_{pu}.
- Preproduction ("disposable," test anchors, typically 1 to 3 in number), Performance and Proof Tests are defined, the latter two covering all production anchors.

- For Performance Testing, the first 2 or 3 anchors plus 2 to 5% of the remainder are selected. The test is a progressive cyclic loading sequence, typically to 1.33 times Working Load. A short (10 or 60 minute) creep test is run at Test Load.
- Proof Tests are simpler, requiring no cycling and are conducted to the same stress limits. The option is provided to return to Alignment Load prior to lock-off (in order to measure the permanent movement at Test Load), otherwise this movement can be estimated from measurements from representative Performance Tests.
- Supplementary Extended Creep Tests are not normally performed on rock anchors, except when installed in very decomposed or argillaceous rocks. A load cell is required and the load steps and reading frequencies are specified.
- Lock-off load shall not exceed 70% F_{pu}, and the wedges will be seated at 50% F_{pu} or more.
- The initial lift-off reading shall be accurate to 2%.
- There are three acceptance criteria for every anchor:
  - Creep: less than 1 mm in the period 1 to 10 minutes, or less than 2 mm in the period 6 to 60 minutes.
  - Movement: there is no criterion on residual movement, but clear criteria are set on the minimum elastic movement (equivalent to at least 80% free length plus jack length) and the maximum elastic movement (equivalent to 100% free length, plus 50% bond length plus jack length).
  - Lift-Off Reading: within 5% of the designed Lock-Off load.
- A decision tree guides practitioners in the event of a failure in any one criterion. The "enhanced" creep criterion is 1 mm in the period 1 to 60 minutes at Test Load.
- The monitoring of service behavior is also addressed. Typically 3 to 10% of the anchors are monitored (or more if desired), by load cells or lift-off tests. Initial monitoring is at 1 to 3-month intervals, stretching to 2 years later.

Epoxy-coated strand (Supplement)

This material and its use was first discussed systematically in 1996, although minor references had been made in 1986. The 2004 document contains a separate supplement dealing with specifications, materials, design, construction and testing, being a condensed and modified version of a document produced by the ADSC Epoxy-Coated Strand Task Force in November 2003. The Scope (Section 1) notes that anchors made from such strand "require experience and techniques beyond those for bare strand anchors." The supplement is a condensed version of the "Supplement for Epoxy-Coated Strand" as prepared by the ADSC Epoxy-Coated Strand Task Force (November 2003). It supplements the recommendations provided in the general recommendations with respect to specifications/responsibilities/ submittals; materials; design; construction; and stressing and testing.

The Issue of Corrosion Protection

1974. Figure 1 illustrates the very simple approach to tendon protection, i.e., cement grout or nothing. "Permanent" is defined as "generally more than a 3-year service life." Sheathing is only discussed as a debonding medium, not a corrosion protection barrier. For permanent anchors "protective corrosion seals over their entire length" are to be provided (but are not defined). For two stage grouted tendons, *sheathing can be omitted*, the implication being that cement grout alone would be acceptable.

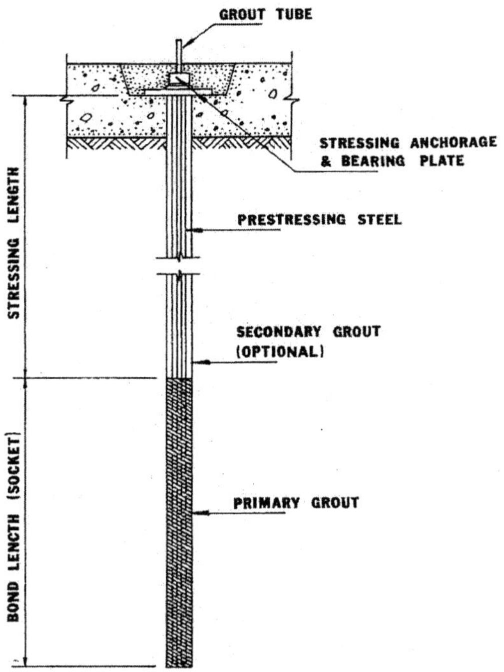

Figure 1. Rock Anchor Components (PCI, 1974) (Note the lack of protection to the steel other than cement grout)

1980. The same Figure 1 is reproduced (as it was also in 1986). The term "permanent" is now reduced to 18 months or more, and growing attention is drawn to the requirements of permanent anchors: sheathing is for debonding "and/or to provide corrosion protection," as is secondary cement grout. Corrugated protection, and epoxy coating for bars, are discussed.

The type and details of corrosion protection are to be based on longevity, anchor environment, consequences of future and in-hole conditions/length of time before grouting. For the bond length, cement grout is considered "the first level of corrosion protection," and plastic corrugated sheathing ("for multiple corrosion protection schemes") or epoxy are permitted. Such protection is to extend at least 2 feet into the free length. The free length is to have, as a minimum, a sheath with cement grout or grease infill. A full length outer sheath is regarded as "good practice."

1986. The emphasis is placed on first investigating the chemical aggressiveness of the soil and ground water: "Permanent anchors placed in environments where any one of these tests indicate critical values must be encapsulated over their full length." Thus, even up until the next set of Recommendations (1996), it was considered acceptable to allow anchors for dams to be installed without any protection for the bond length other than cement grout, depending on the results of laboratory tests on small samples. Encapsulation was not detailed.

1996. Permanence is now defined as a minimum of 24 months in a completely revised set of Recommendations. A wider spectrum of issues than simple chemistry now have to be considered when selecting corrosion protection principles. A major breakthrough was to identify two classes of protection (Class I and II) for permanent anchors to replace the poorly defined and loosely used "double" and "single" corrosion protection systems offered by various tendon manufacturers. The details are summarized in Table 3 and a "decision tree" was provided for the guidance of designers (Nierlich and Bruce, 1997).

Table 3. Corrosion Protection Requirements (PTI, 1996)

CLASS	PROTECTION REQUIREMENTS		
	ANCHORAGE	UNBONDED LENGTH	TENDON BOND LENGTH
I ENCAPSULATED TENDON	1. TRUMPET 2. COVER IF EXPOSED	1. GREASE-FILLED SHEATH, OR 2. GROUT-FILLED SHEATH, OR 3. EPOXY FOR FULLY BONDED ANCHORS	1. GROUT-FILLED ENCAPSULATION, OR 2. EPOXY
II GROUT PROTECTED TENDON	1. TRUMPET 2. COVER IF EXPOSED	1. GREASE-FILLED SHEATH, OR 2. HEAT SHRINK SLEEVE	GROUT

2004. The 1996 Recommendations were revalidated while it is stated that, for permanent anchors, "aggressive conditions shall be assumed if the aggressivity of the ground has not been quantified by testing." Table 3 was revised, as shown in Table 4, mainly to clarify the acceptable Class I status of epoxy protected steel in a "water proofed hole." The sophistication of contemporary tendons is shown in Figure 2. A long supplement is devoted to epoxy protected strand.

Table 4. Corrosion Protection Requirements (PTI, 2004)

CLASS	CORROSION PROTECTION REQUIREMENTS		
	ANCHORAGE	FREE STRESSING LENGTH	TENDON BOND LENGTH
I ENCAPSULATED TENDON	Trumpet Cover if exposed	• Corrosion inhibiting compound-filled sheath encased in grout, or • Grout-filled sheath, or • Grout-encased epoxy-coated strand in a successfully water-pressure tested drill hole	• Grout-filled encapsulation, or • Epoxy-coated strand tendon in a successfully water-pressure tested drill hole
II GROUT PROTECTED TENDON	Trumpet Cover if exposed	• Corrosion inhibiting compound-filled sheath encased in grout, or • Heat shrink sleeve, or • Grout-encased epoxy-coated bar tendon, or • Polyester resin for fully bonded bar tendons in sound rock with non-aggressive ground water	• Grout • Polyester resin in sound rock with non-aggressive ground water

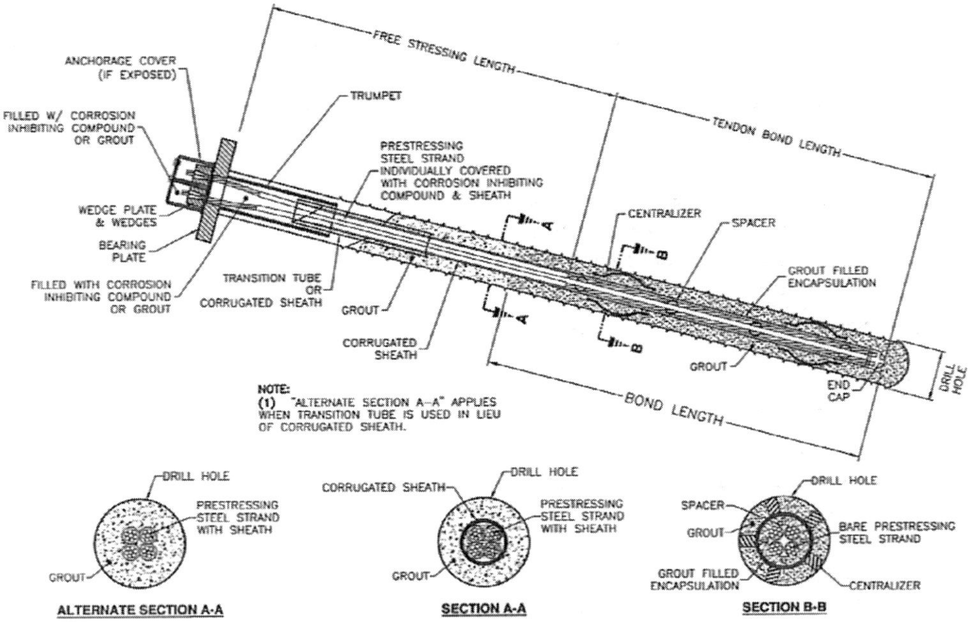

Figure 2. Class I Protection – Encapsulated Strand Anchor (PTI, 2004)

Overall, therefore, one is impressed that between 1974 and 2006 (i) extremely sophisticated corrosion protection systems were developed and (ii) the latitude offered to designers relative to choice of corrosion protection intensity and details was severely restricted: to install a permanent anchor in a dam without Class I protection is now not only impermissible, but unthinkable.

It must also be noted that the philosophy of pregrouting and redrilling the hole ("waterproofing") if it were to fail a permeability test was reaffirmed from 1974 onwards: indeed the early "pass-fail" acceptance criteria were, in fact, very rigorous and led to most anchors on most projects having to be pregrouted and redrilled several times. Although laudable, this was often, in fact, "extra work" since the criterion to achieve grout tightness is really much more lax than the criterion needed to provide the specified degree of water tightness. The saving grace of many of the early anchors was doubtless, therefore, the somewhat erroneous drill hole "waterproofing" criterion under which they were constructed.

Final Observations

The Phase 1 research into the successive "Recommendations" provides a fascinating insight into the evolution of a specialist technology over thirty years of practice. The observations support and often explain the similar trends described in the companion paper dealing with the Program's Bibliography and Database.

Acknowledgements

The "Rock Anchors for Dams National Research Program" was sponsored by The Association of Drilled Shaft Contractors (ADSC), Boart Longyear, Sumitomo (SEI) Steel

Wire Corporation, Sumiden Wire Products Corporation, Freese and Nichols, Inc. and Geosystems, L.P.

References (for companion paper also)

1. ADSC Epoxy-Coated Strand Task Force. (2003). "Supplement for Epoxy-Coated Strand," ADSC: The International Association of Foundation Drilling, November.
2. Aschenbroich, H. (2007). Personal Communication, June 25, 2007.
3. Bogdan L. (2001). "The Use of Epoxy Coated Strand for Post Tensioned Anchors." *Foundation Drilling*, ADSC, September/October, pp 23-34.
4. Bogdan, L. (2001). "Post Tensioned Anchors Using Epoxy Coated Strand for Rehabilitation of Concrete Dams." *The Future of Dams and their Reservoirs*. 21st Annual USSD Lecture Series, United States Society on Dams, Denver, CO, July 30 – August 3, pp 681-692.
5. Bogdan, L. (2007). Personal Communication, July 12, 2007.
6. British Standards Institution. (1989). "Ground Anchorages." BS8081. BSI, London, England.
7. Bruce, D.A. (1988). "Practical Aspects of Rock Anchorages for Dams." Association of State Dam Safety Officials 5th Annual Conference, Manchester, New Hampshire, September 25-28, 26 pp.
8. Bruce, D.A. (2002). "A Historical Review of the Use of Epoxy Protected Strand for Prestressed Rock Anchors." *Dam Safety, 2002*, Annual Conference of the Association of State Dam Safety Officials, Tampa, FL, September 8-11, available on CD, 21 pp.
9. Bruce, D.A. (2003). "Rock Anchors – Then and Now." *Dam Safety 2003*, Annual Conference of the Association of State Dam Safety Officials, Minneapolis, MN, September 7-10, available on CD, 9 pp.
10. Bruce, D.A. and J. Wolfhope. (2006). "Rock Anchors for Dams: National Research Project, The (Semi) Final Results of the Phase 1 Study," ASDSO Dam Safety, Boston, MA, September 10-14.
11. Bruce, D.A. and P.J. Nicholson. (1994). "The Stabilization of Concrete Dams by Post-Tensioned Rock Anchorages." International Symposium on Anchoring and Grouting Techniques, Guangzhou, China, December 7-10, 13 pp.
12. Buro, M. (1972). "Rock Anchoring at Libby Dam," Western Construction, March, pp. 42, 48 and 66.
13. Corns, C.F. and R.B. Jansen (1988). "Concrete Dam Performance and Remedial Measures," Advanced Dam Engineering for Design, Construction, and Rehabilitation, pp 578-608.
14. Crigler, J. (2007). Personal Communication, June 29, 2007.
15. ENR (1971). "Record Tendons Anchor Old Dam to Foundation," Engineering News Record, February 11, pp. 20-21.
16. ENR (1977). "Old Dam Tied Down to Bedrock to Stand up to Greater Flows," Engineering News Record, September 1, Volume 199 Number 9, pp. 18-19.
17. FIP Commission on Practical Construction (1986). "Corrosion and Corrosion Protection of Prestressed Ground Anchorages." Thomas Telford, Ltd., Telford House, London, England, 28 p.
18. Freitas, M.J. and J.M. Smith (1997). "Pardee Dam – South Spillway Interim Modifications, ASDSO Annual Conference Proceedings, Pittsburgh, PA, September 7-10, pp 375-386.
19. Graber F. (1981). "Excavation of a VSL Rock Anchor at Tarbela." VSL Silver Jubilee Symposium, Losinger AG., Berne, Switzerland.

20. Nierlich, H. and D.A. Bruce. (1997). "A Review of the Post-Tensioning Institute's Revised Recommendations for Prestressed Rock and Soil Anchors," *Ground Anchorages and Anchored Structures*, Proceedings of the International Conference, Institution of Civil Engineers, London, U.K., Thomas Telford. Edited by G.S. Littlejohn, March, 20-21, pp. 522-530.
21. Littlejohn, G.S. (1990). "Ground Anchorage Practice." Proc. ASCE Conf., "Design and Performance of Earth Retaining Structures," Cornell, NY, June 18-21, pp 692-733. ASCE Geotechnical Special Publication No. 25, Ed. P.C. Lambe and L.A. Hansen.
22. McWhorter G.H. and M.C. Meeks, Jr. (1990). "Dam Rehabilitation Projects in Northeast Georgia: An Overview," ASDSO Annual Conference Proceedings, New Orleans, LA, October 14-15, pp 133-141.
23. Mishalanie, D.J. (1990). "Rehabilitation of Newton Falls Upper Dam," ASDSO Annual Conference Proceedings, New Orleans, LA, October 14-15, pp 283-290.
24. Paolini, E.M. and M.B. Petrovsky (1991). "Dam Stabilization/Modification at Lighthouse Hill and Schuylerville Hydropower Developments," Waterpower, Denver, CO, July 24-26, pp 1346-1357.
25. Post Tensioning Institute (PTI). (1980). "Recommendations for prestressed rock and soil anchors." First Edition, First Printing, Phoenix, Arizona. 57 p.
26. Post Tensioning Institute (PTI). (1986). "Recommendations for prestressed rock and soil anchors." Second Edition, First Printing, Phoenix, Arizona. 41 p.
27. Post Tensioning Institute (PTI). (1996). "Recommendations for prestressed rock and soil anchors." Third Edition, First Printing, Phoenix, Arizona. 70 p.
28. Post Tensioning Institute (PTI). (2004). "Recommendations for prestressed rock and soil anchors." Fourth Edition, First Printing, Phoenix, Arizona. 98 p.
29. Prestressed Concrete Institute (PCI). (1974). "Tentative Recommendations for prestressed rock and soil anchors." First Printing, Phoenix, Arizona. 32 p.
30. Sharma, R.P. and B.T. Sasaki (1985). "Rehabilitation of Earthquake-Shaken Pacoima Arch Dam," Proceedings of the 15th International Congress on Large Dams, Lausanne, Switzerland, pp 275-296.
31. Standig, K.F. (1984). "Rehabilitation of the Delta Dam." Water Power and Dam Construction, December, pp 21-24.
32. Thompson, F.G. (1969). "Strengthening of John Hollis Bankhead Dam," Civil Engineering, Vol. 39 (12), pp 75-78.
33. Troka, J.R. and J.F. Lane (1988). "Large Capacity Rock Anchors Stabilize Bagnell Dam," Waterpower, pp 1334-1343.

Authors index

Adekunte, A. 427
Adler, S. 302
Al-Mahaidi, R. 321
Altounyan, P. 199
Amano, K. 53
Aoki, K. 188
Aschenbroich, H. 144

Barley, B. 12, 331, 380, 387, 471
Bigby, D. 199
Bluff, C. 321
Bouazza, A. 321
Boyd, L. 132
Bruce, D. 121, 132, 155, 481
Bruce, M. E. 111

Carrigan, L. 321
Chamley, P. 387
Chia-Wei, S. 397
Chihping, K. 312
Ching, J.-J. 397

Daynes, P. 438

Edwards, M. 380
Enomoto, T. 341
Ernst, P. 273

Ferrier, J. 33
Forbes, G. 132
Fuseya, Y. 23

Gomez, J. 111

Hartley, J. 331
Howard, J. 166

Irvin, C. 283
Ivanovic, A. 209, 220
Iwaida, Y. 293
Izakura, M. 238

Jackman, S. 33
John, G. 263, 273
Jones, S. 166
Jordan, J. 99
Jungwirth, D. 302

Kent, L. 199
Kido, T. 351
Klemenc, I. 43
Kline, R. 155
Kubo, H. 407, 451
Kuhar, V. 43

Liao, H.-J. 312, 397
Littlejohn, A. 177
Littlejohn, S. 263, 357
Loidl, D. 321

Madsen, D. 427
Margetson, N. 427
Martin, J. 438
Martin, P. 82
McDonnell, C. 438
Miyatake, H. 451
Mori, T. 231
Mothersille, D. 12, 33, 283, 387, 471

Nakajima, M. 231
Neilson, R. 209, 220
Niki, T. 351

Okada, H. 238
Okubo, K. 92
Okuno, M. 1
Oshita, T. 451

Parrish, S. 63, 368
Pedley, M. 438

Quinn, L. 427

Richards, D. 72
Richards, T. 417
Rodger, A. 209, 220

Schultz, M. 166

Sentry, M. 321
Starkey, A. 209, 220
Sueyoshi, T. 250, 341, 461
Suga, K. 293, 341
Sugisaki, T. 341

Takeda, H. 23
Takemata, T. 1
Takemoto, M. 53, 92
Takemura, J. 407
Takeyama, M. 451
Tanaka, K. 238
Tani, K. 250
Thome, D. 417
Touda, Y. 351
Traylor, R. 111
Tsutsui, T. 231
Turner, M. 72

Urakawa, N. 461

Whitworth, M. 63, 368
Wolfhope, J. 121, 132, 481
Wood, H. 82

Yamada, H. 92
Yamagiwa, T. 351
Yamazaki, J. 407, 461
Yonemura, A. 293
Yoshimura, M. 53

Zicko, K. 155

Discussion sessions

Inspection procedures and physical conditions recorded in service: discussion session 1

Rapporteur's Summary
by Donald Bruce, *Geosystems, L P*

The opening paper by Mothersille and Barley describes observations on anchors in marine conditions installed in Aberdeen, Scotland and at Greenhithe, London. In the former case, the anchors were 22 years old, and in the latter case 35 years upon inspection. In both cases, they were designed and installed with corrosion protection systems that would not be acceptable under contemporary standards. Tests on a strand from an Aberdeen anchor confirmed that its rupture during proof testing was due to combination of corrosive attack in an area where corrosion protection (i.e., grease and sheathing) had been damaged, and "fatigue mechanisms." However, where the protective anchorage cap had remained intact (though outwardly corroded) as an effective retention unit for the corrosion protection compound, the strands and their wedge systems were found to be in "bright steel condition."

At Greenhithe, the anchor heads were graded qualitatively prior to selective lift-off testing. All caps were to some extent damaged by corrosion with 10% being severely damaged. The authors propose a detailed, step-by-step approach for cleaning the exposed anchor heads in situ (including wedges) and reconstituting the protective cap system. Attention is also drawn to the need to evaluate ancillary components such as bearing plates and gusset plates.

A companion paper by Mothersille, Jackman and Ferrier describes the performance and condition assessment of 30-year old Anchorages on the River Clyde at Glasgow. Again, a first phase of visual inspection of the 32-year old marine anchors was followed by selective lift-off testing, and metallurgical examination of certain strands deliberately stressed to failure. The heads were graded, as at Greenhithe, according to the following set of definitions (Table 1).

Table 1. Visual observation grading with definitions.

Grade category	Category definition
A	Good condition with superficial surface rust
B	Satisfactory condition with some corrosion evident. However, future lifespan of unprotected anchor head in aggressive marine environment limited.
C	Poor condition with evidence of deep corroded surfaces. Some section loss and small crevices also visible. Anticipated lifespan of corrosion protection to the anchor steel extremely short. Remedial action recommended.
D	Very poor condition. Severe and significant section loss evident. Immediate remedial action/replacement necessary to prevent further corrosive deterioration of anchor steel.

The authors restate that there is *"currently no method established in the ground anchorage industry for predicting the future performance of corroded anchor components."* As a result, destructive tests on some in situ components were undertaken and the results used to qualitatively assess the performance of the anchored walls (Table 3 in paper).

Table 3. Summary of test categories.

Test type	Test Description
1	Lift-off tests
2	Strand tested to failure through the anchor head
3	Exhumation inspection and removal of strand (minimum 1m length) for remote testing of strand to failure
4	Exhumation inspection and removal of strand (minimum 1m length) through the anchor head with barrel and wedge attached for remote testing of strand head to failure

Tests on the strand proved generally positive in terms of remnant tensile capacity, whereas the gripping efficiency of the wedges proved less satisfactory. The authors concluded that in situ remediation (of the protective cap and top anchorage) would suffice to extend the working life of many anchors. However, the poor condition of certain anchors *"with open exposure of the steel in the strand and barrel and wedge components"* demanded replacement of the entire ground anchorage system.

Representatives of the Japan Anchor Association were extremely active contributors to the Conference, as befits a national market of huge scale and intensity. Their first paper by Okuno and Takemata described a case study of inspection, integrity investigation and repair of anchors used to stabilize slopes around a dam lake. At the time of evaluation, the 358 anchors were already over 20 years old, and had been installed before the establishment in 1988 of the Japanese Anticorrosion Standard. The anchors were visually observed, and subject to *"inspection of soundness by shock and integrity investigation by lift-off test."* The corrosion protection compound had largely been removed and/or deteriorated above and below the head permitting corrosion of the strands. A few strands ruptured during lift-off testing, due to severe pitting corrosion. A method is described to review the corrosion protection above and below the anchor head, the latter involving the placing of a "slide pipe" (Figure 10 in paper).

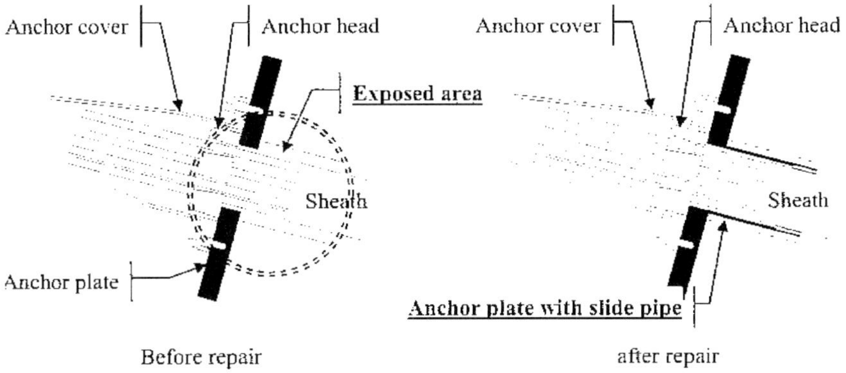

Figure 10. Corrosion protection below the bearing plate.

Despite the refurbishment of the anchors over a 3-year period, *"regular investigation and maintenance"* is recommended for the life of the anchorages, taking special care *"not to overlook the warning signs indicating anchor deterioration."*

In similar vein, Fuseya and Takeda described the "Inspection, Integrity Investigation and Repair of 30-Year Old Anchors," 73 of them, installed in 1975 and 1976 for landslide protection. Their work followed the tasks of Table 1 in the paper.

Table 1. Integrity investigation.

	Subject	Procedure and item	Purpose of investigation
Preliminary investigation	Document research	· Research of basic document at the time of anchor installation · Repair record	· Confirming details of anchors · Confirming details of landslide · Damaged anchors
	Field work	· Geological research in the vicinity	· Confirming of current situation of landslide
Main investigation	Structure survey in the vicinity	· Variation of structure (visual inspection) · Repair record of structure	· Confirming of current situation of landslide
	Anchor head survey	· Detailed investigation by visual inspection	· Propagation of corrosion · Anchors with accident
	Physical investigation	· Investigation of hitting sound · Lift-off test	· Dashing out · Confirming residual anchor strength · Anchors with accident

Although no ruptures were found in the 52 strand anchors, 5 failures were recorded over 8 years in the 20 bar anchors. The visual inspection was conducted of (i) the locking unit and protruding tendons, (ii) "fixed base part" (pedestal steel, supporting unit) and (iii) exposed free length. For each group of components, a simple qualitative grading was used. In summary, *"in all of the anchors investigated, corrosion of the steel was found to have progressed as a result of deterioration of the corrosion protection."* As a result, 35% of the anchors could likely not be refurbished. Lift-off testing confirmed the unacceptably deteriorated situation of the suspect anchors. Repairs were conducted on 45 of the strand anchors, both in the fixed base and locking unit, and in the exposed free length (Figure 8 in paper).

No further deterioration has been observed in the year following the repairs. Thoughtful recommendations are made in support of the evaluation and repair concepts.

The fifth paper by Yoshimura, Takemoto and Amano, colleagues in the NEXCO Expressway Research Institute, continued the theme of the Japanese experience. Their presentation entitled "Maintenance Control of Ground Anchors Constructed in Japanese Expressways" focused on the assessment of "old-type" anchors installed from 17 to 29 years beforehand. Chemical testing of the soil supplemented pullout testing in six discrete sites, and extraction of 3 tendons at each location for metallurgical investigation. Figure 5 in the paper summarizes neatly the loss of tendon cross section with age.

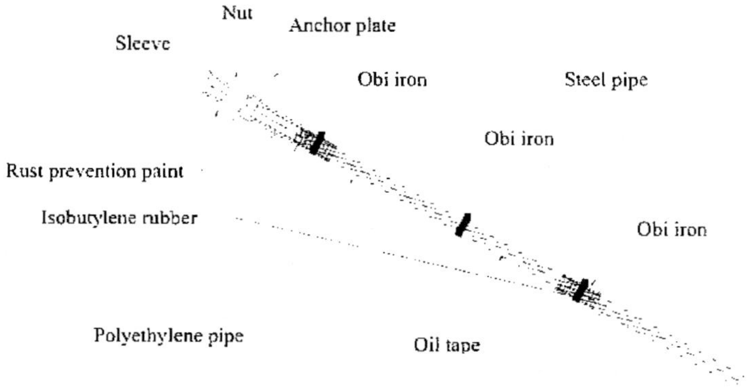

Figure 8. Repair method.

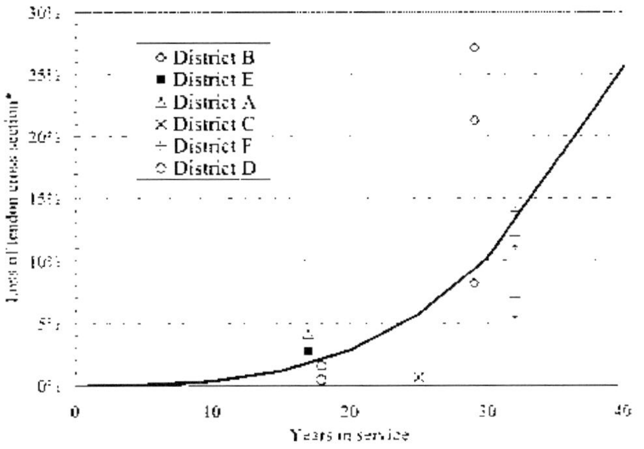

Figure 5. Years passed after construction and loss of tendon cross section.

* Loss of tendon cross section is based on the square of the loss.

It was concluded that the old-type anchors corroded (i.e., lost cross section) at the rate of 0.001 to 0.022 mm per year *"even in low corrosive soils."* Corrosion in the bond length as well as under the anchor head was judged likely.

The final paper by the Slovenians Klemenc and Kuhar provided a fascinating insight into the high quality work being conducted in the region not hitherto widely recognized as being a centre of international excellence. In evaluating anchor conditions at 85 retaining structures built since 1994, the authors highlighted some examples of the poor installation of electrically isolated prestressed ground anchors in Slovenia. Predictably, the likelihood of *"significant corrosion damage to the elements of anchor heads"* was raised, the cause being ascribed to construction defects as opposed to design flaws (the anchors having been conceived along Swiss guidelines). In particular, water was found to have penetrated beneath the anchor caps as a result of gaps in the sealing detail. These "failed anchors" were clearly detected by electrical resistance testing. In the most severe cases, only 2 of the 5 tendon strands were actually observed still to be present and engaged by wedges in the headplate. The paper provides excellent guidance on the chemical changes which had occurred in service to the corrosion protection compounds: the grease in particular now contained *"a considerable amount"* of water, polyurethane foam was classified as *"unsuitable,"* while vaseline was found to have degraded less. Practical recommendations are given concerning details of construction, and principles of service monitoring.

Taken as a group, these papers were consistent in their respective messages:

- There must be a focus on evaluating the current status of any permanent anchor designed and installed under "old type" regulations, which typically means over 20 years old.

- Attention to detail is essential when placing and sealing the protective cap over the anchor head. Even "old" anchors, properly constructed, have been found to have been very well protected, whereas even "new" anchors, if imperfectly constructed, can corrode quickly. This is particularly important since this activity is the last to be conducted in the anchor construction cycle, at a time when attention to and suspension of detail can tend to waiver.

- Good guidance is available to assist practitioners in classifying their visual assessment of anchor components in situ.

- Metallurgical tests of the steel, and chemical tests of the corrosion protection compound, are invaluable in determining rates of material degradation, and hence in helping predict the safe remnant service life.

- The Japanese, in particular, have devised practical methods of refurbishing existing tendons, in situ, to extend their safe service life.

- Regular, routine assessment of permanent anchor installations – if only by visual inspection – is strongly recommended, particularly so when these anchors secure "life line" structures.

John Wasilewski, *Capita Symonds*
We have heard in numerous presentations about the need for full monitoring and I am puzzled as to what is the practical reality. What can full monitoring include? We have also heard many references on the need to study loss of tendon cross section due to corrosion but why has there only been fleeting reference to stress corrosion cracking?

Stuart Littlejohn, *University of Bradford*
We have not used routinely the term "full monitoring" but we have tried to describe the types of physical inspection that should be carried out initially, and then depending on what is discovered, this might then dictate, for example, the need for a more detailed metallurgical examination. Load monitoring is also one of the aspects of the investigation but I do not see any problems today in the concept or procedures associated with the physical inspections of anchor heads. It is the lack of implementation of routine physical inspections that worries me. When you state that there been little discussion on stress corrosion cracking, you are quite right today but you will hear more about the subject tomorrow in Session Six, because stress corrosion cracking is a major concern.

Tony Barley, *Single Bore Multiple Anchor Ltd*
I have a question about electrical inspection. When drawing up the anchor code there were two trends; firstly those who wanted to insist on double corrosion protection, (i.e., two barriers to isolate the anchor from any corrosive materials) and secondly those who favoured carrying out electrical isolation tests on every anchor installed. That would still require a water corrosion protective barrier, so I am interested to know anything from a statistical point of view about the success of electrical corrosion protection monitoring and also the big question that comes to

many people, i.e. when anchors fail a corrosion investigation test from electrical conductivity monitoring, what happens to them? Are they removed or are they left for longer term monitoring?

Iztok Klemenc, *Slovenian National Building and Civil Engineering Institute*
We perform electrical resistance measurements on the completed anchors and according to our experience, about 10% of the installed anchors failed this test. Thereafter, we performed no regular inspections and measurements of this installation, so I cannot give you any data. However, we consider electrical resistance measurements just as a measure to demonstrate the state of the polyethylene sheathing as the corrosion protection measure for the strands of the anchor.

Tony Barley, *Single Bore Multiple Anchor Ltd*
Perhaps a question that those who support very thorough corrosion protection in two barriers could ask is, what happens to the other 10%? In other words, is not this indicative that if you put a single barrier on the tendon during installation and only 90% of them satisfy the test, perhaps you should be looking at double plastic barriers. Hopefully, you would then get a 100% success rate.

Iztok Klemenc, *Slovenian National Building and Civil Engineering Institute*
I agree with your remark. For this 10% of anchors, it is considered that they will not be able to withstand the long term survival of the structure, so we expect that these anchors may fail in one way or another.

Hiroshi Yamada, *Nittoc Construction Co. Ltd.*
If we inspect existing anchors accurately in the first phase then continuing through the maintenance term, it is necessary to inspect routinely to accumulate the data on conditions, e.g. the corrosion state or the tension state. In evaluating the condition, we can progress the further effective maintenance procedure with a helpful index that is created from the accumulated investigation data of existing anchors.

Mike Turner, *Applied Geotechnical Engineering Ltd*
Firstly, I wish to congratulate Minoru Okuno and Hiroshi Yamada on their presentations. I thought the papers were classic examples of the forensic examination of ground anchors. However, my question is addressed to Devon Mothersille in the main because I think that all the examples shown were non-restressable anchors and the problem with non-restressable anchors (even if they are permanent anchors) is that if they go wrong there is nothing that you can do about it. In addition, one problem with the repairs that you completed suggests that although you topped up the corrosion protection on the outer head there is nothing you could do about the inner head protection and that is a bit like papering over the hole in the Titanic.

Chris Irvin, *DYWIDAG-Systems International*
Mike can you just rephrase that question in a little bit more detail in respect of restressable heads?

Mike Turner, *Applied Geotechnical Engineering Ltd*
Yes. When I wrote my paper I was commenting in part on the difference between a restressable, a non-restressable and a detensionable permanent ground anchor. A non-restressable anchor is one where, say in a barrel-wedge type configuration, you cut the strand off so you cannot get back on to it again. A restressable anchor allows you to adjust the load on the anchor by 10% say. If you can grab hold of it, you can change the load a bit if you are losing load, but a detensionable anchor allows you to grab hold of that anchor and take the load off. It enables a full repair or inspection or whatever you want to do and so a lot of the anchors that were shown by Devon were non-restressable; you can do nothing at all with them once you cut them off. When we talk about the anchor codes BS 8081 and DD81 years back, those differences were clearly explained and we tried to emphasise them. It is quite important as an anchor designer or specifier that you understand what you want to do with your structure. If you want to be able to have full control of that anchor they must be detensionable anchors.

Chris Irvin, *DYWIDAG-Systems International*
Your point is a valid one. There are really two systems available, one is a restressable anchor and this is achieved with a threaded headblock or the other option is to leave a tail of strand which you can then connect on to. On the detensionable anchors we have two versions, again one with a long strand on the top or one with a threaded headblock with a set of shims below it, so with either system you can detension the anchor and then inspect it. I really would urge everybody here that where you have the opportunity to specify a restressable anchor with a threaded headblock, it makes life a lot easier.

Chris Irvin, *DYWIDAG-Systems International*
Question for Minoru Okuno and Hiroshi Yamada. It is about the repair of the anchors. What was the bite position for the wedges, was it above or below the original bite position on the strand anchors?

Hiroaki Kubo, *Japan Anchor Association*
The bite position for the wedge was above the original position or at the same position

Horst Aschenbroich, *Con-Tech Systems Ltd*
Every strand anchor is restressable, i.e. every strand anchor can be made restressable, even if you do not have any threaded anchor head. If the anchor head is not externally threaded we now have equipment available on site that can machine a thread on to the anchor head to make it restressable.

Mike Turner, *Applied Geotechnical Engineering Ltd*
Not necessarily. We have experienced scoured anchor heads, so it is quite hard to thread on to that surface. Also, if you take say a barrel and wedge type unit, in some situations that Devon Mothersille has shown with quite heavy corrosion, it is difficult to get back on to that as well, but I take your point.

Jamie Christie, *Aberdeen Harbour Board*
Devon, thank you very much for your excellent presentations, but I do not think in fairness, at the end of the day, it left the role or image of the clients in ports in a particularly good light and I would just like to inform the audience of the type of things that are going on in that context. Aberdeen Harbour Board in fact has 32 quays of which 9 at one time had anchorages. Suspicious of the need to revisit anchorages because we were not informed otherwise at the time, inspection and maintenance were effected on all of our anchorages at each berth from 1990 to 1993. By the way, all examples in Devon's presentation have since been reconstructed due to continuous improvement of our facilities. Finally, Aberdeen Harbour Board are about to partner research with the University of Aberdeen on safe and user-friendly methods of reassessing anchorages, then we plan to effect any further appropriate maintenance measures and inspection regimes as per our asset management policy - just to complete the full picture, otherwise I appreciate that the examples you used were very good.

Devon Mothersille, *Geoserve Global Ltd*
Your points are taken. The main objective of the presentations was to build awareness of similar situations that exist in many other ports and quays across the whole country. So, if some of the details presented for Aberdeen Harbour were lacking, I thank you for filling in the gaps. The important message from this work is that carefully planned programmes of inspection and testing will identify problems with existing anchorages and assist the implementation of maintenance strategies or remedial action.

Jamie Christie, *Aberdeen Harbour Board*
I appreciate that and to be quite frank looking at the delegates list I am amazed to note that I am the only person here representing ports which surprises me and I will spread the word through my own community.

Gerry Robbins, *DYWIDAG-Systems International*
We have seen a lot of old-style anchor heads today and we should state that those are likely to be the ones that require monitoring and testing, as they have been around for a long time. There are a lot of alternatives available to designers today, both for restressable bar and strand tendons, wonderful corrosion protection systems, and on the other side in terms of load testing on site of old anchors, there are a lot of good developments so that check loading can be done even on some of these examples we have seen today with heavy corrosion. From my perspective, if you select the right anchor in the first place, manufacture it correctly, handle it correctly, install it correctly with appropriate inspection, particularly on critical jobs, then there can be real integrity with ground anchors.

Tom Richards, *Nicholson Construction Company*
Question for Devon Mothersille. Regarding the slipped strands, I understand that the wedge becomes corroded and cannot move in anymore. Is the need for the wedge to move because of loss of section on the strand or on the bite marks of the strands or the wedge teeth? In other words, is the reason that the strands slip due to section loss of the wedges at the teeth or because of section loss of the strand or because of some other cause?

Devon Mothersille, *Geoserve Global Ltd*
I think the reason is primarily loss of section of the strand. Usually, if the barrel and wedges are adequately lubricated and you were to have a loss of section of the strand, the wedge would follow that strand and continue gripping, albeit at a slightly different position. If the barrel and wedges are effectively corrosion welded together then the wedges cannot move; once that loss of section occurs and the strand is still under load, it just slips through and the wedge cannot adjust and effectively lose their gripping efficiency.

Written question and answers to session one: Inspection procedures and physical conditions recorded in service

Stuart Littlejohn, *University of Bradford*
The paper by Klemenc and Kuhar provides invaluable case history data on the problems that are encountered with poor installation practice. Bearing in mind that electrical resistance and capacitance measurements were used to assess the integrity of tendon encapsulations, what is your opinion on the reliability of this type of measurement for accepting or rejecting new anchorage installations?

Iztok Klemenc, *Slovenian National Building and Civil Engineering Institute*
In the cases of permanent ground anchors, where a single plastic duct forms the sole protective barrier, the integrity of the plastic duct throughout the length of the anchor can be verified by an in situ test. According to EN 1537, measurement of electrical resistance (ERM I) between each anchor and surrounding soil or structure should be performed in two successive phases to determine the effectiveness of the applied corrosion protection system.

In the first phase the integrity of a plastic sheath along the anchor length after the subsequent stages of the installation of the anchor (from setting the anchor into the borehole, after primary and after possible secondary grouting) and after the testing of load-carrying capacity is tested by the measurements of electrical resistance (ERM I). The overall electrical isolation of the anchor from ground and structure is tested in the second phase, where ERM I is performed after lock-off of the anchor and after the anchor head injection. Moreover, additional measurements of electrical isolation of the anchor accepted for the service from the ground and structure are foreseen in EN 1537 at any time during the service life of the anchor. In both phases the electrical resistance criterion is set as $R_I \geq 0.1$ MΩ, which demonstrates an acceptable level of plastic sheath integrity as well as overall electrical isolation of the anchor from the ground and structure.

Based on the results of the electrical resistance values measured during various aforementioned installation actions and particularly after the lock-off, obtained values on the installed anchors of the anchored structures, which have been constructed in the last decade within the Slovenia's National Motorway Construction Programme, it can be concluded that the electrical resistance (ERM I) of anchor corrosion protection according to EN 1537 is a rather reliable parameter to distinguish between the permanent ground anchors with a sound (watertight) encapsulation and anchors with the deficient protective barrier. Since measured values of resistance can easily be influenced, prerequisites for regular measured values of electrical resistance have to be fulfilled: (1) the points of contact have to be kept clean and dry (which cannot be easily achieved during rainy and winter weather) and (2) faultless earth connection has to be ensured, for which the lengths of reinforcing steel in RC anchored structure in contact with the soil or the moist soil of the site were used in our practice.

In the cases of new anchorage installations where the integrity of the encapsulation has not been proven and the electrical resistance criterion ($R_I \geq 0.1$ MΩ) has not been met, additional information about the state of the anchor corrosion protection can be obtained by measuring the impedance (Z) of the anchor, its polarisation resistance (R_p), its capacitance (C_f) and electrochemical impedance spectroscopy (EIS). These measuring techniques using lower voltage amplitudes of wider frequency ranges could well supplement the standardized method of electrical insulation measurements according to EN 1537. The polarization resistance calculated from measured electrochemical impedance spectra (EIS) is presumably a very good indicator of the corrosion state of an anchor. Results of additional measurements to ERM I can be therefore used as a supplement for the assessment whether the reason for the low value of measured electrical resistance was the damage of the protective PE ducts for the anchor tendons in the form of cracks or perforations or due to the electrical contact between the anchor head and the concrete surface of the anchored structure.

One deficiency of these additional measurements which arises from low voltages and low currents used by these measurements should, however, be noted. Besides the tendons of anchors, there are generally no parallel electrodes in the ground which would enable reliable and stationary conditions during the measurements. Reliable measurements can therefore be made only between two adjoining anchors with accessible steel strands, so only the combination of both of the anchors can be assessed.

Chris Holt, *Jacobs*
With reference to the paper by Mothersille and Barley, the anchor heads appear to be located in the inter-tidal zone and had suffered significant corrosion. Was the depth of corrosion assessed in relation to predicted rates for the given exposure conditions, and was any assessment made to determine whether ALWC was a contributory factor in accelerating the rate of corrosion?

Devon Mothersille, Geoserve Global Ltd
The anchor heads were located between mid tide level and mean high water spring (MHWS). Corrosion was severe whether it was encountered on the exposed strand tendon and the exposed prestressing components, or on the corrosion protection cap.
The question of identification of corrosion rate is of particular interest since it has not previously been assessed on these contracts. In the case of anchor heads at Torry Quay, Aberdeen and at Greenhithe, the rate of corrosion of the protective caps can be simply estimated from the residual thickness of the steel protection caps and the probable time period between installation and total loss of section. Anchor heads at Aberdeen were installed in 1976 and during inspection in 1992 observations established that some caps exhibited corroded holes in the 9.5mm thick steel. At Greenhithe, the caps were also 9.5mm thick and certainly by the time inspection commenced after 30 years some steel caps had fully corroded and been destroyed. Thus in the worst situation at Aberdeen, the rate of corrosion of the steel caps was estimated as 9.5/16 = 0.6mm per annum whilst at Greenhithe, the rate was certainly greater than 9.5/30 = 0.3mm per annum.

At Greenhithe, anchor head plate observation indicated that corrosion pits were as deep as 10mm. Hence in the greater steel section of 40mm thick bearing plates, rate of corrosion appeared also to be 10/30 = 0.33mm per annum. Bearing in mind a general consideration of the average loss of sheet pile section of 0.1 per annum due to traditional marine corrosion in the splash zone then the actual local corrosion of the anchor caps appears to be between 2 and 6 times that value, and corrosion of 40mm plate section approximately 3 times that value. However, these high rates are still consistent with the more extreme observations associated with traditional corrosion as

summarised in Table 21 of the CIRIA Report C 634 "Management of accelerated low water corrosion in steel maritime structures".
Where corrosion of the exposed anchor tendons occurred, no measurement of loss of section was recorded.
With regard to the identification of corrosion mode, there may be some contribution, particularly in the anchor caps, from differential metal corrosion but the possible occurrence of accelerated low water corrosion (ALWC) is unlikely.

Author 2 was given the opportunity by ALWC expert Mike Hodgson to observe "classical" ALWC of sheet piles just above lowest astronomical tide level followed by the removal of the bacterial and corrosion products. All inspection was within a dry cofferdam in the form of a Limpet Dam installed by Nuttall John Martin. This experience in conjunction with data presented in the CIRIA Report C 634 indicate that ALWC generally takes place from approximately 0.7m above low astronomical tide (LAT) level to 0.5m below LAT. Furthermore, the type of corrosion observed on the anchor heads at Aberdeen and Greenhithe did not appear to exhibit any evidence of bacterial attack, adhesion of corrosion products in jelly form or the exhibition of a bright steel surface on product removal. Neither was there any indication of the bright orange colouring nor smells present in the bacterial jelly. Good illustrative photographs of ALWC are presented in the CIRIA Report. In conclusion, it is extremely unlikely that the bacteriological corrosion associated with ALWC made any contribution to the corrosion of the inspected anchor heads.
It is of interest to note that the advanced development of the Limpet Dam System accommodates a drilling mast and drill head and thus allows anchorages to be installed in, and even well below, the level where ALWC may occur.

From the dry environment of the Limpet Dam corrosion protected anchorages were installed in Southampton container terminal 10m below High Water Level. The operation was carried out in order to demonstrate that low level anchorages can safely be installed in the base of an existing quay wall in order to supplement passive toe support and to allow toe level dredging for deep draft container ships. Hence, in the event that production anchorages are installed within ALWC range then special precautions will be required to counter the severe corrosive effects of ALWC. This may require the application of glass fibre reinforced anchor head caps which have been used by Keller Ground Engineering in the marine environment for 15 years.

Jamie Christie, *Aberdeen Harbour Board*
Aberdeen Harbour in fact consists of approximately 30 quays of which 9, at one time, were partially supported by anchorages. Although not known to be common practice at the time inspection and maintenance was carried out on all of these anchorages in the years 1990 to 1993. All examples used in Mr Mothersille's presentation have since been reconstructed as part of Aberdeen Harbour Board's ongoing programme of improvement of berth facilities. Aberdeen Harbour Board are also about to partner research with the University of Aberdeen on user-friendly and safe methods of monitoring existing stresses in anchorages without the need for pull-off tests, and to effect subsequent inspection and maintenance measures as may be necessary thereafter.

Devon Mothersille, Geoserve Global Ltd
Aberdeen Harbour Board must be congratulated on being probably the first authority in the UK to extensively investigate on 9 anchored quays the performance and/or degradation of in situ permanent anchorages in a marine environment. Their link with the University of Aberdeen to investigate the possibility of performance testing without load application must be complimented. The authors look forward to success in the application of impact testing to multi-strand anchorages in order to extend, the capabilities currently believed to be restricted to single tendon anchorages.
It should also be noted that although the anchored quay walls featured in the paper have since been reconstructed, the condition prior to reconstruction is still representative of many existing structures throughout the UK. The authors hope that by increasing awareness of these issues the appropriate authorities and agencies will seize the opportunity to authorise thorough inspection and testing programmes that will lead to appropriate maintenance or remedial action.

Horst Aschenbroich, *Con-Tech Systems Ltd*
The paper by Okuno and Takemata describes interesting repair work. How did they remove the anchor head wedge plate to place a new bearing plate with trumpet? With a long free stressing length the elongation before lock-off must have been longer than the tail. When the anchor is now detensioned to remove the bearing plate and wedge plate, the strand tail will move back into the drill hole?

Minoru Okuno, Japan Anchor Association
The anchor head wedge was removed by connecting and pulling up the strand by coupler.

The strand tail (the protruding length) will move back, but in this project the tail lengths were around 300mm so it was no problem.

For re-stressing, in case of a single strand, the strand tail should be more than 40mm. However, the required tail length could vary depending on the coupler size and number of strands to be re-stressed at the same time.

I have experienced cases where the strand length is insufficient and rather than lose the ground anchor it is often decided not to de-stress the strand and forego replacement of the bearing plate. In such cases, we would usually clean everything up and replace the grease.

I think that a plan should be made to ensure that sufficient tail lengths are maintained to allow maintenance and re-stressing.

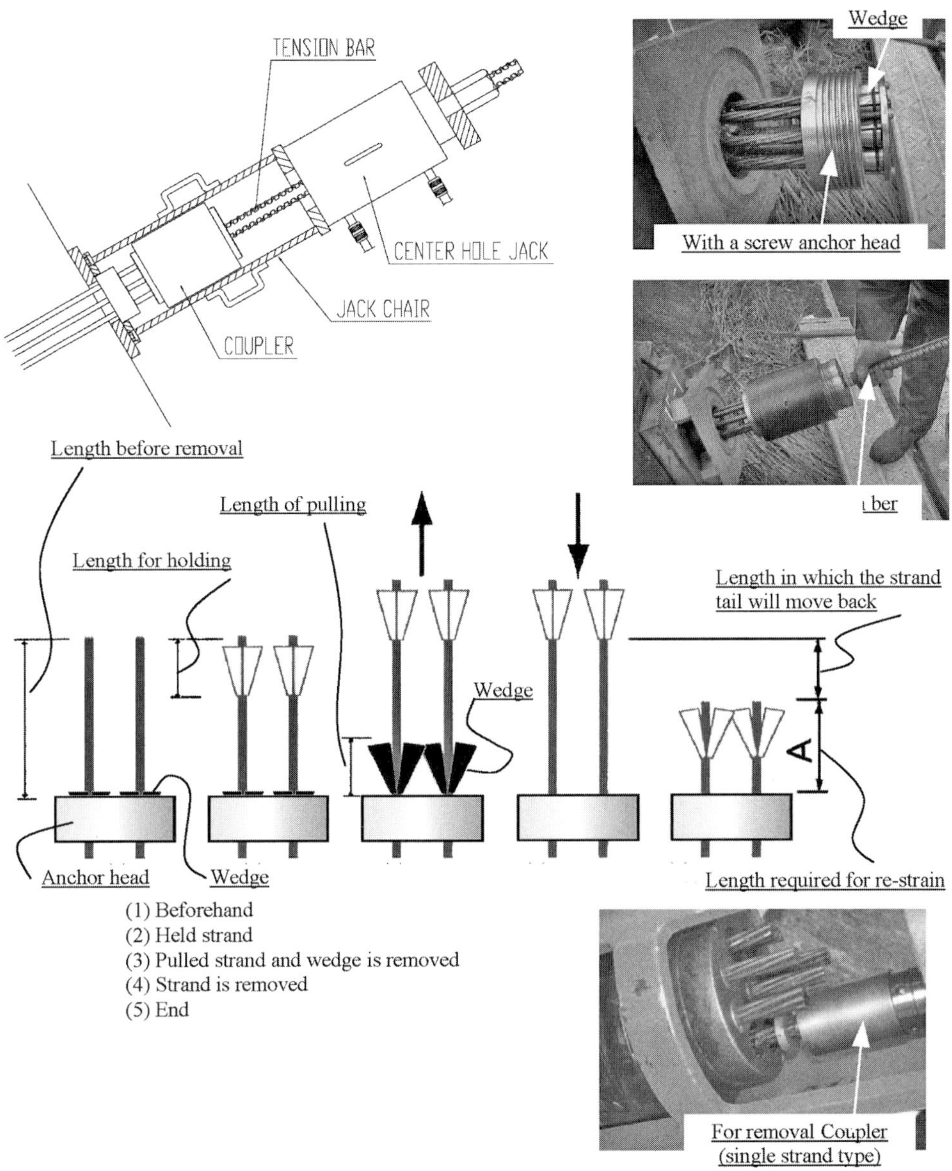

Fig. How to remove a wedge

Horst Aschenbroich, *Con-Tech Systems Ltd*
With reference to the paper by Fuseya and Takeda, how was the threaded bar or sleeve connected to the strand? What is Obi iron?

Hiroshi Yamada, *Nittoc Construction Co. Ltd*
This multiple cable is made at the factory as a rule. First of all, multiple cables are cut as long as designed and ends of them are inserted to allow for proper anchoring to the sleeve. Keeping this condition in mind, they are set with a special jack and pressure is applied at normal temperatures. In this process, the plasticised sleeve bites into the multiple strands and it holds as one body. In turn, cutting of the screw threads at the sleeve can be made.

Obi iron is a description by the Japanese. This is a steel band to tighten a pipe or a sheath.

Service behaviour and monitoring procedure and performance in service: discussion session 2

Rapporteur's summary
by Tony Barley, *Single Bore Multiple Anchor Ltd*

This session specifically addressed papers which recorded information on either long term regular load monitoring (four) or long term infrequent monitoring (one). It always is necessary to consider that unless movement of anchor head and the structure is accurately determined then the load change or extension change in the anchor cannot be realistically attributed to structural movement or to the performance of the anchor and anchor components.
Generally the monitoring of the movement of the anchor head or the structure alone can be carried out more economically and more frequently than monitoring of anchor load change. Furthermore it may be suggested that if the structure does not move then in some instances it provides some indication with regard to the capability of the anchors. For example sheet pile movement at Greenwich quay on the Thames which collapsed in 1990 had been observed by a site operative, some six months before collapse but neither monitored nor reported. Hence annual monitoring of that structure alone could have forewarned of potential collapse and remedial actions could have been carried out.
However measurement of structural movement alone does not provide adequate strength of argument to neglect anchor load monitoring, particularly in structures where anchor failure may result in catastrophic collapse of the structure.
The long term monitoring of anchor load using load cells has long been a contentious issue due to observation of load discrepancies with the jack load and due to suspect long term deterioration in cell performance. At Devonport load discrepancy of approximately 10% was observed and it was acknowledged that the cell monitored load change and not absolute load. The majority of cells were reliable over a 12 year period but some failed due to inadequate waterproofing.
Regular monitoring of the anchors at the Nantgraw Landside Stabilisation, Manchester Stadium, Devonport Dockyard and Hodenpyl Dam along with the intention to regularly monitor anchors in the Japanese Expressways fully supports the anchor monitoring approach.
Investigation at the Silvertown Thames Barrier Project (and other papers relating to Aberdeen Harbour, River Clyde Glasgow, Hull Tidal Barrier and Greenhythe) highlighted that even where anchors have been neglected for a lengthy period load monitoring data and corrosion performance of the anchor heads can still be carried out in order to review satisfactory performance and anchor or corrosion protection replacement. Albeit these papers themselves all draw attention for more regular monitoring of load performance and corrosion being the best approach.
Performance of anchors at the regularly monitored sites were excellent; three sites involving rock anchors and one notable contract involved performance of multiple anchors founded in post grouted stiff clays where previous anchors had lost excessive load. Unfortunately long term monitoring data from soil founded anchors is particularly rare.

Goncalo Veloso, *Keller Gound Engineering*
Question for Lee Jordan. I am aware that the engineers initially specified bent anchors, i.e. anchors working vertically within the ground and then bending to the alignment of the galvanised backstays. I am not aware of any world-wide existing code of practice for these anchors and in the end after several tests and at a cost or around £1 million this scheme was rejected. Would you design in future a scheme using these so called bent anchors and do you still think it is viable and that there will be no problems during its design?

Lee Jordan, *Arup*
I do not understand the question. The anchors were vertical.

Tony Barley, *Single Bore Multiple Anchor Ltd*
I think I can clarify the situation. The initial anchors were to come vertically up the pile and then to be inclined about 20 degrees off vertical to take the tensile load straight from the stays. I think it was a rather contentious issue from the beginning but perhaps Lee would just respond to that. They were anchors that did not comply with any codes, there was a certain specification drawn up for them, the anchors did not perform and maybe the contractor says that anchors should comply with codes and ideas of putting in anchors that do not go in straight lines might not be put forward until there is adequate research and testing carried out.

Lee Jordan, *Arup*
Written answer. The original design was a combination of both existing ground anchor and post-tensioning technologies. It could therefore be seen as the innovative use of existing technology in a new application. The design concept was discussed with several high profile strand and bar anchor manufacturers and was further developed during and after tender in conjunction with well recognised UK anchor experts. The main contractor and their subcontractors signed up to construct the curved anchor detail and were prominent in developing the final details. Therefore, providing a similarly thorough design development exercise takes place, and there is confidence that the relevant contractors can achieve the design intent, a similar detail would be considered again.

Tony Barley, *Single Bore Multiple Anchor Ltd*
Written answer. It is appreciated when liaison takes place prior to the *"innovative use of existing technology in a new application"*. However, in such circumstances it is always appropriate to execute preliminary trials in order to establish known factors such as:
(i) the quantification of frictional forces encountered by a multiple of greased and sheathed strands stressed and de-stressed within a curved steel duct;
(ii) the damage sustained by the individual strand sheathing as a result of lateral displacement of the grease lubricant from the inner plastic sheath face; and

(iii) the long term effect of fluctuating loads (however small) on the friction generated and on corrosion protection sheaths over the length of the bent tendon.

Horst Aschenbroich, *Con-Tech Systems Ltd*
Question for Mr Okubo. I am very interested in this new kind of load cell which you use in Japan. If I read it correctly, it is a pipe support on which you read the different stresses on the outside. Did you find during your testing that you get differential loads on different types of this load cell because we are familiar with the vibrating wire load cells, normally they have 4 – 6 vibrating wire strain gauges inside and if you are not exactly concentric with your load introduction you get differential load readings that can sometimes be very confusing. In North America, we say load cell readings are not the accurate ones, the readings with a hydraulic jack always govern, but in my presentation today you will see that on one project we had 5-strand anchors and the contractor made a mistake in not exactly placing the strands centrally in the 7-strand anchor head, i.e. placing all the strands on one side and we got very different readings. What is your experience of that?

Hiroshi Yamada, *Nittoc Construction Co. Ltd*
For long term monitoring, NEXCO has already tried to establish an effective system using electromagnetics compared with the strain gauge type or hydro loading type. By comparing the data, now accumulated in the field of highway construction, the new system has clearly much merit for long term durability. The existing system has short term durability because it is not well waterproofed. Good waterproofing is essential for long term monitoring.

Tony Barley, *Single Bore Multiple Anchor Ltd*
I think that everybody agrees that load cells and their performance, as compared to lift-off tests, is a contentious issue. Everybody has had some experience with the load cells that work for a certain time and some people prefer lift-off tests.

Stuart Littlejohn, *University of Bradford*
Following Horst's question we are well aware that load cells can suffer from non-linear loading and in fact BS8081 (Appendix L "Load cells") and DD81 have since 1982 recommended calibration tests with eccentric loading, including inclined and uneven bearing plates, that has led to the use of machined distribution plates in practice to try and ensure an adequate seating and reliable readings. Regrettably, load cell manufacturers invariably ignore these BS recommendations so the provision of such calibration records is not standard practice. We should ensure that the load cells are properly calibrated for eccentric loading and that they are properly seated in practice, because otherwise load cells can give erroneous results. For this reason, load cells are often used to record load changes with time, as opposed to absolute loads, bearing in mind that variations in load with time can still be invaluable.

Chris Irvin, *DYWIDAG-Systems International*
Question for Michael Whitworth. With reference to the difference in performance between the bar and strand anchors, are the anchors installed in a different area? Are the bar anchors likely to yield more creep?

Michael Whitworth, *Devonport Royal Dockyard Ltd*
The anchors are installed almost adjacent to each other. There is no real difference in the geology. We have just observed that generally the load loss over time is greater with bar than with strand.

Mike Turner, *Applied Geotechnical Engineering Ltd*
I think we were responsible for the design of the bar anchors at Devonport. Probably the bond, i.e. the friction on the free tendon length is higher in a bar than in a strand, as a generality, that might be part of the explanation. There is a longer degree of load shedding in a bar type anchor.

Chris Irvin, *DYWIDAG-Systems International*
Question for Hugo Wood. What proportion of anchor load can the device for pulling the anchors achieve?

Hugo Wood, *High-Point Rendel*
I am not sure if it was used for the check lifting of the trial anchors but I believe it was used for the proof loading of the working anchors up to 900kN.

Mike Turner, *Applied Geotechnical Engineering Ltd*
Ground Anchors Ltd. installed the anchors at Nantgarw with that system. The destressing system comprised a great bar bolted on to its rectangular plate that would go to at least 80% of the characteristic tensile strength of the tendon.

Caesar Merrifield, *Coffey Geotechnics Ltd*
Just a very quick comment and a request for Stuart, as he is chairing the last session of the conference on recommendations and monitoring. Lee Jordan raised the matter of EN1537 and BS8081 not being as explicit as they should be with respect to monitoring. EN1537 is being reviewed currently and there are two experts in the room here (Tony Barley and Frank Schmidt) who are also on the reviewing committee. I would hope that we would do a better job this time in prevailing upon the European community, who regrettably are not here today in greater numbers, and I think they would have benefited from our deliberations because we are also being faced with ISO22477, i.e. the extraction of all geotechnical testing from existing geotechnical European standards. Lots to talk about with respect to the monitoring of ground anchorages in the future. I think we have a great deal of work to do.

Devon Mothersille, *Geoserve Global Ltd*
Question for Jesús Gomez. I was very interested in your presentation, particularly because I was involved in post grouting anchors in high plasticity clays just a few months ago. From our point of view the post grouting was critical in order to satisfy creep criteria. Please advise how much post grouting was necessary to satisfy the creep criteria.

Jesús Gomez, *Schnabel Engineering*
There are two aspects here as the original pile anchors were also pre-grouted. The reason that the wall kept moving was not due to the potential for creep or the anchors failing - they were just involved in the whole mass movement. During testing of the tiebacks there was creep and there were some tiebacks that did not pass and they were down graded, so those tiebacks I believe were re-grouted once or twice. They were selectively re-grouted and in the case of the SBMAs they were also selectively re-grouted using a packer every two feet. In general, there were two events of re-grouting at every port. I do not know from memory the volumes of grout that were injected. There were some tiebacks that took some good volume of grout, very few, and at some there was hydraulic fracturing of the soil, but that was not the purpose. In general, there were two events of re-grouting and I can give you the exact information if you write to me.

Peter Kingston, *Bachy Soletanche Ltd*
Question for Jesús Gomez. It is in connection with the single bore multi anchor system. Where you have a situation where you have an increasing load, do you run the risk of the shortest free length actually attracting the greatest load and therefore potentially generating a failure situation? When you were checking the load on the anchors did you check the load on each individual strand and was there a significant difference between them?

Jesús Gomez, *Schnabel Engineering*
We checked the load on each individual strand, i.e. the lift-off was recorded for each strand. I could show you a plotted increase of load per strand and it would illustrate similar behaviour following along that 45 degree line so for the shortest strand the stress is more than the longest strand - it was very obvious. In fact, there were a couple of tiebacks where the contractor incorrectly labelled the strands and we were able to find out which one was the shortest by looking at that. We estimate that the tiebacks are going to reach 80% of GUTS in about 2011-12 and they are being monitored yearly up until the point where additional remediation is started, which should happen in the middle of next year.

Tony Barley, *Single Bore Multiple Anchor Ltd*
I could just add a point to that, if it was known that the load is going to increase in SBMAs, let us say you locked off at 60% working load (Tw) and designed at 100%Tw, you would stagger the loads directly proportional to the staggered free lengths. In other words, you would have lower load in the shorter anchor, greater load in the longer anchor such that when the wall movement

took place all the anchor loads would be even, so it is acknowledged that this is a characteristic that people should be aware of and deal with in the design.

Rama Karri, *University of Birmingham*
Question for Mike Turner. You mentioned the anchors going to pull about 14mm at about 100 to 150-200kN. This really worries me because if you have along a longitudinal section of the retaining height a differential wall movement, it means one anchor moves and another anchor does not move, so it is going to cause differential movement between the anchors. Could you suggest when designing the anchors, the differential movements to be considered (as with lateral loads) because if there is movement, it is going to cause lateral loads on the retaining wall and piles.

Mike Turner, *Applied Geotechnical Engineering Ltd*
Basically, we were not doing a research project, we were monitoring a wall during construction work to make sure things were not going wrong and certainly the wall was never in distress from our observations. So the anchor loads probably would be increasing along the whole length of the wall to match that movement outward, say on day one and back on day two, so I am quite confident that the whole thing was acting as a unit with its capping beam. Therefore, the loads measured were typical of the loads expected by the wall movement. Basically, we were looking at items of data that were not necessarily of the same time point. The wall was moving quite surprisingly, I had not met that before myself, and also the anchor loads were fluctuating. Now one bit of data without the other says "have we a problem here?" but we could see that by back analysing our anchor load changes, that they were within the norm of the wall itself, that is not to say that there is not a problem there but overall the assessment of the data from our monitoring indicated that the wall was behaving satisfactorily and not causing problems with the anchors.

Written question and answers session two: service behaviour monitoring procedures and performance in service

Thomas Richards, *Nicholson Construction Co.*
In the paper by Okubo et al, Table 1 Type C explains the cause of sudden decrease in load after lift-off as "wedge and tendon slip after lift-off". At this point in loading, the permanent lock-off wedges should have no load and pulling wedges are the ones with load. I suggest other possible causes of this load drop, as follows:
1. Lift-off of the entire wedgeplate (anchor head), then the wedge plate hitting a restraint in the stressing setup and the wedges freeing from the wedge plate. The load drop being the force required to free the wedges.
2. Grout above the free length sheathing causing load transfer shallow then breaking of this grout, and/or
3. Grout inside the free length sheathing causing load transfer along free length then being lost.

4. Breaking of wires.

All of these causes would have theta1 greater than theta2. For cause 1, the higher theta could occur just before the load drop.

What are the authors thoughts on the above?

Have the authors found the displacement from lift-off to the point of load drop to be similar for a given testing setup? This would give credence to possible cause 1.

Hiroshi Yamada, *Nittoc Construction Co,.Ltd*

Thank you for suggesting the possible causes that are all possible from a general standpoint. However, the evaluation list was made based on findings for old type anchors. Many anchors of this type are in a condition where wedges do not act effectively due to rusting and the spaces between the wedges having become clogged with foreign material. Therefore, slipping between wedges and strands was confirmed after lift-off and this is indicated in Table 1 as Type C.

Stuart Littlejohn, *University of Bradford*

This paper by Wood and Martin describes an excellent example of how permanent anchorages should be inspected and monitored during their service life. When the lift-off load was recorded what was the standard clearance adopted between the stressing head and bearing plate and how was this measured?

Hugo Wood, *High-Point Rendel*

The lift off load was assessed as the load at which a clearance of approximately 1mm was observed. This was measured by visual inspection. The presence of a number of shims between the anchor head and the distribution plate which could become separated means that the distance of lift-off may not be completely uniform around the perimeter of the lowest shim. Because of this, accurate measurement of the lift off distance would, in practice, be difficult hence the reliance on visual observation.

North American Dam rehabilitation:
discussion session three

Rapporteur's summary
by Raymond Coe, *Black and Veatch Ltd.*

Session 3 comprised five papers each of which concerned ground anchoring aspects of dams in North America (Canada and the USA). Their locations are shown on the map below.

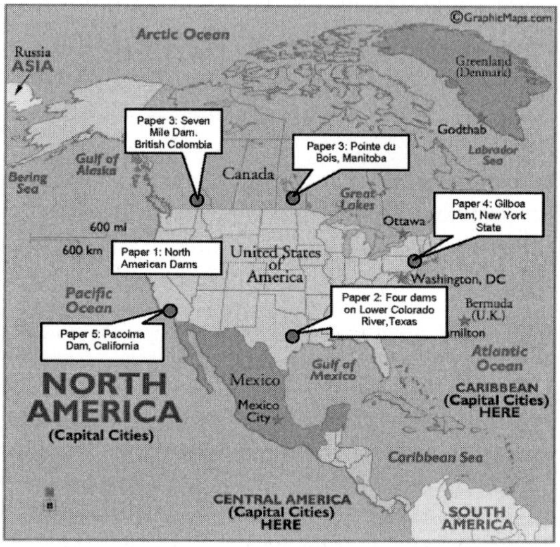

The first paper by Bruce and Wolfhope presents the results of a national research programme conducted in the USA. Given the objectives of the conference, the authors focused on the development of corrosion protection systems from the 1960's to the present time. They reproduce a range of comments from past papers concerning corrosion protection which provide a useful insight into the development of the systems which are now regarded as standard good practice. Up until the early 1970s cement grouting was the normal protection provided. Greased and sheathed tendons then began to be introduced and by the late 1970s early 1980s double corrosion protection systems had been developed. This comprised greased and plastic sheathed tendons inside a grout filled plastic duct which was grouted into the drill hole. An outer corrugated plastic sheath began to be used for the bond length. The authors draw attention to the difference between US and European practice that developed during this period whereby European engineers were not satisfied that cement grout provided a satisfactory barrier to corrosion whilst their US counterparts felt the risks associated with ground water percolating through cracks in the grout was tolerable. During this time, strand sizes increased from 12.7mm up to 15.2mm and the latter is now standard. Techniques for applying the grease and plastic sheath to the strand have also developed together with the arrival of epoxy coating. The authors have provided a valuable summary of their research covering the evolution of ground anchors over some 40 years and interested readers can consult the complete list of references in the companion paper in Session 8.

The second paper by Wolfhope et al describes anchors used to stabilise four of the six hydro power dams on the Lower Colorado River in Texas; they are known as the Highland Lakes Dams. Four of the six dams required stabilisation to withstand probable maximum flood events. Three were founded on igneous rock and one, Tom Miller, on weathered karstic limestone which required considerable grout pre-treatment. The dams typically comprise concrete and earth gravity components. A total of 317 anchors were installed in vertical and inclined holes. All tendons used 15.2mm 7 wire steel strand. On the first dam to be strengthened, Wirtz, corrosion protection was provided through the use of epoxy coated strand plus a minimum 25mm cement grout cover. The bond lengths for the remaining three dams comprised bare strand grouted in corrugated HDPE pipe which met the requirements for a US Class 1 encapsulation. The unbonded anchor strands for Wirtz and Tom Miller dams comprised grease filled polypropylene sheath; this protection was not adopted for the other two dams. It is interesting to reflect that additional corrosion protection would most likely have been specified in Europe. The service life of each of the dams is considered to have been lengthened by more than 50 years.

The third paper by Aschenbroich presents case histories of anchoring for two dams in Canada including long term monitoring. High capacity anchors were required for both facilities to meet dam certification safety requirements. At Pointe du Bois, Manitoba, the spillways, the dam and the powerhouse were all subject to anchoring to resist overturning and uplift forces. The spillway anchors were re-stressable bar anchors. The dam anchors are each 5 N° 15.2mm strand fitted with strain gauges and load cells which allow the loads to be permanently monitored. The power house was provided with additional resistance to uplift and overturning through 48 anchors each 18 N° 15.2mm strand locked off at 3150kN. After five years service the lift off loads were measured slightly less than the original lock off loads and there was no indication of any corrosion. All the anchors can be re-stressed in the future should the need arise. Seven Mile Dam, British Columbia comprises a concrete gravity dam and power house which, due to dam raising in 1988 and higher seismic load provision, required the installation of 57 re-stressable anchors. These were each 92 N° 15.2mm strand with an ultimate load capacity of 23 990 kN and locked off at 17 273 kN. All anchors were fabricated on site due to their large size. They have been tested several times since completion in 2003 using the re-stressing facility and all have exhibited losses less than the anticipated 2%.

The fourth paper by Zicko et al described the installation and testing of 79 high capacity anchors at Gilboa Dam in New York State. After the 1996 flood the sliding stability of the spillway was judged to be marginal and anchors were designed and installed in 12 months. The spillway is underlain by sedimentary rocks including shale which was the predominately weak material. 32 inclined anchors were installed in the downstream face and 47 through the spillway crest. The number of strands ranged from 39 to 58 with lock off loads ranging from 6098 kN to 9069 kN (60% guaranteed ultimate tensile strength). Construction safety must have been quite onerous given the locations of the anchors and the fact that the work was undertaken with the reservoir operational. Corrosion protection was to PTI Class I standard comprising a grout filled polyethylene sheath, corrugated for the bond zone and smooth for the free length, for the full length of the strand. This is not as onerous as European standards where almost certainly the strand itself would have been greased and sheathed. The anchor heads, wedge plates, bearing plates and tails were bitumen coated and encased in concrete which means that re-stressing is not possible but the encasement is felt to provide the best protection particularly in an active spillway. Prior to the main installation a test programme was undertaken on four anchors of smaller size. These were incrementally loaded and unloaded until a bond stress of 1379 kPa was exceeded; they were not loaded to failure. Two of these anchors were equipped with load cells and from 18 months readings the likely lock off load after 100 years was predicted. This confirmed the design and service life of the dam. All 79 production anchors were performance loaded to 133% design load. One anchor had 8 strands break and this was re-tested to a lower load and successfully passed.

The fifth paper by Schultz et al describes the performance of Pacoima Dam, California prior to and post installation of ground anchors following earthquakes in 1971 (San Fernando event – M6.6) and 1994 (Northridge event – M6.7). Surveys following the 1971 earthquake showed a net movement of the dam and left abutment, in particular two distinct block masses. The contraction joint between the dam and the left abutment thrust block opened 10mm. 35 rock anchors were installed in the rock mass underlying the thrust block each comprising 28 N° 12.7mm strand; eight of the tendons were equipped with load cells. In the 1994 earthquake the contraction joint opened 50mm at the crest tapering down to 10mm at the base of the anchored block. Overall, the authors conclude that the 35 anchors were effective in reducing the seismic displacement of the thrust block and improving the seismic performance of the dam. There is some useful data from the load cells but unfortunately none of the original ones function any more. Replacement load cells fitted in 1996 are no longer read because it is too labour intensive. How often do we hear this type of comment and the associated loss of valuable engineering data? After the 1994 earthquake eight additional anchors were installed in the same rock mass. Each one is fitted with a load cell and results are presented for a period of eight years. It is not clear whether or not the dam owner is still reading these cells. Seven of the anchors showed approximately 6% loss of lock off load over the eight years.

In summary, this was a valuable session with plenty of points for future reference. The benefit of installing monitoring gauges is particularly to be noted provided the owner can be persuaded of their value in the long term and will therefore pay for ongoing monitoring.

Mike Turner, *Applied Geotechnical Engineering Ltd*
Question to Donald Bruce. What is a class 1 anchor? How is it defined? How does it differ from European practice?

Donald Bruce, *Geosystems, L.P*
In Class 1 we have at least two layers of inspectable material to protect the tendon and we also pre-grout systematically in rock. You can have a Class 1 protection in soil and of course we do not pre-grout there. I think it is probably equivalent to where you were on tendon protection in the 1980s but it is a big step forward for us and we have been observing Class 1 now for the last 11 years. I think as well what we are trying to do is to compensate for a lack of number of layers by testing each one a lot more

systematically when we are putting them in the ground. So it is a process a much as a protection.

Mike Turner, *Applied Geotechnical Engineering Ltd*
What do you define as the two layers in the protection system?

Donald Bruce, *Geosystems, L.P*
Cement grout inside the corrugated sheath and of course in the free length you also have the greased and sheathed individual strands and then the corrugated sheath is taken over the full length to protect the individual strands during installation.

Chris Irvin, *DYWIDAG-Systems International*
Question to John Wolfhope. For the epoxy coated strand, what creep values do you have when it is subjected to load? I think it is a relatively new characteristic which has come to light that there is some creep on epoxy coated strand. Are you aware of this?

John Wolfhope, *Freese and Nichols, Inc*
I am aware of this and actually the new edition of the PTI recommendations addresses this issue, but when we did this project that information was not available, so we tested the coated strands to a higher standard and actually achieved pretty close to what you would expect for better strand but I do not have the exact numbers with me.

Donald Bruce, *Geosystems, L.P*
John is right and there is a very good supplement in PTI (2004) that addresses this because frankly the earlier materials that were used in the United States, the epoxy coated, epoxy filled strand for different reasons was sub-standard and from one supplier only. Now we have an alternative source in the Japanese material which is just excellent in terms of its quality and its relaxation and creep properties. It is in fact almost equivalent to bare strand, so we tend to be going that way now having gone through the fire with the poorer material in the late 1990s.

Ian Clough
Question to John Wolfhope and Donald Bruce. At the risk of showing that it is 45 years since I last anchored a cable, what was the composition of the anchor grout you used and also the grout that you used around the annulus outside the corrugations?

John Wolfhope, *Freese and Nichols, Inc*
One part sand to one part cement by weight.

Donald Bruce, *Geosystems, L.P*
This grout has a water solids ratio of about 0.6 and is used for pre-grouting of the rock. Tendon grouting and annulus grouting are carried out using a standard mix, somewhere between 0.4 and 0.45 water/cement ratio by weight. As it is terribly hot where John works, typically about 0.5% superplasticiser is added to help it stay fluid with minimal bleed and very high strength.

Gareth John, *CAPCIS Ltd*
With regard to the epoxy coated strands, how actually is the epoxy applied to the strand, fusion bonded or chemical?

Donald Bruce, *Geosystems, L.P*
It is a fusion bonded system and within the conference attendees there are specialists who produce that strand in Japan and who should be consulted. Basically, the strand is heated up, the epoxy is a powder that is fusion bonded and then it is drawn through.

Gareth John, *CAPCIS Ltd*
In terms of the pre-treatment of the strand before the epoxy application what is the pre-treatment?

Donald Bruce, *Geosystems, L.P*
I would not be able to tell you that, and I do not know if I could, even if I knew because of the proprietary nature of the process, but there is a lot of thought goes into it. If you talk to my Japanese colleagues they will give you the scoop.

Edward O'Connor, *ESBI Engineering and Facility Management Ltd*
Would you like to comment on long term creep performance in regard to over tensioning? Is that common practice in the tendons for all the projects or is it specifically designed on a project by project basis?

Donald Bruce, *Geosystems, L.P*
I think it is fair to say that it is quite standard that we automatically design in at least say 10% over, so typically we are locking off at anywhere from 66 to 70% GUTS, knowing that we want say 60% in the long term. We think that it is much more a reflection of relaxation in the steel which we can govern and manage than it is creep. There is very little evidence of creep which is geologically controlled unless you are looking at interfacial aspects with the grout internally, so we tend just to be very conservative. We know that long term loss with the excellent steel that we now have is only 2 or 3%, so we just make the over tension between 6% and 10%.

Edward O'Connor, *ESBI Engineering and Facility Management Ltd*
Actually, what triggered the question was that I am conscious of overhead conductors that are used in the transmission industry and they clearly go for long term creep as well. As far as I know there is a policy of over tension by 10%.

Bede Noonan, *Geotech Pty Ltd*
Question for Donald Bruce. I was interested to see the project where you had the strand break, was there only one strand that broke?

Donald Bruce, *Geosystems, L.P*
No, we lost 8 wires on 8 different strands.

Thomas Richards, *Nicholson Construction Company*
It is my opinion that it was caused by single strand stress.

Donald Bruce, *Geosystems, L.P*
The official position taken by the engineer however - could be that we always pay attention to what Tom says and he is right most of the time but we also think there may have been just a slight misalignment on some of the strands.

Bede Noonan, *Geotech Pty Ltd*
Regarding what Lee Jordan said re the Manchester Stadium project, one of his conclusions was to allow for extra anchors in the future when you design these structures. There is a project we are looking at in Australia where the design proposal at the moment is probably going to limit the ability to put extra anchors in the future if there is a problem with an anchor and it is on a dam. Would your recommendation be that the design should really think about the ability to put in extra anchors in the future?

Thomas Richards, *Nicholson Construction Company*
Our practice is usually to add an extra strand for that unusual case. We very rarely have that happen but typically it may not be every tendon depending on how the load is distributed but we do add an extra strand so that it will allow for that construction issue.

Horst Aschenbroich, *Con-tech Systems Ltd*
The Seven Mile dam was originally designed for 90 strands but then with the strand layout it gave us the opportunity to put 92 in with a little extra cost, to give a 92-strand tendon.

Donald Bruce, *Geosystems, L.P*
I think that is probably the answer to your question. We do not allow really for extra tendons to be installed but we allow extra capacity in the individual tendons.

Bede Noonan, *Geotech Pty Ltd*
Limiting your design in the first place would not seem very smart.

Donald Bruce, *Geosystems, L.P*
In addition, the tendons always group up. You look at parts of the dam and if it includes 43, 44, 45 and 46 strand tendons, you just make it automatically 46 strands for that part of the dam so we try and manage it by design in the initial stages, rather than leave extra pockets because it is tough to get back.

Thomas Richards, *Nicholson Construction Company*
Extra strands are a very small component of the cost compared to extra holes.

Written questions and answers session three: North American dam rehabilitation

Stuart Littlejohn, *University of Bradford*
Compilation of a national bibliography and case history database will be invaluable in assessing the potential condition of anchorages. Those designed and installed before the adoption of PTI (1974) will require particular attention in the future, as the corrosion protection provided may be regarded as inadequate by the standards of PTI.

In the paper, it is stated that no case has been reported of failure resulting from bond length corrosion in a properly grouted anchor. This will invariably be the case, but the key question when you are investigating existing anchors in service is "how do you know when you have a properly or improperly grouted anchor in situ?

Donald Bruce, *Geosystems, LP*

The assessment of the quality and adequacy of the anchor grouting is made in real time during construction, and during stressing and testing: the Engineer of Record must "sign off" on the completed installation, confirming that it is in conformance with the Project Specifications and the Contractor's Method Statement. Other than physical removal of the existing whole anchor (an enterprise which is occasionally proposed by over coring, but is never actually conducted due to the inherent practical difficulties), the authors know of no other method to verify the appropriateness of the grout in situ in an existing anchor.

Rock bolts in tunnels and mines:
discussion session 4

Rapporteur's summary
by Ursula Lawrence, *Cross London Rail Links Ltd*

Session 4 was a relatively short session but contained three fascinating papers. The main aim of the papers was to present experience and techniques from mining operations which could be transferred to more traditional civil engineering applications. All three were well received.

Alison Littlejohn's paper, A review of Glass Fibre Reinforced Polymer (GFRP) Tendon for Rock Bolting in tunnels, was intended as a state of the art review of the use of GFRP in rock bolting practice. In addition comparisons were presented of both the performance and costs of traditional steel anchors. GFRP anchors have several advantages over steel. The flexible and lightweight properties are advantageous where space and access are restricted. Its inherent corrosion resistance is useful in challenging physical environments leading to improvements in durability. The durability of GFRP is affected by five factors, i) sustained long term stress, ii)pH particularly low pH environments produced by cement, iii) presence of moisture, iv)UV radiation and v) temperature. The effects of each of these factors were described and any differences due to resin or glass fibre specification were highlighted.

The properties of both the glass fibres and the resin can be specified individually and as a result the properties of the resulting anchor can vary widely. As with all anchors installation is critical. GFRP rock bolts are weak in transverse shear and problems can arise when using resin capsules. Similarly the locking nut attachment is a point of weakness with the head detail having a tensile capacity or 25-50% of the bar. GFRP rock bolts are widely used in mining applications were the observational method is used. However the installations are often considered temporary and long term monitoring and testing data is not available. In Civil engineering applications ground anchors are considered long term processes typically requiring minimal maintenance. GFRP anchors will need to demonstrate long term performance before it would be widely adopted.

The paper by Kenji Aoki, Design and Construction of Large Rock Caverns Supported by Ground Anchorages, drew on Japan's long history of creating large underground caverns. Approximately 50 underground power generation plants have been constructed in the last 60 years and in recent years large caverns have been created for large volume oil storage. The paper describes the performance of a large underground cavern constructed using shotcrete, rock bolts and pre-stressing anchors.

The whole process of design installation and performance is illustrated with an example, the Kannagawa power plant. The underground cavern is 216m long, 33mwide and 52m high with a volume of 220 000m^3. An investigation was carried out into the geology of the area and the in situ stress field. The caverns are typically designed using finite element analysis supplemented with key block and circular slip analysis. During construction the design process is supplemented with an observational system utilising field measurements and local variations in geological conditions. In addition the stresses during the phased excavation were modelled and checked for accuracy using strain gauges and acoustic emission measurements determined the development of crack growth. The case study showed that even very large excavations could be adequately supported by well designed and monitored ground anchorage systems.

The last paper was presented by Peter Altounyan of Rock Mechanics Technology, Monitoring and Testing of Mine Anchor Systems: Case Studies and Applications to the Civil Engineering Industry, which describes monitoring and testing regimes for anchor systems in mining applications which may be applicable to civil engineering environments.

The large extent of mine workings has resulted in the development of inexpensive instruments for monitoring anchors, which are easy to install, read and maintain. This contrasts with civil engineering tunnels which typically use specialist and expensive systems. Mine support systems typically employ an observational approach based on monitoring data obtained from borehole extensometers and strain gauged rockbolts. These are supplemented by a simple system of telltale extensometers which give a visual measure of roof movements. Both telltale and more recent, electronic extensometer systems were described in detail and typical results presented in case studies.

Traditionally the measurement of strain in rock bolts utilised load cells mounted on the end of the anchor. Unfortunately these load cells provided little information. Recent developments were presented for measuring the load profile along the anchor and determining the integrity of the anchor utilising strain gauges attached in pairs, at a number of points along the bolt, typically 9. This system allows the determination of the maximum load over the any portion of the bolt and any significant bending loads.

Loads in mining operations can exceed the capacity of the bolt or anchor and several methods have been developed to determine whether in situ bolts are intact or broken. These include ultrasonic methods which work on single bolts and a radio frequency technique which works on a pair of bolts in a similar manner to cross-hole tomography. In addition a system has been developed, recently, by RMT, utilising fibre optics which are installed with the bolt/anchor and give a quick indication of the integrity of the system using torch light.

All three papers presented techniques which could be applied to civil engineering applications. However there are a number of significant differences between underground and surface applications. The three cases typically relied on the bolts or anchors purely in tension, whilst the majority of the cases presented in other sessions involved some element of lateral loading, especially those retaining slopes. Underground systems typically reflect less input from detailed design and more influence from observational systems during construction. This is in contrast to traditional civil engineering anchorage systems which designed to a greater level of detail.

Mines and underground caverns provide constant environments for the anchors i.e. temperature, heat, water flow and chemistry. Anchors at the surface exist in seasonally variable environments, involving temperature, heat (sunshine), water flow (rain), and are often installed in dynamic environments e.g. marine environments and earthquake prone areas. The chemistry of near surface groundwater can change depending upon wetting and drying cycles, presence of pollution etc.

Despite the differences in design and environment, the same underlying message came across as for all the other sessions at the conference. This message is that quality, monitoring and maintenance are critical. The quality of the anchor and the quality of the workmanship during installation are critical in determining how long the anchor continues to function. Post construction condition and performance monitoring is fundamental to ensuring that the anchorage system continues to function properly.

Philippe Vincent, *Menard*
Question for Alison Littlejohn. Regarding the glass fibre reinforced polymer, the Young's Modulus is about 20 – 25%, that of steel. Is that the characteristic or isotropic value, or is there a difference depending on the axis you consider?

Alison Littlejohn, *Ramboll Whitbybird*
The Young's Modulus will vary depending on the material constituents used in the composite. Certainly, when you are putting it under stress, if you load it axially in the same direction as you have actually manufactured the fibres, for example, then the creep of the resin matrix is reduced. If you put the bolt under axial stress, it is the axial strain that can cause a problem.

Philippe Vincent, *Menard*
When steel could perhaps tackle problems of a slide in a rock mass, maybe the fibre would be not as strong.

Alison Littlejohn, *Ramboll Whitbybird*
It is widely acknowledged that it is strong in axial tension but weak in transverse shear.

Kitae Chang, *Geotechnical Monitoring Group Europe*
The brittle failure of GFRP worries me. Brittle failure rather than ductile is not good for safety. A ductile type of failure gives us time before it occurs dramatically, i.e. before it fails, but with brittle failure, no time at all.

Alison Littlejohn, *Ramboll Whitbybird*
True. Temperature can also have a significant effect on the behaviour of GFRP, for example, it loses 50% of its tensile strength at 325°C, so for a fire situation in a tunnel where fire temperatures might be much higher, then it has the same problem and after a certain temperature it will fail in a brittle manner. For high tensile loads, GFRP may not be appropriate. For nominal loads and for some of the rock bolting applications where there is lower stress, GFRP may be appropriate if it is supplied in the correct environment. I agree with your concern and in providing a review of the papers in an unbiased way, I am not seeking to sway people for or against its use, but having provided details of the known properties and behaviour of GFRP I am suggesting that

we need further guidance on those kinds of issues, so that GFRP can be used in an appropriate way by designers and installers.

Kitae Chang, *Geotechnical Monitoring Group Europe*
Question for Peter Altounyan. Your monitoring of vertical rock bolts is interesting, but exactly what are you targeting in the detection? In my opinion, as the active zone of the rock mass moves along the rock bolt, you cannot detect the distribution of force acting along the bar. So, for the case of a rock mass deforming along the bar you cannot get anything.

Peter Altounyan, *Rock Mechanics Technology Ltd*
The rock bolt is reinforcing the rock which changes the way it behaves and in order to understand whether you still have a reinforcing system in place or you have got a failed system, you need to monitor the load in the bolts, the strain distribution and how much the rock is moving. By having that information, you will have a fairly good understanding of what is going on. Also, if the rock expands during its failure process by more than a certain amount, it is quite likely that the rock bolts will have failed by that point, so the idea with the tell tales is if the movement is in excess of a certain amount you know that your system of support is no longer effective. That has been proven by a lot of experience.

Kitae Chang, *Geotechnical Monitoring Group Europe*
The central section of the bar may not be deformed at all, but the rock mass surrounding the bar is straining.

Peter Altounyan, *Rock Mechanics Technology Ltd*
The idea with rock bolts used in mining is that you have a very good load transfer so that a very short length of bolt will generate sufficient bond to fail the steel. So, when the rock moves, for example, if you look at the numbers for modern bond strengths, the rock only has to move 0.1mm to generate 10 tons of load within the bolt. You get a reaction very quickly so the rock cannot move without the bolt knowing about it.

Chris Irvin, *DYWIDAG-Systems International*
Question for Peter Altounyan. Do you have much information on the comparison between steel bolts and GFRP bolts?

Peter Altounyan, *Rock Mechanics Technology Ltd*
GRP is used extensively in coal mines because the sides of the roadway are made of coal and they are cut out so GRP is used in very large numbers and there is a lot of technology associated with making those GRP bolts work effectively in coal.

Chris Irvin, *DYWIDAG-Systems International*
I should have been more specific. From a performance point of view, I know the benefit that you can cut through the GFRP with the header but can you summarise the performance characteristics?

Peter Altounyan, *Rock Mechanics Technology Ltd*
With the modern designs profiles, it has equivalent bond strength to steel and equivalent axial strength which Alison mentioned. It is brittle, in other words it does not go into the ductile phase and it is also weak in shear. So, if you have a ribbed coal site that wants to slip thereby generating bending loads in the GRP, then that is the

situation that gives concern. The other problem is the end fitting. In the way we use rock bolts, you start off seeking to reinforce the ground; when that starts to fail you are relying upon the end fittings to carry the weight of broken material. With the steel bolt system the end load characteristics are equivalent and in fact are designed according to the British Standard to be very close to the strength of the bar. The problem with GRP is that it is very hard to fit a good high end load system and it is only recently that those systems have been developed, following a fatality in a coal mine about a year ago.

Jiro Takemura, *Tokyo Institute of Technology*
I have a question about the difference in mechanical properties between steel and GFRP. I suppose a big difference is the stiffness, so if you have deformation predominant in the axial direction maybe you have no difference, but if you have deformations where you expect some bending in your bolt, I suppose steel is better than GFRP. Am I right?

Peter Altounyan, *Rock Mechanics Technology Ltd*
Yes, steel is much better. You can bend a steel bar and you get equivalent strengths to the axial strength. If you take GRP of the same diameter, for example, a 22mm diameter steel rock bolt will fail at about 30 tonnes. For the bending characteristics, there is also a shear test in the standard. If you put a GRP bolt into that test set-up it will fail at about 1.5 tonnes, so it is more than an order of magnitude weaker in bending.

Jiro Takemura, *Tokyo Institute of Technology*
So, if you have a circular cross-section in your tunnel I think you can avoid the deformation but in a rectangular shaped tunnel you might have deformation.

Peter Altounyan, *Rock Mechanics Technology Ltd*
You get a lot of shear occurring in the tunnel. The problem in mining is that you have to provide support in some way so the solution is just to install more bolts and to monitor.

Non-destructive integrity testing: discussion session 5

Rapporteur's summary
by Caesar Merrifield, *Coffey Geotechnics Ltd*

Papers in this session have as a common theme the promotion of the need within the industry to expand the repertoire of non-destructive integrity testing methods of anchorages in service, with special reference to the determination of the state of load in anchorages at any time. The standard direct methods of integrity testing which include the exposure of the anchor head and specifically the lock-off mechanism are both costly and, very often, impractical resulting in many anchorages not receiving the level of monitoring recommended in the international standards.

The five papers in this session present results of research which focus on the ability to monitor the changing levels of load being carried by the anchorages at any time. They do not specifically address the monitoring of the current corrosion protection, although a failure of the corrosion protection system within the anchorage may well be the primary cause of load reduction.

The two papers presented by the researchers from Aberdeen University draw on and develop the concept of a non-destructive testing method based on the dynamic response of an anchorage to an applied external force. This is the basis of the GRANIT© system, under development since 1989. These papers are;-

"Common anchorage issues addressed by numerical modeling" Ana Ivanovic, Richard Neilson, Andrew Starkey and Albert Rodger
"Non-destructive testing of deep mine anchorages" Andrew Starkey, Richard Neilson, Ana Ivanovic and Albert Rodger

The paper dealing with the numerical modeling of the anchorage response to a controlled external dynamic load, builds on the principles of the GRANIT© analysis but introduces a model which is supposed to be able to examine the influence of various elements within the anchorage on the overall dynamic response. The paper deals with issues such as the practice of load estimation, debonding of the fixed anchor length, estimation of free anchor length, the influence of the anchor head and also anchorages installed in sand.

The authors acknowledge that the vibration response measured from the installed anchorages at any time during their working lives indicates changes in the anchorage but the causes of these changes were not readily identified.

The paper concludes that the dynamic numerical model presented in the paper provides a better understanding of how changes in parameters of materials and geometry affect the dynamic response of ground anchorages. The issue of the influence of the head mass and hence it's inertia is addressed and the paper notes that further work is required to clarify this issue. The authors believe that the basic advantage of introducing the model into the GRANIT© system is the ability to use the data from the model to train a neural network to provide a database rather than having to access research anchorages on site to obtain training data.

The second paper from the University of Aberdeen reports the results from a research programme where 264 anchorages in an underground coal mine were subjected to dynamic load testing to undertake a statistically based effective free length analysis on all the anchorages. It describes the research programme, the test procedures and the development of the mathematical model of the anchorage system.

The work concentrated initially on the determination of the effective free lengths of the anchorages tested. Destressing of the anchorage allowed direct assessment of the free length from the impulse tests. The authors conclude that the tests and subsequent analysis using the model allows differences to be observed across the site with respect to installation technique or other factors brought to bear at the time. They also recognize that the control of the effective free length is important in the successful long term monitoring of the anchorage. Their view is that the mathematical model shows that the effective free length is a fundamental physical property of the system – removal from the anchorage design would allow more effective load diagnosis and exception testing. As observed in the other paper from the University of Aberdeen, the authors recognize that control of the anchorage head characteristics during construction is essential for reliable long term condition monitoring using this method.

The three papers presented by delegates from various companies and institutions in Japan are:-

"Ultrasonic technique for rock bolt and rock mass bond quality control" Takayuki Mori, Makoto Nakajima and Tohru Tsutsui
"The development of the method of detecting the anchor tensile force to impact elastic wave" Kuniyasu Tanaka, Hidenobu Okada and Masayasu Izakura
"Development of loading apparatus and test procedure of rapid load test for ground anchors" Tatsuro Sueyoshi and Kazuo Tani

The first two papers focus on the development of theoretical approaches to the integrity testing of very different characteristics of an anchorage. The third paper describes the development of a rapid loading device and associated pull-out load capacity evaluation method.

The first paper deals with the issue of grout integrity post installation. It highlights the possibility of long term reduction in stability of the rock mass due to poor grout filling during installation. It describes the development of a numerical simulation of the wave propagation in a grouted rock bolt and subsequent laboratory and in situ test programmes to validate the method. The system relies on the diagnosis of the differences in the amplitudes of the reflective waves generated from the grout infilled annulus due to emissions from an ultrasonic wave generator. The results from the numerical model suggested that the state of infilling was better observed by the use of low frequency input waves as they were less directional than the higher frequency waves and were therefore more likely to be influenced by the grout boundary conditions. The paper concedes that the inspection method is able to estimate relative quality of the grout infilling.

The second paper describes the development of a method of detecting the anchor tensile force by using an elastic wave provided by an impact. In this case the anchor bar above the lock-off location is struck laterally causing a shock wave to be propagated down through the anchor bar. The paper deals only with anchorages locked off using a nut type fixing. It relies on the assumption that fixity of the nut makes a perfect rigid connection integrated with the anchor plate. Using the generalized equation from free vibration at the anchor head a vibration equation is developed taking into account the length of the bar extending beyond the nut (the anchor extra length). The result is a frequency generated by an impact to the end of the extra length. The proposition is that analysis of the dominant lateral vibration frequency of the anchor relative to a frequency associated with perfect fixity and load level in the anchor provides an indication of any change in the load in the anchorage and hence an indication of the anchorage condition. The authors acknowledge that factors such as head fixity and an air gap in the head influence the interpretation and prior to implementation the method needs to be tested within a wider range of applications.

The final paper presented deals with the development of a loading apparatus and test procedure in order to accelerate the on site load testing of ground anchorages. It is propose that the apparatus provides an effective method of measurement to evaluate the anchorage's pull-out resistance. The main advantage of the system is that the testing time is very short relative to static load tests, hence providing a more economical method of testing.

The paper promotes the advantage of the simple load application and measuring system reporting the development of an improved device which complies with the Japanese Geotechnical Society's Standard (JGS 1815:2002). The essence of the analysis is that the displacement measured either by an accelerometer or rapid reacting displacement transducer, can be related to the anchored free length through the dynamic loading characteristics. The authors report that the device can provide a maximum load of 207kN restricting it to a limited range of anchorage types. Whilst very little background to the analytical method and its relationship to the physical characteristics of the measuring device are presented, the authors appear to have derived sensible data which they acknowledge requires further validation through application to many more sites.

Tony Barley, *Single Bore Multiple Anchor Ltd*
Question for the Aberdeen researchers. The passive rock bolts were installed vertically into the roof of the cavern and you detected de-bonding at the base of the rock bolts, i.e. at the proximal end. What was the loading condition that actually caused the bars to de-bond at the base of the bore hole?
Question to panel. Could any of the monitoring systems presented have been used to detect the catastrophic failures in Boston where there was a failure of bolts at the roof of the tunnel?

Andrew Starkey, *University of Aberdeen*
They were not de-bonded; they were not installed correctly. The bolt is normally a two speed resin installation. It would have a high speed resin at the far end and a low speed resin at the proximal end and they will fill that up until it came out of the borehole, the idea being that the borehole was then completely full of grout, fully encapsulated, fully grouted. Initially, the fast resin would set quickly and then the bolt would be locked-off at a load, normally 50 to 100kN.

Tony Barley, *Single Bore Multiple Anchor Ltd*
Was there any stressing carried out after the lower length resin grouting was done?

Andrew Starkey, *University of Aberdeen*
Once the low speed resin has set, it was just left.

Tony Barley, *Single Bore Multiple Anchor Ltd*
So really it was not caused by the loading but just by poor quality workmanship. I am sure people here are aware of the severity of the Boston failure of the bolts which were quite catastrophic. Would any of the monitoring systems presented have detected the problems with the epoxy prior to the failure occurring, had their systems been tried?

Andrew Starkey, *University of Aberdeen*
In terms of what I was illustrating in the movie, potentially that could show a build-up of loads. I do not know the details of the project but if the load build-up is very quick and if you had something like a GRANIT device in critical areas of mines just continually checking, you would get perhaps early warning and obviously in the mine you do not need very much warning to cover a large space, a large distance as an individual, so it could make a difference if it was deployed in that way.

Horst Aschenbroich, *Con-Tech Systems Ltd*
Question for Mr Sueyoshi. In your rapid load test do you use this as a monitoring device after an anchor has been stressed and come back after half a year to do a rapid test, or do you use this as a test at the time of stressing, because you said a normal test takes up to 1 hour and yours takes only a few seconds. When we test anchors we test them for creep and that takes quite a while and I cannot understand how a rapid load tests could work. How was the rapid rock test done?

Translator
They use dumping rate and spring rate in place of the test of investigating the creep.

Horst Aschenbroich, *Con-Tech Systems Ltd, Canada*
That cannot be done in a few seconds; it takes a while for creep to take place.

Translator
They do not do the creep test.

Caesar Merrifield, *Coffey Geotechnics Ltd*
I think the second part of your question was, is this test done at the time of construction or is it also used in the future to monitor how the anchor will behave in service?

Translator
At the time of construction.

Caesar Merrifield, *Coffey Geotechnics Ltd*
It is a form of stress acceptance test I suspect, i.e. an acceptance test without creep.

Mike Turner, *Applied Geotechnical Engineering Ltd*
Question for Mr Mori. Just in case I missed something within your paper, I presume that the exposed length of the bolt that you strike with a hammer is a uniform length. You say it must be 150mm long. Would that be correct?

Stuart Littlejohn, *University of Bradford*
The answer is that they have done calibration work and they can allow for different lengths. Nevertheless, there are a large number of parameters involved in the test and I suspect that having now established the principles, there is a need to validate this type of testing on real rock bolts where we know that quite a variation in protruding lengths will be encountered, but the authors have carried out initial work to show the influence of protruding length.

Tom Richards, *Nicholson Construction Company*
Question to panel. The Aberdeen Group is focusing on frequency of vibration with GRANIT and it seems to me that Mr Mori's system is similar to GRANIT but he is focusing on amplitude. Is there any reason any reason why one would choose one versus the other?

Andrew Starkey, *University of Aberdeen*
I have done a number of tests in the past and the GRANIT system has been developed over 20 years. If you are going to test multiple bolts, there are lots of characteristics of anchorages on sites and if you just look at amplitude it is very difficult to then diagnose multiple anchorages in our experience. On the other hand, frequency is more tied to the characteristics of the anchorage itself which is the reason we focused on frequency. It means that you can diagnose large numbers of anchorages using the same method without having to retrain, i.e. without having to re-evaluate what each parameter is - you can just do a single test for those anchorages and get a single diagnosis.

Richard Neilson, *University of Aberdeen*
If I could just answer a little bit more about the question you asked. The two systems are completely different, in our case we were applying an impulse and we were measuring the frequency response of the bolts coming back. In the case of Mr Mori, he is actually putting in an ultrasonic signal and he is measuring the size of the response coming back, so the two systems are completely different in terms of ethos for want of a better word.
Question for Mr Mori. How easily can you couple your ultrasonic transducer to the bolt? It seems to me that it is a difficult system to couple high frequency vibration into the bolts effectively.

Takayuki Mori, *Tsutsui Denshi Co. Ltd*
I think it is very easy. It should be easy using grease. Rust is the problem, but roughness of the surface is normal and is no problem.
Written answer. I think it is not so difficult. As you pointed out, it basically matters that the surface of the bolt head is flat and contact is right-angled. However, this technique is little influenced by the contact condition of a probe, compared with other methods. In this technique, echo jelly is applied to the contact surface and the reflective level at the tip of the bolt being the maximum is made into the typical measured value. We use the waveform of high energy and high purity frequency which is generated by burst wave for measurement.

Lee Jordan, *Arup*
Question to panel. Is it possible to use any of these techniques on multi strand anchors or has that not been developed or is it purely for rock bolts and mono strand anchors.

Andrew Starkey, *University of Aberdeen*
We have done tests on multi strand and have shown that it works.

Ana Ivanovic, *University of Aberdeen*
There will be research going on over the next three years so we are trying to establish the relationship within multi strand anchorages in terms of modelling and implementing it into GRANIT system so we are hoping that we will be able to offer much more.

Gareth John, *CAPCIS Ltd*
Following on from the comment about coupling of the ultrasonic system to the bars, we have been involved in work on some large diameter steel bars, we have been using ultrasonic test measurements, in order to look for defects in the bars. One of the main problems there is not the roughness of the face but the lack of normality of the face with respect to the actual bar length. If you position the transducer just off true normal you get completely different results and it is then down to the skill of the operator to actually identify the problems. Have you found this as an issue in the bars that you have been testing?

Takayuki Mori, *Tsutsui Denshi Co. Ltd*
I have no answer at this time. Please give me your email address and I will provide the answer.

Gareth John, *CAPCIS Ltd*
I understand the problem. Do not worry.

Devon Mothersille, *Geoserve Global Ltd*
Question for Ana Ivanovic. Following on from the multi strand question, what challenges do you envisage taking on a multi-strand tendon, as opposes to a bar?

Ana Ivanovic, *University of Aberdeen*
Just the fact that we now have more than one strand means the influence of one strand on another may be important and another major aspect could be the anchor head because with more than one strand we may see differences in response of the anchorage due to the increased stiffness of the anchor head. I have not had a chance to talk in great length but we have done quite a lot of research in terms of sensitivity of the anchor head and the parameters that I am using in the model are actually obtained from laboratory testing. If you see differences in response in multi-strand anchorages this makes the situation much more complex and this is the challenge that I can see in that system.

Caesar Merrifield, *Coffey Geotechnics Ltd*
We have seen during the day dramatic pictures of fairly serious corrosion in anchorages after a number of years in service and again if the philosophy behind this conference is that we should take a positive pro-active role in ongoing monitoring, I did mention earlier about the review of EN1537 and writing into it a greater role or a greater awareness for monitoring. Given the systems that we have seen here, the indirect systems that we have seen from the Japanese delegation and from Aberdeen University, do you believe that there is sufficient confidence in these systems given the lumped parameter basis of your model and how you analyse the system such that you get this variable corrosion right at the position where you are facing the actual GRANIT system. I know the GRANIT system more than I know the other bits but that is where it is all happening. What will be the effect of the actual external condition of the head and of the anchorage and the tendon or the bolt in your case, and multi-strand tendons, hopefully in the future? Will we have confidence in the

results that you come up with, because we saw some pretty ropey anchorages? Would you be able to apply the GRANIT system, could you clean it up basically to a condition where you can actually be confident that what you are reading is something in the anchorage?

Andrew Starkey, *University of Aberdeen*
Yes, it is an interesting question, we have tested on a large number of sites and we tested on sites where there has been extensive corrosion at the anchor head as well, and we have been able to use the model still in those circumstances, but it comes back then to the level of proof that can only really come with time, so from our point of view we have been able to prove the system under those circumstances.

Caesar Merrifield, *Coffey Geotechnics Ltd*
Would it be valid to include indirect systems like this into a European standard for monitoring of anchorage performance in the future?

Andrew Starkey, *University of Aberdeen*
In terms of what we are measuring, e.g. load, this is not a direct measurement of corrosion. If there are load changes then you will see an effect and therefore indirectly, particularly if you are monitoring over a period of time, you can start to see the changes that the corrosion has on the performance of the anchorage and use of the model allows us to investigate the effect of corrosion on various parameters and actually predict what change that would be, so some of the changes may entail us looking for different aspects of the vibration response using the model, but we can then use the model to predict what is going to change and then we look for it and then you can see it changing all the time which was linked to the corrosion.

Stuart Littlejohn, *University of Bradford*
I just wanted to react to Caesar's comment because it is a very important one, namely that we might consider some of these non-destructive testing techniques in a standard. I would not recommend that as step one. What I would recommend is that an independent committee should appraise these non-destructive techniques in an objective way. What do we know about them, how reliable are they, how many case histories are there to justify the approach. If they then come out with recommendations that several techniques are viable in practice that is the time, and only then, that they should go into a code of practice. We already have problems with EN1537 on electrical resistance measurements where you might not get all the anchorages passing at the beginning and then what do you do. In some countries, practitioners are suffering because it is written into the European code, perhaps prematurely, but it is judged acceptable by some engineers simply because it is included in the code.

Written questions and answers session five: Non-destructive integrity testing

Horst Aschenbroich, *Con-Tech Systems Ltd.*
The method shown by Tanaka et al is for bar anchors, how would this work on strand anchors?

Kuniyasu Tanaka, *Shikoku Electric Power Co.*
The method was verified for strand anchors. However, because the principle of this method is simple, we consider this method is applicable for bar anchors.

Paper
"The development of the method of detecting the anchor tensile force to impact elastic wave"
Kuniyasu Tanaka, Shikoku Electric Power Co. Inc, Hidenobu Okada, Ikata Civil & Architectural Engineering JV and Masayasu Izakura, Shikoku Electric Power Co. Inc, Japan
Presenter: Toshiro 'Tiger' Kido

Numbered questions from Richard Neilson, *University of Aberdeen*

Answers in bold from Kuniyasu Tanaka, *Shikoku Electric Power Co.*

1. The units of Equation (1) appear to be inconsistent. I think it should read

$$f = \frac{\lambda^2}{2\pi L^2}\sqrt{\frac{EI}{\gamma A}}$$

g is only required if γ is specific weight (N/m^3) rather than density (kg/m^3).

As indicated in the question, "g" is not needed for Equation (1).

2. For Figure 8 do the authors calculate the frequency of a cantilever beam with variable cross-section?

Yes.

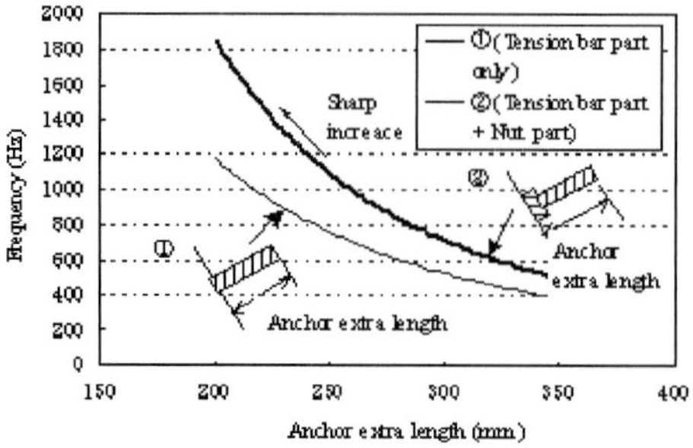

3. In the analytical (FE) model represented in Figure 13 how is the interface/joint between the nut and the plate modelled? If the joint is rigid then the fixity of the joint will be high and it is obvious that no effect of the load or free length will be seen as the joint will effectively reflect any elastic wave.

The joint between the nut and the plate is modelled by dimensional analysis and the degree of fixation of the joint was adequately estimated. So we consider that an elastic wave can propagate.

4. The method appears to be very sensitive to the length of the extra length – in case II, which is presented the method shows no variation in frequency with load. Can the authors comment on this? Would the length of the extra length need to be chosen carefully at installation to ensure the method works?

Certainly there is no variation in frequency with load, however there is variation in length of the extra length with load. The purpose of this test is to get a relation of frequency – load – extra length. In the method, frequency of each anchor is obtained at installation, and change of frequency is detected to check installation load of each anchor. Therefore, extra length is not chosen, frequency to extra length and load of each anchor need to be confirmed.

5. From the correct form of Eq (1) the change in frequency for a length change of 5mm in 250mm, corresponding to 0.1PU is approximately 4% of the frequency of 840Hz. This corresponds to approx 34Hz not 60Hz as quoted. Can the authors comment?

The extra length is shorter, the influence of nut is bigger for frequency. So there is difference between theoretical solution and actual frequency.

6. The figure quoted of 30Hz per 0.1PU would appear to depend on the geometry and the fixity of the nut and is therefore not a general result. Can the authors comment on the general applicability of the results presented?

This result is shown for particular situation of test we examined, and is therefore not general. This result shows that for all anchors, there is not 30Hz reduction per 0.1PU, and there is a relation for reduction of frequency v reduction of load.

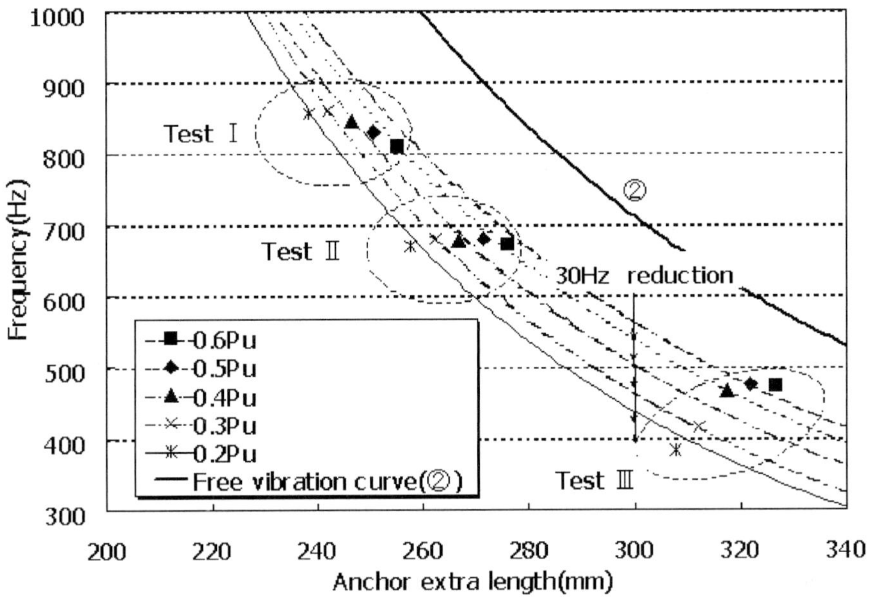

Numbered questions from Ana Ivanovic, *University of Aberdeen*

Answers in bold from Kuniyasu Tanaka, *Shikoku Electric Power Co.*

Regarding the paper by Tanaka et al,

1. The authors do not mention any other non destructive testing currently in use as a reference in the paper. Have the authors undertaken a thorough background research on current non destructive methods available (Boltometer, Ultrasonic technique developed by DMT, GRANIT etc. – see references below)?

Wittenberg, D, Überwachung der Ausbaufunktion eingebauter Gebirgsanker mit Hilfe der Ultraschallmesstechnik, Proceedings of Roofbolting in Mining, RWTH, Aachen, Germany, 6-7 June 2001, p311-322.

Thurner, H.F. Non destructive tests method for rock bolts. Proc.Int.Cong on Rock Mechanics, 3:254-255, 1979.

Ground anchorage testing system; Patent Number: EP1324031; Publication date: 2003-07-02 Applicant: UNIV ABERDEEN (GB); UNIV BRADFORD (GB)

We checked works of Japan only, for example, non detective test using magnetostriction measurement.

2. Some of the findings reported in the paper are comparable to the findings found using axial vibration (see the references below)

Ivanović A, Neilson R.D., Rodger A.A., 2002, Influence of Prestress on the Dynamic Response of Ground Anchorages, *Journal of Geotechnical and Geoenvironmental Engineering, ASCE,* 128(3), 237-249.

Starkey A., Ivanović A., Rodger A.A., Neilson, R.D. 2003 Condition Monitoring of Ground Anchorages by Dynamic Impulses: GRANIT System, *Meccanica* 38:265-282.

Starkey A., Ivanović A., Neilson R.D., Rodger A.A. 2003 Using a Lumped Parameter Dynamic Model of a Rock Bolt to Produce Training Data for a Neural Network for Diagnosis of Real Data *Meccanica* 38:131-142.

We consider that vibration in proposal method involve axial and lateral vibration. Therefore, we will check findings using axial vibration.

3. Can the authors give the reason for use of lateral vibration for the detection of load which presumably gives less propagation of the impulse through the anchor head than the axial vibration used in, for example, GRANIT since 2001?

Starkey A., Ivanović A., Rodger A.A., Neilson, R.D. 2003 Condition Monitoring of Ground Anchorages by Dynamic Impulses: GRANIT System, *Meccanica* 38:265-282.

We consider that vibration in proposal method involves axial and lateral vibration.

4. The authors define a degree of fixity between nut and plate through the constant 'c' (ref. Eq. 2) relating the axial load to the bending stiffness of the interface.

   - Could you clarify the physical interpretation of the equation used
   - How was the constant 'c' obtained from the tests on the actual anchor - are these hammer tests or static bending tests?

$$E \cdot I = \left(\frac{c \cdot A \cdot P}{L'} \right) \cdot I \qquad [c : \text{Constant}]$$

This equation was obtained by dimension analysis using physical elements that influence a degree of fixity. Constant C was obtained by hammering test.

Brian Clifford, *Network Rail*
In regard to the paper by Mori et al, I was surprised to hear that the authors were advocating 3-5MHz frequency for detecting rockbolt grout annulus condition monitoring. In my experience the stress waves generated at these relatively high frequencies are concentrated towards the centre of the rockbolt and have little capability for determining boundary conditions along the length of the rockbolt. This frequency band is more suited to determining the location of defects that intersect the centre of the rockbolt and/or for determining intact rockbolt length.

Lower frequency stress waves, in the range 50kHz to 150kHz, are much broader waveforms and are capable of detecting defects at the bolt/grout and grout/ground interfaces and can be used to determine grout condition. However, the problem with these low frequency waves is that they tend to suffer from a high degree of signal attenuation, represented in the return signal response as "noise".

I was also surprised to hear that the authors had no problems with bolt end coupling condition and its effect on signal transfer between the transducer and rockbolt. Again, in my experience, the condition of the interface between the rockbolt and the probe is critical to the success of launching the signal into the rockbolt and detecting the return wave signature. The contact surface needs to be flat and normal to the bolt axis to give good test results.

During my research into similar techniques I developed a bolt end preparation tool for use in the field that produced a contact surface at the proximal end of the rockbolt suitable for ultrasonic testing (Refs. 1 & 2 below).

I would also caution the authors not to rely on laboratory test work alone to draw significant conclusions about the capability of the systems being developed. Rockbolts installed in a rock mass have completely different attenuation properties to any that can be produced in laboratory test blocks. In service rockbolts that may have been subjected to shear forces and bending and that may be installed in different rock mass types will also give confusing attenuation profiles. Bends in the rockbolt will significantly attenuate ultrasonic stress waves in the proposed frequency ranges, particularly the higher frequency bands (Ref. 1).

Finally, the authors advocate a single shot, or "burst" rather than an "impulse" signal for waveform generation. A system developed by a Swedish company "Geodynamik AB" in the late 1970s and early 1980s called the "Boltometer", also utilised a similar system together with low frequency stress waves. The Boltometer was developed for testing cement grouted rockbolts and was found to give a good indication, through the application of a bolt classification system, of the level/quality of the grout bonding (Ref. 3).

The authors may wish to review references below or are welcome to contact me directly if I can be of assistance to them. I carried out an extensive literature review on NDT systems for testing rockbolts in 2004 and reference to the material is also given in Ref. 1 below.

References:-

1 Clifford B, *The Assessment of Ground Control Risks and Support Integrity in Coal Mine Roadways*, PhD Thesis, Doctor of Philosophy In Geomechanics, University of Exeter, Camborne School of Mines, 2004

2 Clifford B, Bloor A, Bowler J, and Bigby D, *The Use of NDT Methods to Determine the Condition of Rockbolt Support During the Recovery of A Previously Flooded Coal Mine Roadway at Thoresby Colliery*, 21st Conference on Ground Control In Mining, Morgantown USA, August 2002

3 Thurner HF, Geodynamik AB, Stockholm, Sweden, *The Detection of Invisible Faults on Rockbolts In-situ,* Proceedings of International Symposium on Rockbolting, Abisko, 28th August - 2nd September 1983, pp477-480

Takayuki Mori, *Kajima Corporation*
We appreciate your opinion and advice.

Although you claim that 50 to 100 kHz is suitable for grout filling evaluation, it is not exact. Based on sonic wave theory the longitudinal wave of the wavelength of 3.41 times or more of a medium radius will not exist, and only a surface wave will be transmitted. That is, if it is about 24mm in diameter of a rock bolt, at a frequency of less than 100kHz it will not exist as a longitudinal wave but will become a propagation of only a surface wave. Have you noticed this?

Since it is a new proposal, we understand that you are surprised.

It is a basic matter that the surface of the bolt head is flat and contact is right-angled. However, using this technique the contact condition of a probe has little influence compared with other methods. In this technique, echo jelly is applied to the contact surface and the reflective level at the tip of the bolt being the maximum is made into typical measured value. Also, we use the waveform of high energy and high purity frequency generated by burst wave for measurement.

It seems that you have expressed an opinion about vibration of free length of the un-bonded type of rock bolt. It is the bonded type rock bolt which is being considered as our object and it is not vibration but reflection. Therefore, there is very little influence of attenuation as opposed to a reflective wave also in the state where a bolt is subjected to shear and bending forces.

Although you say that you can perform grout-filling evaluation well using "Boltometer", it is a surprise for us. Did you actually use it?

Boltometer attracted the attention of the JTA in the past. However, having tried it good results were not obtained and it was considered useful only as a flaw detector. The purpose of the technique of Boltometer and ours is the same. However, the big difference is that the oscillation is a burst wave and a highly precise log amplifier is used.

Many theoretical verifications and laboratory experiments have been conducted. The proposed technique has performed grout-filling evaluation that was difficult to evaluate using Boltometer.

Gareth John, *CAPCIS Ltd*
With reference to the paper by Mori et al, we have been using compressive wave UT to examine large diameter (75mm Ø) tie bars for defects.

One of the problems encountered was that the ends of the bars are not flat, that is, when you place the probe it is not "true" (normal) to the axis of the bar, and as you move the probe across the end you get different results.

How do you compensate for variable bar ends in your system?

Takayuki Mori, *Kajima Corporation*
Thank you for your opinion. As you pointed out, the reflective waveform is influenced by the contact condition - it is a basic matter that the surface of the bolt head is flat and contact is at right angles.
However, this technique has little influence by the contact condition of a probe compared with other methods. In this technique echo jelly is applied to the contact surface and the reflective level at the tip of the bolt being the maximum is made into a typical measured value.
The second reason is that we use a high energy and high purity frequency waveform, generated by burst wave, for measurement.

Richard Neilson, *University of Aberdeen*
Regarding the paper by Mori et al, can the authors explain how their technique is different from that proposed by Wittenberg? (see reference)

Wittenberg, D, Überwachung der Ausbaufunktion eingebauter Gebirgsanker mit Hilfe der Ultraschallmesstechnik, Proceedings of Roofbolting in Mining, RWTH, Aachen, Germany, 6-7 June 2001, p311-322.

Takayuki Mori, *Kajima Corporation*
Thank you for the question.
Sorry, but we are not familiar with the technique proposed by Wittenberg.

The technique proposed by the authors is based on the principle that the reflection is small when there is no aperture and that the surface diffused reflection amplitude becomes large when there is an aperture.

The features of the proposed technique are as follows: -

- The burst wave is used for the oscillation
- Generating a frequency of high purity is possible at high energy
- There is little noise and long-distance investigation is possible.

Moreover, in order to evaluate the grout filling condition, the signal of "the domain of a small reflective echo" can be faithfully acquired with sufficient accuracy when a LOG amplifier is adopted.

Corrosion monitoring and corrosion protection systems:
discussion session 6 – Part 1

Rapporteur's summary
by Roger Margerison, *Morgan EST plc*

This session comprised a total of ten papers and presented in two parts. The first five papers highlighted the international nature of the conference and included three UK based presentations, one from Japan and one from Germany.

The paper by John and Littlejohn starts with a statement suggesting that the corrosion of steel components in the ground is often given insufficient detailed attention during the initial design and subsequent service period. The authors then set out to demonstrate the factors that need to be considered.

The paper looks at the methods of assessing the aggressivity of the ground from the 1940s through to the current DIN 50 929 Part 3, highlighting the various conditions of the ground that need to be considered during investigation works where ground anchorages are planned. The most important comment by the authors is that aggressivity needs to be considered on a case by case basis which is rare in reality with very few case histories where ground aggressivity and corrosion are quantified. The failure of protection systems is well illustrated by some rather disconcerting photographs of water seepages and corroded anchor components.

The authors then look at the potential for corrosion in the grouted length and that good quality grout provides a very good protective environment. However, the authors point out that the integrity of the grout, the extent of cracking and exposure conditions developed for use in reinforced concrete design could be applied to anchors with reinforcing steel tendons but there is to date little evidence of this in ground anchorage practice.

The final section of the paper presents considerations for the methods of protecting the anchor components and their intended design life. Various methods have been discussed for protecting the anchor elements and head. Particular note has been made about the vulnerability of the void space behind the head and again good photographic examples of corrosion effects due to inadequate protection are shown. Suggested methods are proposed and include particular emphasis on the ability to revisit the anchor head for inspection and future testing.

In conclusion, this paper summarises the importance of the assessment of the aggressivity of the ground in order to determine the likely corrosion rate and therefore the necessary protection measures required and makes a particular reference older installations where information is not available.

The second paper by John and Ernst complements the previous paper by presenting a method of assessing in situ the degree of corrosion of rock anchorages. The paper describes the use linear polarisation resistance on three anchors in an un-named post-tensioned dam.

The paper describes the various techniques adopted, the probe device developed specifically for this purpose and comparative computer modelling used to correlate the site data.

The anchors examined were over 40 years old and required the drilling of boreholes adjacent to the anchors, into which the probe was lowered and the various resistance data gathered. Assuming some familiarity with the formulae used, the polarisation resistance was determined over the length of the anchor and an estimation of the polarised area. Comparative tests were then carried out using computer modelling.

The results and computer models have been plotted against the depth of the anchor and show different corrosion rates within the concrete of the dam to those in the underlying dolerite. An anomaly was attributed to a point in the borehole where water leakage was found.

The resistance testing demonstrated that it is possible to make an assessment of the corrosion condition of the anchor, however the testing and interpretation techniques rely on a level of understanding of the principles of linear polarisation resistance. The method also requires the drilling of adjacent boreholes which could restrict its use in more difficult situations. It does however give hope for the assessment of the long term condition of existing anchors.

The third UK sourced paper by Irvin and Mothersille presents a very concise review of the various methods of providing corrosion protection of ground anchorages. The authors have set out to try to dispel many of the perceived mysteries of these systems and the most common misconceptions.

Initially, the different types of installation are summarised in order to highlight the differences between, for example, ground anchorages and tension piles, and the required level of protection.

The various corrosion protection options are described in sufficient detail so that the performance and limitations of these options can be clearly understood by non-practising engineers. The key aspect covered is the understanding of the purpose and what constitutes double protection as required in BS8081:1989 and EN1537:2000 and that systems need to be integral with the anchorage manufacturing process. The applications of double protection systems in various anchor types are then considered, including an appraisal of the performance of corrugated plastic ducting.

An important section then follows that emphasises the unacceptability of the borehole grout as a means of corrosion protection and its problematic nature with particular reference to self drilled, hollow bar anchors.

Other methods of protection are then considered such as epoxy coatings on rock bolts, duplex coatings for hollow bar anchors and epoxy coated strands. Again the limitations of these systems, particularly the handling damage that can occur are presented. Sacrificial allowances, galvanizing, electrical isolation and cathodic protection are given similar study.

The authors then importantly explain, slightly out of context with the flow of the paper, the key differences between ground anchorages and tension piles.

The final section highlights the criticality of ensuring the void behind the head plate is adequately protected and some of the difficulties that can be encountered whilst trying to achieve this.

This paper provides a very good review of the corrosion protection systems available, illustrated with good photographic examples. It concludes by referring the reader to the appropriate standards, which if followed will ensure a permanent protection for the anchorage.

The fourth paper was prepared by Suga et al, representatives of the Japan Anchor Association. The paper concentrates on the void behind the head plate, an area of an anchor identified previously as being vulnerable to corrosion.

A visual concept of the performance of an anchor over its design life is introduced and used throughout the paper to good effect along with a flow chart to help identify any necessary repair work. Types of repair work are identified, these being durability enhancement, repair enhancement, renewal, emergency countermeasure and life extension countermeasure.

The paper then moves onto specifically addressing issues associated with corrosion below the bearing plate. The causes for the corrosion of this element are described which comprise insufficient initial corrosion protection, deficient construction or deterioration of a corrosion inhibiting compound. An example of an 'old type' anchor is provided.

A system of repair to the anchor head is described where the underhead section of the anchor is cleaned and injected with new corrosion inhibiting compound. The key observation would be that the existing head plate has to be removed. As has been illustrated in the first paper of this section and elsewhere in the conference, this is not always possible. This paper however does emphasise the concern surrounding this element of an anchorage system and offers guidance on an approach that can be adopted.

The fifth paper by Jungwirth and Adler looks to address the performance of steel, normally used as concrete reinforcement, in ground anchorage applications. Specific issues relating to the potential aggressivity and corrosivity of installations in the ground are considered.

Mechanisms of corrosion and the effectiveness of the grout cover to the anchor bar are covered and the rate of corrosion and ultimate performance of the anchor over the design life are presented in a clear series of graphs.

The authors seem to be proposing that an approach to corrosion protection can be adopted that is similar to that used in the design of reinforced concrete. Essentially, protection is provided by the concrete (cementitious grout in an anchor) and that crack widths are limited to 0.2mm, although the thickness of the grout layer would be greater. Only in aggressive soils or for a design life in excess of 50 years would double protection in accordance with EN1537 be necessary. Crack widths would have to be limited by the applied load and a relationship for this is presented.

Presumably, adopting this approach would have to be carried out by carefully considering the risks to the structure and additional material costs. Simpler installation and less sensitivity to materials handling however may give benefits in specific locations. It is unlikely however that this approach would comply with EN1537.

In conclusion, the five papers presented in the first part of session 6 give a good overview of the ground aggressivity and mechanisms for corrosion. Various methods of providing corrosion were presented. The papers by John/Littlejohn and Irvin/Mothersille provide good introductions to less experienced designers and engineers. The paper from the Japan Anchor Association has picked up on a specific area of vulnerability in anchor systems also identified in other papers. The John/Ernst paper is a means of identifying rates of corrosion but is likely to be site specific. However, the technique could be used in other situations and is probably more applicable to specialists in this field. The final paper is probably the most controversial presentation of this session in that it departs from the standards BS8081 and EN1537 for all but the longer design life and more aggressive ground.

Lee Jordan, *Arup*
Question to Gareth John. Given the sight of water spraying out of the top of anchor strand sheaths, how do you think this could occur given that the annulus between the strand and the sheath should be completely filled with grease?

Gareth John, *CAPCIS Ltd*
I agree that it should have been completely full of grease but it was a water soluble grease, so when the water got in, it just washed the grease away which emphasises the point that if you want grease, go and watch the movie, or if you want a protective system, actually specify exactly what you want in terms of inhibitive properties, water rejection and microbiological stability. All these issues are important. As a corrosion materials engineer, I always get very nervous when I hear other people say galvanising, grease, stainless steel because which one do you want because there are so many different types.
I think in that particular case the problem was the use of a water soluble grease. We have no idea what caused the ingress of water; it was not a one-off because there was a large number of tendons on the site that were suffering from the same problem. Something went wrong somewhere, whether it was somebody cutting or not sealing the ends of the sheaths…contractor cutting corners, that would never happen would it, or manufacturers cutting corners, I have no idea but it is what we found.

Lee Jordan, *Arup*
That ties in with what we found during the manufacturing that it is quite difficult to fill the sheaths.

Stuart Littlejohn, *University of Bradford*
Written contribution. The anchorages at a site in Barcelona were temporary and founded in sand under artesian ground water conditions. The multi-strand tendon comprised greased and sheathed strands over the free length and was installed inside a

cased hole. As the casing was withdrawn, the tendon was pressure grouted using neat cement grout. No attempt was made to seal the ends of the sheathed strands at the proximal end of the tendon bond length, thereby giving access for water ingress. As the anchorages were temporary, the ground water was tested and judged to be mildly aggressive, the anchor heads were then sealed to inhibit further water seepage, and thereafter judged fit for purpose, bearing in mind the short service life of less than two years.

Lee Jordan, *Arup*
A comment to Chris Irvin concerning the use of two layers of protection, in the hope that one is going still be in operation. What we have found is that there could be mechanisms that basically unzip both layers of protection, especially in the free length and that could be due to fractures in the rock when the anchors have been homed. It could be high permanent displacements during the stressing, or it could be creep movements over the long term. Is it not better to have one layer of protection that you can check and be confident it still exists rather than have two and sort of keep your fingers crossed that you are going to have one layer over the long term. This leads on to the use of electrical isolation which is quite fundamental and used in Switzerland and they seem to develop the system quite well over there.

Chris Irvin, *Dywidag-Systems International*
These are very interesting points. The first one is if you manufacture and supply an anchorage, a strand anchorage with two layers of plastic sheaths and this is individual strand sheathing within a common external sheath, if that is installed correctly and sealed at the time of installation, experience has shown on hundreds of projects that it is comprehensive protection. It is extremely important to ensure the integrity of the plastic sheathing and in recent times we have introduced thicker wall plastic sheathing, these are extremely robust sheaths and I have brought samples with me, anticipating questions like this so you are welcome to have a look at those afterwards.

In respect of electrical isolation testing, it sounds great in theory. My colleagues in Switzerland know the system very well, but there are a number of practical problems, first of all once the anchorage has been manufactured it needs to have the electrical isolation tests done on it to prove its integrity. This is typically done in a water bath in the factory. It then has to have the same process applied at site when the contractor signs for the anchorages, once they are off-loaded from the truck, so that is another water bath on site, because the last thing the contractor wants is to get the anchorages into the ground and then find that he has got compromised integrity to his sheathing, so for him to be able to take ownership, he has to test the anchorages at the time. Thirdly, the measurements of the resistivity do vary quite widely and if you are in a rainy day or if it is moist, you have to adjust it accordingly. I would invite comments from others on this electrical test. I know you went over to Switzerland to see it.

Lee Jordan, *Arup*
I went to see Ulrik Von Matts the Swiss anchor guru and obviously he is pretty confident in the system. As you know it is part of the Swiss code and it seemed like a reasonable system. It would have solved some of the problems which we have had in terms of things moving during stressing and not being able to know whether the corrosion protection has been damaged in any way.

Chris Irvin, *Dywidag-Systems International*
I think the time when corrosion protection is at highest risk is during installation and there has to be rigorous enforcement and inspection by resident engineers on site. The problem comes when you get a multitude of different trades, you get reinforced concrete people or structural steel people going in there who have little understanding of the importance of the corrosion protection sheathing. It is an extremely well proven system and it is robust when correctly installed.

Lee Jordan, *Arup*
I agree but once the anchor enters the hole there is not really a lot that anyone can do about it, nobody knows what the condition of the rock is that might be effecting the sheathing. You can do anything you want with the anchor head but once it enters the hole and basically these grouts have been stressed, there is very little you can actually do to check what condition it is in and that seems to be the big risk which with electrical isolation you would potentially overcome. I am surprised that it has not really been picked up by the rest of the industry.

Chris Irvin, *Dywidag-Systems International*

That is a fair comment, where there is risk of this. What you are talking about is a side shear of the rock deep down. Why not over drill the borehole and use the American system, re-seal it and re-drill it with grout three or four times so you widen the annulus and once grout is around it that is a fairly robust mechanism to protect it but it is good to have the theory of an electrical isolation system but once it is installed and it does fail what are you going to do about it. I would much rather have two barriers in the hole than one.

Stuart Littlejohn, *University of Bradford*
For the information of Lee Jordan, when we were compiling the draft for development (BS DD81 published in 1982), the United Kingdom Atomic Energy Authority wrote to me and stated that if I did not include a double protection option they would not follow the standard and would write their own manual. If you are in the nuclear industry, double corrosion protection, i.e. two individual barriers are essential. I would challenge anyone on that when it comes to really important structures where the consequences of failure could be absolutely tragic. You are right that the outer barrier can be regarded potentially as sacrificial to protect the inner one. If you do not have that how can you ensure that the first protective barrier has full integrity once installed?

Question for Susanne Adler. Your paper with Dieter Jungwirth gives us a lot of food for thought because you have attempted to rationalise the approach to corrosion and corrosion protection across several geotechnical application, i.e. soil nails, rock bolts, micro piles and ground anchorages. When you look at one of the tables that was presented today for a slightly to moderately aggressive ground water or soil from, there was an option that could also be applied to tension, so therefore ground anchorages, of 35 to 40mm of cement grout cover. That would add 70 to 80mm of diameter to the borehole with extra cost. Would it not just be simpler to have a single sheath with a 5mm cover as required by EN1537?

Dieter Jungwirth, *Technische Universitat Munchen* and **Susanne Adler,** *Stahlwerk Annahutte*

We agree with the three previous presentations of this session. Based on the two extreme systems we have tried to fill the gap with other geotechnical stabilisation applications using the same level of safety and durability, in order to achieve an integrated philosophy.

On one side is the permanent anchor with sensitive pre-stressing steel with a service life up to 100 years and double corrosion protection according to EN1537. On the other side is the unprotected temporary nail or micropile with robust reinforcement steel and corrosion rates according to EN14490 or EN14199. In between, we have tried to fill the gap with different geotechnical systems such as anchors, micropiles, soil nails and rock bolts.

We have considered the following parameters:

- robust reinforcing steel
- galvanised or non-galvanised steel
- corrosion rates (sacrificial corrosion)
- cement grout cover
- permissible crack width of grout
- different aggressive soils
- service lives
- system reserves
- well known elements of construction engineering.
-

All this has been carried out with proper quality of execution and quality control. These very different parameters, e.g. load and service life, result in different costs. Therefore, it is not possible to state in general, what is the most cost-effective solution for an anchor (according to EN1537) with a high load capacity, small borehole and service life of up to 100 years. In our opinion, specialist contractors should be consulted to advise on the different solutions and prices.

Written questions and answers session six part 1: corrosion monitoring and corrosion protection systems

Stuart Littlejohn, *University of Bradford*
Re. paper by Jungwirth and Adler
Question 1
I believe this is a valuable paper that gives us food for thought, but I am concerned that a grout cover of 35 to 40mm can be considered adequate for slightly to moderately aggressive soil conditions, on the basis that the grout cover is ensured by centralisers. Have the authors considered that the cement grout may be compromised by bleeding during the stiffening phase?

Question 2
In addition, a cement grout cover of 40mm increases the diameter of the borehole for the anchorage by 80mm and the drilling cost increases. Would it not be more cost-effective for all permanent anchorages in aggressive ground (no matter how mild) to incorporate a corrugated plastic sheath with an internal grout cover of 5mm, in accordance with EN1537?

Question 3
I would be grateful if the authors would clarify the meaning of "approximately 95% fractile" when describing the crack width at the steel surface.

Susanne Adler, Stahlwerk Annahutte and Dieter Jungwirth,
Technische Universitat Munchen
Question 1:
Considering the illustrated examples of poor execution, especially with pre-stressing steel tendon, the quality of execution works and durability should be our main topic. In our opinion more robust systems should be applied. Without qualified staff the quality of grout injection still implies residual risks. On the other hand, grouting technology has been improved considerably.

In Europe we do have Standards for grouting (EN 445 to EN 447), work instructions, qualified staff as well as the responsibility of system manufacturers according to the European Construction Product Directive to install geotechnical systems in a professional manner, even by a third party (subcontractor).

Insufficient grout conditions of particularly perpendicular load bearing elements can be avoided by post-grouting or removal of the upper grout. For horizontal anchors this problem does not exist, but grout cover should be maintained by centralisers. Shop fabricated double corrosion protected anchors provide higher safety in the execution works. But for a low quantity, shorter service life and robust steel load bearing elements, low-cost cement grouted systems are sufficient on site, if done by qualified staff. For more than 30 years DSI/SAH has been practising this successfully.

Question 2:
Overall costs and profitability are dependent on various parameters:

- choice of suitable system depending on load, soil and service life

- material costs, costs of prefabrication, drilling, installation (not only contractor), quality control
- maintenance
- national conventions or available equipment

Specialised companies do have comparative costing.

Prefabricated double corrosion protected permanent anchors in aggressive soil with high service load and long service life are a more low-priced solution in this case.

For self-drilling systems with small diameter reinforcing steel quality and limited service life, allowance for corrosion rates is more cost-efficient and safer. Furthermore, prefabricated corrosion protection would be damaged during drilling. But in between there are further systems possible and they have been installed successfully already. However, quality and cost-efficiency have to be taken into account.

Question 3:
The average crack distances can be represented by a straight line considering thickness of grout cover ü and rib distance of the thread bar c ($c = 0{,}38 \cdot d_s =$ feature of bond characteristics, see figure 5 of our paper).

It is known from literature and existing crack laws that 5 % of determined crack widths are 1.5 times larger than mean values. This factor was taken into account for determination of crack width $w_{95\%,Steel}$ and named "approximately 95 % fractile". In addition, this formula considers the contribution of the concrete in tension to 0.95 and the fact that crack width at the grout surface decreases to 50 % at the steel surface. Actually, this crack formula even covers the largest crack widths observed in the tests by Wichter.

John Wasilewski, *Capita Symonds*
Many papers presented dealt with corrosion and in particular, the loss of section, increase in stress, loss of grip and/or bond, etc. that corrosion causes.

Most papers dealing with corrosion made little or not reference to stress corrosion cracking ('SCC').

Consistent with the above, the perception of civil/structural engineers generally seems to be that the harm caused to structures by corrosion of steel members and embedded components is limited to,
- loss of section and increase in stress due to loss of section,
- collateral damage due only to corrosive swelling (spalling of stone cladding/protective concrete cover etc.).

Is this a misconception?

Why aren't we more concerned about SCC? Dr John did mention SCC, and I would welcome hearing a little more from him on SCC.

- When is corrosion unlikely to be accompanied by SCC?
- When might SCC happen and how can we detect it?

- Are engineers sufficiently aware of and concerned about SCC in the construction industry?
- How quickly can it occur in the right conditions?

Gareth John, *CAPCIS Ltd*
The quick and simple answer to the question is yes!

By way of background, stress corrosion cracking (SCC) is just one of several mechanisms which form part of the general phenomena of environmental assisted cracking all of which, if not identified, can lead to catastrophic failure of structures and facilities. Stress corrosion cracking can affect most metals & alloys although only in specific environments.

A particular problem with stress corrosion cracking is that it produces fine cracking which can be difficult to find visually, without necessarily producing the more widely appreciated, general corrosion.

While general corrosion of steel structures is undoubtedly of great concern, it is certainly the case that SCC is also a serious potential problem for steel structures and other metal components used in civil engineering. In fact one of the reasons that it is such a potential problem is that it does not necessarily produce the general corrosion (and hence heavy rusting) which engineers are used to looking for during inspection and maintenance.

It is difficult to answer why people are not more concerned about SCC, but there appears in very general terms to be a lack of awareness in the engineering industry about SCC. Perhaps this is related to the fact that it is a much more involved subject than general corrosion and there is no single answer to every possible SCC situation. SCC does not have to be accompanied by general corrosion at all and any corrosion which does occur simultaneously with SCC is likely to be limited. At this point it is worth pointing out that SCC is not confined to steel. It can affect almost all the alloys likely to be used in civil engineering, but only in specific environments which can be different for each alloy, however, in all cases the following combination of conditions must exist together for stress corrosion cracking to occur:

- Susceptible material (metal / alloy)
- Conducive environment
- Stress
- Stress raiser

In many cases, these factors mean that the areas likely to suffer from SCC are localised and not easily accessible.

SCC can occur in a very wide range of situations and there is not sufficient room here to give an exhaustive reply, but to give a few examples.

- failure of dry dock floor, Scaramanga Greece ground anchor in 1977, (stress corrosion cracking of high strength carbon steel ground anchors),
- the collapse of the roof of the Berlin Congress Hall in 1980 (due to hydrogen embrittlement of high strength carbon steel pre-stressing tendons),

- collapse of the Ynys-y-Gwas Bridge in. Wales, UK in 1985 (hydrogen embrittlement of high strength carbon steel pre-stressing tendons),
- failure of stainless steel supports in swimming pool roof , Switzerland, 1985 (http://www.hse.gov.uk/lau/lacs/47-17.htm
- cracking of ferritic stainless steel bridge roller bearings in 2002 (chloride stress corrosion cracking), http://news.bbc.co.uk/1/hi/england/manchester/4242437.stm

As well as major failures, as listed above, SCC can result in failures of small components, a good example being failure of 200 grade stainless steel bandings (used to support road signs and street furniture) or cracking of brasses and copper alloy components in ammonia rich environments. Whilst failures of these items may not be as dramatic as major structure failures, they can still result in increased maintenance/replacement costs and can lead to increase risk to the public.

Detecting SCC often requires the use of non-destructive testing (for example ultrasonic testing, magnetic particle or dye-penetrant inspection) depending on the particular material, geometry, access availability etc. However, once cracks become sufficiently advanced to start to open up they may be seen on visual inspection. As the locations which might suffer SCC can be very localised, it is necessary to know in advance where to look and with what techniques. This requires a specialist to assess the materials involved and to decide on the potential danger areas beforehand.

In some cases SCC can propagate very quickly, however, in other situations propagation can be slow and can take years to manifest itself. There is no single rule which can be applied to cover all situations where SCC might occur.

Horst K. Aschenbroich, *Con-Tech Systems Ltd.*
Under Duplex Coatings - Self Drilled Hollow Bars.

How does the bar drill itself? Also the author must have limited if any experience with corrosion protection of Grout Injection Bored Hollow Bars. In Mr. Ischebeck's short 3 minute presentation, unfortunately without pictures, he mentioned a project in Switzerland, an anchored structure with soil nails,(Dangelstutz), where duplex coated hollow bars are in service and monitored using the electric (potential) resistance measurement for over 5 years. No evidence of corrosion is registered.
The epoxy coating on hollow bars used for slope stabilizations by the California Department of Transportation is a special non brittle coating. A test was conducted on 40 feet long injection bored bars with this special coating. The bars were retracted after drilling and the coating examined, no abrasive damage was noted. Hollow bars are also manufactured and available in INOX, stainless steel for very severe ground conditions.

Chris Irvin, *DYWIDAG-Systems International*
Regarding Duplex coatings for self-drilled hollow bars, the question covers four issues which are
summarised in *italics*, together with the author's responses, as follows:

a) Installation of the hollow bar

Self-drilled hollow bars are installed using rotary percussive drilling in conjunction with simultaneous grouting. The action of rotary percussion ensures efficient drilling in loose or collapsing soils, with the simultaneous injection of grout ensuring that grout is placed at all points as the drill string is advanced.

The authors have 28 years collective experience in the drilling and grouting of bars worldwide, together with further specific experience on hollow bars and an in depth knowledge of the different hollow bar systems offered by other manufacturers. Prior to this, drilling experience has been gained in reverse circulation drilling and down-the-hole (DTH) hammer drilling.

b) Electrical Potential Resistance Test carried out on Duplex coated hollow bars in Switzerland (Dangelstutz), as referred to by Ernst Ischebeck.

It is difficult to comment on this point in the absence of the data from the electrical resistance tests and without knowledge of the aggressivity of the soil. However, some comment on the limitations of the electrical resistance test may be useful:

i) The electrical resistance measurement test (ERM), referred to in EN 1537:1999 requires an electrical resistance of $\geq 0.1 M\Omega$ to demonstrate that the tendon is sufficiently isolated and therefore protected

ii) In practice the ERM test is somewhat meaningless, as the presence of moisture in the air requires the settings on the measuring equipment to be adjusted to compensate for the conductivity within the air, so much so that it is possible to adjust the equipment to produce the desired reading, irrespective of whether the tendon has full isolation or not. In other words, the data are potentially unreliable. The test requires further development before it can credibly be applied to anchor tendons in respect of assessing the level of corrosion protection.

c) Epoxy coating of hollow bars used for slope stabilisation, California Department of Transportation.

There are a number of so-called Duplex/Combination coatings available from hollow bar manufacturers. They all feature a galvanized hollow bar, overlaid with a secondary coating.

i) The galvanized coating should comply with EN 1461:1999 to provide a minimum zinc cover of 85 μm (microns). Electro Plated Galvanizing only offers 5-12 μm and does not comply with EN 1461.

ii) The secondary coating varies depending on manufacturer. The following coatings have been used: Epoxy, Polyester, Rilsan, Accolan. Out of the four, Epoxy was the best, but was still insufficient to withstand the rigours of hollow bar drilling.

There are two principal problems with Duplex coated self-drilled hollow bars:

i) Abrasion damage during rotary installation. Most soils where self-drilled hollow bars are installed have some abrasive materials within them. It is virtually impossible to protect the full length of the bar, or more critically the coupler (which is of a larger diameter), against abrasion damage. The abrasive elements of most soils in conjunction with the slightly spiral nature of the drilling profile inevitably lead to a loss of Duplex on the bar/couplers.

ii) Gripping of the bar for rod release. Once the first hollow bar rod has been drilled in, it is necessary to grip the bar (usually within a set of hydraulic clamps at the base of the drill boom) to enable the drill head to unscrew off the bar, in order that a new rod section may be added on to the drill string. For the hydraulic clamps/jaws to grip the bar; they bite straight through the protective coating, compromising the corrosion protection.

Galvanized coatings by themselves, without a secondary coating, have proved far more practical for corrosion protection of low risk applications (i.e. soil nailing). Where the galvanizing coating complies with EN 1461, it offers supplementary protection to the soil nail. This coating is sufficiently robust to withstand a degree of abrasion. Where this does become compromised, galvanizing offers:

1) A chemically bonded transitional zone comprising a steel/zinc alloy. The make-up of a galvanized bar consist of the base steel, a transition zone of steel/zinc alloy, and a top coating of zinc. Where damage to the outer zinc layer occurs, the transition zone remains intact and therefore maintains a degree of protection.

2) Self-healing to nicks or small areas of damage. At the points where the galvanized coating becomes breached, the zinc oxide expands to cover up areas of damage. This is known as self-healing.

Note. Galvanized coatings are sacrificial coatings with a limited lifespan. They are useful as supplementary protection for soil nails (in conjunction with sacrificial corrosion allowance, for example) but are not applicable to permanent ground anchorages.

d) INOX Stainless Steel for Aggressive Soils

The system referred to here is a stainless steel product. The word INOX roughly translated means "no oxidation", INOX is not a special grade of stainless steel. The use of stainless steel hollow bars in aggressive soils does have some merit, but it is very important that the correct grade of stainless steel is specified, as shown below.

i) Grade 304. This is a low grade stainless steel which is susceptible to corrosion.
ii) Grade 316. This is a higher grade of stainless steel which has much better corrosion resistance.
iii) Duplex Grade (1.4562). This is a superior grade stainless steel which offers increased strength and corrosion resistance over grade 316 in respect of corrosion resistance.

Note. The term Duplex in the context of stainless steel has nothing to do with Duplex Coatings.

It should be pointed out that the costs of Grade 316 or Duplex Stainless Steel hollow bars are hugely expensive, to such an extent, that it is often cheaper to drill a larger hole and install a solid threadbar with double corrosion protection (i.e. within a factory pre-grouted corrugated duct).

Tony Barley, *Single Bore Multiple Anchor Ltd*
Re. paper by Irvin and Mothersille
It is generally accepted that it is not possible to effectively centralise self drilling bars for inclined anchors or for soil nails and thus the bar couplers lie in contact with the borehole. In the case of double protected anchor bars; what are the two recommended protective barriers at the coupler and how should they be applied during installation? What kind of non-corrodible spacers can be strong enough and large enough in diameter to ensure prescribed grout cover to inclined bar couplers and also to prevent barrier damage delivery installation? With the introduction of ultra large diameter heavy bars these appear to be the most critical aspects associated with permanent bar anchors.

Although the paper attempts to clarify which proven barriers may be considered for corrosion protection some doubt or confusion exists universally as to whether cement grout in any form may be acceptable as an impermeable barrier. The concept that cement grout can effectively be considered as a barrier when used in conjunction with a single thread bar tendon has not been proven beyond doubt. The extent of photographic evidence published to establish that tension cracks in the grout within current encapsulation systems are always uniformly distributed at the same pitch as the bar deformation is extremely limited.
The relevant requirement of BS8081 identifies in Clause 11.2.1 "The suitability of all materials, component and methods of construction should be demonstrated to the designer (see Clause 4) before acceptance of any anchorage scheme". Clause 4 puts the responsibility for the corrosion protection system on the Contractor.

Where double plastic barriers are provided their integrity can be easily inspected both before and after encapsulation grouting in conjunction with gun barrel test exhumations. This then puts an emphasis on the demonstration of the integrity and performance of a stressed grout annulus around the deformed bar to ensure "the suitability of all materials". Can the authors provide any documentation to confirm that such proving tests have been carried out to comply with BS 8081?

Prior to the convening of a committee for the drafting EN 1537:2000 (by CEN288 Working Group 9) two tests were carried out, on encapsulated threadbars, the type generally available to the industry. The bar diameters were 25 and 32 mm and cement grouted into the encapsulation ducts of 50 and 65mm diameter respectively. The intent of this test was to allow the provision of a Report (Barley 1993) to the 16 European Anchor Specialists of Working Group 9. The committee members would then take this into consideration when recommending or rejecting the encapsulation grout as an impermeable barrier. The tests carried out did not substantiate that the crack distribution was uniform (Fig 1, Fig 2 and Fig 3).

At the same time a similar test was also carried out using an "elastic polyester resin" (suppliers description) which supposedly deformed with the encapsulated prestressing strands (see BS8081 Clause 8.2.4.3.1 and Fig 19). Subject to the resin remaining uncracked it was possible to consider the resin within the encapsulation as an impermeable barrier. However the resin cracking was from a barrier viewpoint worse than the cement grout in that the crack spacing was greater than 100mm and some visible cracks were in the order of 1mm wide (Fig 4).

The conclusion drawn from the test was that neither cement grout located between the threadbar and the corrugated encapsulation duct nor an "elastic" resin between strands and corrugated encapsulation duct showed evidence of uniformity in distributed of the tension cracks or assurance that crack widths were less than 0.1mm. These limited tests did not support the concept that particular resin or the cement grout within the encapsulation could be considered as an impermeable barrier.

Since this is a major issue in the performance of barriers which may be considered as impermeable to satisfy the British and European Anchor Coding the writer looks forward to the authors' inclusion of update references which are supportive of the crack control concept.

Reference: Barley A.D. 1993 – "The investigation of crack distribution and crack width formed in unrestrained encapsulation ducting during tendon stressing". Submitted to the Working Group 9 of CEN Committee TC 288 at meeting in January 1994. The paper is available for inspection on the website www.theanchorman.com

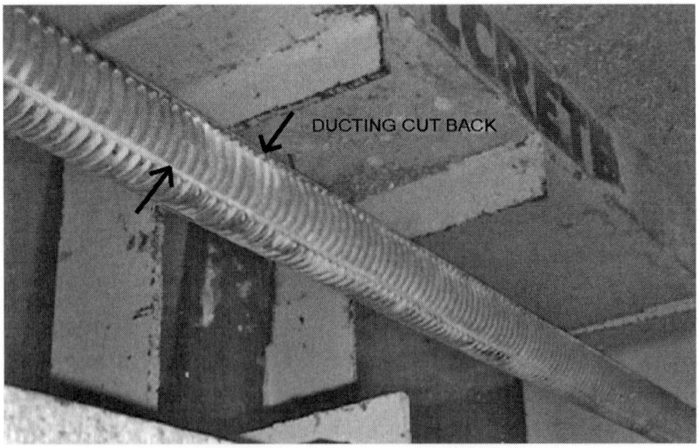

Figure 1 The "window view" provided by the removal of the encapsulation ducting to observe the crack distribution in a stressed unconfined encapsulation. Note that such an unconfined condition can occur in the upper fixed length of a partly grouted borehole – the anchor tendon must cope with the worst possible conditions.

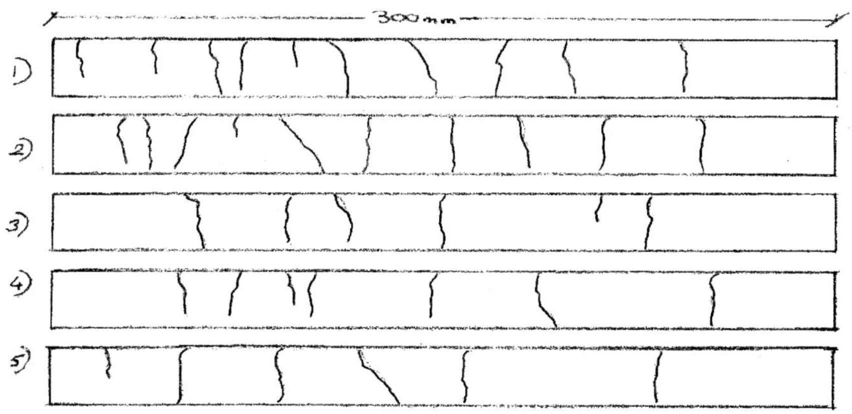

Figure 2 Typical Crack distribution throughout five 300mm inspection lengths of stressed deformed bar and grout. Tendon at 57% Characteristic Strength equivalent to 110% working load.

300mm Observation Length	Cracks including discontinuous Hairline Type		Cracks over Full Section	
	No.	Average Width (mm)	No.	Average Width (mm)
1-2	11	0.124	9	0.152
2-3	9	0.152	9	0.152
3-4	11	0.124	8	0.17
4-5	9	0.152	7	0.19
5-6	12	0.110	6	0.23

Table 1: Crack Distribution in "Crack-Controlled" System A at 57% Characteristic Strength of Tendon

Figure 3 Estimation of crack widths based on elastic extension of a 300mm long deformed bar divided by the number of cracks visible by magnifying glass observation.

Figure 4 Typical Wide Crack visible in the polyester resin alongside a stressed strand tendon.

Chris Irvin, *DYWIDAG-Systems International*

Regarding corrosion protection at couplers and cement grout within a corrugated duct as aneffective corrosion protection barrier, the question covers three issues which are summarised in *italics*, together with the author's responses, as follows:

a) Non-corrodible spacers for support of large diameter pre-grouted bar tendons within the borehole.

With the advent of larger diameter bars (63.5mm and 75mm GEWI and GEWI-*Plus* threadbars) within factory pre-grouted corrugated ducts i.e. double corrosion protection, there is a need to support the bar off the wall of the borehole in horizontally declined installations. This is achieved by the following methods:

i) Bond length in rock - 10mm wrap-around plastic skid spacer
ii) Bond length in soil - lantern spacer (plastic or steel*) to suit borehole diameter, or wrap around plastic centraliser.

* The steel lantern spacer should be used with double corrosion protected bars, such that the spacer shall have no deleterious effect on the tendon (see BS 8081, section 7.3.7).

b) *Corrosion protection of couplers on large diameter pre-grouted bar tendons.*

Pregrouted bar tendons are typically supplied in sections to the jobsite, where they are then joined together by the use of couplers. It is obviously critical that the same level of protection (i.e. double corrosion protection) is provided at the coupler locations as over the pre-grouted lengths of tendon.

The standard protection system for couplers utilizes two individual barriers, comprising:

i) A plastic cover tube over the coupler, with the coupler smeared with an anti-corrosion compound.
ii) An external heatshrink sleeve (with sufficient overlap), over the full length of the cover tube, sealing at each end back on to the tendon sheathing (Figure 1).

Double Corrosion Protection Detail for Couplers on Bar Tendons

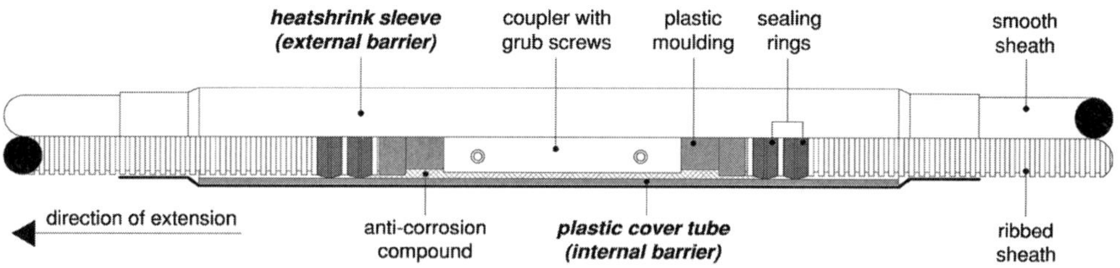

Figure 1

Page 2

c) *Cement grout within a corrugated duct as an effective corrosion protection barrier*

This question relates to the corrosion protection system for bar tendons. To comply with the requirements of double corrosion protection, bar protection comprises:

i) External protection (outer barrier): a corrugated plastic duct of a wall thickness of approximately 1.0mm (previously 0.8mm taking account of manufacturing tolerances),
ii) Internal protection (inner barrier): a cementitious grout placed under factory conditions. For this internal grout to be recognised as an effective barrier against corrosion, it is imperative that cracks induced in the grout under tension do not exceed 0.1mm, in accordance with BS 8081 (1989) and EN 1537 (2000).

The question specifically raised the issue of whether "evidence of uniformity in distribution of the tension cracks or assurance that crack widths were less than 0.1mm" could be substantiated.

Within the Dywidag Group, tests were commissioned from the Technical University of Munich for all the bar diameters within the three steel grades, used for bar tendon production.

The three steel grades for the threadbars are summarised as follows:

Bar Type	Range of Diameters	Steel Grade	Thread Pitch	Rate of extension at working load (60% of U.T.S.) mm/m
Dywidag Prestressing Steel	26.5, 32, 36, 40, 47mm	950 / 1050	Bar Ø / 2	3.07 mm/m
GEWI *Plus*	22, 25, 28, 30, 35, 43, 57.5, 63.5, 75mm	670 / 800	Bar Ø / 3	2.35 mm/m
GEWI Standard	16, 20, 25, 32, 40, 50, 63.5mm	500 / 600	Bar Ø / 2	1.75 mm/m

Figure 2

From the table (figure 2), it can be seen that the most critical bars in respect of rate of extension and thread pitch (bar diameter divided by 2) is the Dywidag Prestressing steel (PST). This bar is worked the hardest, due to its higher strength, yet has a coarser thread pitch than GEWI *Plus* (bar diameter divided by 3). The GEWI Standard, whilst having the same thread pitch as PST, is not worked as hard.

The tests were carried out in accordance with EN 1537 Appendix B for encapsulated bar tendons (i.e. pre-grouted) within a gun barrel. Results from the tests on pre-grouted Dywidag Prestressing Steel (where the rate of extension is greatest and with the coarser thread pitch), together with the less critical GEWI threadbars, showed crack widths below 0.1mm within the encapsulating grout, on all the pre-grouted bars.

In respect of uniformity of crack widths, The Technical University of Munich concluded that the crack widths were uniform.

References:

1. British Standards Institute. BS 8081:1989 British Standard Code of Practice for Ground Anchorages.
2. European Standard. EN 1537:2000 Execution of Special Geotechnical Work - Ground Anchors.
3. Technical University of Munich. Distributiuon of cracks in grouted bodies for double corrosion protection: Prestressing Steel Threadbars 950/1050 N/mm² (GL Reports), GEWI *Plus* Threadbars 670/800 N/mm² (RI Report).

Corrosion monitoring and corrosion protection systems:
discussion session 6 – Part 2

Rapporteur's summary
by John Finnis, *DSI Ltd*

The papers given in this session were broadly relating to the use of newer materials in our rather conservative industry.

The papers generally discussed subjects encompassing carbon fibres, resins and polymers.

The first paper compared the use of cement grout and site mixed and placed resins in borehole grouting operations and the subsequent corrosion protection offered.

The decision to use electrical testing as a means of determining the effectiveness of the two grout types was, I believe, the weak point in the argument. This method of testing, although detailed in EN ISO 1537, is subject to and dependent upon so many environmental variables at a site level that it is difficult to be confident in the results generated.

The second paper compared carbon fibre reinforced polymer elements with steel and discussed the relative advantages of the new material.

Whilst it is accepted that CFRP has greatly enhanced corrosion resistance, it has limitations in general use because of the cost and weakness in shear. The potential for this new material is great and more research is needed to establish it as a geotechnical product.

The third paper further reviews the use of carbon fibre reinforced polymers concentrating on the bond transfer mechanism and gripping methods that are often poor in all FRP elements. By increasing surface roughness, enhanced pull out resistance and bond transfer was achieved.

The forth paper concentrated on the effects of highly corrosive ground conditions on cement and resin grouts and the CFRP and steel strand tensile elements.

In the elevated and acidic conditions reported, the cement grout rapidly deteriorated to a level when bond transfer ceased. The resin however, whilst attacked, maintained sufficient mass to continue to transfer load. The steel, as expected, rapidly corroded to an unusable state.

The fifth paper demonstrated how epoxy coated strand could be modified chemically to achieve the original low relaxation properties of the strand that were degraded during the induction heating as part of the epoxy coating process.

The final paper, an addition to those published, was BBRI Research Project Ground Anchors and it highlighted current multi company development work being undertaken in Belgium to look at, amongst other parameters, how load capacity is influenced by different installation methods.

John Wasilewski, *Capita Symonds*
Question for Tiger Kido. When epoxy coated strand is stressed does the epoxy coating remain within its elastic range or does it go plastic or does it go past the plastic limit and start cracking?

John Finnis, *DYWIDAG-Systems International*
In essence, does the resin material stretch at the same rate as the steel?

Toshiro "Tiger" Kido, *Sumitomo (SEI) Steel Wire Corporation*
Yes.

Mike Turner, *Applied Geotechnical Engineering Ltd*
Question to panel The big problem I think with composite systems is getting hold of the tendon to lock it off, especially in an anchor situation where you often want to overload the anchor, then lock off at its working load, and that often gives you quite long extensions because the fibre tendon is more elastic than steel. How do you get over those problems practically?

Tony Barley, *Single Bore Multiple Anchor Ltd*
Typically, the extension of carbon fibre is about 25 – 30% in excess of steel, so it is in a range that can be accommodated by normal long ram jacks that allow you to overload, relax and lock off. However, there is a problem in developing the final lock-off system and that is the direction that we are looking to develop. I do not think there is any problem with carbon fibre anchors; they are subject to the same acceptance tests, proving tests and suitability tests, as for conventional anchors. We just need jacks with a bit longer ram travel. You get a bit more load loss when you lock the wedges in, but within this extension criterion it is quite acceptable. If you are using glass reinforced plastic which is much more elastic than steel then I think you would get a problem. The carbon fibre is in an appropriate range and I do not see when other developments have been advanced that it is going to be a long term problem.

Mike Turner, *Applied Geotechnical Engineering Ltd*
In practical terms then we can say that the head and lock-off systems are under development. Is that the same with you Matthew?

Matthew Sentry, *Monash University*
What I have found in my research at present is that the most productive form of lock-off or anchorage within the anchor head is a bond type system and I think the Japanese Anchor Association has got some examples out on display. With the resin wedge system, the issues that I have encountered in my research are that it tends to want to slip. The material that I am testing is a similar material to Tony's; it is a flat strand that has a sand coating and the sand coating tends to grind away and prevents the actual gripping of the resin wedge. I have not trialled the soft metal sleeve combination with the steel barrels, purely because through my research it was acknowledged that there were issues with the soft metal biting into the strand

steel as the wedges grip in, and as a result breaking the fibres and causing rupture of the strand at the barrel itself. I think one of the big issues with the bond type system is the length of bond required in the anchor head and then how to accommodate that into a practical design application.

Tony Barley, *Single Bore Multiple Anchor Ltd*
Part of the problem may be solved by having a transitional material in between the carbon fibre and the steel that is more compatible with both materials so that could be a direction that achieves success.

Rama Karri, *University of Birmingham*
My question to Tiger Kido concerns epoxy grout coatings on the ground anchor. Because of the ground stress distribution and also the structure distribution, initially there will be extension in the steel rod. Due to its properties, steel accommodates expansion and contraction, so steel regains its original length after redistribution of the load. What happens to an epoxy coating on a steel rod? Is it going to contract, otherwise it will become slightly elongated or ruptured due to load distributions? In short, is there any limit for expansion and contraction for epoxy resins coated on these tendons? If so, what are the limits?

Toshiro "Tiger" Kido, *Sumitomo (SEI) Steel Wire Corporation*
The elongation of steel strand at its breaking point is usually around 6 to 7%. On the other hand, the elongation of epoxy coatings on epoxy coated strand at breakage is more than 50%. Moreover, the adhesion between the steel and the epoxy coating is so strong that it will not break until the steel wire breaks. Consequently, there is no limit on expansion for the epoxy coatings during stressing of the steel wire strand.
Since the steel wire strand is basically used in tensioning, I am not sure what will happen when the epoxy coating is subject to contraction. However, ASTM specifies that when the epoxy coated strand is bent around a mandrel with 32 times its diameter, there shall not be any crack or damage on the coating. Therefore, there should be no damage due to the contraction experienced on the inner side of the strand in such a test.

Donald Bruce, *Geosystems LP*
I would like to give a bit more of the historical background on the use of epoxy coated strand for anchors in the US in support of Tiger Kido's presentation. The epoxy coated strand had been used in minor applications mainly on the West Coast in the late 1980s and the Bureau of Reclamation had a particular challenge on Stewart Mountain Dam which meant that a very special type of corrosion protection had to be used in the short term. That was when the epoxy coated, but not epoxy filled, strand was on the market. By the grace of God and I think the presence of some very good people in our company at the time, we worked our way through that project without major difficulties using the epoxy coated strand. We then published several papers and a lot of people got interested in the use of the material, but by 1998-99, by which time there had been 20 or 25 dams anchored with the epoxy coded and epoxy filled strand, there were rumours circulating about difficulties with its short and long term performance. One of the jobs that John Wolfhope talked about yesterday was Wirtz

Dam in Texas and we stressed some very high capacity anchors perfectly, but within 4 or 5 hours we had lost several strands on several tendons - they just pulled through the wedges, long after the standard period of lock-off. Everything then came to a grinding halt in terms of the use of material in the US. ADSC sponsored a review team who worked for 2 years with the contractors and the suppliers (like Horst) who were excellent at providing honest information so by the early 2000s we had worked out that in fact it was a problem to do with material itself, including quality control and quality assurance on one hand, but also the techniques of the anchor contractors who were used to employing bare strand and bare strand techniques but were not used to the specialities involved with the use of epoxy coated strands. Then two things happened. We now have an excellent material supplied by Sumiden Wire Products Corporation from the West Coast and we also have a group of about 4 or 5 really experienced and knowledgeable contractors who can use the material properly. In the last 4 or 5 years we have no reports of any unusual behaviour either short term or long term.

John Finnis, *DYWIDAG-Systems International Ltd*
Is that also due to development of the enhancement of the wedges to grip?

Donald Bruce, *Geosystems LP*
Absolutely. That was one of the key things and was mentioned before. The wedges are designed to bite through the epoxy without stripping and one of the things that some contractors did in the early days was to use standard bare strand wedges but strip the epoxy in the head - so what is the point, you are just destroying the corrosion protection that you are paying so much to get in the first place. Now we have these special wedges with big teeth that will bite through the epoxy and will not then rely on the epoxy to carry the load and therefore fail maybe in bond or in creep.

Dieter Jungwirth, *Technische Universitat Munchen*
It was excellent what you mentioned but what happens with varying temperature because the difference in the thermal expansion coefficient between epoxy and strand is very big and if you have additional bending, what happens? In my opinion, small cracks may appear.

John Finnis, *DYWIDAG-Systems International Ltd*
Question to Chihping Kuo. Regarding the electrical isolation testing, when you did your secondary testing you said that the conditions were wet and this is one of the big problems of the electrical isolation tests that I see. We know from anecdotal evidence that even when there is mist, when it is a foggy day, if you are measuring between two electrodes on the surface of a structure or ground anchorage, you run the risk of getting erroneous readings. How much care did you take to prevent any dampness or moisture such that you were reading a short circuit effectively between the one probe on the steel and your other probe in the ground?

Do you accept that if there are wet conditions on the surface of the anchor or the structure then you can get incorrect readings from the electrical tests? In other words, you are not truly testing the electrical resistance between the tendon and the ground, e.g. if there is a short circuit across the surface between your probes.

Chihping Kuo, *National Taiwan University of Science and Technology*
Very true.

Tony Barley, *Single Bore Multiple Anchor Ltd*
When electrical resistance testing was strongly supported in the Eurocode EN1537 by certain continental operators, the people who were not fully acquainted with it suggested that there should be an assessment of the integrity of the test itself because there seems to be various ways of affecting it. What we suggested was that some anchors should be installed with fully intact corrosion protection and some of them deliberately damaged either in the free length or in the fixed length. The integrity of the tests themselves could then be assessed or proven by the results detected. Whilst the concept may appear right, we have got to have confidence that the test is right.

John Finnis, *DYWIDAG-Systems International Ltd*
Absolutely.

Gareth John, *CAPCIS Ltd*
On this issue of electrical resistance testing, I think people should understand that testing component insulation in situ, particularly for large length units, can be extremely difficult and there are several practical issues that have to be addressed For example, the possibility of an electrical short between the anchor head and the test electrode in wet conditions. This should never be overlooked, especially when using a high frequency AC signal, where you can get skin effects coming through. Also, it should be noted that the resistance measured will be proportional to the resistivity of the overall ground and inversely proportional to the square root of the exposed surface area of the steel (i.e. the area of defects). That is why if you just have a small change in the ground resistivity, you can get significant variation in measurement. For example, if you are using your criterion of O.1 Megaohm as a pass or fail criterion, then for a soil with moderate resistivity defects up to $10mm^2$ could exist and the anchor would pass. However, for the same tendon if the resistivity is reduced by an order of magnitude then it would fail with defects of only $0.1mm^2$. The reason for this massive difference in sensitivity is due to the square law effect.

Therefore, if you are going to adopt this type of methodology, first of all you need controls to prevent shorting between the electrodes (e.g. in the case of any misting or surface dampness), and secondly you should normalise the criteria with respect to the overall ground resistivity (this is a standard procedure that is used in rail installation for rail tracks).

In summary, whilst the overall concept is acceptable the actual methodology used in practice is not (in my opinion) sufficiently robust to allow it to be used with any confidence to confirm integrity of the anchor.

Tony Barley, *Single Bore Multiple Anchor Ltd*
Are there any people in the audience with experience in the use of electrical conductivity testing and are completely satisfied that this is a quality test, i.e. a true test of what we are looking for? We are looking for experts who have used the system and are confident with the system.

John Finnis, *DYWIDAG-Systems International Ltd*
Within DSI there has been work carried out to prove that this does in fact give out a satisfactory test result but I think there is still more work needed and I think, as Gareth was saying, you need to be very careful about what you are actually reading and perhaps the second electrode should be taken down the bore hole with the anchor, a second wire that can be attached, say 10 or 20 metres into the bore hole so you truly know that you are measuring the resistance within the anchorage rather than some sort of erroneous surface reading that you could get from all sorts of atmospheric conditions. There is more work to be done but where it has been used successfully people are confident that it works properly but given that, as I understand it, EN1537 is about to be reviewed then perhaps now is the time to ask a few more serious questions about it.

Tom Richards, *Nicholson Construction Company*
Matthew showed a corrugated steel as a corrosion protection layer and I was wondering what other people's thoughts are on that? Or what Matthew's thoughts are?

Matthew Sentry, *Monash University*
I think in Australian conditions we tend to follow the US guide where we do an integrity test on the actual corrugated ducting. In Australia, the primary purpose is just to provide a single corrugated ducting. From the literature and communication that I have had over here in Europe, there is use of either a sacrificial steel corrugated ducting or the double HDPE corrugated ducting. From an Australian point of view, there has been limited or little testing or use of it so either way. Personally, I prefer the double HDPE over the steel purely because you know the properties and the values that can be obtained with HDPE, compared with a material that is going to corrode.

John Finnis, *DYWIDAG-Systems International Ltd*
Where would you use a steel corrugated duct? I cannot imagine many applications

Matthew Sentry, *Monash University*
I do not know of any.

Chris Irvin, *DYWIDAG-Systems International Ltd*
I think it comes from the TAM system (Tube a Manchette) with Bachy Soletanche. Within the European Standard there were five systems proposed, two with a corrugated steel sheath and the other three with a corrugated plastic sheath. I have serious reservations about the corrugated steel sheath solution. The corrugated plastic sheath, yes that would work but I think we have seen that with steel, you cannot guarantee it.

Question to Horst Aschenbroich. Regarding your instrumentation for the testing of the hollow bars, this is very interesting in respect of load performance but there is a key distinction between load performance and long term durability. It is wonderful to know that these self-drilling hollow bars are holding load and performing well in a roughened borehole, but what this low performance testing will not tell you, in my opinion, is the longevity of the self drilled system, perhaps you could give us some comment please.

Horst Aschenbroich, *Con-Tech Systems Ltd*
Very good question and we address it ourselves. On these projects in Las Vegas, we used the sacrificial steel method so the product used was the TITAN 73/53 bar where we actually only needed the 52mm bar, so there was about a 120-year life of sacrificial steel built into the material. The other thing about long term monitoring, especially in the hollow bar, is that these strain gauges are very well protected inside the bar, rather than on reinforced concrete outside of steel bars that can get damaged. If we are putting them inside the hollow bar they stay permanently protected in there. If we encounter corrosion, what does corrosion do? Corrosion causes reduction of cross section. Under a given load, if you have a reduced cross section you get an increased strain so in time you can read an increased strain. Now, whether that is due to corrosion or to something else we do not know, but we know then something is happening.

Frank Schmidt, *DYWIDAG-Systems International*
Just one comment to Horst. You would only recognise the corrosion in strain, the corrosion that happens outside that area where you have got this strain measurement device, if the corrosion is let us say 5 metres above because there is a big crack in the grout above, then you have got bigger strain above your measurement device. This will always remain the weakest point of your pile, anchor or nail but in that area where the strain measurement device is located there is maybe no crack outside and no corrosion at that specific point. Therefore, you would not recognise the corrosion because at that measuring point you have got full cross section of the steel and thus no further strain because of corrosion losses. You can only measure this strain in one specific point and if the corrosion happens anywhere else on the tendon you would not recognise it.

Horst Aschenbroich, *Con-Tech Systems Ltd*
We did not only install one strain gauge, the strain gauge is at the top and then certain distances down over the whole length of the bar.

Gareth John, *CAPCIS Ltd*
Two issues following on from that particular comment. First, you state that your piles had a sacrificial life of 120 years. How did you know that? Do you actually know the corrosivity of the soil you put the piles into? Secondly, again coming onto the issue about using measurements of strain to indicate metal loss, my understanding from investigating other projects over the years, is that the loading on these piles is changing on a regular basis anyway, so how do you know that the change in strain is due to metal loss and not just change in the load?

Horst Aschenbroich, *Con-Tech Systems Ltd*
You are correct as far as piles are concerned, the load is pretty well the same in compression but I am applying it also in soil nails, or in a tension element, or like we have in California, with seismic conditions, where we have compression and tension piles. Now there is another method of measuring (it is mentioned in one of my dam projects in Canada), namely, the SMS gauge (Strain Monitoring System) that we installed immediately beneath the anchor heads in lieu of a load cell and that is a different mechanism. What we found out in the raft slab foundation in Vegas where the piles are primarily in tension, there was one project, filling retaining walls where they used the hollow bars to retro fit it and they also installed these strain gauges there. Unfortunately, you could not see that on a calibration slide with a blue background (normally with a white background you could have seen that curve) and it is a very nice indication of stress transfer. The change of compression to tension is a straight line and the results of the readings which took place are very good. As far as piles are concerned, you are right, it is always a constant load but the interesting part that we could see in Vegas was during construction, the load increased and we had a wonderful opportunity to actually read these loads that were introduced from the outside in. The piles were terminated about a foot below the raft slab, so the loading of the slab to the pile was very well indicated on this instrumentation.

Case histories of satisfactory performance, shortcomings and failures in service: discussion session 7

Rapporteur's summary
by Devon Mothersille, *Geoserve Global Ltd*

Nine contributors cover a wide range of ground anchorage applications with contrasting levels of satisfactory performance, shortcomings and failures in service.

Littlejohn's paper assesses the use of cement grout as sole corrosion protection of permanent tensioned steel tendons. Within the industry there is mixed opinion about the role played by cement grout as a corrosion protection barrier within ground anchor technology. Littlejohn points out that whilst cement grout inhibits corrosion by virtue of its high alkalinity (e.g. pH 9.5 -13.5), any beneficial effects can be undermined by the fact that aggressive ions (e.g. chloride) can intrude through the grout or via cracks in the grout.

A distinction is also made on the influence of grade of steel on corrosion and the effect of crack widths in the surrounding grout. Recognising that high yield prestressing steel is more reactive than low yield reinforcing grade steel with respect to susceptibility to corrosion, Littlejohn advocates relaxation of the permissible crack width from 0.1mm to 0.2mm for reinforcing grade steel in tension.

The paper highlights several case histories which demonstrate the detrimental effects that result from loss of cement grout around steel tendons. Littlejohn points out that when compared to post-tensioned concrete structures there are very few published case histories of the in situ condition of ground anchorages. Despite this, reference was made to studies carried out to investigate the integrity of cement grout as the sole corrosion for steel tendons in post-tensioned dams. Carefully controlled procedures did not prevent gaps in the grout. It was established that bleed within the grout mix was the primary cause of such voids and the fact that grout placed in situ is highly sensitive to quality of workmanship and its integrity cannot be guaranteed.

Littlejohn comments on the agreement between many international codes of practice on the role played by cement grout in protecting steel tendons against corrosion and exposes ambiguity in others. The paper presents a compelling argument that in situ placed cement grout is not appropriate as a corrosion protection barrier for permanent anchorages.

An investigation of the failure of a permanent anchorage comprising a double corrosion protected 75mm diameter bar was the subject of the paper by Whitworth and Parrish. It was pointed that the failure occurred during routine load testing and manifested itself as a large increase in displacement accompanied by a reduction in recorded load on the stressing jack. The anchorage formed part of the support system for a Nuclear Dockyard facility and warranted a thorough investigation.

The investigation required the failed anchorage to be removed from the borehole for inspection and this provided access to the borehole for CCTV equipment to examine the condition of the borehole. It was deduced that rock/grout bond failure was unlikely owing to the high quality rock observed during the surveys. The paper also describes the anchorage extraction process, which was recorded graphically in terms of load/displacement, and further data recorded to assess rock quality by quantify drilling rates through the different of rock conditions encountered during construction.

The authors concluded that failure of the anchorage may have resulted from shearing at the encapsulation/grout interface. The probable causes were linked to features such as contamination of the borehole through detritus material entering the borehole and undermining the quality of the bond. Also cited was the possibility of poor in situ grout quality due to dilution from water entering the bore hole from adjacent operations. It is well documented that ground anchorage technology is highly workmanship sensitive and the probable causes of failure can all be avoided if good procedures are adopted during the anchorage construction phase.

Barley and Edwards describe the satisfactory performance of both permanent and temporary anchorages for the support of a secant pile wall cofferdam. Of particular significance was the nature of the ground which was highly variable, consisting of alternating layers of hard sandy limestone and weakly cemented sandstone interspersed by beds of silts or clays of variable thickness. The authors reported that the use of single bore multiple anchor technology made it possible to design the multiple unit fixed anchor with an extra contingency unit to ensure satisfactory performance in difficult ground.

The permanent anchorages were double corrosion protected with two concentric corrugated ducts with the fixed anchor and double sheaths in the free anchor length in accordance with BS8081:1989. The temporary anchorages were designed such that the tendon could be fully removed from the borehole on completion of the temporary works.

25 out of 100 unit anchor strands were subjected to check lifting after nearly 12 months in service and showed load losses that ranged from zero to 5% of the original lock-off load which is within the permissible limits stipulated in BS8081.

Barley et al reported on the failure of permanent anchorages adjacent to a tidal surge barrier in north east England. The anchorages were constructed in 1979 using the technology available at that time, bearing in mind that there was no code of practice relating to their design, and they were used as the primary support system for a sheet piled wall. When assessed against the standards of corrosion protection stipulated in current practice, the anchorages would be considered unsuitable for permanent structures.

After just 12 years in service the sheet pile wall sustained a sudden lateral deflection of 450mm and in order to investigate this failure a programme of inspections, which included excavation behind the wall, was commissioned. The investigation revealed that during construction unprotected anchor tendons were subjected to aggressive mechanical backfilling that resulted in significant damage in some tendons. Evidence existed of poor attempts to repair damaged free length sheathing using grease impregnated tape which over time proved ineffective.

In addition, the lack of adequate corrosion protection at the inner and outer anchor head location exposed the anchor head components to severe exposure from the marine environment. Observed features included section loss in the steel components and strand slippage, as the heavily corroded barrel and wedges lost gripping capacity. The findings from the investigations led to the recommendation that all the anchorages be replaced with double corrosion protected systems in accordance with BS8081. It is noteworthy that during the mid to late 1970s the upgrading of old Victorian quay walls utilised similar types of construction at various location throughout the UK and serves to strengthen the case for appropriate authorities to commission detailed inspection and testing programmes in order to avoid potential failures in the future.

Liao et al record that almost all the straight shafted anchorages installed in the alluvial soils of the Taipei Basin are drilled with rotary casing incorporating an external flush system. Observations have shown that the process tends to create boreholes with larger diameters than that of the drilling casing and an irregular borehole surface. These features can increase the pull-out resistance of anchorage although the magnitude of this has not been accurately quantified.

To compare the difference between design capacity and actual pullout capacity of straight shaft anchorages, 46 field anchorage tests were carried out. The design capacity was calculated based on the simplified pull-out mechanism based on borehole diameter and the in-situ soil properties. Of the 46 test anchorages, the in-situ pull-out capacities as well as the corresponding borehole information at the test locations were recorded.

By applying a rigorous probabilistic analysis, the authors related the pull-out capacities to a *reliability-based resistance factor*. The results indicate that pull-out capacities obtained from the performance tests are about twice as large as the design capacities based on borehole diameter. The authors also found that even if a factor of safety equal to 1 was applied to calculate the design capacity of an externally flushed anchorage using the in situ soil properties, it was unlikely that the anchorage would fail at the calculated design capacity. Readers should also be aware that for anchorages designed in cohesionless soils, BS8081 acknowledges the potential enhancement of the borehole diameter within the fixed anchor zone that can be created by different grouting techniques and has incorporated this in the recommended design formulae.

The industry has recognised for some time that ground anchorages are resilient to dynamic loading by virtue of the fact that their construction incorporates an elastic free length that can potentially accommodate movements within its elastic range.

Takemura et al report on the fact that anchored slopes subjected to severe earthquake loading (maximum intensity 7 on the 7-grade Japanese intensity scale) showed resilience and were not adversely by the ground motions. Some 1600 failures occurred in non-anchored slopes resulting in 70 million m^3 of soil mass slides. Damaged slopes were substantially restored using ground anchorages and grating cribwork. Interestingly, three years after the initial earthquake the region was subjected to another earthquake of similar magnitude. The intensity of ground motions was less than the previous event since the hypocentre was further away. Despite this, the event presented a unique opportunity to inspect the newly restored structures which together with the original slopes showed no significant signs of distress or damage.

The paper by Richards and Thome describes the restoration of a failed anchored retaining wall in Washington DC. The authors established that the failure resulted from a combination of factors which included poor drainage and inadequate corrosion protection for the ground anchorages used to support the structure. In addition, a slip zone comprising slickensided clay bands was identified in the location of the wall toe which created a failure path within the ground. The paper also describes the unsatisfactory condition of the existing bar anchorages and the process of demolition of a portion of the existing wall and reconstruction of the new anchored wall. Post-grouting proved useful in enhancing the pull-out capacity of the new anchorages together with the use of air flush as a medium for removing drill spoil from the borehole.

The issues that can occur when subjecting anchorages to suitability and acceptance testing were addressed in the paper by Adekunte et al. The authors describe the test programme imposed on 63.5mm diameter bar anchorages used to support a retaining wall. Although the anchorages were designed with free lengths of 4.5m and fixed lengths of 7m, during fabrication of the anchorages it appeared that no attempt was made to debond the free length. The result of this was that subsequent testing yielded free anchor lengths as low as 120mm. With respect to overall stability, this would suggest that no load was effectively transferred beyond the assumed failure surface. Fortunately, during the temporary works period, the wall stiffness and global ground pressures were such that no adverse effects were recorded on the anchored structure.

The final paper in the session, by Martin et al, presented four case histories demonstrating the successful the use of high capacity passive bar anchorages (or anti-flotation tension minipiles - ATMs) to provide restraint against hydrostatic uplift for deep basement structures.

It was emphasised that the retrofitting of measures designed to resist hydrostatic uplift is often not cost effective and greater economies are derived from designing in measures such as the installation of ATMs at earlier stage. It was pointed out that a full understanding of the long-term water regime is necessary in order to produce economic designs. The authors' also held the view that ATMs had many advantages over conventional anchorage solutions, the principal ones being related to cost and ease of construction. However, readers should also be aware that on some projects clients do not wish to permit structural movements necessary to mobilise the loads in passive tensile elements such as ATMs. Under these circumstances conventional anchorages may offer an appropriate solution.

Simple equations expressing the nett uplift force to the basement area and the head of water have proved useful in assessing the demand to the ATMs. The authors recognised the fact that during its service life the grout column around the steel tendon cracks and requires an 'unperforated' corrugated UPVC duct installed along the full length of the steel tendon and extending into the base slab to ensure continuity of protection. Programmes of long-term monitoring using strain gauge instrumented ATMs demonstrated the rapid increase in water pressure that can occur once the dewatering system was decommissioned at the end of the construction phase. Such data serve to improve understanding of the behaviour of deep basements structures subjected to uplift forces and allow the optimal design of structural elements.

Ten years ago the ICE conference entitled *Ground anchorages and anchored structures* covered a broad spectrum of subjects focusing on the design, construction and testing of ground anchorage systems. Ten years later, emphasis at this conference has been placed on the performance of ground anchorage systems in service. The papers presented in this session have provided invaluable insights into the service behaviour of different types of anchorages and anchored structures. The ground anchorage systems presented have been subjected to a variety of loading conditions ranging from the static loading imposed by ground pressures to dynamic loads generated by earthquakes. In addition, some papers have directly addressed issues related to the service environment, providing useful insights into the type of degradation that can occur over time under certain conditions. All these factors are important since ultimately it is the service behaviour of anchorages that provides the most accurate indication of performance under working conditions.

Jan Maertens, *Jan Maertens BVBA*
As a member of the working group on ground anchors for the European Standard EN1537 and as convenor of the working group on micro piles, I have been discussing corrosion protection for a very long time, so I want to take the opportunity to give my personal point of view. When high strength steel is used I fully agree that protective sheaths are necessary and when the risk level is very high I fully agree that double sheaths should be used always, even if it is not easy to install them in all conditions. I do not agree that for ground anchors made with low strength steel, it is always necessary to have protective sheaths. In my opinion, there are a lot of cases where sacrificial steel is much better than corrosion protection and in my personal practice I mostly propose to spend the money that is normally used for corrosion protection to use it to enlarge the thickness and use the sacrificial thickness. I am well aware of the fact when speaking about corrosion protection it is always a discussion between believers and non- believers and even with conferences, we will not solve this problem. I am sorry for that but I think this is the reality. When looking also to corrosion protection we have to keep in mind that when cast in situ piles are used as tension elements for pier walls and also to secure against uplift, it is still a common practice in a lot of countries. There is also the contact between cast in situ concrete piles and the foundation slab, everybody knows there is a layer of quite bad concrete but no one speaks about crack widths of 0.2 mm. There is a big gap between the design of corrosion protection for concrete elements and corrosion protection for micro piles and ground anchors.

Instead of continuing to discuss yes or no, a good approach could be that when anchors are installed with a sacrificial thickness, some additional anchors should be installed, as has been done in the beginning for reinforced earth constructions. There have also been reinforced elements that have been installed and then examined after 5 and 10 years. I think if we want to learn a lot, the proposal should be to install sacrificial anchors and to extract them after 5 and 10 years and then to look to the corrosion and to the capacity of the anchors.

Stuart Littlejohn, *University of Bradford*
I agree with many things that you have said. I am suspicious however when people advocate the use of sacrificial coatings such as galvanisation. How do you then ensure bond if you are allowing for a sacrificial loss. It is all very well in the free length but not necessarily where you are relying on bond between the grout and the tendon. If you are saying we can except sacrificial loss just where you want to mobilise bond that is an issue. I do however support your view at the very end that there is a lack of factual evidence, regarding the performance of anchorages with different types of designed corrosion protection, where the aggressivity of the ground and the ground water are also quantified. Those of us who are geotechnical engineers here will know how little we spend in ground investigations on measuring the aggressivity. We focus on geotechnical properties but not aggressivity and so seldom do you have that quantified. I found in our keynote presentation that we could only find three case histories in the world where the failure descriptions were allied to environmental considerations. This is an important issue and more case histories are required where all the relevant data are quantified, if we are going to resolve these matters.

Jan Maertens, *Jan Maertens BVBA*
I think in highly aggressive soil, corrosion protection is necessary but there are a lot of cases where everybody knows the aggressiveness is very low and it is my personal point of view that corrosion protection is not necessary. On the problem of the bonding of the bond length I believe if you have corrosion of the steel in the bond length the roughness will be larger and I cannot believe that there will be a rupture between the steel and the grout.

Tony Barley, *Single Bore Multiple Anchor Ltd*
I think there is a further consideration associated with loss of steel in a sacrificial steel situation. What happens to the increase in volume, the rust volume, of the corroded material which is typically 6 – 7 times the volume of the original steel. It must completely change the stress field in the cementitious grout and if it is expanding them surely it will aggravate the cracking in the grout.

Gareth John, *CAPCIS Ltd*
When steel corrodes you normally get expansion of anything from 4 – 16 times the volume. This is what you classically see in corrosion of steel in concrete and associated spalling, but that is only when you have full aeration of the condition. If you have an anaerobic condition or limited oxygen that could happen when you have anchorages that are deep in the ground, you will not get the expansive oxide film; at best, you might get

an oxide which is a similar volume to that of the steel it replaces. You can actually get some oxides that are totally soluble and therefore have no expansion at all. In those cases a) you will lose your bond and b) you will actually create more acidity at the surface, as the corrosion product tries to expand, and the only time you will actually know you have got a problem is when you have a total failure. The issue of corrosion and bond is a big one because in steel and concrete, the main bond is from the deformed ribs which react between the steel and the concrete itself and any corrosion can sometimes improve it and can sometimes reduce that bond so again it is not a simple linear effect. This is a major factor and too many people think that a bit of rust on the steel will increase the bond and that it has no real effect.

Bede Noonan, *Geotech Pty Ltd*
Question for Stuart Littlejohn. For the free length poly coating and the grease filling of that, can you comment about whether or not it is really necessary other than for lubrication, or whether you would regard it as an extra form of protection to have the grease completely filling the inside of the free length poly coating.

Stuart Littlejohn, *University of Bradford*
That is a good question and the topic of grease infilling will be addressed in the final session. To answer your question now, aside from the decoupling aspect, it is important to have very good infilling of the annulus with corrosion inhibiting grease, otherwise you allow the possibility of water ingress. Some examples of that were illustrated earlier due to inefficient grease infilling. There is such a variety of techniques for grease infilling in practice around the world that we have encountered figures ranging from zero to 100% infilling, based on cutting and testing greased and sheathed samples, but this will be addressed in greater detail in the final session.

Thomas Richards, *Nicholson Construction Company*
Question for Stuart Littlejohn. Where there were bleed lenses of missing grout, was there a trend for those to be near spacers or something else in the tendon?

Stuart Littlejohn, *University of Bradford*
No, although we expected that where we had spacers or centralisers we would find gaps. What we found, going back to the experimental work in the 1960s, is that we tend to get cracks around spacers and centralisers. When it comes to bleed, for example, in a vertical or sub-vertical column, the majority of the bleed develops over a 2 to 3 hour period during the stiffening phase of the grout. If you imagine bleed developing over the bottom 5m of a 50m deep hole, it has no time to get to the top of the column as it travels upwards; the bleed water develops due of sedimentation of the cement particles when it is fluid (under Stoke's Law) and for water cement ratios of 0.4 to 0.45 (by weight) these bleed pockets tend to be located at intervals of 5 to 7 metres but not necessarily at the locations of centralisers or spacers.

Alistair Henderson, *Ritchies*
In the bleed cavities that you have found, is there any evidence of a fringe of poorer grout that would lend credence to the idea that it is bleed rather than arching, it just seems remarkable to have 300 millimetre gaps as a result of bleed within an anchor column.

Stuart Littlejohn, *University of Bradford*
If you assess the volume of grout in a column between the gaps and take 2 or 3% of that volume of grout, then depending on the borehole volume, the observed gap dimensions work out quite logically.
The 302mm gap was the largest encountered in the field and judged to be caused by a combination of arching and bleed. At the lower interface (bottom of gap), the quality of the grout was of poorer quality with a lower density than that observed at a greater depth within the column. In the absence of arching, 25-75mm wide gaps with a spacing of 5 to 7 metres between gaps were more common for neat cement grouts with water/cement ratios of 0.4 to 0.485 by weight.
In the early full-scale experimental work (1964-65), attempts were made we tried to force the bleed water out of the grout by agitating and re-circulating the grout for an hour, in an attempt to reduce the bleed capacity down to say 1% but voids still appeared within the grout column. For ground anchorage contractors today trying to eliminate bleed, whilst maintaing a high strength, I think we would have to come down to a water/cement ratio of around 0.35 with the use of superplasticisers. If you look at the allied industry of duct grouting in post-tensioned bridges and silos and study the most recent American research (referenced in the paper), that is exactly what they are advocating. In general, we have not moved that way and still retain a fairly simple approach to grouting in the field.

Mike Turner, *Applied Geotechical Engineering Ltd*
I think that is quite right but anchor manufacturers tend to use cable-flow type grouts (as used in post-tensioned concrete ducts) for the grouting of their tendons within the plastic sheathing, so the 0.4 - 0.45 water/cement ratio grouts are typically used for the anchorage grout. Am I correct manufacturers?

Stuart Littlejohn, *University of Bradford*
That is a fair point and I was studying the grout that is placed in situ to bond the tendon to the surrounding ground. I do not think we have the same problems with tendon bond encapsulation units because they tend to be relatively short, and grouted under factory-controlled conditions in a vertical or a sub-vertical position. Bearing in mind what we knew about the bleed issues in the 1960s this directed us towards single and double protection for security. I think people have forgotten about the significance of bleed in relation to the integrity of cement grout and the formation of gaps; there is hardly a mention of it in the PTI recommendations and PTI is silent on cracks even in the 2004 edition.

David Wetherer, *Birkhill & Co*
There have been several references over the last 2 days of anchors in case studies being locked off very low and PTI has a recommendation that anchors should not locked off at less than 50% of their ultimate strength because of anchorage slippage. Is that a correct

procedure for strand, not for bars as you can lock bars off anywhere, but just for strand anchors because of wedge slippage?

Stuart Littlejohn, *University of Bradford*
I tend to advocate designing the anchorage for a given working load (Tw) and then locking off at 110%Tw to allow for steel relaxation and potential consolidation or creep in the ground. However, in the case of a quay is being built where you deposit soil say behind, say, a sheet pile wall during construction and you need the ties, whether they be strand, wire or bars, to hold back the wall from overturning as you backfill, then you cannot apply too high a load initially, otherwise the sheet steel piling is pulled back into the retained soil, so you allow the load in the ties to be mobilised as you backfill. In certain countries they have not checked the residual load on completion of this type of wall construction and I would always advocate carrying out check lifts, and ideally on-site acceptance tests, once the backfilling has been completed and then locking off at 110%Tw, irrespective of how it has been constructed. I do not like the idea of walking away from the permanent anchored structure thinking that you have only locked the anchorages off at some nominal value of 50%Tw. I would rather know precisely what loads have been mobilised and what loads need to be locked off on the wall to stop the wall moving in service and to cater for relaxation or creep. Written contribution. With reference to an initial allowance for wedge pull-in, my rule is to allow for a loss of displacement of 4-6mm, but much depends on the technique used to fix the individual wedges in place. In practice, such losses are overcome by on-site acceptance testing where a proof load is applied (ranging from 1.25Tw up to 1.5Tw) that consolidates the wedge seating and thereafter there is no loss allowance when the anchorage is subsequently locked off at 110%Tw.

Gerry Noonan, *Geotech Pty Ltd*
Question for Stuart Littlejohn. You referred in your talk to finer ground cements. Is there any difference in performance of the finer grained cements compared with normal Portland cement in relation to shrinkage and bleed?

Stuart Littlejohn, *University of Bradford*
In terms of bleed, micro-fine cements are far more attractive and in some countries like Japan these fine grained cements, including ultra-fine cement, are competitively priced. In this country they are not and I think expense may have been an early reason for moving away from very fine grained cements. They hydrate very quickly and you need to employ admixtures to avoid aggregation of the fine particles; they also exhibit a shorter working time, so there were practical issues as well. Nevertheless today, I believe you should consider more the use of micro-fine or ultra-fine cements because of the bleed problem. To answer your question, the finer the grain size the lower the bleed.

Tony Barley, *Single Bore Multiple Anchor Ltd*
Question for Michael Whitworth. With reference to the Devonport anchor that you considered failed at the grout to encapsulation interface. From my experience which is limited in the withdrawal of encapsulated bar anchors, I think it is likely that the failure stems from the failure of the rock to grout interface and that the failure happened when

you were pulling the encapsulation out but it was apparent that it failed at that interface because I have experienced exactly the same situation before. From the observation data, it looks as though it is an encapsulation to grout failure, but it stemmed from the actual failure of the rock to grout interface in the first place. As soon as there is any amount of lateral movement the failure mechanism will transfer itself to the weakest interface that allows you to pull out.

John Finnis, *DYWIDAG-Systems International*
On the point about the micro fine cement, the historical situation has been overtaken in the last 12-18 months as the European Codes of Practice on cement standards have changed which means that the standard OPC cement now supplied in the UK is only a 32N cement at 28 days, so to continue its use would have created problems getting ultimate compressive strengths for encapsulations to comply with the BS8081 code. As a consequence, over the past year or so, we had to move on to effectively what is a micro-fine cement with rapid hardening cement to achieve the same sort of encapsulation strengths that we were getting 2 or 3 years ago.

Alisdair Henderson, *Ritchies*
Responding to John's point, it is a big issue for us. We have to make sure that we are using CEM1 Strength Class 42.5N for the site batched grout that we inject. Coming back to micro-fine cement, at our permeation grouting job at Dounreay, Scotland (although not an anchorage application), we are using grout with a D_{95} of about 16 microns for the cement. It is stabilised with a silica fume additive and has a water/cement ratio well in excess of 1.5 with less than 0.5% bleed after 3 hours, so it is perfectly possible to create a stable grout.

Stuart Littlejohn, *University of Bradford*
So, basically with modern materials we are getting lower bleed capacity grouts. The reference for strength classes of Ordinary Portland cement is BS EN 197-1 (2000).

Devon Mothersille, *Geoserve Global Ltd*
Question to Stuart Littlejohn. Can you just comment again on the influence of lateral confinement on crack widths and how that influences crack widths and orientations in a confined grout?

Stuart Littlejohn, *University of Bradford*
In regard to crack width and orientation, we know particularly following the German use of rib bars in the early days that we tended to get axial cracking opposite the ribs, thereby controlling the crack aperture for a given tensile strain in the bar. We had great debate in the early 1980s about describing systems as double protection, when they only included a single sheath and an internal cement grout, providing the individual crack widths did not exceed 0.1mm. The early field work that confirmed this, before gun barrel tests became popular, was carried out in compact or dense gravels by Ostermayer and Scheele (1977) of the Technical University of Munich. What is important to understand, as we develop anchorages in weaker ground with SBMAs and other systems, is that the lateral restraint of such ground is reduced. If you imagine an encapsulation being pulled through the

outer annulus grout there is a tendency for a bursting mechanism to develop. In that situation, axial cracking no longer dominates, although you cannot ignore it, and you tend to get radial cracks. These cracks do not occur regularly, i.e. conveniently at 25 or 50mm intervals; they have a random and irregular spacing which means that you cannot predict the crack widths, and apertures of 1mm, or more have been reported (Meyer:1977 and Graber:1981). Therefore, the lateral constraint provided by the surrounding ground, particularly soil and compressible weak rock, plays an important role in determining the location, orientation and width of the cracks. The references quoted above are included in BS8081:1989.

Written questions and answers session seven – part 1: case histories of satisfactory performance, shortcomings and failures in service

Tony Barley, *Single Bore Multiple Anchor Ltd*
Similar to Devonport B 160 anchor, after withdrawal of a failed bar encapsulation anchor at Barnsley in 1989 the grout within the bore was inspected by CCTV camera. Initially observation of pull-out of an intact 6m encapsulation supported by the annulus of anchor grout remaining in the borehole (CCTV observation) one may conclude that the failure occurred at the encapsulation/anchor grout interface.

Furthermore inspection of pull-out data and calculation of an increasing value of apparent tendon free length during the load cycles cannot identify whether failure is occurring at the steel/encapsulation grout or encapsulation/anchor grout or anchor grout/rock interface. It is likely that some progressive de-bonding at all three interfaces may have occurred at some stage.

Inspection of the gun barrel tests has been seen to demonstrate that data, limited to load and extension monitored at the top end can be misleading. On dissection of the gun barrel and the grout column, actual failure has been seen to stem from a different interface or even a composite failure mode has been observed (Barley 1997).

At Barnsley there was no evidence of a composite failure which can progress from encapsulation/grout in the upper bond length to the bar/grout. This is generally evident from the existence of elongation or tensile failure of the encapsulated duct.

In order to further investigate the likely anchor failure mechanism an encapsulation identical to the test anchor, but half the length (3m in lieu of 6m) was grouted into a gun barrel test. It was loaded to 80% of the bar capacity and even the half length could not be failed. During the field tests 7 test anchors were installed and 3 failed.

After assessment of all data related to mobilise bond stresses the following was concluded in the Report (Keller Colcrete 1990).
"Jacking of anchor N8 and S8 from the ground resulted in the extraction of the encapsulation from the anchor grout column.

Set grout within the corrugations of the fixed anchor length indicated that this length had been fully bonded at the time of installation.
"Grout paste" on the encapsulation lengths indicated the grinding and crushing of the cured anchor grout around the fixed length during anchor extraction.

Video inspection of the borehole indicated a smooth grout bore. However, any breakdown or cracking of the anchor grout column was most likely masked by the smearing of grout paste over the bore during anchor extraction. Examination of the extracted anchors indicated that failure was not at the bar/encapsulation grout interface and the very low bond stresses developed at the encapsulation/grout interface render failure at this interface extremely unlikely.

The most probable failure mode was at the grout/ground interface. The low ultimate grout/ground bond stress developed was most likely as a result of a reduced in-situ rock strength in the vicinity of the fixed anchor lengths."

We would be interested to hear of further comments from the Author based in this supplementary data.

References:

Barley A.D. (1997) "Ground Anchor Tendon Protected against Corrosion and Damage by a Double Plastic layer" – Proceeding of the International Conference on Ground Anchorages ICE, London.
Keller Colcrete (1989) Report on the Testing of Trial Anchorages at Barnsley Relief Ring road. Nov 1989 – Not published

Michael Whitworth, *Devonport Royal Dockyard Ltd*

Prior to the extraction of B160 the grout to rock interface was thought to be the most likely cause.

However, upon extraction of the anchorage and subsequent inspections detailed below this was rejected.

- Upon extraction of the anchorage, grout was observed within the corrugations throughout the free length, however when the fixed length was extracted no grout was present other than smooth paste similar to "drill slurry"
- This slurry was sent off for testing and found to be a very weak grout mixed with unknown materials.
- A CCTV survey was conducted and found that the grout-rock interface appeared intact and that a very limited sump was drilled.
- A review of drilling records, site investigation data, borehole integrity tests and nearby CCTV surveys showed that the anchorage had been drilled in extremely competent rock

- To supplement this and exclude the grout-rock interface as a failure mode a core of this interface was drilled. This showed that the grout-rock interface was sound and as expected and thus excluded it as a failure mode.
- Having excluded other failure modes i.e. anchorage factory grout and bar-encapsulation failure, the conclusion reached was that the failure mode was grout-encapsulation failure
- This conclusion was reached independently by a specialist contractor, an independent consultant and DRDL.

Session 7 – Part 2: Case Histories Of Satisfactory Performance, Shortcomings And Failures In Service

Syed Ali Raza, *Atkins Ltd*
Question for Thomas Richards. Was your new wall anchor and pile wall designed on the basis of new geotechnical investigations or the previous data that were used in the original design?

Thomas Richards, *Nicholson Construction Company*
We used the previous data but then in investigating the failure the owner drilled some more borings and then before we did the final design of the wall we drilled even more borings. We were searching for slickensides within the hard clay, as evidence of where the failure surfaces were.

Stuart Littlejohn, *University of Bradford*
With Tom Richard's presentation, it was most interesting to hear that their approach now is to encapsulate the whole of the anchorage (i.e. fixed length and free length) in cement grout and it reminded me, that we had gone back to the period from 1934 to the 1960s where nearly all post-tensioned dams were tensioned initially with the fixed anchor grout in place and then the whole of the remaining free length was encapsulated with grout. The view taken was that even if you had the whole of the anchor head rusting away completely, the anchorage would still work. It was only in the early 1970s that some dam owners wanted to know a bit more about how the anchors were performing that we then got decoupled free lengths.

Thomas Richards, *Nicholson Construction Company*
That is not how we did it here. We had greased and sheathed tendons, we grouted them within one metre of the anchor head and flushed the grout out of there so that we could splay the strands out to the anchor head. When I speak of secondary grouting in that application, it is just filling up that bird beak, that last bit for corrosion protection.

Stuart Littlejohn, *University of Bradford*
With regard the paper by Martin et al, I am really pleased that Cementation is still optimising the spacing of the anchorages with respect to slab thickness. I worked formerly with Cementation in the 1960s and when we were designing anchored

diaphragm walls, particularly in the London area, we did exactly the same thing to judge the optimal solution between the spacing of the anchorage and the thickness and reinforcement in the diaphragm walls. The optimum solution in London was a spacings of 2.5 metres, compared with your 3 metres in Dublin. It is nice to see that consistency of approach in Cementation.

I would like finally to comment on the paper by Adekunte et al, presented today by Douglas Madsen. Bearing in mind all the available information, one could conclude that the wall design and the anchorage design, including the overall stability analysis, were sound and arguably conservative. There is no doubt that the recorded performance of the anchored wall in service confirmed satisfactory overall behaviour. However, given the way that the on-site acceptance and on-site suitability tests have been interpreted, the data could be looked at in a different way. I would just like to describe my approach because during the loading tests going from datum load up to the proof load, the displacements measured ranged from 0.12m to 0.3m. The maximum figure represents 6.7% of the apparent free length. The authors acknowledged in the paper that in accordance with BS8081 the apparent free length should have been between 90% and 110% of the design free length and the paper concluded that there was redundancy in the anchorage system. In fact, my view would be different. I believe, perhaps controversially, that the anchorages should have been rejected. I wish you to think about that in relation to the requirements of BS8081. What was actually happening, compared with the drawing that showed a fixed length and a free length, was that the whole of the anchorage was grouted so there was never a decoupled free length. There should have been to ensure that the load travelled through the free length, beyond the potential rupture plane for overall stability reasons. What actually happened was that the load was picked up immediately behind the wall. In my view and with the benefit of hindsight, the stiffness of the secant wall coupled with a greater self-supporting characteristic of the ground actually worked in nature in favour of the project but I would have analysed the on-site acceptance and suitability tests in a different way.

Douglas Madsen, *Parson Brinckerhoff Ltd*
I certainly support the two-stage grouting process. I do agree with you 100% on that, that after the intitial grouting, the bar should have been tensioned up to say 50% working load, locked off and then a second grouting implemented to lock in the tension, so I totally agree with you there. On the measured displacements, are you talking about the displacements with three cycles and suitability testing?

Stuart Littlejohn, *University of Bradford*
Yes.

Tony Barley, *Single Bore Multiple Anchor Ltd*
Question for Douglas Madsen. What attempts were made to de-bond the actual bar in the free length to make sure the load was transferred to the fixed length. The other query is in relation to the optimum angle of inclination of an anchor to retain a horizontal force. I would say you should start at 30 or 35 degrees because that gives you the maximum horizontal component. When you are getting to 45 degrees you are getting a very large

vertical component of force. So there are two points, (i) how was the tendon de-bonded in the free anchor length and (ii) why did you choose 45 degrees for the inclination?

Douglas Madsen, *Parsons Brinckerhoff Ltd*
45 degrees was chosen by the designer. I would agree with your point and I would tend to go a little flatter. We were the designer for a contractor under the design and build contract. I suspect he was trying to get down into the granite as quickly as possible, rather than try and chase it out to some distance. Given the choice I would prefer to go a little bit flatter so you have got a greater horizontal component. On the other point, I support the two-stage grouting, in this case the contractor went with a very simple 0.4 water/cement ratio grout injected in a single pass that did not give the opportunity lock in the load.

Tony Barley, *Single Bore Multiple Anchor Ltd*
Was there a sheath in the free length? Was the bar de-bonded?

Douglas Madsen, *Parsons Brinckerhoff Ltd*
No.

Horst Aschenbroich, *Con-Tech Systems Ltd*
Question to Douglas Madsen. We have seen the bearing plate. It was just bare steel, nothing was protected. How did you protect the anchor head? Did you just fill in the void with concrete?

Douglas Madsen, *Parsons Brinckerhoff Ltd*
I do not believe there was any protection. It was just a straight steel plate using twin nuts for lock-off. Keep in mind that the basement is now completed. It was two-level basement excavation, so now there are floor slabs to support the wall. I suspect that none of that work will be visible on completion of the building, so the anchor heads are being formed in concrete.

Horst Aschenbroich, *Con-Tech Systems Ltd*
Temporary anchors actually.

Douglas Madsen, *Parsons Brinckerhoff Ltd*
Essentially, it is a temporary situation and the anchor heads are not exposed long term.

Written questions and answers session seven part seven – part 2: case histories of satisfactory performance, shortcomings and failures in service

Stuart Littlejohn, *University of Bradford*
Regarding the paper by Adekunte et al, there is no doubt that the initial anchorage design, including overall stability, was sound and that the observed performance of the anchored wall was entirely satisfactory. However, the performance of the individual ground anchorages should not have been accepted, based on the results of on-site suitability tests.

The apparent free lengths ranged from 0.12 to 0.3m, i.e. 6.7% of design free length. As the authors acknowledge, the BS8081 acceptance criterion states that the apparent free length should lie within the range 90% to 110% of the design free length, so the results recorded confirm a major shortcoming. In essence, the results show that applied load was mobilised behind the wall and not beyond the potential rupture plane.

On site the entire length of bar was fully grouted. Importantly, the designed free length of 4.5m should have been fully decoupled to ensure that the applied load was resisted by the fixed anchor located beyond the potential rupture plane.

With the benefit of hindsight, it is considered fortunate that the retaining wall was stiff and that the retained ground was more self-supporting than perhaps assumed.

Douglas Madsen, *Parsons Brinckerhoff Ltd*

We agree that the 4.5m design free length should have been fully decoupled from the 7.0m fixed anchor length (i.e. two-stage grouting). This would have resolved the problem of short apparent free lengths. The decision appears to have been one of construction practicality by the Contractor.

The ground anchorage was provided with a 63mm Dywidag gewi-steel thread bar (1760 kN yield) and pre-stressed to 50% of the 800 kN design safe working load. The measured 2.3mm maximum wall deflection (versus 21mm predicted) for a 7m maximum wall height confirms that the anchorage system was quite robust. Dublin Boulder Clay typically behaves as an undrained material and stands unsupported, and modelling software can over-estimate earth pressures behind walls in Boulder Clay (Ivor et al: 1996). The retained ground on this project probably had greater self-supporting ability than should prudently be assumed by the designer.

References:
Ivor et al. (1996), "The geotechnical aspects of the deep basement for the Jervis Street Shopping Centre". Paper presented to a joint meeting of the civil division of the Institution of Engineers of Ireland and the Geotechnical Society of Ireland, 2 December 1996.

Recommendations and standards for inspection, monitoring and repair:
discussion session eight

Rapporteur's summary
by Stuart Littlejohn, *University of Bradford*

National and international standards provide various levels of guidance on the extent and frequency of monitoring the condition and service behaviour of permanent ground anchorages. In spite of these well-established publications, little attention has been paid in practice to routine inspections and monitoring. In addition, for many existing anchorages, as-built construction records covering design, corrosion protection, installation and stressing are missing or difficult to access as elements are located in different company offices.

The Japanese Manual "Ground anchor inspection, integrity investigation and remedial measures" (described by Miyatake et al) attempts to overcome these historical shortcomings in practice by raising the awareness of owners, particularly those responsible for Public Works, of the importance of maintenance of both individual anchorages and anchored structures as whole.

The requirements are specified in detail but by way of summary, a preliminary survey is required initially that involves a collection of drawings, specifications and maintenance records, where available. It is important to establish if the anchorages are "old", i.e. where the corrosion protection is likely to be insufficient. An initial inspection follows covering the visual condition of the outer head and bearing plate or plinth for all anchorages. Thereafter, a routine "ordinary" inspection is planned and categorised as daily, periodical and emergency.

Daily inspection involves visual observations of the outer anchor heads, bearing plates and ancillary components for any abnormality. Periodical inspection comprises visual examination of all anchorages and close inspection by measurement of say 10% of all anchorages, e.g. by "hammering" (impact acoustics method). A different 10% is examined each year to better understand in time the condition of all anchorages at the site. Emergency visual inspections are carried out immediately after extraordinary events such as earthquakes or typhoons.

Where inspections record any abnormality, an integrity investigation of those individual anchorages is carried out. As described by Yamazaki et al, this involves initially an exposure investigation that includes examination of the condition of the anchor cap, bearing plate, locking unit, tendon and the corrosion inhibiting compound, including efficiency of filling. Within the manual, the scope and content of this type of investigation is described in detail and it is applied to all anchorages demonstrating any abnormality plus 20% of the rest. Load lift-off tests are then applied to the anchorages with an abnormality plus 10% of the rest to determine the residual load and check for compliance with design acceptance criteria. Bearing in mind the known susceptibility of the under head to corrosion, all anchorages exhibiting any abnormality plus 5% of the rest are examined for corrosion and efficiency of grease filling.

Finally, a performance confirmation test can be carried out via incremental loading-unloading cycles up to a proof load, in order to confirm consistent reproducible behaviour, the apparent free length and a margin of safety. Where, the condition or performance of the anchorage is considered inadequate or will become inadequate within the remaining design service life, various options are considered, ranging from removal of the causes to replacement of the anchorages.

It is anticipated that the new Japanese Manual will lead to a widespread improvement in the maintenance of ground anchorages in Japan and potentially extend the service life of existing anchorages by confirming satisfactory condition and behaviour, failing which by the timely implementation of appropriate remedial measures. The Public Works Research Institute and the Japan Anchor Association are to be congratulated for taking the initiative to provide practical guidelines for the benefit of the construction industry and its clients.

Although standards such as BS8081 and EN1537 require a plastic sheath around a strand to be filled completely with a flexible corrosion resistant compound, the paper by Barley and Mothersille, highlights that where greased and sheathed strands are dissected for examination, the efficiency of grease infilling can vary from 0% to 100% in practice. This is an important finding as inefficient greasing can lead to ingress of moisture and even permit the sheath to act as a drainage path.

As a consequence, the processes by which greased and sheathed strand is manufactured are reviewed in the paper. Grease can be applied solely externally or involve internal greasing to fill all the interstices between the wires as well as external greasing. Thereafter, the greased strand can be coated in plastic by heated extrusion followed by cold water immersion to create a high density plastic sheath of a specified diameter and wall thickness. Alternatively, greased strand can be inserted into a preformed plastic sheath, preferably with a leading slug of grease or oil to lubricate the bore of the sheath. In this process, the viscosity of the grease and rate of insertion are critical parameters if the annulus (0.4-0.5mm) between the strand and the sheath is to be filled efficiently.

Simple tests on one metre samples from three manufacturers have confirmed that where specialised equipment and factory-controlled conditions were employed the efficiency of grease infilling ranged from 76% to 100%. On the other hand, tests from one site confirmed results as low as 0%.

A minimal acceptance criterion of 80% for efficiency of grease infilling is proposed and it is recommended that sheathed strand suppliers in future should provide factual data confirming compliance with this recommendation that can also be checked in the field.

The paper by Bruce and Wolfhope describes the key developments in the scope and content of successive North American recommendations over the period 1974 to 2004. In general, the developments highlight the increasing importance attached to testing procedures and corrosion protection, and latterly a supplement covering the specification and use of epoxy-coated strand.

Of particular importance in the planning of anchorage works is the recognition of a range of specific task and responsibilities that have to be allocated for any project. Thereafter, the current recommendations cover the essential elements of anchorage technology, namely, materials, design, corrosion protection, construction and stressing/testing.

It is encouraging to observe that all the components of the anchorage are now addressed under materials and in construction the early watertightness criterion expressed as a flow rate per inch of hole diameter per foot of depth has been replaced sensibly by a standard 10.3 litres in 10 minutes at an excess pressure of 0.035MPa for the entire hole, in order to avoid grout loss.

Bearing in mind the conference theme, the monitoring of service behaviour is also addressed, i.e. typically 3-10% of the anchorages (or more if desired) are recommended for monitoring by load cells or lift-off tests. In general, initial monitoring is at 1 to 3-month intervals, with later monitoring at intervals of not more than 2 years. In spite of these recommendations, there remains a dearth of published information on service behaviour.

Given the importance of corrosion protection, the paper describes in considerable detail how the recommendations have evolved since 1974. It is interesting to note that initially for two-stage grouted tendons, it was judged that sheathing could be omitted on the basis that cement grout alone was acceptable, in spite of the known integrity problems relating to cracks and bleed pockets for neat cement grouts with a water/cement ratio of 0.4-0.5 by weight.

In current North American practice, several options are permitted for permanent anchorages where the service life is 2 years or more. For aggressive conditions or where the aggressivity has not been quantified and where the consequences of failure are serious (even in non-aggressive conditions), a Class 1 encapsulated tendon protection is recommended. For this class, the tendon bond length can comprise a cement grout-filled encapsulation or an epoxy-coated strand tendon in a successfully water pressure tested drill hole. In the free stressing length, options include (i) a corrosion inhibiting compound-filled sheath encased in grout, (ii) a grout-filled sheath, or (iii) a grout encased epoxy coated strand in a successfully water pressure tested drill hole. In North America, Class 1 protection is often referred to as double corrosion protection, where the grout is regarded as one of the physical barriers.

Class II protection is advocated for permanent anchorages where the conditions are non-aggressive and the consequences of failure are not serious, unless the additional in-place cost of Class I protection is inexpensive. For Class II, the tendon bond length can comprise a tendon encased in cement grout, combined with a free length option such as a corrosion inhibiting compound-filled sheath encased in grout. Where cement grout is used to protect the tendon bond length, the installation method should ensure that the grout will remain fully around the tendon, but no guidance is provided in the recommendations on how this achieved in rocks or soils. Class II grout protected tendon is often designated as single corrosion protection in North America.

Overall, it may be observed that the recommended practice in North America is now much closer to that evolved historically in Europe. This augers well for the future, as recommendations for ground anchorages are further developed via international bodies such as the Fédération Internationale du Beton (FIB), bearing in mind the need for consensus.

Stuart Littlejohn, *University of Bradford*
I think we should take up that challenge from John Wolfhope to review the condition and assess the risk associated with our post-tensioned dams. I mentioned in my introduction to the conference that dams had been post-tensioned since 1934 and right through the 40s and 50s post-tensioned dams spread across Europe. French engineers were also extremely active in North Africa, the British in Southern Africa and India. Due to the large number of post-tensioned dams, it would be worthwhile as a starting point looking at all those installed before the 1970s because they may be the ones that are starting to worry us now. It is a challenge that we should address.

Mike Turner, *Applied Geotechnical Engineering Ltd*
I wish to congratulate our two Japanese contributors. Once again the Japanese are ahead of us and what they are proposing is acceptance testing related to the integrity and maintenance of anchors in service, just what we should be doing and should have done a long time ago.

Stuart Littlejohn, *University of Bradford*
Bearing in mind that the new manual is in Japanese, I would like to know if the Japan Anchor Association plans an English version because that would then have much more influence around the world?

Hiroaki Miyatake, *Japan Anchor Association*
We are now planning to publish the Japanese version after which we will start a translation in English. Hiroaki Kubo, the chairman of the publishing committee of the Japan Anchor Association hopes that an English version will be available by December 2008.

Gareth John, *CAPCIS Ltd*
Question for Tony Barley. You demonstrated quite neatly that not all free length strands are greased and it has been a major factor. Have you ever done any tests to establish the efficacy of the grease itself, i.e. how good is the grease in terms of corrosion inhibition and how stable it is over time?

Tony Barley, *Single Bore Multiple Anchor Ltd*
I have been involved in codes and general recommendations and I have done no research into the suitability of the material. I am just trying to establish the point that you need the correct viscosity to get the grease into the annulus, but it is vital to use the right grease as well, whether it is a grease, Vaseline or wax. Whatever it is, the sheaths must be adequately filled, so that they are fully effective as a protective barrier.

Gareth John, *CAPCIS Ltd*
I do know there are ASTM and other standards that relate to the protective behaviour of grease as an inhibitive compound. It would be worth putting these forward and including a suitability test or an efficacy test in relation to these existing standards when specifying the materials, in order to set a minimum basis so you end up with (a) grease in, and (b) grease that works.

Stuart Littlejohn, *University of Bradford*
If you go back to DD81 in 1982 and subsequently BS8081 in 1989, we did in fact investigate the various types of grease that were available, looked at all the standards, and in fact only permitted three types of corrosion inhibiting grease that complied with the required properties. However, with the passage of time and the development of waxes, the recommendations should be reviewed again. We are aware of the ASTM standards but we should be talking to corrosion specialists regularly to get more up-to-date and better advice on what is now available for both corrosion protection and decoupling materials. In addition to specifying what you require in terms of how the grease infilling is carried out and what percentage of infilling is achieved, manufacturers should also record the properties of the grease and the plastics, together with their dimensions, in order to provide a complete package for use by designers.

Chris Irvin, *DYWIDAG-Systems International*
That is a good comment. Following Tony Barley's paper, where you are attempting to push the grease strand into the sheathing, once you get beyond 60 or 70 metres, it becomes incredibly difficult due to friction, even for greased strand which has the interstices or the strand opened up to get the grease around the king wire. For HDPE sheathed strand, the plastic sheathing is extruded on to the greased strand at time of manufacture. There are some very specialist processes that need to be adopted if you use this strand and then de-grease it over the bond length. There have been pullout failures in the past due to strand that was incorrectly de-greased, so you have to be very careful. A further comment relates to good practice, where for example in the UK, the bond length of the strand anchor is often pre-grouted in the factory. When it is handled on site it is extremely important to lift this up correctly. If you have a heavy bond length with an ungrouted free length, you can get a very sharp point or kink and you need to use a spreader beam to avoid damage to the sheathing. Finally on good practice, as soon as the strand tendon is installed in the hole it should be grouted, i.e. both the primary grout around the borehole and the secondary grout within the free length. Quite often you have to do the two stages together in deep anchors to avoid collapse of the sheathing. Once the secondary grout is placed in the free length, it basically will stop water getting in. Then, you should also seal the anchor straight away.

Horst Aschenbroich, *Con-Tech Systems Ltd*
In the United States the extrusion process for strand is well advanced. We do not de-grease the bond lengths and as the HDPE is extruded continuously over the whole pack of strand, e.g. 8,600 feet, the grease (particularly wax) is turned on and off. It is very difficult to do this accurately so we allow additional lengths either way un-greased and greased. When we off-coil and strip the sheath off the bond lengths and we do not see grease where it is supposed to be (actually we can see from the outside as soon as you feel the lay of the strand through the sheath you know there is no grease or very little grease). So we always cut it back to where we do not see the lay of the strand and if we have not sufficient grease there we go a little bit further because we leave enough tolerance on either end to accommodate that. We use both methods, i.e. the "stuffing method" or the "extrusion method". Completely filling the sheath is impossible even if it was filled originally, you haul the anchor, you squeeze that sheath and you squeeze some of that grease out so you never can guarantee a 100% filled sheath. If you go to extrusion, there is a danger that the sheath will be touching the strand and there will be no grease at that point, even if you do not see the lay of the strand and also at that location there is no cushion underneath, so it is susceptible to damage during coiling or uncoiling. Therefore, it is a Catch 22 situation but we test specifically the "stuffed strand" and measure the amount of grease with the water testing method.

Tony Barley, *Single Bore Multiple Anchor Ltd*
If you read the full paper it does talk about the shortcomings and the fact that the strand will not lie in the centre. These points are all raised but we are trying to raise the principle having encountered very poor quality products. We are looking to the specialists to guide us, perhaps it is not your products that are defective but we need to ensure that all products used are appropriate and fit for the purpose. Perhaps, it is none of the companies here, or the better suppliers in the UK or in the States that have a problem, but it is important to make sure that the product that is provided to a contractor (who does not have a vast amount of experience in checking the effectiveness of the corrosion protection) is fit for the purpose, whether it is fully greased and sheathed or not.

Ian Clough
Question for John Wolfhope or Donald Bruce. Have you ever considered in the secondary grouting the use of a mixture of clay and cement or bentonite and cement to cut down on the bleed. It is used in cut-off trenches but admittedly they are about 60cm wide, so it is rather a different situation

Stuart Littlejohn, *University of Bradford*
We used bentonite/cement grout in two-stage grouting, particularly in the late 1960s because bentonite was a very attractive filler to make a cheaper grout. The secondary grout in the free length does not have to be that strong and also bentonite was very good at inhibiting bleed. It is interesting how practice that was quite popular two or three decades ago has lapsed.

Donald Bruce, *Geosystems LP*
We systematically do not use bentonite anywhere near ground anchorages of any sort, except sometimes in post grouting (tube à manchette) work to give stability to higher water/cement ratio grouts. By specifying minimal bleed, you rely on the contractors to produce an appropriate grout using high speed, high shear mixing with low water/cement ratios. In hot conditions you can use a super plasticiser and/or a retarder. We try to control bleed using simple mixes; this helps the contractor on site.

I would also like to repeat John Wolfhope's challenge to all the different nations that are represented here. There are some lessons (bearing in mind that this is going to appear in the conference proceedings) on how to do these national research projects. First of all the key thing is that you need a leader. You need a champion who is interested in doing this and is willing to take up the gauntlet. Secondly, even with a leader supported by his company, you have to wrap the whole study inside an agency that gives its weight, independence and respectability. This agency should be a trade association, or a university, or a professional society. If you try and do it alone as a contractor, for example, you will not gain credibility for the study. The third lesson is that you need money; national studies cost money and you will be looking for at least £20,000 or £30,000 to get a good first phase attack on this work. The fourth thing you need is absolute openness between all the different parties comprising the anchor industry and that means owners, consultants, contractors, and the suppliers of the post-tensioning equipment and materials. Between all of them, they have the whole picture whereas any one group does not have on its own.
Finally, publish early and publish often, even if the study is incomplete. Such papers attract attention amongst interested practitioners who can help "fill the gaps" in the database.

Alisdair Henderson, *Ritchies*
To jump back to bentonite grout, I think the reason that we do not use it for grouting in the free length now is that for multi strand anchors that we are grouting, the primary grouting and secondary grouting are carried out simultaneously, as we wish to maintain the same grout from the same machine and inject at the same time.

John Finnis, *DYWIDAG-Systems International*
Since grease is essentially an engineering lubricant, this is not what we are looking for in the corrosion protection compound around the strand. The standard material used these days is a petrolatum product. Generally speaking, we use the word grease. As Gareth mentioned yesterday water soluble greases can also be a problem, so whilst the word grease is bandied around and used in manuals, we should be referring to either corrosion inhibiting compounds or petrolatum products.

Tom Richards, *Nicholson Construction Co.*
I believe from the Gilboa Dam presentation and from John Wolfhope's presentation, delegates may have the impression that in the US we water test holes and corrugated sheaths routinely. It is my experience that this only occurs on dam anchors and I would like to know if Donald and John have a different opinion.

Donald Bruce, *Geosystems LP*
We have been focusing on our dam anchor practice in the presentations. On the one hand, there has not been a single dam anchor project that I have been involved with that does not have routine water pressure testing and pre-grouting and re-drilling. On the other hand, there are very few other applications of rock anchors, even permanent ones, where I have seen systematic water testing and pre-grouting. Your point is well taken, Tom.

Stuart Littlejohn, *University of Bradford*
Devonport is a classic example where there is an alternative very important type of structure, namely, a nuclear licensed naval dockyard. For this application, you carry out water testing and pre-grouting, just to make sure you have reduced the permeability of the rock mass as low as possible, because of the importance of such structures that are not dissimilar from post-tensioned dams when you consider the high retaining walls tied down to provide stability against overturning and seismic effects.

Raymond Coe, *Black & Veatch*
With reference to the question of having some form of reference or guidance document on physical inspection and service behaviour monitoring, do we wish to take this up through the UK Construction Industry Research and Information Association (CIRIA). If it is something we wish to pursue with them how does this conference wish that to be done because once we have all gone away we will lose the impetus.

Stuart Littlejohn, *University of Bradford*
Given these comments, I will take up the essence of the procedure (the way forward) that Donald has just described and I will write to CIRIA on our behalf and see if we can agree a way forward.

Raymond Coe, *Black & Veatch*
If we do not take some initiative out of this conference we will lose it and historically CIRIA documents, I think most UK people would agree, have been very valuable.

Edward O'Connor, *USB International*
Just a general comment. At the end of the day corrosion is an electro chemical process and a suitably chosen grease acting as a corrosion inhibiting compound can actually prevent the transfer of electrons. When Tony Barley mentioned the importance of viscosity, the suppliers will have no problem in giving you the right grease if you specify it. In fact, they can do a drop test in the laboratory guaranteeing a particular viscosity for a particular product. However, I would suggest that it is a bit late if you are doing the tests in the field; they should be done on factory sample tests to meet your particular requirements. It comes back to the whole chemical process, the classification and the type of soil that you put your anchor into. There could be copper in the soil so you could have dissimilar metals coming in contact; there is a beautiful combination for using electrons.

Stuart Littlejohn, *University of Bradford*
In terms of grease better referred to as corrosion inhibiting compound) there should be factory controls to give a quality assurance (QA/QC) approach but that does not mean that the contractor, or owner or consultant cannot double check the quality of the product actually delivered in the field.

Closing remarks by Stuart Littlejohn, *University of Bradford*
I would like to make a few closing remarks but I have been struggling to review the outcome of our detailed deliberations on the performance of anchorages and anchored structures in service. In a rather light-hearted way, I have picked up two phrases that perhaps reflect the good and the bad aspects of what we have found regarding the condition and performance of anchorages in service. It was Mike Turner who introduced the phrase 'Aha' and then Donald Bruce gave us 'Oh no', so I can now imagine on the one hand the pleasant category 'Aha', where anchorages are performing well and not corroding and then the pessimistic 'Oh no' category, where we encounter severe problems that have to be tackled – such is the variety of conditions that are now being encountered on anchorages in service.

Bearing in mind the popularity of multi-strand tendons where most of our strands comprise seven wires, I use the seven wires just to remind myself of the key elements of ground anchorage technology in practice. These are (i) design, (ii) construction, (iii) testing, (iv) corrosion protection, (v) physical inspection, (vi) service behaviour monitoring and (vii) maintenance. I believe that design, construction and testing procedures have been well established for many years and more recently there has been a growing awareness, perhaps over the last decade in particular, of the importance of corrosion protection with improved requirements and associated standards. However, for reasons that I do not quite understand, there appears to be a general reluctance by owners, including government departments, and the Environment Agency here in Britain, to focus or spend money on mitigation measures. As a result of this conference, I hope you are now alert to the fact that there is insufficient attention being paid to inspection monitoring and maintenance. As anchorages in service grow older there is a need for more effort to address the problem and ensure that practice improves in the future and, hopefully, we can follow the Japanese example.

In closing this conference I hope you have enjoyed the surroundings of the world's oldest engineering institution, meeting old colleagues and making new friends. I wish to thank the sponsors of the conference for their contribution, good will and generosity; the ICE conference office, notably Gemma Doyle, Abbey Clew, Tracy Lebrun and Helen Taylor who have helped to organise the conference and have been serving you at all times; the members of the organising committee and the technical advisory group - they deserve a special thanks from me personally because all their work has been given on a voluntary basis; last, but not least, you the delegates for your participation over the last two days and your excellent company. I now wish you all a very safe journey home and I hope that some of my younger colleagues will organise the next conference when they think it is appropriate. Many thanks indeed.

Donald Bruce, *Geosystems, LP*
We have heard once or twice over the past couple of days that Stuart is thinking about retiring (I will believe that when I see it). I think that we should not let him go today without a special recognition. Many of us here have been directly his students, such as Tony, Devon and myself, but virtually all of us are in fact his disciples. I would go further and say that we also belong to your legion of fans throughout the world. We would like to thank you for

three things in particular and would request that these impromptu remarks are included in the final conference proceedings as a record of the sentiments of the attendees. First, we would like to thank you for developing and sharing your technical knowledge and expertise. We would also like to thank you for your integrity in all our personal and professional dealings; this is one integrity test that has past the test of time. Finally, we would like to thank you for your visionary leadership in the anchor industry, during times which have not always been plain sailing and without controversy. So please stand and give Stuart the recognition he deserves so well. Thank you, Professor Littlejohn.

Stuart Littlejohn, *University of Bradford*
I am overwhelmed by the warmth of your response that leaves me rather lost for words. Thank you Donald for these thoughtful and kind remarks that are very much appreciated.

Written questions and answers session eight: recommendations and standards for inspection, monitoring and repair

Stuart Littlejohn, *University of Bradford*
Bearing in mind that sheath wall thicknesses as low as 0.5mm are mentioned in the paper that would not be acceptable to BS8081 or EN1537, it will also be beneficial if the manufacturers of greased and sheathed strand were to specify their manufacturing procedures, including the materials used, e.g. type of plastic and dimensions of sheathing, together with type and properties of the grease or wax.

Tony Barley, *Single Bore Multiple Anchor Ltd*
The thickness of the strand sheaths that were subjected to the investigation work were 1.1, 1.2 and 1.25 mm, respectively. These satisfy the minimum thickness requirements of the majority of codes. Strand sheaths with wall thicknesses of as little as 0.5mm have only been observed in overseas anchorage works that were for temporary use. However, it is suspected that some of these thin sheaths would also have been used in free anchor lengths of permanent anchorages.

It is agreed that the suppliers of prestressing strand for use in the anchorage industry should offer quality products for both the sheathing and the corrosion inhibiting compound used to fill the annulus. As mentioned in the response to Lee Jordan above, the authors recommend that the essential developments in quality sheathing and corrosion inhibiting compounds for use in the prestressing industry should be considered for use in the anchorage industry, where appropriate.

Stuart Littlejohn, *University of Bradford*
Re paper by Barley and Mothersille
Bearing in mind that sheath wall thicknesses as low as 0.5mm are mentioned in the paper that would not be acceptable to BS8081 or EN1537, it will also be beneficial if the manufacturers of greased and sheathed strand were to specify their manufacturing procedures, including the materials used, e.g. type of plastic and dimensions of sheathing, together with type and properties of the grease or wax.

Thomas D. Richards, Jr, *Nicholson Construction Company*
Have the authors seen water leakage between the sheath and strand with extruded sheathing?

Did the leakage shown occur during grouting? And continue after grout hardened?

What was the grease filling percentage when water leaking observed?

Tony Barley, *Single Bore Multiple Anchor Ltd*
The authors have not seen any water emission direct from the sheaths which have been extruded on to strands.

The authors did not have the opportunity to inspect the sheath placement or greasing method utilised in the overseas fabrication of the leaking strand. The strands were certainly observed to be leaking after the completion of grout injection and setting.
The strands which exhibited grout within the sheaths were fabricated using a poorly controlled greasing and strand pushing system which used no thermostatically controlled method to maintain the correct grease viscosity. The system was however improved by the tendon fabricator upon receipt of written complaints from the authors.

Lee Jordan, *Arup Manchester*
The sight of water being emitted from inside the strands of freshly installed anchorages, as seen in Tony Barleys paper should be very concerning to the industry. We have seen similar examples ourselves and were indeed very concerned.

The paper confirms that at present, the strand manufacturers can not readily comply with the requirements of the BS8081, EN1537 and FIP in respect of grease content and this will need to be addressed or otherwise strands displaying evidence of voiding should continue to be rejected from projects.

Arup believe, from experience, now backed up by the testing by Tony Barley, that strands manufactured using the ‘push in’ method have a high risk of being incompletely filled with grease and therefore have a high susceptibility to water ingress. If water is allowed to enter the top of strands fabricated using such a method, for example by rainfall infiltration, there is a possibility it will track along this interface until it meets a full annulus of grease. This could be at any point in the anchor, possibly in the stressed length. Once it enters the strand there is no method of which we are aware, using current technology, of tracking its migration, nor any method of removing it. If water is held in this moist air pocket it is possible to attack the strand and result in pitting corrosion, which could cause the strand to fail prematurely.

The purpose of corrosion protection is to prevent a moist gaseous environment from developing around the tendon. These examples of water in the strand are clear breaches of this. Once it is agreed that there is a susceptibility of water ingress via the tops of the strands, the number of corrosion protection layers around the anchor become irrelevant as this appears to be the primary risk. What does become critical is the protection of the tops of the strands from water ingress in between fabrication and final placement of head protection.

It should also be noted that electrical testing is not necessarily the solution to this problem since if water has entered the top of the strand by rainfall, the anchor could still be isolated from the ground and therefore pass the test.

It is concerning that the anchor industry has taken so long to identify this as a potential problem and the many photographs of unprotected strands hanging out of anchor installations before head placement, during the conference and on trade literature, should be concerning to the owners of these installations who may be relying on there being dry weather at time of installation as their primary corrosion protection barrier.

Does the author concur that additional work is required on the development of greasing techniques and grease types for use in anchor installations to ensure a fully greased strand. Also, in the knowledge of these findings, should any critical structures already supported by anchors manufactured by less reliable methods be subject to more frequent monitoring, unless rainfall records are available from the installation period?

Tony Barley, *Single Bore Multiple Anchor Ltd*
Beyond the intent of the paper to demand production of quality greased and sheathed tendon manufacture, Lee Jordan's comments are of great value and draw attention to several points of weakness in anchorage construction technologies that are rarely fully addressed. Not only should the delivered product be fit for the purpose, the construction operations must ensure that the quality of the protection system is fully maintained prior to, during, and after installation. Part of the workmanship problem stems from the vastly different expertise required from operatives carrying out anchorage installation and stressing.

It is rare that even the most specialist anchorage contractors have foremen or supervisors fully acquainted with all the high demands of drilling, grouting, stressing and corrosion protection technology. Thus, in the majority of cases the "stressing team" is a separate entity from the "installation team" and this discontinuity can result in the application of a differing rationale in attention to critical phases in the protection of tendons. For example, the strands (or bars) may be left unprotected, simply projecting out of the structure after installation. One reasonable precaution is the fitment of brightly coloured tape or ducts over the protruding tendons, in order to reduce the risk of other contractors' operatives causing damage. Furthermore, it is then a logical step to fit end caps over each strand (or bar) to prevent both the ingress of water and the egress of grease which may become fluid in the event of direct exposure to the sun. Physical protection measures should also be fitted at points where the flexible strands leave the structure such that the sheaths cannot be subjected to abrasion damage, particularly from wave action in a marine environment.

It is recommended that each stage of the subsequent operations also requires close attention. For example:

- Does the protective sheath terminate below or pass through the bearing plate?
- Should the sheath be in contact with the base of the stressing head?
- Will the sheath crimp and be damaged on proof loading?
- Is it reasonable to expect the strand sheath end to remain in the same place during load cycles?
- When is it most appropriate to apply the underhead protection? Can its performance be demonstrated in advance?
- Is it better to remove all the grease from the strand above the sheath or leave a smearing on the strands? The authors recommend leaving a smearing on the strand to provide some isolation from immediate corrosion.
- After acceptance testing, what protection should be offered prior to the cap fitting? The authors recommend refitting of the removed greased sheath complete with end cap.
- How long should the temporary condition exist, with particular attention to severity of weather?
- In the case of marine anchorages, is it necessary to address wave wash action and evolve a more sophisticated approach to temporary protection.

With reference to the remaining queries from Lee Jordan, we agree that additional investigation work may be required with regard to the choice of "grease" or more appropriately "corrosion inhibiting compounds". This is particularly relevant since many of the suppliers of strand tendons to the anchorage industry also supply to the prestressing industry for use in structures. It is acknowledged that up to about 1995 the structural prestressing industry left many tendons protected solely by cement grout of variable quality. As a consequence, some structural failures occurred and some structures required demolition. With the enormous advancement in corrosion protection technologies applied to prestressing strand for structural works, it is appropriate that the prestressing industry should ensure the provision of updated guidelines with regard to the types of corrosion inhibiting compound, including wax, for use in the ground anchorage industry.

With regard to structures that are supported by "old" anchorages that may have limited protection against corrosion, attention should be paid initially to anchorages installed prior to the demands for corrosion protection introduced by the draft anchorage standard BS DD81 in 1982. Anchorages installed in a marine environment may be particularly vulnerable to degradation of the steel components and the corrosion protection. Although investigation works have recently been carried out by authorities such as the Aberdeen Harbour Board and Glasgow City Council (River Clyde) there are over 30 coastal towns and harbours with old anchored structures where the condition and service performance of the anchorages should be investigated (Barley, 1992).

Barley A.D. (1992). The use of ground anchorages in Waterfront Structures. Presented at the International Shipping and Marine Technology Market with Congress, Hamburg 29th September to 3rd October 1992.

Lucky Wehalle, *Environment Agency*
(1) Are there recent advances made with regard to locating accurately the actual position of ground anchors embedded in the ground?. This is important for subsequently installing piles, diaphragm walls, or other substructures within the area between the retaining wall and the landward end of the ground anchors without damaging the integrity of the ground anchors.

If such techniques are not available then please advise me what global angle of deviation need to be allowed for the position of the anchors in order to define an exclusion zone within which no substructure works like piling can take place. For a site at Erith, Kent on the banks of the River Thames an angle of deviation of 2.5 degrees from the cone axis was taken as the angle of the cone of deviation with the ground anchor head as the apex of the cone. However this angle of 2.5 degrees was found to be insufficient as three anchors were ruptured during the piling operations. I shall be pleased to receive any suggestions regarding the angle of the cone of deviation for future substructure works in the vicinity of ground anchors?. As you are no doubt aware there are still several sites with long ground anchors adjacent to the Thames where it is important to locate the actual position of the ground anchors accurately prior to development of the site. The 2.5 degree was plucked out from the air in consultation with the developer's representative and I considered this was a reasonable assumption.

(2) Please advise me on the optimum levels of monitoring for the ground anchors supporting a retaining wall when substructure works are carried out in the vicinity of ground anchors? The inspection and monitoring of the ground anchors involved visual inspection & photographic record of the anchor head and check lifting of the ground anchors. I have placed conditions when granting approval for works in rivers requiring that ground anchors are monitored (a) before starting any substructure works to establish base readings (b) immediately after any substructure works (c) 3 months after completion of the substructure works. Some selected anchors (which indicated widely varying loads during the monitoring phases a,b and c above) were required to be further tested after 6 months and 12 months for a fewer. The retaining walls were required to be monitored daily while piling was in progress then at weekly intervals while the works were progressing and at 6 monthly intervals after completion of the substructure works for 2 years.

Is the above monitoring regime about right? over the top or not enough as far as the ground anchors were concerned? Or was the retaining wall monitoring about right?

Regarding Professor Littlejohn's remarks in the closing speech in relation to the lack of monitoring of ground anchors by the Environment Agency I wish to clarify the situation as follows:
** Ground Anchors were monitored at several sites e.g.** North Woolwich near the Woolwich ferry / ** Erith (near the Morrisons Supermarket) and ** Woolwich Arsenal area in the mid--1980's with check lifts and the loads released down to design loads where these were greater than the designed loads.
** At present, generally every 6 months, visual inspections of the anchor heads and any visible supporting brackets, anchor caps, walings and the adjoining sheet piles are carried out. This inspection is part of the routine when inspecting the condition of flood defence structures along the tidal river Thames and its tidal tributaries.
** When a site is redeveloped extensive monitoring including check lifting as in (2) above is carried out e.g Chandlers Wharf Erith (1999/ 2001) and Royal Arsenal DLR Woolwich Extension (2005/06), Manor Way Erith (1990's).

However there is no programme for routine check lifting of ground anchors or inspection for corrosion of the anchor head etc where there is no new development. This needs to be addressed.

(3) I shall try to improve this situation by proposing the following measures are these ADEQUATE?

** visual survey of the retaining wall every 6 months for movement --done at present
** removal of the caps and inspecting the condition of the anchor pad, bearing plate every 5 years
** checking whether the wedges in strand anchors or nuts in the Mcalloy bars are in good condition within the anchor head every 5 years. Any corrosion to be removed and corrosion resisting paint applied.
** cleaning out and refilling caps with grease where required,
** check lifting selected representative number of anchors every 10 years for up to 10% of the anchors.
** excavation behind the retaining wall to check for corrosion below the anchor bearing plate and of the anchor free length at selected locations
** investigate presence of voids below anchor head plates by probing (is endoscope probing necessary?)

Stuart Littlejohn, *University of Bradford*
Locating the actual position of ground anchorages
With reference to the accurate location and geometry of buried ground anchorage tendons by remote methods, I am not aware of any relevant case history. Ground penetrating radar is worthy of consideration. However, if the water is saline, the signal will be highly attenuated and may not penetrate more than 5-6m. It may also be necessary to use a lower frequency antenna (e.g. 500 -100MHz), in order yield a lower resolution. For a more informed opinion on the application of radar, contact should be made with Professor Michael Forde or Dr Antonis Giannopoulos at the University of Edinburgh.

Where saltwater conditions are present, the electromagnetic induction method may be applicable. It will also be appropriate to investigate down-the-hole geophysical tools where boreholes are drilled within 0.5 to 1m of a representative number of ground anchorages, in order to establish the range of tendon geometries for the project. The Journal of Applied Geophysics will be a useful first contact point to locate experts in this field, in order to evaluate the potential of current geophysical techniques.

Where such techniques are not available, guidelines on initial alignment and drill hole deviation are provided in sub-section 10.2.4 "Hole geometry" and Figure 28 of BS8081 (1989), so the $\pm 2.5^0$ should not have been plucked out of the air for application at Erith. The global angle of deviation should take account of the maximum acceptable alignment (set-up error) of $\pm 2.5^0$ at the entry point of the anchorage plus a typical acceptable drill hole deviation of 1 in 30. Thus, in the absence of recorded tolerances on an anchorage project, a deviation angle of $\pm 4.5^0$ would be considered reasonable. In current practice, ground anchorage specifications should require that the alignment and drill hole deviation be measured and recorded during construction, so that the range of global angles of deviation are known for a particular project.

Optimum levels of monitoring ground anchorages in close vicinity to sub-structure works
Where substructure works are carried out that could influence adjacent ground anchorages tying back a retaining wall, the conditions described for granting approval are basically sound, particularly during the period of the works.

On completion of the works, the original service loads of all anchorages should be restored where variations greater than ±10% design working load are recorded. Given a record of locked-off service loads on completion of the project, the loads should be checked after 3 and

6 months to confirm via three sets of readings that the service loads have stabilised, bearing in mind load fluctuations due to temperature and tidal conditions, if applicable. As part of the forward monitoring strategy, acceptance criteria for magnitude and rate of load loss should be established as part of the policy (Section 11.5 of BS8081).

In the event of untoward readings, e.g. significant load loss with time, the situation should be re-assessed by restoring the required service loads and extending the period of monitoring for a further 3 and 6 months to confirm a stable situation. Overall, and bearing in mind that the monitoring is not focused on corrosion performance, I would seek to approve the performance of the anchorages within 6 to 12 months, rather than specify at the outset, readings at 6 monthly intervals for two years, but this is a matter of individual judgement.

Lack of inspection and monitoring of ground anchorages
With reference to the closing remarks concerning the lack of inspection and monitoring of anchored structures by owners and government departments, including the Environment Agency (EA), here in the UK, Ms Wehalle (EA) has clarified the situation and in so doing has highlighted that the Environment Agency carries out routinely a visual inspection of the outer covers protecting the anchor heads, any visible brackets, walings and the adjoining sheet piles, generally every six months.

These inspections are valuable in recording external mechanical damage, general external corrosion and in severe cases holing of the sheet piling, but such observations provide no insight into the physical condition of (i) the locking unit, e.g. stressing head with wedges for strand tendon or nut for a bar, (ii) the steel tendon at and beneath the locking unit, and (iii) the service load of the individual anchorages. Based on the information provided transparently by Ms Wehalle, routine load lift-off checks have not been carried out for over 20 years. The only exceptions to the current six-monthly visual surveys occur where a specific site is to be redeveloped.

In other words, there is no routine inspection of the anchor head components or load monitoring, in spite of the importance of the anchorages in maintaining the overall stability of the river walls. This highlights a fundamental flaw in the asset management system that has been implemented historically for the exposed river wall structure in that there is no routine inspection of the critical components of the ground anchorages that are hidden from general view.

For critical infrastructure such as ground anchored retaining walls along the tidal river Thames and its tidal tributaries, it is necessary to produce an appropriate asset management policy and strategy that enables risk identification, assessment and control, as recommended, for example, in British Standards Institution PAS 55 -1 (2004) “Asset Management Part 1: Specification for the optimised management of physical infrastructure assets”.

CIRIA Report C634 (2005) “Management of accelerated low water corrosion in steel maritime structures” is also relevant for the anchored Thames river walls as inadequately protected anchorages over 20 years old are at greater risk of showing ALWC. In marine conditions, localised corrosion of the stressed anchorage components is a particular concern where there is inadequate protection, and localised corrosion rates are typically 1.5 to 3 times general uniform rates.

In order to establish the risk of river wall failure due to anchorage failure, it is necessary to establish for each distinct section of river wall (involving a specific anchorage design) the effect of failure of 1, 2 and 3 adjacent anchorages on the overall stability of the retained structure. This analysis will determine whether or not isolated failures can be tolerated (usually the case in conservative designs) and the critical number of failed neighbouring anchorages that will lead to progressive failure of the river wall.

Ms Wehalle clearly recognises the need to address the current situation and comments on his proposed measures for routine inspection and service behaviour monitoring are provided below.

The Keynote Presentation at this conference provides detailed guidance on maintenance testing and service behaviour monitoring (pages 5-21) and should be studied in order to assist in the preparation of a long-term asset management strategy that include an inspection/monitoring specification, condition and performance requirements, and a risk assessment for each type of anchored wall.

Bearing in mind that no routine rigorous inspections have been carried out for some 20 years at the river Thames, it is first necessary to inspect the inner and outer anchor heads of a representative sample of the anchorage types in service that will include both bar and multi-strand tendons. The different types of anchorages in service should be available from the asset register containing the historical design, construction and testing records.

Given the current dearth of relevant information on the condition of the anchorages retaining the river walls, this exercise should be implemented as soon as possible. Bearing in mind the experience gained from published case histories to date, it is particularly important to investigate the condition of the tendon below the stressing head. In the absence of de-stressing to remove the stressing head/locking nut and bearing plate to permit this examination, the use of an endoscope or fibrescope camera is essential.

Assessment of the results will indicate the probable condition of the remaining anchorages and the need for additional inspections, should the conditions recorded, e.g. corrosion of the stressing head components and tendon, or losses and/or gains in service loads, give concern. Early intervention with appropriate preventative or remedial measures makes economic sense.

In the event that there is an absence of corrosion, a frequency of five years is deemed appropriate for both physical inspections and service behaviour monitoring, as proposed by Ms Wehalle. This frequency also complies with the recommendations of BS8081 and should not be extended. The degree of corrosion or mechanical damage on any anchorage will dictate what additional measures are required, ranging from the restoration of appropriate corrosion protection to anchorage replacement.

Given the circumstances described above, the EA should carry out an asset management review and thereby address the need for changes to policy, strategy and objectives that take account of the additional elements of physical inspection and service behaviour monitoring required for anchored retaining walls. By this means, the management review process will reflect a commitment to improvement and ensure that the necessary information is collected to allow the EA to effectively manage this asset, as recommended in BS PAS 55 -2 (2004) "Asset Management Part 2: Guidelines for the application of PAS 55-1".

Horst Aschenbroich, *Con-Tech Systems Ltd.*
I thank you for the successful and interesting Conference in London. It was a real eye-opener to hear from other parts of the world about anchored structures in service and corrosion failures experienced, also in the UK. I am surprised to learn about the different multiple corrosion protection systems used in England in the bond length (two layers of corrugated HDPE), but only one layer of smooth HDPE in the free length of the anchors. I was not surprised to see pictures of failed anchors where anchor heads are without corrosion protection, i.e. no trumpet transitions under the bearing plates, discontinued HDPE sheath under the strand wedge plates with only little ore no protection by corrosion inhibiting grease or cement grout, and also poor head cover details. Also internally corroded grippers (wedges) have shown no evidence of any grease protection cover. Surprisingly, I heard statements that there are no standards or recommendations for the corrosion inhibiting greases or compounds in Europe. I would like to draw your attention to the North American PTI Recommendations for Rock and Soil Anchors, Table 4-1 Requirements for corrosion inhibiting compounds, Grease Wax and Gel. No other materials are used.

How the Class I protection is detailed in North America you can see in the attached Figures 5.2a 5.2b (PTI) and FHWA details.

Stuart Littlejohn, *University of Bradford*
Your positive response to the conference is much appreciated.

Given your comments on corrosion protection in England, there is a need for some clarification. Double corrosion protection in the UK implies two physical barriers, where the purpose of the outer second barrier is to protect the inner barrier against the possibility of damage during tendon handling and installation (in accordance with BS DD82:1982 and its successor BS8081:1989). The second barrier therefore provides additional insurance, given the distinction between the degree of protection of the tendon once installed in the ground, and that of the tendon, as supplied. The free length comprises, for example in a multi-strand tendon, greased and sheathed individual strands supplemented by a tendon "group" sheath encapsulating all the strands in the free length. Your quote of one layer of smooth HDPE over the free length represents "single protection" in the UK. Your photographs taken from PTI and FHWA are well known and were in fact derived over a period of time from European practice.

For your interest, greased and sheathed strands, first developed commercially by British strand manufacturers, e.g. Richard, Johnson & Nephew, and British Ropes Ltd. (now Bridon Wire) were employed for permanent anchorages in London Clay to resist hydrostatic uplift at an underground car park in London in 1968 and, interestingly after a meeting with New York consultants in the UK, the same solution was specified shortly afterwards by Bob White (President of Spencer, White and Prentiss Inc.) for the anchored diaphragm walls at the World Trade Centre in New York.

The photographs of examples of corroded tendons in various presentations where the anchor heads did not include trumpet transitions under the bearing plates were taken from old anchorages installed in the 1970s or earlier, i.e. before the adoption of national standards in Europe. I agree that it is not surprising that such anchorages have suffered severe corrosion.

The statements during the discussion in Session Six that there were no standards or recommendations covering corrosion inhibiting greases is incorrect. By way of example, recommended properties, acceptance criteria and associated test procedures are stipulated in BS DD81:1982, FIP:1986, BS8081:1989 and EN1537:1999.

Jamie Christie, *Aberdeen Harbour Board*
The common thread that I detected throughout was the emphasis on good, sound, basic inspection and maintenance regimes with respect to most types of anchorage. In other words, asset management.

I've been involved in developing asset management principals and strategies for a number of years now through my involvement with accelerated corrosion of maritime steel piles. With CIRIA we produced the following book:

Management of Accelerated Low Water Corrosion in Steel Maritime Structures" Ref. CIRIA C634 ISBN 0-86017-634-7

The difficulties and practicalities of managing steel piles as well as anchorages as "assets" are very similar and there is a lot of good advice contained within.
Furthermore I am currently on a CIRIA workgroup just finishing off a project called:

Whole Life Infrastructure Asset Management Across Industries: Good Practice Guide (Research Project 737)

We have examined asset management techniques across civil engineering infrastructure assets AS WELL AS components, to try to identify common elements in the approach to the asset management of each. We've found those and are just finalising the book which is due for publication by CIRIA mid 2008.

Stuart Littlejohn, *University of Bradford*
CIRIA Report C634 (2005) provides a comprehensive guide to the phenomena of accelerated low water corrosion (ALWC) in steel and its management, condition appraisal, repair, protection and monitoring. The recommendations are applicable to many anchored river and sea walls and I fully support the adage referenced, namely “a stitch in time saves nine”.

Given the guidance that is provided, asset managers should be aware that failure to implement measures for the monitoring and control of corrosion can expose an organisation and its management to the full severity of Heath and Safety and Environmental Legislation if a fatal incident, pollution or harm to the public occurs.

The new CIRIA Good Practice Guide “Whole-Life Infrastructure Asset Management Across Industries” focuses on the management of long life civil engineering infrastructure assets that include dams, flood defences, highways, ports and harbours. These structures may incorporate anchorages and consequently the principles and key steps required to successfully implement whole life infrastructure asset management that are described in the CIRIA Guide will be of considerable interest to delegates.

Stuart Littlejohn, *University of Bradford*
Re Paper by Miyatake et al
In my experience, ultrasonic testing for the presence of cracks is limited to 2-3 metres of tendon from the anchor head, what is the Japanese experience?

Hiroaki Miyatake, *Public Works Research Institute, Japan*
According to the results of our experiments, ultrasonic testing is limited to within 2m of the anchor head for steel bar tendons and to within 5m for strand tendons. So, there is no difference between your experience and ours in Japan.

Nevertheless, we think ultrasonic testing is a valuable method to find very small cracks in the anchor bar near to the anchor head for such shallow investigations.

Various methods are presented in our new manual that can be used as a discussion tool between administrators and specialists. We introduce ultrasonic testing as one of the various methods.

Stuart Littlejohn, *University of Bradford*
In figure 5 of the paper by Yamazaki et al, a head displacement of around 5mm is indicated for lift-off that is quite a high figure. Does this figure include extension of the tension rod? For information, in the UK we monitor the load lift-off at a stressing head clear displacement of 0.5 or 1mm, using feeler or dial gauges, as a standard approach for consistency.

Junichi Yamazaki, *Japan Anchor Association*
The picture of lift-off test of Figure 5 in paper is re-shown below.
A displacement gauge is fixed to the hydraulic jack in this test. The amount of displacement at a point in the tension rod is measured and so a "familiar" lift off and the expansion of tension rod are included in the measured value.
As the purpose of the test is just to find the variation point on the load/displacement curve and get residual load, we usually use this method so that the displacement gauge can be easily set up.

Figure of lift-off test

Stuart Littlejohn, *University of Bradford*
With reference to the paper by Bruce and Wolfhope, what cement grouts are typically used for Class II grout protected tendons? PTI (2004) states "Cement grout can be used to protect the tendon bond length in non-aggressive ground when installation methods ensure that the grout will remain fully around the tendon. How is this achieved in practice, bearing in mind that PTI (2004) recognises bleed but is silent on cracks?

John Wolfhope, *Freese and Nichols, Inc.*
Standard practice in North America is to use neat cement grout mixtures for Class II grout protected tendons. Grout mixtures are designed according to Chapter 6 of the PTI Recommendations (2004), typically using Type I cement and a water/cement ratio of 0:45:1 by weight. The grout mixtures are designed considering strength, bleed and pumpability. It is not standard practice to design grout mixtures specifically to prevent cracking of the grout around the tendon. Type II grout protected tendons are only used for:

1) temporary anchors with a service life less than 24 months, or

2) permanent anchors in non-aggressive ground conditions having insignificant consequences of failure and high incremental costs associated with providing an additional corrosion protection mechanism.

Class I Encapsulated Tendon anchors shall be provided for all permanent anchors regardless of ground aggressivity if the failure of the anchors could result in serious consequences (such as loss of life or significant economic impact) or if providing additional corrosion protection does not significantly increase the costs of the work. Accordingly, anchors in dams and other important infrastructure are routinely installed with Class I Encapsulated Tendon corrosion protection.

Stuart Littlejohn, *University of Bradford*
Compilation of a national bibliography and case history database will be invaluable in assessing the potential condition of anchorages. These designed and installed before the adoption of PTI (1974) will require particular attention in the future, as the corrosion protection provided may be regarded as inadequate by the standards of PTI.

In the paper, it is stated that no case has been reported of failure resulting from bond length corrosion in a properly grouted anchor. This will invariably be the case, but the key question when you are investigating existing anchors in service is "how do you know when you have a properly or improperly grouted anchor in situ?

Donald Bruce, *Geosystems, L.P.*
The assessment of the quality and adequacy of the anchor grouting is made in real time during construction, and during stressing and testing: the Engineer of Record must "sign off" on the completed installation, confirming that it is in conformance with the Project Specifications and the Contractor's Method Statement. Other than physical removal of the existing whole anchor (an enterprise which is occasionally proposed by over coring, but is never actually conducted due to the inherent practical difficulties), the authors know of no other method to verify the appropriateness of the grout in situ in an existing anchor.